Progress in Colloid and Polymer Science · Volume 115 · 2000

Springer-Verlag Berlin Heidelberg GmbH

Progress in Colloid and Polymer Science

Editors: F. Kremer, Leipzig and G. Lagaly, Kiel

Volume 115 · 2000

Trends in Colloid and Interface Science XIV

Volume Editor:
V. Buckin

Springer

The series Progress in Colloid and Polymer Science is also available electronically (ISSN 1437-8027)

- Access to tables of contents and abstracts is *free* for everybody.
- Scientists affiliated with departments/institutes subscribing to Progress in Colloid and Polymer Science as a whole also have full access to all papers in PDF form. Point your librarian to the LINK access registration form at http://link.springer.de/series/pcps/reg-form.htm

ISBN 978-3-662-16016-9   ISBN 978-3-540-46545-4 (eBook)
DOI 10.1007/978-3-540-46545-4

Springer-Verlag is a company in the BertelsmannSpringer publishing group.
© Springer-Verlag
Berlin Heidelberg 2000
Originally published by Springer-Verlag Berlin Heidelberg New York in 2000
Softcover reprint of the hardcover 1st edition 2000

Production: PROEDIT GmbH, 69121 Heidelberg, Germany

Typesetting: SPS, Madras, India

Cover: Estudio Calamar, F. Steinen-Broo, Pau/Girona, Spain

SPIN: 10757065

Printed on acid-free paper

Progr Colloid Polym Sci (2000) 115 : V
© Springer-Verlag 2000

PREFACE

The 13th Conference of the European Colloid and Interface Society (ECIS 99) was held from the 12th to the 17th of September 1999 in Trinity College, Dublin, Ireland. 245 participants from Europe, North and South America, Asia and Australia, 26 countries overall, attended the Conference. The scientific program included 77 oral and 145 poster presentations and consisted of eight topics:

- Surfactant colloids
- Polymer colloids and solid particles
- Food colloids
- Soft matter interfaces
- Biosystems
- Rheology
- Methods
- New in colloid and surfactant science.

This volume contains a selection of the contributions presented at the Conference, which outlines the key theoretical and experimental aspects of modern colloid and interface science.

The Conference was organised by the Irish Centre for Colloid and Interface Science and Biomaterials, which includes research groups in the University College Dublin and Queen's University Belfast. The Organising Committee (V. Buckin, J. Earnshaw, K. Dawson, R. Lynden-Bell and H. Hoffman) wishes to thank all participants of the Conference for their contributions and discussions resulting in an exciting scientific program. We are especially grateful to the members of the Scientific Committee: M. Corti, E. Dickinson, K. Dawson, H. Hoffmann, U. Kaatze, B. Lindman, D. Langevin and W. Poon for the success of the Conference. We also gratefully acknowledge the help of the postgraduate students and postdoctoral fellows of University College Dublin (S. Morrissey, C. Smyth, C. Dwyer, N. Hunt, E. Kudriashov, M. Hayes and D. McCloughlin) for their contribution to the organisation of the Conference.

We would like to thank the European Colloid and Interface Society, University College Dublin, Queen's University Belfast, Zeneca, Guinness and Radionics for their financial support of the Conference.

We remember with gratitude Prof. John Earnshaw, who died tragically in 1999. He was President of the European Colloid and Interface Society in 1998/99. He contributed greatly to the Society and did a great deal of work preparing for this Conference.

On behalf of the Organising Committee:
Vitaly Buckin,
Chairman of the Organising Committee,
University College Dublin, Ireland

Progr Colloid Polym Sci (2000) 115: VII–XI
© Springer-Verlag 2000

CONTENTS

VIII

Contents of Volume 115

Progr Colloid Polym Sci (2000) 115: 1–4
© Springer-Verlag 2000

M. S. Romero-Cano
A. Martín-Rodríguez
F. J. de las Nieves

Colloidal stability of a cationic latex covered with Triton X-100

M. S. Romero-Cano
F. J. de las Nieves (✉)
Complex Fluids Physics Group
Department of Applied Physics
Faculty of Experimental Sciences
University of Almería
04120 Almería, Spain
e-mail: fjnieves@filabres.ualm.es
Tel.: +34-50-215434
Fax: +34-50-215434

A. Martín-Rodríguez
Biocolloid and Fluids Physics Group
Faculty of Sciences
University of Granada
18071 Granada Spain

Abstract The stability behaviour of a cationic latex after adsorption of Triton X-100 has been studied with respect to the surfactant coverage, with pH being an important variable due to the weak character of the surface groups. It is possible to find an adequate coverage of surfactant that causes high stabilization. Moreover, the colloidal stability of some latex–surfactant complexes are insensitive to the addition of electrolyte. The chemical characteristics of the particle surface are an important factor that should be considered, because a specific interaction between the surface and the ions that determine the solution pH (H^+ or OH^-) may appear at extreme pH.

Key words Polymer colloids · Nonionic surfactant · Steric stabilization

Introduction

Nonionic surfactants adsorbed onto colloidal particles can act as steric stabilizers [1]. In this case the colloidal stability is provided by the combination of electrostatic and steric effects. The classical colloidal stability theory, Derjaguin–Landan–Verwey–Overbeek (DLVO) theory [2–3], has to be modified by the addition of a steric repulsive interaction energy [4–5] that considers the osmotic and elastic effects in the interaction between the surfactant layer. The colloidal stability theory including steric repulsion is usually referred to as extended DLVO theory, which gives the net interaction energy as the sum of all three potential energies: $V = V_R + V_A + V_S$.

The presence of nonionic surfactant adsorbed onto the surface of the colloidal particles causes a stabilization of the system when the coverage is adequate. At low coverage, the complex should have the same colloidal stability as the bare particles; however, at low coverage, systems with a variable surface charge present anomalous behaviour (destabilization) which is not explained by extended DLVO theory [6]. In this respect, the physicochemical characteristics of the surface groups are an important factor that should be considered. This new experimental variable enables colloidal systems to be made whose colloidal stability can be controlled in different ways.

In order to find universality of the anomalous behaviour that appears in a carboxyl latex [6], an other colloidal system with variable charge has been selected. In this work attention is focused on the influence of the surfactant coverage over the colloidal stability of a cationic latex. In particular, we have studied the influence of the surfactant desorption on the colloidal stability of the latex–surfactant complexes. The pH of the medium is also an interesting variable that can modify the colloidal stability of the complexes due to the change in the surface charge density of the cationic latex.

Experimental

Chemicals

All chemicals were of analytical grade and were used without further purification. The water in all experiments was ultrapure with a specific electrical conductivity lower than 1 μS/cm (ATAPA, Spain).

2

Latex characterization

The latex used in this work was synthesized in our laboratories. Styrene (Merck) was previously distilled under low pressure (10 mm Hg, 40 °C). Positively charged polystyrene latex (PS-CAT) was prepared using the emulsifier-free method, with N,N'-azobis(dimethyl-isobutylamide hydrochloride) as initiator, in a discontinuous reaction, following the method described in the pioneering work of Hidalgo-Alvarez et al. [7]. The maximum surface charge density and the diameter from electron microscopy were 9.2 ± 2.4 μC/cm^2 and 192 ± 10 nm, respectively, with a polydispersity index of 1.0066.

Surfactant characterization

Triton X-100 (TX-100) [p-(1,1,3,3-tetramethylbutyl) phenyl poly(ethylene glycol)], a gas chromatography grade material from Merck, was used without further purification. Surfactant concentrations during adsorption and desorption experiments were determined by UV spectrophotometry at 275 nm [8] using a Spectronic Genesys 5 spectrophotometer (Milton Roy, USA). The extinction coefficient obtained was 1.33×10^3 M^{-1} cm^{-1}. The critical micelle concentrations were determined by measuring the absorbance at different concentrations due to the change in slope of the absorbance versus concentration curve that appears during the micellization process. The value of this magnitude was of $(5.1 \pm 0.1) \times 10^{-4}$ M. In order to gain more information about the purity of the surfactant sample, the mass spectrum was obtained (University of Granada). The result showed that the surfactant is of good purity with a polydispersity index of 1.031.

Preparation of complexes

Latex–surfactant complexes were obtained by surfactant adsorption [9, 10] in batteries for 24 h at 25.0 ± 0.1 °C with various amounts of surfactant added onto a 0.25 m^2 latex surface. The surfactant–latex mixtures were gently shaken during the adsorption experiment.

Colloidal stability

The stability of the dispersions was evaluated with a Spectronic Genesys 5 spectrophotometer (Milton Roy, USA) by measuring the absorbance ($\lambda_0 = 600$ nm) as a function of time for different electrolyte concentrations. In a typical coagulation experiment, 2.4 ml buffered latex solution was put into the spectrophotometer cell and the optical absorbance was measured. Then, 0.6 ml NaCl solution at a given concentration was quickly added and mixed automatically [11]. The final particle concentration in the cell was 5.0×10^{16} particles/m^3. The optical absorbance was measured immediately and recorded continuously via computer for a period of 30 s. The initial slope of these curves, $(dAb/dt)_0$, is directly proportional to the initial coagulation rate (Fig. 1).

Results and discussion

The colloidal stability experiments of the latex–surfactant complexes were performed at two extreme pHs and different surfactant coverages. The control of these variables permits the colloidal system to be studied in a region characterized by the combination of electrostatic and steric repulsive interaction energies.

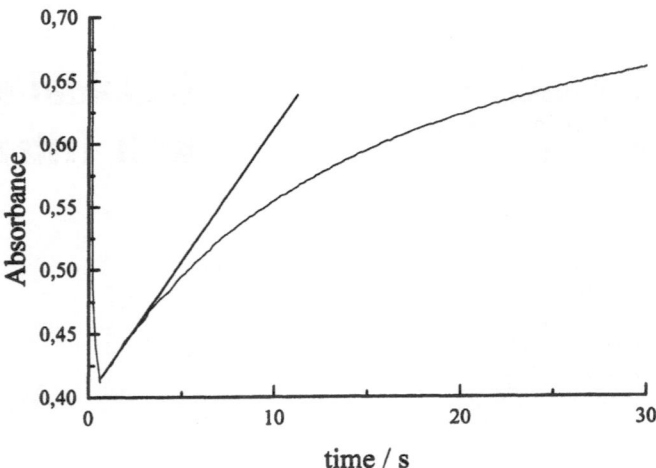

Fig. 1 Optical absorbance variation versus time in an aggregation reaction

Figures 2 and 3 show the absorbance change versus the NaCl concentration for different PS-CAT–TX100 complexes, with different amounts of surfactant adsorbed, at pH 6 and 10. The weak character of the amidine groups yields an electrostatic repulsion that is a maximum at pH 6, where the amidine groups are protonated, and that is much lower at pH 10, where most of the weak acid groups are deprotonated.

The coverage that causes total stabilization (around 2.5–3.0 μmol/m^2) can be determined from the curves. The complexes with this coverage are insensitive to the addition of electrolyte due to the high repulsive steric barrier that does not depend on electrolyte concentration. On the other hand, at pH 6 and 10 and for a low

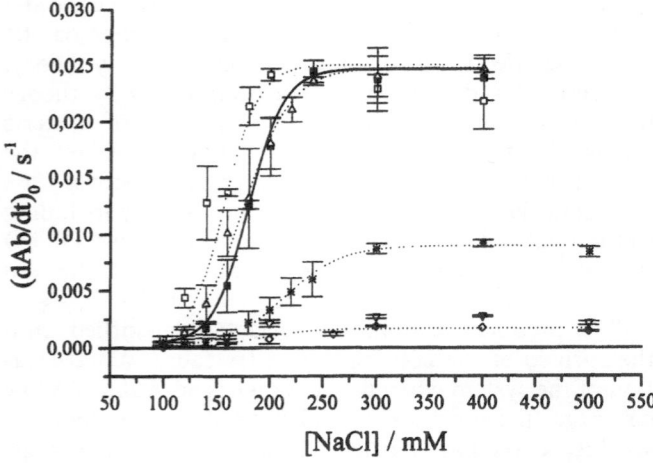

Fig. 2 Aggregation kinetics of the positively charged polystyrene latex and Triton X-100 (PS-CAT–TX100) system at pH 6 with different coverages: bare (■), 1.0 μmol/m^2 (□), 1.5 μmol/m^2 (△), 2.0 μmol/m^2 (∗), 2.5 μmol/m^2 (◇), 3.0 μmol/m^2 (▽)

3

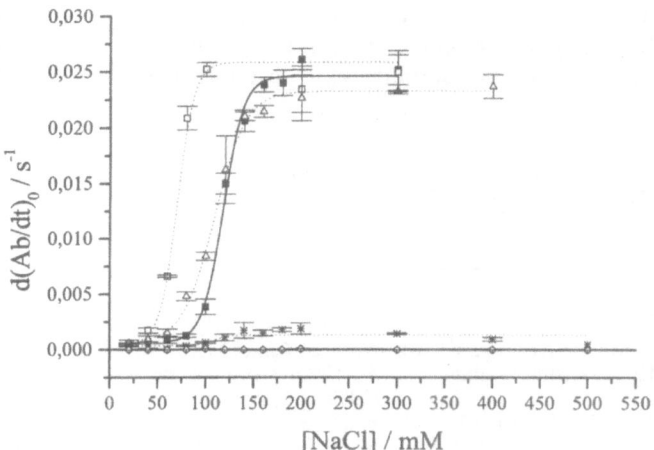

Fig. 3 Aggregation kinetics of the PS–CAT–TX100 system at pH 10 with different coverages: bare (■), 1.0 μmol/m^2 (□), 1.5 μmol/m^2 (△), 2.0 μmol/m^2 (∗), 2.5–3.0 μmol/m^2 (◇)

surfactant coverage the colloidal stability of the complexes is similar or surprisingly relatively lower (significant change in dAb/dt) than that of the bare particles. This anomalous behaviour is in agreement with that previously found for the carboxyl latex [6]. This result cannot be explained within the frame of current colloidal stability theories and may be accounted for by looking at the chemical nature of the molecules present on the particle surface.

At pH 6 the interaction between the ethylene oxide and the amidine groups could be possible through hydrogen bonding. This interaction will rarely occur in complexes with a highly packed monolayer of TX-100, while at low coverages and because of the great flexibility of the hydrophilic moieties the interaction could be favoured. A direct consequence of this interaction is the change in the surfactant layer configuration from extended to flat, thus resulting in the screening of the steric potential. This fact explains why the observed stability for this complex is similar to that of the bare particles but does not explain the previously mentioned destabilization phenomenon. A possible explanation could be the electrostatic screening of the particle surface charge by the flat layer of surfactant.

On the other hand, the destabilization that was found for complexes with low coverage at pH 10 could be explained as a consequence of the specific adsorption of OH$^-$ ions. The electrokinetic behaviour of the complexes [12] confirms this fact because a relative decrease in surface charge at high pH seems to indicate the possible adsorption of negative charge (OH$^-$ ions).

The explanation for these unexpected results may then be attributed to a decrease in electrostatic repulsive interaction energy due to a screening of the charge by the flat layer of surfactant, in the first case, or by the OH$^-$ ions, in the second one. In recent work, Gana-

chaud et al. [13] found experimental evidence for the destabilization of a cationic latex partially covered with Triton X-405.

The colloidal stability results for the same system at pH 6 and 10 after centrifugation are shown in Figs. 4 and 5. The colloidal stability behaviour is considerably different from that shown in Figs. 2 and 3. In this case, all complexes show a clear destabilization phenomenon. As found in others work [12], an important desorption phenomenon occurs when a precipitated complex is redispersed in a new medium. If we take this fact into account the results shown in these figures are explained by the previous arguments.

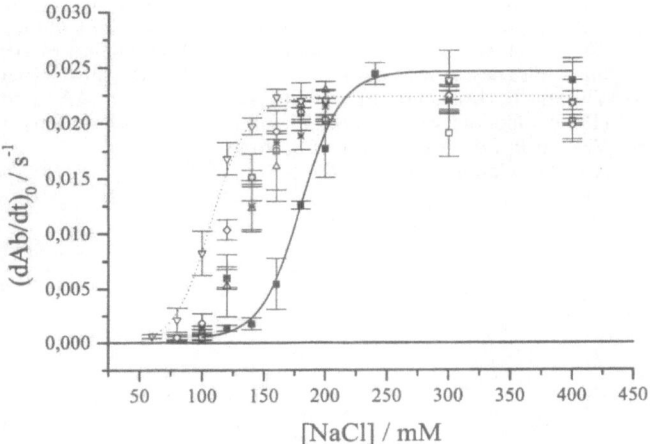

Fig. 4 Aggregation kinetics of the PS–CAT–TX100 system at pH 6 with different coverages after centrifugation: bare (■), 1.0 μmol/m^2 (□), 1.5 μmol/m^2 (△), 2.0 μmol/m^2 (∗), 2.5 μmol/m^2 (◇), 3.0 μmol/m^2 (▽)

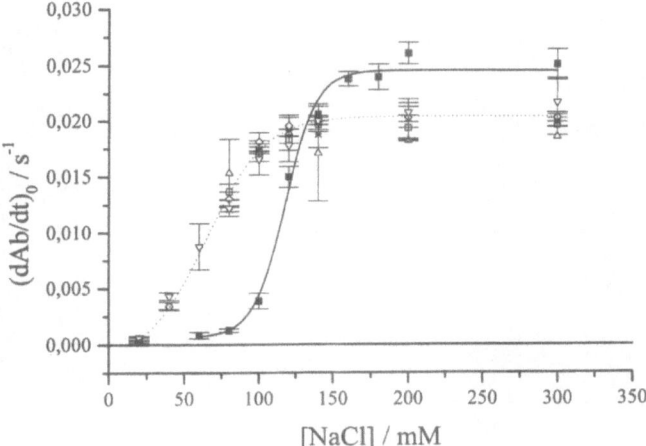

Fig. 5 Aggregation kinetics of the PS–CAT–TX100 system at pH 10 with different coverages after centrifugation: bare (■), 1.0 μmol/m^2 (□), 1.5 μmol/m^2 (△), 2.0 μmol/m^2 (∗), 2.5 μmol/m^2 (◇), 3.0 μmol/m^2 (▽)

4

Conclusions

It is well-known that nonionic surfactants are excellent tools for controlling the stability of a colloidal system. This stability is improved with an adequate surface coverage. However, the physicochemical characteristics of the surface groups are also an important factor that should be considered as a new experimental variable that enables colloidal systems to be obtained whose colloidal stability can be controlled in different ways.

Acknowledgements The financial support provided by the CICYT (under project MAT 96-1035-C03-03) is greatly appreciated.

References

1. Napper DH (1983) Polymeric stabilization of colloidal dispersion. Academic, London
2. Derjaguin BV, Landau LD (1941) Acta Physicochim URSS 14:633
3. Verwey EJW, Overbeek JThG (1948) Theory of stability of lyophobic colloids. Elsevier, Amsterdam
4. Vincent B, Luckham PF, Waite FA (1980) J Colloid Interface Sci 73:508
5. Vincent B, Edwards J, Emmett S, Jones A (1986) Colloids Surf 18:261
6. Romero-Cano MS, Martín-Rodríguez A, de las Nieves FJ. Macromolecular Symposia, in press
7. Hidalgo-Álvarez R, de las Nieves FJ, van der Linde AJ, Bijsterbosch BH (1986) Colloids Surf 21:259
8. Chiming M (1987) Colloids Surf 28:1
9. Martín-Rodríguez A, Cabrerizo-Vílchez MA, Hidalgo-Álvarez R (1994) Colloids Surf A 92:113
10. Romero-Cano MS, Martín-Rodríguez A, Chauveteau G, de las Nieves FJ (1998) J Colloid Interface Sci 198:266
11. Rubio-Hernández FJ, de las Nieves FJ, Hidalgo-Álvarez R, Bijsterbosch BH (1994) J Dispersion Sci Technol 15:1
12. Romero-Cano MS (1998) PhD thesis. University of Granada
13. Ganachaud F, Elaïssari A, Pichot C (1997) Langmuir 13:7021

Progr Colloid Polym Sci (2000) 115: 5–9
© Springer-Verlag 2000

D. Letellier
V. Cabuil

Dilution of 6-lauryl poly(oxyethylene ether) lamellar phases doped with magnetic nanoparticles

D. Letellier · V. Cabuil (✉)
Laboratoire des Liquides Ioniques et
Interfaces Chargées – Colloïdes Magnétiques
Université Paris 6, Bât. F, Case 63
75252 Paris Cedex 05, France

Abstract 6-Lauryl poly(oxyethylene ether) is a nonionic surfactant, the phase diagram of which exhibits a large lamellar domain. Negatively charged magnetic particles are incorporated inside the lamellae, the average diameter of the particles being about 7 nm. The volume fraction of particles that can be included is determined. The dilution of such lamellar phases leads to the formation of onions. Upon dilution, doped lamellar phases give rise to onions stuffed with magnetic particles. These onions move in a magnetic field gradient. During the dissolution of the lamellar phase, "myelin figures" are observed. Such nonequilibrium pattern formations have been observed in other cases, namely in the case of dissolution of almost insoluble lamellar phases. Myelin patterns formed with loaded lamellar phases align in the direction of an applied magnetic field.

Key words 6-Lauryl poly(oxyethylene ether) · Lyotropic lamellar phase · Myelin · Magnetic particles · Ferrofluid

Introduction

The inclusion of magnetic nanoparticles in self-organized phases of surfactants [1–5] or polymers [6] has been widely described in recent last years. For example, positively charged nanoparticles made of a magnetic ferric oxide (namely maghemite, γ-Fe_2O_3) have been incorporated into electrostatistically stabilized lamellar phases [didodecyldimethylammonium bromide (DDAB) water system] [1], but it has been shown that the incorporation of solid nanoparticles in a lamellar lyotropic phase is favored when the interactions between the membranes of the lamellar system are dominated by the fluctuations of the membranes as described in the Helfrich model [7]. That is the reason why we include here magnetic nanoparticles in the L_α lamellar phase of the nonionic surfactant 6-lauryl poly(oxyethylene ether) ($C_{12}E_4$)/water system [8]. Upon addition of water, the lamellar phase swells and below 20 wt% $C_{12}E_4$ it dissolves. Then a W + L_α phase constituted of multilamellar vesicles (onions) dispersed in water is observed. We were especially interested in incorporating magnetic oxide nanoparticles in these onions as it has proved possible to incorporate large (0.4–1 μm) particles in such systems [9]. Moreover, onions are often described as promising carriers for drug delivery or other applications and making them partly magnetic can improve their ability to move or to be recovered.

Experimental

Materials

$C_{12}E_4$ was obtained from Sigma and was used as received.

The γ-Fe_2O_3 magnetic nanoparticles were synthesized as described elsewhere [10] by chemical precipitation of metallic salts in an alkaline medium and further oxidation in acidic medium. The particle size distribution follows a log–normal law:

$$P(d) = \frac{1}{\sqrt{2\pi}\sigma d} \exp\left[-\frac{1}{2\sigma^2}\left(\ln\frac{d}{d_0}\right)^2\right] ,$$

where d_0 is the mean diameter and σ the standard deviation. The particles obtained had a mean diameter ranging between 5 and 12 nm. They were coated by citrate species (trisodium citrate) as described in Ref. [11] in order to have negative surface charges for pH higher than 3. They were then dispersed in water and a stable dispersion of magnetic nanoparticles was obtained. The ionic strength of the dispersion was rather high due to unadsorbed trisodium citrate species in solution. This ionic strength was deduced from conductivity experiments and was kept constant, for example a citrate concentration of about 5×10^{-3} mol/l.

Methods

The inclusion of particles in $C_{12}E_4$ lamellar phases was performed by dispersing the $C_{12}E_4$ surfactant into the aqueous dispersion of nanoparticles at room temperature. The mixture was stirred vigorously and then centrifuged for 15 min at 10,000g. Magnetic onions were obtained on diluting the magnetic lamellar phases obtained with water. The dilution was performed either in glass tubes for a preparative purpose or in capillaries for microscopic observation of the dilution process.

Samples were observed with an Ortholux optical microscope using a Leitz ×40 objective. For usual observations, the samples were placed in homemade cells constituted of two glass slides separated by 100-μm spacers. For the study of the dilution process, the samples were set in glass capillary tubes and were observed between crossed polarizers to see the specific confocal patterns of the lamellar phase.

Small-angle X-ray scattering (SAXS) experiments were carried out at the Laboratoire pour l'Utilisation du Rayonnement Electromagnétique (Orsay, France). The samples were set between two Milar sheets and the q range was 0.0045–0.19 Å^{-1}.

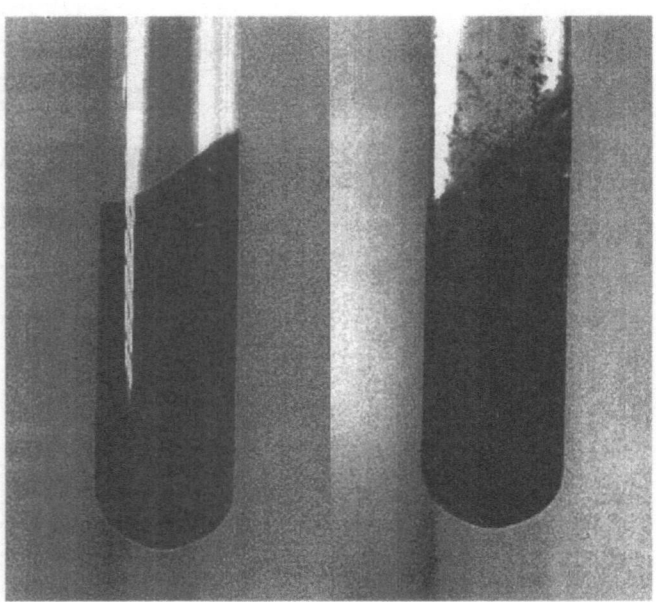

Fig. 1 Pictures of macroscopic samples of 6-lauryl poly(oxyethylene ether) ($C_{12}E_4$) and ferrofluid mixtures after centrifugation. *Left*: stable doped lamellar phase; *right*: flocculated sample

Results

Macroscopic observations

Several mixtures were synthesized with the amount of surfactant ranging between 0 and 75 wt% and the particle content ranging between 0 and 2 vol% of the aqueous phase. The ionic strength was kept constant. All the samples were centrifuged to probe their stability. Stable lamellar phases containing magnetic nanoparticles appeared as nonflowing, homogenous orange dispersions (Fig. 1 left). Unstable ones separated between a colorless viscous liquid upper phase and a flocculent magnetic orange subphase (Fig. 1 right). For samples near the phase-separation threshold, the upper phase was still a stable magnetic lamellar phase but a drop of dense magnetic liquid phase was found at the bottom of the tube. The observations are summarized in Fig. 2, in which the volume fraction of magnetic particles included inside the lamellae is plotted against the swelling of the lamellar phase.

Microscopic observations

Direct observation of the lamellar samples revealed a homogenous orange background. Between crossed polarizers typical so-called maltese cross patterns, characteristic of the lamellar order were observed. After dilution, dark spherical objects were observed in a

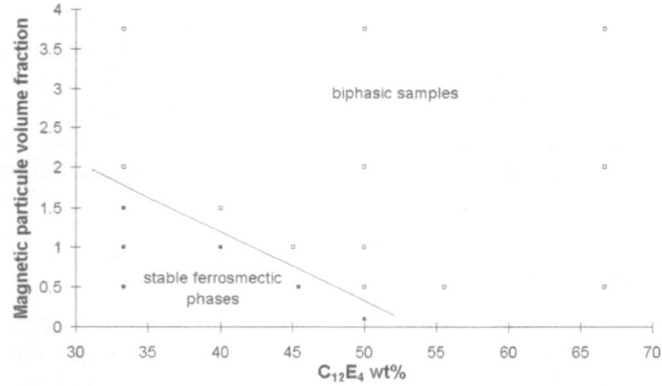

Fig. 2 Stability diagram of $C_{12}E_4$ lamellar phases doped with magnetic nanoparticles. The volume fraction of particles is plotted as a function of $C_{12}E_4$ weight percent. *Open squares* are unstable biphasic samples; *filled points* are stable doped lamellar phases

colorless background (Fig. 3a). They were polydisperse in size and their diameter ranged from less than 1 μm to 5 μm. They aligned when a magnetic field was applied and exhibited magnetophoretic mobility (Fig. 3b). Between crossed polarizers, malt crosses were observed, indicating that these objects present a local lamellar order and are multilamellar vesicles or onions.

Some contact experiments were performed in order to follow the dilution of the lamellar phase. Between a $C_{12}E_4$ lamellar phase and pure water, the flow of water in the lamellar phase is balanced by a backflow of

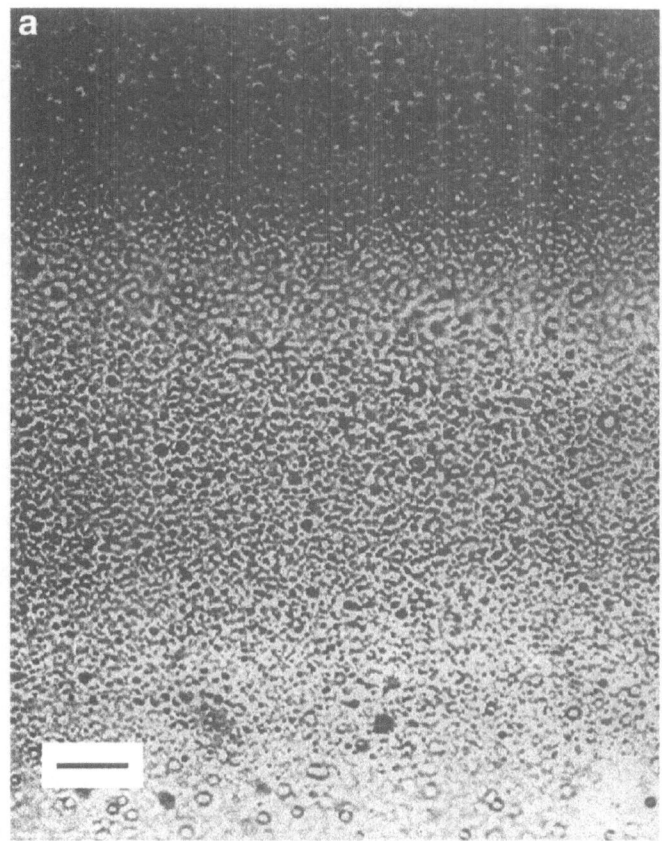

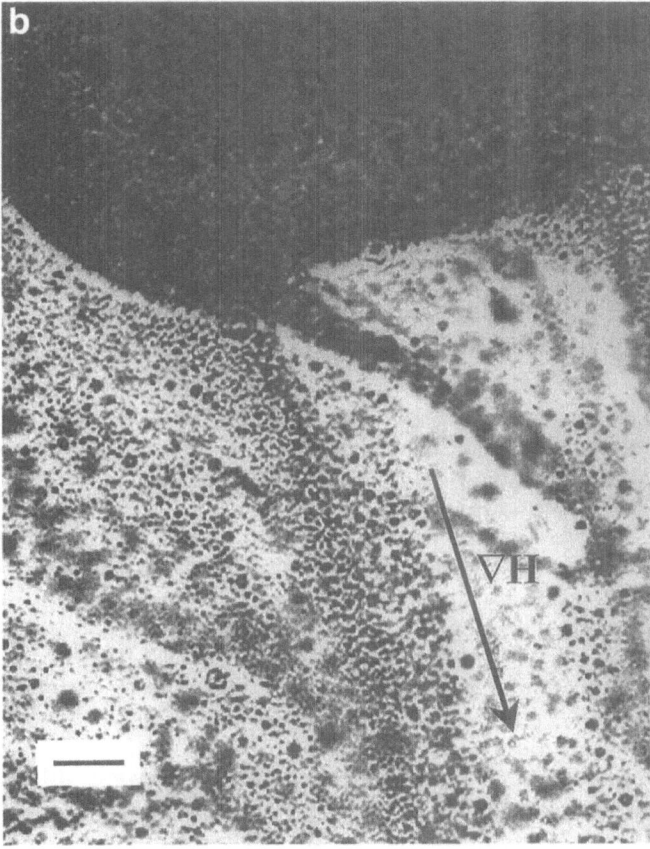

Fig. 3a, b Optical microscopy picture of magnetic onions resulting of the dissolution of a doped lamellar phase in a contact experiment. **a** Onions falling from the lamellar phase at the L_α/water interface. **b** Flow of magnetic onions under the action of a magnetic force. The *bar* is 2.5 μm

surfactant towards the aqueous phase consisting of growing myelin figures [12–16] (Fig. 4). Between magnetic $C_{12}E_4$ lamellar phases and water, the lamellar phase dissolves, releasing a cascade of magnetic onions into the aqueous phase (Fig. 3a). Magnetic myelin figures are observed when the lamellar phase is less soluble, i.e. for $C_{12}E_4$ content above 45 wt% (Fig. 5a). In this case, as for the onions, the myelin is sensitive to a magnetic field. While ordinary myelin grows in random directions, magnetic myelin aligns in the direction of an applied external magnetic field (Fig. 5b, c). It should be noted that, as well as onions, the diameter of the myelin can vary over a wide range (about 2 μm in Fig. 5a and about 0.25 μm in Fig. 5b, c).

Nanoscopic observations

Preliminary SAXS experiments were performed for compositions ranging from 75 to 20 wt% $C_{12}E_4$. The spectra obtained for a lamellar sample $C_{12}E_4$ (40 wt%)/ trisodium citrate aqueous solution (5×10^{-3} mol/l) without and with magnetic nanoparticles (0.1%) are compared in Fig. 6. The position of the Bragg peak is the same for both samples. The periodicity of the lamellar structure is about 85 Å and is not altered by the incorporation of the particles (Fig. 7).

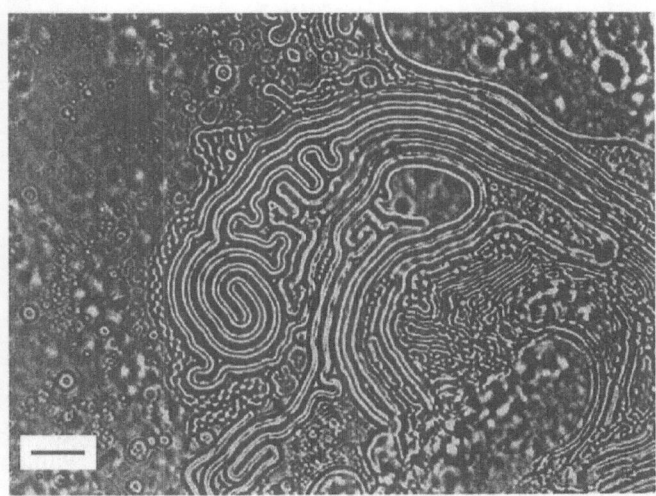

Fig. 4 Optical microscopy pictures of myelin figures in a $C_{12}E_4$ lamellar phase–pure water contact experiment. The *bar* is 10 μm

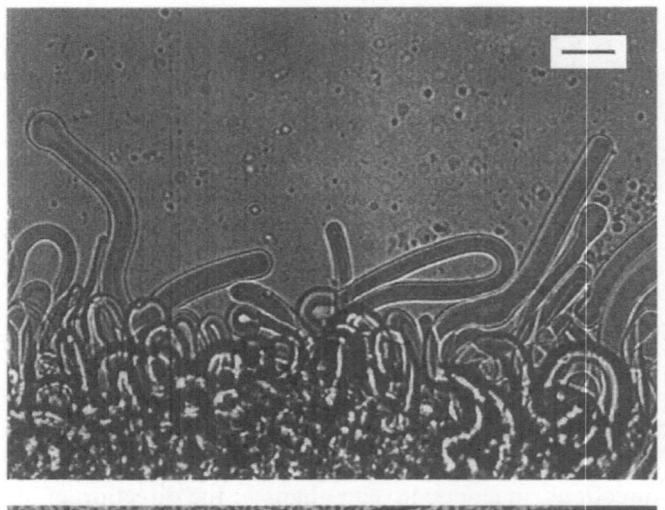

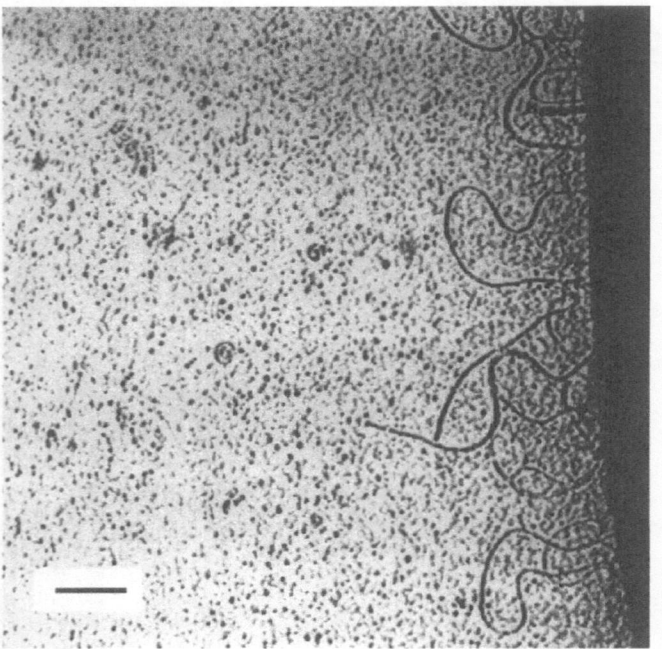

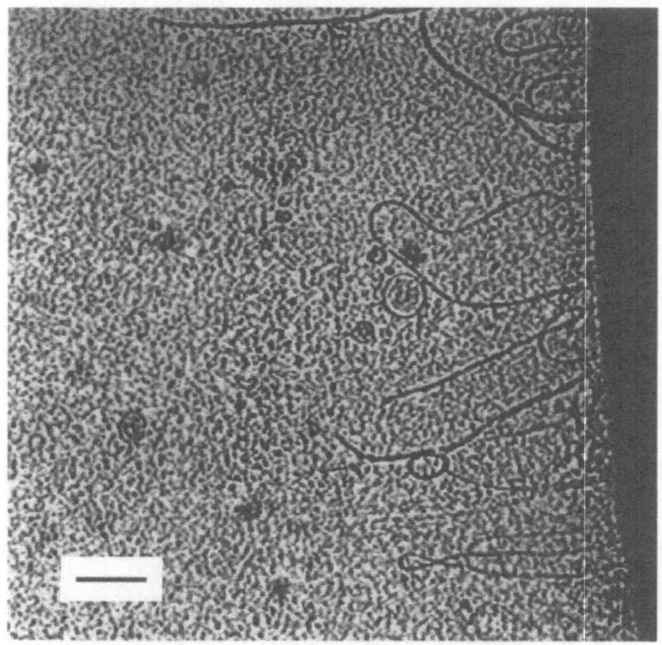

Fig. 5a–c Optical microscopy picture of a contact experiment between pure water and a magnetic $C_{12}E_4$ lamellar phase. a Typical myelin figures ($C_{12}E_4$ 50 wt%, magnetic particles 0.1%). b Randomly growing myelin figures. c Same myelin figures as in b aligned in an applied external magnetic field. The *bar* is 2.5 μm

Discussion and conclusion

These results show that it is possible to incorporate magnetic nanoparticles having surface charges in an aqueous lamellar phase. In contrast to the case of the DDAB/water phases, which were purely electrostatistically stabilized systems, $C_{12}E_4$/water phases are able to include volume fractions of negatively charged particles as high as 1.5%. The preliminary SAXS results obtained indicate that the interlamellar spacing is not modified by the presence of magnetic inclusions. When the volume fraction of particles is higher than a given value, particles are excluded from the lamellae as a dense magnetic phase or as a precipitate. When the interlamellae spacing is too small (e.g. smaller than 65 Å),

i.e. for an amount of $C_{12}E_4$ greater than 50 wt%, the systems become biphasic as well. All these results are in good agreement with previous results obtained for other ferrosmectics [17].

Magnetic onions have also been obtained by dilution of the magnetic lamellar phase. The particles seem to stay inside the lamellae and so are trapped inside the onions. The next step in order to use the magnetic properties of these stuffed onions would be to coencapsulate magnetic nanoparticles with other chemical species, making the magnetic onions new possible vectorization tools.

Dilution of $C_{12}E_4$ lamellar phases with water allows nice myelin figures to be observed. Such structures have been observed in other systems when the lamellar phase

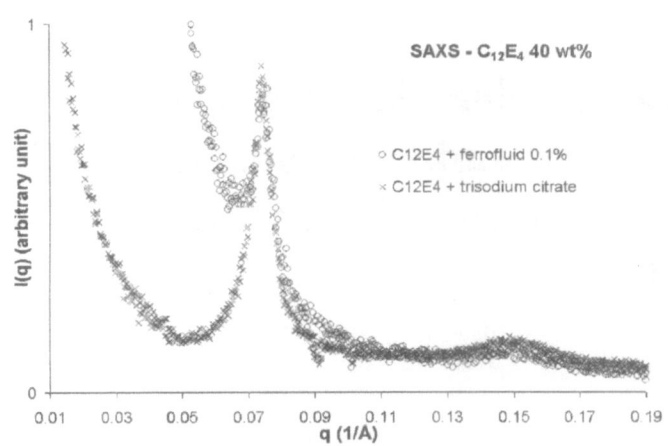

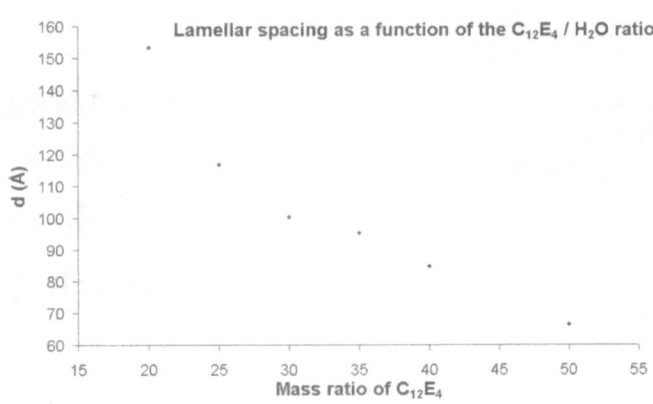

Fig. 6 Small-angle X-ray scattering (*SAXS*) spectra of $C_{12}E_4$ (40 wt%)/trisodium citrate 5×10^{-3} mol/l) aqueous solution without (×) and with (O) magnetic nanoparticles (0.1%). The diffusion intensity (in arbitrary units) is plotted as a function of the wave vector, q (Å^{-1})

Fig. 7 Evolution of the lamellar spacing. The periodicity (calculated from SAXS) is plotted as a function of the composition of the sample ($C_{12}E_4$ wt%)

is insoluble and they are still observed in the case of the dilution of a magnetic lamellar phase. Studying the growth of doped myelin figures under an applied magnetic field could key to measure the driving forces and to elucidate the mechanisms of their formation of the myclins. The encapsulated particles could also act as local probes of the swelling of the lamellae.

Acknowledgements The authors deeply indebled to P. Lesieur for his help with the SAXS experiments.

References

1. Ménager C, Belloni V, Cabuil V, Dubois M, Gulik-Krzywicki T, Zemb T (1996) Langmuir 12:3516–3522
2. Fabre P, Quilliet C, Veyssié M, Nallet F, Roux D, Cabuil V, Massart R (1992) Europhys Lett 20:229
3. Ménager C, Cabuil V (1994) Colloid Polym Sci 272:1295–1299
4. (a) Berejnov V, Raikher Y, Cabuil V, Bacri J-C, Perzynski R (1998) J Colloid Interface Sci 199:215–217; (b) Berejnov V, Bacri J-C, Cabuil V, Perzynski R, Raikher Y (1998) Europhys Lett 41:507–512
5. Letellier D, Sandre O, Ménager C, Cabuil V, Lavergne M (1997) Mater Sci Eng C 5:153–162
6. Lefebure S, Cabuil V, Ausserré D, Paris F, Gallot Y, Lauter-Pasyuk V (1998) Prog Colloid Polym Sci 110:94–98
7. Ramos L, Fabre P, Dubois E (1996) J Phys Chem 100:4533–4537
8. Mitchell J, Tiddy G, Waring L, Bostock T, McDonald M (1983) J Chem Soc Faraday Trans 1 79:975–1000
9. Arrault J, Grand C, Poon W, Cates M (1997) Europhys Lett 38:625–630
10. Massart R (1981) IEEE Trans Magn 17:131
11. Bacri J-C, Perzynski R, Salin D, Cabuil V, Massart R (1990) J Magn Magn Mater 85:27–32
12. Buchanan M, Arrault J, Cates M (1998) Langmuir 14:7371–7377
13. Tsao Y, Evans F, Rand R, Parsegian V (1993) Langmuir 9:233–241
14. Mishima K, Yoshiyama K (1987) Biochim Biophys Acta 904:149–153
15. Sakurai I (1985) Biochim Biophys Acta 815:149–152
16. Sakurai I, Kawamura Y (1984) Biochim Biophys Acta (1984) 777:347–351
17. Fabre P, Casagrande C, Veyssié M, Cabuil V, Massart R (1990) Phys Rev Lett 5:64

Progr Colloid Polym Sci (2000) 115:10–14
© Springer-Verlag 2000

A. Stradner
O. Glatter
P. Schurtenberger

Hexanol-induced sphere-to-flexible-coil transition in sugar surfactant solutions: a small-angle neutron and static light scattering study

A. Stradner · O. Glatter
Institut für Physikalische Chemie
Universität Graz
8010 Graz, Austria

A. Stradner (✉) · P. Schurtenberger
Department of Physics
University of Fribourg
1700 Fribourg, Switzerland
e-mail: anna.stradner@unifr.ch
Tel.: +41-26-300 9120
Fax: +41-26-300 9747

Abstract Sugar surfactants have recently attracted considerable interest. We report systematic light and neutron scattering experiments with aqueous solutions of the alkyl polyglucoside $C_{8/10}G_{1.5}$. The system exhibits a transition from globular to polymer-like micelles upon adding hexanol. We can finely tune the extent of micellar growth and reach conditions where solutions of relatively short cylinders exist; these cylinders grow dramatically into giant wormlike micelles upon the addition of small amounts of hexanol. This system provides us with an ideal data set for analyzing micellar growth and interactions in solutions with cylindrical particles of moderate axial ratios.

Key words Sphere-to-rod transition · Polymer-like micelles · Equilibrium polymers · Light scattering · Small-angle neutron scattering

Introduction

Alkyl glucosides or alkyl polyglucosides are nonionic surfactants produced from renewable raw materials – especially starch and fats and their derivatives [1]. The hydrophilic headgroup consists of one or more glucose molecules, while the tail is a hydrocarbon chain. Notations such as AG, APG, or sugar-based surfactants are often used when referring to alkyl glucosides. The industrial processes generally do not yield the pure alkyl monoglucosides, but a complex mixture of alkyl mono-, di-, tri-, and oligoglucosides, as well as α- and β-anomers are produced; therefore, the industrial products are called alkyl polyglucosides, C_iG_j, and are characterized by the length of the alkyl chain, i, and the average number of glucose units, j, linked to it.

At present there is great interest in alkyl glucosides due to the fact that these molecules may be synthesized from renewable resources. In addition, the ecological and toxicological properties of this class of surfactants are extremely favorable, which is not that surprising considering their structural similarity to glycolipids or other biological surfactants which are very environmentally compatible. They are extremely mild, display dermatological safety, very good biodegradability, and interesting surface-active properties, such as good foaming behavior; therefore, these surfactants are becoming increasingly important for use in detergents and in cosmetic products.

It has been demonstrated that it is possible to find conditions where micelles grow dramatically with increasing concentrations into giant cylindrical aggregates [2, 3]. The phenomenon of the formation of locally cylindrical giant micelles has also been observed in aqueous solutions of the sugar surfactant $C_{8/10}G_{1.5}$, where the addition of small amounts of hexanol induces one-dimensional micellar growth [4]. A qualitative explanation of the hexanol-induced growth can be given based on the cosurfactant dependence of the spontaneous curvature or packing parameter of the surfactant molecule. Hexanol acts as a cosurfactant that dissolves mainly in the interfacial layer of the sugar surfactant aggregates, thereby changing its packing properties and curvature. Hexanol thus plays the same role as temperature for nonionic surfactants of the C_iE_j type or salt for ionic surfactants. Under these circumstances the molecules now have a preference for locally cylindrical packing, which provides them with a possibility for almost unlimited uniaxial growth.

Here we try to understand the structural properties of alkyl polyglucoside micelles as a function of hexanol content based on previous work on giant micelles (Ref. [5] and references therein). They normally have a high degree of flexibility, and their overall structure is generally well described by polymer theory. The postulate of an analogy between classical polymers and polymer-like micelles is an important step towards a quantitative understanding of surfactant systems. While the close analogy between polymer and micelle structure has indeed been demonstrated successfully, there remain major differences between polymers and micelles. In contrast to classical polymers, which are covalently bound, micelles are dynamic entities which constantly exchange monomers and can break spontaneously at a random point along their contour and later recombine with another micelle. The reversibility allows the aggregates to exchange material: a given monomer could belong to different chains at different times. This transient nature of the micelles not only has important consequences for the dynamic properties of micellar solutions on time scales long compared to the average micellar life time, it also means that the micellar size and size distribution are equilibrium properties that strongly depend upon parameters such as surfactant and cosurfactant concentration or temperature [2]. Therefore, wormlike micelles serve as a prime example of equilibrium polymers, where the term equilibrium polymer is used for linear aggregates that can break and recombine, which have recently attracted considerable attention in the physics community.

The equilibrium nature of the micellar size also has important consequences in the analysis of interaction effects when performing scattering experiments. Once the micellar size and concentration are large enough, the polymer-like structures overlap and start to entangle. At even higher concentrations, the micelles can form an entanglement network, and the solution becomes viscoelastic with properties analogous to semidilute polymers. The entanglement threshold, c^*, where the transition from the dilute to the semidilute regime occurs, is in reality not a sharp boundary but rather a concentration range which starts when the coils touch and ends when the coils are completely entangled [6]. For polymers, enormous progress has been made in the understanding of the results from scattering experiments; however, for micelles the situation is more complicated due to the fact that the concentration dependence of the experimental results reflects contributions from the equilibrium size distribution, i.e. micellar growth, as well as from intermicellar interactions. It is only recently that it has been possible to combine both micellar growth and interaction effects and to apply polymer renormalization group theory and Monte Carlo simulations in order to quantitatively analyze the scattering results from micellar solutions [7–9].

Here we demonstrate that through a combination of static light scattering (SLS) and small-angle neutron scattering (SANS) information can be obtained about the overall size, the flexibility, and the local cross-section of the wormlike micelles. While SLS experiments are ideally suitable for the determination of the overall size and apparent molar mass of these giant aggregates, information on the flexibility and local structure is obtained from SANS experiments.

Materials and methods

Materials and preparation

The technical grade alkyl polyglucoside $C_{8/10}G_{1.5}$ (Glucopon 215 CSUP) was kindly provided by Henkel, Germany. This commercial alkyl polyglucoside represents a mixture of α- and β-glucosides. It is supplied as a 62–65 wt% solution in water with high pH value to avoid microbial attack. The surfactant was used without further purification. 1-Hexanol (purity greater than 99%) was purchased from Fluka (Buchs, Switzerland) and was used without further purification. D_2O with an isotopic purity greater than 99.75% was obtained from Merck (Darmstadt, Germany). The buffer 0.1 M tris(hydroxymethyl)aminomethane/HCl in D_2O (pH 7.4) was used as a solvent. A highly concentrated stock solution of alkyl polyglucoside in buffer solution was prepared by weighing the surfactant into a glass bottle and diluting it to the desired concentration. Next, hexanol was weighed into glass bottles and diluted to the appropriate hexanol weight percent with the alkyl polyglucoside solution. Stock solutions with four different hexanol-to-alkyl polyglucoside ratios were prepared, namely, a pure alkyl polyglucoside solution and solutions with additional small, medium, and high amounts of hexanol (concentration ratios of C_6E_0 to alkyl polyglucoside of 0.08, 0.12, and 0.15, respectively). The concentration series were prepared by adding appropriate amounts of buffer to the stock solutions. All experiments were performed at 30 °C.

Methods

All samples were measured at a temperature of 30 °C. SANS measurements were performed at the SINQ small-angle scattering facilities of the Paul Scherrer Institute, Switzerland. The initial data treatment and the data analysis were performed as described in Ref. [10]. SLS experiments were carried out and analyzed as described in detail previously [10]. Measurements were performed with a commercial goniometer system (ALV/DLS/SLS-5000F monomode fiber compact goniometer system with ALV-5000 fast correlator). From the normalized scattering intensity an apparent aggregation number, N_{app}, and a static correlation length were determined using a Lorentzian scattering law for the initial q dependence of the intensity. The "surfactant molecule" was assumed to consist of one alkyl polyglucoside molecule and an additional contribution from the respective number of hexanol molecules per alkyl polyglucoside molecule.

Results and discussion

Hexanol-induced micellar growth

N_{app} obtained from SLS measurements are summarized in Fig. 1, where the data at different total surfactant

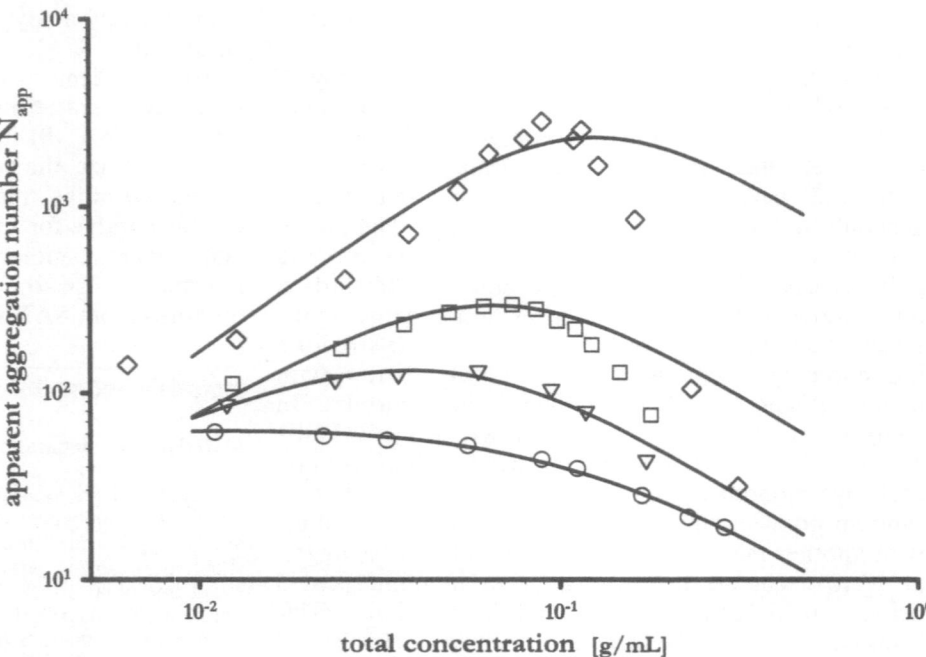

Fig. 1 Apparent aggregation number, N_{app}, obtained from static light scattering data as a function of the total concentration of alkyl polyglucoside (*APG*) solutions in buffer with different amounts of hexanol: pure APG (○); low hexanol content (▽, [hexanol]/[APG] = 0.08); medium hexanol content (□, [hexanol]/[APG] = 0.12); high hexanol content (◇, [hexanol]/[APG] = 0.15) (see text for details)

concentrations and varying hexanol content are shown. Hexanol has a dramatic influence on the resulting micellar size. At low surfactant concentration one can observe the hexanol-induced micellar growth. Adding hexanol not only produces larger micelles, it also has an enormous effect on the concentration dependence of the micellar size distribution. For the pure alkyl polyglucoside samples the micelles remain relatively small with increasing concentration, and at the highest concentrations the intermicellar interaction effects lead to decreasing N_{app}. For the samples with hexanol added, however, we find a dramatic concentration-induced micellar growth; this results in a strong initial increase in N_{app} followed by a decrease at higher surfactant concentrations, where intermicellar interaction effects start to dominate.

Figure 2 further illustrates the effect of hexanol on the micellar size and shape at the much higher spatial resolution provided by SANS. The forward intensity increases dramatically with increasing hexanol content, which is in agreement with the findings of the SLS experiments. On the other hand, the high-q data completely superimpose, which demonstrates that micellar growth occurs in a one-dimensional fashion along the micellar contour, leaving the local cylindrical cross-section of the micelles unchanged. For the samples with higher hexanol content, we can observe all the typical features of giant polymer-like aggregates at low and intermediate q values, for example, the asymptotic power-law dependence of $I(q) \sim q^{-1.66}$. The excellent agreement between the corresponding predictions from polymer theory and the experimental data is demon-

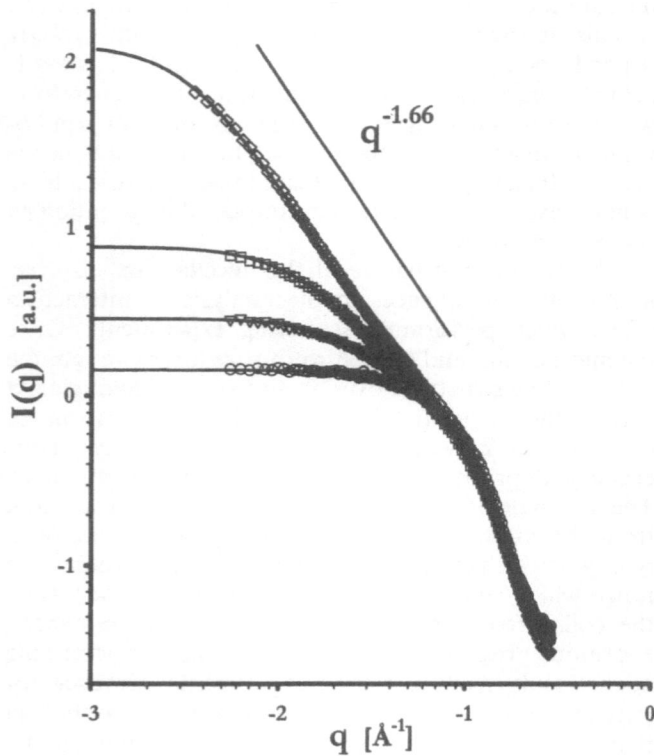

Fig. 2 The effect of hexanol on the q dependence of the scattered intensity for samples with constant APG concentration (10 wt%) and four different ratios of hexanol to APG: pure APG (○); low hexanol content (▽, [hexanol]/[APG] = 0.08); medium hexanol content (□, [hexanol]/[APG] = 0.12); high hexanol content (◇, [hexanol]/[APG] = 0.15) (see text for details)

strated in Fig. 2 by the solid lines, which are theoretical curves based upon a least-squares fitting procedure of a wormlike chain scattering function with excluded-volume effects [11].

Micellar growth and interaction effects

The analysis of the detailed structure and size distribution of anisotropic micelles is hampered by the difficult task of distinguishing between the contributions from intermicellar interactions and concentration-dependent micellar growth to the scattering data. The concentration dependence of N_{app} shown in Fig. 1 reflects contributions from the equilibrium size distribution as well as from intermicellar interactions. One can combine both micellar growth and interaction effects and apply renormalization group theory and Monte Carlo simulations to analyze scattering results from micellar solutions [9]. The fits shown as solid lines in Fig. 1 are based on the combination of two assumptions. First we allow for anisotropic particles showing concentration-induced micellar growth, where we assume the well-established fact that for micelles exhibiting one-dimensional growth the molar mass depends with a power law on the surfactant concentration, and where the respective exponent is the so-called growth exponent; for the structure factor $S(0)$ we assume an expression derived from renormalization group theory for polymers [7, 8]. The growth exponents used for the fits are 0.1 (pure alkyl polyglucoside micelles), 0.8 (low hexanol content), 1.1 (medium hexanol content), and 1.2 (high hexanol content), reflecting the dramatically enhanced concentration dependence of the micellar growth in samples where hexanol is added. Figure 1 demonstrates that the combination of these expressions is able to reproduce the experimental data over the entire range of concentration both for the solutions with pure alkyl polyglucosides as well as for the solutions with additional small amounts of hexanol. However, the situation completely changes for the two series with higher hexanol content, where we find much more extended micellar growth and the formation of giant polymer-like micelles. While the scattering data at low concentrations are fully consistent with the presence of polymer-like micelles, the theoretical framework obtained from polymer analogy is not capable of reproducing the data at higher concentrations. The extremely strong decay of N_{app} with increasing concentration is completely inconsistent with polymer theory for semiflexible chains in a good solvent. One possible explanation for the surprising behavior at high surfactant concentration is micellar branching, where the micelles would start to build star polymers rather than linear polymers, with a subsequently modified effective intermicellar interaction behavior. However, to understand this behavior at higher concentrations

additional experimental and theoretical effort is required.

In the next step we try to calculate the full normalized scattering intensity for the samples without and with small amounts of hexanol. Here we use the results from a previously performed systematic Monte Carlo simulation study [12]. Details of these calculations are given elsewhere [9]. The close correspondence between calculation and experimental data shown in Fig. 3 is quite remarkable as it is performed on an absolute scale and relies on the perfect agreement between the SLS and the SANS data. It also demonstrates that the phenomenological expression for the full static structure factor of semiflexible polymers with excluded-volume effects derived from systematic Monte Carlo simulations works even in the limit of very short chains, where we previously lacked an appropriate form that could reproduce the effect of interparticle interactions on the scattering data.

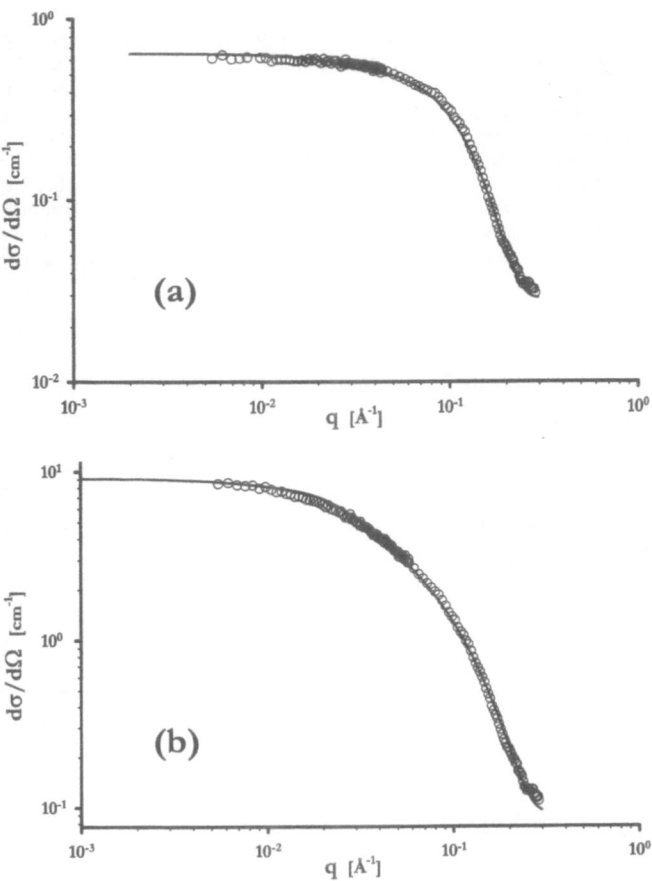

Fig. 3a, b Scattered intensity as a function of q from small-angle neutron scattering experiments. The *open symbols* represent the experimental data and the *full lines* are theoretical calculations based on many-chain Monte Carlo simulations for semiflexible polymers [9, 12]. **a** Pure APG sample (15 wt% APG). **b** APG sample with small amounts of hexanol (10 wt% APG, [hexanol]/[APG] = 0.08)

Conclusion

The present study of the structural properties of these sugar-based surfactants as a function of surfactant concentration and added hexanol demonstrates the enormous effect of hexanol on the resulting micellar size and shape. We have been able to demonstrate that the application of theoretical expressions for the static structure factor derived from Monte Carlo computer simulations of semiflexible polymers with excluded-volume interactions allow us to self-consistently incorporate micellar interactions and growth in a quantitative analysis of the light and neutron scattering data on an absolute scale. In aqueous solutions of the sugar surfactant $C_{8/10}G_{1.5}$ one can finely tune the extent of micellar growth and in particular reach conditions where solutions of relatively short cylinders exist, which grow dramatically into wormlike micelles upon the addition of small amounts of hexanol. While considerable progress has been made in the understanding of interaction effects in solutions of spheres or large polymer-like objects, limited information exists for short flexible cylinders due to the lack of good model systems. The alkyl polyglucoside system provides us now with an ideal set of data for an analysis of micellar growth and interactions in solutions with cylindrical particles of moderate axial ratios.

Acknowledgements We acknowledge the Paul Scherrer Institute Villigen, for providing the neutron research facilities, and we gratefully acknowledge the expert help of our local contact Joachim Kohlbrecher. We are grateful to Wolfgang von Rybinski from Henkel, who supplied us with the alkyl polyglucoside $C_{8/10}G_{1.5}$ used in this study. This work was supported by the Österreichischer Fonds zur Förderung der wissenschaftlichen Forschung under grant P11778-CHE.

References

1. Hill K, von Rybinski W, Stoll G (1997) Alkylglycosides. VCH, Weinheim
2. Cates ME, Candau SJ (1990) J Phys Condens Matter 2:6869
3. Jerke G (1997) PhD thesis. Swiss Federal Institute of Technology, Zurich
4. Stradner A, Mayer B, Sottmann T, Hermetter A, Glatter O (1999) J Phys Chem B 103:6680
5. Schurtenberger P (1996) Curr Opin Colloid Interface Sci 1:773
6. Schurtenberger P, Cavaco C (1994) Langmuir 10:100
7. Schurtenberger P, Cavaco C (1993) J Phys II 3:1279
8. Schurtenberger P, Cavaco C (1993) J Phys II 4:305
9. Jerke G, Pedersen JS, Egelhaaf SU, Schurtenberger P (1997) Phys Rev E 56:5772
10. Jerke G, Pedersen JS, Egelhaaf SU, Schurtenberger P (1998) Langmuir 14:6013
11. Pedersen JS, Schurtenberger P (1996) Macromolecules 29:7602
12. Pedersen JS, Schurtenberger P (1999) Europhys Lett 45:666

Progr Colloid Polym Sci (2000) 115:15–19
© Springer-Verlag 2000

M. Freda
G. Onori
A. Santucci

Hydration and dynamics of bis(2-ethylhexyl)sulfosuccinate microemulsions: effect of change in counterion

M. Freda
Dipartimento di Fisica and Istituto
per la Fisica della Materia
Unità di Tor Vergata
Via della Ricerca Scientifica 1
00133 Rome, Italy

G. Onori (✉) · A. Santucci
Istituto per la Fisica della Materia
Unità di Perugia and Dipartimento
di Fisica, Università di Perugia
Via Pascoli, 06100 Perugia, Italy
e-mail: symbio@pg.infn.it

Abstract We report results of a detailed dielectric and IR study of bis(2-ethylhexyl)sulfosuccinate based micellar systems in CCl_4, where the Na^+ counterion of the surfactant was replaced by the divalent cations Ca^{2+}, Cu^{2+} and Mg^{2+}. In all these micellar systems a relaxation phenomenon was found, whose behaviour strongly depends on the degree of hydration of the micellar aggregate. This relaxation process reflects complex intrinsic dynamics of the reverse micelles, which can be activated above a threshold in the degree of hydration of the same aggregates. The connection between hydration and dynamical properties of the reverse micelles can be considered to be a possible model to explain more complex biological phenomena, for example, the starting of protein functionality above a defined degree of hydration.

Key words Dielectric · Infrared · Dynamics · Bis(2-ethylhexyl)sulfo-succinate

Introduction

The water dissolved in reverse micelles is similar, in many respects, to the interfacial water near biological membranes or protein surfaces. For this reason, it is interesting to study the characteristics of water confined inside a reverse micelle as a model of specific water in biological systems. Among the surfactants capable of forming reverse micelles in apolar solvents, the most widely investigated in the literature is sodium bis(2-ethylhexyl)sulfosuccinate (NaAOT). This surfactant can form quasispherical, nanometre-sized reverse micelles in a large variety of nonpolar solvents.

Recently, we reported results [1–3] of a detailed study concerning dielectric and IR properties of diluted samples of NaAOT reverse micelles as a function of the water content inside their internal core. A relaxation process was observed, reflecting some properties inherent of the single particle dynamics, whose behaviour strongly depends on the micellar degree of hydration.

More recently [4–6], we investigated the effect of changing the Na^+ counterion of NaAOT for divalent cations (Ca^{2+}, Cu^{2+}, Mg^{2+}) on the hydration mechanisms and dynamical properties of AOT-based reverse micelles. A close connection between the behaviour of the relaxation time versus W and the progressive hydration of AOT polar head groups was found, similar to the case of the NaAOT micellar system. These results have been successfully interpreted in terms of two coexisting diffusion mechanisms: the reorientation of the whole micellar aggregate and the rotational diffusion of the completely hydrated AOT head groups.

The connection between the degree of hydration and the dynamical properties of the micelles could represent a model to explain some biologically relevant phenomena, such as the role of the hydration water in the starting of protein functionality.

The main results of these investigations are reviewed and are discussed with reference to possible biological aspects.

Results and discussion

The complex dielectric function of AOT reverse micelles in CCl₄ samples was measured by a frequency-domain coaxial technique in the range 0.02–3 GHz. Measurements were performed at a fixed volume fraction of the dispersed phase ($\Phi = 0.1$) by varying the ratio, W, between water and AOT concentrations. The sample preparation and the measuring techniques were described in Refs. [7, 8].

The dielectric spectrum of a NaAOT/H₂O/CCl₄ sample at $\Phi = 0.1$ and $W = 10$ is shown in Fig. 1. Two distinct relaxation phenomena are present in the spectrum and the frequency dependence of the complex dielectric function $[\varepsilon^*(\omega) = \varepsilon'(\omega) - i\varepsilon''(\omega)]$ can be described in terms of Cole–Cole and Debye-type relaxation processes according to the equation

$$\varepsilon^*(\omega) = \varepsilon_\infty + \frac{\Delta\varepsilon_1}{1 + (i\omega\tau_1)^{1-\alpha}} + \frac{\Delta\varepsilon_2}{1 + i\omega\tau_2} \ , \qquad (1)$$

where ε_∞ is the high-frequency dielectric constant, ω is the angular frequency of the applied electric field, $\Delta\varepsilon_1$ and $\Delta\varepsilon_2$ are the low- and high-frequency dielectric increments, respectively, τ_1 and τ_2 are the relaxation times of the two processes and α is a parameter characterising the width of the relaxation time distribution around τ_1.

The best-fit curves of the experimental spectra are reported in Fig. 1 together with the Cole–Cole and Debye-type contributions. The Debye dispersion is located at higher frequencies in the region of the relaxation of bulk water . In our experimental range of frequency it makes only a little contribution and this contribution was found to increase with the degree of hydration. On these grounds this relaxation process was attributed to the reorientation of bulk water inside the reverse micelles. Attention was directed to the low-frequency relaxation phenomenon characterised by the dielectric parameters τ_1, $\Delta\varepsilon_1$ and α.

The trends of the relaxation time τ_1 for NaAOT, Cu(AOT)₂ and Ca(AOT)₂ reverse micelles in CCl₄, are shown in Fig. 2 as a function of water content at $\Phi = 0.1$.

For each series the experimental values of τ_1 exhibit similar trends as a function of the degree of hydration, i.e. τ_1 shows an initial decrease followed by a region in which it is nearly constant. The values of τ_1 for Cu(AOT)₂ and Ca(AOT)₂ are systematically higher than the corresponding ones for NaAOT and are practically coincident, except for the fact that the solubility limit for water in Cu(AOT)₂/CCl₄ micelles is higher than that in the Ca(AOT)₂ system. However, both NaAOT and Cu(AOT)₂ systems shows the same asymptotic behaviour for τ_1 versus W, i.e. the experimental values of τ_1 tend to a value of about 0.2 ns at the solubility limit of water (larger W). The trend of τ_1 for Mg(AOT)₂ samples (not reported here) is very different from those of the other samples [4]. In fact, the values of τ_1 for the Mg(AOT)₂ system are systematically higher than those of NaAOT and of the other surfactants, and a marked

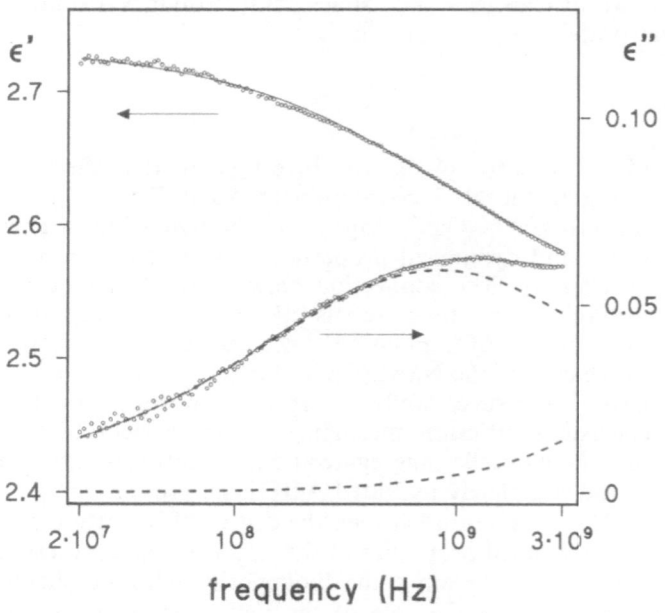

Fig. 1 Real (ε') and imaginary (ε'') parts of the dielectric function of a sodium bis(2-ethylhexyl)sulfosuccinate (*NaAOT*)H₂O/CCl₄ mixture versus frequency. $W = 10$, $\Phi = 0.1$. Experimental points (o); best-fit curve according to Eq. (1) (—). Cole–Cole and Debye-type contributions (- - -) to the best fit of $\varepsilon''(\omega)$ are also shown

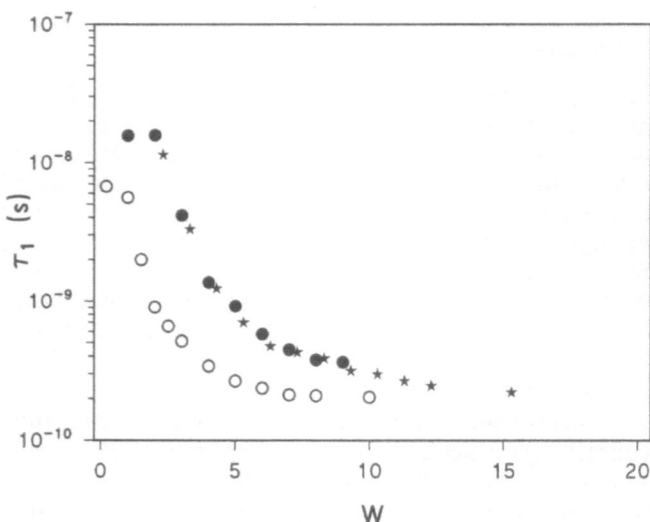

Fig. 2 Relaxation time (τ_1) versus W for M^x (AOT)$_x$/micellar systems ($\Phi = 0.1$). NaAOT/H₂O/CCl₄ (O); Ca(AOT)₂/H₂O/CCl₄ (●); Cu (AOT)₂/ H₂O/CCl₄ (★)

effect of particle volume fraction is observed. This effect can be ascribed to the presence of attractive interactions between reverse micelles.

Referring to Fig. 2, the analogies between the trends of τ_1 versus W for NaAOT, Ca(AOT)$_2$ and Cu(AOT)$_2$ samples allow the dielectric data to be treated with the same quantitative model first reported for NaAOT [1–3] and successively applied to Ca(AOT)$_2$ [5] and Cu(AOT)$_2$ [6] reverse micelles. In this model, one supposes that at the lowest water content the almost dehydrated reverse micelle has a nearly rigid structure and, therefore, the observed dielectric relaxation process is ascribed to the reorientation of the whole micelle in the applied electric field. The relaxation time τ_1 is then related to the radius, R, of the aggregate by the Debye–Stokes formula:

$$\tau = \frac{4\pi\eta R^3}{k_B T} \quad , \tag{2}$$

where $k_B T$ is the thermal energy and η is the viscosity of solution. From Eq. (2) the estimated radii of the NaAOT, Ca(AOT)$_2$ and Cu(AOT)$_2$ micellar aggregates are in agreement with the values found in the literature [9].

With increasing water content inside the micelle, R increases almost linearly with W, so τ_1 is expected to increase approximately with W^3, according to the Debye–Stokes law (see Eq. 2). In contrast, for all micellar systems τ_1 initially decreases and then it is nearly constant as a function of W (Fig. 2). The quantitative model applied to NaAOT, Cu(AOT)$_2$ and Ca(AOT)$_2$/H$_2$O/CCl$_4$ dielectric data states that for each value of W there exists a fraction $[1-X(W)]$ of AOT head groups which reorientate rigidly with the whole micelle with the Debye diffusion time, $\tau_D(W)$, calculated from Eq. (2), and a fraction $X(W)$ of "free" AOT head groups which reorientate in the applied electric field independently with respect to the micellar aggregate. This reorientation is characterised by a relaxation time $\tau_0 \ll \tau_D(W)$, corresponding to the limiting value of τ_1 at the highest W. The configuration depicted has an intrinsic dynamical nature, because each AOT group experiences fluctuations in its local aqueous environment, the characteristic time of which is comparable to the mean residence time of a water molecule in the hydration shell of each ion [10]. Since it is found experimentally that this fluctuation rate is higher with respect to the intrinsic rates of the observed process [10] (τ_D and τ_0) we applied the so-called fast-exchange condition [11] to express the experimental relaxation time τ_1 in terms of τ_0 and τ_D:

$$\frac{1}{\tau_1} = \frac{X(W)}{\tau_0} + \frac{1-X(W)}{\tau_D} \quad . \tag{3}$$

Starting from Eq. (3) the fraction $X(W)$ of the free AOT head groups for NaAOT, Ca(AOT)$_2$ and Cu(AOT)$_2$ reverse micelles was calculated by inserting into Eq. (3)

the experimental values of τ_1, the values of τ_D calculated according to the Debye formula (Eq. 2) and, finally, a value of τ_0 of 1.9 ns (the same for the three systems). The values of $X(W)$ obtained from Eq. (3) are reported in Fig. 3 for NaAOT, Ca(AOT)$_2$ and Cu(AOT)$_2$. In all the samples examined, $X(W) = 0$ at the lowest values of W

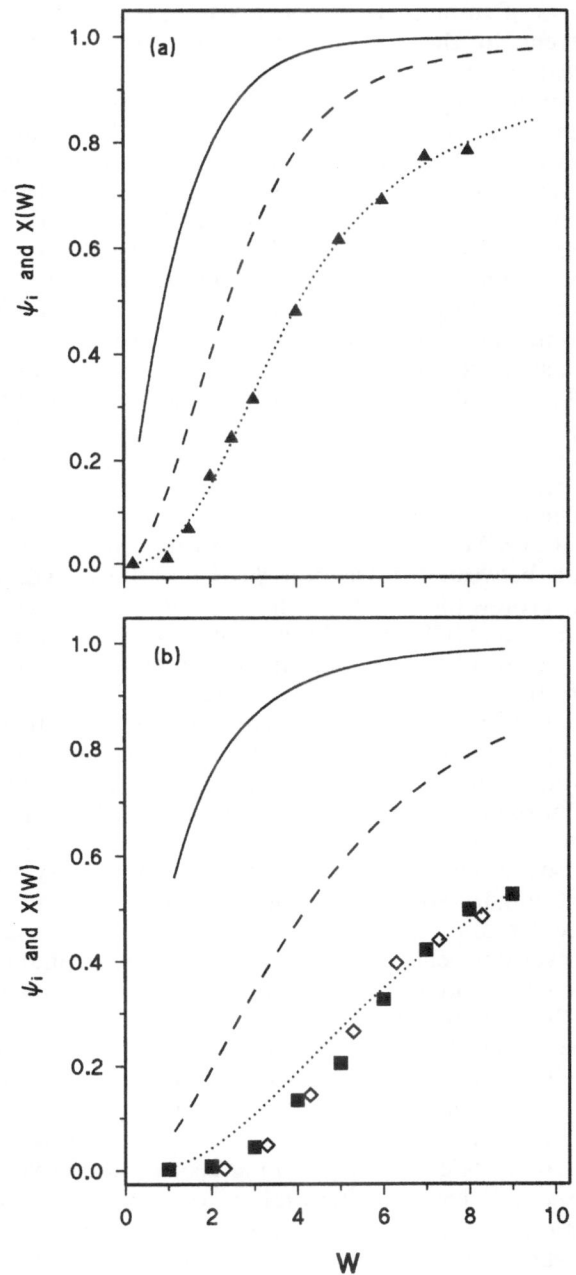

Fig. 3a, b Plots of the fraction, ψ_i, of AOT head groups with at least one (—), two (- - -) or three (· · ·) water molecules. **a** Fraction $X(W)$ of free AOT head groups of the NaAOT system (▲) calculated from dielectric data (Eq. 3) **b** Fraction $X(W)$ of free AOT head groups of the Ca(AOT)$_2$ (■) and Cu(AOT)$_2$ (◇) systems

and tends to increase above a threshold value of W, which depends on the nature of the counterion [$W \cong 1$ for NaAOT, $W \cong 2$ for Ca(AOT)$_2$ and Cu(AOT)$_2$]. It is evident that the values of $X(W)$ for Ca(AOT)$_2$ and Cu(AOT)$_2$ are practically coincident and are lower than that calculated for NaAOT.

The hydration properties of the reverse micelles were also studied by means of IR spectroscopy. The IR spectra of surfactant-entrapped water [5, 6, 12] are very different to the corresponding one for bulk water, indicating that the water in the reverse micelles lacks the normal hydrogen-bond network present in pure water. The spectra of water encapsulated in NaAOT and Ca(AOT)$_2$ reverse micelles were analysed assuming the existence of a continuous equilibrium between two spectroscopically distinguishable water species, i.e. the "bulk" and the "hydration" water. The IR data for these systems were interpreted in terms of the existence of three binding sites per AOT group that could be filled by water molecules and they were treated assuming a three-stage equilibrium model to construct a hydrated shell around the surfactant polar heads. From this three-stage equilibrium model the fractions of AOT head groups hydrated with at least one (ψ_1), two (ψ_2) or three (ψ_3) water molecules were calculated.

The fractions ψ_1, ψ_2 and ψ_3 are plotted for NaAOT and Ca(AOT)$_2$ in Fig. 3. A surprising coincidence is found between the fraction $\psi_3(W)$ of fully hydrated AOT groups (derived from the analysis of IR data) and the fraction $X(W)$ of AOT head groups which reorientate independently in the applied electric field (from dielectric measurements). This coincidence indicates that AOT head groups with sufficient mobility to contribute separately to the observed relaxation process can be identified with the totally hydrated AOT head groups.

Our data suggest that the effect of the hydration is to dilute the interactions between charged groups in the reverse micelle, enhancing their individual mobility. This result might have a relevance that goes beyond the particular case of dynamics–hydration connections in the reverse micelles. In fact, water plays an essential role in starting and maintaining the protein functionality, but this occurrence is still not clear, especially at the molecular level. Experimental evidence that water affects protein functionality arises, for example, from measurements of enzyme activity of hydrated lysozyme powders [13, 14]. Figure 4a shows the trend of the logarithm of the activity rate of lysozyme in aqueous solution versus the ratio, α, between the total amount of water in the sample and the amount of water (0.38 g protein) that corresponds to the formation of a complete hydration monolayer around the protein $\left(\alpha = \frac{\text{grams of water}}{0.38 \text{ g protein}}\right)$. The enzyme activity is negligible at the lowest degree of hydration ($0 < \alpha < 0.5$); the activity increases on increasing the water content ($0.5 < \alpha < 1.5$) and at $\alpha > 1.5$

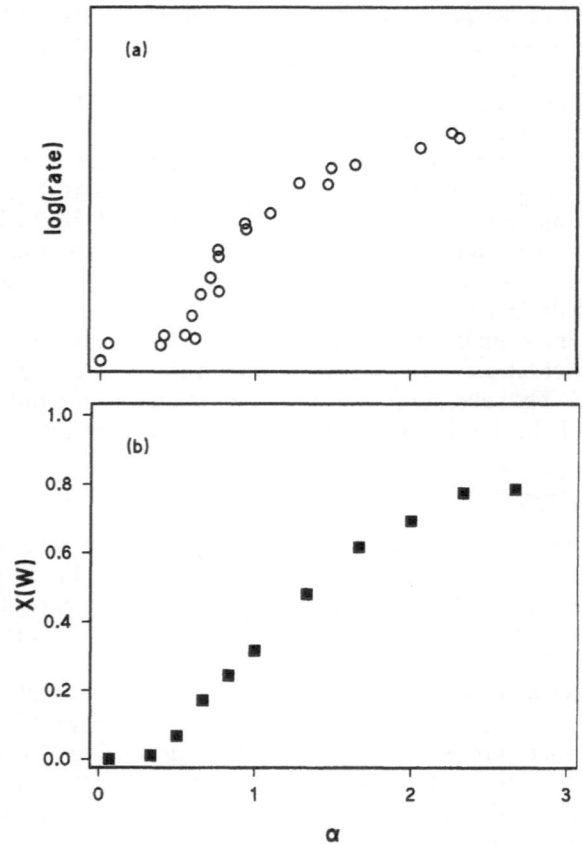

Fig. 4 a Enzymatic activity for lysozyme in the solid phase as a function of α (see text). Reproduced from Rupley et al. [14]. **b** Plot of the fraction $X(W)$ for NaAOT/H$_2$O/CCl$_4$ samples versus $\alpha = W/(W_{\text{Max}} = 3)$

it increases again, approaching the value corresponding to dilute lysozyme–water solution.

It is noteworthy that this behaviour of lysozyme activity versus α closely resembles that observed for the fraction $X(W)$ of AOT head groups which reorientate independently in an external electric field. For comparison, in Fig. 4b the trend of $X(W)$ calculated for NaAOT micelles is plotted versus α, where $\alpha = W/W_{\text{max}}$ is now the ratio between the amount of water dissolved in the micelles and the amount of water that corresponds to the complete hydration of an AOT head group ($W_{\text{max}} = 3$). For both $X(W)$ and the logarithm of activity rate there is a threshold of hydration ($\alpha \sim 0.5$) beyond which the two quantities becomes significantly different from 0. These analogies support the hypothesis that the presence of water in protein systems may have a major effect on the protein dynamical flexibility, which is in turn essential for their functionality. This intrinsic dynamics activates only when the hydration process is completed and it can explain the starting of lysozyme activity with a threshold degree of hydration.

References

1. D'Angelo M, Fioretto D, Onori G, Palmieri L, Santucci A (1995) Phys Rev E 52:R4620
2. D'Angelo M, Fioretto D, Onori G, Palmieri L, Santucci A (1996) Phys Rev E 54:993
3. D'Angelo M, Fioretto D, Onori G, Santucci A (1998) Phys Rev E 58:7657
4. Fioretto D, Freda M, Onori G, Santucci A (1997) Prog Colloid Polym Sci 105:256
5. Fioretto D, Freda M, Mannaioli S, Onori G, Santucci A (1999) J Phys Chem B 103:2631
6. Fioretto D, Freda M, Onori G, Santucci A (1999) J Phys Chem B 103:8216
7. Fioretto D, Marini A, Massarotti M, Onori G, Palmieri L, Santucci A, Socino G (1993) J Chem Phys 99:8115
8. Eastoe J, Fragneto G, Robinson BH, Towey TF, Heenan RK, Lang FJ (1992) J Chem Soc Faraday Trans I 88:461
9. Maitra A (1984) J Phys Chem 88:5122
10. Giese K (1972) Ber Bunsenges Phys Chem 76:495
11. Anderson JE (1967) J Chem Phys 47:4879
12. Onori G, Santucci A (1993) J Phys Chem 97:5430
13. Rupley JA, Yang PH, Tollin G (1980) In: Rowland SP (ed) Water in polymers: thermodynamic and related studies of water interactions with proteins. ACS Symposium Series 127. American Chemical Society, Washington, D.C., pp 111–132
14. Rupley JA, Gratton E, Careri G (1983) TIBS 18

Progr Colloid Polym Sci (2000) 115: 20–24
© Springer-Verlag 2000

M. Freda
G. La Penna
V. Minicozzi
S. Morante
G. Salina

Molecular dynamics and hybrid Monte Carlo simulations of a sodium bis(2-ethylhexyl)-sulfosuccinate reverse micelle

M. Freda · V. Minicozzi
S. Morante (✉) · G. Salina
Dipartimento di Fisica
Università di Roma "Tor Vergata"
Via della Ricerca Scientifica
00133 Rome, Italy
e-mail: morante@roma2.infn.it
Tel.: +39-6-72594554
Fax: +39-6-2023507

G. La Penna
Istituto di Studi Chimico-Fisici di
Macromolecole Sintetiche e Naturali
CNR, Via De Marini 6
16149 Genoa, Italy

M. Freda · V. Minicozzi · S. Morante
INFM, Unità di Roma 2

G. Salina
INFN, Sezione di Roma 2

Abstract The bis(2-ethylhexyl)sulfosuccinate (AOT) reverse micelle is a nanometer-sized quasispherical aggregate that can dissolve water in its interior. This feature makes a reverse micelle a valuable model system for the study of "hydration water", i.e. water confined in the proximity of polar and ionic groups. We present a first study of the structural and thermodynamical properties of a system formed by 2048 CCl_4 molecules in which an AOT reverse micelle, including its internal hydration water and Na^+ counterions, is contained. CCl_4 equilibrium configurations are obtained from a very long hybrid Monte Carlo simulation of a cubic box of 256 CCl_4 molecules which is then duplicated in all three dimensions. The whole system is successively equilibrated using a standard molecular dynamics simulation algorithm.

Key words Reverse Micelle · Bis(2-ethylhexyl)sulfosuccinate · Molecular dynamics · Hybrid Monte Carlo

Introduction

Among the surfactants capable of forming micellar aggregates in apolar solvents the most widely studied is sodium bis(2-ethylhexyl) sulfosuccinate (AOT). The AOT reverse micelle is a nanometer-sized, quasi-spherical aggregate that is capable of dissolving water inside itself, forming the so-called "water pool" in its interior. This peculiar property makes reverse micelles a unique model for the study of the physicochemical properties of "hydration water", i.e. of water confined in the proximity of polar and ionic centers. There is almost universal agreement concerning the "qualitative" difference between "hydration" and "bulk" water, but many quantitative aspects of both the formation processes and the structural properties of the hydration region still remain to be clarified [1, 2]. In this context, the atomic structural detail reachable by numerical simulations appears to yield information of no value. To our knowledge, in the few studies based on numerical simulations available in the literature, reverse micelles have never been described at the level of atomic detail [3].

We have performed numerical simulations of an AOT reverse micelle (of realistic dimensions) in a CCl_4 solution, where the hydrophilic head of the AOT molecule has been modeled in its full atomic detail. We present the results of a hybrid Monte Carlo (HMC) simulation of 256 CCl_4 molecules in a cubic box with periodic boundary conditions (PBCs). We decided to use the HMC algorithm as it is expected to be able to explore the system phase space in a much better way than standard molecular dynamics (MD). We have implemented a new version of the HMC algorithm that employs the multiple time step (MTS) [4, 5] integration algorithm in generating trial configurations via MD moves. Bulk properties (such as specific heat, internal energy, etc.) of the CCl_4 solvent, as well as its structural properties (for example, its radial distribution function (RDF)), are correctly reproduced. Structural data and thermodynamical quantities extracted from our simulations are found to be in good agreement with previous simulations and with experimental results. We also present the construction of a system consisting of a cubic box of CCl₄ of volume 70 Å³ in which an AOT reverse micelle is immersed. In the interior of the micelle, water molecules and an appropriate number of Na⁺ counterions have been inserted. The whole aggregate has been equilibrated using standard MD simulation techniques. We have checked that, during equilibration, all the structural properties of the aggregate remain unaltered. This proves the reliability of the force field we have employed and of our simulation procedure. The next step of this investigation will be to perform an adequately long simulation of the whole reverse micelle–CCl₄ system with the purpose of studying in detail its physicochemical properties.

The code

The problem of simulating a complex system with a standard Monte Carlo algorithm is that trying to update more than one particle in a single move usually results in a prohibitively low acceptance rate. The HMC algorithm [6, 7] overcomes this problem as it combines the ability of the Monte Carlo method in generating canonically distributed configurations with the possibility of performing, in an efficient way, nonlocal collective moves typical of MD. As is well known, a sufficient condition to produce configurations with the required Maxwell–Boltzmann probability distribution is that the acceptance probability of the Metropolis criterion obeys the "detailed balance" condition. The latter is easily implemented in HMC algorithms as it is enough to have a MD integration scheme which is time-reversible and area-preserving.

A HMC move consists of the following steps:

1. Given the initial positions, $\mathbf{r}$, of the particles, their initial momenta, $\mathbf{p}$, are taken as Gaussian-distributed variables at the working temperature.
2. n steps of MD are performed and the last configuration, $(\mathbf{r'}, \mathbf{p'})$, is taken as the new trial configuration.
3. The new configuration, $(\mathbf{r'}, \mathbf{p'})$, is accepted or rejected on the basis of the standard Metropolis criterion.
4. Go back to 1.

It has been shown [8] that HMC algorithms allow fairly fast collection of large sets of uncorrelated, canonically distributed configurations, thus substantially reducing, with respect to a straight MD simulation, the problem of an incomplete sampling of the system "ensemble". In our version the global moves are performed by a MD code in which the classical equations of motion are integrated using the so-called MTS algorithm. The MTS algorithm is a flexible integration method based on a procedure in which the evolution of the system is

followed with time steps of different magnitude, according to how rapidly a given type of interaction makes coordinates and momenta change in time.

To explain how the MTS algorithm works, we recall that the potential $V(\mathbf{r})$ can be naturally subdivided into two parts:

$$V(\mathbf{r}) = V_I(\mathbf{r}) + V_E(\mathbf{r}) \ , \qquad (1)$$

where $V_I(\mathbf{r})$ and $V_E(\mathbf{r})$ are the intramolecular ("fast") and intermolecular ("slow") part of the potential, respectively. Following this subdivision, the classical Liouville operator can be split into a "fast" and a "slow" part by the repeated use of the Trotter formula, as discussed in detail in Ref. [9]. The "slow" degrees of freedom are updated with time steps of the order of a few femtoseconds, while the "fast" degrees of freedom are updated with smaller time steps (of the order of a few tenths of a femtosecond). For completeness, we note that short-range potentials are driven to zero beyond $R_{ij} > 9$ Å, while the long-range Coulomb potential is computed by using the so-called smooth particle mesh Ewald (SPME) [10] summation method. The SPME tolerance, which is estimated as in Ref. [10] by evaluating the difference between actual and approximated Coulomb energies, was set to 10^{-4}.

Results and discussion

HMC simulations of CCl₄

We started by simulating a system of 256 CCl₄ molecules at 298 K contained in a cubic box with PBCs using a HMC of the type described in the Introduction and expressly developed by us for the purpose of this investigation. The side of the box was 35 Å; in this way we obtained a system with the density of the liquid CCl₄ [11] at the chosen simulation temperature. MD global moves were performed by solving the equations of motion with the force field taken from Chang et al. [12]. The whole CCl₄ simulation consists of two consecutive parts in which different choices of the time steps are used in the MTS integration algorithm. The first part of the simulation can be considered as a kind of equilibration, which is not included in the following analysis. Structural and thermodynamical data are extracted from the subsequent 540,000 configurations generated by our HMC algorithm.

The RDFs of the C—C, C—Cl and Cl—Cl pairs obtained from the HMC simulation data are shown in Fig. 1

In all the RDFs we obtained, both the position and the magnitude of the peaks are consistent with similar results reported in Refs. [12, 13]. The simulation data are also in good agreement with experimental measurements [14]. Looking in particular at the C—C pair distribution

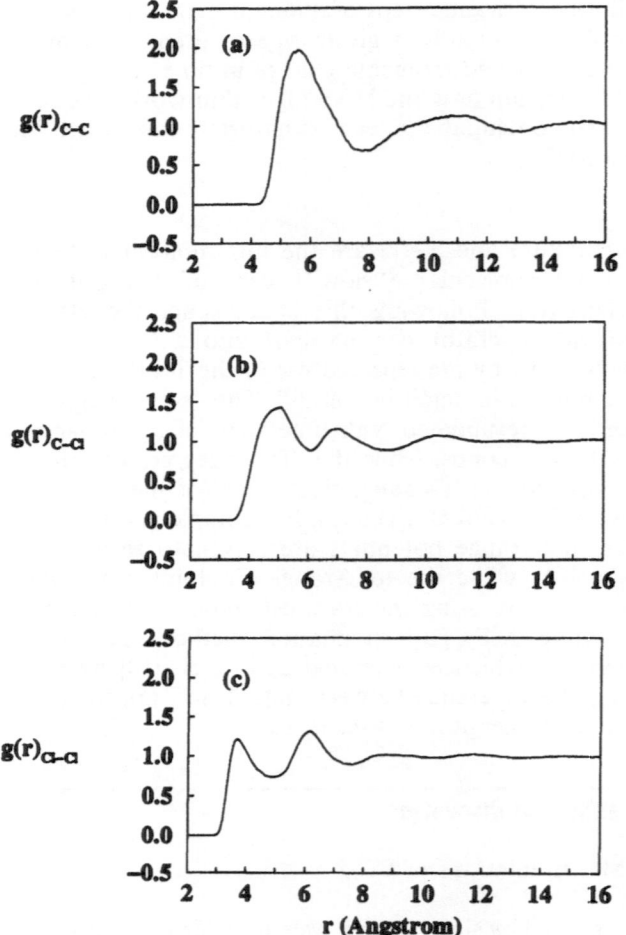

Fig. 1a–c Radial distribution function $g(r)$, computed from CCl$_4$ hybrid Monte Carlo simulation data. **a** $g(r)_{C-C}$; **b** $g(r)_{C-Cl}$; **c** $g(r)_{Cl-Cl}$

Table 1 Molar specific heat and enthalpy of vaporization from the hybrid Monte Carlo (*HMC*) simulation and comparison with experimental results

Specific heat, C (cal/mol K)		Vaporization enthalpy, ΔH_{vap} (kcal/mol)	
HMC	Exp. [11]	HMC	Exp. [15, 16]
28 ± 4	31.49 ± 0.01	7.53 ± 0.03	7.79 ± 0.01

$$\Delta H_{vap}^{sim} = -E_{pot} - E_{corr} + RT \ , \qquad (2)$$

where R is the gas constant, E_{pot} is the computed total intermolecular potential energy at 298 K and E_{corr} is the correction coming from our cutting off the Lennard-Jones potential at distances of 9 Å and greater. We find $E_{pot} = -6.92 \pm 0.03$ kcal/mol, $E_{corr} = -0.018 \pm 0.001$ kcal/mol and $RT = 0.59$ kcal/mol. ΔH_{vap}^{sim} from Eq. (2) compares very nicely with the experimental value of the vaporization enthalpy [15, 16].

MD simulations of an AOT reverse micelle

A very important novelty of the present approach lies in the way we have modeled the AOT surfactant molecule. The hydrocarbon tails are modeled, as usual, by using the "united atoms" approximation, while the detailed atomic structure of the sulfonic polar head has been fully implemented in our code for the first time. The force field describing the interactions of the hydrocarbon tails is obtained by combining the (OPLS) potentials [17] with the potentials developed by Wilson and Pohorille [18] for phospholipidic molecules. The head group force field was obtained by appropriately modifying the potential developed by Schweighofer et al. [19] for the sulfate head group of sodium dodecyl sulfate surfactant. Improper dihedrals are used to maintain the AOT molecule in a well-defined chirality state and to preserve the planar structure of the C=OO groups. Furthermore, since the sulfonic head group must be allowed to rotate freely around the CH—S bond, the torsional potentials for dihedrals, including S, are set to zero. The Lennard-Jones and Coulomb parameters used in the force field of the sulfonic head group are reported in Table 2.

Once the force field of the AOT molecule has been chosen, the next step is to build up the micelle aggregate. This is done more or less manually by using, in a strongly interactive way, the INSIGHT [20] package.

The micelle itself is made of 20 AOT molecules assembled in a quasi-spherical configuration (Fig. 2) with the heads facing the interior of the aggregate. The distance between two opposite sulfur atoms is between 12.5 and 13 Å.

In order to neutralize the negative charge of the AOT molecule, a Na$^+$ counterion is put, at a distance of about 3 Å, opposite each sulfonic group. The water

(Fig. 1a), we notice that, as in Ref. [12], the first peak is centered at about 5.8 Å and the area under the peak corresponds to a coordination number of 12. This is the result expected for a close-packed monoatomic liquid [12, 13]. The peak that is known to be present around 20–25 Å is not visible in our data because the use of PBCs limits the range of explored distances to half the side of the box, i.e. to about 17 Å in our case. Peaks corresponding to the Cl—C and Cl—Cl pair correlation, which are expected to be seen at shorter distances [12, 13], are very visible in Fig. 1b and c, respectively.

The values of the two thermodynamic quantities we computed, C_v and ΔH_{vap}, are reported together with the corresponding experimental data [11, 15, 16] in Table 1. The computed specific heat at constant volume, C_v^{sim}, is compared with the measured specific heat at constant pressure, C_p^{exp}, assuming, as in the case of a non-compressible liquid, that $C_p = C_v$. C_v^{sim} and C_p^{exp} coincide within the error limit. ΔH_{vap}^{sim} is evaluated using the relation

Table 2 The constants used in the sulfonic polar head force field. σ and ε are the constants appearing in the Lennard-Jones potential; q are the partial charges of the listed groups in units of electron charge

Group	σ (Å)	ε (kcal/mol)	$q(e)$
CH_3	3.906	0.175	0.000
CH_2	3.906	0.118	0.000
O (ester)	3.000	0.170	−0.400
S	3.550	0.250	1.562
O (carbonyl)	2.960	0.210	−0.450
O (SO_3)	2.960	0.210	−0.654
CH	3.850	0.080	0.000
C	3.750	0.105	0.550
IP(Na^+)	3.328	0.00277	1.000

inside the reverse micelle is finally added by the following procedure:

1. A HMC simulation (about 3×10^5 configurations were collected) of 512 TIP3P [21] water molecules is performed at 298 K and at a density of 0.976 g/cm^3.
2. A few tens of water molecules, whose relative positions are taken from the HMC simulation, are inserted inside the aggregate.
3. Water molecules whose distance either from any AOT atom or from any Na$^+$ ion happens to be less than 1 Å are eliminated. At the end of this operation, the number of water molecules left over is 22. This leads to a ratio between water and surfactant concentration of about 1.
4. The CCl$_4$ configuration that comes out from the simulation described in the previous sections is

replicated in all three directions and a new, larger box with a volume of 70 Å^3, containing 2048 CCl$_4$ molecules, is obtained.

5. The whole micellar aggregate (AOT + Na$^+$ + water) is inserted in the larger CCl$_4$ box by digging a hole in the liquid and throwing away all CCl$_4$ molecules whose distance from any of the aggregate atoms is less than 2.5 Å.

A delicate issue is the equilibration of the very complicated system we have assembled. This was done by a standard MD simulation using a single (as opposed to a multiple) time step integration algorithm with $\delta t = 0.5$ fs. In order to keep the system at a fixed temperature (298 K) we appropriately rescaled the velocities of the particles at each time step. The equilibration of the system was carried out for 20 ps. During the equilibration the structure of the aggregate was continuously monitored. We checked that, remarkably, the system remains structurally stable, giving us confidence that our choice of the force field and all our procedures is very reliable.

Conclusions and outlook

The efficiency of the Monte Carlo method in exploring the configurational space of a system can be greatly enhanced in the difficult case of very complex systems (such as micelles) if classical MD is used to generate the trial configurations that are then subjected to the Metropolis test. This variant of the Monte Carlo method goes under the name of HMC. We have developed a new HMC code in which the global MD moves are performed by exploiting the widely appreciated virtues of the MTS algorithm for the integration of the classical Newton equations of motion.

We have successfully tested our HMC code by performing a large-scale simulation of a system of 256 CCl$_4$ molecules in which we generated up to 540,000 configurations. Thermodynamical and structural quantities extracted from this configuration sample are found to be in very good agreement with results obtained in previous simulations and with experimental data. As a second step of this investigation, a system made of a reverse micelle (of realistic dimensions) with inclusion of hydration water and counterions, immersed in a box of 2048 CCl$_4$ molecules, was prepared and fully equilibrated. After this very long preparation the final step of our program will be to perform adequately long simulations of the whole aggregate by both using classical MD and the HMC algorithm.

Acknowledgements This work was partially supported by the Istituto Nazionale per la Fisica della Materia. We thank G.C. Rossi for discussion and for reading the manuscript.

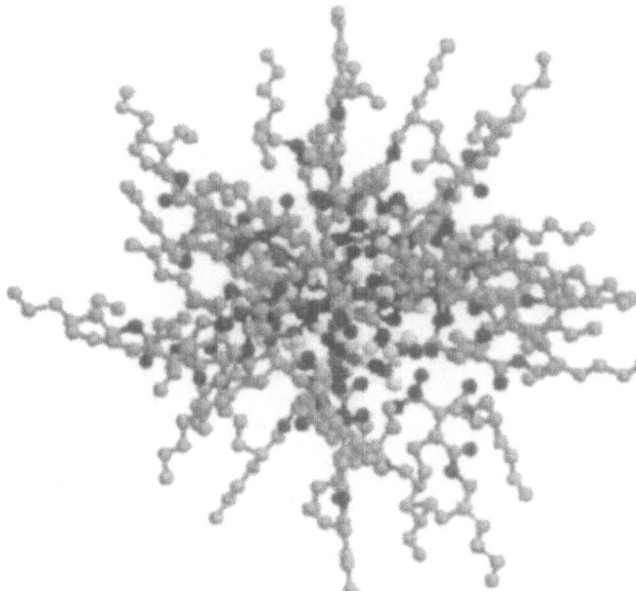

Fig. 2 Picture of the assembled bis(2-ethylhexyl)sulfosuccinate reverse micelle

References

1. Wong M, Thomas JK, Gratzel M (1976) J Am Chem Soc 98:2391
2. Hauser H, Haering G, Pande A, Luisi PL (1989) J Phys Chem 93:7869
3. Brown D, Clarke JHR (1988) J Phys Chem 92:2881
4. Tuckermann M, Martyna GJ, Berne BJ (1992) J Chem Phys 97:1990
5. Procacci P, Berne BJ (1994) J Chem Phys 101:2421
6. Duane S, Kennedy BJ (1987) Phys Lett B 195:216
7. Mehlig B, Heermann DW, Forrest BM (1992) Phys Rev B 45:679
8. Zhou RH, Berne BJ (1997) J Chem Phys 107:9185
9. La Penna G, Minicozzi V, Morante S, Rossi GC, Salina G (1997) Comput Phys Commun 106:53
10. Essmann U, Perera L, Berkowitz ML, Darden T, Lee H, Pedersen LG (1995) J Chem Phys 103:8577
11. Weast RC (1982–83) (ed) Handbook of chemistry and physics. CRC, Boca Raton
12. Chang TM, Peterson KA, Dang LX (1995) J Chem Phys 103:7502
13. McDonald I, Bounds DG, Klein ML (1982) Mol Phys 45:521
14. Narten AH (1976) J Chem Phys 65:573
15. Timmermans J (1975) Physico-chemical constants of pure organic compounds, Elsevier, Amsterdam
16. Majer V, Svab L, Svibida V (1980) J Chem Thermodyn 12:843
17. Jorgensen WL, Maxwell DS, Tirado-Reves J (1996) J Am Chem Soc 118:11225
18. Wilson MA, Pohorille A (1994) J Am Chem Soc 116:1490
19. Schweighofer KJ, Essmann U, Berkowitz ML (1997) J Phys Chem B 101:3793
20. MSI (1998) INSIGHT. MSI, San Diego
21. Jorgensen WL, Chandrasekhar J, Madura JD (1983) J Chem Phys 79:926

Progr Colloid Polym Sci (2000) 115 : 25–30
© Springer-Verlag 2000

M. Giustini
G. Palazzo
A. Ceglie
J. Eastoe
A. Bumujdad
R. K. Heenan

Studies of cationic and nonionic surfactant mixed microemulsions by small-angle neutron scattering and pulsed field gradient NMR

M. Giustini (✉)
Dipartimento di Chimica
Università "La Sapienza"
P.le A. Moro 5, 00185 Rome, Italy
e-mail: giustini@axrma.uniroma1.it
Tel.: +39-6-49913336
Fax: +39-6-490321

G. Palazzo
Dipartimento di Chimica,
Università di Bari, Bari, Italy

A. Ceglie
C.S.G.I., c/o Dept. Food Technology
(DISTAAM), Università del Molise
Campobasso, Italy

J. Eastoe · A. Bumujdad
School of Chemistry
University of Bristol, Bristol, UK

R. K. Heenan
ISIS Facility
Rutherford Appelton Laboratory
Chilton, UK

Abstract For mixed surfactant films in water-in-oil microemulsions, insight into the interfacial structures and compositions has been obtained by combining data from pulsed field gradient NMR (PFG-NMR) and small-angle neutron scattering (SANS). As an example of this approach we report a preliminary study on water in n-heptane microemulsions stabilised by mixtures of didodecyldimethylammonium bromide and pentaethylene glycol monododecyl ether. The results deduced from both methods are consistent with spherical microemulsion droplets stabilised by a mixed surfactant film. It is shown that the combination of SANS and PFG-NMR is extremely powerful for investigating these multi-component systems. In particular, dynamic studies by PFG-NMR are useful for validating models for SANS data.

Key words Small-angle neutron scattering · Pulsed field gradient NMR · Didodecyldimethylammonium bromide · Pentaethylene glycol monodecyl ether · Microemulsions

Introduction

Surfactant mixtures often give rise to enhanced performance over the individual components and so blends are employed in many commercial applications [1]. In aqueous systems mixtures have received much attention [1–4]. In emulsions and microemulsions it is well known that surfactant mixtures can improve solubilisation [5–7]. To understand how surfactant mixtures affect behaviour of curved interfaces direct measurements of film compositions are needed: these can be made with small-angle neutron scattering (SANS) [3–5] and pulsed field gradient NMR (PFG-NMR) [8]. PFG-NMR is a dynamic method which is useful for studying molecular diffusion of individual components in a mixed system and it is particularly useful for discriminating between discrete droplets and bicontinuous structures. On the other hand, SANS is an equilibrium technique which can provide detailed information about internal particle structure and interparticle interactions. Contrast-variation SANS experiments were used to determine the compositions of the interfacial films via the simultaneous analysis of multiple data sets. Here, data from SANS and PFG-NMR are combined to study model mixed surfactant water-in-heptane microemulsions. The main surfactants used were a two-tailed cationic didodecyldimethylammonium bromide (DDAB) and pentaethylene glycol monododecyl ether ($C_{12}E_5$): both have identical C_{12} hydrophobic groups.

The actual composition of the mixed film, in terms of the mole fraction of single-tail surfactant at the interface (X_{film}^1), has also been measured directly by means of both SANS and PFG-NMR in an independent way. It is shown that the combination of the two techniques is extremely powerful for investigating structure and properties in mixed surfactant, water and oil systems. In particular, dynamic studies by NMR are useful for validating models for SANS data, which are strictly equilibrium measurements.

Experimental

The source and the purification procedures of the microemulsion components as well as the sample preparation are thoroughly described in Ref. [5]. Each sample is characterised by the mole fraction of $C_{12}E_5$ in the surfactant mixture [$X^1 = n_{C12E5}/(n_{C12E5} + n_{DDAB})$] and by the volume fractions of water, surfactant and oil, which are ϕ_w, ϕ_s and ϕ_o, respectively. Other useful quantities are the water-to-surfactant molar ratio (W_o) and the dispersed-phase volume fraction ($\phi = \phi_w + \phi_s$).

The NMR self-diffusion experiments (PFG-NMR)[9] were performed at 298 K using the classical Stejskal–Tanner sequence, 90°-τ-180°-τ-echo, with two rectangular field gradient pulses of strength G and duration δ applied within the diffusion time, Δ, on the apparatus described elsewhere [10]. For isotropic Brownian motion the echo amplitude is given by

$$A(\delta) = A(0)\exp^{-[\gamma^2 G^2 D\delta^2(\Delta-\delta/3)]} , \qquad (1)$$

where γ is the magnetogyric ratio of the nucleus under investigation (^{1}H in this instance) and D is the self-diffusion coefficient of the species responsible of the spin-echo decay (D_w, D_{C12E5} and D_{oil} for water, $C_{12}E_5$ and n-heptane, respectively). The G value was checked before each set of experiments by a separate calibration with neat dimethyl sulfoxide (DMSO) ($D_{DMSO}^{25°C} = 7.3 \times 10^{-10}$ m^{2}s^{-1}). The accuracy of the experimentally determined self-diffusion coefficients was always better than 5% (the accuracy of the gradient calibration, however, was always better than 0.5%).

SANS measurements were carried out on the D22 diffractometer at ILL (Grenoble, France), using neutrons with a wavelength of 10 Å, or on the time-of-flight LOQ instrument at ISIS, UK, where incident wavelengths are $2.2 \leq \lambda \leq 10$ Å.[1] The scattering length densities are from previous work [5] and the momentum transfer, Q, ranges were 0.0041–0.362 Å^{-1} on the D22 diffractometere and 0.009–0.22 Å^{-1} at LOQ. Absolute intensities for $I(Q)$ were determined to within ± 5% by standard calibration methods. Detailed accounts of scattering from core–shell particles and application of the FISH SANS analysis programme to multicontrast microemulsion samples have been given elsewhere [5, 11]. For polydisperse spherical particles the general law is

$$I(Q) = n_p \Delta\rho^2[P(Q,R)p(R)]S(Q) , \qquad (2)$$

where n_p is the number density, $P(Q,R)$ is a particle form factor, $p(R)$ is a normalised distribution function and $S(Q)$ is a structure factor. The term factor, $\Delta\rho^2$, represents a scattering length density difference [12, 13]. As shown before [5, 12, 13], detailed structural and compositional information can be obtained from contrast-variation SANS experiments by employing a simultaneous (partial-structure-factor type) analysis. For example, with D-DDAB and H-$C_{12}E_5$, the following contrasts are used: core (D-water/H-DDAB:H-$C_{12}E_5$/H-heptane or D/H:H/H), shell 1 (D-water/H-DDAB:H-$C_{12}E_5$/D-heptane or D/H:H/D) and shell 2 (H-water/D-DDAB:H-$C_{12}E_5$/H-heptane or H/D:H/H). The fitted value of the scattering length density, ρ_{film}, for shell 2 is linked to the mole fraction of single-chain surfactant in the film, X_{film}^1, via molecular volumes of the single- and double-chain surfactants, V^1 and V^2.

---

[1] For SANS data processing information see http://isise.rl.ac.uk/LargeScale/LOQ/loq.htm and http://www.ill.fr.

These volumes can be estimated from mass density measurements. Literature values for nuclear scattering lengths, b and mass densities [5] were used to calculate scattering length densities of solvents and surfactants. Since scattering length densities, ρ, ϕ_{oil} and ϕ_{water} were known they were input as constants in the modeling. For mixtures of H-DDAB with H-$C_{12}E_5$ in shell 1 samples the ρ_{shell1} value can be estimated using

$$\rho_{shell1} = X^1\left(\frac{\sum_i b_i^1}{V^1}\right) + (1 - X^1)\left(\frac{\sum_j b_j^2}{V^2}\right) , \qquad (3)$$

where the superscripts 1 and 2 refer to single- and double-tailed surfactant, respectively; hence, ρ_{shell1} is also known. This leaves four adjustable parameters: the average radius, R_c^{ay}, the polydispersity, p, and for the film an apparent thickness, t, as well as the scattering length density, ρ_{shell2}. This last value relates to the D-DDAB/H-single-chain surfactant mixes in shell 2 samples, and because ρ_{shell2} is most sensitive to film composition, X_{film}^1, [5] it has also been called ρ_{film} in this report. A similar expression to Eq. (3) defines ρ_{film} (ρ_{shell2}), where Σb^1 is for H-$C_{12}E_5$ and Σb^2 is for chain-deuterated DDAB.

Results and discussion

In principle, in a PFG-NMR experiment one can follow the decay of all the chemical groups giving an NMR signal; however, due to the strict analogy of the radio frequency pulse sequence employed in a PFG-NMR experiment with the classical Hahn spin-echo sequence, the relaxation phenomena taking place during the diffusion time, Δ, can strongly influence the amount of information which can be obtained, starting from the mere appearance of the echo NMR spectrum of the surfactant mixture. This is demonstrated by the spectra

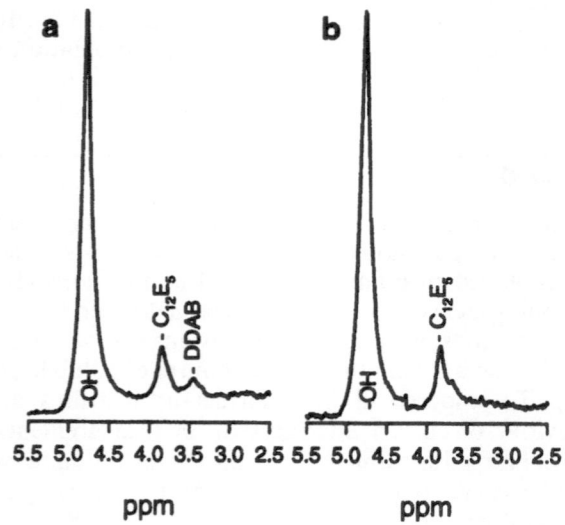

Fig. 1a, b ^{1}H spin-echo NMR spectra of a 0.1 mol/l, $W_o = 15$, $X^1 = 0.2$ pentaethylene glycol monododecyl ether ($C_{12}E_5$) didodecyl-dimethylammonium bromide (DDAB)/water in n-heptane microemulsion at different echo times. **a** $2\tau = \Delta = 60$ ms; **b** $2\tau = \Delta = 140$ ms

shown in Fig. 1. For an experiment at $\Delta = 60$ ms (Fig. 1a), the resonances of DDAB and $C_{12}E_5$ are partially resolved, though the DDAB trimethylammonium group appears extremely broad (probably due to ^1H T_2 effects). An attempt to measure the self-diffusion coefficient of both surfactants at this Δ was made but the results were affected by a high uncertainty (around 15%); therefore, the surfactant self-diffusion coefficient determination was made for $\Delta = 140$ ms, and it was limited to the nonionic one (Fig. 1b). At this Δ, in fact, the contribution of almost all the solvent is lost within the first 2–3 δ, and even the hydroxyl signal is appreciably reduced. In this way, a more convenient set of instrumental parameters can be used, leading to a better signal-to-noise ratio of the spectra (Fig. 2). With this experimental approach, the error in the self-diffusion coefficient is lower than 5%. From the experiments performed at $\Delta = 60$ ms, however, some information on the DDAB diffusional behaviour has also been obtained which points to its complete presence at the interface (data not shown) as expected because DDAB is virtually insoluble in both oil and water.[2]

From the self-diffusion data collected in Table 1, some preliminary conclusions can be drawn. It is evident that the oil diffusion is higher than that found for water and $C_{12}E_5$, thus indicating a water-in-oil structure of the microemulsion with a dispersed phase made of water and surfactants. A decrease in both D_W and D_{C12E5} with increasing W_o is clearly observed. Such experimental evidence suggests the existence of a water core which increases in size with added dispersed component. The significance is that SANS can be used to full effect, as described later, to obtain detailed information on internal droplet structures and interfacial composition.

The NMR data clearly show that the nonionic surfactant diffuses faster than the water, a strong indication of partitioning between the interface and the organic bulk. Assuming a fast exchange between two sites, namely the interfacial film and the bulk, evaluating the apparent partition coefficient (P_{mic}) is possible via Lindman's relationship [14]:

$$D_{C12E5} = P_{mic}D_{mic} + (1 - P_{mic})D_{free} \ , \tag{4}$$

where D_{mic} and D_{free} are the self-diffusion coefficient of $C_{12}E_5$ in the aggregates and free in n-heptane, respectively. The contribution of water molecularly dispersed in the oil bulk to the measured D_W can be significant at low water content [10, 15]; however, in our samples the water concentration was always higher than 1.5 M and the $C_{12}E_5$ solubility in n-heptane is much higher than that of water [5]. Thus we can assume, to a first

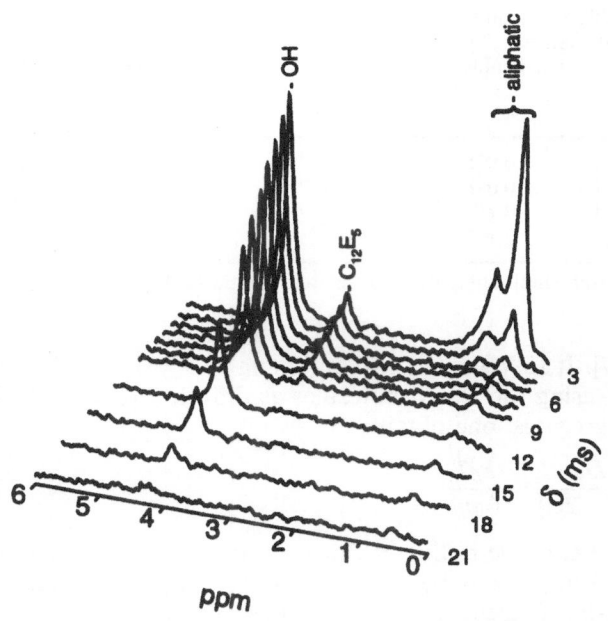

Fig. 2 Stacked plot of a typical ^1H pulsed field gradient NMR self-diffusion experiment performed on a 0.1 mol/l, $W_o = 15$, $X^1 = 0.2$, $C_{12}E_5$/DDAB/water in n-heptane microemulsion

approximation, that $D_{mic} = D_W$ (this is substantiated by SANS data, see later). As long as D_{free} is known, it is possible to apply Eq. (4) to calculate P_{mic}. D_{C12E5} in a solution in n-heptane at the same average composition of the microemulsion under investigation was measured and was found to be 1.22×10^{-9} m^2 s^{-1}; therefore, the P_{mic} values reported in Table 1 at each W_o indicate that almost 90% of $C_{12}E_5$ is present at the interface.

The limiting diffusion coefficient at infinite dilution is related to the micelle hydrodynamic radius, r via the Stokes–Einstein relationship:

$$D_{mic}^0 = \frac{k_B T}{6\pi\eta r} \ . \tag{5}$$

In this study the self-diffusion coefficients were obtained on a time scale of hundreds of milliseconds and correspond to the macroscopic self-diffusion coefficients. With typical diffusion constants ranging from 10^{-11} to 10^{-9} m^2 s^{-1}, the mean squared displacement is in the range 1–15 μm. For solutions of finite concentration, obstruction effects may come into play and would give rise to a decrease in the observed micellar self-diffusion coefficient. For spherical particles the first term in a virial expansion [16] is

$$D_{mic} = D_{mic}^0(1 - k\phi) \ . \tag{6}$$

Theoretical and experimental studies find different values for the interaction constant, k, depending on the nature of the interdroplet interactions. For hard spheres without any hydrodynamic interactions, $k = 2.0$

---

[2] An indirect confirmation of this assumption comes from Fig. 2, where the aliphatic peak decay reveals a component (safely assignable to DDAB) which follows the hydroxyl peak decay.

Table 1 ^{1}H pulsed field gradient NMR self-diffusion coefficients and geometrical parameters as a function of ϕ and W_o for 0.1 mol/l, $X^1 = 0.2$, pentaethylene glycol monododecyl ether ($C_{12}E_5$) dido- decyldimethylammonium bromide ($DDAB$)/water in n-heptane microemulsions

W_o	ϕ	D_W (m^{2}s^{-1})	D_{C12E5} (m^{2}s^{-1})	D_{oil} (m^{2}s^{-1})	r (Å)	t (Å)	P_{mic}	a_s (Å^2)
15.5	0.078	9.1×10^{-11}	2.0×10^{-10}	2.2×10^{-9}	51	–	0.90	37
20.7	0.086	7.2×10^{-11}	2.2×10^{-10}	2.1×10^{-9}	63	–	0.87	38
30.4	0.102	7.0×10^{-11}	2.2×10^{-10}	2.2×10^{-9}	63	–	0.87	56
35.0	0.113	–	–	–	65	12.5[a]	0.87	62

[a] From small-angle neutron scattering data of Table 2

[17]. It follows that by combining Eqs. (5) and (6), and by using the water molecules as diffusion probes of the aggregates, one obtains

$$\frac{D_W}{(1-2\phi)} = \frac{k_B T}{6\pi\eta r} . \tag{7}$$

The effective radius, r, can be calculated by making use of Eq. (7) and by taking into account the amount of $C_{12}E_5$ present at the interface ($\phi = \phi_W + \phi_{DDAB} + P_{mic}\phi_{C12E5}$).

If the system consists of spherical droplets film, geometrical considerations give

$$r = t + \frac{3v_{water}}{a_s} W_{o,mic} , \tag{8}$$

where t is the interfacial film thickness, v_{water} the volume of one water molecule, $W_{o,mic}$ the molar ratio of water to interfacial surfactants and a_s the surfactant mean polar head area.

Knowing the interfacial film composition ($W_{o,mic}$), the a_s values in Table 1 can be obtained by making use of the film thickness directly measured with SANS (see later) and of the hydrodynamic radii obtained from PFG-NMR.

Since DDAB-based microemulsions usually show a wide variety of aggregate shapes upon changing the water content [18], the SANS experiments were performed at the emulsification–failure phase boundary, where only spherical droplets are expected. Typical SANS data and simultaneous analysis fits for samples with a DDAB/$C_{12}E_5$ mixture at $X^1 = 0.20$ are shown in Fig. 3. Owing to constraints of data input and output with the fitting package it was only possible to simultaneously analyse three $I(Q)$ curves in one go (core + shell 1 + shell 2 sets, in the present case; more extensive data are given elsewhere [5]). The fitting gives R_c^{av}, p, t, and ρ_{film}, which in this case is for the mixtures of D-DDAB with H-$C_{12}E_5$ surfactant in the shell 2 samples. Any small differences in the fitted answers were within the absolute errors (caption to Table 2) For shell contrast, at a given droplet size and with fixed solvent scattering length densities, the actual value of ρ_{film} affects the absolute SANS intensity [5, 13]. Since the scattering length density is related to X^1 (Eq. 3), the

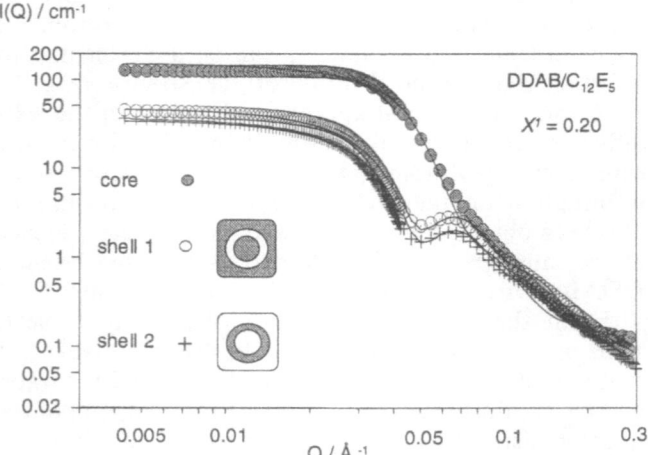

Fig. 3 Example small-angle neutron scattering ($SANS$) data from D22 for microemulsions at $W_o = 33$ and $X^1 = 0.20$ for DDAB/ $C_{12}E_5$ mixtures with contrasts core–shell 1–shell 2. The total surfactant is 0.10 mol/l. The *lines* are simultaneous fits to the model described in the text. Systems composition and fitted parameters are given in Table 2. *Error bars* are shown, but they are well within the data points

Table 2 Compositions of $C_{12}E_5$/DDAB/water in n-heptane microemulsions and values derived from analyses of small-angle neutron scattering data. Uncertainties: R_c^{av} and $t \pm 1$ Å, $p \pm 0.02$ and $\rho_{film} \pm 0.15 \times 10^{10}$ cm^{-2}

X^1	W_o	R_c^{av} (Å)	p	t (Å)	ρ_{film} (10^{10} cm^{-2})
0.08	29	43.7	0.21	12.5	5.49
0.16	33	50.0	0.20	12.5	5.08
0.20	35	52.4	0.20	12.5	4.86

changes in measured intensities may be related to the film composition. The intensity from the mixed D-DDAB/H-$C_{12}E_5$ film (shell 2) is lower than the corresponding H-DDAB/H-$C_{12}E_5$ sample (shell 1), as clearly shown in Fig. 3. This difference is direct evidence for interfacial mixing. In the low Q range (below 0.06 Å^{-1}), the average ratios $I(Q)$ shell 2/$I(Q)$ shell 1 are 0.76, 0.68 and 0.60 for $X^1 = 0.08$, 0.16 and 0.20, respectively. The values of ρ_{film} (which is the scattering

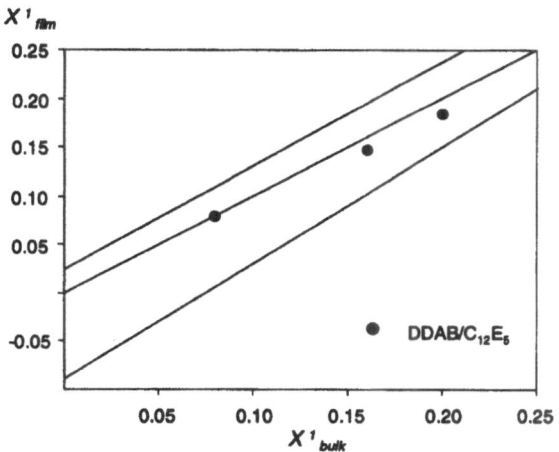

Fig. 4 Film compositions versus bulk composition, X^1_{bulk}, as determined by simultaneous analyses of contrast-variation SANS data and Eq. (5). The *solid line* is for a film density of 0.87 g cm^{-3} and the *dotted lines* are for 0.80 (*lower*) and 0.90 g cm^{-3} (*upper*)

length density for the shell 2 samples) given in Table 2 decrease consistently with increasing proportion of H-C$_{12}$E$_5$: this shows the film is becoming richer in the H-C$_{12}$E$_5$. Figure 4 is a plot of film composition versus the bulk composition derived from the SANS data and analyses using Eq. (3). The observed gradient of unity is consistent with strong binding of both surfactants at the interface. The small deviations seen for C$_{12}$E$_5$ are within the uncertainties of the calculations, based on the accuracies of the measured scattering length density and mass density.

From shell-contrast SANS experiments, P_{mic} can also be evaluated in an independent way by comparing data obtained from samples made of H-DDAB/H-C$_{12}$E$_5$ and D-DDAB/H-C$_{12}$E$_5$ [5]. From the scattering length density of the interfacial film obtained, a film composition $X^1_{\text{film}} = 0.18$ is obtained. From this value, P_{mic} can be easily evaluated using the following relationship

$$P_{\text{mic}} = \frac{(X^1 - 1)X^1_{\text{film}}}{(X^1_{\text{film}} - 1)X^1} , \qquad (9)$$

thus obtaining a partition coefficient of 0.87, in excellent agreement with the PFG-NMR data.

Conclusions

SANS and PFG-NMR have been applied to investigate effects of surfactant mixing in model water–heptane interfaces in water-in-oil microemulsions. Self-diffusion coefficients, measured by PFG-NMR, were consistent with a spherical droplet structure and a mixed surfactant film. By applying extensive neutron contrast variation in partial-structure-factor type experiments (Figs. 3, 4) the droplet structures were determined, and fitted parameters are given in Table 2.

More importantly it has been possible to measure compositions of mixed cationic–nonionic films. For mole fractions of single–chain surfactant X^1 up to 0.20 it was found that both DDAB and C$_{12}$E$_5$ partition strongly into the film. These results show the effects of surfactant blending on phase behaviour ($W_{\text{o,max}}$) and the properties are a direct consequence of mixing in the films. Furthermore, the value of combining information from PFG-NMR and SANS has been clearly demonstrated.

Acknowledgements A.B. was awarded a Kuwaiti Government studentship as well as a grant for travel and consumables. The EPSRC funded neutron beamtime and travel costs (J.E.).

M.G., G.P. and A.C. wish to acknowledge the MURST Progr. Naz. Cofin. 1998 for financial support.

References

1. (a) Scamehorn JF (1986) Phenomena in mixed surfactant systems. American Chemical Society, Washington, DC; (b) Scamehorn JF (1993) In: Ogino K, Abe M (eds) Mixed surfactant systems. Dekker, New York
2. Hines JD, Thomas RK, Garrett PR, Rennie GK, Penfold J (1997) J Phys Chem B 101:9215, and references therein
3. Penfold J, Staples EJ, Tucker I, Thompson L (1997) Langmuir 13:6638
4. (a) Lusvardi KM, Full AP, Kaler EW (1995) Langmuir 11:487; (b) Lu JR, Purcell IP, Lee EM, Simister EA, Thomas RK, Rennie AR, Penfold J (1995) J Colloid Interface Sci 174:441; (c) Brasher LL, Kaler EW (1996) Langmuir 12:6270
5. Bumajdad A, Eastoe J, Heenan RK, Lu JR, Steytler DC, Egelhaaf SU (1998) J Chem Soc Faraday Trans 94:2143
6. Bumajdad A, Eastoe J, Griffiths P, Steytler DC, Heenan RK, Lu JR, Timmins P (1999) Langmuir 15:5271, and references therein
7. (a) Johnson KA, Shah DO (1985) J Colloid Interface Sci 107:269; (b) Gradzielski M, Hoffmann H (1994) J Phys Chem 98:2613, and references therein; (c) Binks BP, Fletcher PDI, Taylor DJF (1998) Langmuir 14:5324, and references therein; (d) Yamaguchi S (1998) Langmuir 14:7183, and references therein
8. Lindman B, Olsson U (1996) Ber Bunsenges Phys Chem 100:344, and references therein
9. Stilbs P (1987) Prog Nucl Magn Resen Spectrosc 19:1–45
10. Giustini G, Palazzo G, Colafemmina G, Della Monica M, Giomini M, Ceglie A (1996) J Phys Chem 100:3190–3198
11. Eastoe J, Hetherington KJ, Dalton JS, Sharpe D, Lu JR, Steyler DC, Heenan RK (1997) J Colloid Interface Sci 190:449
12. Eastoe J, Dong J, Hetherington KJ, Sharpe D, Steytler DC, Heenanoid RK (1996) Langmuir 12:3876

30

13. Markovic I, Ottewill RH, Cebula DJ, Field I, Marsh J (1984) Colloid Polym Sci 262:648
14. Nilsson PG, Lindman B (1983) J Phys Chem 87:4756–4761; (b) Nilsson PG, Lindman B (1984) J Phys Chem 88:4764–4769
15. Angelico R, Balinov B, Ceglie A, Olsson U, Palazzo G, Soderman O (1999) Langmuir 15:1679–1684
16. Ohtsuki T, Okano K (1982) J Chem Phys 77:1443–1451
17. Lekkerkerker HNW, Dhont JKG (1984) J Chem Phys 80:5790–5792
18. Chen SJ, Evans DF, Ninham BW, Mitchell DJ, Blum FD, Pickup S (1986) J Phys Chem 90:842–847

Progr Colloid Polym Sci (2000) 115:31–35
© Springer-Verlag 2000

L. Grosmaire
M. Chorro
C. Chorro
S. Partyka
R. Zana

Thermodynamics of micellization of cationic dimeric (gemini) surfactants

L. Grosmaire · M. Chorro · C. Chorro
S. Partyka (✉)
LAMMI, University of Montpellier II
Place E. Bataillon
34095 Montpellier Cedex 5, France
e-mail: lgrosmaire@crit.univ-montp2.fr

R. Zana
ICS-CRM (CNRS, ULP)
6 rue Boussingault
67083 Strasbourg Cedex, France

Abstract In the present work, the thermodynamics of the micellization process of three alkanediyl-α,ω-bis(dodecyldimethylammonium bromide) dimeric (gemini) surfactants with the alkanediyl spacer groups C_2H_4, C_6H_{12} and $C_{10}H_{20}$ has been investigated by microcalorimetry and conductometry. This aim of the study was to understand better the behavior of cationic gemini surfactants in aqueous solution with particular emphasis on the effect of the length of the spacer group on the structure of the aggregates. This parameter appears capable of influencing the surfactant properties in water. The variations of the differential and cumulative molar enthalpies of dilution of the three gemini surfactants in water solution have been evaluated and the conductivity variations in aqueous solution were measured. The effect of the length of the spacer on the surfactant micellization process is evidenced and different micellar structures are proposed.

Key words Calorimetry · Cationic gemini surfactants · Conductometry

Introduction

Dimeric (gemini) surfactants are made up of two identical amphiphilic moieties connected at the level of the head groups by a spacer group. They constitute a new class of surfactants that is arousing considerable interest in some of their properties [1]. Indeed, they offer some advantages, such as lower critical micellar concentrations (cmc), higher surface activity and better wetting properties than the corresponding monomeric surfactants [2, 3]. As for conventional surfactant molecules, gemini surfactants tend to self-associate in water, where they produce micellar solutions. The change in the cmc with the spacer carbon number S, is complex and is attributed both to the conformational change in the surfactant ion at low S and to the progressive penetration of the spacer into the micelle hydrophobic core for longer spacers ($S \geq 10$) [4]. The aim of this work was to study, using microcalorimetry, the thermodynamics of the micellization of three alkanediyl-α,ω-bis(dodecyldimethylammonium bromide) gemini cationic surfactants with different spacer groups: C_2H_4, C_6H_{12} and

$C_{10}H_{20}$ (the corresponding surfactants are called 12-2-12, 12-6-12 and 12-10-12, respectively). This experimental method affords the direct determination of the differential molar enthalpies of dilution ($\Delta_{dil}h$) of the three surfactants in aqueous solution. These results allow us to compare the variations in enthalpy required for the micellization of the three surfactants according to the characteristics of the spacer group (length, rigidity). In addition, the values of the three cmcs were determined by conductivity measurements. Finally, the cumulative enthalpies ($\Sigma\Delta_{dil}h$) are reported: they exhibit the transition between monomeric solutions and micellar structures. Moreover, the determination of the cmcs by this method permits the reliability of the calorimetric experiments to be verified.

Experimental

Material

The alkanediyl-α,ω-bis(dodecyldimethylammonium bromide) dimeric surfactants are referred to as 12-S-12, with 12 being the carbon number of the amphiphilic moieties and S the carbon number of the alkanediyl spacer group. The three surfactants used, 12-2-12, 12-6-12 and 12-10-12, were synthesized as in previous studies [2]. Their cmcs have been reported to be 0.81, 1.03 and 0.63 mM in pure water at 25 °C [2].

The water used throughout the experiments was deionized and purified with a Millipore "Super Q" system. It had a conductivity ranging between 5 and 10 μS m^{-1} and a pH around 6.

Methods

The calorimetric measurements were performed with a batch microcalorimeter "Montcal 3" schematized in Fig. 1 [5]. The calorimeter contains the calorimetric cell, the agitation system and the syringe pump. The calorimetric device is placed in a calorimetric metallic block, where the temperature is controlled with a precision of $\pm 5 \times 10^{-4}$ °C. The time of return to the baseline (after one aliquot injection) was about 15 min, the detector (thermistors) sensitivity was 0.05 μV μW^{-1}, the volume of water in the cell was about 8 g and one aliquot injection was about 60 μg. The temperature of the calorimetric investigation was 35 °C. The enthalpic change, Δh_{exp}, due to the micelle destruction is calculated from

$$\Delta h_{exp} = \frac{\Delta H_{c_e} - \Delta H_{c_0}}{n_2} \, ,$$

where c_e is the equilibrium surfactant concentration after injection of a given mass of stock solution corresponding to n_2 moles of surfactant and c_0 is the initial concentration of the stock solution (10^{-2} mol kg^{-1}).

The conductivities were measured using a Tacussel CDM 210 conductimeter at 25 °C.

Discussion

The experimental curves of the differential molar enthalpies of dilution ($\Delta_{dil}h$) of the three surfactants investigated show differences both for micellization enthalpic values and for the demicellization process (Fig. 2). The values of the micellization enthalpy for 12-2-12, 12-6-12 and 12-10-12 surfactants are -21 kJ mol^{-1}, -12 kJ mol^{-1} and -15 kJ mol^{-1}, respectively, (Fig. 3). The difference between the values relating to 12-2-12 (-21 kJ mol^{-1}) and to 12-6-12 (-12 kJ mol^{-1}) is certainly assigned to the shape of micelles. Indeed, the 12-2-12 surfactant gives rise to homogeneous and compact micelles with an aggregation number $N = 40$ and a micelle ionization degree $\alpha = 0.2$ [1]. On the other hand, the 12-6-12 surfactant micellizes in more reduced and ionized entities: $N = 27$, $\alpha = 0.3$ [1]. This difference can be related to the effect of the spacer length. In fact, the rigidity of the spacer group of the 12-6-12 surfactant (six CH$_2$ groups) increases the distance between the polar heads, thus allowing the penetration of water molecules into the palisade layer.

The profiles of the differential molar enthalpic curves of dilution (Fig. 2) of the 12-2-12 and 12-6-12 surfactants are quite similar and show a rapid transition

Fig. 1 Calorimeter "Montcal 3"

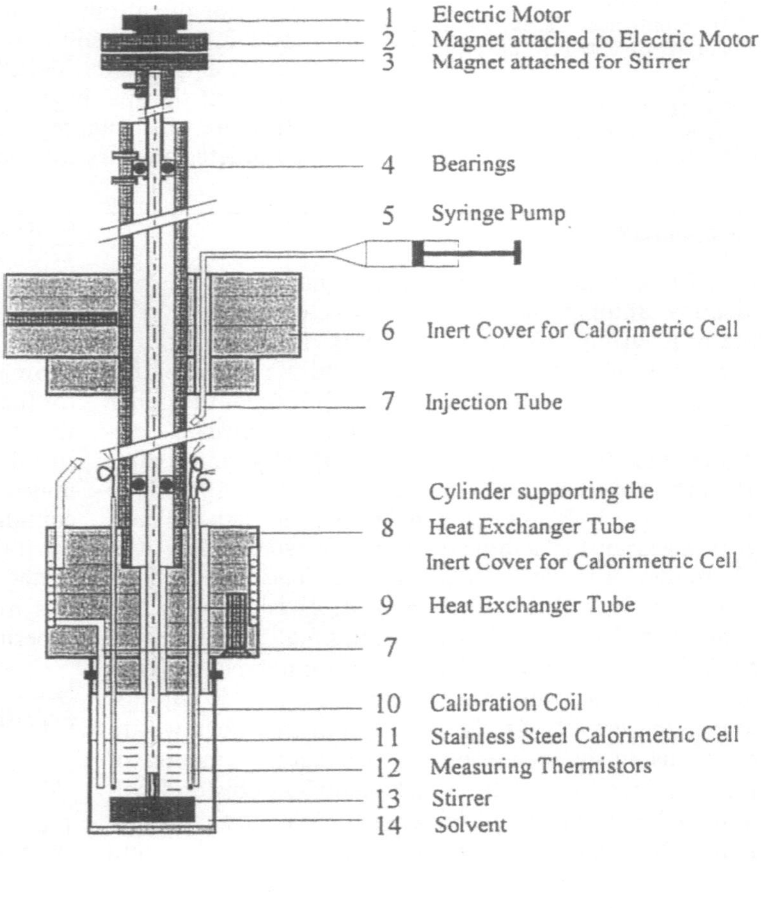

1	Electric Motor
2	Magnet attached to Electric Motor
3	Magnet attached for Stirrer
4	Bearings
5	Syringe Pump
6	Inert Cover for Calorimetric Cell
7	Injection Tube
8	Cylinder supporting the Heat Exchanger Tube / Inert Cover for Calorimetric Cell
9	Heat Exchanger Tube
7	
10	Calibration Coil
11	Stainless Steel Calorimetric Cell
12	Measuring Thermistors
13	Stirrer
14	Solvent

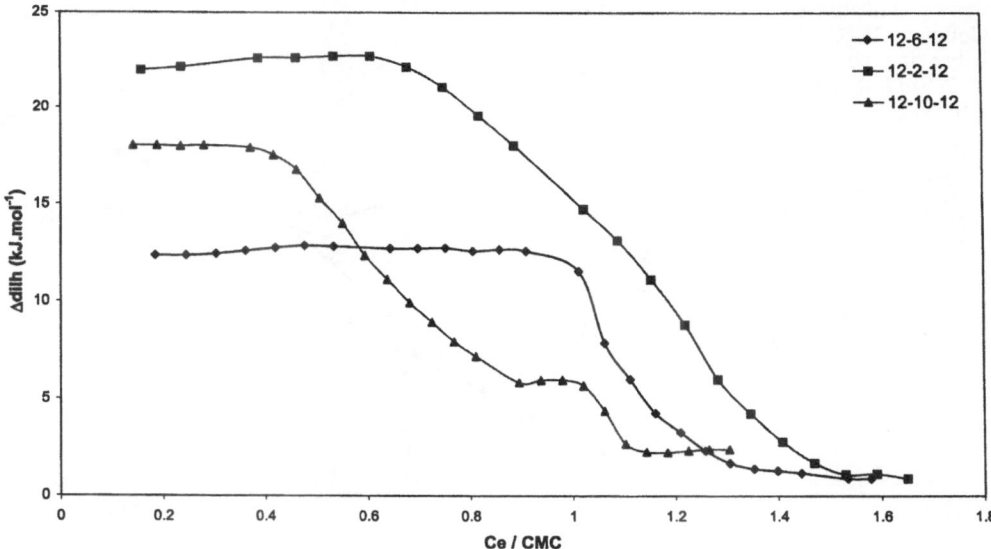

Fig. 2 Differential molar enthalpy of 12-2-12, 12-6-12 and 12-10-12 dilution versus reduced scale of concentration at 35 °C. See text for an explanation of the notation

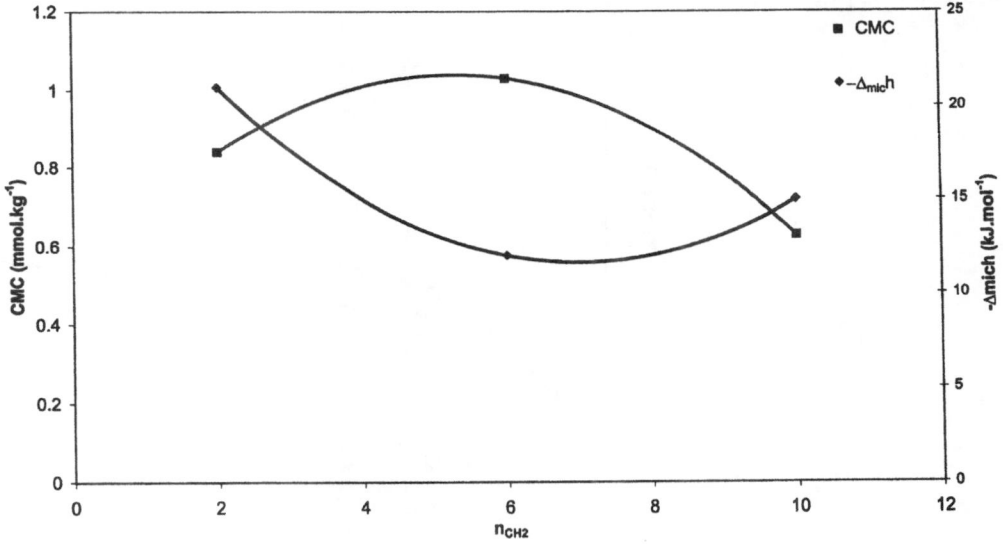

Fig. 3 Enthalpy of micellization and critical micellar concentration (*cmc*) versus spacer length

between monomeric species and micelles. On the other hand, the analysis of the results obtained for the 12-10-12 surfactant is more complex. The enthalpic value of micellization of the 12-10-12 surfactant is intermediate between the 12-2-12 and 12-6-12 surfactants (Fig. 3), suggesting a different behavior of the spacer. Its length renders it flexible to penetrate the core of the micelle through hydrophobic interaction, inducing a more compact organization of the palisade layer, thereby limiting the penetration of water molecules inside this layer. This fact justifies a more exothermic value of the micellization enthalpy than that of the 12-6-12 surfactant in spite of the formation of little more reduced and ionized micelles $N = 23$ and $\alpha = 0.55$ [1]. The curve of the differential molar enthalpy of dilution of the 12-10-

12 surfactant shows a different profile compared to the 12-2-12 and 12-6-12 surfactants: the transition between the monomeric and micellar solutions stretches over a wider range of concentration (Fig. 2) and foreshadows a more complex transition. The conductivity and cumulative heat curves support these expectations: both exhibit three different regions (Fig. 4). It seems that the breakdown of micelles into monomeric surfactants occurs at very low concentration below 0.4 cmc. At higher concentration, one observes a second region, which can be ascribed to the destruction of micelles into premicellar aggregates such as dimers, trimers and little oligomers.

On the other hand, in the case of the 12-2-12 and 12-6-12 surfactants, the conductivity and cumulative heat

34

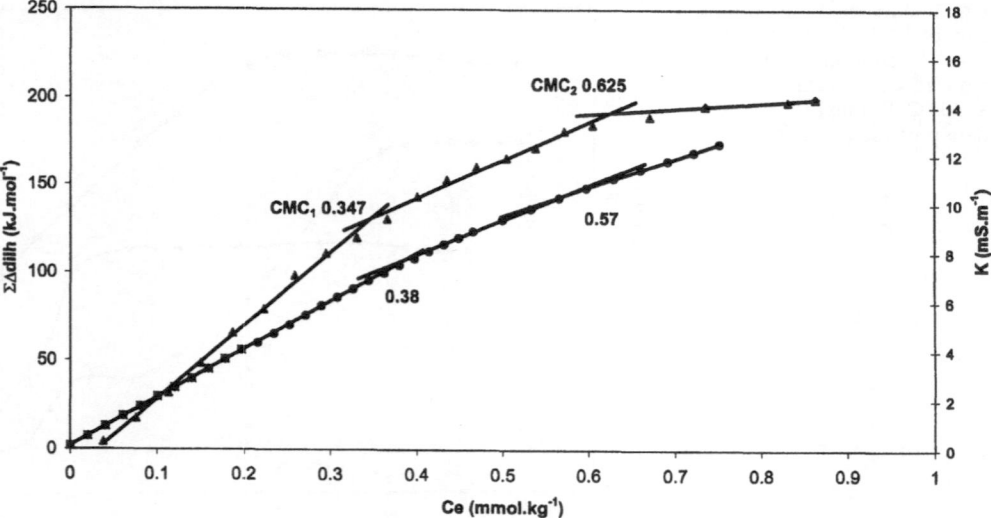

Fig. 4 Cumulative enthalpy of dilution at 35 °C and conductivity versus concentration of 12-10-12: determination of the cmc

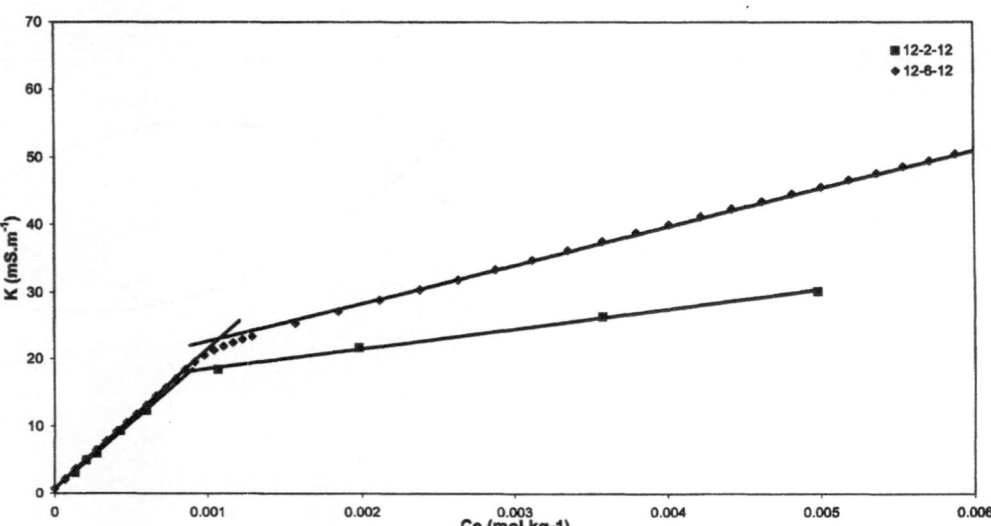

Fig. 5 Conductivity of 12-2-12 and 12-6-12 in pure water

curves (Figs. 5, 6) exhibit only two regions, indicating a single type of demicellization. The Van Oss points clearly indicate the cmcs of the two surfactants in each case and allow the validation of the calorimetric experiments.

Conclusion

The experimental determination of differential molar enthalpies of dilution of surfactants by calorimetry allows, on one hand, the molar enthalpies of micellization for the three surfactants investigated to be determined and, on the other hand, the different demicellization processes to be discerned. Indeed, the less exothermic value of the micellization enthalpy

obtained for the 12-6-12 surfactant suggests the hydration of the palisade layer due to the rigidity of the spacer group, which is longer than the 12-2-12 spacer group; however, the 12-10-12 surfactant shows a more exothermic micellization enthalpy than the 12-6-12 surfactant due to the hydration of the palisade layer being limited by a more flexible spacer group, which can penetrate into the hydrophobic core of the micelle. Moreover, the comparison of the conductivity and cumulative enthalpy curves for the three surfactants shows a difference in the demicellization process for the 12-10-12 surfactant. In fact, for $S = 2, 6$ the curves show two regions indicating the cmc, while for $S = 10$ the curves exhibit three regions and two cmcs. It seems that for the 12-10-12 surfactant, the destruction of micelles follows two processes: formation of monomers at low concentration (0.35 mmol kg^{-1}) and formation of premicellar aggre-

Fig. 6 Cumulative enthalpy of
dilution at 35 °C versus concentration of 12-2-12 and 12-6-
12: determination of the cmc

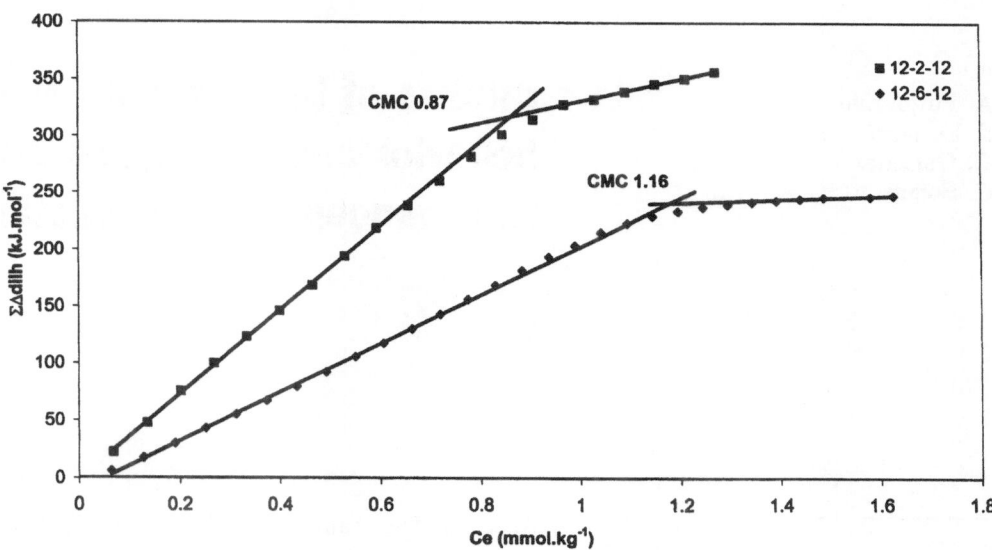

gates at higher concentration (0.62 mmol kg^{-1}). The
calorimetry of dilution in aqueous solution of gemini
cationic surfactant with different spacer groups is a

relevant method to obtain information on the thermodynamics of the micellization process.

References

1. Zana R (1997) Spec Surf 4:81
2. Zana R, Benrraou M, Rueff R (1991)
Langmuir 7:6
3. Chorro C, Chorro M, Dolladille O,
Partyka S, Zana R (1998) J Colloid
Interface Sci 199:169
4. Zana R, Talmon Y (1993) Nature 362
5. Zajac J, Chorro M, Chorro C, Partyka S
(1995) J Therm Anal 45:781

Progr Colloid Polym Sci (2000) 115:36–39
© Springer-Verlag 2000

SURFACTANT COLLOIDS

A. Forgiarini
J. Esquena
C. González
C. Solans

Studies of the relation between phase behavior and emulsification methods with nanoemulsion formation

A. Forgiarini · J. Esquena · C. Solans (✉)
Dept. Tecnologia de Tensioactius
Instituto de Investigaciones Químicas y
Ambientales de Barcelona, CSIC
Jordi Girona 18-26
08034 Barcelona, Spain
e-mail: csmqci@cid.csic.es
Tel.: +34-93-4006159
Fax: +34-93-2045904

C. González
Dept. Ingeniería Química y Metalurgia
Universidad de Barcelona
Martì i Franquès 1
08028 Barcelona, Spain

A. Forgiarini
Fac. de Ingeniería
Universidad de Los Andes, Venezuela

Abstract The main aim of this work was to study the relationship between the type of phases present during the emulsification process, the order of addition of components and the droplet size of the resulting emulsions. In this study, a pseudo-ternary water/poly(oxyethylene) nonionic surfactant/decane system was chosen as a model system to form oil-in-water emulsions at 25 °C. The phase behavior of the model system was determined at constant temperature in order to know the equilibrium phases and also those involved in the emulsification process. The low-energy emulsification methods studied were

A. Addition of oil to an aqueous surfactant dispersion.
B. Addition of water to a surfactant solution in oil.
C. Mixing preequilibrated samples of the components.

Emulsion droplet size distributions were obtained by means of laser diffraction and scattering methods as well as by optical microscopy. The emulsions obtained with methods A and C were more polydisperse than those obtained by method B. Furthermore, with method B very small droplet size emulsions, nano-emulsions, could be obtained. The results have been interpreted according to the changes in the natural curvature of the surfactants during the emulsification process.

Key words Emulsification · Nanoemulsion · Ethoxylated nonionic surfactant · Phase behavior

Introduction

Emulsions are thermodynamically unstable liquid/liquid dispersions stabilised, generally, by surfactants, polymers or solids particles [1]. The sizes of the droplets, which constitute the dispersed phase, are in the range from 0.5 to 100 μm, approximately. In this range, gravity forces attract emulsion droplets. As nonequilibrium systems, emulsion properties depend not only on physicochemical variables (nature of components, composition, temperature and pressure) but also on preparation methods and the order of addition of the components [1–4].

On the other hand, microemulsions are thermodynamically stable, isotropic liquid/liquid dispersions with characteristic sizes of the order of 0.01 μm [5–12]. They form spontaneously when the aqueous, oily and amphiphilic components are brought into contact. A major disadvantage of microemulsions for practical applications lies in the fact that microemulsion formation requires higher amounts of surfactant than emulsions, typically over 10 wt%.

In this context, a new class of emulsions with droplet sizes in the range of nanometers, similar to those of microemulsions, has been reported in recent years [13–15]. These emulsions, termed nanoemulsions, miniemulsions or ultrafine emulsions, are transparent or translucent and show high kinetic stability. Due to their characteristic properties, namely extremely small droplet size (between 20 and 500 nm), kinetic stability and transparency, nanoemulsions are attracting increasing theoretical and practical interest.

The main objective of this work was to study the relationship between the type of phases present during the emulsification process, the order of addition of components and nanoemulsion formation.

Experimental

Products

The surfactant was a technical grade polytetraoxyethylene dodecyl ether with four oxyethylene groups purchased from Sigma, abbreviated as $C_{12}(EO)_4$. The oil component was n-decane and water was deionized by Milli-Q filtration.

Methods

Phase diagram

The phase behavior of the water/$C_{12}(EO)_4$/decane system was determined at constant temperature (25 °C). All components were weighed, sealed in ampoules and homogenized with a vibromixer. These samples were equilibrated at 25 °C. The phase boundaries were determined by visual inspection. The type of liquid crystal was identified using a polarising microscope.

Hydrophile–lipophile balance temperature determination

The hydrophile–lipophile balance (HLB) temperature, T_{HLB}, was determined by conductivity with a Crison model 525 conductimeter with a Pt/platinized electrode with a cell constant of 0.960 cm^{-1}. The samples were prepared with an electrolyte solution (NaCl 10^{-2} M) instead of water.

Emulsion formation

Emulsions were prepared at 700 rpm with a magnetic stirrer and the speed of addition of the components was kept constant. The final surfactant concentration in all the emulsions was 5 wt% and the temperature was 25 °C.

Droplet size

The droplet size of the emulsions was determined by optical microscopy (Reichter Polyvar 2, Leica), laser light refraction (Mastersizer-S, Malvern) or laser light scattering Photon Correlator Spectrometer 4700 PS/MV Malvern depending on the size range.

Results and discussion

To determine the relationship between nanoemulsion formation, the type of phases present during the emulsification process and the order of addition of the components, a model system, water/$C_{12}(EO)_4$/decane, was chosen. The phase behavior of the system was first determined in order to know the equilibrium phases and also those involved in the emulsification process. Then, oil-in-water (O/W) emulsions were prepared according

to different methods. Finally, emulsion droplet sizes were determined in emulsions with constant surfactant concentration as a function of the oil weight fraction, R = oil/(oil + water).

Phase behavior of the water/$C_{12}(EO)_4$/decane system

The phase diagram of the water/$C_{12}(EO)_4$/decane system at 25 °C is shown in Fig. 1. The behavior of this system conforms to that of similar nonionic surfactant systems [16–18]. The main characteristics are as follows. A one-phase region of lamellar liquid crystal (I_{CL}). A zone of one liquid phase (I), considered to be formed by inverse micelles or water-in-oil (W/O) microemulsions, with a subregion (I′) characterized by its birefringence under shear. There is a multiphase region of lamellar liquid crystal (II_{LC}), a two-phase region (II_V) composed of a dispersion of vesicles in water and a region of two liquid phases (II). The water–surfactant binary system is a dispersion of lamellar liquid crystal in water (vesicles) at surfactant concentrations below 17 wt%.

The nonionic surfactant used in this work was of technical grade; therefore, the HLB temperature is not a system property, but depends on the water/oil ratio and surfactant concentration [16]. The HLB temperature at a constant surfactant concentration (5 wt%) and different oil-weight fraction, R, was determined by conductivity, using aqueous NaCl solution instead of water, as indicated in the Experimental section. The conductivity is shown as a function of R in Fig. 2. Independent of the R ratio, initially, the conductivity is high and increases slightly with temperature. At a certain temperature (T_{HLB}), the conductivity decreases suddenly, reaching

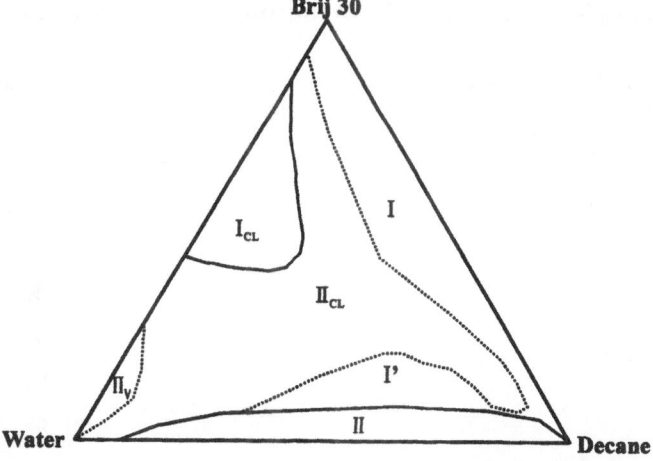

Fig. 1 Phase behavior of the water/poly(oxyethylene lauryl ether) $C_{12}(EO)_4$/decane system at 25 °C. *I*: isotropic liquid phase; *I′*: shear birefringence liquid phase; *I_{CL}*: lamellar liquid-crystalline phase; *II_{CL}*: multiphase region including lamellar liquid crystal; *II_v*: lamellar liquid-crystal dispersion (vesicles); *II*: two liquid phases

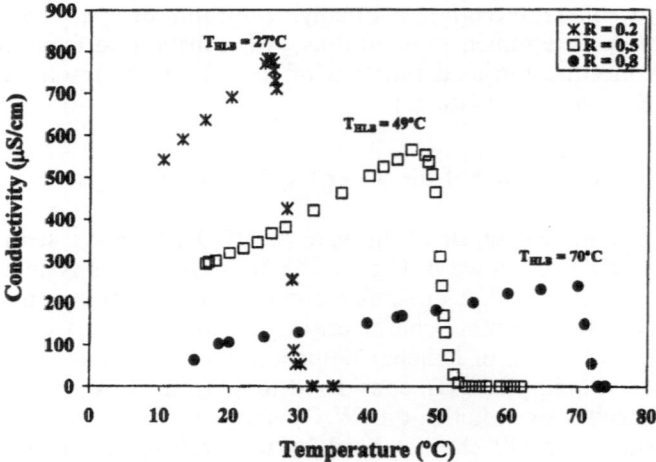

Fig. 2 Conductivity as a function of temperature for the system of an aqueous solution of 10^{-2} M NaCl/C$_{12}$(EO)$_4$/decane. R = oil/(oil + water), S = 5 wt%

Table 1 Interfacial tensions for equilibrated samples of water/C$_{12}$(EO)$_4$/decane as a function of the oil-to-water weight ratio, R. The surfactant concentration was 5 wt% and the temperature was 25 °C

R = oil/(oil + water)	Interfacial tension (mN m^{-1})
0.25	6.0×10^{-3}
0.50	7.1×10^{-2}
0.80	2.7×10^{-1}

concentration of 5 wt% at 25 °C. Table 1 shows that the interfacial tensions increase with R. The lower measurable interfacial tensions were obtained at R values close to 0.25, in the region where nanoemulsions can only be formed by method B (see later).

Emulsification and emulsion droplet size

In this study, emulsions were prepared using low-energy emulsification methods. The surfactant concentration was kept constant at 5 wt%. The emulsification methods used were the following:

A. Adding, slowly, hydrocarbon to mixtures of water and surfactant.
B. Adding, slowly, water to surfactant and hydrocarbon solutions.
C. Weighing all components and letting the samples equilibrate prior to emulsification.

Figure 3 shows that droplet sizes decreases with an increase in R when emulsions are obtained by methods A and C. Nanoemulsion formation was not observed using these methods. The lowest sizes are of the order of 3 μm. It is worth noting that using a high-energy input (ultraturrax at 10000 rpm) nanoemulsions could not be

very low values, an indication that surfactant molecules change their affinity. O/W (high conductivity) emulsions are formed when $T < T_{HLB}$ and W/O (low conductivity) emulsions are formed when $T > T_{HLB}$. As expected, at constant surfactant concentration, T_{HLB} diminishes when the oil weight fraction increases due to the different partition of the surfactant homologues between the water and the oil phases.

The requirement of low values of interfacial tension for nanoemulsion formation has been a subject of debate [13, 15, 19]. The formation of nanoemulsions has been related to the order of addition of the components. In order to find out whether a relationship between interfacial tension and nanoemulsion formation could be established, the interfacial tensions for equilibrated samples were measured as a function of R at a surfactant

Fig. 3 Droplet size as a function of R for emulsions obtained by emulsification methods A and C

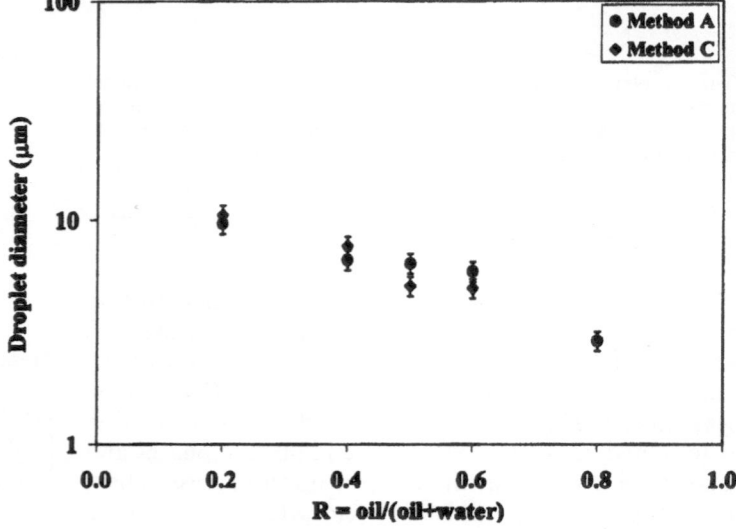

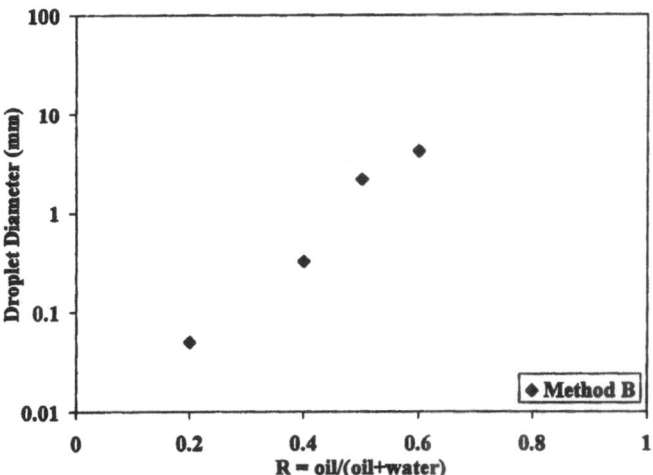

Fig. 4 Droplet size as a function of R in emulsions obtained by emulsification method B

formed by these methods. Moreover, all emulsions obtained were highly polydisperse.

On the other hand, droplet size increases with an increase in R in emulsions obtained with method B (Fig. 4). All the emulsions show narrow size distributions. It should be noted that small droplet sizes (of the order of 50 nm) were obtained in emulsions with R lower than 0.35. These emulsions, which have a transparent appearance and are bluish, can be considered as nanoemulsions.

These results clearly show that nanoemulsions could be formed when the interfacial tension was very low; however, low equilibrium interfacial tension is not the only requirement. For instance, at $R = 0.25$, the equilibrium interfacial tension is 6×10^{-2} mN m^{-1}, and depending on the order of addition of the components, emulsions of 10 μm (method A, C) or 50 nm (method B) are obtained; therefore, the key to nanoemulsion forma-

tion could be attributed to the phase transitions that occur during the emulsification process. With emulsification method A, initially the system is formed by a dispersion of liquid crystal and water (vesicles). Addition of decane to the system leads to O/W emulsions via a multiphase region with lamellar liquid crystal (Fig. 1). In emulsification method B, the transitions during emulsification are the following (Fig. 1): an isotropic phase, I, (a W/O microemulsion); a multiphase region including lamellar liquid crystal (II$_{LC}$); a shear birefringent (bicontinuous) phase (I'); an O/W emulsion. Similar transitions can be produced at constant compositions by changes in temperature [15–17]. In emulsification method B there is a more pronounced change of curvature of the surfactant during the emulsification process from W/O to O/W than in method A. Consequently, the interfacial tensions achieved during the emulsification process are probably not as low as in method B.

Conclusions

Nanoemulsions have been obtained with emulsification method B (addition of water to surfactant in oil solution) at certain oil weight fractions where very low values of interfacial tensions are achieved; however, low values of equilibrium interfacial tensions are not the only requirement to obtain nanoemulsions. The change in the natural curvature of the surfactant during the emulsification process could play a major role. Independent of the oil weight fraction, emulsions obtained by method B have lower polydispersity than those obtained by method A (addition of oil to surfactant in water mixture).

Acknowledgements Financial support from CICYT (QUI 96-0454) and Generalitat de Catalunya (1997SGR-00105) is gratefully acknowledged. A.F. acknowledges CONICIT-ULA for a Ph.D. grant. J. Puebla from Optilas Ibérica is also gratefully acknowledged for facilitating the use of a Malvern Mastersizer-S.

References

1. Becher P (1965) Encyclopedia of emulsion technology, vol 1.
2. Forster T (1997) Surfactants in cosmetics. Surfactant Science Series 68, pp 105–125
3. Lin TJ, Haruki K, Hideaki Ohta BSJ (1975) Cosmet Chem 26:121–139
4. Esquena J, Solans C (1998) Prog Colloid Polym Sci 110:235–239
5. Friberg SE, Vesable R (1985) In: Becher P (ed) Microemulsions, encyclopedia of emulsion technology, vol 1, pp 287–336
6. Langevin D (1991) Adv Colloid Interface Sci 34:583–595
7. Solans C, García-Celma MJ (1997) Curr Opin Colloid Interface Sci 2:464–471

8. Kahlweit M (1995) J Phys Chem 99:1281–1284
9. Schubert KV, Kaler EW (1996) Ber Bunsenges Phys Chem 100:190–205
10. Lekkerkerker HNW, Kegel WK, Overbeek JThG (1996) Ber Bunsenges Phys Chem 100:206–217
11. Sjöblom J, Lindberg R, Friberg SE (1996) Adv Colloid Interface Sci 65:125–287
12. Friberg S (1982) Colloids Surf 4:201
13. El-Aasser M, Lack C, Vanderhoff J, Fowkes F (1988) Colloids Surf 29: 103–118
14. Tomomasa S, Kochi M, Nakajima H (1988) J Jpn Oil Chem 37:1012

15. Nakajima H (1997) Industrial application of macroemulsion. Surfactant Science Series 66, pp 175–197
16. Kunieda H, Shinoda K (1985) J Colloid Interface Sci 107:107–121
17. Kunieda H, Shinoda K (1982) J Dispersion Sci Technol 3:233–244
18. Buzier M, Ravey JC (1983) J Colloid Interface Sci 91:20–33
19. Rosano HI, Lan T, Weiss A, Whittam JH, Gerbacia WEF (1981) J Phys Chem 85:468

Progr Colloid Polym Sci (2000) 115:40–43
© Springer-Verlag 2000

SURFACTANT COLLOIDS

K. Aramaki
A. Akahane
H. Kunieda

Oil-induced phase transition from lamellar to reverse hexagonal liquid crystals in the Aerosol OT system

K. Aramaki
Faculty of Engineering
Yokohama National University,
Tokiwadai 79-5, Hodogaya-ku,
Yokohama 240-8501, Japan

A. Akahane · H. Kunieda (✉)
Graduate School of Engineering
Yokohama National University
Tokiwadai 79-5, Hodogaya-ku
Yokohama 240-8501, Japan
e-mail: kunieda@ynu.ac.jp
Tel.: +81-45-3394190
Fax: +81-45-3394190

Abstract The effect of added n-decane or m-xylene on the lamellar liquid crystal (L_α) phase in a water/purified Aerosol OT system was investigated by small-angle X-ray scattering (SAXS) at 25 °C. Upon addition of n-decane at constant water/surfactant weight ratio, the reverse micellar solution phase (water-in-oil microemulsion) separates beyond the phase boundary of the single L_α phase, whereas the L_α–hexagonal liquid crystal (H_2) phase transition takes place by adding m-xylene. In the m-xylene system, the effective cross-sectional area per surfactant molecule at the interface, a_s, calculated from SAXS data suddenly drops off at the L_α–H_2 phase transition. In the poly(oxyethylene)-type nonionic surfactant system, however, a_s monotonically increases at the same transition upon addition of n-decane. It is considered that the difference in the size of surfactant head group and the penetration behavior of oil affect the change in a_s at the L_α–H_2 transition.

Key words Aerosol OT · Effective cross-sectional area · Small-angle X-ray scattering · Liquid crystal · Effect of added oil

Introduction

It is known that added oil induces structural change of surfactant aggregates or changes in the phase transition in many surfactant systems [1–4]. One of the factors to dictate the change is the penetration of oil into the surfactant palisade layer, which forces the surfactant layer curvature to be negative due to the increase in the effective hydrophobic volume of the surfactant. The degree of penetration depends on the size of the oil molecule and/or the polarity of the oil [5]. In a previous study [6], it was found that the effective cross-sectional area per surfactant molecule at the interface, a_s, is not changed at the transition of lamellar (L_α)-to-reverse hexagonal (H_2) liquid crystals when n-decane is solubilized in the water/trioxyethylene dodecyl ether ($C_{12}EO_3$) system. When the surfactant layer curvature is changed from 0 (layer structure) to negative (reverse cylindrical structure), the size of the hydrophilic group of the surfactant is crucial for the structural change. Since the

head groups of ionic surfactants are considerably smaller than the poly(oxyethylene) chains of nonionics, the steric hindrance of the head group may be almost negligible.

In this context, we investigated the effect of head group size on the molecular packing of surfactant at the interface in water/Aerosol OT (AOT)/hydrocarbon systems around the phase transition from a L_α to a H_2 phase and the results are compared with the same transition in a poly(oxyethylene)-type nonionic surfactant system.

Experimental

Materials

Aerosol OT [sodium bis(2-ethylhexyl) sulfosuccinate] was purchased from Sigma and was used after being purified in accordance with the process described in a previous paper [7]. n-decane (>99.0%) and m-xylene (purity above 99.0%) were purchased from Tokyo Kasei Kogyo and were used without further purification. Water distilled from deionized water was used.

Methods

All chemicals were weighed and sealed in test tubes. The samples were mixed by shaking to obtain homogeneity. They were kept in a water bath with a thermostat (controlled within ± 0.05 °C). The liquid crystals were distinguished by the observation of optical birefringence under crossed polarizers and these structures were distinguished by polarized microscopy and small-angle X-ray scattering (SAXS). The interlayer spacing, d, of the liquid crystal was measured by SAXS. The effective cross-sectional area per surfactant molecule at the interface of the water-lipophilic part of the surfactant, a_s, was calculated using the following equations [6].

$$a_s = \frac{2v_L}{d\phi_L} \quad \text{(for the L}_\alpha \text{ phase)} \tag{1}$$

$$a_s = \frac{\{2\sqrt{3}\pi(1 - \phi_L - \phi_o)\}^{1/2}v_L}{d\phi_L} \quad \text{(for the H}_2 \text{ phase)} , \tag{2}$$

where v_L is the molecular volume of the lipophilic part, and ϕ_L and ϕ_o are the volume fractions of the lipophilic part and the added oil in the system, respectively. Molecular volumes and volume fractions were evaluated from the densities of the surfactant solutions measured using a density meter (Anton Paar DMA 40) and the molecular weights. The densities of AOT, water, n-decane and m-xylene at 25 °C are 1.13 g cm^3, 0.997 g cm^3, 0.726 g cm^3 and 0.860 g cm^3, respectively. We considered the hydrophilic part of AOT to be the sodium sulfonate unit and the volume of the lipophilic part was estimated by subtracting the molecular volume of disodium sulfonate, 0.08 nm^3 [8], from that of AOT, 0.65 nm^3.

Results and discussion

Phase Behavior

The phase behavior in the water/purified AOT system at 25 °C was determined. Up to 77.6 wt% surfactant, a L$_\alpha$ liquid-crystalline phase is formed, except in an extremely dilute region in which an isotropic solution appears. A H$_2$ liquid-crystalline phase is formed above 81.5 wt% up to 100 wt%. A reverse bicontinuous cubic phase (V$_2$) is formed in-between these two phases. Upon addition of n-decane at constant water-to-surfactant weight ratio (2/3) at 25 °C, the L$_\alpha$ phase is maintained up to $\phi_o/\phi_L = 0.22$. A two-phase equilibrium of the L$_\alpha$ and the reverse micellar solution phase (water-in-oil, w/o, microemulsion or O$_m$ phase) is observed with further addition of n-decane and a single phase (O$_m$) is formed above $\phi_o/\phi_L = 0.40$. On the other hand, with the addition of m-xylene, the L$_\alpha$ phase is terminated at $\phi_o/\phi_L = 0.10$ and the H$_2$ phase is observed above $\phi_o/\phi_L = 0.28$. The phase then changes to the O$_m$ phase above $\phi_o/\phi_L = 0.47$.

In contrast, in the case of the L$_\alpha$ phase of a C$_{12}$EO$_3$ system [6], the L$_\alpha$–H$_2$ transition takes place upon addition of n-decane, whereas the L$_\alpha$–O$_m$ transition occurs upon addition of m-xylene at 25 °C. Hence, the

effect of the type of oil on the stability of the H$_2$ phase is opposite in both the AOT and the C$_{12}$EO$_3$ systems.

Oil-induced structural change of liquid crystals

The interlayer spacings of the liquid crystals, d, measured by SAXS are plotted as a function of ϕ_o/ϕ_L in Fig. 1a. The a_s values calculated using Eq. (1) or Eq. (2) using the measured d values are plotted in Fig. 1b. In the L$_\alpha$ phase, d increases on addition of saturated hydrocarbons, whereas it does not increase greatly on addition of m-xylene. Aromatic hydrocarbons tend to be solubilized in the surfactant palisade layer (penetration), whereas long hydrocarbons such as n-decane make an oil core in the L$_\alpha$ phase (swelling) [3, 6]. If the penetration tendency is strong, the interlayer spacing does not increase much and a_s increases greatly. If the swelling tendency were strong, the opposite behavior for

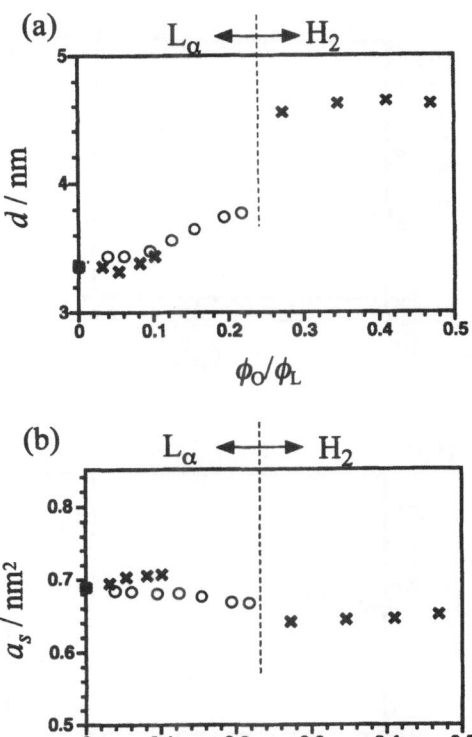

Fig. 1 a The change in interlayer spacing, d, measured by small-angle X-ray scattering as a function of volume ratio of oil to lipophilic moiety of Aerosol OT (*AOT*), ϕ_o/ϕ_L, in the water/AOT/oil system at 25 °C. The oils used are n-decane (*circles*) and m-xylene (*crosses*). The water/AOT weight ratio was kept at 2/3. **b** The change in effective cross-sectional area per surfactant molecule at the interface, a_s, as a function of ϕ_o/ϕ_L, which is evaluated from d values. Note that the L$_\alpha$–H$_2$ transition does not take place in the n-decane system

the changes in d and a_s values would be observed. As shown in Fig. 1, m-xylene exhibits penetration behavior, whereas n-decane exhibits swelling behavior when they are solubilized in the L_α phase of the AOT system.

On passing through the L_α–H_2 phase transition in the m-xylene system, a_s suddenly drops off. Note that the a_s values in the H_2 phase, which were measured by small-angle neutron scattering [9], are in good agreement with those in w/o-type microemulsions in the water/AOT/hexamethyldisilane system (approximately 0.6 nm^2). In the water/$C_{12}EO_3$/n-decane system [6], the phase transition from the L_α phase to the H_2 phase also takes place. In this case, however, a_s increases continuously with increasing oil content even after the L_α phase is changed to the H_2 phase as is shown in Fig. 2.

The EO chain length of $C_{12}EO_3$ is approximately 1 nm in its extended form [10], whereas the -SO_3^- group in AOT is about 0.2 nm because the length of the S—O bond is 0.175 nm [11]. Since AOT has double hydrocarbon chains, a_s is large and the hydrocarbon chains

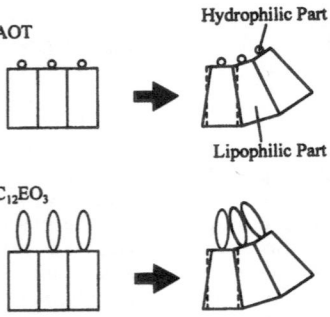

Fig. 3 The schematic representation of the geometrical change of distance between neighboring head groups at the phase transition from L_α to H_2. In the case of the short head group (AOT), the distance does not change very much, whereas it becomes small in the case of the long head group ($C_{12}EO_3$)

experience considerable contact with water because the actual cross-sectional area of sulfonate is much smaller than a_s. Note that the actual area of sulfate is 0.27 nm^2 [11]. Hence, if oil greatly penetrates in the AOT palisade layer and a_s is expanded, eventually, the repulsion of the head groups is reduced, whereas the attraction due to the interfacial tension between water and the hydrocarbon chain increases: this causes the L_α–H_2 phase transition. Then a_s sharply drops off as shown schematically in Fig. 3 because the steric hindrance of the head group is small.

On the other hand, the hydrophilic group of $C_{12}EO_3$ is approximately 5 times longer than that of AOT. As shown in Fig. 3, if a_s decreases at the L_α–H_2 transition point, it causes a large increase in repulsion due to the steric hindrance of the hydrophilic moieties; therefore, the decrease in a_s is not observed in the H_2 phase.

As mentioned in the former section, the effect of the type of oil on the stability of the H_2 phase is different for both systems. m-xylene stabilizes the phase compared with n-decane in the AOT system, whereas the opposite behavior is observed in the $C_{12}EO_3$ system. This will be investigated in the near future.

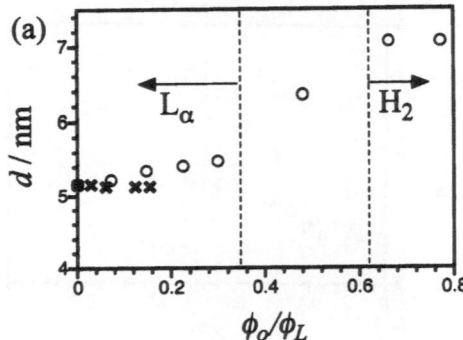

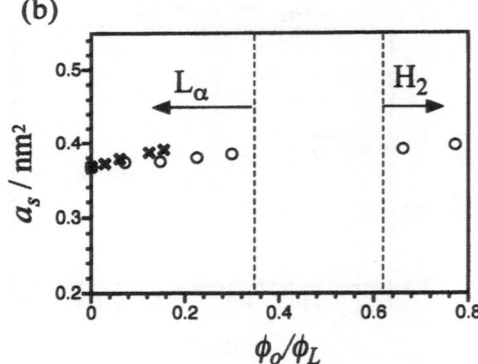

Fig. 2 a The change in d and **b** the change in a_s in the water/trioxyethylene dodecyl ether ($C_{12}EO_3$)/oil system at 25 °C as a function of ϕ_o/ϕ_L. The data are reproduced from Ref. [6]. The oils used were n-decane (*circles*) and m-xylene (*crosses*). The volume ratio, $\phi_L/(\phi_H + \phi_W)$, was kept at 0.376/0.624, where the subscripts H and W indicate the hydrophilic moiety of surfactant and water, respectively. Note that the L_α–H_2 transition does not take place in the m-xylene system

Conclusion

In the water/AOT/m-xylene system, the phase transition from a L_α phase to a H_2 phase occurs, whereas the L_α phase is directly changed to a O_m in the n-decane system. In the water/$C_{12}EO_3$/oil systems, however, the L_α–H_2 transition takes place in the n-decane system, whereas the L_α–O_m transition occurs upon addition of m-xylene at 25 °C. a_s is almost constant or increases slightly within a liquid-crystalline phase upon addition of these oils; however, at the L_α–H_2 phase transition, a_s shrinks suddenly, which is opposite to the continuous increase in a_s upon addition of n-decane in a poly(oxyethylene)-type nonionic surfactant system. Aromatic hydrocarbons such as *m*-xylene have a strong tendency to penetrate in the surfactant palisade layer and make the surfactant layer curvature negative. In the AOT system, the repulsion does not increase when the curvature becomes negative because the hydrophilic head group is small and short. On the other hand, the EO head group is long and the repulsion would be large at the negative curved interface due to the steric hindrance of the hydrophilic moieties.

References

1. Hoffmann H, Ulbricht W (1989) J Colloid Interface Sci 129:38
2. Chen SJ, Evans DF, Ninham BW, Mitchell DJ, Blum FD, Pickup S (1986) J Phys Chem 90:842
3. Aramaki K, Kunieda H (1999) Colloid Polym Sci 277:34
4. Kunieda H, Taoka H, Iwanaga T, Harashima A (1998) Langmuir 14:5113
5. Evans DF, Wennerström H (1999) The colloidal domain, 2nd edn. Wiley, New York, p 548
6. Kunieda H, Ozawa K, Huang KL (1998) J Phys Chem B 102:831
7. Kunieda H, Shinoda K (1979) J Colloid Interface Sci 70:577
8. Hodgman CD, Weast RC, Selby SM (1955) Handbook of chemistry and physics 37th edn. CRC, Cleveland, p 601
9. Steytler DC, Dowding PJ, Robinson BH, Hague JD, Rennie JHS, Leng CA, Eastoe J, Heenan RK (1998) Langmuir 14:3517
10. Rösch M (1966) In: Shick MJ (ed) Nonionic surfactants. Dekker, New York, pp 760–761
11. Evans DF, Wennerström H (1999) The colloidal domain, 2nd edn. Wiley, New York, p 184

Progr Colloid Polym Sci (2000) 115:44–49
© Springer-Verlag 2000

F. Bordi
C. Cametti
A. Di Biasio

High-frequency dielectric polarization mechanism in water-in-oil microemulsions below percolation

F. Bordi
Dipartimento di Medicina Interna
Università di Tor Vergata
Rome, Italy

C. Cametti (✉)
Dipartimento di Fisica, Università di Roma
"La Sapienza", Rome, Italy
e-mail: cesare.cametti@roma1.infn.it
Fax: +39-06-4463158

A. Di Biasio
Dipartimento di Matematica e Fisica
Universita' degli Studi di Camerino
Camerino, Italy

F. Bordi · C. Cametti · A. Di Biasio
Istituto Nazionale di Fisica della Materia
Unita' di Roma I, Rome, Italy

Abstract In this note, we report high-frequency dielectric relaxations of water-in-oil microemulsion systems occurring in the frequency range from 1 MHz to 1.8 GHz. In the composition range and temperature interval investigated (water-to-surfactant molar ratio of 40.8 and fractional volumes from 0.05 to 0.30), the system behaves as a collection of isolated water droplets, well below the percolation threshold, where a more complex structural arrangement prevails. The observed dielectric spectra evidence a polarization mechanism occurring within the single water droplet, characterized by a dielectric strength of several units and a relaxation time of the order of 1–2 ns. The results presented here give support to the identification of the dielectric relaxation as being due to the orientational correlation of the surfactant head groups at the water–surfactant interface resulting in the formation of an apparent relaxing dipole moment. This polarization mechanism constitutes a novel contribution to the overall microemulsion dielectric spectra, whose features should be accurately investigated to gain a complete description of the electrical behaviour of these systems.

Key words Microemulsions · dielectric properties · dielectric polarization

Introduction

Water-in-oil microemulsions are homogeneous, isotropic and thermodynamically stable solutions composed of water and oil domains separated by surfactant monolayers [1–3]. Within certain ranges of composition and temperature, the system consists of water droplets covered by a surfactant layer and dispersed in an apolar continuous phase. Owing to their ionic nature, the surfactant molecules dissociate, causing a partially fixed charge distribution at the water droplet surface and leaving movable oppositely charge counterions in the aqueous droplet core.

Colloidal systems like the one investigated here should present, in principle, at least three different relaxation regions, indicating that different polarization mechanisms contribute to the overall response of the system to an applied electric field [4]. At low frequencies, of the order of 1–10 kHz, a surface polarization mechanism is caused by counterion tangential flows at the inner surface water droplets. These motions result in very large dielectric increments, such as those observed in aqueous suspensions of latex particles [5, 6]. Here, counterions move in the external aqueous medium and the nonconducting latex particles are bounded by a fixed (or semimobile) charge distribution. In the case of microemulsions, in contrast, the roles of the different media are completely exchanged, the interior is substituted with the exterior medium and vice versa, but the essence of the surface polarization mechanism remains unchanged, even if a stronger constraint to the counterion motion is expected.

At higher frequencies, the usual interfacial Maxwell–Wagner polarization mechanism is to be taken into account [7]; however, this contribution in the case of water particles (permittivity $\varepsilon_p = 80$) in nonconductive media (permittivity $\varepsilon_m = 2-3$) at low volume fractions ($\Phi < 0.30$) results in a dielectric dispersion of very small dielectric strength, falling at frequencies of about 100 kHz–1 MHz. Finally, in the very high frequency region, the orientational polarization of the water molecules dominates, the maximum loss occurring at about 17 GHz (at 20 °C) [8]. This dispersion extends over more than two frequency decades and consequently this effect also contributes to the higher limit of the frequency range investigated here.

This analysis seems to exclude the possible existence of other relaxation regions in the frequency range investigated (1 MHz–1.8 GHz). In contrast, the experimental investigation in this "unusual" frequency region evidences the presence of a marked dielectric dispersion located between the expected region of the Maxwell–Wagner effect and that of the bulk water.

Two different mechanisms could, in principle, be responsible for the observed effect. The first involves the distribution of counterions inside each water droplet. Owing to the peculiar arrangement of the system, the bulk aqueous solution does not satisfy the electroneutrality condition, since bulk counterions (in this case Na^+ ions) derived from the surfactant ionization are not neutralized by an equivalent number of anions. Under the influence of an external electric field, an asymmetrical distribution of these counterions inside the water core could produce an apparent dipole moment, whose relaxation could cause the observed dispersion. The second possibility comes from the orientational correlation of the head groups of the ionized surfactant molecules at the boundary of the water droplets. This correlation induces, also in this case, an apparent dipole moment and, consequently, a relaxation process.

Here, we present some high-frequency dielectric spectra of water-in-oil microemulsions below the percolation threshold which evidence this further relaxation process and we briefly discuss the reliability of the two mechanisms.

Experimental

The microemulsion investigated here was composed of water, n-decane and sodium bis(2-ethylhexyl) sulfosuccinate (AOT) as an ionic surfactant. AOT, purity 99%, was purchased from Sigma Chemical Co. (St. Louis, Mo) and was used as received without further purification. The alkane n-decane was obtained from Aldrich, USA. The water used was distilled and deionized with an electrical conductivity lower than 10^{-6} Ω^{-1} cm^{-1} at room temperature.

The composition of the microemulsion is characterized by two parameters: the water to surfactant molar ratio, $W = [H_2O]/[AOT]$, and the fractional volume of the dispersed phase (water + surfactant per unit volume), Φ. We studied microemulsions with the same W (40.8) and different Φ (0.10, 0.15, 0.20, 0.30) at different temperatures between 5 and 25 °C.

The dielectric spectra, showing a well-defined dielectric dispersion, were measured in the frequency range from 1 MHz to 1.8 GHz by means of a Hewlett–Packard model 4291A high-frequency impedance meter. The modulus of the reflection coefficient, ρ, and the phase angle, ϕ, produced by the mismatch of a section of a coaxial line filled with the sample at the input of the meter, were converted into the permittivity, ε', and the total dielectric loss, ε''_{tot} by means of an interpolation procedure described in detail elsewhere [9, 10]. Three different liquids of different dielectric constants and conductivities were employed as reference liquids. The overall accuracy of the experimental setup was within 0.2 dielectric units for ε' and within 0.1 dielectric units for ε''_{tot}. The dielectric loss, ε''_{diel}, is evaluated from the relation

$$\varepsilon''_{diel} = \varepsilon''_{tot} - \frac{\sigma_0}{\varepsilon_0 \omega} , \tag{1}$$

where σ_0 is the low-frequency limit of the electrical conductivity, ε_0 the dielectric constant of free space and ω the angular frequency of the applied electric field.

At the composition and temperatures investigated, the system behaves as a collection of noninteracting spherical water droplets (or interacting by a short-range attractive potential) whose size is governed by the W according to

$$R_H = \left[\left(\frac{3V}{a_0} \right) W + \left(\frac{3V_H}{a_0} \right) \right] + \delta , \tag{2}$$

where V is the volume of an individual water molecule, V_H and a_0 are the average volume and the area occupied by the surfactant head group and δ is the length of the surfactant molecule ($\delta = 1.05$ nm in the case of AOT without the SO_3^- group). In the present case, the hydrodynamic radius, R_H, of the surfactant-coated water particles is estimated to be about 75–80 Å.

We studied this system below the percolation threshold, i.e., for each value of Φ, at temperatures where the water droplets are essentially isolated from each other and the complex phenomena resulting in the formation of large-sized bicontinuous clusters do not occur. In the region of the phase diagram investigated, the system behaves as a typical colloidal system of conducting spheres uniformly distributed in a nonconducting medium and should be analysed following the typical methods of highly dispersed heterogeneous systems [11].

Results and discussion

Typical dielectric spectra of a microemulsion at $\Phi = 0.30$ and $W = 40.8$ at some selected temperatures between 5 and 15 °C are shown in Fig. 1. As can be seen, as the temperature is increased, by approaching the percolation threshold, the dielectric increment increases. Nevertheless, also at temperatures far from percolation, a well-defined relaxation process occurs, characterized by a dielectric strength of some units and a relaxation time of the order of 5 ns. The same data in the typical Cole–Cole semicircle representation are shown in Fig. 1.

As previously stated, we confined our investigation to microemulsions below percolation. Figure 2 shows the electrical conductivity behaviour at temperatures close to and above percolation, where the very large conductivity increase, over 4 orders of magnitude, is the index

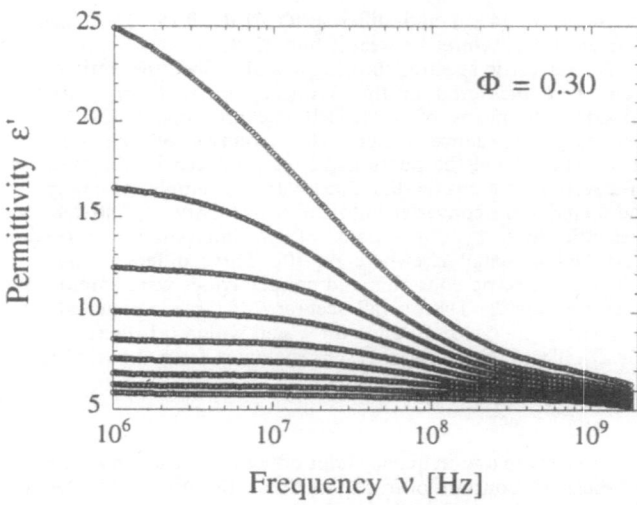

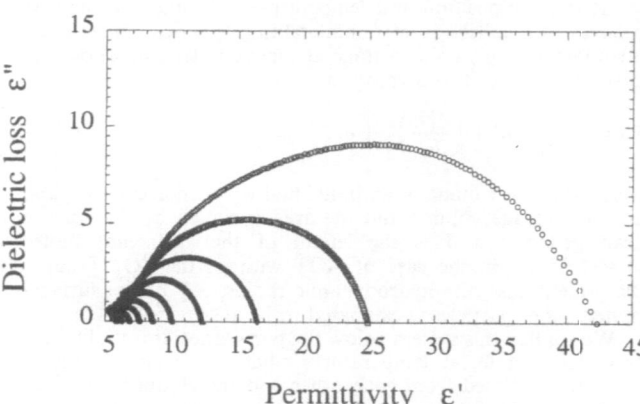

Fig. 1 The dielectric relaxation of a water-in-oil microemulsion system at the water-to-surfactant molar ratio $W = 40.8$ and volume fraction $\Phi = 0.30$ at different temperatures from 5 to 15 °C. The frequency range investigated covers the interval from 1 MHz to 1.8 GHz. The same data are also plotted in a typical Cole–Cole semicircle representation (*bottom*)

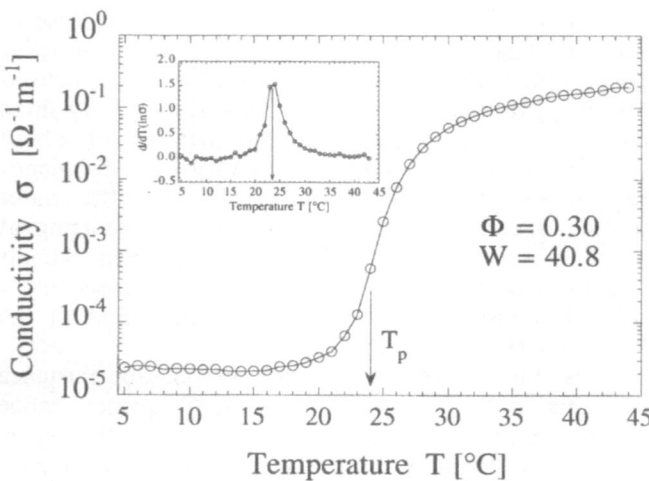

Fig. 2 Typical electrical conductivity behaviour of a water-in-oil micro-emulsion system ($W = 40.8$, $\Phi = 0.30$) over an extended temperature interval, showing the occurrence of percolation. In the present case, the percolation temperature, T_p, is 24 °C

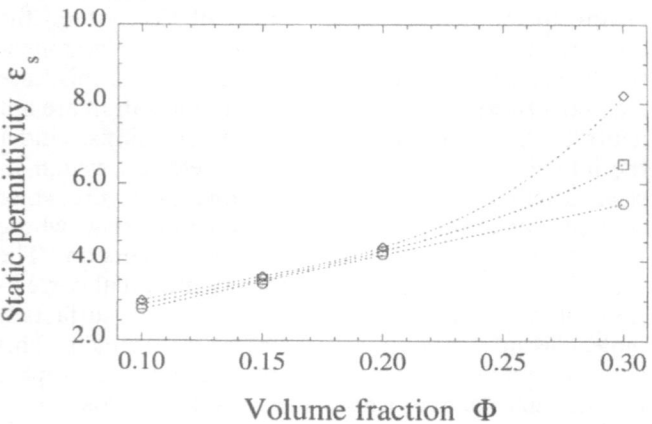

Fig. 3 The static permittivity, ε_s, of a water-in-oil microemulsion system as a function of Φ at three different temperatures: 5 °C (○); 10 °C (□); 15 °C (◇). The deviation from a straight line at higher temperatures and higher volume fractions evidences the beginning of percolation

that a new structural arrangement modifies the whole transport mechanism occurring in the system. The dependence of the static dielectric constant on Φ at three selected temperatures is shown in Fig. 3. Deviation from a straight line at higher temperatures and higher Φ evidences the beginning of percolation.

We are not aware of any other colloidal system exhibiting dielectric relaxations with dielectric strengths as large as 10 dielectric units at frequencies around 10^7–10^8 Hz, with the exception of aqueous solutions of zwitterionic phospholipids [12–14].

This behaviour is an indication of a new polarization mechanism and demands a physical explanation. Here we make a first attempt at this. In what follows, we analyse the two mechanisms separately and derive the main expressions for the dielectric increment of this relaxation process.

As far as the bulk counterion distribution within the water core is concerned, the electrical polarization of a conducting spherical particle of permittivity ε_p^* and radius R (the radius of the water core) in a nonconducting medium of permittivity ε_m^* results in an apparent polarization given by

$$\vec{P^*} = \varepsilon_m^* R \frac{1 - \frac{\varepsilon_m^*}{\varepsilon_p^*}}{1 + 2\frac{\varepsilon_m^*}{\varepsilon_p^*}} \vec{E} \qquad (3)$$

and the effective dielectric constant of the whole system can be written as

$$\varepsilon^* = \varepsilon_m^* + 4\pi \frac{N\vec{P}}{\vec{E}} \quad , \tag{4}$$

where N is the number of droplets per unit volume and $\vec{E}$ the external electric field. If the motion of counterions within each water droplet gives rise to a current density according to the following expression

$$J(\vec{r}) = ze\left[-n(\vec{r})u\nabla\phi - D\nabla n(\vec{r})\right] \quad , \tag{5}$$

where $n(\vec{r})$ is the concentration of ions of charge ze, mobility u and diffusional coefficient D and ϕ is the electrical potential, ε_p^* of the water droplet in Eq. (3) must be replaced with [15]

$$\varepsilon_p^* \to \frac{\varepsilon_p^*}{\left[1 - \frac{K_D^2}{\gamma^2}\left(\frac{n-m}{n}\right)\right]} \quad , \tag{6}$$

where K_D is the inverse of the Debye screening length, $\gamma^2 = K_D^2 + i\omega/D$ and the quantities m and n are defined according to

$$n = (\gamma^2 R^2 + 2)\sinh\gamma R - 2\gamma R\cosh\gamma R$$
$$m = \gamma R\cosh\gamma R - \sinh\gamma R \quad . \tag{7}$$

This substitution yields the effective permittivity of the microemulsion system, whose static value (in the limit $\omega \to 0$) is

$$\varepsilon = \varepsilon_m + 3\Phi\frac{\varepsilon_m\left[\left(\frac{\sigma_p}{\sigma_m}\right)^2 - 2a\left(\frac{\sigma_p}{\sigma_m}\right) + 2(a-2)\right] + 3\varepsilon_p}{\left[2a + \left(\frac{\sigma_p}{\sigma_m}\right)\right]^2} \quad , \tag{8}$$

where a is the real part of the denominator in Eq. (6), i.e., $a = Re\left[\frac{K_D^2}{\gamma^2}\left(\frac{n-m}{n}\right)\right]$.

The dielectric increment normalized to Φ as a function of the ratio σ_p/σ_m in the two extreme cases $a = 1$ and $a = 0$ is shown in Fig. 4. As can be seen, a dielectric increment comparable with those observed experimentally occurs for values of σ_p/σ_m lower than 10. In the present case, the electrical conductivity of the external medium is extremely low (of the order of 10^{-7}–10^{-9} Ω^{-1} m^{-1}), whereas the conductivity of the aqueous core is of the order of 10^{-2}–10^{-3} Ω^{-1} m^{-1}). At these values of σ_p/σ_m it corresponds to a completely negligible dielectric increment, independent of the ion polarization contribution governed by the parameter $K_D R$, despite the dielectric increment of several units, as found experimentally. The above considerations rule out the mechanism of bulk counterion polarization as being responsible for the observed dielectric increment.

As already stated, a further mechanism derives from motion of the anionic head groups with respect to the apolar part of the surfactant molecules at the water

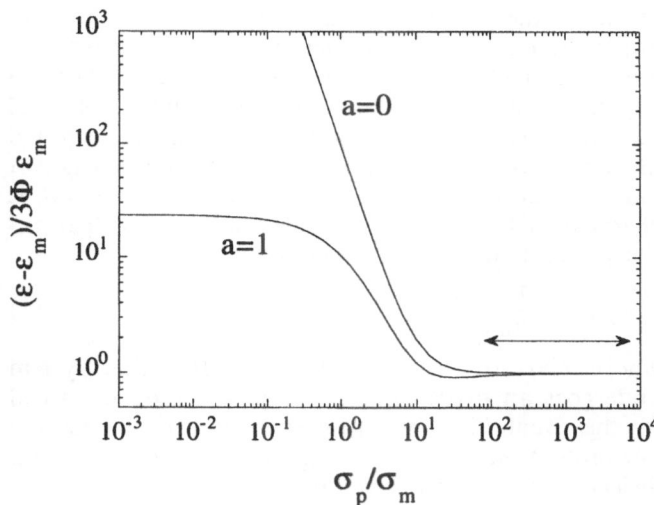

Fig. 4 The reduced dielectric increment due to the polarization of each water droplet induced by counterion motion within the water core (Eq. 8), normalized to Φ, as a function of the ratio σ_p/σ_m. A measurable effect occurs only for σ_p/σ_m lower than 10. In the case of a microemulsion, σ_p/σ_m exceeds this limit and reasonable values are marked by the *arrow*

droplet boundary. These motions, correlated over a distance of the order of 1–5 Å, impart a fluctuating electric dipole moment to the whole water droplet which, in turn, gives rise to a dielectric relaxation.

Dielectric studies on aqueous solutions of zwitterionic phospholipids [12–14] have shown the occurrence of a dispersion of strength of a few dielectric units centred at about 80–100 MHz between that due to counterions at lower frequencies and that attributed to the rotational diffusion of water molecules at microwave frequencies. Here, the model postulates that cationic groups, characterized by a mobility u, perform diffuse motion relative to the anionic groups around an axis perpendicular to the permanent electric dipole. In the case of AOT, an anionic double-chain surfactant, the anionic head group motion with respect to the apolar part of the molecule occurring within a distance ξ causes an increase in the surface droplet polarizability given by

$$\alpha = \frac{g\bar{n}(\xi e)^2}{2k_B T} \quad , \tag{9}$$

where g denotes an orientational correlation factor taking into account the existence of neighbouring head groups with identical orientation and $\bar{n}$ is the surface number density of correlated ionic groups. The surface polarizability, α, can be converted into the bulk polarizability of the surface head group layer coating the water droplet, whose dielectric constant ε_s can be written as

$$\varepsilon_s = \frac{4\pi\alpha}{3R} \quad . \tag{10}$$

Owing to the presence of this polarized layer (the head group region) at the interface between the water core and the surfactant chains, the whole system behaves as a triphasic system, being composed of an inner spherical aqueous medium disjoined from the apolar external medium (surfactant chains and oil) by the surfactant ionized head group layer. This layer, of thickness d, is characterized by a complex dielectric constant (permittivity ε_s and electrical conductivity σ_s)

$$\varepsilon_s^* = \varepsilon_s - i\frac{\sigma_s}{\omega\varepsilon_0} \tag{11}$$

whose real part is given by Eq. (10). The system undergoes an overall interfacial polarization described by the usual Maxwell–Wagner effect [11]. Under the assumption $\sigma_s < \sigma_p$, $\varepsilon_s < \varepsilon_p$ and $\sigma_m \approx 0$, the total dielectric increment is given by

$$\Delta\varepsilon = 9\Phi\frac{d}{R}\left(\frac{\varepsilon_m}{(1-\Phi)^2\varepsilon_s + (1-\Phi)(2+\Phi)(d/R)\varepsilon_m}\right) . \tag{12}$$

Fitting Eq. (12) to the measured dielectric increment yields the values of the model parameter $(g\xi^2)$ related to the cooperativity factor through ξ. In this analysis, we used $R = R_H - \delta = 67.5$ Å, $d = 5$ Å and $n = 1/65$ Å^{-2}. The diffusion path length, ξ, of the anionic groups is not known, so the correlation factor cannot be explicitly calculated; however, if ξ is assumed to be at most 5 Å, the minimum value of g is of the order of 30. This means that a head group correlation induced by the external electric field exists at the water–surfactant interface, causing a new polarization mechanism in these systems. A more accurate analysis on the basis of this model, taking into account that in a triphasic system more than two different dielectric dispersions could occur, is necessary for a full description of the orientational correlation of the anionic groups of the surfactant molecules.

Finally, associated with the dielectric dispersion, a well-pronounced electrical conductivity dispersion occurs in the same frequency range. The typical behaviour of the electrical conductivity of a water-in-oil microemulsion with $W = 40.8$ and $\Phi = 0.30$ is shown in Fig. 5. This dispersion will be analysed on the basis of appropriate triphasic mixture equations elsewhere.

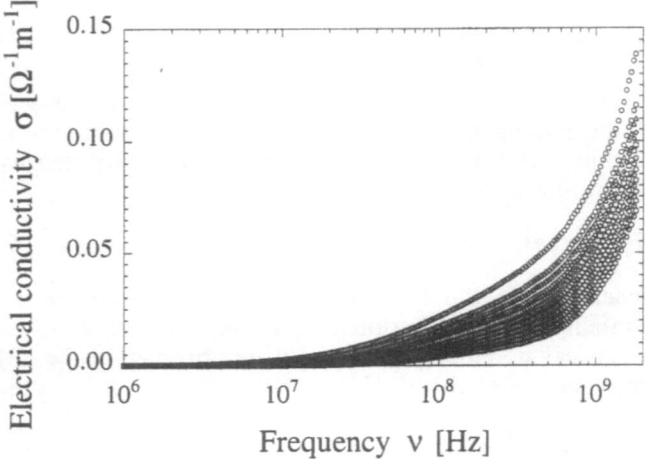

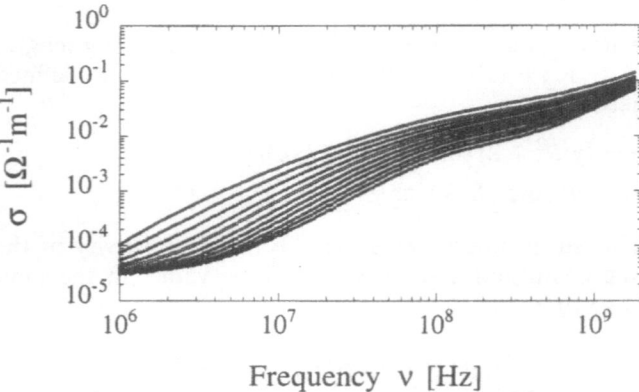

Fig. 5 The electrical conductivity behaviour of a water-in-oil microemulsion system ($W = 40.8$, $\Phi = 0.30$) as a function of frequency from 1 MHz to 1.8 GHz at different temperatures from 5 to 15 °C. The same date are plotted (*bottom*) on a log–log scale to evidence possible behaviour described by scaling law

The polarization mechanism associated with an induced correlation of the surfactant head groups is able to take into account the observed high-frequency dielectric dispersions in the water-in-oil microemulsion system. A detailed analysis of this relaxation region on the basis of the previously stated polarization mechanism, in a wider concentration range and a more extended temperature interval, is in progress.

References

1. Shah DO (ed) (1998) Micelles, microemulsions and monolayers: science and technology. Dekker, New york
2. Chen SH, Huang JS, Tartaglia P (eds) (1991) Structure and dynamics of strongly interacting colloids and supramolecular aggregates in solution. NATO ASI Series 369, Kluwer Academic Press, Boston
3. Degiorgio V, Corti M (eds) (1985) Physics of amphiphiles: micelles, vesicles and microemulsions. Proceedings of the International School of Physics "E. Fermi" course XC, North Holland, Amsterdam
4. Havriliak S, Havriliak SJ (1997) Dielectric and mechanical relaxation in materials. Hanser, Munich
5. Schwan HP, Schwarz G, Maczuk J, Pauly H (1962) J Phys Chem 66:2626
6. Schwarz G (1962) J Phys Chem 66:2636
7. Takashima S (1989) Electrical properties of biopolymers and membranes. Hilger, Bristol
8. Hasted JB (1973) Aqueous dielectrics. Chapman & Hall, London

9. Takashima S, Casaleggio A, Giuliano F, Morando M, Arrigo P, Ridella S (1986) Biophys J 49:1003

10. Bordi F, Cametti C, Paradossi G (1996) Biopolymers 40:485

11. Clausse M (1983) In: Becher P (ed) Encyclopedia of emulsion technology, vol 1. Dekker, New York, pp 481–715

12. Trukhan EM (1967) Sov Phys Solid State 4:2560

13. Pottel R, Gopel KD, Henze R, Kaatze U, Uhlendorf U (1984) Biophys Chem 19:233

14. Uhlendorf U (1984) Biophys Chem 20:261

15. Kaatze U, Muller SC, Eibl H (1980) Chem Phys Lipid 27:263

Progr Colloid Polym Sci (2000) 115:50–54
© Springer-Verlag 2000

A. Terreros
P. Galera Gomez
E. Lopez-Cabarcos

Influence of the surfactant chain length and the molecular weight of poly(oxyethylene) on the stability of oil-in-water concentrated emulsions

A. Terreros · P. Galera Gomez
E. Lopez-Cabarcos (✉)
Departamento de Quimica Fisica II
Facultad de Farmacia
Universidad Complutense, 28040
Madrid, Spain
e-mail: cabarcos@farm.ucm.es
Tel.: +34-91-3941751
Fax: +34-91-3942032

Abstract The effect of n-alkylammonium chloride surfactants on the stability of concentrated oil-in-water emulsions has been investigated as a function of the surfactant chain length. Surfactants with longer - hydrocarbon chains were found to form droplets with smaller sizes and to enhance the emulsion stability.

An additional increase in the stability of the oil-in-water concentrated emulsions was achieved by using poly(oxyethylene) with $M_w = 10000$.

Key words Concentrated emulsions · Surfactants · Interfacial free energy · Poly(oxyethylene)

Introduction

A concentrated emulsion is defined as an emulsion in which the volume fraction, Φ, of the dispersed phase is greater than 0.74. This number represents the maximum volume fraction corresponding to a compact arrangement of spheres of equal size; however, emulsions containing a higher volume fraction of the dispersed phase have been attained [1–5] and their properties have been studied extensively [5–10]. The study of highly concentrated emulsions is of both scientific [11–14] and practical importance, and the applications of these gel-like emulsions gained increased attention over the last decade [15–19]. It has been demonstrated that the polymerisation of monomers included in the dispersed phase of concentrated emulsions is rapider than polymerisation in bulk and provides polymers with higher molecular weights [16]. Simultaneous polymerisation of monomers included in both the dispersed and the continuous phase of a concentrated emulsion lead to hydrophobic–hydrophilic composite polymers with a wide range of possibilities [20–22].

It is well established that emulsions may undergo a number of breakdown processes determined by the interaction forces between droplets, namely creaming, sedimentation, flocculation, Ostwald ripening and coalescence. The stability of concentrated emulsions has been investigated extensively from the theoretical and experimental point of view; however, the question of whether the surfactant hydrocarbon chain length influences the stability of concentrated emulsions has, to the best of our knowledge, never been addressed. In this paper we report an experimental study of the stability of oil-in-water (o/w) concentrated emulsions prepared with ionic surfactants of different chain length. On the other hand, polymers have given excellent results when they have been used to stabilise emulsions. The influence of the molecular weight of poly(oxyethylene) (POE) on the stability of the o/w concentrated emulsions is also presented.

Materials and methods

Materials

Dodecane (Fluka) was used as supplied and water was super-Q-Millipore grade. The surfactants were synthesised in our laboratory by passing gaseous hydrogen chloride through a solution of the n-alkylamine in ether. The resulting precipitate was washed several times with ethyl ether and then crystallised in an acetone/ethanol/ethyl ether mixture with solvent evaporation in vacuum. Commercial POE of different molecular weights (POEM_w) and poly(vinylpyrrolidone) (PVPM_w) from Sigma and dextran from Fluka were employed as received.

Emulsion preparation

The emulsions were prepared according to procedures outlined in the literature [5, 9]. The o/w concentrated emulsions were produced by dropwise addition, with a syringe, of the dispersed phase (dodecane) to a water solution of 0.48 M *n*-alkylammonium chloride surfactant. The homogenisation used was 10000 rpm for 5 min at room temperature (22 °C). This emulsion had a $\Phi \approx 0.91$; the surfactant was considered part of the continuous-phase volume. Concentrated emulsions prepared using *n*-alkylammonium chloride surfactants with 9, 10, 11 and 12 (henceforth referred to as C9, C10, C11 and C12) carbon atoms in the hydrocarbon chain were investigated. The surfactant concentration was in all cases above the critical micelle concentration (cmc), which was determined from surface tension measurements as is illustrated in Fig. 1. The values obtained for the cmc were 0.11 M for C9, 0.055 M for C10, 0.028 M for C11 and 0.012 M for C12.

Instrumentation

The particle size measurements in the range from 0.5 μm to about 150 μm were performed with a Galai Cis-1 particle analyser system, which combined a laser-based size analyser and a video-based shape analyser. Samples for analysis were prepared by diluting the concentrated emulsion with water. Dilution of a concentrated emulsion to lower values of Φ did not significantly alter the size distribution profile [18]. The maximum concentration of particles detected was 10^8 particles cm^{-3}. The laser-based size analysis employs the time-of-transition theory. According to this theory the time it takes a particle to cross a laser beam depends on the particle's diameter. Particle–laser interactions can thus be processed to yield diameter size. Aggregation and contamination problems are immediately identified using an image monitor which is connected to a charge-coupled-device camera in order to provide a live image of the material under study. For shape analysis the video camera transfers an image of the measurement area to the computer. Particles caught by the camera out of focus are rejected from the analysis. The frequency distribution is expressed in terms of the surface volume mean diameter, d_{av},

$$d_{av} = \frac{\sum nd^3}{\sum nd^2} , \qquad (1)$$

where *n* is the number of particles in a size range of 2 μm whose midpoint is *d*. Freeze-fracture scanning electron microscopy (SEM) studies were conduced on a Jeol model JSM-6400 microscope operating at 20 kV.

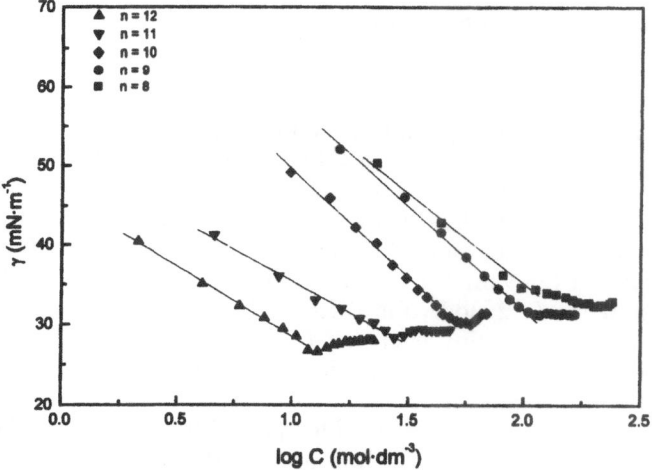

Fig. 1 Surface tension versus *n*-alkylammonium surfactant concentration in water at 20 °C. The hydrocarbon chain length of the surfactant was varied between 8 and 12 as indicated

Results and discussion

Influence of the hydrocarbon chain length of the surfactant

We use the Galai Cis particle analyser for measuring the size distribution of the $\Phi \approx 0.91$ emulsions prepared with surfactants C9, C10, C11 and C12. Droplet size profiles of these emulsions are given in Fig. 2. The values of d_{av}, the standard deviation and the mode of the distribution are presented in Table 1. The droplets are polydisperse and d_{av} decreases with increasing surfactant chain length.

The coalescence of the emulsions was followed by measuring the time evolution of d_{av}, which is a consequence of the aggregation of the emulsion droplets [23]. The increase in d_{av} is directly related to interactions between the emulsion droplets that promote flocculation and coalescence. Coalescence leads to phase-separation after 5 days for the concentrated emulsion prepared with surfactant C9. The time needed for phase-separation increases with the surfactant chain length and reaches a maximum of 22 days for the concentrated emulsions prepared with C12 (Fig. 3). This seems to indicate that the longer the hydrocarbon tail of the surfactant (smaller droplets) the more stable the concentrated emulsion.

Correlation between the hydrocarbon chain length of the surfactant and the interfacial free energy

Chen and Ruckenstein [9] have reported the correlation between the stability of concentrated emulsions and the interfacial free energy between the water and oil phases. According to Ruckenstein, the energy of interaction between the surfactant molecules and the two phases can be decomposed into two terms: the interaction between the hydrocarbon tails and the hydrophobic phase, and the interaction between the polar head groups and the water. The excess interaction energy per unit area between the two phases of the emulsion can be written as

$$\Delta U = U_{sw} + U_{so} - U_{wo} , \qquad (2)$$

U being the energy of interaction per unit area between the hydrophilic head groups of the surfactant and the water, U_{sw}, the hydrophobic tails of the surfactant and the oil phase, U_{so}, and the oil and the water phases, U_{wo}. The evaluation of the energy of interaction considers that the tails and the head groups constitute bulk phases and the excess interaction energy can be expressed as [9]

$$\Delta U = U_{sw} - \gamma_w + \gamma_{wo} + 2\phi(\gamma_s\gamma_o)^{1/2} - \gamma_o , \qquad (3)$$

where γ_s is the surface tension of the hydrocarbon tails, γ_o is the surface tension of the oil phase, γ_{wo} is the

52

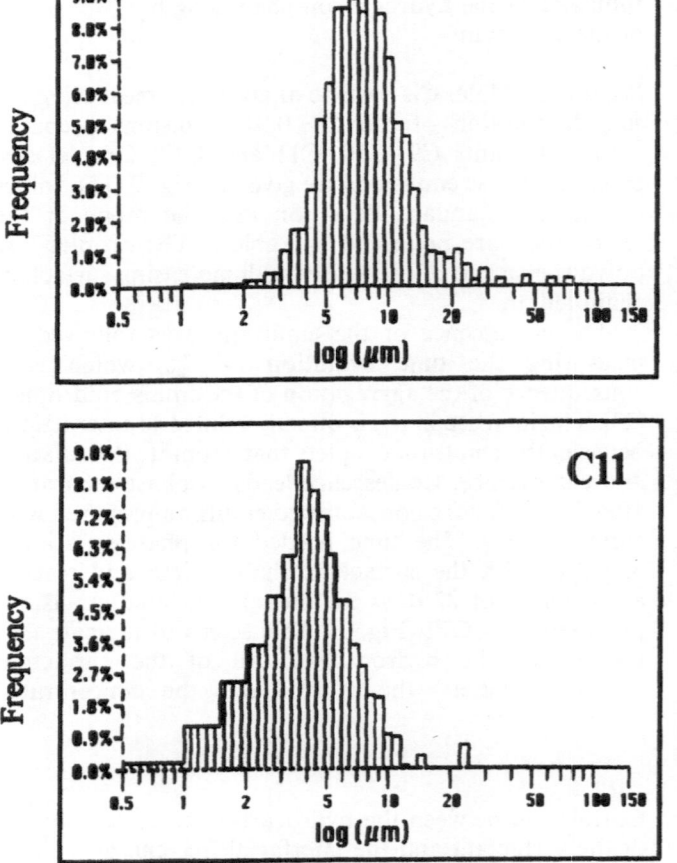

Fig. 2 Droplet size histograms for the dodecane/alkylammonium/water concentrated emulsions

Table 1 Values of the surface volume mean diameter, d_{av}, the standard deviation, σ, and the mode, m, of the emulsions whose size distribution profiles are shown in Fig. 2

	d_{av} (μm)	σ (μm)	m (μm)
C9	8.8	6.0	8.3
C10	5.2	3.1	4.8
C11	4.3	2.7	3.8
C12	2.9	1.7	4.1

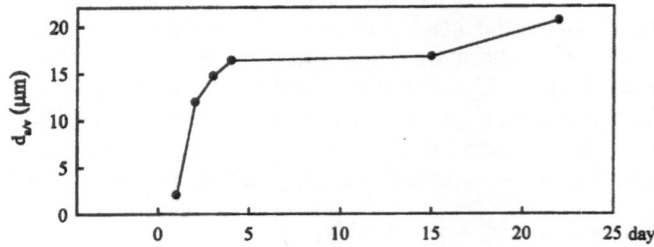

Fig. 3 Time evolution of the surface volume mean diameter, d_{av} for the oil-in-water concentrated emulsion

interfacial energy between the oil phase and the water and ϕ is the interaction parameter. Surfactants belonging to a homologous series have the same U_{sw} and, therefore, the quantity $U_{sw} - \gamma_w$ is constant and can be considered as a reference state [9]. In such a case, emulsions can be compared using the expression,

$$\Delta U_c = \gamma_{wo} + 2\phi(\gamma_s\gamma_o)^{1/2} - \gamma_o \ . \tag{4}$$

The values of ϕ, γ_o and γ_{wo} can be found in the literature [24]. The value of γ_s is taken to be the value of the surface tension of the alkane with the same number of carbons in the hydrocarbon chain; thus, for C9 we take the surface tension of nonane (22.8 mN m^{-1}), for C10 decane (23.9 mN m^{-1}), for C11 undecane (24.5 mN m^{-1}) and for C12 dodecane (25.4 mN m^{-1}). The calculated excess interaction energy, ΔU_c, is given in Table 2 and increases with increasing hydrocarbon chain length of the surfactant.

Chen and Ruckenstein [9] reported that the higher the interfacial free energy the greater the stability of the concentrated emulsion. Our results support this conclusion as illustrated in Fig. 4, where we plotted d_{av} as a function of ΔU_c for the concentrated emulsions prepared

Table 2 Calculated excess interaction energy per unit area, ΔU_c, between the two phases of the emulsion as a function of the surfactant chain length

	C9	C10	C11	C12
ΔU_c (m Nm^{-1})	62.8	64.0	64.5	65.5

Table 3 Numerical values of d_{av} and the stability of the concentrated emulsions prepared with the polymers indicated

	d_{av} (μm)	Stability
No polymer	2.3 ± 0.2	22 days
POE600	1.1 ± 0.4	32 days
POE4600	1.0 ± 0.4	32 days
POE10000	0.7 ± 0.2	>1 year
POE35000	No gel	
POE100000	No gel	
PVP49000	No gel	
Dextran35000	0.9 ± 0.3	4 months

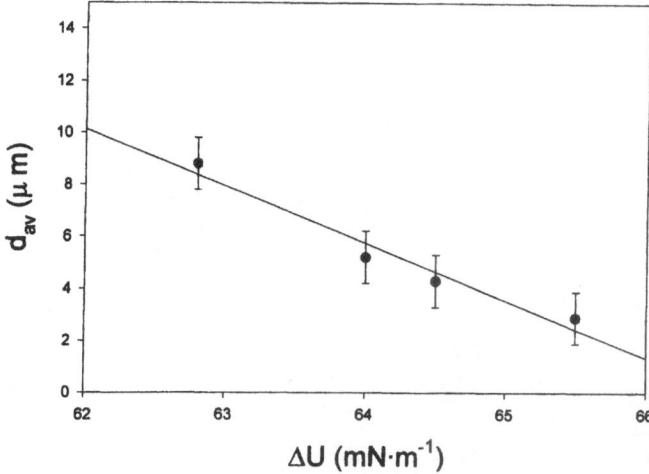

Fig. 4 d_{av} as a function of the interfacial free energy

with surfactants of varying chain lengths. Furthermore, the presence of micelles in the continuous phase cannot be ruled out when discussing the stability of concentrated emulsions [25]. Micelles are thermodynamically stable structures that can contain dissolved oil and the surfactant molecules would be partitioned between the micelles and the droplets. The micelle aggregation number for pure C12 in water is 80 [26] and for C9 is 28 [27]. It is known that the large micelles solubilise greater amounts of oil than the small micelles. The higher oil solubility of the C12 micelles as well as a steric factor could also effectively contribute to the increase in the stabilisation when the hydrocarbon chain length is increased.

Influence of the POE molecular weight

With the aim of increasing the half-life of the emulsion, we investigated the stabilisation by polymers of the concentrated o/w emulsions prepared with C12. The value of 22 days must be increased in order to obtain an emulsion with a half-life of 1 year, which is an acceptable level of stability for most practical applications. One possibility to increase the half-life of the emulsion is to enhance the viscosity; a second possibility is to introduce an energy barrier, B, between the droplets. The flocculation (and coalescence) is governed

by the Schmoluchowski rate equation [28], which provides the half-life

$$t_{1/2} = \frac{3B\eta}{8k_B Tn} \tag{5}$$

where k_B is Boltzmann's constant and η is the viscosity. There are two kinds of barriers: the electric double layer and adsorbed polymers. We investigated the stabilisation of the emulsions, prepared with C12 POE of different molecular weights (POE600, POE4600, POE10000, POE35000, POE100000) and also by PVP49000 and dextran ($M_w = 35000$). The stabilities of various o/w emulsions made with POE of different molecular weights are given in Table 3. The results in Table 3 indicate that concentrated emulsions do not form after the addition of high-molecular-weight polymers such as POE35000, POE100000 and PVP49000. In contrast, emulsions prepared with low-molecular-weight POE had droplet diameters in the range of about 0.7–1.1 μm, which are considerably smaller than the droplet diameter of the emulsion without POE.

Even though the role of the molecular weight of the polymer seems important it is not the only factor. Thus, POE and dextran of similar molecular weight show a completely different result and in one case (POE35000) the concentrated emulsion cannot be prepared whereas in the other case (dextran35000) the emulsion is stable for 4 months. The emulsion made with POE10000, which has a smaller droplet diameter, reaches the highest stability. This sample was stored at room temperature and after 9 months we took the SEM micrograph which is shown in Fig. 5. In this figure we can see the polyhedral cells of the dispersed phase separated by thin films of the continuous phase. The incorporation of the hydrophilic polymer within the thin aqueous film inhibits flocculation and increases the stability of the concentrated emulsion.

Conclusion

The stability of the concentrated emulsions increases following the increase in the hydrocarbon chain length of

54

Fig. 5 Scanning electron micrograph, the *magnification* is indicated at the *bottom right*, showing the polyhedral cells characteristic of the dispersed phase of concentrated emulsion ($\phi \cong 0.91$). The cells are separated by a thin film which was stabilised by the incorporation of POE10000. The micrograph was taken after about 9 months of ambient storage of the sample prepared with C12

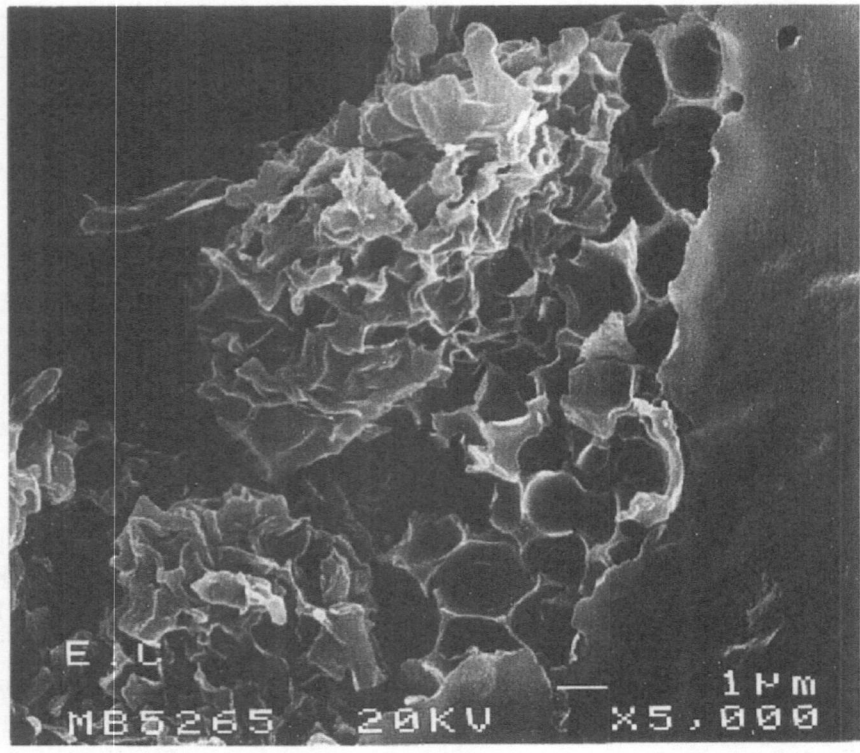

the surfactant, which produces an increase in the excess interaction energy. The stabilisation of concentrated emulsion with POE of different molecular weights seems to indicate that there is a value of the molecular weight ($M_w = 10000$) which gives the maximum stability to the emulsion and beyond which concentrated emulsions do not form.

Acknowledgements Grateful acknowledgements are due to DGI-CYT (grants PB95-0397 and PB96-0594) and to CICYT (Mat 97-1811 E) for the generous support of this investigation.

References

1. Lissant KJ (1966) J Colloid Interface Sci 22:462
2. Lissant KJ, Mayhan KG (1973) J Colloid Interface Sci 42:201
3. Lissant KJ, Peace BW, Mayhan KG (1974) J Colloid Interface Sci 47:416
4. Nixon J, Beerbower A (1969) J Am Chem Soc Div Petrol Chem 14:49
5. Princen HM, Aronson MD, Moser JC (1980) J Colloid Interface Sci 75:246
6. Princen HM (1983) J Colloid Interface Sci 91:160
7. Princen HM, Kiss AD (1986) J Colloid Interface Sci 112:427
8. Ruckenstein E, Ebert G, Platz G (1989) J Colloid Interface Sci 133:432
9. Chen HH, Ruckenstein E (1990) J Colloid Interface Sci 138:473
10. Bhakta A, Ruckenstein E (1995) Langmuir 11:1486
11. Kunieda H, Evans DF, Solans C, Yoshida M (1990) Colloids Surf 47:35
12. Solans C, Pons R, Zhu S, Davis HY, Evans DF, Nakamura K, Kunieda H (1993) Langmuir 9:171
13. Ozawa K, Solans C, Kunieda H (1997) J Colloid Interface Sci 188(2):275
14. Park JS, Ruckenstein E (1990) Polymer 31:175
15. Ruckenstein E, Li H (1996) Polymer 37:3373
16. Ruckenstein E (1997) Adv Polym Sci 127:1
17. Ruckenstein E, Li H (1997) Polym Compos 18(3):320
18. Das AK, Mukesh D, Swayambunathan V, Kotkar DD, Ghost PK (1992) Langmuir 8:2427
19. Paraskevopolou A, Kiosseoglou V, Alevisopoulos S, Kasapis S (1999) Colloids Surf B12:107
20. Kim KJ, Ruckenstein E (1988) Makromol Chem Rapid Commun 9:285
21. Park JS, Ruckenstein E (1989) J Appl Polym Sci 38:453
22. Ruckenstein E, Park JS (1990) J Appl Polym Sci 40:213
23. Egusa S (1982) J Colloid Interface Sci 86:135
24. Gardon JT (1965) Encyclopedia of polymer science technology, vol 3. Wiley, New York, p 838
25. Aronson MP (1992) In: Emulsions – A Fundamental and Practical Approach Sjöblom J (ed), NATO ASI Series 363, Kluwer Academic Publishers, Holland, p 75
26. Kushner LM, Hubbard WD, Parker RA (1957) J Res Nat Bur Stand US 59:113
27. Lopez-Cabarcos E, Galera Gomez PA (1994) Nuovo Cimento D16:1515
28. Friberg SE (1992) In: Emulsions – A Fundamental and Practical Approach Sjöblom J (ed), NATO ASI Series 363, Kluwer Academic Publishers, Holland, p 1

Progr Colloid Polym Sci (2000) 115: 55–58
© Springer-Verlag 2000

A. M. Puertas
A. Fernández-Barbero
F. J. De las Nieves

Aggregation between oppositely charged colloidal particles

A. M. Puertas · A. Fernández-Barbero
F. J. De las Nieves (✉)
Group of Complex Fluids Physics
Department of Applied Physics
University of Almería
04120 Almería, Spain

Abstract The aggregate structure and long-time kinetics of aggregation between oppositely charged particles is studied. Experimental and simulation results show that heteroaggregation produces more branched aggregates than diffusive aggregation. The kinetics is shown to slow down as time proceeds. Both results are connected by considering the friction of the aggregates with the suspension medium.

Key words Polymer colloids · Colloidal aggregation · Heterocoagulation

Introduction

Aggregation between particles of different characteristics (sizes, charges, materials, etc.) is generally known as heteroaggregation [1]. This type of process is quite common in nature, as well as in industry [2, 3]. Heteroaggregation between two systems of hard spheres bearing opposite charges is of paramount interest since long-range attractive forces control the aggregation.

In this work, we study the long-time behaviour of aggregation between oppositely charged particles. Aggregation experiments using polystyrene latexes and Brownian dynamics (BD) simulations are presented, covering the cluster structure as well as the aggregation kinetics. The agreement between simulation and experiment indicates that the presence of electrical attractive long-range forces in the system is responsible for the behaviour. The initial stages of this aggregation have been studied previously [4].

Materials and methods

Two polystyrene latexes with different sign of charge were used as hard spheres. The presence of weak groups in the surfaces of the latexes, carboxyl and amidine for the anionic and cationic latexes, respectively, produced a pH-dependent surface charge. The latexes were synthesized following the methods and recipes described in Ref. [5] and in Ref. [6] for the anionic and cationic latexes, respectively.

The particle sizes were obtained by transmission electron microscopy: (187 ± 7) nm for the anionic latex and (185 ± 9) nm for the cationic one. Both latexes presented low polydispersity indexes: 1.004 and 1.006, respectively.

The aggregation experiments were carried out at pH 4.5, where both latexes presented very similar stabilities [7]. The pH was adjusted using only HCl, which ensures a very low bulk ionic concentration and, thus, that the attraction is not screened. All of the aggregations took place at 25 °C.

Static and dynamic light scattering were employed for accessing the cluster structure and kinetics of aggregation, respectively. In static light scattering, an angular scan of the scattered intensity allows the calculation of the aggregate fractal dimension (d_f) by means of [8]

$$I(q) \approx q^{-d_f} \quad R^{-1} \ll q \ll a^{-1} \ , \tag{1}$$

where $q = 4\pi/\lambda \sin \theta/2$ is the wavevector (λ being the wavelength of light in the suspension medium and θ the dispersion angle). R and a are the aggregate and particle radii, respectively.

The mean hydrodynamic radius of the aggregates was measured as a function of time in order to study the aggregation kinetics. This magnitude not only depends on the aggregation kinetics, but also on the structure of the aggregates. Thus, it is more convenient to study the evolution of the mean number of particles per aggregate, which contains only kinetics information, and can be calculated from the hydrodynamic radius, once the fractal dimension is known:

$$\langle n \rangle \sim R_H^{d_f} \ . \tag{2}$$

The number-mean cluster size is directly related to the mean cluster mass, which obeys a power law in time $\langle n \rangle \sim t^z$, as shown by experiments, simulations and some kinetics models [8, 9]. The value of the kinetics exponent, z, is representative of constant velocity aggregation $(z = 1)$, accelerating reaction $(z > 1)$ or aggregation slowing down $(z < 1)$.

The results of BD simulations were compared with the experimental results. The simulation details are described elsewhere [4, 10]. The structure of the aggregates was studied by means of the integrated pair correlation function, $c(r)$, which is calculated simply by counting the number of pairs of particles separated by a distance less than r. The structure was proven to be fractal following the scaling behaviour

$$c(r) \sim r^{d_f} , \tag{3}$$

thus allowing the determination of d_f.

Results and discussion

The temporal evolution of the mean hydrodynamic radius is presented in Fig. 1. The salt concentration was kept very low in the homo- and heteroaggregations in order to make the interaction range as long as possible. In contrast, diffusive aggregation was obtained at high salt concentration. As expected, the attractive forces between oppositely charged particles make heteroaggregation the fastest reaction. In addition, homoaggregation is very slow, since repulsive interactions control the process.

In order to obtain information on the kinetics, the effects of cluster structure on the hydrodynamic size have to be taken into account; therefore, the structures of the aggregates must be determined. The scattered intensity is plotted versus the scattering vector in Fig. 2. It is important to notice the power law behaviour of I versus q for heteroaggregation, which implies a uniform internal fractal structure. Its fractal dimension, as shown in the plot, is much lower than the diffusive one. For the other two cases, the fractal dimensions are within the reported values in the literature for DLCA (Diffusion Limited Cluster Aggregation) and the tran-

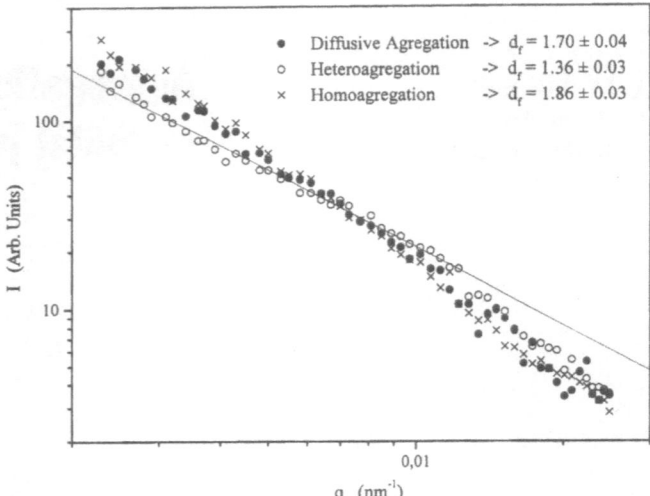

Fig. 2 Scattered intensity versus scattering vector. The linear fit is presented for the case of heteroaggregation to show the power-law behaviour

sition to RLCA (Reaction Limited Cluster Aggregation).

The low fractal dimension indicates that the structures of the clusters growing in heteroaggregation are more open than those formed under diffusive conditions. BD simulations were carried out to compare the results with those of experiments. The structure of the aggregates was studied by means of the integrated pair correlation function, $c(r)$.

$c(r)$ is shown for hetero- and homoaggregation and for diffusive aggregation in Fig. 3. In this plot, a linear behaviour in the log $c(r)$ versus log r plot is clearly observed, which implies uniform internal growing structures. The values for diffusive aggregation and for

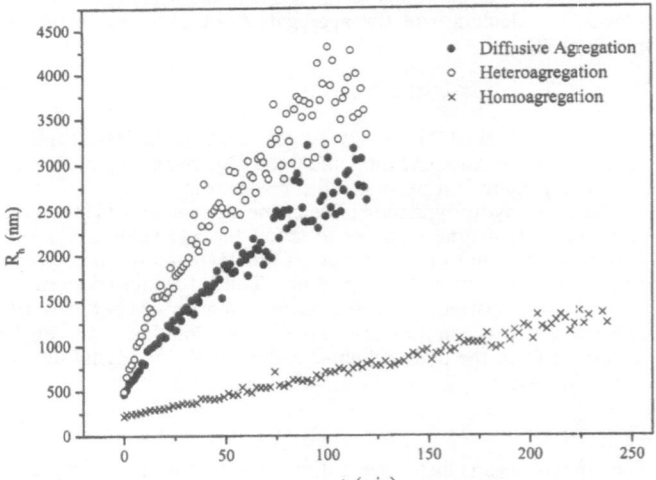

Fig. 1 Temporal evolution of the mean hydrodynamic radius

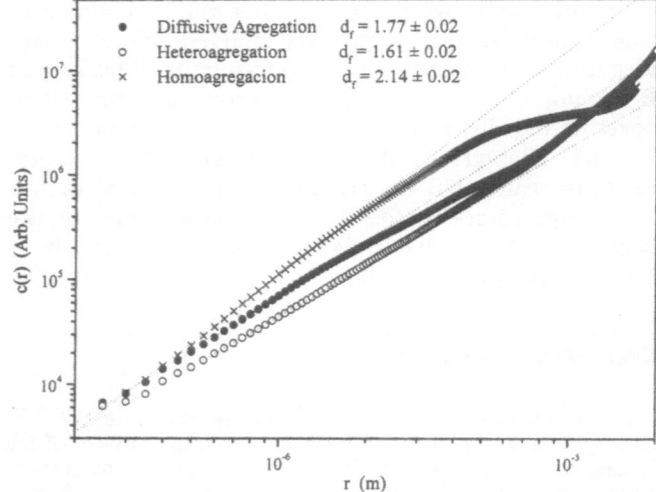

Fig. 3 Integrated pair correlation function from simulation

homoaggregation are typical of DLCA and RLCA, respectively. In contrast, more branched aggregates are produced in heteroaggregation, in agreement with the experimental results. However, it should be pointed out that the decrease in the experimental fractal dimension is larger than that from simulations. This difference arises from the fact that κ cannot be taken to be too low, since the potential range has to be kept within the simulation box size. This problem is not present in experiments.

The agreement between experiment and simulation concerning the decrease in the fractal dimension indicates that this effect is due to the presence of attractive forces in the system. This attraction is the driving force for aggregation, in contrast to the RLCA regime, where repulsive forces slow down the reaction.

Therefore, it is likely that the differences in the aggregation mechanism affect not only the cluster structure, but also the kinetics of aggregation. This features is dealt with in Figs. 4 and 5. In Fig. 4, the mean cluster mass, $\langle n \rangle$, is obtained from Fig. 1 and by taking into account the fractal dimensions of the aggregates. The kinetic exponent, z, is also shown in the figure.

The aggregation kinetics is tackled in Fig. 5 via simulations where the mean number of particles per aggregate is plotted for the three cases.

Comparison between Figs. 4 and 5 leads to some important conclusions. In all cases, the evolution of the mean cluster mass obeys a power law. The kinetics exponent is 1 for diffusive aggregation and is larger than 1 for homoaggregations, as expected. These features are observed both experimentally and with simulations and are within the values reported in the literature [8]; however, the kinetics exponent for heteroaggregation is lower than the diffusion one.

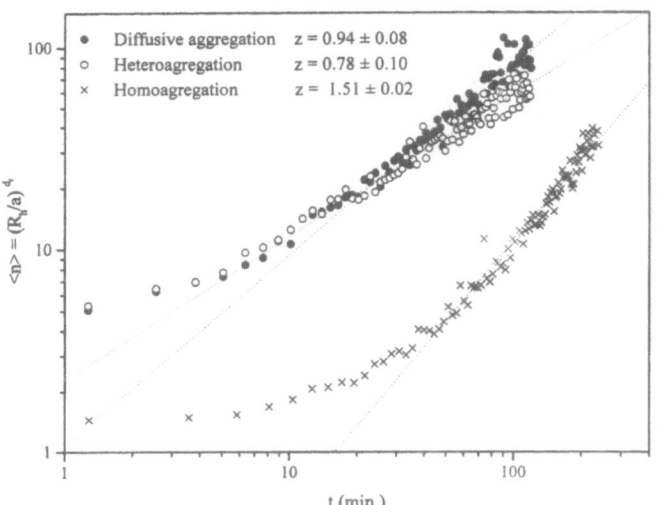

Fig. 4 Experimental temporal evolution of the average number of particles per aggregate

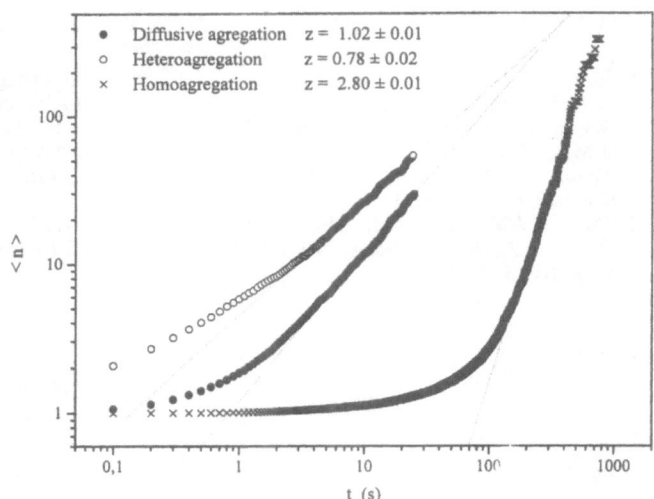

Fig. 5 Temporal evolution of the average number of particles per aggregate from simulation

This kinetics exponent is representative of a reaction that slows down as time proceeds. This fact can be explained by taking into account that as clusters grow by addition of both positive and negative particles their charge-to-mass ratio decreases. This feature leads to very fast aggregation in the initial stages, but to a slowing down when the aggregates grow and diffusion is the main driving force for aggregation; however, at this time, the process is mainly diffusive and DLCA results should be recovered.

Thus, a further effect must be present to obtain slower than diffusive aggregations. Since the aggregates have such open structures, their drag coefficient is larger than for aggregates formed in DLCA or RLCA. This effect will result in a much lower diffusion coefficient for heteroaggregates than for DLCA aggregates. Thus, heteroaggregation will be slower than diffusive aggregation.

Finally, we would like to remark that the presence of long-range attractive forces produces branched aggregates and an aggregation kinetics that slows down as time proceeds. These results have been obtained both experimentally and by computer simulation, which indicates that they are independent of the particle nature and that they are only related to the attractive long-range forces. The aggregation kinetics slows down since the aggregates reduce their charge-to-mass ratio as aggregation proceeds. Moreover, it becomes even slower than diffusive aggregation because the friction of the aggregates is larger than in diffusive aggregation.

Acknowledgements The financial support provided by CICYT (under project MAT 96-1035-C03-03) is greatly appreciated. The authors are also grateful to M.S. Romero-Cano for synthesizing the latexes used throughout this work.

58

References

1. Hogg R, Healy TW, Fuerstenau RW (1966) Trans Faraday Soc 62:1638
2. Wang Q, Heiskanen K (1992) Int J Miner Process 35:121
3. Wang Q, Heiskanen K (1992) Int J Miner Process 35:133
4. Puertas AM, Maroto JA, Fernández-Barbero A, de las Nieves FJ (1999) Phys Rev E 59:1943
5. Bastos D, Ortega JL, de las Nieves FJ, Hidalgo-Alvarez R (1996) J Colloid Interface Sci 176:232
6. Hidalgo-Alvarez R, de las Nieves FJ, van der Linde AJ, Bijsterbosch BH (1986) Colloids Surf 21:259
7. Puertas AM, de las Nieves FJ (1999) J Colloid Interface Sci 216:221
8. Vicsek T (1992) Fractal growth phenomena. World Scientific, Singapore
9. Botet R, Jullien R (1984) J Phys A 17:2517
10. Puertas AM, Fernandez-Barbero A, de las Nieves FJ (1999) Comput Phys Comm 121:353

Progr Colloid Polym Sci (2000) 115:59–65
© Springer-Verlag 2000

A. Stipp
C. Sinn
T. Palberg
I. Weber
E. Bartsch

Crystallizing polystyrene microgel colloids

A. Stipp (✉) · C. Sinn · T. Palberg
Johannes-Gutenberg-Universität
Institut für Physik, Staudingerweg 7
55099 Mainz, Germany
e-mail: andreas.stipp@uni-mainz.de
Tel.: +49-6131-3923636
Fax: +49-6131-3922991

I. Weber · E. Bartsch
Johannes-Gutenberg-Universität
Institut für physikalische Chemie
Jakob-Welder-Weg 11
55099 Mainz, Germany

Abstract Spherical microgel particles of sufficiently high degree of internal cross-linking and swollen in a good solvent in many respects behave quite similarly to hard-sphere colloids. Due to solvent uptake they can be refractive-index-matched and density-matched in suitable organic solvents. We present preliminary measurements of the crystallization kinetics of 1:10 cross-linked polystyrene microgel particles. We measured Bragg and small-angle light scattering of the solidifying shear melt. Two different scattering patterns, a set of Debye–Scherrer rings and a second ring pattern at small angles could be observed. We check for similarities and differences compared to previously investigated colloidal systems.

Key words Colloidal suspensions · Microgel particles · Hard spheres · Crystallization kinetics · Light scattering

Introduction

Colloidal suspensions of spherical particles with hard-sphere (HS)-like interactions have become a well-recognized experimental model system of soft condensed matter physics over the last few years. A large body of work has been carried out on systems of poly(methyl methacrylate) (PMMA) particles [1] and silica particles [2] in organic solvents and their structural and dynamic properties have been comprehensively studied. Another important aspect was their phase behaviour and the formation of crystalline long-ranged order under certain conditions [3]. The HS system shows a first-order freezing transition (the freezing point) at a volume fraction φ_f of 0.494; above the melting point ($\varphi_m = 0.545$) the sample is fully crystallized [4]. In-between fluid and crystal coexist. Also the kinetics of the crystallization kinetics has been investigated extensively and with various techniques [5, 6]. For example, Harland et al. [7] used Bragg light scattering (BLS) to monitor the temporal evolution of the crystal order on the length scale of inter particle distances. Small-angle light scattering (SALS) detects long-ranged density fluctuations, such as growing crystals with surrounding depletion zones and remaining fluid [8, 9]. From both, nucleation rates, growth velocities and crystal densities were extracted and analysed in terms of classical nucleation and growth theories [7, 9, 10].

Microgel particles, also known as micronetwork particles, on the other hand are spherical networks of cross-linked polymer chains which are swollen in a good solvent. They may be synthesized with high average cross-link density, which guarantees that free polymer chains on the particle surface are very short relative to the particle radius, R. Second, the volume swelling ratio, $S = R^3_{\text{swollen}}/R^3_{\text{unswollen}}$, can be tuned to be very small and then elastic deformation of the particles at high volume fractions can be avoided. Recent investigations show that under these conditions microgel particles may be regarded as well-defined model suspensions with HS-like potential [11, 12]. In addition, however, due to solvent uptake they can be density- (and index-) matched in suitable organic solvents; thus, sedimentation effects can be either introduced or avoided. Further, in principle, they offer the possibility of introducing subtle deviations from a pure HS interaction potential. In fact, for a smaller cross-link density, deviations towards more soft-sphere-like behaviour are clearly detectable [13, 14].

For these reasons there has been growing interest in the behaviour of microgel particles compared to the HS reference system. Up to now mainly states of short-range order both below and above the glass transition (located at $\varphi_f \approx 0.58$) have been investigated. There structural and diffusive data of highly cross-linked samples were observed to closely follow predictions for HS suspensions [12, 15, 16]. While it is known that monodisperse microgel particles do show crystallization, to the best of our knowledge, no measurements of the kinetics of solidification have been presented so far. Such data would be very desirable to have a further check of HS behaviour.

We here present the first measurements of the crystallization kinetics of a suspension of highly cross-linked microgels. The particles were synthesized and prepared to yield crystallizing systems. Microscopy allowed the selection of a suitable sample. We use combined BLS and SALS experiments to analyse the temporal evolution of crystal sizes and structural details. We observed many features very similar to those known from HS systems but we also observed specific differences.

Experimental

Polystyrene microgel particles were synthesized by surfactant-free emulsion polymerization with the cross-linker diisopropenyl benzene as co-monomer, using $K_2S_2O_8$ as the radical initiator and bidistilled water as the solvent. The average cross-link density was 1:10. Details of the synthesis and the subsequent freeze-drying process can be found elsewhere [17]. The particles were redispersed in 2-ethylnaphthalene (EN), which is an index- and density-matching, good solvent for the microgels.[1]

Samples with different volume fractions were made by filling the desired amount of stock suspension into rectangular light scattering cuvettes with dimensions 10×5 mm^2 and adding weighed amounts of solvent. The particles were left undisturbed for some months to equilibrate the swelling process. In the final state, they consisted of about 50% solvent. Dilute samples were characterized by SLS and dynamic light scattering, from which a particle radius of $R_{\text{swollen}} = 360$ nm and a polydispersity $\sigma \approx 6\%$ resulted.

Volume fractions of the more concentrated samples were calculated from the weight fractions [14] using a swelling ratio of $S = 2.1 \pm 0.1$ [18].[2] As this estimate introduces a nonnegligible uncertainty, the volume fractions given later are referred to as nominal volume fractions.

After equilibration samples of nominal volume fraction, φ, larger than φ_f, showed more or less complete crystallization but no

---

[1] $n_D^{PS} = 1.598, n_D^{DIPB} = 1.556, n_D^{EN} = 1.599, \rho^{PS} = 1.05\,\text{g cm}^{-3}, \rho^{DIPB} = 0.952\,\text{g cm}^{-3}, \rho^{EN} = 0.998\,\text{g cm}^{-3}$ (all values given for ambient temperature).

[2] The complete phase diagram of the particles under study here is not yet available. Therefore we used a similar batch of the same cross-linking density to estimate the swelling ratio leading to typical values of $S = 2.1 \pm 0.1$.

detectable sedimentation effects. Fully crystalline samples recrystallized within roughly 1 day after shear-melting. Thus, their solidification kinetics is conveniently accessible in both the early stage of nucleation and growth and in the later stage of ripening.

Under white light illumination beautiful iridescence indicated the presence of ordered structures with interparticle spacings on a length scale comparable to the wavelength of visible light. We note that not all samples showed a homogeneously crystalline appearance. One of the exceptions is shown in Fig. 1. In this sample of $\varphi = 0.55$, both crystalline regions (as identified by the brilliant opalescence of the crystallites) and noncrystalline regions (showing a continuous rainbowlike change in colours) appeared. The boundary between the two regions was observed to be rather sharp: the reason for this is not yet known, but it possibly originates from a transient "polydispersity" caused by still incomplete swelling. Other samples at volume fractions well above φ with very small crystallites showed regions of preferred orientations as identified by a still specular scattering pattern but a more continuous colour change. These possibly result from the shear-melting process applied before. Similar observations were made on HS and are shown and discussed in Ref. [19].

To avoid artefacts such as structural or orientational inhomogenieties we chose the sample of $\varphi \approx 0.54$, shown in Fig. 2. It shows randomly oriented, homogeneously nucleated crystals over the whole sample volume. After several days the average crystal diameter was about 50–100 μm.

We note the appearance of "streaky" crystals, which indicates twinning. This is also well known from crystals of charged spheres [20]. For PMMA spheres–crystallizing in either face-centred-cubic (fcc) or hexagonal-close-packed lattices–the formation of a twin boundary corresponds to the introduction of a stacking fault [21]. While in our case the number of stacking faults seems to be rather small at late stages, in some HS cases even complete random stacking has been observed [22].

The crystallization process was investigated by BLS and SALS. To avoid parasitic reflections the sample cuvette was placed in an index-matching bath containing EN and the detection optics were chosen to be of similar refractive index. For illumination we used a widened (diameter 8 mm) He-Ne laser beam ($\lambda_0 = 632.8$ nm). This was focused onto the centre of the small-angle detector (distance from sample about 1.5 m). To avoid mechanical disturbances, the sample was fixed and the Bragg detector, which was an array of photodiodes mounted on an arm, was rotated around the optical axis. Thus, either the full two-dimensional scattering pattern or

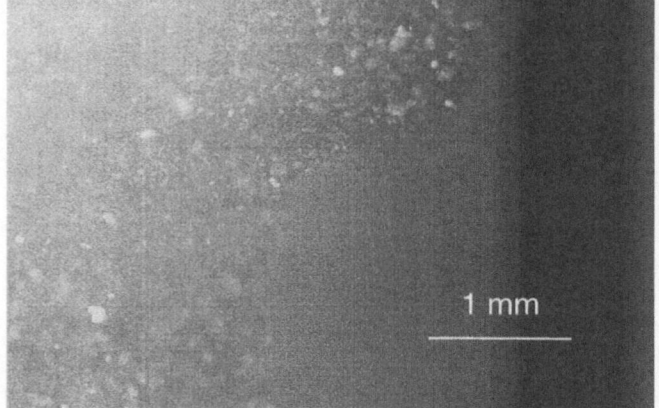

1 mm

Fig. 1 Incomplete crystallization in a sample of nominal volume fraction of $\varphi = 0.55$. See text for details

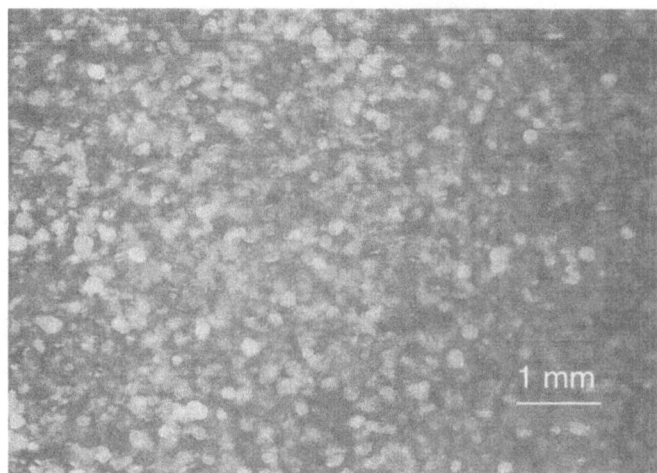

Fig. 2 Complete and homogeneous crystallization in a sample of nominal volume fraction of $\varphi = 0.54$

after angular averaging the usual Debye–Scherrer pattern was recorded. A similar construction was used for the small-angle regime, where a ring pattern was observed to grow in time. This light scattering apparatus was described in more detail before [23]. Due to technical reasons we performed the BLS and SALS measurements subsequently on each sample.

Before each measurement, the sample was slowly rotated for some hours to destroy any crystalline order. Abortion of shear and insertion of the sample into the experiment defines $t = 0$ min. A single measurement (one rotation of the detector arms with a stepper motor and detection of the scattered intensity in about 6000 steps) lasted about 30 s. Since the crystallization process occurred on times scales of hours to days it could be monitored with high temporal resolution.

Results

BLS results

BLS detects the time- and q-dependent scattered intensity, $I(q, t)$, on the length scale of the interparticle distances. Here $q = (4\pi n_S/\lambda_0)\sin(\theta/2)$ is the magnitude of the scattering vector, with n_S being the refractive index of the solvent, λ_0 the laser wavelength and θ the scattering angle. For better statistical accuracy, we summed up the scattered intensity over the whole ring. To correct for the form factor, $P(q)$, of the particles and the slightly different sensitivities of our photodiodes, we divided $I(q, t)$ by the intensity, $I(q, t = 0$ min$)$, at which only the shear molten state was present in the sample. We are aware that this procedure is somewhat problematic since we not only divide by $P(q)$, but also by other contributions, such as scattering from the remaining fluid. Further, after the appearance of crystallites an additional background may be present. In particular, completely random stacking may lead to a very pro-

nounced broadening of selected peaks, thus affecting neighbouring reflections [22]. We note that more accurate correction procedures than the ones used here are available. For details see Refs. [7, 24].

In Fig. 3 the normalized scattering intensity, $I(q, t)/I(q, 0)$, for three selected times from $t = 10$ min up to $t = 3380$ min shows different growing Bragg peaks. They can be indexed $\{111\}$, $\{200\}$, $\{220\}$, $\{311\}$ and $\{222\}$ using the Miller indices of a fcc structure. We note that in the full two-dimensional scattering both $\{111\}$ and $\{311\}$ showed a sixfold symmetric scattering pattern superimposed on a ring pattern. This has been observed before in HS systems and originates from the presence of wall-nucleated, shear-oriented crystals [24]. The kinetics measured on angular-averaged data of both peaks shows two contributions and is not easy to interpret. Further, a distortion in the vicinity of the $\{111\}$ reflection is present, probably due to the division by the fluid structure factor. $\{200\}$ and $\{222\}$ are very weak in comparison to the other peaks. $\{200\}$, which is expected to occur for a pure fcc lattice only, in particular may be subject to stacking-fault broadening and may contribute a nonnegligible background to the interlayer $\{111\}$ reflection. Its weakness in the present case indicates a considerable degree of random stacking present in our sample at early times.

For these reasons we refrained from further evaluating $\{111\}$, $\{200\}$, $\{311\}$ and $\{222\}$. In contrast, the fluid background at the position of the $\{220\}$ reflection can be well approximated by a straight line [24] and no significant additional background due to random stacking or contributions from wall crystals should be present at this position. To isolate the peak we chose two limiting q values, q_1 and q_2, and subtracted a linear

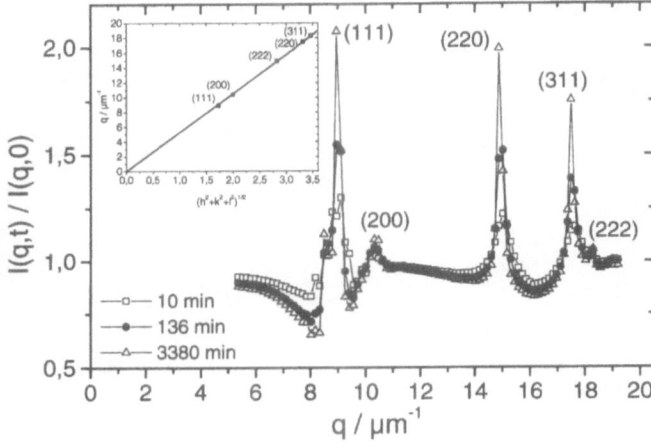

Fig. 3 Temporal evolution of the Bragg light scattering signal normalized as $I(q, t)/I(q, 0)$ for $t = 10$ min ($\square$), 136 min ($\bullet$) and 3380 min ($\triangle$). The indexing uses Miller indices and the insert shows a face-centred-cubic structure to be present

function $I_U(q, t)$ through $I(q_1, t)$ and $I(q_2, t)$, resulting in the structure factor in the vicinity of the reflection:

$$S_{220}(q, t) = \frac{I(q, t)}{I(q, t = 0)} - I_U(q, t) \ . \tag{1}$$

The result for selected times covering more than 3 orders of magnitude is shown in Fig. 4. From $S_{220}(q, t)$ the area under the peak, $A_{220}(t)$, the full width at half maximum, $\Delta q_{220}(t)$, and the position of the maximum, $q_{220}(t)$, can be determined. An increase in $A_{220}(t)$ indicates an increase in the amount of crystalline material through both nucleation and growth. We fitted a Lorentzian function to $S_{220}(q, t)$ and integrated it between q_1 and q_2. We note that there is no meaning behind this choice other than a convenient parameterization of our data. A decrease in $\Delta q_{220}(t)$ indicates increasing crystal sizes. In fact, we may further calculate the radii, $R_C(t)$, from

$$R_C(t) = \frac{2\pi K}{\Delta q_{220}(t)} \ , \tag{2}$$

with the constant $K = 1.107$ resulting from the assumption of crystallites of spherical shape [25].

The results for $A_{220}(t)$ and $R_C(t)$ are shown in Fig. 5 in a log-log plot. $A_{220}(t)$ shows a steep increase only at short times up to 10 min and then remains constant up to the longest times investigated. Since $A_{220}(t)$ is proportional to the crystallized volume in the sample, our data seem to indicate that after 10 min the processes of nucleation and growth have already finished and that the sample has completely solidified. On the other hand, $R_C(t)$ increases continuously. $R_C(t)$ is well described by a single power law of t^α, with $\alpha \approx 0.28$, until $t \approx 400$ min. Note that for our microgel particles, no transition regime is visible in $R_C(t)$ at $t = 10$ min. This is markedly

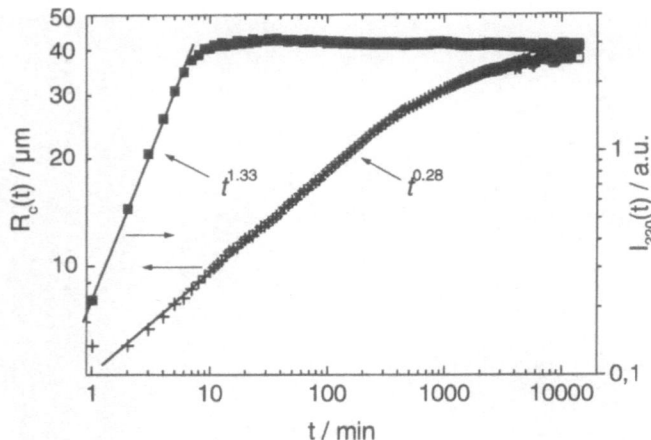

Fig. 5 *Upper curve, right scale*: integrated intensity $A_{220}(t)$ (■); *lower curve, left scale*: crystal size $R_C(t)$ (+), determined from the width of the reflection. *Solid lines*: fits using indicated power laws

different to previously investigated HS systems, where saturation in $A(t)$ corresponded to a change in the power law for $R_C(t)$, indicating the transition from nucleation and growth to ripening [7, 8].

Finally shifts in the peak position indicated compression or expansion of the crystal lattice. The results for $q_{220}(t)$ are shown in Fig. 6. We observed a shift in the peak position towards smaller scattering vectors in time. This is known from previous work on HS systems [7]. It has been interpreted as a slow expansion of crystallites from an initially compressed state. We calculated the corresponding crystal density decrease to obtain

$$\frac{\varphi_C(t = 0) - \varphi_C(t \to \infty)}{\varphi_C(t \to \infty)} \approx 6\% \ . \tag{3}$$

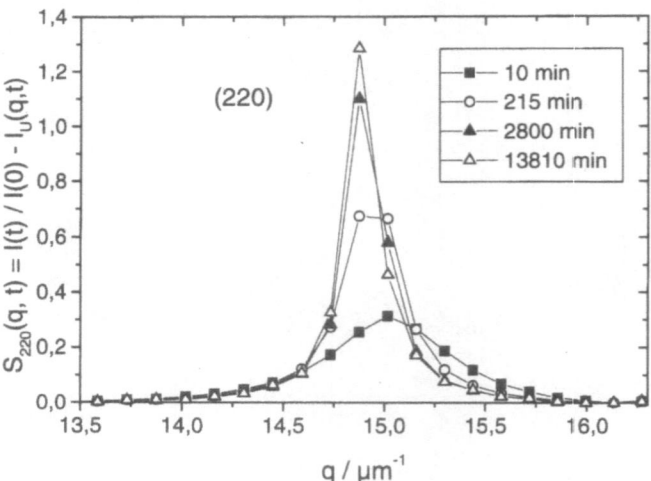

Fig. 4 Temporal evolution of the background-corrected intensity $I(t)/I(0) - I_U(q, t)$ of the {220} reflection for $t = 10$ min (■), 215 min (○), 2800 min (▲) and 13810 min (△)

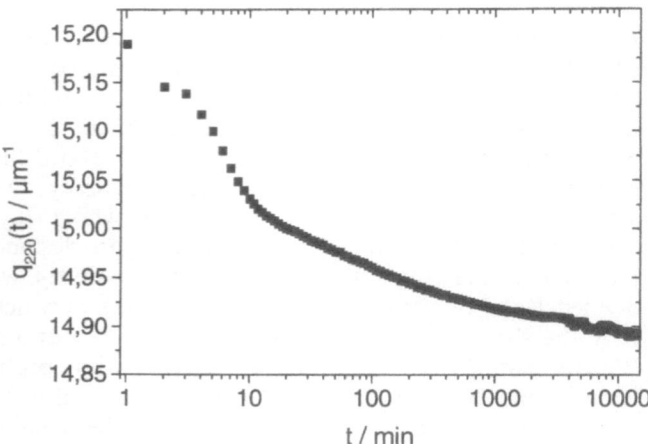

Fig. 6 Position of the maximum of the {220} reflection $q_{220}(t)$ versus time

This is in good agreement with the results in Ref. [7]. We note that while a change in the temporal behaviour of $q_{220}(t)$ seems to be present around $t = 10$ min, the data could not be fitted by power laws or other simple functions and thus no quantitative assertion can be given here.

SALS results

The SALS intensity was observed to remain practically constant, or even to slightly decrease, until $t_i \approx 100$ min. After this time a strong ringlike scattering signal i.e. a small-angle peak with a maximum at nonzero wave vector, evolves. To correct for stray light at small scattering vectors we followed Schätzel and Ackerson [8] and He et al. [9] and subtracted an early measurement from all the following measurements. We chose a correction time of $t = 93$ min, just before the peak appeared. The corrected small-angle intensity was then calculated as $I_{SA}(q, t) = I(q, t) - I(q, 93 \text{ min})$. The corrected data are shown in Fig. 7. The peak gains intensity and its maximum shifts towards smaller q.

Long induction times for SALS patterns have been reported before by van Duijneveldt [26]. For a system of slightly charged colloidal silica spheres with added electrolyte he reported t_i to vary between 610 and 2800 s. This was attributed to the formation of crystals with critical size, which then can grow further. Since our samples are already completely solidified at $t = 10$ min, we can rule out this explanation. Ring patterns have also been observed for HS systems of PMMA particles [8, 9]. They were interpreted as the "form factor" of growing crystals surrounded by a depletion zone of lower density and immersed in the shear molten fluid; however, there the patterns were already observed during nucleation and growth. Note that here the SALS pattern appears

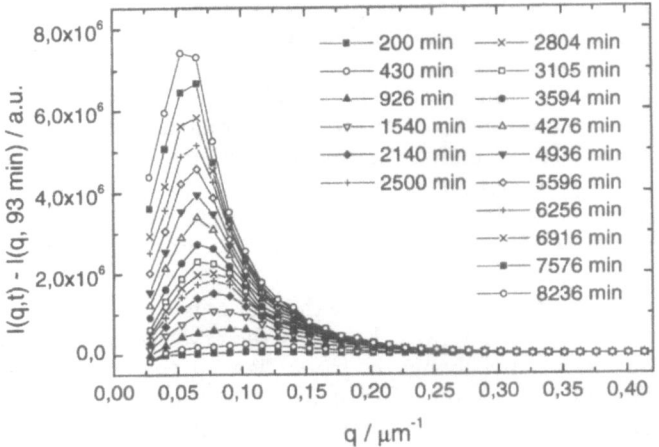

Fig. 7 Corrected small-angle data $I(q, t) - I(q, 93 \text{ min})$ for $t = 200$ min (*lowest curve*) to $t = 8236$ min (*uppermost curve*)

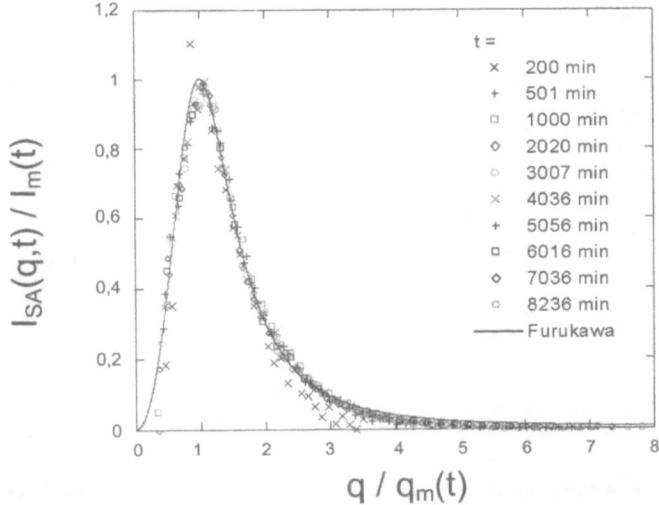

Fig. 8 Scaled small-angle light scattering data. The *solid line* is a plot of the Furukawa function calculated for $\delta = 3.0$

for times significantly longer than 10 min. This indicates a different scattering mechanism is present for our system.

To analyse the time dependence and check for scaling laws we plotted $I_{SA}(q, t)/I_m(t)$ versus $q/q_m(t)$. Here I_m is the intensity at the position of the maximum q_m. All the data collapse nearly on one single curve for times longer than 200 min. This is shown for times up to 8236 min in Fig. 8. Scaling has been observed before, but rarely over the whole range of times $t > t_i$ [9].

We tried to find a suitable scaling function $P(Q)$ with $Q = q/q_m$, such that the following relation holds:

$$I_{SA}(q, t) = I_m(t) P[q/q_m(t)] . \qquad (4)$$

Schätzel and Ackerson [8] used the function $P(Q) = 27Q^2/(2 + Q^2)^3$ to describe their data on PMMA HS. This function does not fit our data. Instead, we found the well-known Furukawa function [27] described the data well for intermediate and late times:

$$P(Q) = \frac{(1 + \delta/2)Q^2}{\delta/2 + Q^{\delta+2}} . \qquad (5)$$

The Furukawa function is known to apply to late-stage spinodal decomposition after off-critical quenches. It may be regarded as an interpolation between the limits $P(q) \sim q^2$ for small q and $P(q) \sim q^{-\delta}$ for large q. δ is related to the fractal dimension, d_f, of the scattering objects by $\delta = d_f + 1$. From the visual inspection of the sample at late times (Fig. 3) we first assumed compact and homogeneous scattering objects of smooth surface ($\delta = 4.0$). This does not describe our data. However, using $\delta = 3.0$ the solid curve in Fig. 8 results. As can be seen only for times up to $t \approx 200$ min there are visible

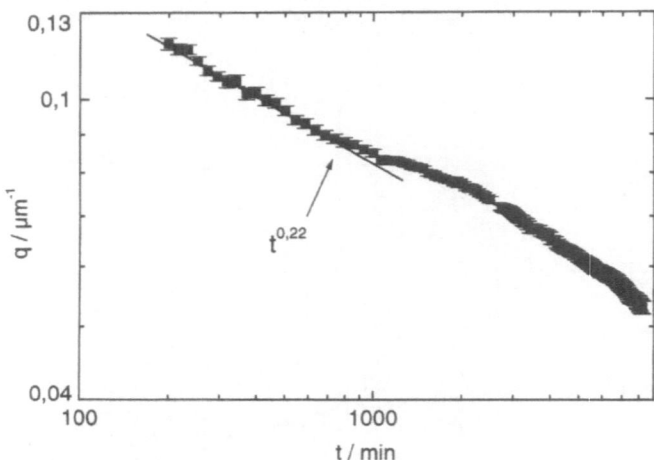

Fig. 9 Position of the small-angle peak $q_m(t)$ versus time. The fitted *solid line* indicates a $t^{0.22}$ behaviour for the first 1000 min

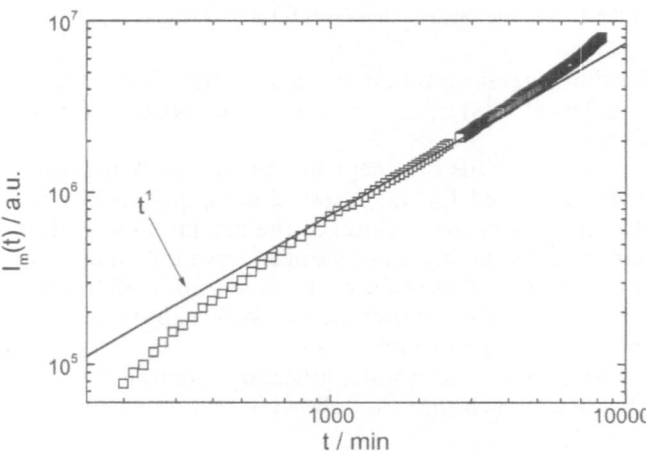

Fig. 10 Intensity of the small-angle peak $I_m(t)$ versus time. The *solid line* for comparison indicates a linear time law

deviations and the data fall slightly below the Furukawa curve.

The series of fits yield nearly identical values for δ, which slightly decreases from 3.2 for $t = 200$ min to about 2.7 for $t \geq 2000$ min. It is interesting to note that van Duijneveldt [26] found even smaller values of the large q exponent of 1.8–2.9. He suggests the form factors of fractal or highly asymmetric scattering objects as an explanation. It is not clear how to explain the deviation from Porod behaviour in our case.

The time evolution of $q_m(t)$ and $I_m(t)$ is shown in log-log plots in Figs. 9 and 10, respectively. $q_m(t)$ decreases in the first 1000 min with the power law $q_m(t) \sim t^{0.22}$. We conclude that the typical dimension of the objects giving rise to the SALS also increases with the same power law. Note that $\alpha = 0.22$ is very close to the

exponent observed for $R_C(t)$ of 0.28. After a transition regime of slightly lower exponent the same or even a larger exponent is observed. $I_m(t)$ shows a continuous increase which is only approximately described by a power law. For comparison, the solid line shows a linear increase in $I_m(t)$ with time. Agreement is best around $t \approx 2000$ min. Deviations towards larger exponents are clearly visible for early and late times. Both exponents seem to show related temporal behaviour. The correspondence between the two exponents, however, remains unclear.

Discussion

We have investigated the crystallization kinetics in a sample of 1:10 cross-linked polystyrene microgel particles by combined SALS and BLS. Like in samples of HS colloids, we identified an early period of nucleation and growth, which is followed by a ripening region. The structure of the resulting solids was identified to be stacking-faulted fcc. The lattice constants show a slight increase indicating an expansion of the growing or ripening crystals. Further, a SALS peak could be observed to grow in time and shift towards smaller values of q, indicating sample inhomogeneity on length scales comparable to the crystal size. Relevant derived quantities could be shown to follow power-law behaviour for both BLS and SLS. In all these features our microgel samples very much resemble the behaviour known from crystallizing HS systems [3, 6–9, 19, 24–26] and are quite different to low-volume-fraction charged-sphere suspensions [10].

Most notably, however there are also specific differences. First of all, the SALS peak appears only after completed solidification. This rules out depletion zone scattering as a possible origin. Further, the values of the power-law exponents for $R_C(t)$ and $q_m(t)$ are nearly identical, indicating the possibility of SALS originating from the crystals alone; however, we are not aware of any form factor of crystallites giving rise to a maximum at finite q. Finally, this is the first time that Furukawa scaling was observed to hold for practically the whole range of times and scattering vectors; however, the value of $\delta = 3$ obtained would indicate two-dimensional objects as possible scattering objects. We therefore conclude that the origin of SALS is not identified in our case.

Also concerning the growth mechanism there seem to be marked differences. For BLS the crystallite size does not show any change in power law at the completion time of $t = 10$ min. This is in contrast to observations on HS, where nearly always a significant change in growth exponents occurred upon mutual touching of the crystallites or in depletion zones. Furthermore, the exponents of the power laws in our case are much smaller than those reported before both for BLS and for SLS. The crystal-size-growth exponent, for example, is even slightly lower than that expected theoretically for Ostwald ripening ($\alpha = 1/3$). The mechanistic origin of such behaviour is not understood.

Our data pose a number of fundamental questions for the possibilities and mechanisms of solidification via crystallization in soft condensed matter systems. They clearly show that microgel colloids behave quite, but not completely, unlike crystallizing HS systems. Further research is clearly needed to clarify why these differences to HS systems appear, while the short-range ordered states are much more alike.

Acknowledgements We gratefully acknowledge the financial support of the following institutions: the Deutsche Forschungsgemeinschaft (DFG: Pa459/6-3), the Materialwissenschaftliches Forschungszentrum Mainz and the Sonderforschungsbereich SFB 262, Mainz.

References

1. Phan SE, Russel WB, Cheng Z, Zhu J, Chaikin PM, Dunsmuir JH, Ottewill RH (1996) Phys Rev E 54:6633
2. Vrij A (1983) Faraday Discuss Chem Soc 76:19
3. Pusey PN, van Megen W (1986) Nature 320:340
4. Hoover WG, Ree FH (1968) J Chem Phys 27:1208
5. Bartlett P, van Megen W (1994) In: Mehta A (ed) Granular matter. Springer, Berlin Heidelberg New York, pp 195–257
6. Palberg T (1997) Curr Opin Colloid Interface Sci 2:607
7. Harland JL, Henderson SI, Underwood SM, van Megen W (1996) Phys Rev Lett 75:3572
8. Schätzel K, Ackerson BJ (1992) Phys Rev E 48:3766
9. He Y, Ackerson BJ, van Megen W, Underwood SM, Schätzel K (1996) Phys Rev E 54:5286
10. Palberg T (1999) J Phys Condens Matter 11:R323
11. Saunders BR, Vincent B (1999) Adv Colloid Interface Sci 80:1
12. Bartsch E, Kirsch S, Lindner P, Scherer T, Stölken S (1998) Ber Bunsenges Phys Chem 102:1597
13. Bartsch E (1995) In: Yip E (ed) Relaxation kinetics in supercooled liquids – mode coupling theory and its experimental test, Transport Theory and Statistical Physics 24:1125
14. Bartsch E, Frenz V, Baschnagel J, Schärtl W, Sillescu H (1997) J Chem Phys 106:3743
15. Kasper A, Kirsch S, Renth F, Bartsch E, Sillescu H (1996) Prog Colloid Polym Sci 100:151
16. Kasper A, Bartsch E, Sillescu H (1998) Langmuir 14:5004
17. Kirsch S, Doerk A, Bartsch E, Sillescu H, Landfester K, Spiess HW, Maechtle W (1999) Macromolecules 32:4508
18. Kirsch S (1996) Ph D thesis. Mainz
19. Henderson SI, Mortensen TC, Underwood SM, van Megen W (1996) Physica A 233:102
20. Monovaukas Y, Gast A (1990) Phase Transitions 21:217
21. Elliot MS, Bristol BTF, Poon WCK (1997) Physica A 235:216
22. Pusey PN, van Megen W, Bartlett P, Ackerson BJ, Rarity JG, Underwood SM (1989) Phys Rev Lett 63:2753
23. Schätzel K (1996) In: Arora AK, Tata BVR (eds) Ordering and phase transitions in charged colloids. VCH, New York, pp 17–40
24. Heymann A, Stipp A, Sinn C, Palberg T (1998) J Colloid Interface Sci 206:119
25. Heymann A (1997) Ph D thesis. Mainz
26. van Duijneveldt JS (1994) PhD thesis. Utrecht
27. Furukawa H (1984) Physica A123:497

Progr Colloid Polym Sci (2000) 115: 66–71
© Springer-Verlag 2000

C. Charnay
S. Lagerge
S. Partyka

Determination of surface heterogeneity of talc materials

C. Charnay · S. Lagerge (✉) · S. Partyka
Laboratoire des Agrégats Moléculaires et
Matériaux Inorganiques, UPRESA5072
Université Montpellier II
Place E. Bataillon
34095 Montpellier Cedex 5, France
e-mail: slagerge@univ-montp2.fr

Abstract The adsorption of the cationic molecule benzyltrimethyl-ammonium bromide (BTMAB) from aqueous solutions on negatively charged surfaces of talc materials at 298 K and free pH has been studied using adsorption microcalorimetry. The most important features of the adsorption of this cationic compound on the talc surface are the constancy of the pH, the increase in the divalent ion concentration in the bulk phase and the decrease in the exothermic displacement enthalpy with increasing amount adsorbed. The comparison of these results with those obtained for the adsorption of cationic molecules onto silica surfaces suggests the same mechanism of adsorption. As a consequence, adsorption in the first stage is mainly governed by electrostatic interactions involved through ion exchange. Because of the lack of a hydrophobic moiety, it is evident that $BTMA^+$ molecules can only adsorb on the hydrophilic parts of talc particles. Assuming that the surface area occupied by one $BTMA^+$ molecule is around 60 nm^2, the hydrophilic area of the talc surface should be estimated by measuring the maximum amount adsorbed.

Key words Talc · Surfactant adsorption · Surface heterogeneity · Hydrophobic–hydrophilic balance · Microcalorimetry

Introduction

Talc is a pigment or mineral filler widely used in paint, polypropylene, cosmetics and at different stages of paper-making processes, etc. When mixing talc with other components, the solid–liquid interface properties should be modified to improve the dispersion process and to avoid particle flocculation due to the natural hydrophobicity of talc. For this purpose, surfactants or polymers, selected according to the desired surface solid properties, are added to the dispersion medium [1–3]. The surface heterogeneity of talc particles in aqueous solution is a crucial parameter for predicting interactions between the different molecules used in industrial formulations and the solid surface [4]. Particularly, quantification of the surface polarity is of vital impor-

tance and remains a topic of current debate. However, there is no general agreement about a standard method of determining the hydrophilic–hydrophobic balance of a surface [5]; the method used depends on the industrial application. In the case of talc materials a flow adsorption microcalorimetric method has been used to quantify the hydrophilic/polar surface sites or contribution on the solid. The method of determining the surface area of the hydrophilic sites in talc materials is based on the determination of the heat of adsorption of 1-butanol from dilute solution in n-heptane, capable of forming a close-packed monolayer on the hydrophilic surface sites of talc (SiOH and MgOH). The comparison of $\Delta_{ads}H_{(\text{butanol/heptane})}$ obtained for the material studied with $\Delta_{ads}H_{(\text{butanol/heptane})}$ of a reference material allows the hydrophilic surface area of the solid studied to be

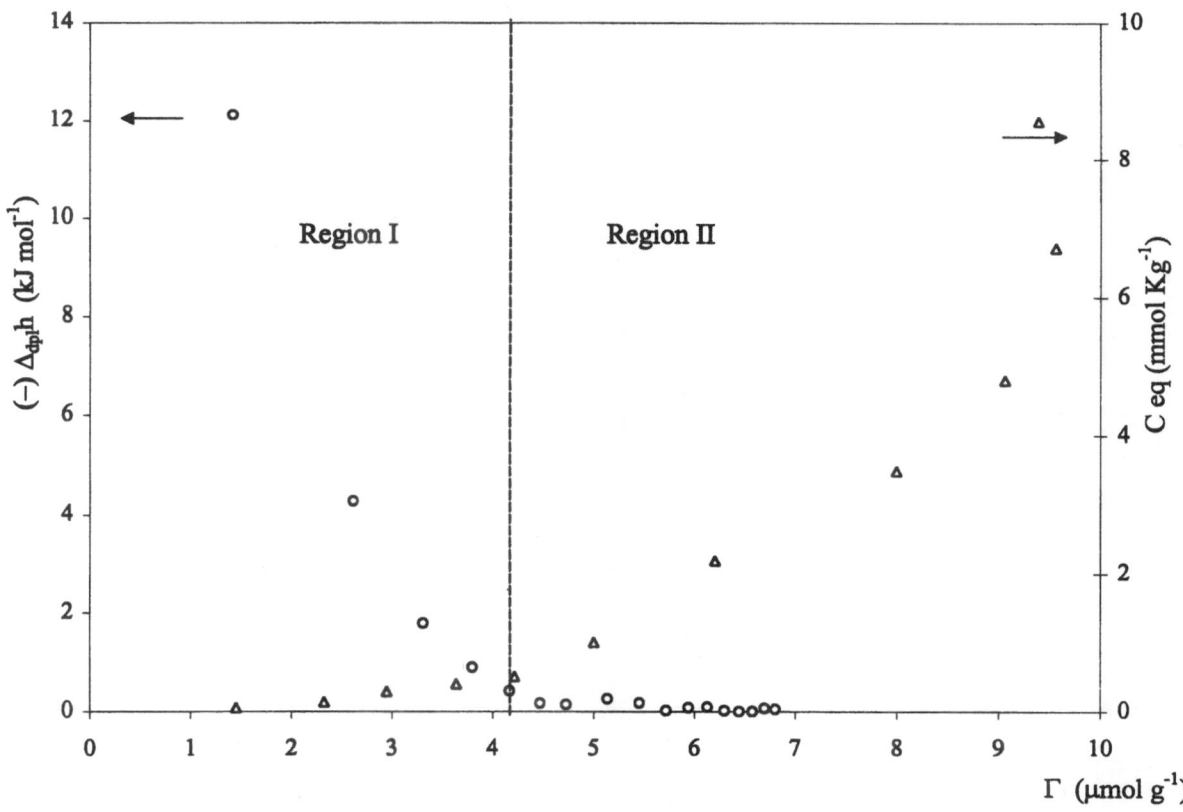

Fig. 1 Adsorption isotherm and differential molar enthalpies of displacement of benzyltrimethylammonium bromide (*BTMAB*) onto talc sample C from aqueous solutions at 298 K and free pH. The amount adsorbed is plotted on the *abscissa*

determined [5]. However, the experimental results obtained in this way [6] reveal very high hydrophilic surfaces, which is in disagreement with the hydrophobic properties of talc materials described by a weak value of the Lewis-base component of the surface tension [7].

This aim of the present study was to quantify the hydrophilic surface sites of talc materials by a conventional adsorption method. We investigated the adsorption of benzyltrimethylammonium cation (BTMA$^+$) from water solution onto four different talc samples with different chlorite contents. Our choice was guided by the fact that the chlorite content appeared capable of

significantly affecting the hydrophilic–hydrophobic balance of the talc surfaces [8, 9].

For this purpose, in order to obtain a good description of the geometric part of the hydrophilic surface of talc materials in aqueous solution, adsorption of an ionic molecule used as a surface property probe was performed. On account of the negative charge of the talc particles [10, 11], BTMA bromide (BTMAB) may be used as a "molecular probe", allowing the hydrophilic area of the talc surface to be assessed.

This study aims to follow the adsorption mechanisms and the structure of BTMA$^+$ on the talc surface by a calorimetric method and by titration of the divalent ions Ca^{2+} and Mg^{2+}.

Materials and methods

Materials

Talc samples were supplied by Talc de Luzenac (France) and were used as received without further treatment. These materials are composed of two minerals: talc and chlorite. Both are phyllosilicate-type clays and occur in the form of lamellae. Talc and chlorite have no layer charge.

Both talc and chlorite are trioctahedral minerals with a sheet structure; the theoretical structural formula are Mg$_3$Si$_4$O$_{10}$(OH)$_2$ and (Mg$_{6-x-y}$Fe$_y$Al$_x$)(Si$_{4-x}$Al$_x$)O$_{10}$(OH)$_8$, respectively. An individual sheet is made of a brucite layer between two external silica layers in the case of talc [12, 13] and in the case of chlorite, an

Table 1 Some characteristics of talc samples used in this study

Sample	Mineralogy		Brunauer–Emmett–Teller surface area (m^2 g^{-1})
	Talc %	Chlorite %	
A	94	3	10
B	77.5	21	10
C	64	33	8.5
D	36	53	11

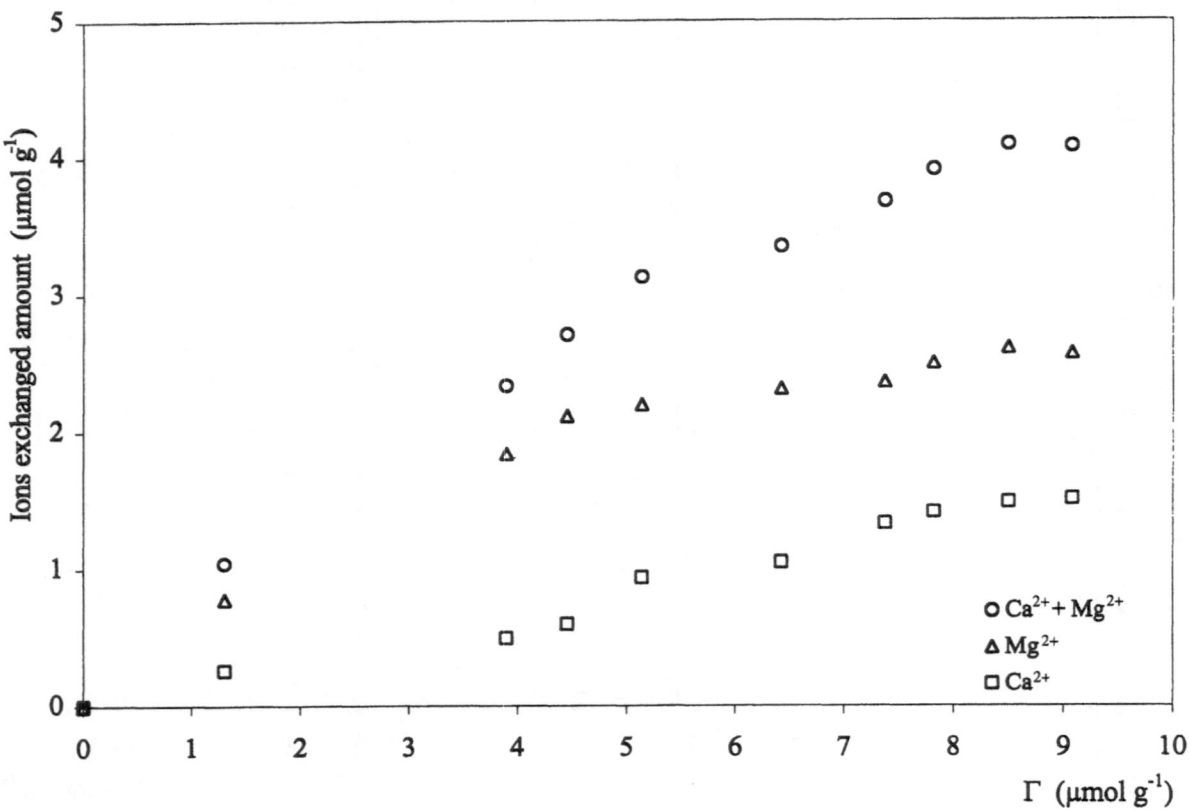

Fig. 2 Amount of divalent Ca^{2+} and Mg^{2+} exchanged following the adsorption of BTMAB

individual sheet is made of a talc layer with an external brucite layer [12].

The mineralogical composition and the N_2 Brunauer–Emmett–Teller (BET) specific surface area ($\sigma_{N2} = 16.2 \text{ Å}^2$) for each sample are reported in Table 1.

The pH of a natural talc dispersion is about 9 and therefore the particles carry negative charge. Moreover, from previous studies [10], only the edge surface of a talc particle is known to exhibit a negative charge. These lateral surfaces are indeed composed of ionised and neutral hydroxyl groups created by the rupture of covalent bonds. The basal surfaces, composed of fully compensated oxygen atoms, are considered to be nonpolar and therefore are expected to show hydrophobic behaviour in water.

The cationic compound, BTMAB, was supplied by Fluka (France) with a purity greater than 99%.

Methods

Individual BTMAB adsorption isotherms were obtained using the classical solution depletion method. Samples (2 g) of talc adsorbent were mixed with 20 g BTMAB solutions at different concentrations in stoppered Pyrex tubes and were equilibrated by gentle rotation at a constant temperature of 298 K for 12 h. Subsequently, the solid samples were separated from the supernatant by filtration (through 0.45-μm cellulose-free acetate membrane filters, Millipore France). The clear supernatants collected from each tube were analysed for BTMAB, Ca^{2+} and Mg^{2+} contents using a total organic carbon analyser (TOC-5000 A, Shimadzu, Japan) and flame spectroscopy, giving the corresponding concentration in the

equilibrium bulk phase, after the attainment of the adsorption equilibrium.

The adsorption measurements were carried out at free pH and in the absence of extra salt at 298 K. The pH value of the talc particle dispersions in deionised water was 9.4 and did not change during adsorption of BTMAB.

Calorimetric measurements of the enthalpy of dilution and displacement accompanying the adsorption of cations onto talc provide detailed knowledge of the interactions involved in the different adsorption stages. Thermal effects were measured using a microcalorimetric batch technique. A detailed description of the "Montcal" microcalorimeter used for this study and its functional parts can be found in previous papers [14, 15].

Results and discussion

Adsorption of $BTMA^+$ species from water

The adsorption isotherm of BTMAB onto talc sample C from deionised water at 298 K and free pH is displayed in Fig. 1. The corresponding differential molar enthalpies of displacement of water by BTMAB on the talc surface are also reported.

The BTMAB concentrations, in the equilibrium bulk solution after the attainment of the equilibrium of adsorption are plotted against the amount adsorbed per unit mass of talc (Γ) and the corresponding enthalpies of displacement ($\Delta_{dpl}h$). The amount of adsorption is plotted on the abscissa to shed more light on the adsorption mechanism of BTMAB on the talc surface.

The adsorption isotherm consists of two regions. The first region is a steeply rising part (region I), where all the BTMAB adsorbs. This initial part of the isotherm, corresponding to a very low equilibrium concentrations range (c_e < 0.1 mmol kg^{-1}) shows a quasiinfinite initial slope and suggests a very strong affinity of the BTMAB molecules for some of the talc surface sites. The second region above c_e = 0.1 mmol kg^{-1} (region II), where first the adsorption of BTMAB gradually decreases and then the isotherm reaches a limiting value, Γ_m, of 9.3 μmol g^{-1}, is characteristic of the system and corresponds to the saturation of the surface. The microcalorimetric experiments were performed with a high sensitivity (1 mJ) at which the molar enthalpy of dilution of BTMAB in the calorimetric cell was found to be close to zero over the whole concentration range [16]; therefore, no correction term arising from the dilution of solute injected into the calorimetric cell should be subtracted from the total enthalpic effects. The corresponding differential molar enthalpies are exothermic over the whole concentration range. The enthalpic

effects related to the initial adsorption (region I) of cationic species (0 < Γ < 4.5 μmol g^{-1} $\approx$ Γ_c) decline and are exothermic with a value of $\Delta_{1,2}h$ ranging from -12 kJ mol^{-1} to about 0 kJ mol^{-1}. Then, from an amount adsorbed of 4.5 μmol g^{-1}, the enthalpy of displacement becomes quite constant and athermal. This regular decrease of the experimental enthalpy of adsorption up to 4.5 μmol g^{-1} suggests the occurrence of significant site heterogeneity on the talc surfaces.

Following the adsorption of BTMAB, some divalent ions (Ca^{2+} and Mg^{2+}) were detected in the equilibrium bulk solution. This clearly indicates that Ca^{2+} and Mg^{2+}, initially present on the talc surface, are removed from the surface to the bulk solution upon the adsorption of BTMAB.

The evolution of the amount of Ca^{2+} and Mg^{2+} exchanged with the adsorption of BTMA$^+$ is displayed in Figs. 2 and 3. The curves consist of three different regions:

1. Initially the number of cations exchanged increases strongly and linearly up to Γ_{BTMAB} $\approx$ 2.5 μmol g^{-1}. In this initial part, the equilibrium concentration of BTMAB in the bulk solution is close to zero (Fig. 1), which indicates a very strong affinity of the BTMAB molecules for the talc surface. Moreover the ratio of the number of divalent ions exchanged to the amount of BTMAB adsorbed (r) is above 1.

Fig. 3 Ratio of the amount of Ca^{2+} and Mg^{2+} exchanged to that of the BTMA$^+$ adsorbed against BTMA$^+$ adsorption

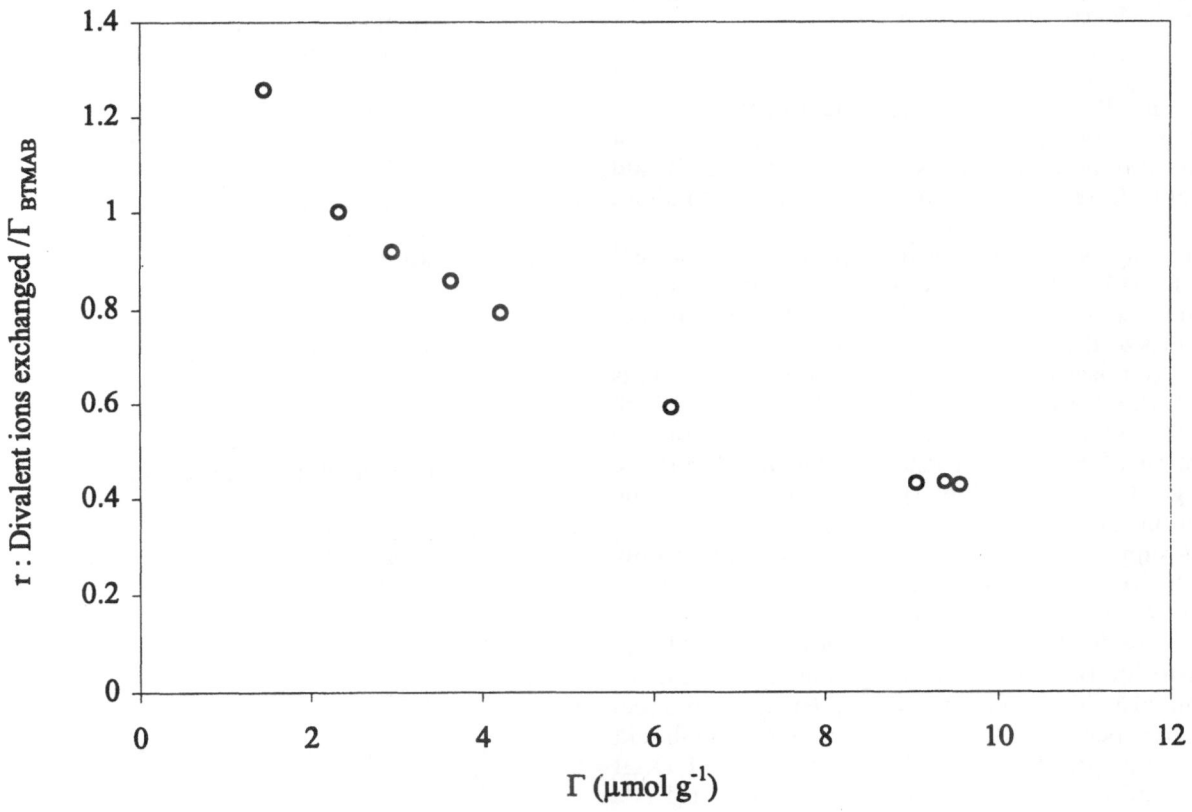

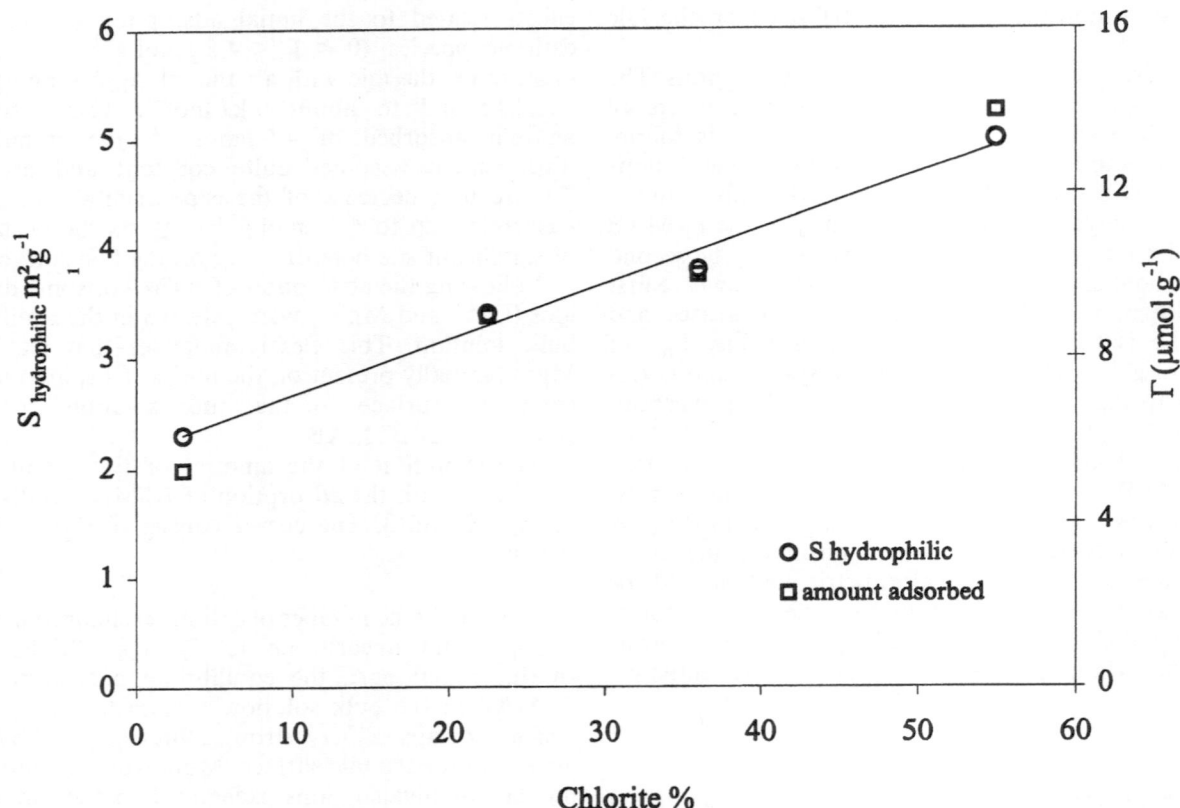

Fig. 4 Relationships between the hydrophilic surface or the amount of BTMA$^+$ adsorbed with the chlorite content

2. A region in which $2.5 < \Gamma_{BTMAB} < 8 \; \mu mol \; g^{-1}$ where some BTMAB molecules remain in the supernatant (Fig. 1). In this range of adsorption the number of divalent cations exchanged still increases (Fig. 2) and the r value decreases slowly and linearly from 1 to about 0.5.

3. Finally a region above $\Gamma_{BTMAB} > 8 \; \mu mol \; g^{-1}$ where BTMAB still adsorbs and its amount attains a plateau value of 9.3 $\mu mol \; g^{-1}$ (Fig. 1), while the number of desorbed divalent ions saturates at 4.2 $\mu mol \; g^{-1}$ (Fig. 2). In this range of surface saturation, the amounts of BTMAB adsorbed and Mg^{2+} and Ca^{2+} desorbed are equal to the stoichiometric values required for the adsorption of BTMA$^+$ species through a 1:2 cationic exchange ($r \approx 0.5$ in Fig. 3) involving electrostatic interactions at oppositely charged surface sites. This supports the idea of BTMA$^+$ adsorption only on the hydrophilic surface sites of talc–chlorite particles because of the lack of hydrophobic moiety.

On the basis of the BTMA$^+$ chemical structure being similar to the polar head of benzyldimethyldodecylammonium bromide surfactant (BDDAB), the cross-sectional area per BTMA$^+$ adsorbed at the solid–liquid interface is assumed to be equal to the area of a BDDAB molecule adsorbed at the liquid–gas interface (71 Å^2)

[16]. However, in the absence of an alkyl chain, it is probable that BTMA$^+$ ions are adsorbed with the aromatic part perpendicular to the surface. In such a configuration, the cross-sectional area is close to that of tetramethylammonium, i.e. 60 Å^2 [15].

From the plateau value ($\Gamma_{max} = 9.3 \; \mu mol \; g^{-1}$) and on the basis of the value of the cross-sectional area ($\sigma = 60$ Å^2) per adsorbed molecule, the hydrophilic surface sites of the talc could be calculated according the following relationship:

$$S_{hydrophilic} = \Gamma_{max} N \sigma \; ,$$

where N is the Avogadro number.

For sample C, the hydrophilic surface area was found to be 3.4 m^2 g^{-1}, which represents 40% of the BET surface area. The $S_{hydrophilic}$ values obtained in the same way, for samples A, B and D, are reported in Table 2.

Table 2 Relationships between the amount of benzyltrimethylammonium ($BTMA^+$) adsorbed for the four talc samples studied with different chlorite content

Sample	A	B	C	D
Chlorite content (%)	3	21	33	53
Amount of BTMA$^+$ adsorbed ($\mu mol \; g^{-1}$)	5.3	8.6	9.3	14
Hydrophilic surface (m^2 g^{-1})	1.9	3.1	3.4	5
Polarity (%)	19	31	40	45

Hydrophilic character of the talc samples

As previously stated, it is known that the hydrophilic properties of talc increase with the chlorite content [9]. In fact, the immersion enthalpy in water for a talc sample without impurities is near 300 mJ m^{-2} and for a chlorite sample the immersion enthalpy is near 900 mJ m^{-2} [9]. The hydrophilic character of chlorite, which is much higher than that of talc, can be explained by the presence of a polar group (hydroxyl) on the basal surface [9]. Indeed the results in Fig. 4 show a linear progression of the hydrophilic surface with the chlorite content.

Conclusion

The adsorption of BTMA$^+$ from deionised water appears particularly useful and reliable for the quantification of the hydrophilic surface sites of talc particles. The experimental results of adsorption (isotherm and calorimetry) show that the adsorption of BTMA$^+$ occurs only onto the polar fraction of the total talc surface through a cationic exchange involving electrostatic interactions. The hydrophobic part of the talc surface is fully ineffective for the adsorption of BTMAB because of the lack of specific adsorption. This makes the adsorption method reliable for determining the hydrophilic/polar surface contribution in different talc materials. However, the values of the hydrophilic surfaces obtained from BTMAB adsorption are far from those determined from geometric considerations (30%) [17]. This is probably due to the fact that these latter values do not take into account all the parameters responsible for the hydrophilicity of the material, particularly the chlorite content.

References

1. Li Z, Giese RF, Van Oss CJ, Yvon J, Cases J (1993) J Colloid Interface Sci 156:279
2. Krysztafkiewicz A, Domka L (1987) Tenside Surfactants Deterg 24:227
3. Allbee N (1984) Plast Compd (Jan/Feb) 65
4. Malhammar G (1990) Colloids Surf 44:61
5. Groszek AJ, Partyka S (1993) Langmuir 9:2721
6. Malandrini H, Clauss F, Partyka S, Douillard JM (1997) J Colloid Interface Sci 194:183
7. Giese RF, Costanzo PM, Van Oss CJ (1991) Phys Chem Miner 17:611
8. Douillard JM, Malandrini H, Zoungrana T, Clauss F, Partyka S (1994) J Therm Anal 41:1205
9. Malandrini H (1995) PhD thesis. University of Montpellier II, Montpellier
10. Fuerstenau MC, Lopez-Valdievieso A, Fuerstenau W (1988) Int J Miner Process 23:161
11. Chander S, Wie JM, Fuerstenau W (1975) AIChE Symp Ser 71 150:183
12. Evans BW, Guggenheim S (1988) Rev Miner 19:225
13. Rayner JH, Brown G (1973) Clays Clay Miner 21:103
14. Zajac J, Lindheimer M, Partyka S (1995) Colloids Surf A 98:197
15. Zajac J, Partyka S (1996) In: Dabrowski A, Tertykh VA (eds) Series studies in surface science and catalysis, vol 99. Elsevier Amsterdam, pp 797–829
16. Trompette JL, Zajac J, Keh E, Partyka S (1994) Langmuir 10:812
17. Pugh RJ, Tjus K (1990) Collloids Surf 47:179

Progr Colloid Polym Sci (2000) 115:72–76
© Springer-Verlag 2000

M. J. Meziani
H. Benalla
J. Zajac
S. Partyka

Effect of the pore size on the adsorption of dimeric cationic surfactants from aqueous solution onto aluminosilicate adsorbents with a hexagonal pore arrangement

Presented at the 13th Conference of the European Colloid and Interface Society, Dublin 1999

M. J. Meziani · H. Benalla
J. Zajac (⊠) · S. Partyka
Laboratoire des Agrégats Moléculaires et Matériaux Inorganiques
UPRESA 5072, Université Montpellier 2
Place E. Bataillon
34095 Montpellier Cedex 5, France
e-mail zajac@univ-montp2.fr
Tel.: +33-4-67143340
Fax: +33-4-67143304

Abstract The effect of the pore size of the adsorbent on the adsorption of dimeric alkanediyl-α,ω-bis (dodecyldimethylammonium bromides) with alkanediyl spacer groups C_2H_4, C_6H_{12}, and $C_{12}H_{20}$ has been studied in aqueous solution at 298 K and free pH. Three mesoporous aluminosilicate materials of the MCM-41 type were used as solid supports with homogeneous pore size distributions. The analysis was based on the experimental adsorption isotherms, supplemented with measurements of the sodium and hydronium ion concentrations in the equilibrated supernatant liquid as well as electrophoretic curves. The adsorption and interfacial aggregation of the dimeric surfactants were found to occur both on the external surface and on the pore walls. The ultimate structure of the adsorbed phase appeared to depend not only on the pore size but also on the spacer group length and flexibility.

Key words Surfactant adsorption · Dimeric (gemini) cationic surfactants · MCM-41 aluminosilicates

Introduction

Gemini or dimeric surfactants have raised considerable interest with respect to their high surface activity, low critical micelle concentration (cmc) values, and remarkable rheological properties [1–3]. This new class of surfactants includes molecules that contain two hydrophilic groups connected with a flexible hydrophilic [4–7], flexible hydrophobic [8, 9], or rigid hydrophobic [10, 11] spacer.

The behaviour of dimeric surfactants in aqueous solution and at the air–solution interface is relatively well recognised [1, 2, 12, 13]. In aqueous solution, they can form a variety of aggregates depending mainly on the spacer length and flexibility. In contrast to the classical, monomeric surfactants, which form mostly globular micelles, dimerics with a single spacer form linear threadlike aggregates and those having a longer spacer produce a mixture of spheroids and treelike micelles [14, 15].

Some attempts have been made to study the mechanism of adsorption of these surfactants at the solid–solution interface [3, 16–18]. Manne et al. [17] used atomic force microscopy to monitor the structure of interfacial aggregates formed by dimeric surfactants with varying tail and spacer length. The most important conclusion of this work was that, due to the possibility of double bonding to the charged surface of mica, dimeric aggregates showed a closer orientation relative to the underlying mica lattice than the conventional surfactants did [17]. The interfacial aggregates on the graphite surface were predominantly half-cylindrical, regardless of the surfactant structure [17]. For dimeric surfactants belonging to the family of alkanediyl-α,β-bis(dodecyldimethylammonium bromide) adsorbed onto silica [18], the first adsorption step was ascribed to ion exchange between the dimeric ions and the residual sodium cations. As the adsorption progressed, the growth of surfactant aggregates was accompanied by the retention of some bromide counterions. At this stage, the maximum amount of the surfactant adsorbed increased with decreasing length of the spacer group. The trend was explained by the increasing aggregate size.

The intention of the present report is to direct attention to the effect of the adsorbent porosity on the adsorption of cationic dimeric surfactants on solid surfaces. For this purpose, model mesoporous solids, characterised by the uniform porous structure, were used as adsorbents. Three alkanediyl-α, ω-bis(dodecyldimethylammonium bromide) dimeric surfactants were adsorbed on aluminosilicates of the MCM-41 type having different pore sizes. The discussion is based mainly on the results of surfactant adsorption measurements, supplemented by the determination of the changes in the sodium and hydronium ion concentrations during adsorption and electrokinetic experiments.

Experimental

Materials

MCM-41 aluminosilicate materials were prepared and characterised following the methods described in a previous article [19]. They are designated SiAlxCn, where x is the Si:Al molar ratio and n the surfactant template chain length. To recall the most important surface characteristics of the calcined samples, surface area and pore structure parameters have been collected in Table 1.

The alkanediyl-α, ω-bis(dodecyldimethylammonium bromide) dimeric surfactants are referred to as 12-s-12, where 12 represents the number of carbon atoms in the hydrophobic moiety and s is the number of carbon atoms within the alkanediyl spacer group. The surfactants used were 12-2-12, 12-6-12, and 12-12-12. The synthesis and purification procedures were described previously [9]. They all are soluble in water at very low concentrations and their cmc values at 298 K are as follows: 0.81, 1.03, and 0.37 mmol kg^{-1}. The cross-sectional areas of the polar head group at the air–solution interface are 1, 1.45, and 2.25, respectively.

Methods

Adsorption isotherms were obtained using the solution depletion method. Surfactant solutions were equilibrated with a solid sample at room temperature for 24 h by slow rotation in glass joint stoppered tubes. Each suspension was then filtered through 0.2-μm cellulose-free acetate membrane filters and the final concentration of the supernatant was determined by the combustion/nondispersive IR gas analysis method with a total organic carbon analyser (TOC-5000A, Shimadzu, Japan). The amount adsorbed was

calculated from the difference between the initial and the final bulk molality. Simultaneously, samples of solid suspension were collected for electrophoretic measurements performed with the use of a Rank Brothers microelectrophoresis apparatus having a rectangular cell [20].

The measurements of pH in the equilibrated supernatant solution were carried out with a Tacussel PHM 240 ionometer in conjunction with a pH electrode (Tacussel XC 601).

The sodium ion content was determined using a Varian atomic absorption spectrometer. The water used throughout all the experiments was deionised and purified with a Millipore Super Q system. It had a pH value of 6 and a conductivity which varied between 0.05 and 0.1 μS cm^{-1}.

Results and discussion

The adsorption isotherms of the dimeric 12-2-12, 12-6-12, and 12-12-12 surfactants onto SiAl32C8 and SiAl32C18 are presented in Figs. 1 and 2. Since the cmc value depends markedly on the spacer group, the amount adsorbed, Γ_2, has been plotted as a function of the relative molality of the equilibrium bulk solution, i.e., m_2^b/cmc. Such a presentation makes the comparison between the different adsorption curves easier.

Two-step isotherms were found in most cases. For cationic surfactant adsorption on negatively charged, flat solid surfaces [20], the first step is usually attributed to the compensation of the negative surface charge by the adsorbed surfactant cations. The second adsorption step is dominated by hydrophobic association of the surfactant tails between themselves at the solid–solution interface, leading to the formation of interfacial

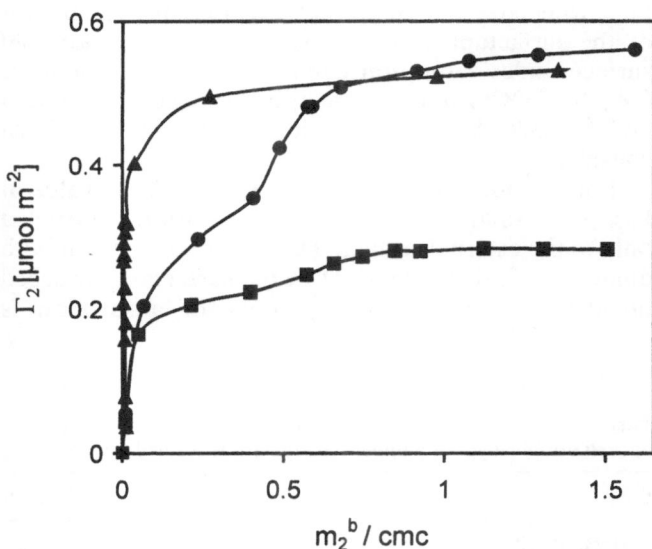

Fig. 1 Adsorption isotherms of dimeric 12-2-12 (●), 12-6-12 (■), and 12-12-12 (▲) surfactants onto SiAl32C8 from aqueous solution at 298 K and free pH

Table 1 Total surface area, S_t, external surface area, S_{ext}, and pore size, d_m, of the MCM-41 materials studied [19]

Solid	SiAl32C8	SiAl32C14	SiAl32C18
S_t (m^2 g^{-1})	608	803	607
S_{ext} (m^2 g^{-1})	126	28	62
d_m (nm)	1.8	2.7	3.7

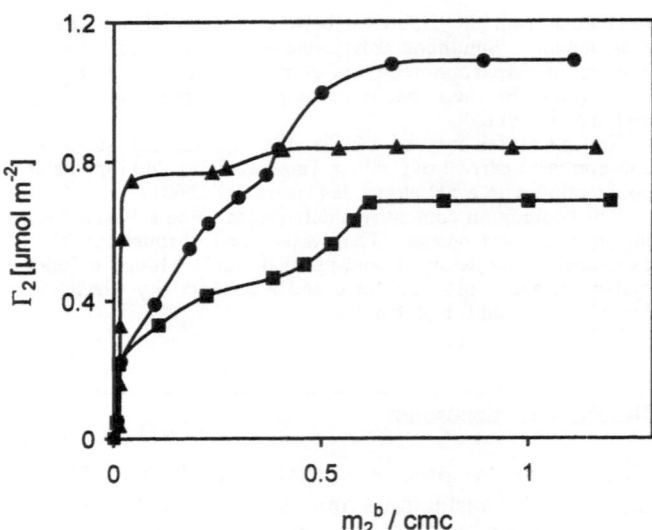

Fig. 2 Adsorption isotherms of dimeric 12-2-12 (●), 12-6-12 (■), and 12-12-12 (▲) surfactants onto SiAl32C18 from aqueous solution at 298 K and free pH

aggregates with varying size and structure. There arises a question as to whether the dimeric surfactant ions adsorb only on the external surface or whether they can penetrate into the pore space. The following geometrical consideration will allow the question to be decided.

In Table 2, the experimentally measured effectiveness of adsorption, i.e., the amount of the surfactant adsorbed on the solid surface at saturation, Γ_{max}, is compared with the hypothetical quantity of adsorption, Γ_{hyp}, calculated assuming that a densely packed bilayer of the surfactant adsorbate is formed on the external surface under saturation conditions. One microporous (i.e., SiAl32C8) and two mesoporous (i.e., SiAl32C14 and SiAl32C18) MCM-41 aluminosilicates have been considered.

For mesoporous samples, it is clear that the value of Γ_{hyp} is too small to account for the adsorption restricted only to the external surface. The difference between both quantities yields the amount of the surfactant adsorbed on the pore walls. This quantity of adsorption is consistent with a fractional monolayer coverage of the surface included within the pores. In the case of the 12-12-12 surfactant molecule, this monolayer coverage is even complete.

For the microporous SiAl32C8 sample, only the 12-12-12 dimeric surfactant can easily penetrate into the pore space. The adsorption of two other molecules on the pore walls is questionable. This cannot be ascribed to the kinetic effects in adsorption, since the amounts adsorbed measured at an equilibration time of 48 h have not changed at all. It is more likely that the increased flexibility of the 12-12-12 spacer group enables this molecule to enter the pore space and to adopt a favourable conformation on the pore walls.

Effect of the spacer length

The pore space of the SiAl32C18 aluminosilicate is thus available to the three dimeric surfactants. The maximum amount of adsorption decreases in the following order: 12-2-12 > 12-12-12 > 12-6-12. Note that a different trend has been obtained for the adsorption of the same molecules on a flat silica surface [18]. It is also important to know that the amount of a monomeric cationic surfactant adsorbed on the same sample is twice that obtained with 12-2-12 and 3 times that measured for 12-6-12 [19]; therefore, the adsorption of dimeric surfactant ions cannot be regarded as the adsorption of two independent monomers.

It is obvious that the capacity of double attachment of a dimeric ion to two neighbouring surface sites results in extended monolayered aggregates. If a particular spatial arrangement of active sites prevents such strong bonding, the formation of small aggregates with a high degree of curvature is more favourable. As a consequence, the effectiveness of adsorption should diminish. The dimeric 12-2-12 ion has a short spacer group, which matches well with active sites densely distributed on the solid surface. For a rigid C_6H_{12} spacer, the number of possible configurations is markedly limited. Some of the neighbouring active sites are not available if one ionic head is already fixed on the surface. The spacer of 12-12-12 is much longer than that of 12-2-12, but is much more flexible than the C_6H_{12}

Table 2 Comparison of the maximum amount adsorbed, Γ_{max}, with the hypothetical amount, Γ_{hyp}, calculated assuming that a densely packed double layer of a dimeric surfactant is formed only on the external surface of the mesoporous adsorbents studied (298 K, free pH)

Solid	SiAl32C14		SiAl32C8			SiAl32C18	
Surfactant: 12-s-12	$s = 12$	$s = 2$	$s = 6$	$s = 12$	$s = 2$	$s = 6$	$s = 12$
Γ_{max} (μmol g^{-1})	643	333	169	330	639	403	492
Γ_{hyp} (μmol g^{-1})	41	418	289	186	206	142	91

one. For this reason, the maximum amount of 12-12-12 is greater than that of 12-6-12, in spite of the greater size of the former.

Effect of the pore size

The effect of the pore size on the adsorption of dimeric surfactants may be discussed following the example of 12-12-12, which can penetrate into the micropores of the SiAl32C8 sample. The effectiveness of adsorption per unit surface area of the adsorbent attains the following values: 0.54 μmol m^{-2}, SiAl32C8; 0.8 μmol m^{-2}, SiAl32C14; 0.81 μmol m^{-2}, SiAl32C18. When the pore diameter passes from the micopore to the mesopore size range, the maximum amount adsorbed increases on account of the increased adsorption space available to the interfacial aggregates. If the distance between the opposite pore walls is too small for bilayered aggregates to be formed on each side, the interfacial aggregates can only grow in the direction parallel to the pore walls. Alternatively, the formation of internal aggregates with interpenetrating hydrophobic tails may be envisaged. In both cases, the growth of interfacial aggregates in the direction perpendicular to the pore walls is significantly restricted. This is why the values of Γ_{max} measured for SiAl32C14 and SiAl32C18 differ only a little.

The evolution of the electrophoretic mobility of the finely divided solid particles during the adsorption of 12-12-12 onto SiAl32C8, SiAl32C14, and SiAl32C18 is very similar. In all cases, the electrokinetic potential changes sign at an amount adsorbed of 0.25 μmol m^{-2}, then increases and reaches a saturated positive value. Similarly to monomeric surfactants [20], the adsorption of dimeric cations in excess over the negatively charged surface sites provides a convincing argument for the formation of interfacial aggregates. However, the location of the sliping plane outside the pore space is very probable in the present case; therefore, the changes in the electrophoretic mobility take account of the 12-12-12 ions adsorbed only on the external surface.

Changes in the surface charge density during cationic surfactant adsorption are usually accompanied by a release of hydronium ions to the bulk phase and a consequent decrease in the pH of the equilibrated supernatant [20]. For 12-12-12 ions adsorbed onto SiAl32C14, the pH diminishes from 7.5 to 4.2; note that the pH of the supernatant during the adsorption of monomeric surfactant decreased only from 7.5 to 5 [19]. This difference may be explained by a stronger specific adsorption of dimeric ions. Moreover, some exchangeable sodium cations may be displaced from the surface by the adsorbing surfactant cations [20]. The sodium cation concentration in the supernatant hardly changes during the adsorption of the dimeric ion; the amount of sodium released to the supernatant by a unit mass of the solid is 0.75 mmol g^{-1} for SiAl32C8, 0.82 mmol g^{-1} for SiAl32C14, and 0.91 mmol g^{-1} for SiAl32C18. In this case, the electric compensation of the negatively charged groups on the solid surface by the adsorbing surfactant cations is not necessarily accompanied by a stoichiometric ion exchange with sodium ions.

Conclusion

The adsorption of dimeric cationic surfactants on the negatively charged surface of mesoporous MCM-41 aluminosilicates depends both on the pore diameter and on the effective size of the surfactant spacer group. The attachment of dimeric ions to the solid surface with densely populated negative sites is favoured if the spacer is short or flexible enough. This in turn, determines the formation of extended interfacial aggregates on the pore walls, leading to the increased packing of the surfactant adsorbate on the solid surface.

References

1. Zana R (1996) Curr Opin Colloid Interface Sci 1:566
2. Rosen MJ (1993) Chem Tech 23:30
3. Esumi K, Takeda Y, Goino M, Ishiduki Y, Koide Y (1997) Langmuir 13:2585
4. Okahara M, Masuyama A, Sumida Y, Zhu YP (1988) J Jpn Oil Chem Soc 37:716
5. Zhu YP, Masuyama A, Okahara M (1991) J Am Oil Chem Soc 68:268
6. Rosen MJ, Zhu ZH, Hua XY (1992) J Am Oil Chem Soc 69:30
7. Masuyama A, Hirono T, Zhu YP, Okahara M, Rosen MJ (1992) J Jpn Oil Chem Soc 41:301
8. Devinsky F, Lacko I, Bittererova F, Tomeckova L (1986) Colloid Interface Sci 114:314
9. Zana R, Benrraou M, Rueff R (1991) Langmuir 7:1072
10. Menger FM, Littau CA (1991) J Am Chem Soc 113:1451
11. Menger FM, Littau CA (1993) J Am Chem Soc 115:10083
12. Zana R (1997) In: Robb ID (ed) Specialist surfactant. Chapman and Hall, London, p 81
13. Alami E, Beinert G, Marie P, Zana R (1993) Langmuir 9:1465
14. Zana R, Talmon Y (1993) Nature 362:228
15. Kern F, Lequeux F, Zana R, Candau SJ (1994) Langmuir 10:1714

16. Esumi K, Goino M, Koide Y (1996) J Colloid Interface Sci 183:539

17. Manne S, Schäffer TE, Huo Q, Hansma PK, Morse DE, Stucky GD, Aksay IA (1997) Langmuir 13:6382

18. Chorro C, Chorro M, Dolladille O, Partyka S, Zana R (1998) J Colloid Interface Sci 199:169

19. Meziani MJ, Zajac J, Jones DJ, Rozière J, Partyka S (1997) Langmuir 13:5409

20. Zajac J, Partyka S (1996) In: Dabrowski A, Tertykh VA (eds) Series studies in surface science and catalysis, vol 99. Elsevier, Amsterdam, pp 797–829

Progr Colloid Polym Sci (2000) 115 : 77–83
© Springer-Verlag 2000

F. Cousin
V. Cabuil

Fluid–solid transitions in aqueous ferrofluids

F. Cousin · V. Cabuil (✉)
Laboratoire des Liquides Ioniques
et Interfaces Chargées
Equipe Colloïdes Magnétiques
Université Paris 6, Case 63
75252 Paris Cedex 5
France
e-mail: cabuil@ccr.jussieu.fr
Tel.: +33-1-44273174
Fax: +33-1-44273675

F. Cousin
Centre de Recherche sur la Matière Divisée
Centre National de la Recherche
Scientifique, 45071 Orléans Cedex 2
France

Abstract We report the phase behaviour of aqueous dispersions of
magnetic nanoparticles in the high-
volume-fraction regime. Osmotic
compression experiments are used to
obtain high particle volume fractions
in the range of low ionic strength.
The suspensions exhibit a reversible
rheological liquid–solid transition
for a given particle volume fraction.
The threshold of the transition is
shifted towards low volume fractions by a decrease in the ionic
strength. The structure factor of the
suspensions obtained from small-
angle neutron scattering measurements shows that the sum of the
interactions in the system is dominated by strong electrostatic repulsions and that the suspensions
exhibit a liquidlike structure even in
the solid phase. The polydispersity of
the system does not allow the crystal
phase to be reached; the solid phase
is thus a Wigner glass. The threshold
of the transition is consistent with the
theoretical threshold of a solid transition for a system of hard spheres,
taking for the radius of the particle
an effective radius: the sum of the
radius and the Debye length. Repulsions are so efficient that the sum of
the interactions remains isotropic
under the appliance of a magnetic
field. It is nevertheless possible to
induce a solid–liquid transition by
the appliance of a magnetic field.

Key words Magnetic fluids ·
Osmotic compression · Structure
factor · Phase transition ·
Wigner glass

Introduction

We report the phase behaviour of aqueous dispersions
of nanometer-sized magnetic spherical particles (ferrofluids) [1]. Considering the solvent as a continuous
medium, an analogy between such suspensions and
atomic systems is usually made. Indeed the shape of the
interaction potential between objects is the same in both
systems [2]. Fluid, liquid, gas and solid phases are thus
observed in colloids, but, in contrast to atomic systems,
the interparticle interactions in colloidal suspensions can
be tuned using several experimental parameters, such as
salinity, temperature and, in the case of ferrofluids, the
magnetic field. Because of the variety of the potentials
specific behaviours, predicted by numerous theoretical
studies [3], are induced. The polydispersity in the
colloidal suspensions also strongly influences their phase
behaviour [1, 4].

Few experimental studies concern aqueous suspensions of nanoparticles. Recent studies have reported the
stability of aqueous magnetic fluids in the low-particle-
volume-fraction regime and have pointed out the
astonishing stability of these suspensions – Derjaguin–
Landau–Verwey–Overbeek (DLVO) theory fails to
describe such systems [5, 6]. In this low-volume-fraction
regime and for the particle size under consideration
($d \leq 12$ nm), the dipolar interactions are sufficiently
weak compared to electrostatic repulsions so the magnetic fluids can be considered as isotropic dispersions.
An increase in ionic strength lowers the strength of
repulsions and turns the fluid phase into a biphasic
"gas–liquid" suspension where droplets of a dense phase

spontaneously nucleate in a dilute one. Such gas–liquid-like transitions have been described elsewhere [6–8]. The effect of temperature has always been found to be negligible compared to the effect of ionic strength, except near the phase-transition line.

In the present work, we investigate the regime of high particle volume fractions and low ionic strength, i.e. the regime of strong repulsions. We show that in this case a fluid–solid transition is observed and we discuss the nature of this transition.

Experimental

Materials

The spherical nanoparticles under consideration are made of a magnetic ferric oxide, maghemite (γ-Fe$_2$O$_3$). They were synthesized by condensation of metallic salts in an alkaline medium according to Massart's method [9] and can be dispersed either in acidic or in alkaline media when they are uncoated. Nevertheless, the effective charge at the surface of the particles depends on the pH [5, 10]. In order to get particles with a constant surface charge density for all pH $\geq$ 3, the particles are coated with citrate species and dispersed at pH 7. They bear negative charges due to the acidic-basic behaviour of the citrate species. Above a concentration of citrate in solution of 2×10^{-3} mol/l, the plateau of adsorption is reached [6]. The ionic strength in such suspensions is thus ensured by the unadsorbed electrolyte (3:1), i.e. trisodium citrate.

The suspensions are usually polydisperse systems. The particle size distribution is described by a log–normal law [1]:

$$P(d) = \frac{1}{\sqrt{2\pi}\sigma d} \exp\left[-\frac{1}{2\sigma^2}\left(\ln\frac{d}{d_0}\right)^2\right],$$

where d_0 is the mean diameter and σ the standard deviation. The particles obtained have a mean diameter ranging between 5 and 12 nm according to the experimental conditions used for their synthesis. It is possible to reduce the polydispersity by a size-sorting process [11] but it is not possible to get suspensions with a σ inferior to 0.1. The samples under consideration here have a mean diameter of 8 nm and a standard deviation of 0.35 (d and σ are deduced from the analysis of the shape of the magnetization curve [12]). The mean volume, V_W, is estimated by taking into account the polydispersity through $\langle r^3 \rangle = \langle r_0^3 \rangle \exp\left(\frac{3^2\sigma^2}{2}\right)$. It allows an average diameter, d_{mean} of 9.6 nm to be deduced for the samples described here.

Methods

Osmotic compression experiment

The ionic strength and the osmotic pressure are imposed by osmotic compression [13], which is a suitable way to get suspensions with a high volume fraction of particles: the suspensions after synthesis are dialysed against a bath, called a reservoir, which has the required ionic strength. A dextran polymer ($M_w = 110000$ g/mol) is added in the bath and imposes its own osmotic pressure, which is neither dependent on temperature nor dependent on ionic strength or pH. The phenomenological law followed by the osmotic pressure as a function of the concentration of polymer has been established elsewhere [14]: $\log_{10}(\Pi_{dyn/cm^2}) = 1.826 + 1.715 w^{0.297}$ ($100w$ is the mass fraction of polymer in solution). The reservoir is replaced as many times as necessary to reach equilibrium, which usually takes about 3 weeks.

The volume fraction of the suspensions after the osmotic compression is determined either by chemical titration [15] or from the value of the saturation magnetization [12].

Small-angle neutron scattering experiments

We investigated the structure of the concentrated suspensions by small-angle neutrons scattering (SANS) experiments. The measurements were performed at LLB (CEA-Saclay) on the two-dimensional PAXY spectrometer. The neutron wavelength was 10 Å and two sample–multidetector distances were used: 3.20 and 1.05 m. The scattering vector ranged from 8×10^{-3} to 7×10^{-2} Å^{-1} in the first configuration and from 5×10^{-2} to 4×10^{-1} Å^{-1} in the second one.

In order to evaluate the effect of the appliance of a magnetic field on the suspensions, we performed SANS measurements under a constant magnetic field. The magnetic field was parallel to the plane of the multidetector. The magnetic field had a constant value of 0.03 T. The sample–multidetector distance was 3.20 m and the neutron wavelength was 10 Å (8×10^{-3} Å$^{-1} < q < 7 \times 10^{-2}$ Å^{-1}).

Results

Establishment of the rheological phase diagram of the concentrated suspensions

Osmotic compression of liquid samples allows solid samples to be obtained. The "liquid" and "solid" phases are here rheological definitions: a sample is considered to be a liquid if it flows and to be a solid if it does not. The osmotic decompression of the solid sample allows the liquid to be recovered: the liquid–solid transition is reversible.

The threshold of the "liquid–solid" transition is shifted towards the low volume of particles when the ionic strength decreases.

The experimental phase diagram at room temperature of the liquid–solid transition is presented in Fig. 1 as a function of the ionic strength. All the samples were obtained by osmotic compression.

Equation of state of the system

The equation of state of the system of concentrated suspensions for a citrate ionic strength of 2.5×10^{-3} mol/l is presented in Fig. 2. The regime of the low volume fraction is compared in Fig. 2a to previous results reported in Ref. [5]. The high-volume-fraction regime is investigated in Fig. 2b. It was obtained using the osmotic compression experiment. It has revealed that it is impossible to determine the volume fraction of samples for which the imposed osmotic pressure is about 18000 Pa.

Structure of the suspensions

The structure of the suspensions was determined using SANS. The scattered intensity of a colloidal suspension

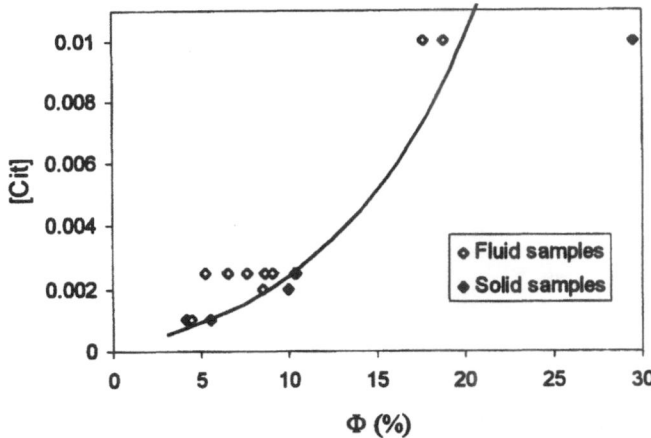

Fig. 1 Experimental phase diagram of suspensions as a function of the citrate concentration. The *line* represents the theoretical value of a liquid–solid transition for a system of hard spheres supposed to occur for a renormalized volume fraction $\Phi_{eff} = 50\%$: renormalization is performed for each ionic strength taking into account for the particle volume the range of the repulsions

in a SANS experiment is $I(q) = n^*(\Delta\rho)^{2*}P(q)^*S(q)$, where q is the wave factor, n the density of the system, $(\Delta\rho)^2$ the contrast, $P(q)$ the form factor of the particles and $S(q)$ the structure factor of the suspension.

For low concentrations of colloids, a suspension behaves as a perfect gas: there are no interactions between particles and the structure factor $S(q)$ is 1 for every q. As the contrast and $P(q)$ are similar for concentrated and dilute suspensions, the structure factor of the concentrated suspensions was calculated as follows: $S(q) = [I_{conc}(q)/\Phi_{conc}]/[I_{dil}(q)/\Phi_{dil}]$.

Fig. 2a, b Experimental equation of state of a suspension of maghemite particles for a citrate concentration of 2.5×10^{-3} mol/l. **a** Low-volume-fraction regime: experimental results are compared with results from Ref. [5]. **b** High-volume-fraction regime: *open symbols* are related to liquid samples and *filled symbols* to solid samples

The curves obtained for a dilute suspension ($\Phi = 1\%$) and for three concentrated suspensions, two liquid samples (A and B) and a solid sample (C) are presented in Fig. 3a. The citrate ionic strength was set to 10^{-2} mol/l. Despite anisotropic dipolar interactions, the 2D spectra are isotropic and are thus presented in an integrated form. As it is very difficult to measure the volume fraction of maghemite in concentrated suspensions, the volume fraction was deduced from the calculation of $I/I_{1\%}$, which is linear to $S(q)^*\Phi$. As $S(q) \rightarrow 1$ at high q, Φ can be determined. We found the following A: $\Phi = 17.6\%$; B: $\Phi = 18.8\%$; C: $\Phi = 29.5\%$ (Fig. 3b).

The structure factor of suspensions A, B and C is presented in Fig. 4. It is obtained by dividing the curves in Fig. 3b by Φ.

Effect of the appliance of a magnetic field on the samples

Macroscopically, a magnetic field has no effect on the solid samples, except for the ones which are just above the liquid–solid transition threshold. For these samples the appliance of a magnetic field turns the solid phase into a liquid one.

Evolution of the structure of the suspensions under the appliance of a magnetic field

The scattered intensity of a suspension with $\Phi = 17.6\%$ (citrate ionic strength of 10^{-2} mol/l) in the parallel and perpendicular directions to the magnetic field is represented in Fig. 5. The two spectra are similar, indicating that the scattered intensity remains globally isotropic.

As the experimental spectrum of a concentrated suspension remains isotropic under the appliance of a magnetic field, it is possible to integrate the measured signal over the surface of the detector in order to get the

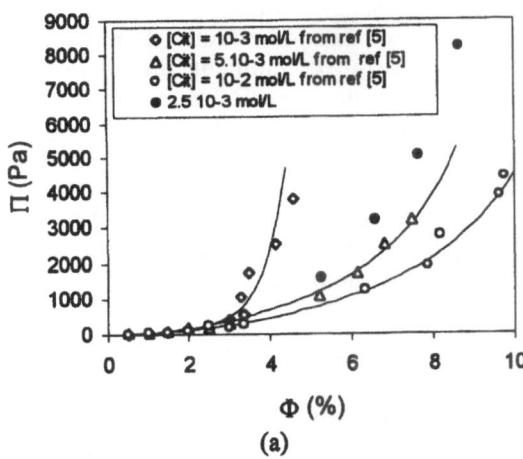

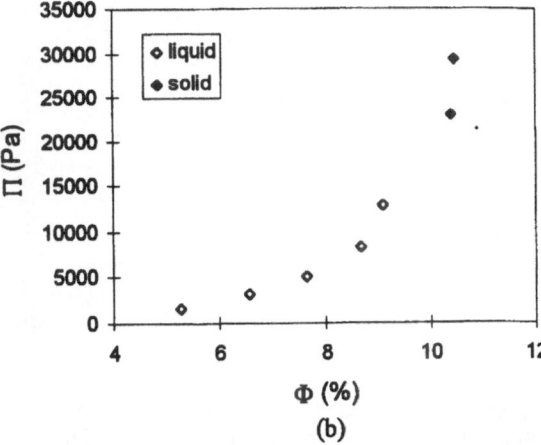

(a)   (b)

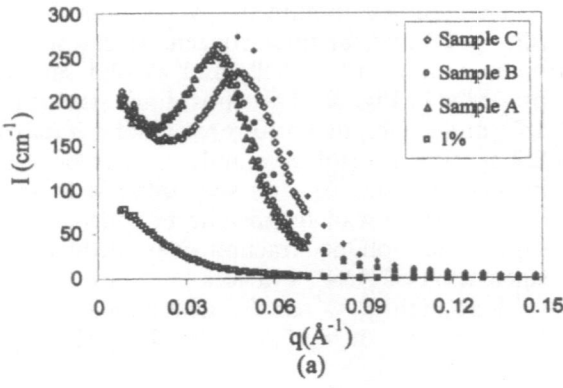

Fig. 3 a Small-angle neutron scattering spectra obtained for the high-volume-fraction samples A, B and C and for a dilute suspension of maghemite particles ($\Phi = 1\%$). **b** Renormalization of the spectra for samples A, B and C by the form factor $P(q)$ and the contrast $(\Delta\rho)^2$. *Open symbols* represent the data obtained from the first configuration of the experimental setup and *filled symbols* represent the data obtained from the second configuration

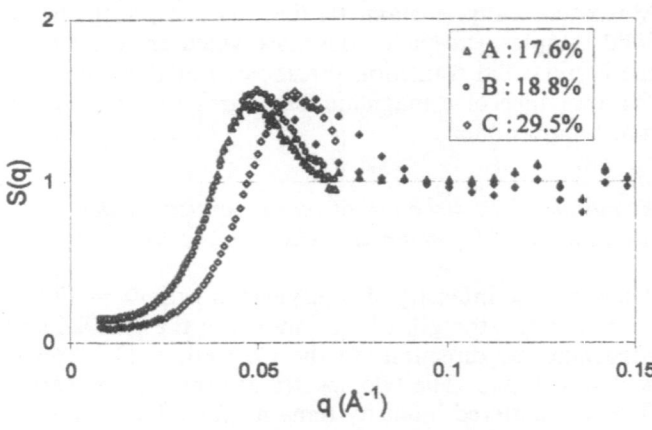

Fig. 4 Structure factor of samples A ($\Phi = 17.6\%$), B ($\Phi = 18.8\%$) and C ($\Phi = 29.5\%$). *Open symbols* represent data obtained from the first configuration of the experimental setup and *filled symbols* represent data obtained from the second configuration

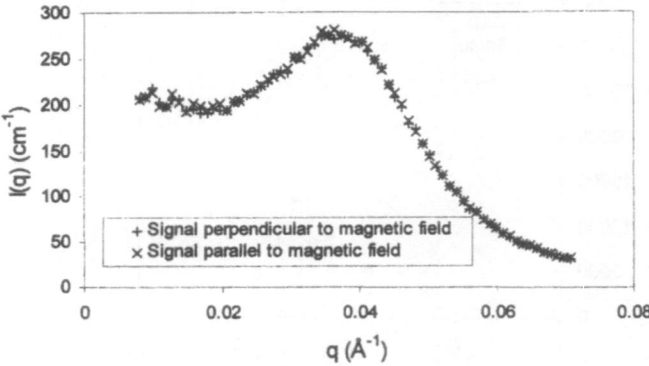

Fig. 5 Scattered intensity of sample A ($\Phi = 17.6\%$) under a 0.03 T magnetic field in parallel and perpendicular directions to the field

isotropic structure factor of the suspensions submitted to a magnetic field. The structure factor of the suspension with $\Phi = 17.6\%$ (citrate ionic strength of 10^{-2} mol/l) in zero magnetic field is compared with that in a field of 0.03 T.

Discussion

Interactions in the suspensions

The interparticle interactions in the aqueous dispersions under consideration here are van der Waals attractions, magnetic dipolar interactions and electrostatic repulsions.

For a dispersion of spherical particles of diameter d_0, the reduced potential due to van der Waals attractions can be written

$$\frac{U_{\text{vdW}}(s)}{kT} = -\frac{A}{6kT}\left(\frac{2}{s^2-4} + \frac{2}{s^2} + \ln\frac{s^2}{s^2-4}\right) ,$$

where A is the Hamaker constant and $s = 2r/d_0$, for particles with $r \gg d_0$, r being the distance between the centres of the two particles ($A = 10^{-19}$ J for the maghemite particles) [16].

In the dispersions under consideration, each particle is a magnetic monodomain, i.e. each particle has a permanent magnetic moment, the intensity, μ, of which depends on the specific magnetization, m_s, of the material and on the particle diameter, d_0, $\mu = m_s\pi d_0^3/6$ ($m_s = 3.1 \times 10^5$ A/m for these maghemite particles) [12]. If $\vec{\mu}_i$ is the dipolar moment of particle i and $\vec{r}$ is the vector joining the centres of the particles, the anisotropic potential relative to the dipolar interactions between two dipoles in zero magnetic field is

$$\frac{U_{\text{dip}}(r)}{kT} = \frac{\gamma}{4\pi}(2\cos\theta_1\cos\theta_2 - \sin\theta_1\sin\theta_2\cos\varphi) ,$$

where θ_i is the angle between $\vec{\mu}_i$ and $\vec{r}$ and φ is the azimuthal angle between both dipoles.

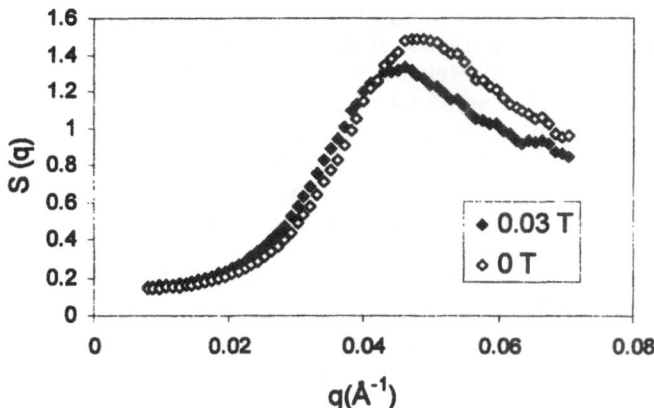

Fig. 6 Structure factor of sample A ($\Phi = 17.6\%$) in zero magnetic field (*open symbols*) and in a 0.03 T magnetic field (*filled symbols*)

A reduced parameter, $\gamma = \frac{\mu_0 m_s^2 V^2}{kTr^3}$ (V is the volume of a particle), allows the magnetic coupling between the particles to be quantified:

- If $\gamma/4\pi \ll 1$, the dipoles rotate independently and the magnetic potential in zero field, averaged over all directions, may be simplified to $\frac{\langle U_M \rangle}{kT} = -\frac{\gamma^2}{48\pi^2}$ (low magnetic coupling).
- If $\gamma/4\pi \gg 1$, there are correlations between the dipoles and the potential becomes anisotropic. Nevertheless, it is possible to integrate over the space to obtain a mean value of the dipolar interaction: $\frac{\langle U_M \rangle}{kT} = -\frac{\gamma}{2\pi}$ (high magnetic coupling) [10].

When a magnetic field is applied, the expansion of the potential around the most probable configuration leads to the mean potential decreasing as $1/r^3$.

Concerning repulsions, such aqueous dispersions of particles with surface charges are usually described using the celebrated DLVO formalism [17], which takes into account the van der Waals attractions and the Coulombic repulsions; however, this DLVO theory fails to describe the astonishing stability of aqueous suspensions of maghemite particles coated by citrate ions at high ionic strength as reported in Ref. [6]. The repulsive Coulombic potential is nevertheless well described by a Yukawa potential:

$$\frac{U_{Coul}}{kT} = K \frac{d_0}{r} \exp\left[-\frac{(r - d_0)}{D}\right],$$

where K corresponds to the contact value and D characterizes the interaction range, with r being the distance from the interparticle centre and d_0 the particle diameter. The value of K is much higher than the one used in the DLVO potential that takes into account the effective surface charge. These strong repulsions at contact can be due to the finite size of the trivalent citrate ions which induce an additional steric repulsion

at the surface of the particles. The range of the interactions, D, is of the order of the Debye length, $1/\kappa$, defined by $\kappa = \left(\frac{e^2}{\varepsilon_0 \varepsilon kT} \sum_i c_i z_i^2\right)^{1/2}$.

Structure of the suspensions in the high-volume-fraction regime

The structure factors obtained from SANS for liquid samples A and B clearly indicate that the structure of the dispersions of the particles is that of a liquid (or a fluid). The isotropy of the spectra and the value of the maximum of $S(q)$ indicate that the sum of the interactions is dominated by the electrostatic repulsions despite dipolar interactions. The shape of the structure factor of solid sample C is similar to that of liquid samples A and B and no second Bragg peak is observed, indicating that the structure stays liquidlike.

The mean distance between particles, r_{mean}, is calculated from the abscissa q_{max} corresponding to the maximum of the structure factor by $r_{mean} = 2\pi/q_{max}$. r_{mean} is perfectly linear to $(\pi/6\Phi)^{1/3} d_{mean}$ (the values are reported in Table 1), proving that the particles are homogeneously dispersed. The mean diameter d_{mean} of the particles deduced from r_{mean} (9.5 nm) is in good accordance with the mean diameter obtained from the magnetization curve (9.6 nm).

Nature of the transition

Figure 2a indicates that for the low ionic strength under consideration in this study the regime of the low volume fractions is dominated by strong electrostatic repulsions. The suspensions remain fluid in those experimental conditions, far from the regime of lower strength repulsions where the transition from a fluid phase to a biphasic liquid–gas system occurs.

According to Fig. 2b, it is very difficult to conclude if the equation of state presents a plateau. Moreover the evidence of the biphasic samples has not been established.

As shown by the structure factor, the regime of the high volume fractions is also dominated by strong electrostatic repulsions. The strength of these electrostatic repulsions is so high that it governs the structure of the system, which remains isotropic. Nevertheless, in

Table 1 Mean interparticle distance, r_{mean}, deduced from q_{max} as a function of the volume fraction, Φ, of the samples

Φ (%)	r_{mean} (nm)
17.6	12.93
18.8	12.58
29.5	10.58

contrast to the case of the low volume fractions, for which the dipolar interactions remain in the low-coupling regime (i.e. $\gamma/4\pi \ll 1$), an intermediate magnetic coupling regime is reached for the high volume fractions as the interparticle distance is strongly reduced (we get $\gamma/4\pi \approx 0.4$ for the most concentrated suspension under consideration here). The dipolar interactions are thus anisotropic.

As the SANS spectra indicate that the liquid and the solid phases obtained are both fluid phases (no crystalline order is observed), the liquid–solid transition observed can be described as a vitreous transition. As it is driven by long-range electrostatic repulsions, the solid phase is a Wigner glass [18]. The solid transition for a system of hard spheres occurs theoretically for a given particle volume fraction of 50%. In Fig. 1 we compare this theoretical threshold to our experimental results for different ionic strength and renormalize the volume of the particles by taking into account the range of the electrostatic repulsions: the radius of the particle is replaced by an effective radius: the sum of the particle's radius, the citrate shell (6 Å [19]) and the Debye length. The experimental data are in good agreement with this theoretical value of the threshold.

The polydispersity of our suspensions ($\sigma = 0.35$) explains why it is not possible to make fluid–crystal transitions [4]. When a significant volume fraction of particles has a diameter greater than the mean interparticle distance, too many defects in the lattice are created and the formation of a crystal is not allowed. An effective polydispersity, inferior to the real one, can be evaluated in our experimental system by taking into account the range of the repulsions, but it remains too important to allow the crystal phase to be reached.

Dynamic measurements usually allow the ergodicity of the system to be discussed, in order to determine if the nature of a transition is vitreous [20]. Unfortunately the long time which is necessary to reach equilibrium during an osmotic compression experiment does not allow dynamic measurements to be made on our suspensions. Our conclusion concerning the vitreous nature of the transition is thus only based on static arguments: the liquidlike structure of the solid and the high polydispersity of the system.

Effect of a magnetic field

The structure factor of the suspensions for which a magnetic field was applied shows that the interactions in the system continue to be dominated by repulsions and thus stay globally isotropic. Nevertheless, the sum of the interactions is less repulsive when a magnetic field is applied (the first peak of the structure factor is lowered). This has to be related to the rheological solid–liquid transition observed when a magnetic field is applied to solid samples. The sum of the interactions is lowered and the change in the effective volume fraction of the spheres through the lowering of the range of the interactions can be sufficient for the threshold of the solid transition to be crossed.

It is astonishing that at the same time the mean distance between the particles increases: the value of q_{max} of the structure factor is shifted towards lower q, meaning that the density of the objects changes through $r_{mean} = (\pi/6\Phi)^{1/3}d_{mean}$. The decrease in q_{max} corresponds to a decrease in the volume fraction from 17.6 to 16.5%. This has to be related to the appearance of a small peak in the structure factor at $q = 0.0665$ Å^{-1} (9.45 nm in real space). This distance is of the order of two particle radii, which means that some doublets of particles are formed under the appliance of the magnetic field. For a field of 0.03 T, the decrease in the volume fraction corresponds to a proportion of particles involved in doublets of about 12%. As the polydispersity of the interactions is very important in the system (dipolar interactions are function of d^3), it is reasonable to suppose that the particles involved in the creation of the doublets are the biggest. Nevertheless, the system continues to be dominated by repulsions.

Conclusion

This experimental work concerning the phase behavior of aqueous dispersions of magnetic nanoparticles in the high-volume-fraction regime completes previous studies on the same systems performed for low volume fractions and illustrates the fact that magnetic fluids are a suitable experimental system for the study of the phase behaviour of suspensions of nanometric particles (it is possible to reach the fluid, solid, gas and liquid phases). The structure of concentrated suspensions is governed by strong electrostatic repulsions and the suspensions exhibit a vitreous transition. The threshold of the transition can be monitored by the ionic strength in the suspension. The suspensions keep their vitreous structure when a magnetic field is applied, but the sum of the interactions is reduced and it appears possible to induce solid–liquid transitions, which give specific magnetorheological properties to the suspensions.

Acknowledgements We thank François Boué for his help during the SANS experiments and Emmanuelle Dubois for helpful discussions.

References

1. Bacri JC, Perzynski R, Salin D, Cabuil V, Massart R (1990) J Magn Magn Mater 85:27
2. Pusey PN (1991) In: Hansen JP, Levesque D, Zinn-Justin J (eds) Liquids, freezing and glass transitions. North-Holland, Amsterdam, pp 765–942
3. (a) Tejero CF, Daanoun A, Lekkerkerker HNW, Baus M (1994) Phys Rev Lett 73:752; (b) Pusey PN, Poon WCK, Ilett SM, Bartlett P (1994) J Phys Condens Matter 6:A29–A24
4. Bartlett P (1997) J Chem Phys 107:188
5. Cousin F, Cabuil V J Mol Liq 83:203
6. Dubois E, Cabuil V, Boué F, Perzynski R (1999) J Chem Phys 111:7147
7. Dubois E, Cabuil V, Boué F, Bacri JC, Perzynski R (1997) Prog Colloid Polym Sci 104:173
8. Bacri J-C, Perzynski R, Salin D, Cabuil V, Massart R (1988) J Colloid Interface Sci 132:43
9. Massart R (1981) IEEE Trans Magn 17:1247
10. Ménager C, Belloni L, Cabuil V, Dubois M, Gulik-Krzywicki T, Zemb T (1996) Langmuir 12:3519
11. Massart R, Dubois E, Cabuil V, Hasmonay E (1995) J Magn Magn Mater 149:1
12. Bacri JC, Cabuil V, Massart R, Perzynski R, Salin D (1986) J Magn Magn Mater 62:36
13. Parsegian VA, Fuller N, Rand RP (1979) Proc Natl Acad Sci USA 76:2750
14. Mourchid A, Delville A, Lambard J, Lécollier E, Levitz P (1995) Langmuir 11:1942
15. Charlot G (1966) In: Les méthodes de la chimie analytique. Masson et Cie (ed), p 737
16. Sholten PC (1978) In: Berkovsky B (ed) Thermomechanics of the magnetic fluids. Hemisphere, Bristol, Pa, 1
17. Verwey EJW, Overbeek JTG (1948) Theory of the stability of lyophobic colloids. Elsevier, Amsterdam
18. (a) Wigner E (1938) Trans Faraday Soc 34:678; (b) Bosse J, Wilke SD (1998) Phys Rev Lett 80:1260
19. Dubois E (1997) Thesis. Paris VI University
20. (a) Van Megen W, Underwood SM, Pusey PN (1991) Phys Rev Lett 67:1586; (b) Bonn D, Tanaka H, Wegdam G, Kellay H, Meunier J (1998) Europhys Lett 45:52

Progr Colloid Polym Sci (2000) 115:84–87
© Springer-Verlag 2000

A. Fernández-Barbero
A. Loxley
B. Vincent

Heteroaggregation between charged hard and soft particles

A. Fernández-Barbero
Complex Fluid Physics Group
Department of Applied Physics
University of Almería
Cañada de San Urbano s/n
04120 Almería, Spain

A. Loxley
E Ink Corp, 45 Spinelli Place, Cambridge
Massachusetts, MA 02138, USA

B. Vincent (✉)
School of Chemistry, University of Bristol
Cantock's Close, Bristol BS8 1TS, UK

Abstract The heteroaggregation of hard and soft particles with opposite sign of charge was studied by static and dynamic light scattering. The cluster structure as well as the aggregation kinetics were investigated. The clusters were observed to be highly branched in the presence of attractive and repulsive forces, although the mass evolution does not indicate relevant differences with respect to diffusion aggregation; however, at high ionic concentration, where diffusion aggregation is expected, very compact clusters were found as well as very fast aggregation. These findings support the idea that the reactivity of the larger clusters is especially relevant.

Key words Fractal structure · Aggregation · Heteroaggregation · Light scattering · Mesoscopic systems

Introduction

Colloidal aggregates have attracted much research interest since it was shown that the concept of a fractal can be used to characterize the structure of a disordered system, such as a random cluster growing under non-equilibrium conditions [1]. Under most circumstances, only certain limiting values of the fractal dimension have been found with gold, silica and polystyrene colloidal particles: $d_f \approx 1.75$ under diffusive conditions and $d_f \approx 2.0$ when the aggregation is controlled by long-range repulsive forces [2]. These two universal behaviors are known as diffusion-limited cluster aggregation and reaction-limited cluster aggregation (RLCA).

In the present article, the aggregation of positively and negatively charged particles is studied. A repulsive long-range electrical force is considered within the aggregates, which does not compete against the van der Waals force due to its different range. This force provokes repulsion among some particles, modifying the fractal structure (particles with different sign of charge will try to join, but those with similar sign will be located as far apart as possible). This introduces a different energy balance and more branched clusters are expected. Moreover, the soft character of the positively charged particles makes the short-range attractive van der Waals force weaker. At high salt concentration only the short-range force should be relevant and the soft nature should become apparent; however, at low salt concentration, the long-range electrical forces have to be responsible for the system behavior.

Theory

Colloidal clusters exhibit a structure characterized by a fractal dimension, d_f, directly related to the cluster compactness. They show open, self-similar, branched structures described by a scaling law relating the characteristic cluster radius, R, and its volume V [3]:

$$V(R) \sim R^{d_f} . \tag{1}$$

In addition, the growth mechanism may be characterized by a kinetic exponent, z, modeling the mean mass growth for long aggregation times:

$$\langle M \rangle \sim t^z . \tag{2}$$

This exponent is unity for diffusion aggregation (linear behavior) and is greater than unity when repulsive forces modify the particle trajectories. In RLCA, for which larger clusters are more reactive than smaller ones, the kinetics exponent must increase.

Experimental

Materials

Two systems of spherical particles were used for the heteroaggregation experiments. ANTO⁻ is an aqueous suspension of charged surfactant-free microspheres, synthesized by emulsion polymerization of styrene, using sodium persulfate as the initiator [6]. The negative particle charge arises from sulfate groups on the surface. ANDY⁺ is an aqueous cationic microgel suspension synthesized by polymerization of poly(2-vinylpyridine), cross-linked with divinylbenzene (0.25 wt%). The initiator used was 2,2′-azobis(2-amidinopropane) dihydrochloride (Wako) [7]. Two types of groups are able to confer charge to the colloidal particles: amidininum groups arising from the initiator, located essentially at the periphery of the particles and the constituent monomer 2-vinylpyridine (2VP), uniformly distributed within the particles. The pK_a of 2VP is 5 and that of the amidinium groups is around 10; therefore, for pH higher than 5, the 2VP residues are uncharged, the microgel particles are hydrophobic and the particles are in their collapsed state. The particle charge is dominated by the surface groups since the particle interior is essentially uncharged.

The sign of charge was tested by electrophoresis and the particle size was determined by transmission electron microscopy to be 225 ± 10 and 205 ± 14 for the anionic and cationic particles, respectively. The salt employed was NaCl (Merck, analytical grade), and the temperature was always set to (20 ± 1)°C.

Methods

Static light scattering (SLS) allows d_f to be determined from the angular dependence of the mean scattered intensity. For elastic scattering, the intensity from a system of clusters under the condition $qR \gg 1$ is expressed as [4]

$$I(q) \sim q^{-d_f} , \tag{3}$$

where $q = 4\pi/\lambda \sin(\theta/2)$ is the scattering wave vector, with λ the wavelength of the light in the solvent and θ the scattering angle. For higher q values, the length scale corresponds to individual spheres within the cluster and the intensity is related to the particle form factor.

Information about the aggregation kinetics can be extracted from the time evolution of the cluster mass or, equivalently, from the evolution of the number-average cluster size, $\langle n_n \rangle$. In order to determine the latter, the relationship for fractal aggregates between the mean cluster mass and the mean hydrodynamic radius, $\langle R_h \rangle(t)$, was employed:

$$\langle n_n \rangle = \frac{\langle M \rangle}{m_0} = \left(\frac{\langle R_h \rangle}{R_0} \right)^{d_f} . \tag{4}$$

In this equation m_0 and R_0 are the monomer mass and radius, respectively. In order to measure the hydrodynamic radius, dynamic light scattering (DLS) was employed to determine the intensity autocorrelation function. Data analysis was performed by cumulative analysis [5].

Light scattering experiments were performed using a slightly modified Malvern 4700 system working with a 632-nm-wavelength He-Ne laser. The intensity autocorrelation functions were determined at different times during the aggregation experiments. After

data analysis, the mean hydrodynamic radius was obtained as a function of time. The mean scattered intensity was also obtained for different angles in the range 20°–140°. $I(q)$ exhibited an asymptotic time-independent power law related to the cluster fractal structure. From these curves, the fractal dimensions were determined.

Results and discussion

Particle aggregation was performed under the influence of two long-range forces: an attractive electrical force between particles of different sign of charge and a repulsive one between those with the same sign of charge. Moreover, a weaker short-range force due to the softness of the positively charged particles is also involved. The ionic concentration was used for controlling the strength of the interactions. As a first approximation, we investigated the problem of heteroaggregation of equal numbers of particles of each type. Under these conditions, any extra asymmetry is taken into account.

Aggregations were monitored by DLS. Figure 1 shows the time evolution of the mean diameter for heteroaggregation at low and high salt concentration as well as for diffusion aggregation. Every curve scales according to a power law in time (in the inset, one of the aggregations is plotted on a double-logarithmic scale as an example). From the slope, a size-related kinetics exponent may be obtained.

It is essential to emphasize that the size-related kinetics exponent not only contains information on the kinetics of aggregation, but also on the cluster structure. Thus, for proper data interpretation, SLS was employed

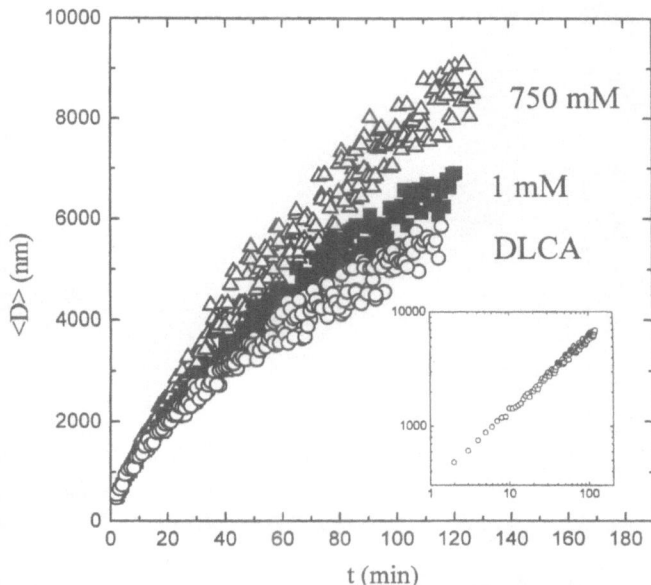

Fig. 1 Evolution of the mean cluster size. The *inset* shows a log–log plot for heteroaggregation at 1 mM salt

for determining the cluster fractal dimension. The angular dependence of the mean scattered intensity for different aggregation processes is shown in Fig. 2 (inset). A decreasing power law in q is observed as theory predicts for fractal aggregates. From the slopes, the fractal dimensions were obtained and are plotted in the main graphic as a function of the salt concentration. At low salt concentration, where the long-range forces dominate, the aggregates are more branched than those growing under diffusion-controlled conditions. This result is quite intuitive by taking into account the repulsion between particles the same sign of charge within the clusters. For high salt concentration, where the short-range forces are especially relevant, a strong increase in the fractal dimension is exhibited in spite of the fact that diffusion aggregation was expected. This result could be related to the fact that the van der Waals forces involved in the particle aggregation are now weaker due to the soft character of the positive particles and, thus, a rearrangement within the clusters is possible.

The time evolution of the cluster mass (proportional to $\langle n_n \rangle$ was studied in order to extract information regarding the aggregation kinetics. $\langle n_n \rangle$) was calculated by applying Eq. (4). A power law at long aggregation times was observed for every case (Fig. 3, inset). The mass-related kinetics exponent was determined and is plotted in Fig. 3 as a function of the salt concentration. At low salt concentration, no relevant differences were detected with respect to the diffusion value. This means that the larger cluster size detected by DLS (Fig. 1) is a

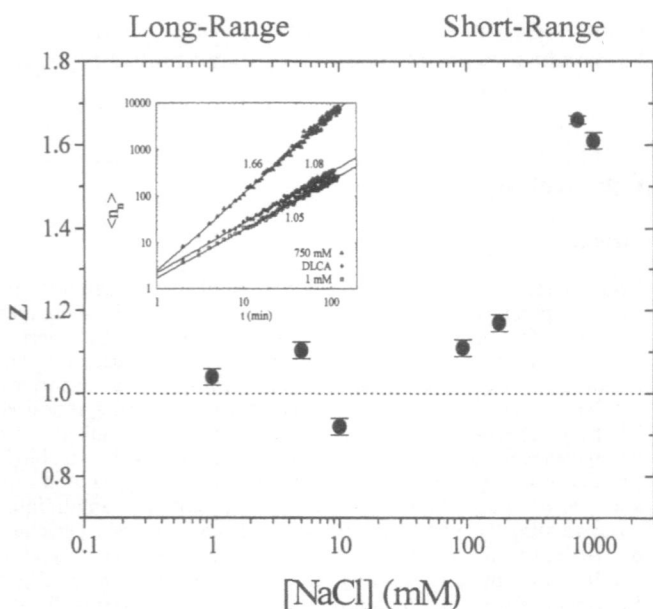

Fig. 3 Evolution of the mass-based kinetics exponent with salt concentration. The number-average mean cluster size is plotted in the *inset* at different salt concentrations. Different slopes indicate changes in the fractal dimensions

consequence only of the more branched structure of the aggregate growing under the influence of long-range attraction forces. However, for high salt concentration the kinetics exponent strongly increases, indicating that the reactivity of larger clusters is higher that of small clusters. This result is also unexpected since diffusion aggregation should become apparent. The aggregation becomes so fast that the mean size grows over the rest of the process studied, despite the fact that the clusters are more compact (Fig. 1).

Conclusions

The clusters were observed to be more branched in the presence of attractive and repulsive forces, although the mass evolution indicates that the process is not very different from the diffusion-controlled case. However, at high ionic concentration, where diffusion aggregation is expected, very compact clusters were found as well as a very fast aggregation process. These findings support the idea that a rearrangement within the clusters due to the weakness of the short-range attraction force is possible. Moreover, the reactivity of the larger clusters is especially relevant.

Acknowledgements A.F.B. acknowledges financial support through the University of Almería (Plan Propio de Investigación) and the Andalusian Government for a period of study at the University of Bristol. This work was supported by the Acción Integrada Hispano-Británica (HB 1998-0225).

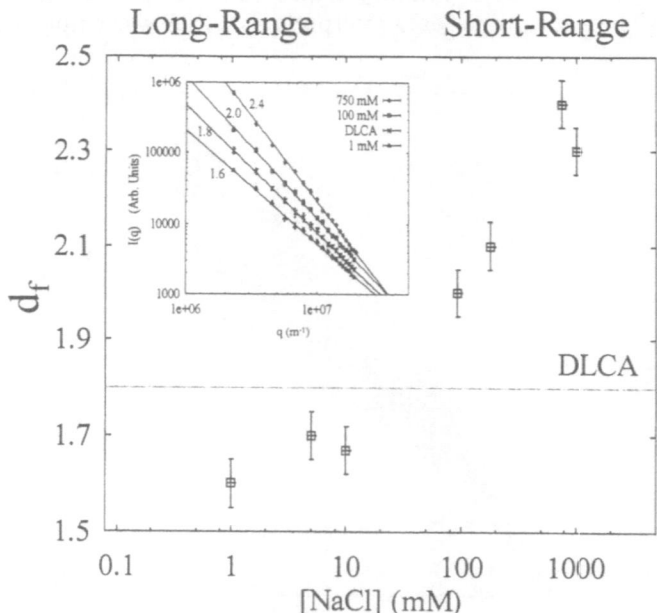

Fig. 2 Fractal dimension as a function of the salt concentration for the heteroaggregation processes. The region of diffusion-limited cluster aggregation (*DLCA*) is indicated. The mean scattered light intensity is shown in the *inset*

References

1. Witten TA, Sander LM (1981) Phys Rev Lett 47:1400
2. Carpineti M, Giglio M (1990) Adv Colloid Interface Sci 46:13
3. Julien R, Botet R (1987) Aggregation and fractal aggregates. World Scientific, Singapore
4. Dhont JKG (1996) An introduction to dynamics of colloids. Elsevier, Amsterdam
5. Chu B (1991) Laser light scattering. Academic Press, San Diego 247
6. Goodwin JW, Hearn J, Hu CC, Ottewill RH (1974) Colloid Polym Sci 259:464
7. Loxley A, Vincent B (1997) Colloid Polym Sci 275:1108

Progr Colloid Polym Sci (2000) 115:88–92
© Springer-Verlag 2000

J. Stellbrink
J. Allgaier
M. Monkenbusch
D. Richter
A. Lang
C. N. Likos
M. Watzlawek
H. Löwen
G. Ehlers
P. Schleger

Neither Gaussian chains nor hard spheres – star polymers seen as ultrasoft colloids

J. Stellbrink (✉) · J. Allgaier
M. Monkenbusch · D. Richter
IFF-Neutronenstreuung
Forschungszentrum Jülich
52425 Jülich, Germany

A. Lang* · C. N. Likos · M. Watzlawek
H. Löwen
Institut für Theoretische Physik II
Heinrich-Heine-Universität Düsseldorf
Universitätsstrasse 1
40225 Düsseldorf, Germany

G. Ehlers · P. Schleger
Institute Laue-Langevin, BP156
38042 Grenoble, France

Permanent address:
* Inst. Theor. Phys.
Tu Wien, Wiedner Hauptstr. 8-10
A-1040 Wien, Austria

Abstract In dense solution high functionality star polymers show ordering phenomena which give rise to a well-pronounced peak in the static structure factor, $S_{exp}(Q)$, observed by small-angle neutron scattering. The concentration dependence of $S_{exp}(Q)$ gives evidence for unusual phase behaviour as predicted by theory. In addition, the dynamics of the star polymer solutions is dominated by an increasing amount of structural arrest with increasing concentration. The mean square displacement obtained from neutron spin-echo spectroscopy is compared to the blob size obtained from dynamic light scattering. Thermal energy enables each star core to perform restricted motion over a spatial extent equal to the blob size of the surrounding dense star polymer solution.

Key words Star polymers · Small-angle neutron scattering · Neutron spin-echo spectroscopy · Liquid-state theory · Pair potential

Introduction

Recently we introduced star polymers as a new class of ultrasoft colloids showing both polymer-like and colloidal characteristics. This hybrid character gives rise to some unusual phenomena in the structural properties [1–4] as well as in the dynamics of dense star polymer solution [5, 6]. However, all these studies were limited to low functionality star polymers and/or concentrations below the overlap concentration, $c^* = 3/(4\pi R_g^3) \times (M_w/N_A)$.

Here, we present new results concerning the structure and dynamics of star polymer solutions which have overcome these limitations. The article is divided as follows. In the first part, we present small-angle neutron scattering (SANS) data of a 57-arm polybutadiene (PB) star polymer in a good solvent up to a concentration well above c^*. The experimental data are analysed using the microscopic pair potential introduced by Likos et al. [1] and are finally compared to the phase diagram of Watzlawek et al. [3]. In the second part, we discuss the microscopic dynamics of an 18-arm polyisoprene (PI) star polymer as investigated by neutron spin-echo (NSE) spectroscopy. To cover the slowed-down dynamics at the peak position in the static structure factor, Q_m, [6], the dynamic time range of NSE is extended for the first time up to 350 ns using a long wavelength of 19 Å at IN15(ILL). Finally, we draw some conclusions emphasizing the exceptional position of star polymers as a connecting link between polymer physics and colloidal science.

Experimental

The molecular characteristics of the polymers investigated are summarised in Table 1. The partially labelled high functionality PB star was prepared by anionic polymerization following an established procedure [7, 8]. The synthesis of the arms started with deuterated PB and secondary butyl lithium as initiator and proceeded with protonated PB. The still-living polymer chains were coupled to the linking agent, a 4G-chlorosilan dendrimer. The result of the synthesis is a labelled multiarm star which has a protonated core and a deuterated shell. Due to steric hindrance during the coupling [9], the synthesis results in a small polydisper-

Table 1 Molecular characteristics of the star polymers investigated

Sample	Monomer	f	M_w (g/mol)	M_w/M_n[a]	R_g (Å)	c^* (g/cm^3)	R_h^b (Å)
S18	Isoprene	18	139,700[c]	1.01	72.0[d]	0.149	80.5
S64	Butadiene	57	467,800[e]	1.01	105.5	0.158	124.5

[a] Gel permeation chromatography
[b] Dynamic light scattering
[c] Low angle laser light scattering
[d] Interpolated from literature data [17, 18]
[e] Small-angle neutron scattering

sity in f, which gives a mean functionality, f_{mean} of 57. The partially labelled 18-arm PI star could be prepared nearly monodisperse and is the same as that described in Refs. [1, 5]. Using fully deuterated methylcyclohexane, the solvent and the deuterated shell scatter neutrons in the same way and in the experiments only the protonated core is visible.

The SANS experiments were performed using the KWS1 instrument at Forschungszentrum Jülich and the PAXY instrument at LLB, Sacclay. NSE spectroscopy data were obtained using the NSE instrument at FRJ-2, Jülich, and at IN15, ILL, Grenoble.

Results and discussion

Small-angle neutron scattering

For analysing our SANS data the same procedure as described in detail in Ref. [1] was applied. That is, starting from a pair potential, $V(r)$, for star polymers which reads as follows

$$\frac{V(r)}{k_B T} = \begin{cases} (5/18)f^{3/2}[-\ln(r/\sigma) + (1+\sqrt{f}/2)^{-1}] & (r \le \sigma); \\ (5/18)f^{3/2}(1+\sqrt{f}/2)^{-1}(\sigma/r) \\ \times \exp[-\sqrt{f}(r-\sigma)/2\sigma] & (r > \sigma) \end{cases}$$

(1)

and applying the Rogers–Young closure [10] and associated Monte Carlo simulations we obtain information about the pair structure of the liquid, in particular the centre-to-centre structure factor, $S(Q)$, of the stars. In attempting to fit the experimental data for the total scattering intensity, $I(Q)$, with the theoretical predictions based on an analytic pair potential, we must take into consideration the fact that the star size itself has a dependence on the concentration. Whereas in a previous study [1] this could be done using experimental data for $\sigma(\Phi)$, in the present study we have to use σ as (the only) adjustable parameter [11].[1]

A representative fit for a volume fraction, Φ, of 18.75%, i.e. approximately 1.2Φ^*, is shown in Fig. 1. It can be seen that the fit is quite satisfactory for the whole Q range. The insert of Fig. 1 shows how σ varies with increasing Φ. Obviously, there are three different regions:

1. Up to $\Phi \approx 0.5\Phi^*$, σ stays nearly constant at about 150 Å.
2. A steep decrease up to $\Phi \approx 1.2\Phi^*$.
3. A much less pronounced Φ dependence is reached for the highest volume fractions under study.

The Daoud–Cotton [12] scaling approach for star polymers predicts exactly these three different concentration regimes for high functionality star polymers. The proposed power laws $\sim \Phi^0$, $\sim \Phi^{-3/4}$ and $\Phi^{-1/8}$ are also shown in Fig. 1. Although not completely convincing, the observed Φ dependence agrees reasonably with these power laws.

For discussing in more detail the observed structures it is recommended to refer directly to $S(Q)$ rather than to $I(Q)$. The experimental static structure factor, $S_{exp}(Q)$, of the high functionality PB star polymer at Φ^* is compared to that of the 18-arm PI star polymer used in the previous study in Fig. 2. $S_{exp}(Q)$ is obtained assuming a decoupling between the form factor and the structure factor, i.e. $I(q) = V_W P(Q) S(Q)$, where $P(Q)$ is the intramolecular form factor.[2] For $f = 57$, $S_{exp}(Q)$ shows two well-developed maxima, indicating the high degree of structural order. From the peak height of the first maximum, we can estimate that we are already close to a phase transition to a crystalline phase [13]. The insert of Fig. 2 shows the dependence of the peak height, S_{max}, of the first maximum on the volume fraction. S_{max} itself shows a first maximum exactly at Φ^* as predicted by theory [14]. After a short monotonous decrease S_{max} starts to fluctuate strongly between $1.2\Phi^* \le \Phi \le 2\Phi^*$ until it finally increases again monotonically. These fluctuations can be interpreted as a precursor of a reentrant melting as predicted in the phase diagram of dense star polymer solutions [3]. This reentrant melting should take place exactly in the Φ region where S_{max} shows its unusual behaviour. The fact that no Bragg peaks are observed in the SANS data, as expected for a crystalline phase, can be attributed to the small polydispersity of our star polymers, which was also not taken into account in the phase diagram. To summarize this part, the agreement obtained between our SANS data

---

[1] Zero-average contrast experiments on similar star polymers have shown that R_g stays constant with increasing concentration, but some changes took place an smaller length scales. Probably the size of the outermost blob changes, which could not be resolved within experimental errors.

[2] This assumption is still valid in our case at high volume fractions. Due to the fact that only the inner quarter of each star is protonated, i.e. visible for the neutrons, the form factor of this part is not affected by any overlap of the outer regions of the stars.

Fig. 1 Experimental (*points*) versus theoretical results (*solid line*) for the total scattering intensity, $I(Q)/\Phi$, of 57-arm polybutadiene stars at a volume fraction, Φ, of $18.75 \approx 1.2\Phi^*$. Also shown is the star form factor obtained from extrapolation to zero concentration (*dashed line*). *Insert*: concentration dependence of the star radius, σ, used as an adjustable parameter in the fit. The *solid line* shows the proposed scaling laws of Ref. [12] (see text)

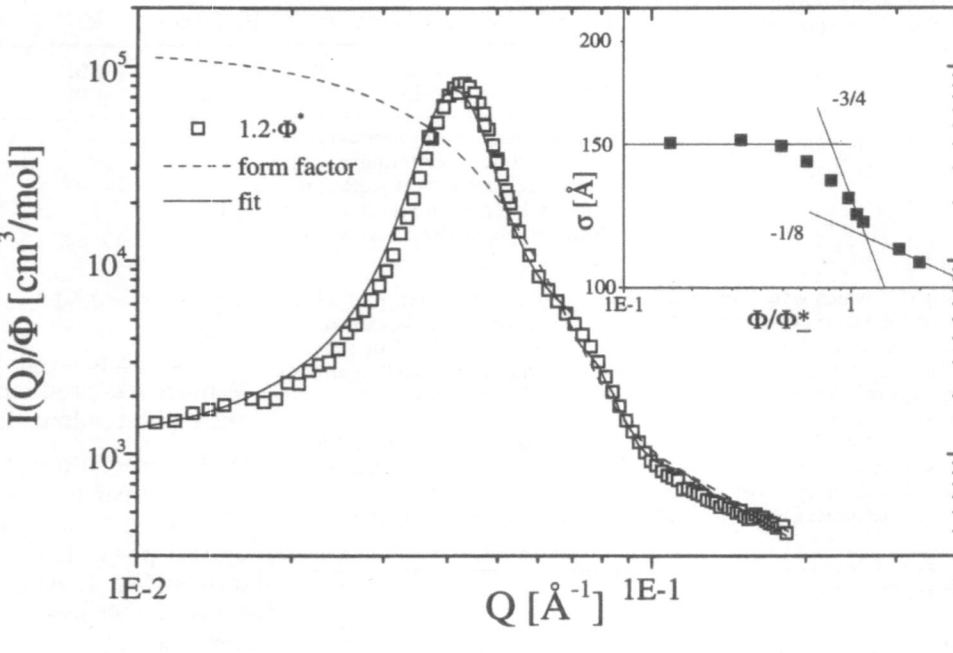

Fig. 2 The static structure factor, $S_{exp}(Q)$, obtained by small-angle neutron scattering is shown for star polymers of varying functionality, f, at their overlap concentration, c^*. With increasing functionality the degree of order increases substantially. From the height of the first maximum we can estimate that we are already close to a phase transition to a crystalline order for $f_{mean} = 57$. The *insert* shows the unusual concentration dependence of the peak height of the first maximum, S_{max}. The *dotted lines* indicate the region where reentrant melting is expected in the phase diagram [3]

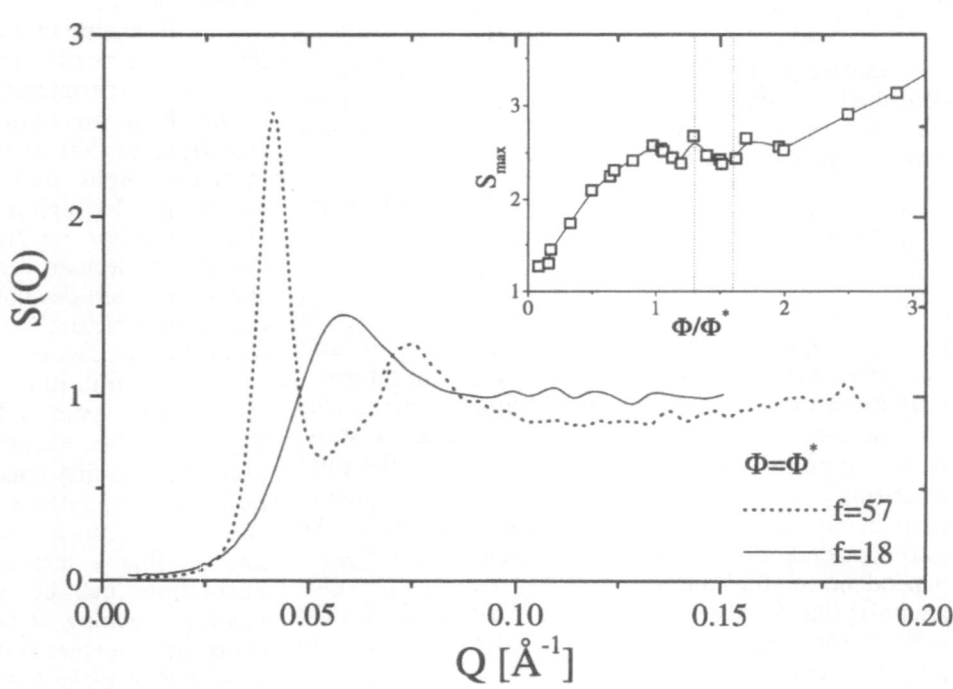

and predictions from polymer scaling theory as well as from a colloidal approach emphasizes the hybrid character of star polymers as a connecting link between polymer physics and colloidal science.

NSE spectroscopy

Due to the applied matching conditions NSE experiments observe the collective motion of the star centres if we measure close to the structure factor peak [15] (for our star polymer this occurs around $Q_m \approx 0.075$ Å^{-1} in the concentration regime of interest, see previous section).[3] The intermediate scattering function, $S(Q,t)$, is strongly modulated in the region around Q_m. To cover

[3] In a very recent light scattering study Semenov et al. [15] observed precursors of the structural influence on star dynamics, but using light they could not achieve Q_m

Fig. 3 Normalized intermediate scattering functions, $S(Q,t)/S(Q,0)$, as obtained by neutron spin-echo (*NSE*) spectroscopy for an 18-arm polyisoprene star polymer at different Φ^*. From *top* and *bottom* $\Phi = 30$, 25 and 20%. Above Φ^* a concentration-dependent plateau is developed in $S(Q,t)$ near Q_m, the peak position in the static structure factor. *Solid line*: fit to Eq. (2). The dynamic time range of about 350 ns was achieved for the very first time using a long wavelength, 19 Å, at IN15(ILL)

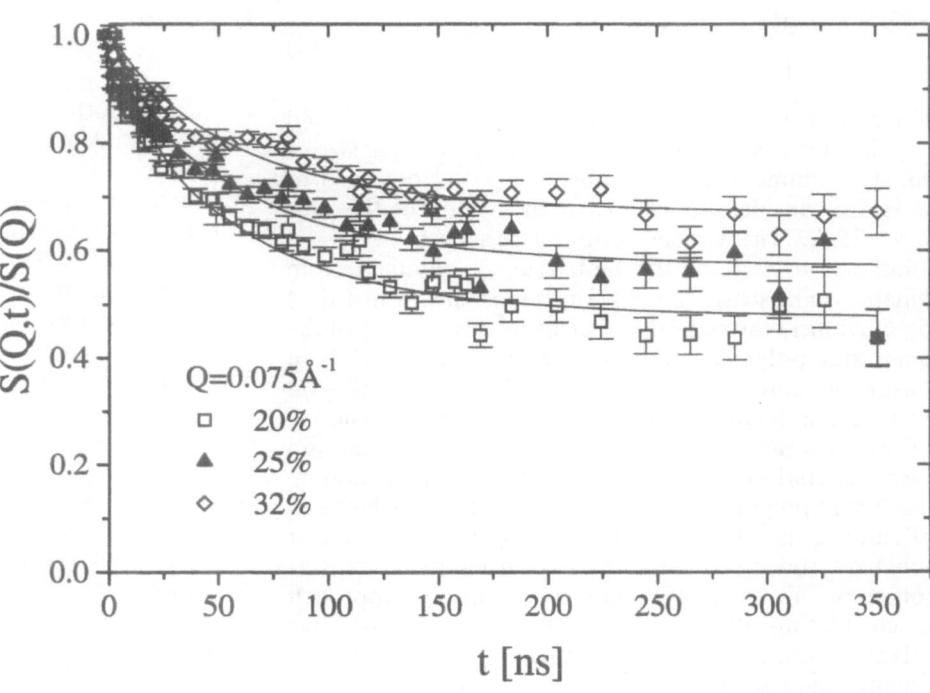

Fig. 4 Concentration dependence of the mean square displacement, $\langle r^2 \rangle^{1/2}$ of star cores obtained by NSE and of the blob size of the surrounding entangled star polymer solution, ξ, as obtained by dynamic light scattering (data taken from Ref. [5])

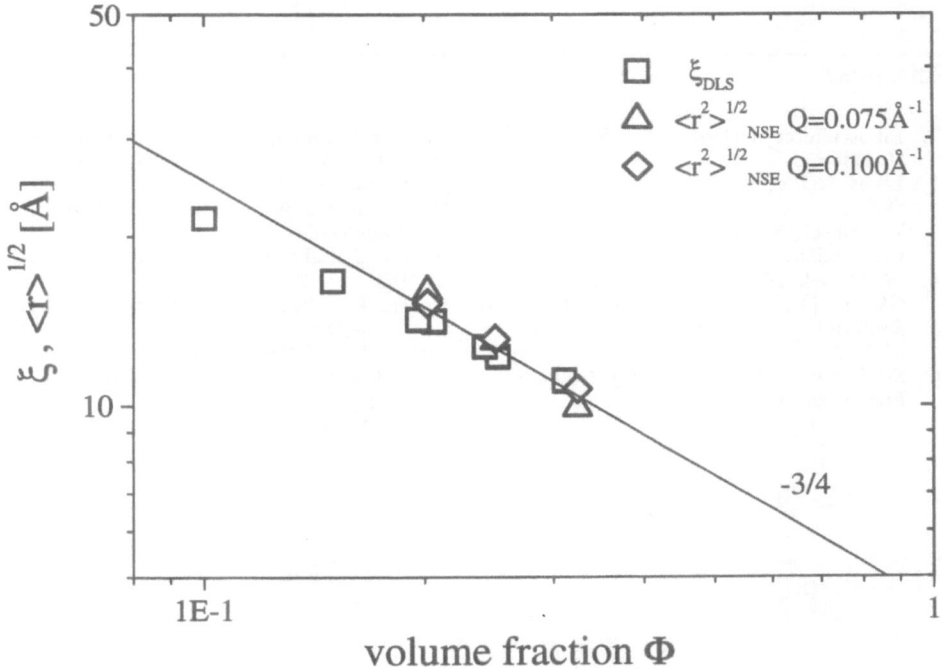

the slowed-down dynamics at Q_m, the dynamic time range of NSE was extended up to 350 ns using a long wavelength, 19 Å, at In15(ILL).

The normalized intermediate scattering functions, $S(Q,t)/S(Q,0)$, at $Q_m = 0.075$ are shown in Fig. 3 at three different volume fractions: $\Phi = 20, 25$ and 30% ($\Phi^* = 15\%$). It can clearly be seen that at each volume fraction a final plateau is reached at about 180 ns. Due to the increasing amount of structural arrest with increasing concentration, the height of the plateau depends on concentration. For quantitative analysis, the restricted motion of the star polymers can be approximated by the following intermediate scattering function (solid lines in Fig. 3):

$$S(Q,t) = \exp(-1/3\ Q^2 \langle r^2 \rangle)$$
$$+ [1 - \exp(-1/3\ Q^2 \langle r^2 \rangle)]H(Q,t)\ , \tag{2}$$

with $H(Q,t)$ the Zimm dynamic structure factor [16] and $\langle r^2 \rangle$ the mean square displacement or the average size of the star confinement. The values of $\langle r^2 \rangle^{1/2}$ obtained of 15 Å are considerably smaller than the size of the star, $R_g = 75$ Å. The concentration dependence of $\langle r^2 \rangle^{1/2}$ compared to that of the blob size, ξ, obtained from dynamic light scattering experiments is shown in Fig. 4 [5]. Obviously the star confinement relates to the ξ of the dense star polymer solution. The slope obtained is in reasonable agreement with the predicted value of $-3/4$ from scaling theory. Thus, the available thermal energy, kT, enables each star core to perform restricted motion over a spatial extent equal to the ξ of the surrounding dense star polymer solution. This relationship between kT and ξ is also known from solutions of linear polymers and emphasizes the hybrid character of star polymers. These results are a promising approach to elucidating the complex dynamics of dense star polymer solutions by combining SANS, NSE and dynamic light scattering.

Conclusions

The exceptional hybrid character of star polymers influences both their structural properties and their dynamics in dense solution. The colloidal aspect is more pronounced in the intermolecular properties of star polymers, for example, the centre-to-centre structure factor. $S_{exp}(Q)$ obtained by SANS can be described by a colloidal approach combining a new pair potential and liquid-state theory as known from hard-sphere systems. The observed agreement is, in particular, convincing due to the fact that we are starting from a microscopic level. The polymer aspect, on the other hand, is more pronounced in the single star properties. For example, the dependence of star size on concentration and the relationship between the dynamical confinement of a single star core and the blob size of the already entangled outer regions of different stars. All these results underline the importance of star polymers as a new class of colloids, which we would like to call ultra soft colloids.

Acknowledgement J.S. acknowledges the Deutsche Forschungsgemeinschaft for financial support. A.L. likes to acknowledge the "Osterreisehische Forschungsfond" for financial support under Proj. No P13062.

References

1. Likos CN et al (1998) Phys Rev Lett 80:4450
2. Likos CN et al (1998) Phys Rev E 58:6299
3. Watzlawek M et al (1999) Phys Rev Lett 82:5289
4. Watzlawek M, Löwen H, Likos CN (1998) J Phys Condens Matter 10:8189
5. Stellbrink J et al (1997) Phys Rev E 57:3772
6. Stellbrink J et al (1998) Prog Colloid Polym Sci 110:25
7. Allgaier J, Young RN, Efstratiadis V, Hadjichristidis N (1996) Macromolecules 29:1794
8. Hadjichristidis N, Fetters LJ (1980) Macromolecules 13:191
9. Allgaier J et al (1999) Macromolecules 32:3190
10. Rogers FA, Young DA (1984) Phys Rev A 30:999
11. Jucknischke O (1995) Doctoral thesis. Westfälische Wilhelms-Universität Münster
12. Daoud M, Cotton JP (1982) J Phys Paris 43:531
13. Hansen J-P, Verlet L (1969) Phys Rev 184:151
14. Witten TA (1986) Europhys Lett 2:137
15. Semenov AN et al (1999) Langmuir 15:358
16. Dubois-Violette E, de Gennes PG (1968) Physics 3:181
17. Willner L et al (1994) Macromolecules 27:3821
18. Fetters LJ, Hadjichristidis N, Lindner JS, Mays JW (1994) J Phys Chem Ref Data 23:619

Progr Colloid Polym Sci (2000) 115:93–96
© Springer-Verlag 2000

Phase behaviour of aqueous mixtures of sodium dodecyl sulfate with a weakly cationically charged acrylamide-based copolymer

G. Mylonas
G. Bokias
G. Staikos

G. Mylonas · G. Bokias (✉) · G. Staikos
Department of Chemical Engineering
University of Patras and Institute
of Chemical Engineering
and High-Temperature Processes
ICE/HT-FORTH
P.O. Box 1414, 26500 Patras, Greece
e-mail: bokias@chemeng.upatras.gr
Fax: +30-61-997266

Abstract The phase behaviour at constant polymer concentration of aqueous mixtures of sodium dodecyl sulfate (SDS) with a weakly charged cationic polymer (PAM6) based on an acrylamide backbone, containing 6 mol% cationic sites, is presented. Upon increasing the SDS concentration, the known strongly hydrophilic character of the acrylamide polymer turns gradually to moderately hydrophobic. Thus, although at room temperature the aqueous PAM6/SDS mixtures present the expected phase behaviour (fully homogeneous or phase-separated, depending on the polymer/surfactant concentration ratio), they exhibit upon heating either upper critical solution temperature or lower critical solution temperature characteristics.

Key words Weakly charged polyacrylamide · Sodium dodecyl sulfate · Upper critical solution temperature · Lower critical solution temperature

Introduction

Polyacrylamide (PAM) is one of the best studied water-soluble polymers, due mainly to its large use for practical applications [1]. Its especially pronounced hydrophilic character is revealed by the good solubility of the homopolymer in water over the whole temperature range and by the large Mark–Houwink–Sakurada exponents [1, 2]. Moreover, it is estimated to present upper critical solution temperature (UCST) behaviour, a behaviour common to most hydrophilic polymers, at −38 °C [3]. In copolymers, acrylamide (AM) is used as the hydrophilic component in order to preserve water solubility or to moderate the special solution or phase behaviour of the comonomer. Such copolymers are hydrophobically modified PAM samples [4–7] or AM/N-isopropylacrylamide copolymers [8–11]. Obviously, in both cases, the special solution or phase behaviour of the copolymer in water is only due to the comonomer.

Another consequence of the strong hydrophilic character of PAM is the absence of any important interaction with surfactants. To induce the interaction of such mixtures, AM-based copolymers with appropriate comonomers can been used, exploiting, for instance, the hydrophobic or electrostatic interactions of the comonomer with the surfactant.

Here, we present some preliminary results concerning the solution and phase behaviour of dilute aqueous mixtures of sodium dodecyl sulfate (SDS) with a weakly cationically charged copolymer based on AM. We focus mainly on the temperature dependence of the phase behaviour of such mixtures. The chemical structure of the copolymer used is presented in Scheme 1. This polymer, notated as PAM6, contains 6 mol% of the charged comonomer (3-(methacryloylamino) propyl) trimethylammonium chloride (MAPTMAC).

Note that copolymers of a similar structure but with a much higher charge content, 30–60 mol%, associate strongly with SDS [12] and their phase and solution behaviour is similar to that of aqueous mixtures of charged homopolymers, polycations or polyions with oppositely charged surfactants [13]. Namely

1. The association starts for very low surfactant concentrations.
2. On approaching charge neutralisation, associative phase-separation takes place.

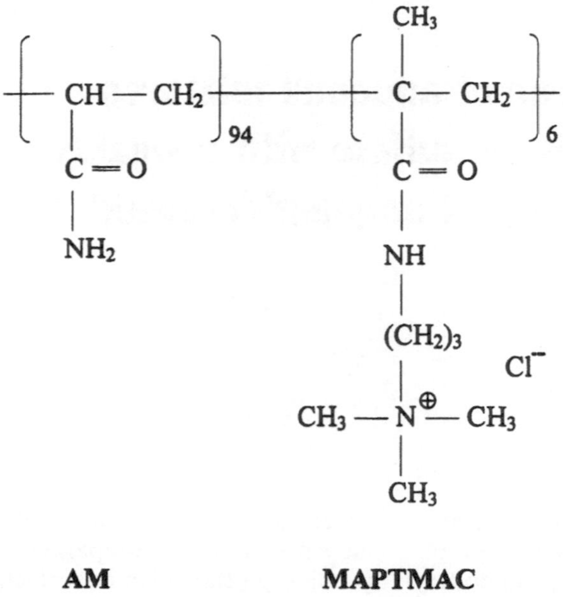

$$\left[\begin{array}{c} CH - CH_2 \\ | \\ C=O \\ | \\ NH_2 \end{array}\right]_{94} \left[\begin{array}{c} CH_3 \\ | \\ C - CH_2 \\ | \\ C=O \\ | \\ NH \\ | \\ (CH_2)_3 \\ | \\ CH_3 - N^{\oplus} - CH_3 \\ | \\ CH_3 \end{array}\right]_{6} Cl^-$$

AM   **MAPTMAC**

Scheme 1.

3. The mixture is redissolved for high surfactant concentrations.
4. If the polymer concentration is rather high, thickening properties may be observed just before or after the two-phase region.

However, in these highly charged systems, either the temperature dependence of the two-phase region is not considered at all or it is considered as not to be significant. In contrast, in our system, the polymer–surfactant interactions should not be so strong, as PAM6 is only weakly charged, and the observation of temperature-sensitive phenomena is more probable.

Experimental

Materials

SDS, used with no further purification, was a product of Aldrich. PAM6 was prepared by copolymerisation in water of AM and MAPTMAC at 28 °C, initiated by the redox couple $(NH_4)_2S_2O_8$/$Na_2S_2O_5$. The product was dialysed and then recovered by freeze-drying. The MAPMAC content of the copolymer was determined by potentiometric titration of the Cl^- counterions with $AgNO_3$ and by ^1H NMR. The intrinsic viscosity of the copolymer was determined in an aqueous 0.5 M NaCl solution at 25 °C [14] and its molar mass was estimated to be 150000. All reagents used for the synthesis of PAM6 and the subsequent studies were of analytical grade. The water used was reagent grade from a Seralpur Pro 90C apparatus.

Methods

Turbidimetric measurements

A U2001 (Hitachi) UV/vis spectrophotometer equipped with a water-circulating cell holder was used for the turbidimetric measurements. The temperature was changed manually at a rate of about 0.5 °C/min. The absorbance of a light beam at 500 nm was measured. The cloud points, determined as the onset of the absorbance–temperature curves, were generally in good agreement with those determined by simple visual observation of the samples.

Viscosity measurements

The reduced viscosity of PAM6/SDS aqueous solutions at various temperatures was determined using a Schott-Gerate AVS 300 automated viscosity measuring system equipped with an Ubbelohde type viscometer. The polymer concentration for both turbidimetric and viscosity measurements was always 2.5×10^{-3} g cm^{-3}.

Results and discussion

The phase behaviour at room temperature of the PAM6/SDS mixtures in water with increasing SDS concentration is drawn schematically in Fig. 1. We observed the phase behaviour usually expected for an aqueous system of a charged polymer with an oppositely charged surfactant: after a homogeneous region at low SDS concentrations, an associative two-phase region is observed, starting below, but not very far from, the charge neutralisation (indicated with the vertical dotted line in Fig. 1). This two-phase region is rather large, as the system becomes homogeneous again only when the SDS concentration exceeds about 6 mM. We should note that with the exception of the region very close to the charge neutralisation point the biphasic region is not characterised by two well-defined phases, but we only observe a uniform, more or less turbid, solution. Some of these turbid solutions were tested for several days and no changes in their appearance were observed. The stabilisation of the polymer-rich phase in this suspension-like form is probably due to the excess charge borne by the polymer as we go far from charge neutralisation.

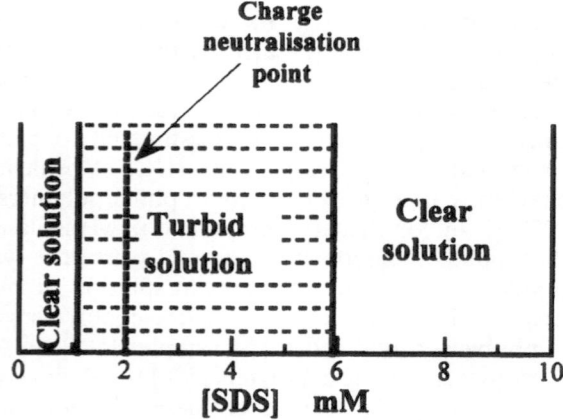

Fig. 1 A schematic illustration of the phase behaviour at 25 °C of aqueous mixtures of sodium dodecyl sulfate (SDS) and the acrylamide-based copolymer (PAM6), containing 6 mol% [(3-methacryloylamino)propyl]trimethylammonium chloride. The concentration of PAM6, c, is 2.5×10^{-3} g cm^{-3}

In addition, the high affinity of the main component of the backbone (AM) to water disfavours the formation of two well-separated phases.

Despite the, more or less, expected behaviour of the PAM6/SDS mixtures at room temperature, the dependence of their phase behaviour with temperature is rather surprising. Some turbidimetric measurements of mixtures containing SDS at low (before the charge neutralisation point) or rather high (after the charge neutralisation point) concentration are presented in Fig. 2a and b, respectively. First of all, we observe that at a given temperature the solutions get less and less turbid as we go far from the charge neutralisation point (to lower SDS concentrations in Fig. 2a or to higher SDS concentrations in Fig. 2b). In fact, this decreasing turbidity, that could be the result of a lower quantity or the smaller size of the phase-separated particles, supports the argument presented earlier that these suspension-like structures are stabilised by the excess charge (arising from the polymer chain when SDS is deficient, Fig. 2a, or from the charged, bound SDS molecules when SDS is in excess, Fig. 2b). However,

the most important observation in Fig. 2 concerns the temperature dependence of the turbidity of these solutions. Thus, for low SDS concentrations (Fig. 2a), the initially turbid solutions turn less and less turbid with increasing temperature and at a given temperature they become transparent. In contrast, for relatively high SDS concentrations (Fig. 2b), the initially transparent solutions turn turbid upon heating. Depending, therefore, on the SDS concentration, this aqueous system (PAM6/SDS) exhibits properties of both UCST and lower critical solution temperature (LCST) behaviour.

This is more clearly illustrated in Fig. 3, where the cloud-point temperature, i.e. the temperature of the turbidity onset upon heating or cooling, is presented as a function of the SDS concentration. At low SDS concentration, the system does not exhibit any temperature sensitivity and it is homogeneous over the whole temperature range: from the freezing point of water up to about 90 °C. When approaching the charge neutralisation point, the UCST behaviour is evidenced, as the system is turbid at low temperatures and turns transparent upon heating. The cloud-point temperature is very sensitive to the SDS concentration, as it increases rapidly within a short SDS concentration region, so very soon the two-phase region is extended to the whole temperature range studied. Furthermore, the system remains biphasic for a rather large SDS concentration region independently of the temperature, until LCST phase behaviour is observed: the system is homogeneous at low temperature and it turns cloudy upon heating.

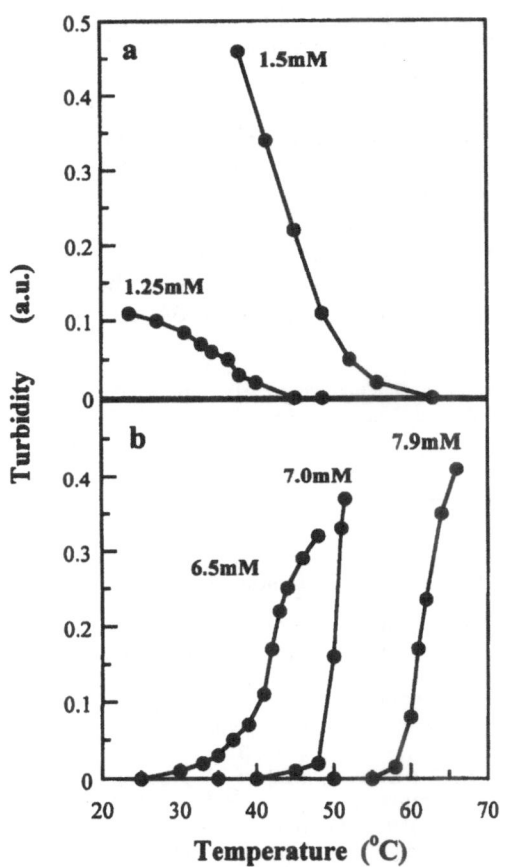

Fig. 2a, b The turbidimetric behaviour of aqueous PAM6/SDS mixtures with temperature. $c = 2.5 \times 10^{-3}$ g cm^{-3}. The SDS concentration is indicated next to each curve

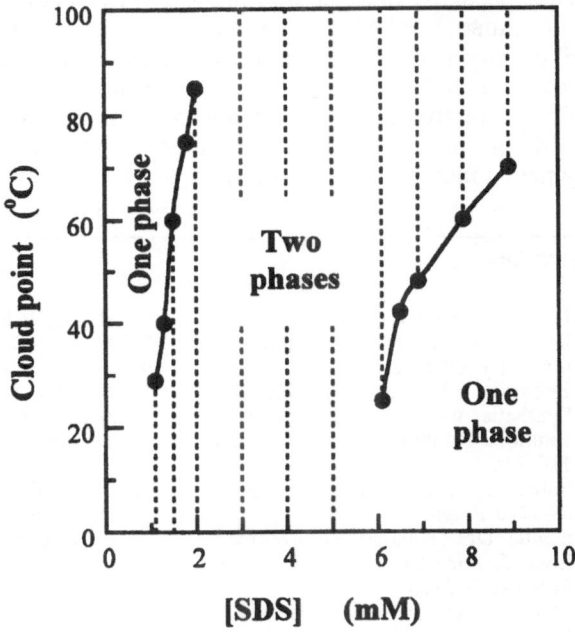

Fig. 3 The influence of the SDS concentration on the cloud point of aqueous PAM6/SDS mixtures. $c = 2.5 \times 10^{-3}$ g cm^{-3}

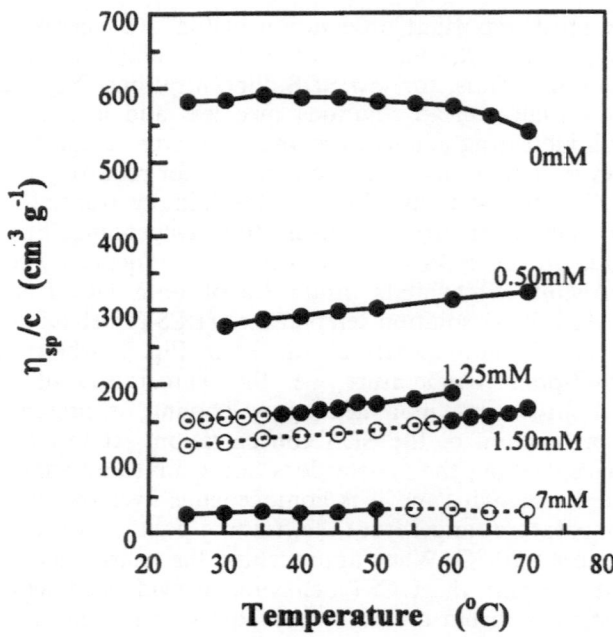

Fig. 4 The influence of temperature on the reduced viscosity, η_{sp}/c, of aqueous PAM6/SDS mixtures. $c = 2.5 \times 10^{-3}$ g cm^{-3}. The SDS concentration is indicated next to each curve

The one-phase region is extended gradually to higher temperatures upon addition of SDS, so the cloud-point temperature is shifted to temperatures higher than the boiling point of water when the SDS concentration exceeds 10 mM and the system is again homogeneous over the whole temperature range.

Unexpectedly, this rich and original phase behaviour of the aqueous PAM6/SDS mixtures is not accompanied by significant conformational changes, as far as we can conclude from the viscometric data presented in Fig. 4. In this figure, the reduced viscosity of pure PAM6 or of PAM6/SDS mixtures at a constant polymer concentration is plotted as a function of temperature. Some of the mixtures are chosen to be in the UCST or LCST regions (turbid regions are indicated by the dotted lines, open circles). For all mixtures presented, the reduced viscosity–temperature patterns are rather flat, even if the curve crosses the two-phase region. This hard to explain behaviour is in contrast with other known thermosensitive systems, for instance, aqueous poly(N-isopropylacrylamide)/SDS mixtures [15], where a conformational transition from a coil to a compact conformation is observed when approaching the two-phase region.

Conclusion

The introduction of positive charges on a PAM backbone enhances, as expected, the interaction of this polymer with SDS, as in most charged polymer/oppositely charged surfactant systems. Moreover, if the polymer is only weakly charged, like PAM6, association with SDS leads to remarkable phase behaviour: depending on the SDS concentration, fully homogeneous or two-phase aqueous mixtures are formed and, more importantly, thermally sensitive systems (comprising both UCST and LCST characteristics) can be obtained. These first results presented here show how the increasing binding of SDS molecules onto the PAM6 backbone changes progressively the strong hydrophilic character of the polymer to moderately hydrophilic (observation of UCST properties) and, then, to moderately hydrophobic (observation of LCST properties). However, to fully understand this behaviour detailed studies on the binding mechanism or the influence of several parameters, such as the charge density of the backbone, the polymer concentration or the molecular weight of the polymer, are needed.

Acknowledgements This work was supported by the European Union under contract ERBFMBICT983240 (G.B.'s return grant to Greece).

References

1. Molyneux P (1987) Water-soluble synthetic polymers: properties and behavior, vol 1. CRC, New York
2. Bekturov EA, Bakauova ZKh (1986) Synthetic water-soluble polymers in solution. Huethig & Wepf, Basel pp. 126–135
3. Silberberg A, Eliassaf J, Katchalsky A (1957) J Polym Sci 23:259
4. Schulz DN, Kaladas JJ, Maurer JJ, Bock J, Pace SJ, Schulz WW (1987) Polymer 28:2110
5. McCormick C, Nonaka T, Johnson CB (1988) Polymer 29:731
6. Hill A, Candau F, Selb J (1993) Macromolecules 26:4521
7. Hwang FS, Hogen-Esch TE (1995) Macromolecules 28:3328
8. Chiklis CK, Grasshoff JM (1970) J Polym Sci Part A Poly Chem Part A-2 8:L6L7
9. Taylor LD, Ceranowski LD (1975) J Polym Sci Polym Chem Ed 13:2551
10. Priest JH, Murray SL, Nelson RJ, Hoffman AS (1987) In: Russo P (ed) Reversible polymeric gels and related systems. ACS Symposium Series 350. American Chemical Society, Washington, DC, pp. 255–262
11. Feil H, Bae YH, Feijen J, Kim SW (1993) Macromolecules 26:2496
12. Fundin J, Brown W, Iliopoulos I, Claesson PM (1999) Colloid Polym Sci 277:25
13. Goddard ED (1993) In: Goddard ED, Ananthapadmanabhan KP (eds) Interactions of surfactants with polymers and proteins. CRC, New York, pp 171–201
14. Mabire F, Audebert R, Quivoron C (1984) Polymer 25:1318
15. Staikos G (1995) Macromol Rapid Commun 16:913

Progr Colloid Polym Sci (2000) 115:97–99
© Springer-Verlag 2000

M. Bele
S. Pejovnik
K. Kočevar
I. Muševič
J. O. Besenhard

Substrate-induced deposition: molecular bridging between dispersed particles and gelatine-modified surfaces

M. Bele · S. Pejovnik
National Institute of Chemistry
Hajdrihova 19, 1000 Ljubljana
Slovenia

K. Kočevar · I. Muševič
J. Stefan Institute, Jamova 39
1000 Ljubljana, Slovenia

I. Muševič
Faculty of Mathematics and Physics
University of Ljubljana, Jadranska 19
1000 Ljubljana, Slovenia

J. O. Besenhard
Institute of Chemical Technology
of Inorganic Materials
Technical University of Graz
Stremayrgasse 16/III, 8010 Graz, Austria

Abstract In the process of substrate-induced deposition of particles from a dispersion (e.g. carbon black) the surface of the substrate is first covered with a water-soluble polymer film (e.g. gelatine). The subsequent immersion into a colloid dispersion leads to deposition of dispersed particles on the surface of the substrate. An atomic force microscope has been used to measure the distance dependence of the forces between microscopic-sized glass spheres coated with gelatine and a solid surface in the presence of an aqueous solution.

Key words Atomic force microscope · Force spectroscopy · Gelatine · Particle deposition · Substrate-induced deposition

Introduction

Interactions of macromolecules and colloid particles with interfaces leading to their adsorption, deposition, and adhesion are of great practical interest, for example, for coating surfaces with fine particulate materials (as a part of the metallization processes of through holes in printed wiring boards [1–7] or the deposition of microporous particles [8]). The substrate is first covered with a water-soluble polymer film as an adhesion agent (e.g. proteins, cellulose derivatives, vinyl polymers). The subsequent immersion of such substrates into a colloid dispersion leads to deposition of dispersed particles on the surface of the substrate. Because the deposition is induced by the polymer adsorbed on the substrate, this technique is usually called a substrate-induced deposition. However, aqueous colloids are charge-stabilized, and it is difficult to prepare "close-packed layers" of charged particles. It is clear that an understanding of the mechanism of particle deposition from dispersions and the binding to a substrate can improve coating technology significantly.

Materials and methods

Freshly cleaved mica (Balzers) was used as the substrate and the flatness was checked on a nanometer scale. Glass spheres of 15–25 μm diameter (Duke Scientific Corporation) made of soda lime glass were attached to the cantilever of an atomic force microscope (AFM) microscope. Gelatine no. 48722 obtained from Fluka was used as the deposition initiator (adhesive substance). The cationic surfactant cetyltrimethylammonium bromide (CTAB), 95% Aldrich no. 85,582-0 was used.

The forces between the gelatine-covered glass particle and the mica-covered or carbon black covered surface were measured in solution using a Nanoscope III AFM equipped with a liquid cell. The AFM was operated in the force-plot mode [9]. In this mode of operation, the surface of the sample and the cantilever approach each other periodically and one measures the deflection of the cantilever, which is proportional to the force, exerted on the cantilever. Usually, a spherical particle is attached to the cantilever and the forces are measured between this particle and a flat surface.

In our experiments, a 20-μm glass sphere was attached to the standard AFM cantilever with a spring constant of 0.05 Nm^{-1}.

Results and discussion

The deposition of particles onto gelatine-coated substrates from dispersions is controlled by the pretreatment of surface-adsorbed gelatine, as was demonstrated in our previous work [7]. There, the experiments indicated that the main binding mechanism which was responsible for the deposition of carbon black particles onto gelatine-covered surfaces could be molecular bridging between the solid surfaces. In the present work, this hypothesis is confirmed using the AFM. A typical measurement is presented in Fig. 1, which shows the distance dependence of the force between a glass sphere coated with a thin layer of gelatine and the bare mica surface in the presence of pure water. First, we performed a full approach as illustrated by the full line in Fig. 1. In the second step we performed a force scan, where the approach of the two surfaces was stopped at a separation of approximately 50 nm. Although the two surfaces obviously did not come into hard contact, there is some attractive force when the sphere is retracting away from the surface and we clearly observe a very weak and fluctuating binding force between the glass sphere and the mica surface. This force increases slightly at a separation of 350 nm, where it suddenly disappears. We explain this observation as follows. When the "hairy" interface of gelatine comes to

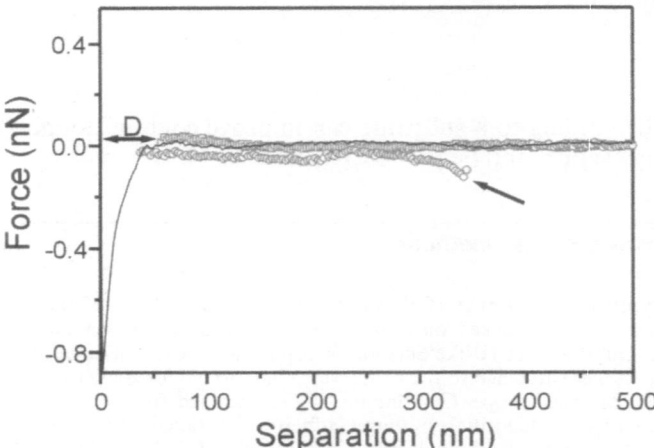

Fig. 1 Force between a 25-μm diameter glass sphere coated with gelatine and the bare surface of mica, measured in pure water. The *full line* is a force plot, where the approach has been made to the hard contact. *Squares* represent the force when approaching the mica surface from a distance of 500 nm to a distance of 55 nm. *Circles* represent the force when retracting from this distance. Note that there are no forces during approach (*open squares*). During the withdrawal (*open circles*) attractive bridging forces are effective up to the pull-off distance of 340 ± 30 nm. The pull-off point is indicated with an *arrow*. The energy needed to pull the probe from the surface is several 10^3 $k_b T$

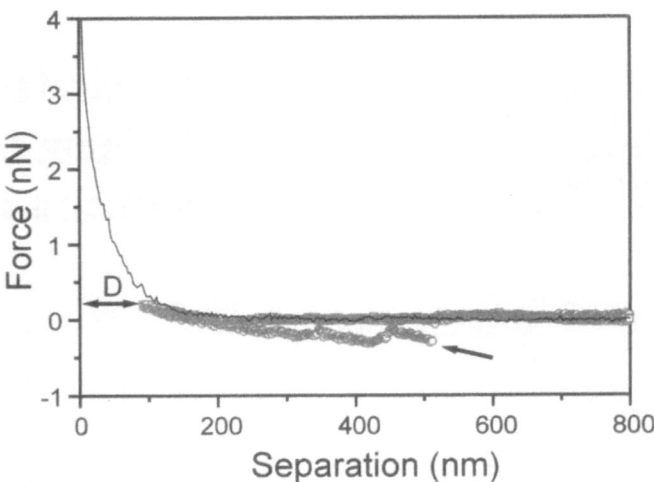

Fig. 2 Forces between a 25-μm glass sphere coated with gelatine and the bare mica surface in 0.98 mM cetyltrimethylammonium bromide and 0.05 M CH$_3$COONa water solution (pH 8.8). The *full line* is a force plot, where the approach has been made to the hard contact. The decay length of the exponential repulsion is 50 nm. *Squares* and *circles* represent a force scan, where we stopped approaching the sphere at a distance of 90 ± 10 nm from the surface. Only a weak repulsion force is observed during the approach (*open squares*). During withdrawal (*open circles*) the adhesive bridging forces are effective till a pull-off distance of 500 ± 30 nm. The pull-off point is indicated with an *arrow*

a minimum distance from the mica surface, gelatine molecules are tethered to the mica surface and create bonds between a sphere and the surface. Upon retracting, these bonds are broken and the heat is dissipated at the interface. The gelatine molecules can be extended to the full length of several hundred nanometers, where the bonds are broken and the sphere is detached from the surface.

On replacing pure water by a solution of 0.98 mM CTAB and 0.05 M CH$_3$COONa, the force curve appears quite different, as shown in Fig. 2. Again, we first made a full approach and detected the repulsion with a screening length of 50 nm. After that, the two surfaces approached only to a separation of 90 nm, where practically no repulsive electrostatic force could be detected; however, when the two surfaces were separating, we observed a characteristic molecular bridging force. The binding force is of the order of 100 pN, it is far-reaching, and shows saw-tooth behavior with sudden jumps which terminate at a distance of 500 nm. The work needed to separate the sphere from the surface is again several 10^3 $k_B T$ at ambient temperature. The last molecular bridges are broken at a separation of 0.5 μm.

Conclusions

For successful deposition of close-packed particles on a substrate, it is necessary to satisfy a number of criteria.

Firstly, water-soluble polyelectrolyte (e.g. proteins, cellulose derivatives, vinyl polymers) should be adsorbed on the substrate. This provides a "hairy" interface, with polymer loops and tails sticking out into solution.

Secondly, the surfactant should be used in the particle dispersion. On the one hand, the dispersion has to be stable. Due to dissociation of adsorbed surfactant onto the surfaces of the particles the surfactants become charged in aqueous solution and the resulting electrostatic repulsive forces between equaly charged surfaces leads to stabilization of the dispersion against van der Waals attraction. On the other hand, the surfactant is not only needed to keep the particle suspension stable, but it also swells the polyelectrolyte layer to a greater distance from the surface, which results in more effective particle trapping and thicker layers adsorbed onto the polyelectrolyte.

Thirdly, the electrostatic properties of the particles in the dispersion have to be near the point where the particles can approach a certain minimum distance so that deposition can occur. This point is controlled by the appropriate addition of a salt, which screens the electric field of the surfaces and allows the particles to approach to shorter distances during their Brownian motion. Under such conditions, coagulation can only occur on a modified substrate surface, but not on other surfaces and, of course, not in the bulk dispersion.

Acknowledgement The authors thank M. Gaberšcek for his helpful comments during the preparation of the manuscript.

References

1. Besenhard JO, Claussen O, Gausmann HP, Meyer H (1991) German Patent DE 4141416 and related European and USA applications
2. Besenhard JO, Meyer H, Gausman HP, Mahlkow H (1991) German Patent DE 4141744
3. Besenhard JO, Meyer H, Gausmann HP (1991) German Patent DE 4113407
4. Pendleton P (1988) J Adhes Sci Technol 2:137
5. Pendleton P (1989) IPC Inst. Interconnecting and Packaging Electron. Circuits, Lincolnwood, Il, USA, Technical Papers. IPC 32nd Annual Meeting 776/1–20, Lake Buena Vista, F (1989) USA p 610
6. Besenhard JO, Hanna S, Haag Ch, Fiedler DA, Bele M, Pejovnik S, Meyer H (1998) Proceedings of the Symposium on Interconnect and Contact Metallization (Rathore MS, Mathad GS, Plougonven C, Schuckert CC, Eds.) Electrochemical Society, Pennington, NY, USA, p 96
7. Bele M, Pejovnik S, Besenhard JO, Ribitsch V (1998) Colloids Surf. A 143:17
8. Bele M, Pejovnik S (1999) J Mater Sci Lett 18:1841
9. Ducker WA, Senden TJ, Pashley RM (1991) Nature 353:239

Progr Colloid Polym Sci (2000) 115:100–105
© Springer-Verlag 2000

Y. Chevalier
A. Chivé
B. Delfort
M. Born
L. Barré
R. Gallo

The emulsification of inorganic colloidal particles in an organic medium

Y. Chevalier (✉)
Laboratoire des Matériaux Organiques à
Propriétés Spécifiques
UMR 5041 CNRS-Université de Savoie
BP 24, 69390 Vernaison
France
e-mail: yves.chevalier@lmops.cnrs.fr
Tel.: +33-4-78022271
Fax: +33-4-78027187

A. Chivé · B. Delfort · M. Born · L. Barré
Institut Français du Pétrole
BP 311, 1-4 av.de Bois-Préau
92506 Rueil-Malmaison Cedex
France

R. Gallo
ENSSPICAM, Faculté Saint-Jérôme
13397 Marseille Cedex 20, France

Abstract Colloidal suspensions of inorganic particles in xylene could be prepared by means of reactions between two solid powders. In all cases, the reagents were transported to the reaction sites by means of a microemulsion of the reverse type containing a surfactant, water and tetrahydrofuran as a cosurfactant. This process, where colloidal particles are formed and emulsified from a chemical reaction site at the surface of a solid, is illustrated with examples of the synthesis of particles containing calcium thiophosphate, calcium hydroxide or sodium phosphate. The role of the cosurfactant was to speed up the transport rate of the reagents, allowing a high yield of colloidal particles and the control of side reactions in the case of calcium thiophosphate synthesis. The mechanism by which the inorganic particles were emulsified is still an open question and this is discussed. The particle sizes did not depend on the amount of surfactant as in the case of classical emulsions. The type of inorganic product formed inside the reverse micelles and the spontaneous curvature of the surfactant film appear as the most relevant parameters which control the particle size.

Key words Colloidal suspension · Calcium thiophosphate · Calcium hydroxide · Phosphate · Microemulsion

Introduction

The preparation of colloidal dispersions of inorganic particles in organic media is an important technological challenge [1]. There are two well-known processes, neither of these can give satisfactory results. The direct dispersion by mechanical grinding of a coarse powder in the presence of a dispersant additive in the solvent results in large particles, which settle too fast because of their large size. The in-situ synthesis of the particles as a colloidal dispersion in the reaction medium is more efficient for attaining particles of small size [2]. The most widely used process consists of precipitation of the inorganic solid by mixing two inorganic precursors solubilized in reverse micelles. Scaling up this process is limited by the low solubility of reagents in reverse micelles, which impedes the direct preparation of concentrated dispersions.

The synthesis of the so-called "overbased calcium carbonate micelles" in an example of a successful preparation of a concentrated colloidal dispersion of inorganic particles by means of a simple process [3]. Dispersions of calcium carbonate particles in an organic medium could be obtained by reacting CO_2 with CaO or $Ca(OH)_2$ in the presence of a calcium alkylarylsulfonate surfactant. The synthesis of such inorganic particles has been known for a long time and the mechanism of their formation has been studied extensively [4–7]. The micellization process was described as a nucleation and growth mechanism, where the rate of growth was controlled by the transport of $Ca(OH)_2$ to the growing colloidal particles by means of reverse micelles. The particles grew in size as the reaction proceeded because the chemical reaction of CO_2 with $Ca(OH)_2$ took place inside the colloidal particles.

In the present article, we report a process leading to the formation of colloidal inorganic particles dispersed in an organic medium by the reaction between two solid powders. This process involves the transport of reagents from one solid to the other by means of a microemulsion. The direct formation of $Ca(OH)_2$ particles by reaction of water with CaH_2 solid powder is also reported since it involves similar mechanisms. Three examples relating to the successful synthesis of stable concentrated dispersions of small inorganic particles in xylene with high yields are given. The role of the microemulsion medium is emphasized. Finally, the mechanism of the emulsification of the inorganic particles and the parameters which control the particle size are discussed.

Synthesis of calcium thiophosphate particles

The reaction of CaO and P_4S_{10} leads to the micellization of the products of the reaction, namely calcium thiophosphates $Ca_3(PO_xS_{4-x})_4$, where x ranges from zero (thiophosphate) to 4 (calcium orthophosphate). This system involves the reaction of two powdered solids, CaO and P_4S_{10}, in an organic solution of a calcium alkylarylsulfonate surfactant, upon progressive addition of water. The synthesis of calcium thiophosphate particles, avoiding the formation of calcium orthophosphate, was successful when a microemulsion of the reverse micelle type was used as the reaction medium [8, 9]. The following mechanism was proposed according to detailed studies of the reaction [9]. Because the specific area of CaO is much larger than that of P_4S_{10}, the reaction starts by the hydrolysis of CaO into $Ca(OH)_2$, which solubilizes inside the reverse micelles. No formation of colloidal particles occurs at this stage

when water is slowly added to the reaction medium. The micelles containing water and CaO reach the P_4S_{10} surface where the chemical reaction takes place. The micellar core contains an aqueous solution of $Ca(OH)_2$. The reaction with P_4S_{10} is a basic hydrolysis leading to thiophosphate anions followed by the precipitation of the insoluble calcium salt.

$$P_4S_{10} \xrightarrow{OH^-} PO_xS_{(4-x)}{}^{3-} \xrightarrow{Ca^{2+}} Ca_3(PO_xS_{(4-x)})_4 \qquad (1)$$

At this stage, particles of solid calcium thiophosphate leave the reaction site and become emulsified as a stable colloidal suspension. Kinetic measurements performed at low temperature have shown that the particle sizes were small and independent of the conversion. Thus, as the reaction proceeds, the colloidal suspension is enriched with colloidal calcium thiophosphate at constant particle size, new particles are emulsified at the surface of P_4S_{10} and the number of particles increases. The mechanism of the process is summarized in Fig. 1; it is quite different from the nucleation and growth mechanism involved in the formation of colloidal suspensions of calcium carbonate by reaction of CaO and CO_2 in a microemulsion [6, 7].

When water reaches the surface of P_4S_{10}, the hydrolysis yields thiophosphoric acids, which are not chemically stable in acidic water, and there hydrolyse rapidly into orthophosphoric acid and sulfhydric acid (H_2S).

$$P_4S_{10} \xrightarrow{H_2O} H_3PO_xS_{(4-x)} \xrightarrow{H_2O} H_3PO_4 + H_2S \qquad (2)$$

Particles of calcium phosphate (instead of thiophosphates) are formed after the neutralization by $Ca(OH)_2$. This side reaction occurs when the transport rate of $Ca(OH)_2$ is not fast enough with respect to the rate of water transport. The formation of phosphate could then

Fig. 1 Sketch of the reaction path which leads to the formation of colloidal calcium thiophosphate particles

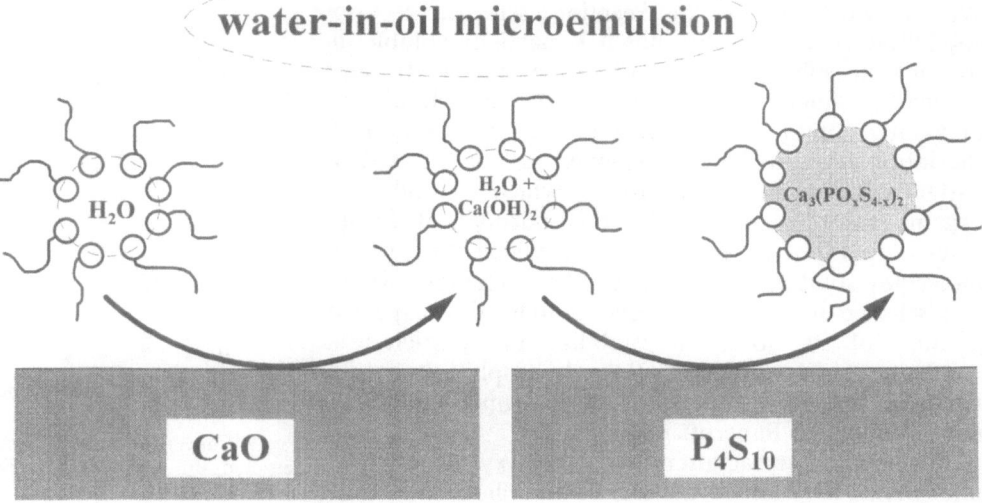

be used as an indicator for the transport rate of the basic species to the P_4S_{10} surface. Thus, large enough concentrations of surfactant, water and tetrahydrofuran (THF) are required for the formation of thiophosphate species with $\langle\chi\rangle = 1.5$, according to the stoichiometry of reaction 1. Reverse micelles are necessary for the control of the hydrolysis into phosphate. $Ca(OH)_2$ is transported inside the reverse micelles, but water which solubilizes in the micellar cores and in the organic solvent (xylene: THF 2:1 mixture) can be transported to the reaction site, even in the absence of reverse micelles. Because water can reach the reaction site through the organic solvent path, the reaction is possible in the absence of surfactant, but calcium phosphate species (orthophosphate, pyrophosphate and tripolyphosphate) are produced instead of thiophosphates (but the suspension is not stable because of the lack of surfactant).

The particles have a core–shell structure, with an inorganic core containing either calcium thiophosphates or calcium phosphates surrounded by a shell of surfactant ensuring the colloidal stability. The inorganic materials are in an amorphous state, as shown by means of wide-angle X-ray scattering (WAXS). The particles are small and nearly spherical, their mean radii range between 10 and 20 Å, whatever the details of the process. The yield of the whole process reaches high values (80%). The yields of the chemical reaction itself and of the dispersion process into a colloidal suspension contribute to the overall yield; it is calculated as the phosphorus content of the colloidal particles with respect to the phosphorus feed (P_4S_{10}).

Role of the microemulsion as a reaction medium

The liquid reaction medium ensures the transport of the reagents to the reaction site at the solid surface. Since the ionic species are not soluble in the organic medium, reverse micelles are required for their solubilization. Water, which enters the reaction scheme, is also solubilized in the micelles but it is partially soluble in the organic solvent mixture of xylene and THF. The surfactant which is used for the formulation of the microemulsion is also necessary for the stabilization of the inorganic particles. The calcium salt of the AS117 surfactant is an alkylarylsulfonate which is soluble in organic solvents suitable for this purpose; it forms reverse micelles of small size in which few water molecules solubilize ($w = $ [water]/[surfactant] $= 8$). It was selected in spite of its low solubilization capacity because of its ability to stabilize the particles: a "hydrophobic" surfactant (low hydrophile–lipophile balance) stabilizes dispersions in hydrophobic media according to the Bancroft rule.

When such a surfactant is used in pure xylene, calcium phosphates are formed instead of thiophosphates and

the yield of the emulsification is poor. Such micelles are inefficient for supplying the P_4S_{10} surface with water and $Ca(OH)_2$ simultaneously. Acidic hydrolysis into phosphates takes place and the inorganic materials formed are not emulsified. In order to increase the solubilization capacity for ionic species and to accelerate both the transport and the exchange rates of materials through the interfaces, a cosurfactant is necessary. THF was used for that purpose since conventional short-chain alcohols can react with P_4S_{10}. Thus, the reaction medium is a microemulsion of the reverse micelle type where the dynamic phenomena are fast. The structure of the reaction medium and its role in the control of the reaction have been studied in detail [9]. According to the well-established knowledge of dynamic phenomena in microemulsions [10–11], large amounts of THF, which acts as cosurfactant, loosen the surfactant film at the interfaces, a larger solubilization capacity of water and CaO results and the exchange phenomena are speeded up.

The acceleration of the chemical processes by the presence of THF can be observed in kinetic experiments. Because the water was added in one shot at the beginning of the reaction, some differences resulted from this modification with respect to the optimized process, where water was added progressively. In particular, the yield of colloidal particles was lower. The conversion of P_4S_{10} into colloidal calcium (thio) phosphates is clearly faster with THF present (Fig 2): the maximum (60%) conversion level is reached within 0.5 h, while the reaction lasts 6 h without THF and only reaches 40% conversion.

The efficiency of the CaO transport can be monitored by the electrical conductivity of the microemulsion (Fig. 3). This is a direct measurement of the transport rate of ionic species, namely $Ca(OH)_2$, which are the basic species preventing the acidic hydrolysis into phosphates. In the absence of THF, the electrical

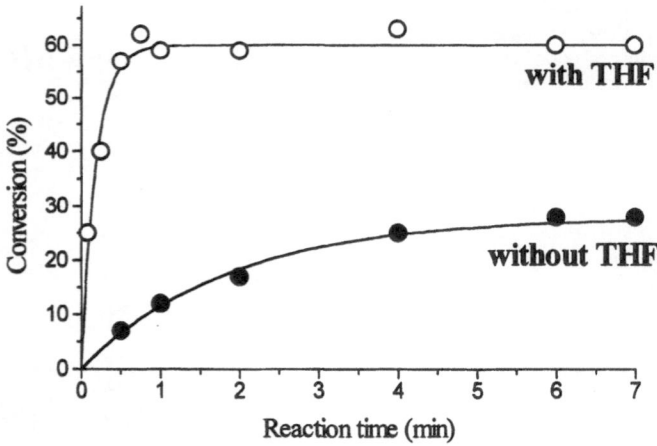

Fig. 2 Conversion of P_4S_{10} into colloidal particles at 30 °C with tetrahydrofuran (THF) according to the recipe of Ref. [8] and without THF

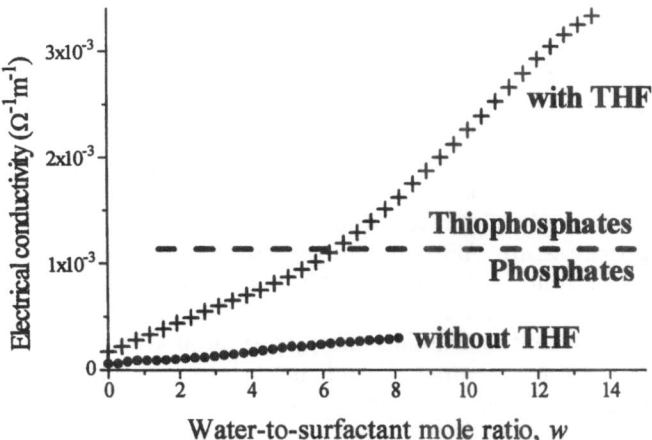

Fig. 3 Electrical conductivity of microemulsions containing 5% surfactant in pure xylene and in a 2/1 (v/v) mixture of xylene and THF as a function of their water content

conductivity is low and the water solubilization capacity is poor ($W_{max} = 8$); phosphates are formed in such conditions and the yield of the emulsification process is low (39%). When THF is added, the electrical conductivity increases by 1 order of magnitude at a fixed water content. The solubilization capacity of water and Ca(OH)$_2$ also increases upon addition of THF. Large electrical conductivities (thus large transport rates) are attained with high THF and water contents. The steep increase in the electrical conductivity with the water content is to be noted since it indicates that the CaO transport is accelerated by the presence of large amounts of water. As a consequence, the quenching of the hydrolysis of P$_4$S$_{10}$ by Ca(OH)$_2$ basic species which leads to calcium thiophosphates is more efficient when the water content is high. Indeed calcium thiophosphates do form with large yields at high water contents, while calcium phosphates form predominantly at low water contents, even in the presence of THF cosurfactant. This is a paradoxical result, where the hydrolysis of thiophosphates into phosphates is prevented when large amounts of water are used. This side reaction is controlled by physicochemical processes in the microemulsion.

It clearly appears that a microemulsion with a reverse micelle structure allows the fast materials transport and exchanges required for the present chemical process. A different point which deserves discussion is the emulsification itself and the control of the size of the particles. Obviously, no reaction is possible when two insoluble solid powders are immersed in a nonpolar organic solvent (xylene), where water is not soluble. No emulsification is possible in the absence of surfactant, but the hydrolysis takes place at high THF contents when water can be transported in the xylene–THF mixture. The phosphates formed in this case remain in the solid powder. In the presence of surfactant in pure xylene (no

THF), the reaction takes place because of the presence of reverse micelles and colloidal particles are formed because of the presence of the surfactant; however the yield of the emulsification process is low (and phosphates are formed) in the absence of THF cosurfactant. As the THF content increases, the yield of the emulsification increases (from 39 to 80%). The process by which reverse micelles containing water and Ca(OH)$_2$ react with P$_4$S$_{10}$ at the surface of the solid and leave the solid as a colloidal particle is poorly understood; it is considered in the Discussion. The present study shows that a cosurfactant is necessary.

Synthesis of sodium and potassium phosphate particles

NaOH or KOH as aqueous solutions could be used in place of CaO for the synthesis of calcium thiophosphate particles; however, when NaOH or KOH was used, the hydrolysis of P$_4$S$_{10}$ into phosphates was complete. Colloidal particles of sodium (potassium) phosphates dispersed in xylene could be prepared in this way [13]. A detailed study of the mechanism was not carried out. According to the discussion presented earlier, the reverse micelles are not able to ensure fast transport of NaOH (KOH) in their aqueous core. One can suspect that the solubilization capacity for NaOH is less than that for Ca(OH)$_2$; interactions of Ca^{2+} ions with the sulfonate headgroups of the surfactant may be the origin of the higher solubilization of Ca(OH)$_2$ in the micellar cores.

The results for sodium and potassium phosphates are rather similar: a mixture of orthophosphate, pyrophosphate and polyphosphates is formed. The process is identical to that used for the synthesis of the calcium thiophosphate particles described earlier. In particular, the yields of the colloidal particles approach 100%. There are two main differences, however. The inorganic content of the colloidal particles is crystalline, at least partially. The Bragg peak patterns could not be ascribed to a known crystalline structure (which could be found in a database); crystalline sodium pyrophosphate was present in the case of the synthesis with sodium hydroxide, but together with other unidentified compounds. The second difference with respect to calcium thiophosphate is the larger size of the alkaline phosphate particles. Both small-angle X-ray scattering (SAXS) and transmission electron microscopy (TEM) observations indicate spherical particles with a broad size distribution and mean radius of 50–100 Å.

Synthesis of Ca(OH)$_2$ particles

Ca(OH)$_2$ particles could be prepared by the reaction of CaH$_2$ solid powder with water solubilized in the same microemulsion [14]. This process differs from the preceding ones because water solubilized in the liquid directly reacts at the surface of the solid: There is no transport of materials from one solid to another; however, the same type of emulsification of the inorganic particles into the microemulsion takes place during the chemical reaction. This emulsification was of very low yield when the same reaction was carried out with CaO upon a massive addition of water in one shot [10].

$Ca(OH)_2$ particles are emulsified according to this process with a yield of 40%. Part of the $Ca(OH)_2$ remains as a solid powder, which also retains a significant fraction (50%) of the surfactant adsorbed. The colloidal fraction is made of pure $Ca(OH)_2$ particles surrounded by a surfactant shell. WAXS shows that the inorganic core is crystalline with the structure of Portlandite. The particles are large and disklike in shape as shown by TEM and SAXS. The thickness of the platelets as estimated from SAXS or from the broadening of the 00l reflections is WAXS is 30 Å and their diameter is 180 ± 20 Å.

Discussion

There are common features in these synthesis processes, several of which remain quite obscure. In particular, two points deserve discussion: What is the mechanism of the emulsification phenomenon? What controls the particle sizes?

In every case, the inorganic particles are emulsified at the surface of the solid where the hydrolysis reaction takes place. The inorganic materials are formed as the reagents reach the reaction site. The reagents (water or water + basic species) are brought to the surface inside reverse micelles. Once the reaction products are formed, either particles leave the reaction site into the solution or the inorganic materials stay in the solid powder. In the first case, where particles are emulsified, a colloidal suspension is obtained. At full conversion of the chemical reaction, the yield of the overall process is identical to the yield of the emulsification process, which expresses the extent of the former event.

Once the particles are emulsified, the colloidal dispersion has to be stabilized. The presence of the surfactant at the surface of the particles provides large enough steric repulsions between the particles, preventing their coagulation. The AS117 surfactant was selected and is efficient for that purpose. The main difficulty is the emulsification itself, where particles come unstuck from the surface. When the adhesion energy is negative, this process takes place spontaneously; however, this situation is unlikely since two polar surfaces usually adhere strongly when immersed in a nonpolar organic medium. When the emulsification phenomenon is not associated with a decrease in the free energy, a state out of thermodynamic equilibrium is aimed at. Schematically,

there are two ways to perform a successful emulsification: either lower the energy barrier for separating the surfaces or launch a large energy input which overcomes this energy barrier. The latter process is more common; the technology of emulsions is based on this idea, where the liquid to be dispersed is "ground" in the liquid-dispersing medium by means of a high-mechanical-energy input.

The chemical reactions involved in the present processes are quite exothermic, so intense local heating at the solid surface occurs. The chemical transformation is also associated with volume changes. The large thermal and mechanical energy produced at the interface has to be dissipated and may lead to instabilities responsible for the emulsification of small particles. The emulsification has obviously something to do with the strength of the chemical reaction. Thus, the reaction of CaH_2 with water was very exothermic (it should be carried out at 0 °C under slow addition of water) and its yield was 50% [14], whereas the reaction of water with CaO was sluggish in the microemulsion medium and led to very poor emulsification yields. The cosurfactant, which accelerates the transport phenomena, also increases the yield of emulsification since it speeds up the heat release and materials transfer.

This mechanism is still speculative, but the Marangoni effect is predicted to generate droplets at interfaces where heat or mass transfer take place [15]. Different phenomena may also operate. Examples of similar processes can be found in various domains. Spontaneous emulsification of droplets was observed at the slag–steel interface at high temperature (1550 °C) while the chemical reaction of Al (contained in steel) and CaO (in slag) took place [16]. In this case, the interfacial tension at thermodynamic equilibrium is high, so an increase in the interfacial area is difficult, but the chemical reaction is vigorous. Droplet formation could be observed by means of X-ray photography and an apparent lowering of the interfacial tension was deduced from the shapes of the droplets observed during the chemical reaction [17].

In contrast, smooth emulsification is observed when a thin film of phospholipid cast from an organic solution is hydrated with water [18, 19]. In this case, the interfacial tension between the lamellar phase of the partially hydrated lipid and pure water is low and the penetration of water into the lamellar phase gives rise to the growth of cylindrical excrescences of the lamellar phase into the water; these finally break into droplets of multilamellar vesicles dispersed in water. The myelin figures resulting from this smooth process at room temperature can be observed with a microscope.

In the present process, the size of the particles does not vary with the conversion as would be the case for a nucleation and growth mechanism [20]. When particles are formed by nucleation and growth in the presence of surfactant, the final particle size depends on the surfactant concentration; the most typical example is emulsion polymerization and the overbased calcium carbonate micelles enter this class [3–7]. The particle size does not depend on the concentration of surfactant in the present case. It is also quite different for systems at thermodynamic equilibrium, such as microemulsions, where the particle size decreases as a function of the surfactant concentration because the amount of surfactant sets the total interfacial area [21].

The particle sizes presented in this paper depended on the type of inorganic material, although the surfactant was the same in every case. Thus, in the case of the calcium thiophosphates particles, the particle radii were identical to the radii of the reverse micelles present in the microemulsion [9]. The radius of the inorganic particles was set by the surfactant optimum curvature as in the microemulsion [21]. According to this observation, the inorganic particles would be able to leave the P_4S_{10} surface when their curvature reached the optimum curvature of the surfactant. The adsorption of the surfactant is a maximum in this case, the energy barrier for emulsification is lower and the colloidal stability higher. This situation, where the particle size is controlled by the surfactant optimum curvature, is encountered when the inorganic core of the particles can accommodate such high curvatures. This is possible with calcium thiophosphates which are in an amorphous state; however, particles containing crystalline materials such as sodium (potassium) phosphates are of larger size and the $Ca(OH)_2$ particles which are fully crystalline are even larger and are anisotropic.

The size of the particles is the result of competition between the surfactant, which tends to impose its optimum curvature on the particle, and the bulk energy of the inorganic core, which dictates the crystal habit. Finally, the emulsification yield is lower when the particle curvature departs too much from the optimum curvature of the surfactant because the surfactant free energy of adsorption to the particle is lower by its bending free energy. Colloidal suspensions of highly anisotropic crystalline particles of $BaCO_3$ (nanowires) could be prepared in a similar way in a microemulsion of reverse micelles [22]. In a theoretical calculation of the stability of small inorganic particles stabilized by surfactant, the most stable size (with lower energy) depended on the surfactant concentration [23]. The present case is different: the particles which are formed are those which are able to leave the surface where the exothermic reaction takes place; their energy has to be low but not necessarily at the minimum.

Conclusions

A microemulsion where dynamic phenomena of transport and materials transfer across interfaces are very fast allows the emulsification of particles generated during chemical reactions at the surface of solids. In the same way, it also allows reactions between materials contained in different solids. Colloidal suspensions of particles in hydrocarbon media can be synthesized according to this process. The composition of the microemulsion has to be carefully chosen, especially in complex systems where the rates of the side reactions have to be controlled. In particular, large amounts of cosurfactant are required for efficient emulsification.

References

1. Eicke H-F, Parfitt GD (eds) (1987) Interfacial phenomena in apolar media, Surfactant Science Series 21. Dekker, New York
2. Pileni M-P (ed) (1989) Structure and reactivity in reverse micelles. Elsevier, Amsterdam
3. Bray UB, Dickey CR, Voorhees V (1975) Ind Eng Chem Prod Res Dev 14:295
4. Glavati OL, Fialkovskii RV, Marchenko AI, Premyslov VKh, Alekseev OL (1980) Kolloidn Zh 42:26
5. Roman J-P, Hoornaert P, Faure D, Biver C, Jacquet F, Martin J-M (1991) J Colloid Interface Sci 144:324
6. Belle C, Beraud C, Faure D, Gallo R, Hornaert P, Martin J-M, Rey C (1990) J Chim Phys 87:93
7. Gallo R, Jacquet F, Hoornaert P, Roman J-P (1991) Rev Inst Fr Pet 46:251
8. Delfort B, Chivé A, Barré L (1997) J Colloid Interface Sci 186:300
9. Chivé A, Delfort B, Born M, Barré L, Chevalier Y, Gallo R (1998) Langmuir 14:5355
10. (a) Leung R, Shah DO (1987) J Colloid Interface Sci 120:320; (b) Leung R, Shah DO (1987) J Colloid Interface Sci 120:330
11. Jada A, Lang J, Zana R (1990) J Phys Chem 94:381
12. Jada A, Lang J, Zana R, Makhloufi R, Hirsch E, Candau SJ (1990) J Phys Chem 94:387
13. Delfort B, Normand L, Dascotte P, Barré L (1998) J Colloid Interface Sci 207:218
14. Delfort B, Born M, Chivé A, Barré L (1997) J Colloid Interface Sci 189:151
15. Velarde MG (1998) Philos Trans R Soc London Ser A 356:829
16. Riboud PV, Lucas LD (1981) Can Metall Q 20:199
17. Chung Y, Cramb AW (1998) Philos Trans R Soc London Ser A 356:981
18. Bangham AD, Standish MM, Watkins JC (1965) J Mol Biol 13:238
19. New RRC (ed) (1990) Liposomes, a practical approach. Oxford University Press, Oxford
20. La Mer VK, Dinegar RH (1950) J Am Chem Soc 72:4847
21. Chevalier Y, Zemb T (1990) Rep Prog Phys 53:279
22. Qi L, Ma J, Cheng H, Zhao Z (1997) J Phys Chem B 101:3460
23. Whetten RL, Gelbart WM (1994) J Phys Chem 98:3544

Progr Colloid Polym Sci (2000) 115:106–111
© Springer-Verlag 2000

P. G. Klepetsanis
A. Kladi
P. G. Koutsoukos
Z. Amjad

The interaction of water-soluble polymers with solid substrates. Implications on the kinetics of crystal growth

P. G. Klepetsanis · A. Kladi
P. G. Koutsoukos
Institute of Chemical Engineering and
High Temperature Chemical Processes
265 00 Patras, Greece

P. G. Klepetsanis
Department of Pharmacy
University of Patras, 265 00 Patras, Greece

P. G. Koutsoukos (✉)
Department of Chemical Engineering
University of Patras, 265 00 Patras, Greece

Z. Amjad
The BF Goodrich Company
Advanced Technology Group
9921 Brecksville Road
Brecksville, Ohio, USA

Abstract The precipitation of calcium carbonate was investigated at pH 8.5 and at temperatures of 25 and 80 °C in stable, supersaturated, synthetic high-salinity solutions (0.62 M NaCl) seeded with calcite, in the presence of fulvic acid (FA), humic acid (HA) and poly(acrylic acid) (PAA), which acted as inhibitors. The rates of crystal growth and the effect of the inhibitors on the mineral formation kinetics were evaluated under conditions of sustained supersaturation. Concentrations of the inhibitors tested in the range 0.02–0.1 ppm for solutions with a supersaturation value (SR) of 8.39 with respect to calcite suppressed the rates of crystal growth by up to 95%. For 0.1 ppm PAA as well as for 0.1 ppm HA the rates of crystal growth showed a linear increase with $SR_{calcite}$ of the working solution ($SR_{calcite}$ = 7.31–13.9) at 80 °C. The second-order dependence of the rate on the solution supersaturation suggested surface diffusion control. The inhibitory activity of the compounds tested was ascribed to their interaction with the solid surface. Kinetics analysis of the results obtained for the formation of calcium carbonate in the presence of the compounds tested, based on the assumption of Langmuir-type adsorption on the calcite seed crystals, fitted satisfactorily the data for PAA and FA, while for HA the interaction did not follow this model. In the latter case aggregation phenomena may occur during the growth of crystals, resulting in rate reduction.

Key words Calcite · Fulvic acid · Humic acid · Poly(acrylic acid) · Adsorption

Introduction

The formation of deposits of alkaline-earth insoluble salts is a serious problem in installations where untreated natural waters are used, such as in geothermal energy exploitation, in secondary oil production by waterflooding, in cooling towers, in the production of potable water by reverse osmosis, etc. [1, 2]. The formation of these deposits reduces heat transfer and the internal diameter of pipes, increases the operating pressure of pumps and enhances the probability of corrosion damage of metallic constructions. In many cases, the removal of deposits leads to discontinuous operation of the installations, resulting in higher operation costs.

Calcium carbonate is one of the most commonly encountered scale deposits. It is found in different crystalline forms in the following order of increasing solubility: calcite, aragonite, vaterite, calcium carbonate monohydrate and calcium carbonate hexahydrate. Calcite, the thermodynamically most stable polymorph of calcium carbonate, forms tenaciously adhering, hard mineral deposits. The precipitation and stabilization of the calcium carbonate polymorphs depends on the precipitation conditions, i.e. level of supersaturation, pH, temperature, pressure and the concentration and chemical structure of the additives [1, 3]. A number of methods have been applied for the prevention of calcium carbonate polymorph formation [4], the most promising being the addition of water-soluble polyelectrolytes [5–8]. Water-soluble compounds (polyelectrolytes, phosphonates) give satisfactory results even at very low concentrations (lower than 1 ppm) and reduce significantly the cost-effectiveness for the prevention of deposits. In several cases, the presence of these water-soluble compounds may cause significant modifications of the crystal habit, reducing their ability to adhere on the surfaces. The main disadvantage of polyelectrolytes (polyacrylates, polymaleates and their copolymers) and phosphonates is their low biodegradiability. As a result these compounds contribute to environmental pollution. Moreover, phosphonates at high temperatures decompose and the phosphate ions released may cause the formation of calcium phosphates. In the last decade there was a tendency to use "green inhibitors" to control deposit formation. Humic and fulvic acids, commonly found in a natural environment, are polymeric molecules with molecular weights ranging from a few hundreds to several thousands. These acids mostly contain phenolic and carboxylic acid functional groups and behave as negatively charged colloids or anionic polyelectrolytes in natural waters. Although there is no general agreement on the exact chemical structure of these compounds, hypotheses concerning the structure of humic acid have been reported [9]. Recently, fulvic and humic acids have been reported to be good inhibitors for hydroxyapatite and calcium phosphate dihydrate formation [10, 11]. In the present work, we studied the effect of two natural polyelectrolytes, humic and fulvic acids (Suwannee River) and one synthetic polyelectrolyte, poly(acrylic acid), on the crystallization of calcium carbonate from metastable supersaturated aqueous solutions.

Experimental

Seeded growth experiments of calcium carbonate on calcite seed crystals at sustained supersaturation conditions were done in the absence and it the presence of additives. All experiments were done in a batch-type, magnetically stirred glass reactor, thermostated by water at constant temperature circulated through a water jacket. The temperature in all experiments was kept at 25.0 ± 0.1 °C. The stock solutions were prepared from crystalline, reagent grade chemicals (Merck, *pro analisi*) dissolved in triply distilled, CO_2-free water and were filtered through membrane filters (0.2 μm, Millipore). Calcium chloride and sodium chloride stock solutions were standardized by atomic absorption spectroscopy (Perkin Elmer, AAnalyst 300). The sodium bicarbonate and sodium carbonate solutions were prepared fresh for each experiment by weighing the exact amounts of the respective solids (dried at 105 °C overnight) followed by dissolution in triply distilled water. The fulvic and humic acid stock solutions were prepared by weighing the exact amounts of the respective solids. Poly(acrylic acid) stock solution was prepared by dilution of the appropriate volume of its concentrated solution. The molecular weight of the poly(acrylic acid) used for the experiments was 2100.

The calcite seed crystals were prepared by slow mixing of calcium chloride and sodium carbonate solutions at 70 °C [12]. The crystalline solid was aged for 1 week under continuous stirring and it was filtered, washed with saturated calcium carbonate solution and dried. The final solid was characterized by physicochemical methods including powder X-ray diffraction (XRD, Philips PW 1840), scanning electron microscopy (SEM, JEOL JSM 5200) and by specific surface area measurements (GEMINI, Micromeritics). The powder XRD pattern of the crystalline calcite preparation matched that of the respective reference material [13]. The specific surface area of the seed crystals was determined by a multiple-point-method Brunauer–Emmett–Teller adsorption isotherm and was found to be 0.30 m^2/g.

The total volume of the supersaturated solutions was 500 ml. The supersaturated working solutions were prepared in the reaction vessel by careful and rapid mixing of equal volumes (250 ml) of calcium chloride and sodium bicarbonate solutions under continuous stirring. In all experiments the total calcium was equal to the total carbonate concentration. The ionic strength of the supersaturated solutions was adjusted by the addition of an appropriate volume of concentrated sodium chloride stock solution. For the crystal growth experiments in the presence of fulvic and humic acids the appropriate volume of the respective additive stock solution was added to the working solution. Next, the solution pH was adjusted to 8.50 by the slow addition of 0.10 M standard sodium hydroxide solution (Merck, Titrisol). The solution pH was measured by a combination glass/Ag–AgCl pair of electrodes (Ingold), standardized before and after each experiment with N-bromosuccinimide buffer solutions (7.413 and 9.180 at 25 °C) [14]. The working solutions were stirred by a magnetic stirrer with a Teflon-coated stirring bar. A schematic representation of the constant composition experimental setup is shown in Fig. 1.

In all experiments the precipitation of calcium carbonate resulted in a decrease in the solution pH due to the proton release concomitant with the solid formation according to the reaction

$$Ca^{2+} + xH_2CO_3 + yHCO_3^- + zCO_3^{2-} \Leftrightarrow CaCO_{3(s)} + uH^+ \ ,$$

where x, y, z and u are stoichiometric coefficients.

The supersaturated solutions remained stable after pH adjustment, indicating the absence of any precipitation process. Under the experimental conditions of the seeded-growth experiments reported here, the working solutions were stable for periods exceeding 24 h and were supersaturated with respect to all calcium carbonate polymorphs. Following pH adjustment in the working solutions and after 1–2 hours, a precisely weighed amount of calcite (about 20 mg) was introduced in the supersaturated solutions and the precipitation started immediately. A pH drop as small as 0.005 pH units triggered the addition of titrant solutions from two mechanically coupled syringes of a computer-controlled titrator through the appropriate software. Throughout the course of the seeded growth experiments the pH of the working solution and the added volume of the titrants as a function of time were recorded and stored in the computer for further analysis. The titrant solutions in the two burettes consisted of calcium chloride (titrant A) and of a mixture of sodium carbonate, sodium bicarbonate and additive as appropriate (titrant B). The supersaturated working solution contained sufficient sodium chloride (inert electrolyte) so as to maintain the solution ionic strength during addition of the titrants [15].

During the course of the seeded growth experiments, samples were withdrawn and filtered through membrane filters (0.2 μm). The filtrates were analyzed for total calcium by atomic absorption spectroscopy in order to confirm the constancy of the solution composition. In all the seeded growth experiments the analysis showed that the total calcium concentration remained constant to within $\pm 2\%$. The working solutions at the end of the experiments were filtered and the solids were dried for further analysis. Crystal growth rates were determined from the traces of the titrant volume added as a function of time, using curve-fitting software. The rates of calcium carbonate formation were normalized for the seed crystal total surface area as follows:

Fig. 1 Experimental apparatus for the investigation of the seeded crystal growth kinetics of calcium carbonate under conditions of constant supersaturation

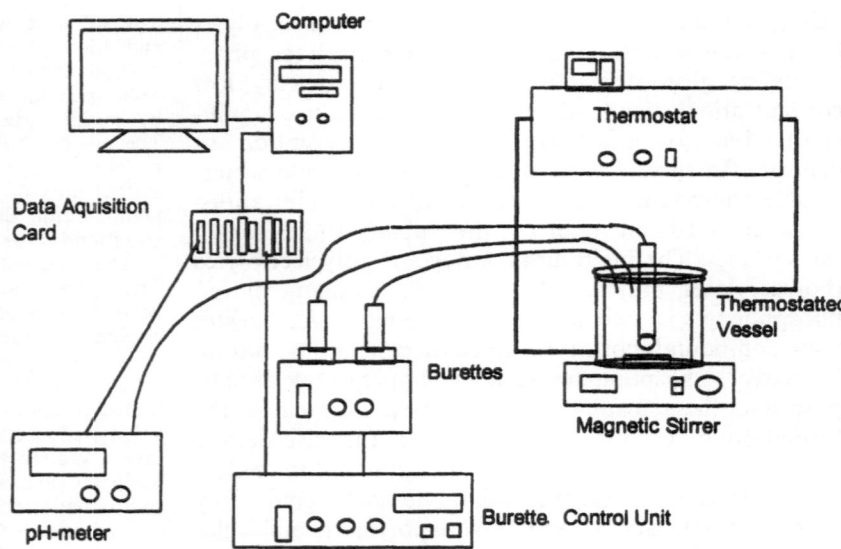

$$R_g = \frac{dV}{dt}\frac{c_t}{A_t} \ (\text{mol min}^{-1}\,\text{m}^{-2}) \ , \tag{1}$$

where dV/dt is the rate of titrant addition of concentration $c_t(\text{mol/l})(= 10c_1)$ and A_t is the total surface area of the added calcite seed crystals.

Results and discussion

Characterization of the precipitated solids

In all experiments calcite was the only calcium carbonate phase found. Calcite is the thermodynamically most stable polymorph of calcium carbonate and its formation was favored in our experimental conditions. The experimental conditions and the kinetics results obtained are summarized in Table 1.

The identification of solid precipitates was done by XRD and SEM analysis. The XRD spectrum of the precipitated solid in the presence of the additives tested and the respective reference spectral lines of calcite from the JCPDS database [15] are shown in Fig. 2. As may be seen, there is a complete match between the reflections of the spectra.

Morphological examination by SEM of the calcite crystals grown in the presence of the additives tested showed that the presence of these compounds in the supersaturated solutions did not cause any appreciable morphological changes in the characteristic rhombohedral shape of the calcite crystals formed in the absence of additives. In all experiments, no induction times preceded the crystal growth of calcite at all concentrations of the additives tested. The relative inhibition of calcium carbonate overgrowth on calcite seed crystals in the presence of humic acid, fulvic acid and poly(acrylic acid) was found to depend strongly on their concentrations

Table 1 Crystal growth of calcium carbonate on calcite seed crystals in the presence of humic acid, fulvic acid and poly(acrylic acid) (*PAA*) at 25 °C and pH 8.50. Initial conditions and relative inhibition

Exp. no.	Additive	Concentration (ppm)	Relative inhibition
1	Blank[a]	0	–
2	Humic acid	0.05	25.9
3	Humic acid	0.1	56.4
4	Humic acid	0.2	79.1
5	Humic acid	0.3	89.3
6	Humic acid	0.4	95.7
7	Humic acid	0.5	96.9
8	Humic acid	0.6	98.7
9	Humic acid	0.8	99.1
10	Fulvic acid	0.1	12.5
11	Fulvic acid	0.2	39.4
12	Fulvic acid	0.3	59.3
13	Fulvic acid	0.5	78.1
14	Fulvic acid	1.0	85.9
15	Fulvic acid	1.5	94.2
16	Fulvic acid	2.0	94.4
17	PAA	0.05	90.0
18	PAA	0.1	97.5
19	PAA	0.15	96.9
20	PAA	0.2	98.5
21	PAA	0.4	100

[a] Blank: total calcium = total carbonate = 2.00×10^{-3} M, total sodium chloride = 2.00×10^{-2} M

but to a different extent. The dependence of the percentage relative inhibition on the concentration of the additive tested is shown in Fig. 3. As may be seen, poly(acrylic acid) gave higher relative inhibition than humic and fulvic acids for the same concentration. Also, the presence of poly(acrylic acid) resulted in the complete inhibition of calcium carbonate precipitation at concentrations exceeding 0.2 ppm. Similar results were obtained for humic acid, which was found to be a

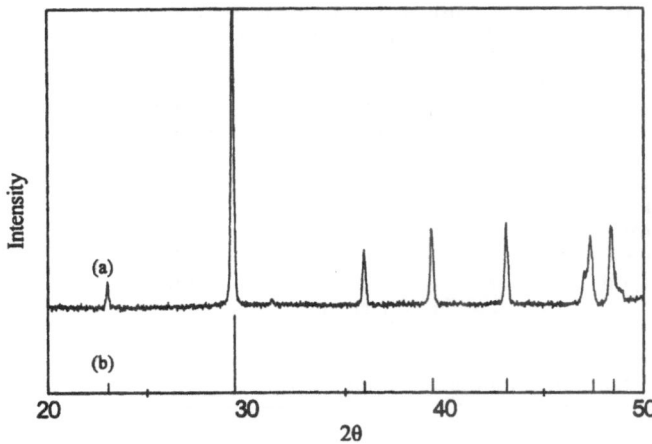

Fig. 2 Powder X-ray diffraction spectrum for (*a*) calcium carbonate formed in the presence of humic acid at decreasing supersaturation and (*b*) the reference spectrum of calcite

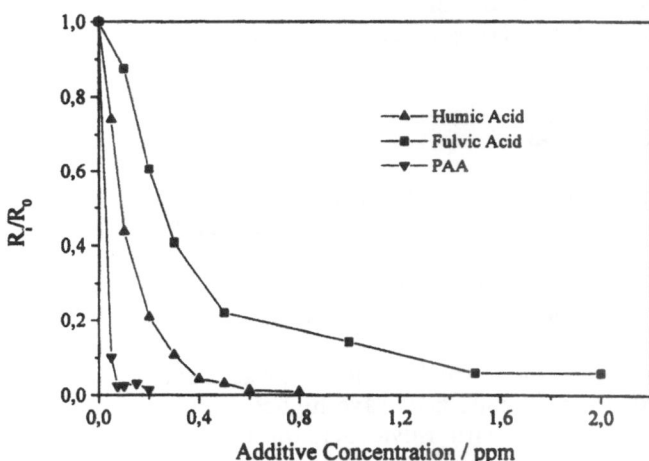

Fig. 3 Dependence of the ratio R_i/R_0 on the humic acid, fulvic acid and poly(acrylic acid) (*PAA*) concentration for the calcium carbonate overgrowth on calcite seed crystals at pH 8.50, 25 °C

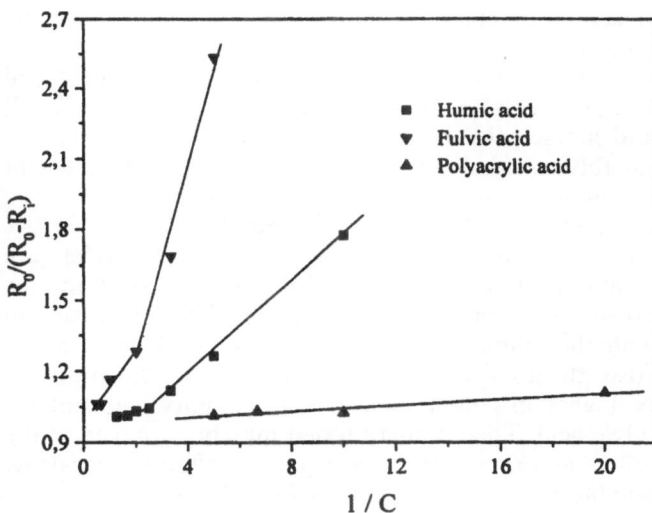

Fig. 4 Fit of the kinetics results for the overgrowth of calcium carbonate on calcite seed crystals in the presence of humic acid, fulvic acid and PAA at pH 8.50 and 25 °C to the Langmuir-type kinetic model

more efficient inhibitor in comparison with fulvic acid. Humic acid completely inhibited calcium carbonate precipitation at concentrations higher than 0.8 ppm.

In contrast to poly(acrylic acid) and humic acids fulvic acid did not yield complete inhibition of calcite crystal growth even at concentrations of 2 ppm, significantly higher than the concentration at which humic acid resulted in complete inhibition. The relative inhibition for fulvic acid reached a plateau. This was probably due to the stronger adsorption of poly(acrylic acid) and humic acid molecules on the active growth sites of the calcite seed crystals in comparison with fulvic acid at pH 8.50. A linear dependence of the relative inhibition on the additive concentration was observed at concentrations higher than 0.4 ppm for humic acid and

0.5 ppm for fulvic acid. Thus, from the kinetics analysis according to the Langmuir model [16]

$$\frac{R_0}{R_0 - R_i} = \frac{1}{1 - b} + \frac{1}{K(1 - b)c_i} \ , \tag{2}$$

where R_0 and R_i are the rates of crystal growth in the absence and in the presence of additives of concentration c_i and b is the limiting rate ($0 \leq b < 1$). The limiting rate is the rate of crystal growth corresponding to monolayer formation of the additive onto the calcite seed crystals. In Eq. (2) K is a constant related to the adsorption/ desorption rate constants and is defined as the affinity constant. This constant is considered to be a measure of the affinity of the adsorbent for the adsorbate. A linear relationship between the ratio $R_0/(R_0 - R_i)$ and the inverse of c_i is predicted for concentrations higher than 0.4 ppm for humic acid and 0.5 ppm for fulvic acid. Poly(acrylic acid) gave a linear dependence between the ratio $R_0/(R_0 - R_i)$ and the inverse of the additive concentration, $1/c_i$, according to the Langmuir model over the entire range of the concentrations tested. The analysis of the kinetics data obtained from the overgrowth of calcium carbonate on calcite seed crystals is shown in Fig. 4. At concentrations lower than 0.4 ppm for humic acid and 0.5 ppm for fulvic acid a second linear dependence was observed. This may be ascribed either to the presence of active growth sites with different energy at the calcite surface or to changes in the mode of adsorption with increasing surface concentration. At higher surface coverage the structure of the molecules adsorbed on the mineral surfaces may change. Poly(acrylic acid) molecules are linear and simpler in

110

structure in comparison with humic and fulvic acid molecules.

From the kinetics analysis according to the model of Eq. (2) the value of the intercept for poly(acrylic acid) and humic acid was found to be lower than unity and for fulvic acid slightly higher than unity. The values of the affinity constants for humic acid, fulvic acid and poly(acrylic acid), as calculated from the kinetics analysis according to the Langmuir-type model, are summarized in Table 2. The affinity constant for poly(acrylic acid) was significantly higher in comparison with the affinity constants for humic and fulvic acids. Also, the affinity constant for humic acid was found to be higher in comparison with the affinity constant for fulvic acid. The additives tested may be arranged in the following order with respect to their affinity for calcite: poly(acrylic acid) ≫ humic Acid > fulvic Acid.

Poly(acrylic acid) and humic acid, which caused complete inhibition at lower concentrations in comparison with fulvic acid, yielded higher affinity constants. The affinity constant of poly(acrylic acid) was significantly higher in comparison with the affinity constants of the organophosphorus compounds, 1,3-bis[(1-phenyl-1-dihydroxyphosphonyl)methyl]-2-imidazolidinone (BPDMI) and Ethylenediaminetetrabis(methylene phosphonic acid) (ETMPA) which were reported to inhibit effectively the precipitation of calcium carbonate. Poly(acrylic acid) is more effective than these organophosphorus compounds because it gave the same results but at lower concentrations in comparison with these compounds [17–19]. The affinity constant for humic acid was of the same order of magnitude as the affinity constants found for the organophosphorus compounds. Moreover BPDMI and ETMPA gave the same relative inhibition at almost the same concentration and inhibited completely the precipitation of calcium carbonate at the same low concentration in which humic acid was tested [19]. In contrast, the affinity constant of fulvic acid was lower comparable to mellitic acid and its inhibition efficiency was poorer in comparison with poly(acrylic acid) and humic acid and BPDMI and ETMPA. The difference in effectiveness as precipitation inhibitors between poly(acrylic acid) humic acid and fulvic acid may be attributed to the characteristics of their molecules, such as the degree of ionization of their functional groups, their charge density at pH 8.50, their molecular size, their geometry, their flexibility and to the ability for binding to more than one active growth site by one molecule. The poly(acrylic acid) molecule is linear, fully ionizable at pH 8.50 and more flexible for adsorption on crystal surfaces in comparison with humic

Table 2 Affinity constants for the additive–calcite interface calculated from the kinetics of crystal growth data according to the Langmuir model

Additive	Affinity constant
Humic acid	1.9×10^7
Fulvic acid	4.6×10^6
PAA	1.6×10^9
Mellitic acid	2.0×10^6 [17]
Phosphate	5.9×10^7 [18]
1,3-Bis[(1-phenyl-1-dihydroxyphosphonyl)methyl]-2-imidazolidinone	1.6×10^7 [19]
Ethylenediaminetetrabis (methylene phosphonic acid)	1.0×10^7 [19]

and fulvic acids. The exact formulae for humic and fulvic acids are not known and it is therefore very difficult to obtain an insight into the etiology for the superior performance of humic acid as a precipitation inhibitor in comparison with fulvic acid.

Conclusions

In the present work it has been shown that the presence of humic acid, fulvic acid and poly(acrylic acid) at very low concentrations can strongly inhibit the crystal growth of calcium carbonate from supersaturated aqueous solutions. Humic acid, fulvic acid and poly(acrylic acid) did not affect the nature of the calcium carbonate phase formed, which in all cases was calcite. Poly(acrylic acid) and humic acid gave higher relative inhibition in comparison with fulvic acid at the same conditions. Poly(acrylic acid) and humic acid were found to completely inhibit the crystallization of calcium carbonate at very low concentrations, while fulvic acid failed to yield complete inhibition even at significantly higher concentrations. The higher inhibition efficiency of poly(acrylic acid) in comparison with humic and fulvic acids is due to the larger extent of adsorption of poly(acrylic acid) molecules on calcite surface in our experimental conditions. This suggestion was corroborated by the calculated higher affinity constant of poly(acrylic acid) molecules for the calcite surface. Also, humic acid showed better behavior as a precipitation inhibitor in comparison with fulvic acid. The difference in inhibition efficiency between poly(acrylic acid) humic acid and fulvic acid may be attributed to their molecular characteristics.

References

1. Cowan JC, Weintritt DJ (1976) Water formed scale deposits. Gulf, Houston, Texa
2. Amjad Z (1987) Langmuir 3:224
3. Wilson D (1994) Corrosion 94 paper no. 48
4. Glater J, York JL, Campbell KS (1980) In: Spiegler KS, Laird ADK (eds) Principles of desalination. Academic, New York, pp 627–650
5. Weijnen MPC, Van Rosmalen GM (1985) Desalination 54:239
6. Vetter OJ (1972) J Pet Technol 997
7. Amjad Z (1988) J Colloid Interface Sci 123:523
8. Carrier AM, Standish ML (1995) In: Amjad Z (ed) Mineral scale formation and prevention. Plenum, New York, p 63
9. Mortvedt JJ, Giordano PM, Lindsay WL (1972) Micronutrients in agriculture. American Society of Agronomy, Madison, Wis
10. Lacout JL, Koutsoukos PG, Rouquet N, Freche M (1992) Agrochimica 36:500
11. Amjad Z, Reddy MM (1998) In: Amjad Z (ed) Water-soluble polymers: solution, properties and applications, Plenum, New York, pp 77–90
12. Reddy MM, Nancollas GH (1971) J Colloid Interface Sci 37:824
13. JCPDS ASTM Card No. 05-0586
14. Bates RG (1973) Determination of pH: theory and practice, 2nd edn. Wiley, New York, p 479
15. Sabbides Th, Koutsoukos PG (1993) J Cryst Growth 133:13
16. Nancollas GH, Zawacki SJ (1984) Ind Cryst 84:51
17. Leung WH, Nancollas GH (1978) J Cryst Growth 44:163
18. Giannimaras EK, Koutsoukos PG (1987) J Colloid Interface Sci 116:423
19. Xyla AG, Mikroyannidis J, Koutsoukos PG (1992) J Colloid Interface Sci 153:537

Progr Colloid Polym Sci (2000) 115:112–116
© Springer-Verlag 2000

POLYMER COLLOID AND SOLID PARTICLES

M. J. García-Salinas
M. S. Romero-Cano
F. J. de las Nieves

Zeta potential study of a polystyrene latex with variable surface charge: influence on the electroviscous coefficient

M. J. García-Salinas · M. S. Romero-Cano
F. J. de las Nieves (✉)
Complex Fluids Physics Group
Department of Applied Physics
University of Almería
04120 Almería, Spain
e-mail: fjnieves@ualm.es
Tel.: +34-50-215434
Fax: +34-50-215434

Abstract Electrophoretic mobilities of a carboxyl polystyrene latex have been measured against electrolyte (NaCl) concentration for different pH values. The aim of this study is to know how the changes in the surface charge density affect the electrokinetic behaviour of this latex and, particularly, the presence and position of the maximum in the mobility versus electrolyte concentration curves. Mobilities become greater with increasing surface charge and the maximum mobility becomes greater and is reached at higher electrolyte concentration. This behaviour is discussed by analysing the classical explanations (hairy layer, coion adsorption, surface conductivity) proposed. Zeta potentials have also been calculated using different mobility–zeta potential conversion theories and the results have been compared. These mobility–zeta potential conversion theories have also been tested with an experimental study that is different from the usual electrokinetic techniques: the primary electroviscous effect.

Key words Polystyrene latex · Electrophoretic mobility–zeta potential maximum · Primary electroviscous effect

Introduction

Among the electrokinetic phenomena used to characterize colloids, this article deals with electrophoresis and the electroviscous effect.

In electrophoresis, the usual experimental quantity is the electrophoretic mobility, μ, the velocity of the particle per unit field. The theory of electrophoresis involves theoretical treatment to convert μ into the ζ potential, which is the potential at the slipping or shear plane (the plane of the transition between zero and bulk fluidity). This theory has been dealt with extensively [1, 2] and even recently some of the basic assumptions of the classical treatment have been removed from the hypothesis and their possible occurrence included in the equations [3, 4].

The primary electroviscous effect occurs, for a dilute system, when the complex fluid is sheared and the electrical double layers around the particles are distorted by the shear field. The viscosity increases as a result of an extra dissipation of energy, which is taken into account as a correction factor "p" to the Einstein equation [5, 6]:

$$\eta_r = 1 + k(1 + p)\phi \; ,$$

where η_r is the relative viscosity (relation between the viscosity of the suspension and the viscosity of the solvent), ϕ is the volume fraction of solid and $k = 2.5$ for spherical particles. The theoretical determinations of the coefficient p began with a simple formula given by Smoluchowski [7] in 1916. A more general analysis of the primary electroviscous effect was carried out by Booth [8], who found a dependence on ζ^2 for the primary electroviscous effect coefficient. Later, Watterson and White [9] obtained the solution of the problem by solving a system of coupled differential equations.

All these theories use the ζ potential of the particles to calculate the electroviscous effect and they have been tested numerous times, but always using the classical theories (Smoluchowski [10], O'Brien–White [11]) to obtain the ζ potential.

In this work, electrophoretic mobilities of a carboxyl polystyrene latex were measured against electrolyte (NaCl) concentration for different pH values. The aim of this study was to know how the changes in the surface charge density affect the electrokinetic behaviour of this latex. From the mobility values the ζ potentials were obtained from different mobility–ζ potential conversion theories: Smoluchowski [10], O'Brien–White [11] and Dukhin–Semenikhin [3]. Furthermore, this ζ potential calculation was also taken into account for a study of the primary electroviscous effect for dilute suspensions of the same latex. The calculations of the electroviscous coefficient were made by using different ζ potentials. In such a way we can test the mobility–ζ potential conversion theories from a new point of view.

Materials and methods

All the chemicals in this study were of analytical grade and were used without further purification. Ultrapure water with an electrical conductivity less than 1 μS/cm was used in all experiments.

The carboxylated polystyrene latex used in this work was synthesized by our group using an emulsifier-free polymerization method in a discontinuous reaction [12]. The carboxyl groups were provided by the initiator: 4,4'-azobis(4-cianopentanoic acid), with the advantage that the surface characteristics are those of the polystyrene and that there is no emulsifying agent present. Styrene (Merck) was previously distilled under low pressure (10 mmHg and 40 °C). The latex was cleaned by serum replacement until the conductivity of the supernatant was similar to that of the water. The particle diameter (from transmission electron microscopy) was 187 ± 7 nm, and the surface charge density was determined by conductimetric titration, being the highest ($\sigma_0 = 22.2 \pm 1.2$ μC/cm^2) at basic pH [13].

The electrophoretic mobility measurements were carried out with a Malvern Zetamaster S device. A series of μ against electrolyte concentration measurements were carried out for different pHs. pH 5 was adjusted without buffer solution, a pH around 6 was reached spontaneously, and the rest of the pH values were obtained using buffer solutions. The mobilities were taken as the average of at least ten measurements and the standard deviation of these measurements was considered to be the experimental error. The values that seemed to deviate from the expected trend were measured again in order to confirm them. The ζ potentials were calculated using the Dukhin–Semenikhin [3], O'Brien–White [11], and Smoluchowski [10] theories.

For the viscosity measurements the concentration of the stock latex solution was determined by evaporating to dryness at about 90 °C and the volume fractions of the suspensions were carefully calculated and prepared. The viscosity of the samples was measured with a Schott-Geräte apparatus using Ubbelohde capillary viscometers in a thermostatic bath (with refrigeration and agitation) keeping a constant temperature of 25.0 ± 0.1 °C [13]. For fixed pH and salt concentration, the viscosities of several suspensions with different volume fractions, ϕ, were determined. The value of the p coefficient was obtained from the slope of the η_r versus ϕ plot.

Results

The study of this system was carried out for different pHs (5, 6, 7, 8, 10). The surface charge density changed, but there was no other effect on size or any surface property because carboxyl groups were provided by the initiator with no other monomer than polystyrene present in the polymerization reaction. The measurements of mobility against electrolyte concentration for the five pH values are shown in Fig. 1. First of all, the mobilities increase with increasing pH and, thus, surface charge density. Previous studies comparing sulfonated latexes of similar size but very different surface charge reported similar mobility values for these systems [14, 15] or even lower mobilities for the systems with higher surface charge density [16, 17]. This was attributed to a water-soluble polymer layer, appearing because of the shot-injection step in the synthesis [15], which covered the particles and shifted the shear plane, thus decreasing the ζ potential. In our case the system was prepared by conventional emulsion polymerization (i.e. there was not a second injection) and its behaviour was as expected, so that layer may be absent. Moreover, we are also comparing the same system in different conditions, not different systems prepared by adding different amounts of the ionic comonomer. These results confirm the importance of the preparation method on the final behaviour of the latex particles. Another possible explanation could be related to ion–condensation phenomena [18].

Another remarkable fact is the appearance of a maximum in all the curves; this has been reported in most of the mobility studies and has also been widely discussed [2]. Several explanations have been proposed (hairy layer, coion adsorption, or surface conductivity) [19–28]. There is strong controversy and, depending on

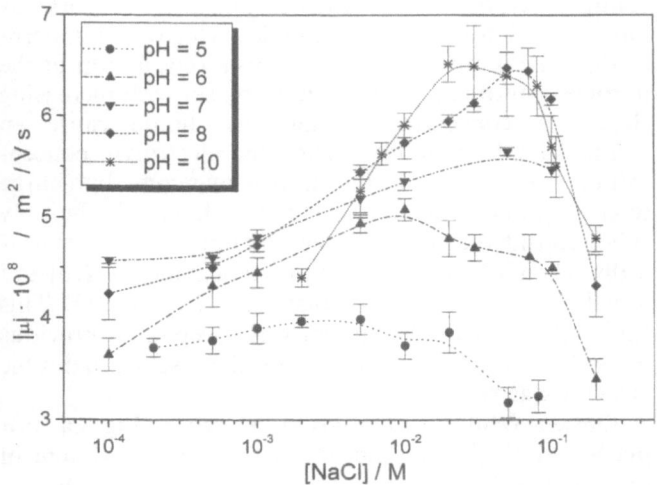

Fig. 1 Electrophoretic mobility against NaCl concentration for different pHs

the experimental system and the method of preparation, some of the explanations for the origin of the maximum may be accepted or ruled out.

The possibility of a hairy layer covering the particles has been considered. Heat treatment has been applied to several systems with different results [19–22], but for carboxylated latexes the presence of a hairy layer cannot be concluded from these studies [21]. We also carried out various size measurements for different electrolyte concentrations but found no trend in the results, so no compression of a hairy layer can be postulated.

The approach of coions to the surface of the particles, thus increasing the electrokinetic potential, has been considered as an explanation for the increasing mobility before the maximum [23–25]. Elimelech and O'Melia [23] found that the mobility maximum appeared at higher electrolyte concentration for systems with higher surface charge, but no explanation was given. Here Fig. 1 shows the same trend clearly. As the electrolyte concentration is increased, coions approach the particle, increasing the negative electrokinetic potential; on the other hand, counterions neutralize the charge and the double layers are compressed. These opposing mechanisms cause the maximum, with the former effect being dominant in the ascending leg of the curve before the maximum. If the maximum is reached when counterions and neutralization of charge dominate, it seems logical that the higher the initial surface charge, the more counterions are needed and, thus, the maximum shifts to higher electrolyte concentration. In Fig. 1 this is seen to happen except for at pH 10, but in this case the surface charge is so high ($\sigma \cong 22\ \mu C/cm^2$) that coions experience greater repulsion on approaching the particle. A recent study of the colloidal stability of this latex [26] also raises the possibility of coion adsorption, which is favoured when the particles discharge.

Polarization of the double layer, related to surface conductivity, has been considered in a series of works by Dukhin and coworkers [1]. For low electrolyte concentration, the induced dipole leads to a retardation of the particle movement. This effect decreases with increasing electrolyte concentration and this should cause an increase in the mobility. When the electrolyte concentration is high enough this effect is overcome by charge screening, and a maximum is reached. This is the only effect considered to cause the maximum that is specifically taken into account in a theoretical model [2, 3] and it will be used later to obtain the ζ potential. This explanation for the maximum was accepted in previous works [14, 27–29], but in other specific cases (thin double layer) it was ruled out [30].

Obrien–White ζ potentials are shown in Fig. 2a. For pH 8 and 10 (highest surface charge) the maximum in the curves still remains and is in the same position.

Dukhin–Semenikhin ζ potentials (Fig. 2b) show a clear descending trend over the whole range studied,

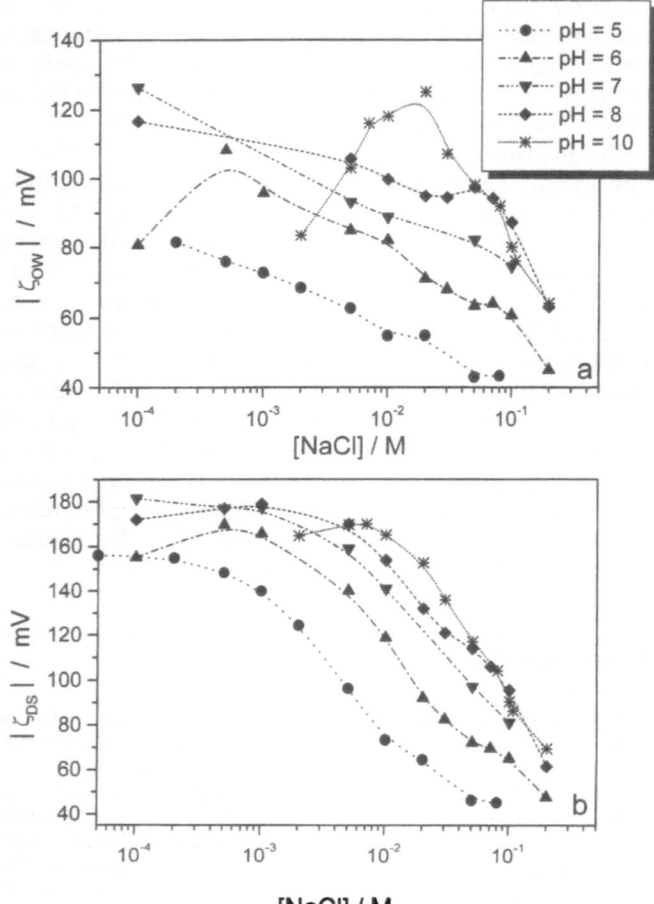

Fig. 2a, b ζ potential against NaCl concentration for different pHs. **a** O'Brien–White (OW) ζ potential; **b** Dukhin–Semenikhin (DS) ζ potential

even for pH 8 and 10. This suggests that the steeper increase in the mobility values for these pHs is due to surface conduction, which is more important for higher surface charge. When this effect is taken into account to find the ζ potential (Dukhin–Semenikhin theory), the maximum is removed. It must also be noted that for these pHs and for low electrolyte concentration, the mobilities are lower than they are for lower pHs, when the system is assumed to be less charged. This contradiction almost disappears when converting the data to Dukhin–Semenikhin potentials. This means that the low mobility values were due to high surface conduction (and thus a retardation in the particle movement).

So to sum up, we can conclude from this analysis of the mobility and ζ potential data that in addition to the classical electrokinetic theory, a coion approximation to the particle and surface conductance have to be taken into account in order to understand the electrokinetic behaviour of colloidal particles. For the latex studied in this work, coion adsorption is likely to occur, and

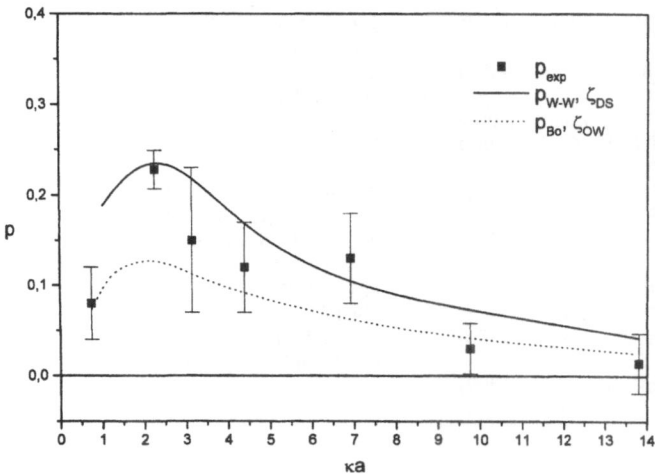

Fig. 3 Electroviscous effect coefficient: experimental values (*squares*); Watterson–White (*W-W*) theoretical values using the *DS* ζ potential (*solid line*); Booth (*Bo*) theoretical values using the (*OW*) ζ potential (*dotted line*)

surface conductance is also present and is considered when calculating ζ potentials (Dukhin–Semenikhin theory). The use of this theory and the mechanism explaining electrokinetic behaviour is corroborated, as will be shown next, by considering another electrokinetic phenomenon: the electroviscous effect.

A study of the primary electroviscous effect was also made for this system. By comparing the experimental results with the theoretical predictions from different ζ potentials, we have a consistent argument to support the validity of the conversion theories. In previous work [13, 31] the electroviscous effect for low pHs was determined and, in general, good results were obtained. As an example, Fig. 3 shows the electroviscous coefficient, p, for this system measured at pH 5 and the theoretical values. When the most accurate theory (Watterson–White, [9]) for the electroviscous effect is used, the best agreement is obtained with the Dukhin–Semenikhin ζ potential; however, a good description of the experimental values was also found using Booth's theory [8] and the O'Brien–White ζ potential. In Watterson and White's paper [9], Booth's theory was

analysed and it was found to be valid for low ζ potentials, otherwise it overestimated the electroviscous coefficient. The use of both the O'brien–White ζ potential (below the real potential) and the Booth electroviscous coefficient (which is higher than the more exact value from Watterson–White) leads to an apparently good agreement, since two opposite deviations from the exact values cancel each other.

In a study of the primary electroviscous effect for a sulfonated latex very good agreement using Watterson–White theory and the Dukhin–Semenikhin ζ potential was also found. Thus, for higher surface charge densities the surface conductance becomes more important and Dukhin–Semenikhin theory provides the best ζ potential data. This result is confirmed by the best fitting of the electroviscous coefficient when the ζ potential is calculated by this theory.

Conclusions

The electrokinetic behaviour of a carboxyl polystyrene latex, particularly the presence and position of a maximum in the mobility versus electrolyte concentration curves, was studied at different pHs. For this latex, prepared by conventional emulsion polymerization, a direct correlation between surface charge and mobility was found.

By analysing these results and those of a complementary study of the primary electroviscous effect for this latex, the mobility–ζ potential conversion theories and the mechanisms involved in electrophoresis were discussed. Although coion adsorption may occur, polarization of the double layer is reported as the main effect causing the nonclassical behaviour of the mobility. The Dukhin–Semenikhin theory seems suitable to explain these mobility data, giving a ζ potential which gives the best experimental–theoretical agreement for electroviscous effect data.

Acknowledgements The financial support provided by the Comisión Interministerial de Ciencia y Tecnología under project MAT96-1035-C03-03 is greatly appreciated.

References

1. Hunter RJ (1981) Zeta potential in colloid science. Principles and applications. Academic, London
2. Hidalgo-Alvarez R (1991) Adv Colloid Interface Sci 34:217
3. Semenikhin NM, Dukhin SS (1975) Kolloidn Zh 37:1127
4. Mangelsdorf CS, White LR (1990) J Chem Soc Faraday Trans 86:2859
5. Hiemenz PC, Rajagopalan R (1997) In: Principles of colloid and surface chemistry, 3rd edn. Dekker, New York, pp 145–192
6. Einstein A (1906) Ann Phys (Leipzig) 19:298
7. Smoluchowski M (1916) Kolloid-z 18:190
8. Booth F (1950) Proc R Soc Lond Ser A 203:533
9. Watterson IG, White LR (1981) J Chem Soc Faraday Trans 2 77:1115
10. Smoluchouski M (1918) Z Phys Chem 92:129
11. O'Brien RW, White LR (1978) J Chem Soc Faraday Trans 2 74:1607

12. Bastos-González D, Ortega-Vinuesa JL, de las Nieves FJ, Hidalgo-Alvarez R (1999) J Colloid Interface Sci 176:232
13. García-Salinas MJ, de las Nieves FJ (1998) Prog Colloid Polym Sci 110:134
14. Hidalgo-Alvarez R, de las Nieves FJ, van der Linde AJ, Bijsterbosch BH (1986) Colloids Surf 21:259
15. de las Nieves FJ, Daniels ES, EL-Aasser MS (1991) Colloids Surf 60:107
16. Bastos-González D, de las Nieves (1993) Colloid Polym Sci 271:860
17. Peula-Garcia JM, Hidalgo-Alvarez R, de las Nieves FJ (1997) Colloids Surf 127:19
18. Fernández-Nieves A, Fernández-Barbero A, de las Nieves FJ. Langmuir (in press)
19. Seeberg JE, Berg JC (1995) Colloids Surf A 100:139
20. Wu X, van de Ven TGM (1996) Langmuir 12:3859
21. Bastos-Gonzalez D, Hidalgo-Alvarez R, de las Nieves FJ (1996) J Colloid Interface Sci 177:372
22. Dunstan DE (1993) J Chem Soc Faraday Trans 89:521
23. Elimelech M, O'Melia CR (1990) Colloids Surf 44:165
24. Voegtli LP, Zukoski CF (1991) J Colloid Interface Sci 141:92
25. Zukoski CF, Saville DA (1986) J Colloid Interface Sci 114:32
26. Puertas AM, de las Nives FJ (1999) J Colloid Interface Sci 216:221
27. Baran AA, Dudkina LM, Soboleva NM, Chechik OS (1981) Kolloid-Z 43:211
28. Moleon-Baca JA, Rubio-Hernandez FJ, de las Nieves Lopez FJ, Hidalgo-Alvarez R (1991) J Non-Equilib Thermodyn 16:187
29. Hidalgo-Alvarez R, Moleon JA, de las Nieves Lopez FJ, Bijsterbosch BH (1992) J Colloid Interface Sci 149:23
30. Midmore BR, Pratt GV, Herrington TM (1996) J Colloid Interface Sci 184:170
31. García-Salinas MJ, de las Nieves FJ, Macromol Symp (in press)

Progr Colloid Polym Sci (2000) 115:117–120
© Springer-Verlag 2000

E. G. Timoshenko
Yu. A. Kuznetsov

Study of mesoglobules in solutions of amphiphilic heteropolymers

E. G. Timoshenko (✉)
Yu. A. Kuznetsov
Theory and Computation Group
Department of Chemistry
University College Dublin, Belfield
Dublin 4, Ireland
e-mail: edward.timoshenko@ucd.de
Fax: +353-1-7062127

Abstract We study the conformational states in solutions of amphiphilic copolymers by means of the Gaussian variational theory and lattice Monte Carlo simulation. We find that due to a delicate balance of microphase separation and entropic effects under the connectivity constraints macromolecular clusters consisting of a few distinct chains may become thermodynamically stable in some narrow regions of the phase diagram within the conventional two-phase coexistence region. These are characterised by a relatively monodispersed size distribution, with the mean size of these mesoscopic globules related to a characteristic scale of the micro-phase separation. Thus, in contrast to the homopolymer case, here collapse and aggregation can compete with each other, producing thermodynamically stable particles with a predominantly hydrophobic core and a hydrophilic shell. These were recently observed experimentally as, at least, rather long lived final states in the kinetics after quenching beyond the spinodal in aqueous solutions of poly(N-isopropylacrylamide) with some of the monomers hydrophobically modified.

Key words Mesoglobule ·
Heteropolymer · Solution · Phase
separation

Introduction

The physical properties of polymers in dilute solutions have been a matter of intensive study using various experimental techniques. Much of the previous experimental work has been carried out near the upper critical solubility temperature of polystyrene in cyclohexane. In recent years there were also numerous works devoted to water-soluble polymers near the lower critical solubility temperature. Popular systems include poly(oxyethylene)–poly(oxypropylene) block copolymers and poly(N-isopropylacrylamide) (PNIPAM) homopolymer. Generally, experiments in this area are quite difficult as they are obscured by interchain aggregation phenomena. Although contraction of chains in the poor solvent regime has been observed many times, compact isolated globules have probably never been seen at equilibrium for homopolymers in pure solvent [1].

Recently, it has been discovered experimentally that at temperatures above the transition point dense spherical mesoscopic globules composed of a number of polymer molecules are formed [2] in dilute aqueous solution of PNIPAM. Clearly, the process of formation of these rather stable and monodispersed particles with sizes in the 50–500 nm range involves some kind of competition between the collapse of single globules and aggregation, which thus take place simultaneously. We called these particles mesoglobules because of their size being intermediate between that of a single globule and a macroscopic aggregate. To make this notion more precise we shall use the term mesoglobules only when talking about equally (or nearly equally) sized globules composed of several distinct chains.

It is our intention here to explain by theory and simulation this experimental observation. In Ref. [3] we conjectured that for copolymers the mesoglobules are stabilised by the microphase separation and could really become thermodynamically stable. A number of recent experimental works [4], in which, however, ionomers were also included, seem to indicate that this might be indeed the case.

Here, we shall concentrate on trying to understand the issue of the monodispersity of the mesoglobules. First, we shall use the Gaussian variational theory to investigate which of the clusters obtained by association of distinct chains are most stable and thus possess the lowest free energy among other possible local minima. Second, to understand the fluctuations beyond the mean-field approximation, we shall use lattice Monte Carlo simulations to obtain the histograms of the cluster size and mass distributions directly. Knowing the widths of these distributions would allow us to characterise quantitatively the monodispersity property of the mesoglobules.

We should also mention perhaps that block and random heteropolymers in solutions and melts have been attracting a great deal of interest as they exhibit ordered microphase-separated and disordered glassy phases [5]. Block copolymers are often used as surfactants in ternary mixtures of two otherwise immiscible liquids, such as water and oil, and these mixtures produce many sophisticated structures, such as micelles and lamellae [6].

Numerical results from the variational theory

First of all, we have to perform a careful numerical analysis of the free energy obtained from the Gibbs–Bogoliubov variational principle with a generic quadratic Hamiltonian [7].

It is reasonable to expect that in the simple case of diblock copolymers mesoglobules are nothing but ordinary polymer micelles. Clearly, in a poor solution, hydrophobic units would tend to escape from unfavourable contacts with the solvent, but the connectivity of each chain seriously restricts their freedom to do so. However, for arbitrary heteropolymer sequences simple micellar structures are not possible due to the connectivity of the chains. For heteropolymers with an essential heterogeneity along the chain there are many competing interactions and the free-energy profile is very rugged; therefore, some new free-energy minima may appear due to a specific compensation of the interaction terms and the entropy.

The free energy, $\mathscr{A} = \mathscr{E} - T\mathscr{S}$, obtained from the Gibbs–Bogoliubov variational principle, $\mathscr{A} = \mathscr{A}_0 + \langle H - H_0 \rangle_0$, as described in Ref. [7] has the following conformational "entropy" part $\mathscr{A}_0$,

$$\mathscr{S} = \frac{3}{2}k_B \ln \det' R, \quad R_{AA'} = \frac{1}{N^2 M^2} \sum_{BB'} D_{AB,A'B'}, \quad (1)$$

$$D_{AA',BB'} = -(1/2)(D_{AB} + D_{A'B'} - D_{AB'} - D_{A'B}),$$

$$D_{AA'} = (1/3)\left\langle (\mathbf{X}_A - \mathbf{X}_{A'})^2 \right\rangle, \quad (2)$$

where we have denoted by $\mathbf{X}_n^a$ the coordinates of the nth monomer in the ath chain, multi-index $A \equiv (a, n)$, and N and M are the number of monomers in a chain and the total number of chains, respectively. The mean energy part, $\mathscr{E} = \langle H \rangle_0$, is then written as follows [7]

$$\mathscr{E} = \frac{3k_B T}{4L^2 N^2 M^2} \sum_{AA'} D_{AA'}\left(\frac{1}{M} - \delta_{aa'}\right) + \frac{3k_B T}{2l^2} \sum_{n,a} D_{n\,n-1}^a$$
$$+ \frac{1}{(2\pi)^{3/2}} \sum_{AA'} \frac{\bar{u}^{(2)} + \Delta(\sigma_A + \sigma_{A'})/2}{D_{AA'}^{3/2}} + \frac{3u^{(3)}}{(2\pi)^3} \sum_{AA'} D_{AA'}^{-3}$$
$$+ \frac{u^{(3)}}{(2\pi)^3} \sum_{AA'A''} \left(D_{AA'}D_{A''A'} - D_{AA',A''A'}^2\right)^{-3/2}, \quad (3)$$

where L is the box size, l is the statistical segment length, $\bar{u}^{(2)}$ is associated with the quality of the solvent, the amphiphilicity, Δ, characterises the difference in interactions of hydrophilic and hydrophobic units with the solvent and $u^{(3)}$ is the third virial coefficient[1]. The set $\{\sigma_n^a\}$ expresses the chemical composition, or the primary sequence of a chain. Here the variables σ_A take only two values: -1 and 1, corresponding to the hydrophobic a and hydrophilic b monomers, respectively. The free energy then has to be minimised with respect to the full set of mean-squared distances, $D_{AA'}$.

We see from numerical analysis that there is indeed a large number of local free-energy minima. These correspond to conformations in which polymers form one or several clusters, each consisting of one or more chains. Let us consider the system composed of $M = 12$ chains of $N = 12$ monomers each for two different sequences, one periodic and the other random. Values of the free energy at local minima corresponding to equally sized (symmetric) clusters are presented in Table 1 for different values of $\bar{u}^{(2)}$ at a fixed sufficiently high Δ. All the asymmetric clusters considered were found to possess a higher value of the free energy than these symmetric ones. One can see that in some range of $\bar{u}^{(2)}$ the main (deepest) minimum is reached at a state corresponding to mesoglobules – clusters of equal size, such as 6×2, 4×3, 3×4 or 2×6. This means that such a state becomes thermodynamically stable there. We may also note from the table that, with all other parameters being fixed, the size of the stable mesoglobules increases with the concentration and with $|\bar{u}^{(2)}|$, but it is also very sensitive to the heteropolymer sequence. It is important to emphasise that mesoglobules can have the lowest free energy not only for heteropolymers with a periodic

---

[1] We chose the system of units such that $l = 1$, $k_B T = 1$ and fix $u^{(3)} = 10\,k_B T l^6$.

Table 1 Values of the specific free energy, $a = \mathscr{A}/MN$, at its various minima for the system of $M = 12$ heteropolymers of sequences $(a_3b_3)_2$ (*top half*) and $b_3a_2ba_2baba$ (*bottom half*) versus the mean second virial coefficient, $\bar{u}^{(2)}$. Here $L = 20$ and $\Delta = 30$. The value for the global (deepest) minimum is printed in **boldface**. Note that all rows except the first and last for each sequence correspond to the region where one of the mesoglobular states is stable

$\bar{u}^{(2)}$	12×1	6×2	4×3	3×4	2×6	1×12
$(a_3b_3)_2$						
0	**−3.02**	−2.81	−2.61	−2.45	−2.18	−1.65
−10	−8.12	**−8.37**	−8.34	−8.26	−8.08	−7.63
−15	−11.23	−11.78	**−11.85**	−11.83	−11.71	−11.33
−20	−14.74	−15.60	−15.80	**−15.84**	−15.79	−15.51
−30	−23.05	−24.61	−25.07	−25.26	**−25.39**	−25.32
−35	−28.00	−29.89	−30.48	−30.75	−30.97	**−31.03**
$b_3a_2ba_2baba$						
−10	**−7.78**	−7.77	−7.61	−7.43	−7.10	−6.40
−15	−10.93	**−11.23**	−11.16	−11.03	−10.77	−10.14
−20	−14.49	−15.11	**−15.17**	−15.11	−14.92	−14.37
−25	−18.47	−19.44	−19.63	**−19.65**	−19.54	−19.11
−35	−27.91	−29.60	−30.08	−30.27	**−30.36**	−30.18
−45	−40.47	−42.75	−43.26	−43.31	−43.59	**−43.67**

(block) structure, but essentially for many random sequences as well.

Results from lattice Monte Carlo simulation

It is interesting to check these predictions of the Gaussian variational method by Monte Carlo simulation on a lattice. We adopt the Metropolis technique in the lattice model of Ref. [8]. This model, apart from the connectiv-

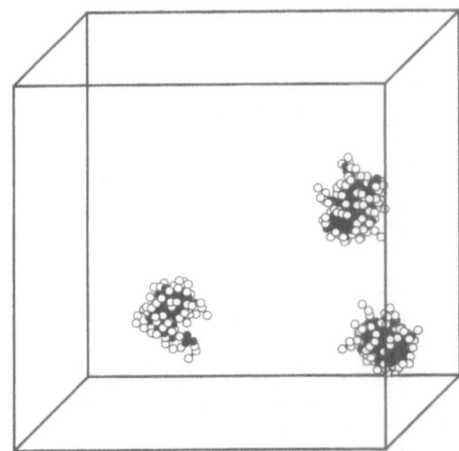

Fig. 1 Snapshot of a typical mesoglobular conformation of heteropolymer sequence $(a_3b_3)_4$. Here the equilibration time was 1.92×10^9 of attempted Monte Carlo moves and the other parameters were $L = 60$, $N = 24$, $M = 20$, $\chi_{aa} = 1$, $\chi_{ab} = 0.4$ and $\chi_{bb} = -0.2$. *Black circles* correspond to hydrophobic monomers and *white circles* correspond to hydrophilic monomers

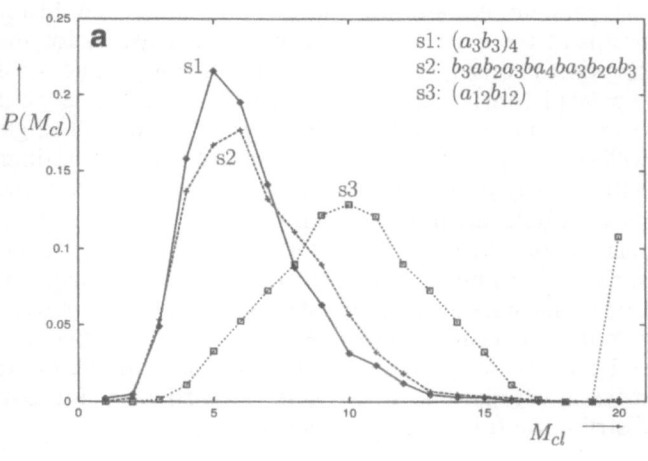

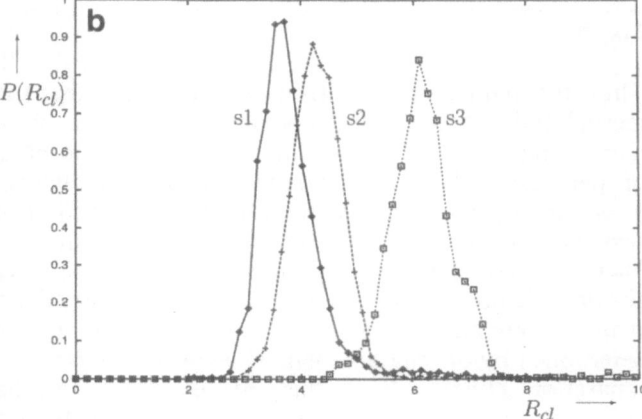

Fig. 2 Probability densities of **a** the number of chains (mass), M_{cl}, and **b** the radius of gyration (size), R_{cl}, for mesoglobules of different sequences. These results were obtained by analysing data for the ensemble size $Q = 1000$

ity and excluded-volume constraints, includes short-ranged pairwise interactions between lattice sites. The system is completely characterised by three Flory interaction parameters, χ_{aa}, χ_{ab} and χ_{bb}, along with N, M and the linear lattice size, L. In addition to local monomer moves [8], we include translational moves representing diffusion of chains. The latter moves are applied to all clusters of chains with a probability inversely proportional to the number of monomers within (Stokes law).

We considered several concrete sequences consisting of strongly hydrophobic and slightly hydrophilic units. From the snapshot in Fig. 1 we can clearly see a number of distinct clusters there, each consisting of several chains. These tend to have a larger amount of hydrophilic (white) material on the outside. Strikingly, these clusters are of nearly equal size, something we have already noted from the above theory.

To address the question about the size polydispersity of the mesoglobules we obtained a large ensemble of independent equilibrium states. The calculated histograms of the mass and size probability densities for the sequences considered are presented in Fig. 2. The most typical picture is seen for sequences s1 (intermediate-sized blocks) and s2 (a random sequence). These have a single well-distinguished peak in the mass and size distributions with a fairly narrow Gaussian-like shape; therefore, the mesoglobules are quite monodisperse with about 10–15% relative dispersion in size. Other sequences, however, such as s3 (diblock copolymer) have different distributions. The mass distribution of s3, in addition, has a large population of single aggregates, $M_{cl} = 20$. This is believed to be a finite-size effect as the characteristic mesoglobule mass is comparable to M here. Nevertheless, the size distribution for s3 still has a single peak.

Conclusion

Thus, the main conclusion from our analysis is that the mesoglobules become thermodynamically stable in a narrow region of the phase diagram for a wide class of periodic and random amphiphilic heteropolymers possessing hydrophobic and hydrophilic monomers. The size of the mesoglobules varies somewhat due to fluctuations transforming symmetric clusters into slightly asymmetric ones, although the barriers separating these structures are high. The mean size of the mesoglobules is determined by the characteristic scale of the microphase separation. Thus, the larger this scale (e.g. block size) the larger the mesoglobules produced. The size distribution of the mesoglobules is sufficiently monodisperse due to the thermodynamic preference for clusters to be of equal size. The two methods employed here give the upper and lower bounds on the variations in mesoglobule size. The variational method, being merely an optimised mean-field theory, predicts perfectly monodispersed mesoglobules. The Monte Carlo method, on the other hand, tends to overemphasise their polydispersity. This is because breaking away from an already-formed dense cluster is harder than joining it. It seems quite reasonable that even weak electrostatic repulsion may play a crucial role for further stabilisation of the mesoglobules and for improving their monodispersity.

We believe that these observations may shed some light on understanding the problems of competition aggregation versus folding in protein solutions and self-organisation of the quaternary structure in multimeric proteins. It appears that for the latter process to take place a considerable number of "sticky" hydrophobic amino acid residues should be exposed on the exterior of each of the folded subunits. The resulting association produces a well-defined quaternary structure, which is biologically functional as in hemoglobin, for example, rather than a disordered aggregate. In our view, this feature looks similar to the process of mesoglobule formation, but of course one should bear in mind the complexity of real proteins not represented in the current oversimplified model of heteropolymers. Hopefully, work in this direction would also help unravel the mechanism of association of so-called Bence–Jones proteins believed to be related to amyloid fibril formation.

Speaking of other possible applications, water-soluble polymers are widely used in chemical, pharmaceutical and food industries. Therefore, detailed understanding of the mechanism of mesoglobule formation is important for being able to select polymers from which nanoparticles of the required size and polydispersity can be prepared. The main advantage of this new approach is that nanoparticles can be formed reversibly in mild conditions from a prepurified polymer. Also, nanoparticles carrying functional groups properly located in the mesoglobule can be obtained and these could be further modified. Such nanoparticles with a well-controlled size distribution can find a broad range of applications in pharmaceutical, biotechnological and cosmetic industries, where technological control is one of the major problems.

Acknowledgements The authors acknowledge interesting discussions with F. Ganazzoli, G. Allegra, G. Raos and our colleague A.V. Gorelov. This work was supported by grant SC/99/186 from Enterprise Ireland.

References

1. (a) Chu B, Ying Q, Grosberg AY (1995) Macromolecules 28:180; (b) Ganazzoli F, Raos G, Allegra G (1999) Macromol Theory Simul 8:65
2. (a) Gorelov AV, Vasil'eva LN, Du Chesne A, Timoshenko EG, Kuznetsov YA, Dawson KA (1994) Nuovo Cimento D 16:711; (b) Gorelov AV, Du Chesne A, Dawson KA (1997) Physica A 240:443
3. Timoshenko EG, Kuznetsov YA (1998) Nuovo Cimento D20:(12BIS) 2359–2364
4. (a) Deng Y, Pelton R (1995) Macromolecules 28:4617; (b) Qiu X, Kwan C, Wu C (1997) Macromolecules 30:6090
5. Miles IS, Rostami S (eds) (1992) Multicomponent polymer systems. Longman, London
6. Mittal K (ed) (1977) Micellization, solubilization and microemulsion. Plenum, New York
7. (a) Timoshenko EG, Kuznetsov YA, Dawson KA (1997) Physica A 240:432; (b) Timoshenko EG, Kuznetsov YA, Dawson KA (1998) Phys Rev E 57:6801
8. Kuznetsov YA, Timoshenko EG, Dawson KA (1995) J Chem Phys 103:4807

Progr Colloid Polym Sci (2000) 115:121–127
© Springer-Verlag 2000

I. Lynch
A. V. Gorelov
K. A. Dawson

Systematic comparison of effect of structural and architectural changes to the network structure on the kinetics of collapse of *N*-isopropyloacrylamide gels

I. Lynch · A. V. Gorelov
K. A. Dawson (✉)
Irish Centre for Colloid Science
and Biomaterials
Department of Chemistry
University College Dublin
Belfield, Dublin 4, Ireland
e-mail: kenneth@fiachra.ucd.ie
Tel.: +353-1-7062300
Fax: +353-1-7062415

Abstract The shrinking kinetics of submillimetre spherical gels of composition 10:20:70 mol% *N-tert*-butylacrylamide: *N,N'*-dimethylacrylamide: *N*-isopropylacrylamide have been studied in detail. A description of the shrinking kinetics throughout the transition region (34–38 °C) is given as, in the size range studied, the gels showed no shape distortion during temperature jumps to these temperatures. All gels relaxed to their equilibrium size exponentially after temperature jumps to temperatures in the transition region. Using this unique molar composition, we systematically studied the effect of changing the network structure on the shrinking kinetics. The aim was to achieve rapid controlled collapse with potential application in medicine and medical devices. The approaches used included grafting of freely mobile polymer chains onto the polymer backbone, introduction of poly(acrylamide) and incorporation of a second component to form a composite gel matrix.

Key words *N*-Isopropylacrylamide gels · Shrinking kinetics · Network structure

Introduction

Much attention has been focused in recent years on improving the shrinking time of *N*-isopropylacrylamide (NIPAM) gels in order to utilise them in practical devices such as drug-delivery systems and optical switching devices, where response times are critical. The shrinking process is controlled mainly by the diffusion of the network through the solvent medium, but there is also a contribution from the nonzero shear modulus, which works to maintain the gel's shape as it collapses. Moreover, in these gels there is often an additional factor to consider, which is the critical slowing down of the shrinking and the formation of a "metastable" phase, which is thought to be due to the formation of a dense collapsed layer on the surface of the gel [1]. While in some applications this effect has been beneficial, such as "on–off" switching devices, in cases where maximum shrinkage is required in the minimum time, this slowing down is a major disadvantage.

Several novel gel structures have been described in the literature which have much faster collapse transition times than conventional NIPAM gels. One of the earliest architectural changes to the network structure of gels was reported by Kaneko et al. [2], who grafted freely mobile poly(NIPAM) (PNIPAM) chains into NIPAM gels. Since the chains were tethered to the network backbone at one end only, they are considered to behave as freely mobile chains and, as such, they respond to changes in their environment much faster than the more constrained chains of the network. These tethered chains speeded up the overall collapse, possibly by pulling the network with them as they collapsed. More recently, Hirotsu [3] chemically constrained poly(acrylamide) (PAM) chains into an NIPAM gel network. When the temperature is increased the NIPAM chains shrink, whereas the PAM chains swell and absorb the water expelled by the NIPAM sections. Very quickly the PAM sections meet up and provide effective channels to release the water from the gel, dramatically speeding up the shrinking time.

We recently reported a terpolymer gel composed of 10:20:70 mol% *N-tert*-butylacrylamide (BAM): *N,N'*-dimethylacrylamide (DAM) NIPAM which has a conventional gel structure, but which has considerably faster shrinking kinetics than reported by others for similar-sized gels [1]. We further improved the shrinking times by changing the network architecture to a two-component gel structure.

Thus, we now have access to a range of gels, all with faster than normal shrinking times. Until now these gels have all been prepared differently and studied using different approaches. Thus, there is no absolute way of determining which gels really have the fastest shrinking combined with the simplest preparations, both of which are important factors when designing an application of gel systems. In this work, we prepared all the gels in exactly the same way and here we compare their shrinking kinetics systematically to obtain absolute kinetics comparisons. The 10:20:70 BAM:DAM:NIPAM gel mentioned previously has been described in detail [1] and allows us to study the shrinking process throughout the transition region. Using this as a model gel, we determined the effect of grafting PNIPAM chains onto the backbone of the network, incorporating PAM chains and incorporating a second component, which results in a composite matrix structure, on the shrinking kinetics of the gel. Thus, we can quantitatively state the most efficient gel for drug-delivery applications in terms of both ease of preparation and speed of the shrinking process.

Experimental

Materials

NIPAM monomer (purity above 99%) from Acros Organics (Geel, Belgium) was recrystallised twice from hexane. BAM (purity 97.0% or above), DAM (purity above 98%), and *N,N'*-methylene-bis(acrylamide) (purity above 99.5%) from Fluka (Dorset, UK) were all used as supplied. Acrylamide (purity 99 + %) and ammonium peroxydisulfate (APS) (purity 99.99%) from Aldrich (Dorset, UK) were used as supplied. *N,N'*-Azobis(isobutyronitrile)(AIBN) from Phase Separations (Clwyd, UK) was recrystallised from methanol. 2-Aminoethanethiol (AESH) from Fluka (Dorset, UK) and *N*-acryloxysuccinimide (NAS) from Acros Organics (Geel, Belgium) were used as supplied. *N,N,N',N'*-Tetraethylenediamine (TEMED) from Sigma (Dorset, UK) was used as supplied. Paraffin oil from BDH (Dublin, Ireland) was washed with deionised water before use. *N,N*-Dimethylformamide (DMF) from Aldrich (Dorset, UK) was distilled under vacuum and obtained as the fraction which boiled at 76 °C at 39 mmHg. All water used was of Milli-Q quality (Millipore, Hertfordshire, UK) and was degassed under vacuum before use. All organic solvents used were of analytical grade from Sigma–Aldrich (Dorset, UK).

Synthesis of PNIPAM macromonomers

The synthesis of the PNIPAM macromonomers was carried out in two steps, according to the method of Kaneko et al. [2]. Firstly, a semitelechelic PNIPAM with a terminal amino end group was synthesised by radical telomerisation of NIPAM monomer (2.6814 g) with (AESH) (0.02558 g) as a chain-transfer agent and AIBN (0.00394 g) in 10 ml dry DMF. The flask containing the reactants was degassed by three freeze–thaw cycles and sealed, and the reaction was carried out at 70 °C for 15 h. The product was concentrated by evaporation of DMF, and semitelechelic PNIPAM was precipitated in diethyl ether, filtered, and cleaned by repeated precipitation from DMF into diethyl ether.

In the second step, a polymerisable end group was introduced into the amino semitelechelic PNIPAM using an amide condensation reaction between the amino groups of PNIPAM and NAS with a molar ratio of 1:10 in DMF at 4 °C for 2 days. The macromonomer was purified using exactly the same procedure as for PNIPAM with terminal amino groups. The viscosity-average molecular weight of the PNIPAM polymer was determined by viscosity measurements using a Ubbelohde viscometer to be 9000.

Synthesis of PAM

PAM was synthesised by conventional free-radical polymerisation, according to the method of Hirotsu [3]. Acrylamide (0.994 g, 700 mmol) was dissolved in 19 ml water. The solution temperature was raised to 70 °C and the solution was degassed by bubbling with N_2. APS (8 mg) was dissolved in 1 ml water. The APS solution was degassed under vacuum and added to the acrylamide solution and the reaction proceeded at 70 °C for 2 h. The viscosity-average molecular weight of the PAM polymer was determined by viscosity measurements using a Ubbelohde viscometer [4] to be 2×10^5.

Synthesis of spherical gel beads

Submillimetre spherical gel beads were prepared by inverse polymerisation, according to the method of Matsuo et al. [5]. Paraffin oil was washed with Milli-Q water, heated to 100 °C for 1 h, cooled to room temperature, and degassed by bubbling with N_2. All the gels had 0.8 wt% cross-linking, and the pregel concentration was 700 mM monomer (490 mM NIPAM, 70 mM BAM, 140 mM DMA) in 4 g water. Additional network components (i.e. PAM, amino-terminal PNIPAM chains or the second component of the composite gel matrix) were added to the pregel solution. The concentration of the bulk network was always 700 mmol, and the additional components were added at a concentration of 140 mmol (20 wt% of the bulk gel monomer concentration). The pregel solution was degassed under vacuum and 15 μl TEMED was added. The initiator concentration was 40 mg in 1 g water, of which a 70-μl aliquot was added. A 1-ml aliquot of this solution was added via a syringe to the vigorously stirred oil. The reaction was purged after 1 h by adding water. The oil was removed and unreacted materials were removed by washing with copious amounts of distilled water.

Temperature jumps

The temperature-jump method is the same as described previously [1]. Gel beads were inserted onto a "thermoslide" and heated by applying a current. The degree of shrinkage was measured by monitoring the diameter of the gels, D_t, and the change in diameter [either the diameter at time t divided by the initial diameter, D_t/D_0, or the renormalised diameter, $(D_t/D_0 - D_f/D_0)/(1 - D_f/D_0)$] was plotted as a function of time. The relaxation times, or the times taken for the gels to reach their equilibrium sizes at the new temperature, were determined from these plots by curve-fitting.

Results and discussion

The composition of the gels used (10:20:70 BAM: DAM: NIPAM) was selected to have a similar transition temperature to that of the NIPAM gel, but to have a continuous volume phase transition, which reduces the instability associated with the transition region. Using this composition we can study the temperature-induced volume phase transition throughout the transition region [1]. The gels undergo no shape distortion unless the temperature jump is into the fully collapsed region, which is above 38 °C. Based on this we divide the equilibrium curve into two regions – the transition region (34–38 °C) and the collapsed region (above 38 °C) as shown in Fig. 1. We define T^*, the fully collapsed temperature, as being above 38 °C. This particular gel has a conventional network structure, but we have shown that its shrinking kinetics is much faster than reported by others for similarly sized gels [1]. All the gels have the same bulk composition and, thus, are directly comparable in terms of shrinking kinetics, since any change in the shrinking kinetics must be a direct consequence of changes to the network structure. The equilibrium collapse curves of the four gels used in this study are compared in Fig. 2, showing that none of the structural modifications alter the transition temperature.

A sketch of the various gel structures used in this study is shown in Fig. 3. Figure 3a represents an idealised homogenous or conventional gel network. Figure 3b shows PNIPAM chains grafted onto the

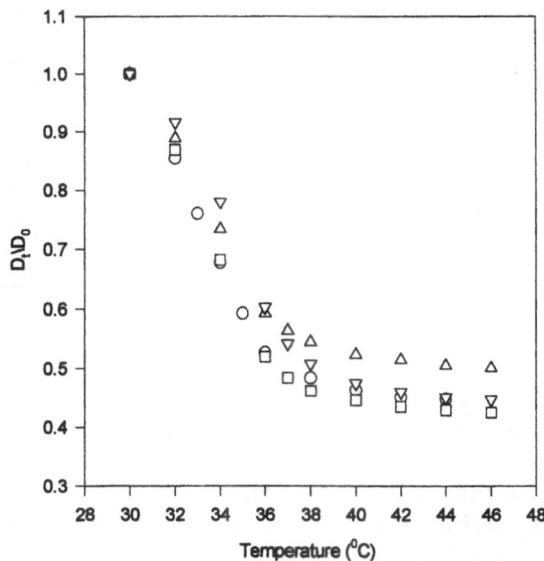

Fig. 2 Comparison of the equilibrium collapse curves for the four gel types used. Model gel (O), graft poly(NIPAM) (*PNIPAM*) gel (□), polylacrylamide (*PAM*) gel (△) and composite gel (▽)

backbone of the gel network, as proposed by Kaneko et al. [2]. The PNIPAM chains are constrained at only one point (where they are attached to the network) as opposed to the network backbone polymers, which are constrained at several places along their length into a three-dimensional network by the bifunctional cross-linking molecules. Thus, the PNIPAM chains are considered to be freely mobile and extend into the pores of the gel network in a "Flory-coil"-type arrangement. Figure 3c shows the structure of the gel containing long-chain PAM polymers which was proposed by Hirotsu [3]. As the gel forms the terminal acrylamide groups of the PAM chains are activated and the polymers become chemically linked into the gel network.

With all four gels used in this study, the shrinking process occurs in two distinct ways depending on the depth of the temperature jump. The first shrinking process results from small temperature jumps, where the gel network is able to rearrange and thus the gel shrinks rapidly and exponentially, with no shape distortion and no appearance of opacity. The second shrinking process is a two-stage collapse which results from large temperature jumps, where the gel is taken far out of equilibrium. In this case, an initial burst of shrinking is followed by a slow relaxation to the final size. The appearance of bubbles and opacity often accompany the two-stage collapse, due to the formation of a "skin-layer" of collapsed gel on the surface. Such a skin layer was first described by Yoshida et al. [6]. The temperature at which the shrinking process becomes two-stage is dependent on the gel structure and the size of the gel bead.

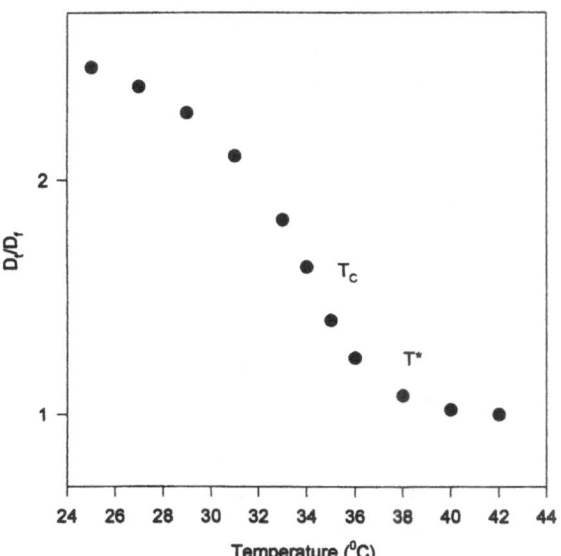

Fig. 1 Equilibrium collapse curve for the 10:20:70 *N-tert*-butylacryla-mide: *N,N′*-dimethylacrylamide: *N*-isopropylacrylamide (*NIPAM*) mol% gel. T_C is the critical temperature and T^* is the fully collapsed temperature

124

Fig. 3a–c Schematic representations of the gels used in this study. **a** Unmodified network, **b** modified with freely mobile PNIPAM chains, and **c** modified with PAM chains

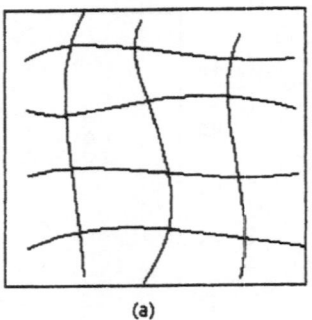
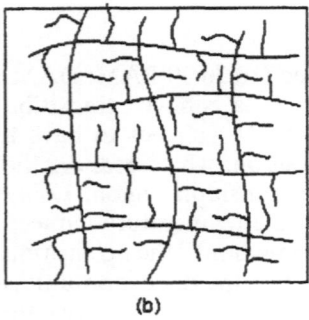
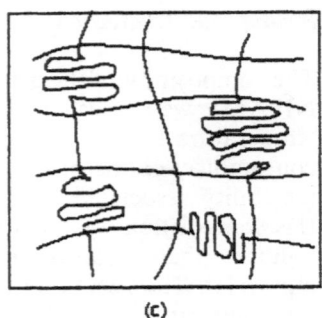

(a)　　　　　　　　　　　(b)　　　　　　　　　　　(c)

As we reported previously [1], the model (unmodified) gel collapses exponentially throughout the transition region and up to the temperature at which the gel becomes fully collapsed (T^*), which is 38 °C for the model gel (Fig. 1) in the size range studied. Above T^*, the collapse becomes two-stage. The effect of the "skin-layer" becomes more pronounced with increasing gel size, as the surface area to volume ratio decreases, resulting in more water being trapped inside the gel at the time of skin formation. In the case of the graft PNIPAM gel the shrinking process is exponential after temperature jumps to temperatures up to and including, T_C, the collapse transition temperature, and two-stage after temperature jumps to temperatures above T_C; however, the exact onset temperature of two-stage collapse behaviour is size-dependent. The larger gel (260 μm) began to show two-stage kinetics at a lower temperature than the smaller gel (185 μm). The onset temperatures for two-stage collapse behaviour were 36 and 37 °C, respectively, for these gels. Shrinking times were faster for the graft PNIPAM gel than the model gel in all cases, and in the case of two-stage collapse, the macroscopic effects of the skin layer were less pronounced. There was no evidence of bubbling except at the very highest temperatures in the largest gel. The gels did turn opaque where collapse was two-stage. In the case of the PAM gels the situation is slightly different. Here the shrinking is exponential only up to T_C and becomes very noticeably two-stage above T_C. Where collapse is two-stage, the initial burst of shrinking is so pronounced that the water being forced out of the network causes a disturbance in the surrounding water, which can be seen in the images collected using the microscope. Also, this initial shrinking burst is shorter than in the graft PNIPAM gel or the model gel, but more water is ejected, and thus more shrinkage has occurred. This means that there is less water trapped inside the gel network when the second stage of the collapse begins, and thus less pressure builds up inside the gel. Consequently, there is no appearance of bubbles on the surface in any of the PAM gels at any temperature. The gel does turn opaque however, when collapse is two-stage. In the case of the composite gel matrix the shrinking process is similar to that of the model gel – shrinking is exponential up to T^* and two-stage above T^*. The shrinking times are faster, however, and as with the PAM and graft PNIPAM gels, "skin-layer" formation has less of an influence on the shrinking. Again, there is the appearance of opacity after temperature-jumps to $T > T^*$. The shrinking times (τ) for the various gels and the onset temperatures for two-stage kinetics are shown in Table 1. A comparison of the shrinking profiles after temperature jumps to 34 °C (below T_C) and 37 °C (above T_C) are shown in Figs. 4 and 5, showing clearly that the PAM gel has the fastest shrinking kinetics in the temperature range studied.

It is important to notice that the shrinking times given in Table 1 are the times for exponential shrinking where this is the process occurring. The values were obtained by fitting the experimental data to a single exponential. This process is described in a previous article [1]. In all cases where the collapse is two-stage, the second stage was removed before fitting the data to the exponential fitting equation [1], and thus the values obtained are approximate.

In many drug-delivery applications, it is not necessary for the gel to reach its fully collapsed state. Where speed of release is the important factor, a gel which shrinks rapidly and releases a large proportion of its water in an initial rapid burst may be more effective than one which releases almost all its water content more slowly. Thus, information about the degree of shrinking in the first few seconds would also be useful when deciding which gel to use. Since the aim of this paper is to compare a range of rapid-shrinking gels for their effectiveness and suitability for rapid-release applications, we compare the amount of shrinkage attained and thus the amount of water released by the initial burst of shrinking. The degree of shrinking attained in the first burst of shrinking for each of the four gels studied after temperature jumps to 34 and 38 °C is shown in Figs. 6 and 7. It can be clearly seen that the amount of water released by each of the gels is quite different even at temperatures below the transition temperature. This trend becomes more pronounced at temperatures above

Table 1 Shrinking times, τ(s), calculated from experimental data using curve-fitting. The shrinking times in *italics* indicate two-stage shrinking kinetics

Gel type	Temperature after temperature jump (°C)					
	34	35	36	37	38	40
Model						
178 μm	27	28	29	28	29	*16*
212 μm	38	43	41	48	42	*21*
Graft poly(*N*-isopropylacrylamide)						
185 μm	31	25	21	*17*	*13*	*10*
260 μm	42	34	*29*	*20*	*14*	9
Poly(acrylamide)						
210 μm	25	20	*16*	*13*	9	6
290 μm	35	33	*19*	*14*	9	7
Composite						
200 μm	26	29	38	24	23	*12*
260 μm	26	44	42	35	30	*21*

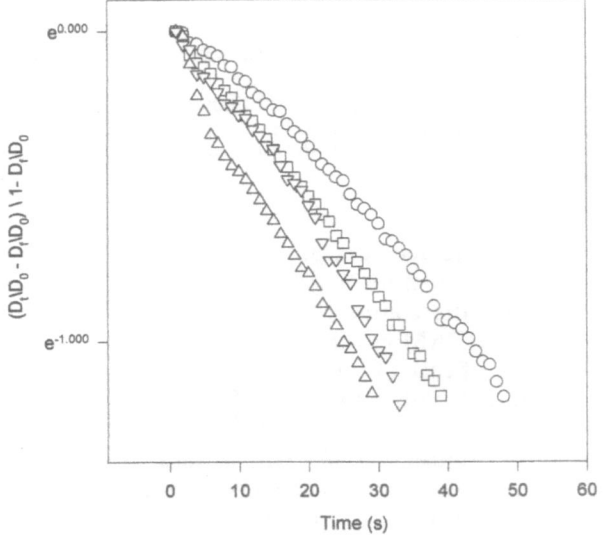

Fig. 4 Comparison of the shrinking times of the various gels (about 200 μM) after temperature jumps to 34 °C. Model gel (○), graft PNIPAM gel (□), PAM gel (△) and composite gel (▽)

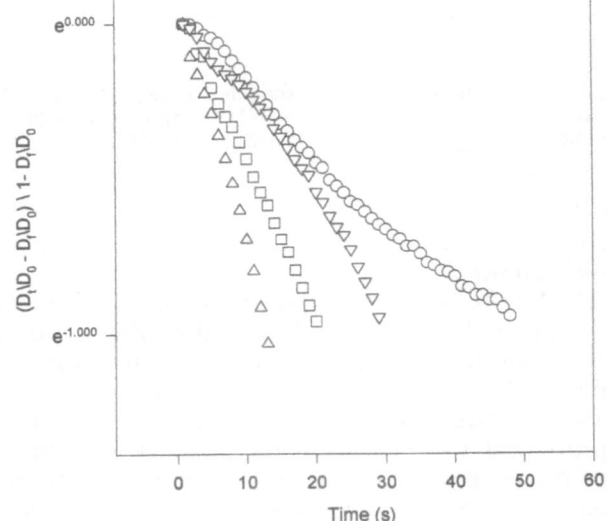

Fig. 5 Comparison of the shrinking times of the various gels (about 200 μM) after temperature jumps to 38 °C. Model gel (○), graft PNIPAM gel (□), PAM gel (△) and composite gel (▽)

T_C, where the PAM gel clearly has the fastest release of water in the initial 15 s following the temperature jump followed by the graft PNIPAM gel. The composite and model gels release approximately the same amount of water above and below T_C. Thus, if extremely rapid release of a large quantity of the network water is required, clearly the PAM gel is the most efficient for this purpose. The degree of shrinkage in the initial 15 s following temperature jumps for the various gels are shown in Table 2.

Conclusion

A series of gels, reported in the literature to have faster shrinking times than conventional gel networks, were prepared and their shrinking kinetics determined. The effect of changes to the network structure on the shrinking kinetics was systematically compared, and the effectiveness of each of the gels for use in drug-delivery systems was evaluated. The parameters considered included speed of collapse, ease of preparation, and amount of water released in the initial shrinking period. The gels studied were all basic modifications of a conventional gel network. The model gel was composed of 10:20:70 mol% BAM:DAM:NIPAM. The network modifications studied were PNIPAM chains grafted onto the network backbone, long chains of PAM incorporated into the network, and addition of a second component to produce a composite gel network. Each of the modified gels had a bulk gel composition identical to that of the model gel, with

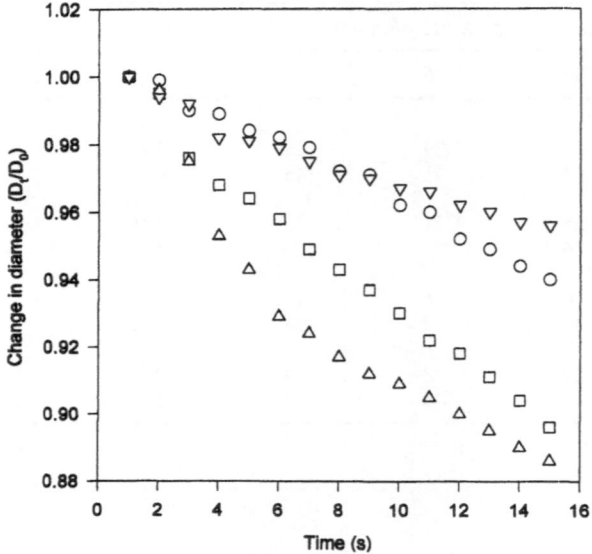

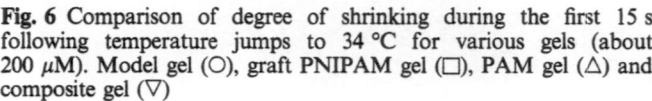

Fig. 6 Comparison of degree of shrinking during the first 15 s following temperature jumps to 34 °C for various gels (about 200 μM). Model gel (O), graft PNIPAM gel (□), PAM gel (△) and composite gel (▽)

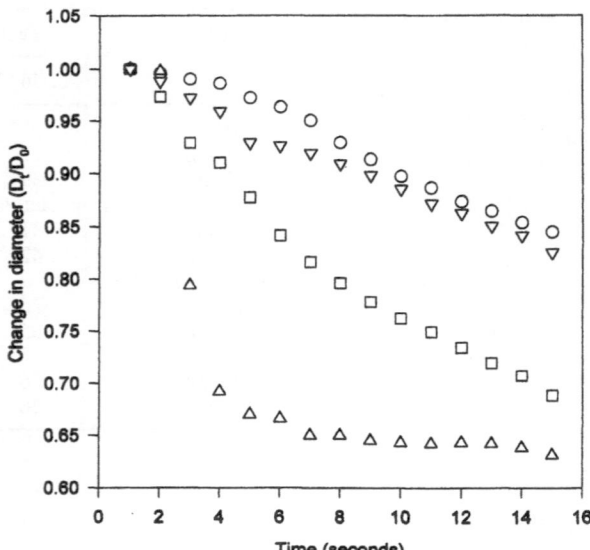

Fig. 7 Comparison of degree of shrinking during the first 15 s following temperature jumps to 38 °C for various gels (about 200 μM). Model gel (O), graft PNIPAM gel (□), PAM gel (△) and composite gel (▽)

140 mmol of the additional component added to the pregel solution.

The kinetics of the collapse transition of the model gel was described in detail [1], and the shrinking times were much faster than any reported previously for unmodified gels. Each of the three structural changes further enhanced the shrinking times, and the extent of improvement was dependent on the final temperature after the temperature jump. Below the transition temperature, T_C, the graft PNIPAM gel showed the least improvement, with the PAM gel showing the most. Above T_C the composite gel showed the least improvement and again the PAM gel had the fastest shrinking times. Thus, in terms of overall speed of collapse, the PAM-modified gel clearly has the fastest shrinking rate

over the entire temperature range studied. The graft PNIPAM and composite gels vied with each other to be fastest, with the composite gel winning at $T < T_C$ and the graft PNIPAM gel shrinking faster at $T > T_C$.

In terms of ease of preparation, the PAM gel and the composite gel both involve an extra step prior to the gelation reaction. The graft PNIPAM gel involves two additional reactions, along with several precipitations, and is therefore more troublesome to prepare.

In several proposed applications of rapid-shrinking gels release of a large portion of the network water is sufficient (attainment of the fully collapsed state is not necessary). Here, it is the amount of water released in the initial burst of shrinking that is important and not the time required to reach the equilibrium size. Having compared

Table 2 Percentage shrinking during the first 15 s of shrinking following the temperature jumps

Gel type	Temperature after temperature jump (°C)					
	34	35	36	37	38	40
Model						
178 μm	10	14	18	20	21	22
212 μm	6	8	12	14	16	15
Graft poly(N-isopropylacrylamide)						
185 μm	10	17	22	26	31	43
260 μm	8	12	18	21	31	39
Poly(acrylamide)						
210 μm	10	18	24	29	37	42
290 μm	7	11	17	28	35	42
Composite						
200 μm	4	7	9	11	18	36
260 μm	5	7	9	12	14	23

the degree of shrinkage in the first 15 s following a temperature jump, the PAM gel is clearly the most efficient at releasing large quantities of water rapidly over the entire temperature range studied, although it actually has the smallest overall degree of shrinking.

No attempt has been made here to explain the shrinking mechanisms in operation in any of these gels.

More work is required in order to quantitatively state the effect of the network structure on the shrinking mechanism.

Acknowledgements This work was funded by INCO-Copernicus grant no. IC15CT96-0756. The authors thank J. C. Jacquier for his careful reading of this manuscript.

References

1. Lynch I, Gorelov AV, Dawson KA (1999) Phys Chem Chem Phys 1:2103
2. Kaneko Y, Sakai K, Kikuchi A, Yoshida R, Sakurai Y, Okano T (1995) Macromol 28:7717
3. Hirotsu S (1998) Jpn J Appl Phys 37:L284
4. Ubbelohde L (1933) J Inst Petr Technol 19:376
5. Matsuo ES, Tanaka T (1988) J Chem Phys 89:1695
6. Yoshida R, Sakai T, Okano T, Sakurai Y (1992) J Biomater Sci Polym Ed 3:243

Progr Colloid Polym Sci (2000) 115:128–133
© Springer-Verlag 2000

M. Casagrande
C. Heldmann
U. Pawelzik
G. Meier
M. Stamm

Influence of composition on the interdiffusion of poly(vinyl acetate) latex particles

M. Casagrande · U. Pawelzik · G. Meier
Max-Planck-Institut für
Polymerforschung, Ackermannweg 10
55128 Mainz
Germany

C. Heldmann
Clariant GmbH, 65926 Frankfurt am
Main, Germany

M. Stamm (✉)
Institut für Polymerforschung Dresden e.V.
Hohe Strasse 6, 01069 Dresden
Germany
e-mail: stamm@ipfdd.de
Tel.: +49-351-4658224
Fax: +49-351-4658281

Abstract The influence of the co-monomer sodium vinyl sulfonate (SVS) on the interdiffusion of poly(vinyl acetate) latex particles during film formation was studied. Poly(vinyl acetate) latices with contents of 0, 0.5, 1.5 and 3 wt% SVS were investigated utilizing small-angle neutron scattering. For each SVS content pairs of identical particles differing only by deuteration were synthesized by emulsion polymerization. The measurements were performed at 55 and 60 °C with samples containing 5 wt% of deuterated and 95 wt% of protonated particles, respectively. The hydrophilic shell formed by SVS and vinyl acetate copolymers at the particle surface hinders interdiffusion partially. The addition of only 0.5 wt% SVS significantly lowers the value of the diffusion coefficient with respect to the SVS-free sample. Higher content of SVS leads to further retardation of the interdiffusion. In conjunction with NMR measurements of comparable samples it is concluded that the decreasing mobility of the hydrophilic surface layer with increasing SVS content is the determining factor for the interdiffusion process.

Key words Interdiffusion · Polymer latex · Film formation · vinyl acetate

Introduction

Films prepared from dispersions of poly(vinyl acetate) (PVAc) latex particles are widely used in paint applications. To stabilize the particles sodium vinyl sulfonate (SVS) is added during synthesis. The vinyl acetate (VAc) and SVS react to form a polyelectrolyte copolymer which forms a hydrophilic layer at the surface of the latex particles and therefore, provides, electrostatic and steric stabilization of the particles in the dispersion [1]. Especially, the stability in high ionic strength environments is very important for the addition of pigments in paints. While the effect of stabilization has already been investigated [1], less is known about the influence of surface-bound acid groups and their salts on interdiffusion during film formation from latices. This article addresses the influence of the sulfonate group at the surface on the interdiffusion of PVAc latices. Since VAc is widely used for latex synthesis, such an investigation is important not only from a scientific point of view but also from a technological perspective in order to understand interdiffusion in films formed from latex particles.

If the hydrophilic polymer is copolymerized on top of the core polymer, the copolymer forms a separate interconnected phase at the surface of the particles during film formation [2]. The hydrophilic membranes which separate the core particles may act as a barrier to prevent or retard polymer interdiffusion between neighboring particles [2]. Kim and Winnik [3, 4] measured the kinetics of polymer diffusion in latex films prepared from differently labeled pairs of poly(butyl methacrylate) latex particles using direct nonradiative energy transfer. They investigated the effect of carboxylic acid groups at the surface of the particles on the interdiffusion rate and showed that there is a progressive decrease in the diffusion coefficient with increasing carboxylic acid group content in the hydrophilic shell. A linear relationship was found between the logarithm of the mean apparent diffusion coefficient and the temperature difference, $T - T_g$, where T is the annealing temperature of the film and T_g is the estimated glass-transition temperature of the ionomer phase in each film.

Experimental

PVAc latices with contents of 0, 0.5, 1.5 and 3 wt% SVS were prepared by emulsion polymerization. For these measurements, we employed pairs of almost identical particles which differed only in their labeling by deuteration. We used commercial deuterated VAc for the polymerization of the deuterated latices. A blend of the latex component containing 100% of the deuterated species and the second latex component with 100% of the protonated species was prepared for each SVS concentration. Extensive screening showed that best results concerning reproducibility of the synthesis were obtained utilizing batch emulsion polymerization. Semicontinuous synthesis [5] leads to a poorer reproducibility of the particle diameter and other material constants.

Synthesis

In a reactor especially constructed for emulsion polymerization, 15 ml degased water, 1.5 g distilled VAc (protonated or deuterated), 15 g 0.5% solution of dodecyl benzyl sulfonate and sodium vinyl sulfonate (0–3 wt%) were heated under vigorous stirring under nitrogen atmosphere. At 70 °C 100 mg of the starter potassium peroxodisulfate was added. The latex mixture was maintained at 72 °C for 2 h and was used without further purification.

Sample preparation

The deuterated latex was mixed with the corresponding protonated latex of the same SVS content. The resulting mixed latices contained 5 wt% of deuterated particles and were freeze-dried in order to remove residual solvent. Subsequently the dry PVAc particle mixtures were sintered in a vacuum hot press at 43 °C for 30 min under a pressure of 10 MPa.

Small-angle neutron scattering measurement and data correction

The small-angle neutron scattering (SANS) experiments were carried out at the Forschungszentrum Jülich, Germany. A sample detector distance of 20 m and a neutron wavelength of 9.5 Å ($\Delta\lambda/\lambda = 0.16$) were used. The data were collected on a two-dimensional position-sensitive detector (64 × 64 channels, 8-mm spatial resolution). The two-dimensional scattering patterns were converted to absolute intensities and were radially averaged to obtain the one-dimensional scattering curves.

Data evaluation

The interdiffusion distance, $d(t)$, of the polymer chains across the particle boundaries was estimated by subtracting the original radius, $R(0)$, of the deuterated latex particles in the dispersed state from the radius, $R(t)$, of the expanded deuterated particles

$$d(t) = R(t) - R(0) \ , \tag{1}$$

where t stands for the annealing time. The radius of the deuterated was calculated from

$$R^2 = (5/3) \, R_g^2 \ , \tag{2}$$

with the radius of gyration, R_g, determined from the slope of the plot $I(Q)^{-1}$ versus Q^2 (Zimm plot). Q is the magnitude of the scattering vector, $Q = 4\pi/\lambda \sin\Theta$, where 2Θ is the scattering angle. For small values of Q ($Q^2 R_g^2 < 1$) the differential scattering cross-section per unit volume, $d\Sigma/d\Omega$, can be written as

$$\left[\frac{d\Sigma}{d\Omega}(Q)\right]^{-1} = \frac{1}{C \, M_w}\left(1 + \frac{Q^2 R_g^2}{3}\right) \ , \tag{3}$$

where the contrast factor, C, is given by

$$C = \frac{N_A}{M_D} \, \rho(1-x)x(b_H - b_D)^2 \ . \tag{4}$$

b_H and b_D are the scattering lengths of protonated and deuterated structural units of the polymer, respectively, N_A is Avogadro's number, ρ stands for the polymer density, x is the mole fraction of deuterated polymer and M_D is the deuterated polymer molecular weight.

Calculating the diffusion coefficient

Summerfield and Ullman [6, 7] derived an equation to obtain the diffusion coefficient from SANS data of polymers composed of randomly mixed particles of protonated and deuterated polymers of the same molecular weight. A more generalized diffusion and scattering theory based on this equation was developed by Eu and Ullman [8]. This generalized equation extends the applicability to polymers with differences in molecular weight or chemical composition and has the following form:

$$S(q,t) = S(Q,0)\exp(-2Q^2 Dt) + S(Q,\infty)$$
$$\times [a(t) - a(0)\exp(-2Q^2 Dt)] \ , \tag{5}$$

where $S(Q,t)$ is the scattering intensity from the interdiffused latex particles at time t, and $S(Q,0)$ and $S(Q,\infty)$ are the scattering intensities of the initial and completely intermixed latices, respectively. D is the diffusion coefficient. In Eq. (5) it is assumed that the interdiffusion is of Fickian nature. $a(t)$ corresponds to the space–time correlation function, which increases with annealing time. If the mixed latices contain a low concentration of deuterated particles, $S(Q,\infty)$ is much smaller than $S(Q,0)$ and Eq. (5) reduces to

$$S(Q,t) = S(Q,0)\exp(-2Q^2 Dt) \ . \tag{6}$$

In the Guinier region, where $QR_g \ll 1$, it follows that $S(Q,0) = S(0,0)\exp[-2QR_g^2(0)/3]$. $R_g(0)$ is the radius of gyration of the initial deuterated particles. Equation 6 can be rewritten as

$$S(Q,t) = S(0,0)\exp\left\{-Q^2\left[R_g^2(0) + 6Dt\right]/3\right\}$$
$$= S(0,0)\exp\left[-Q^2 R_g^2(t)/3\right] \ , \tag{7}$$

leading to a simple relation between the measured radii of gyration and the diffusion coefficient, D,

$$R_g^2(t) = R_g^2(0) + 6Dt \ . \tag{8}$$

Temperature dependence of the diffusion coefficient

The temperature dependence of the diffusion coefficient can be described by the Williams–Landel–Ferry (WLF) equation. The corresponding temperature shift factor, α_τ, may be represented by [12]

$$\log(\alpha_\tau) = \log\left(\frac{D \, T_0}{D_0 \, T}\right) = \frac{C_1(T - T_0)}{C_2 + T - T_0} \ , \tag{9}$$

Fig. 1a, b Zimm plot of the small-angle neutron scattering curves at a temperature of 55 °C. **a** poly (vinyl acetate) (*PVAc*) without sodium vinyl sulfonate (*SVS*) and **b** PVAc with a content of 0.5 wt% SVS.

a)

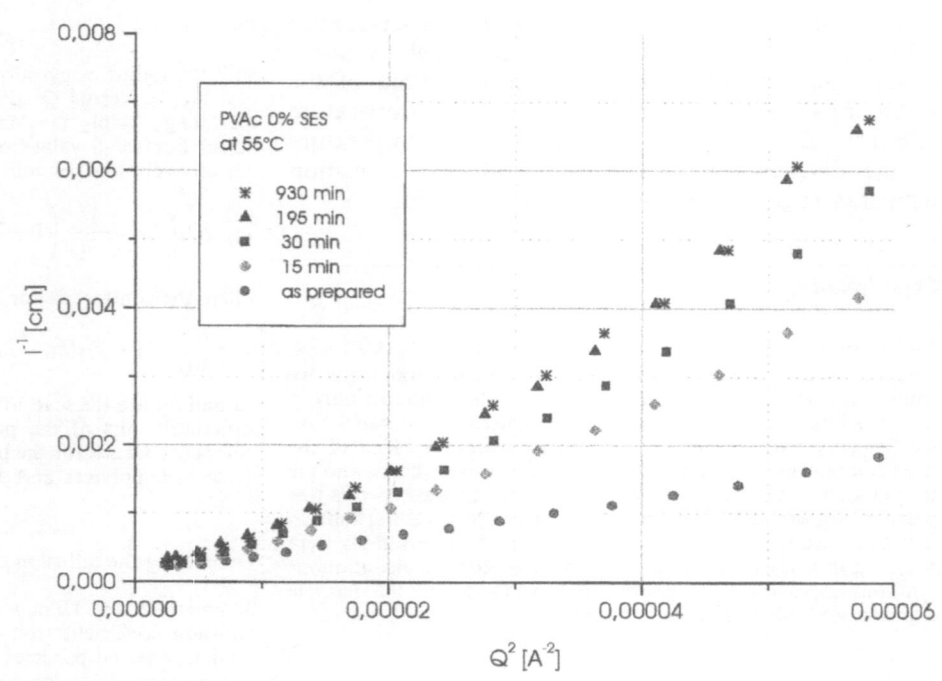

b)

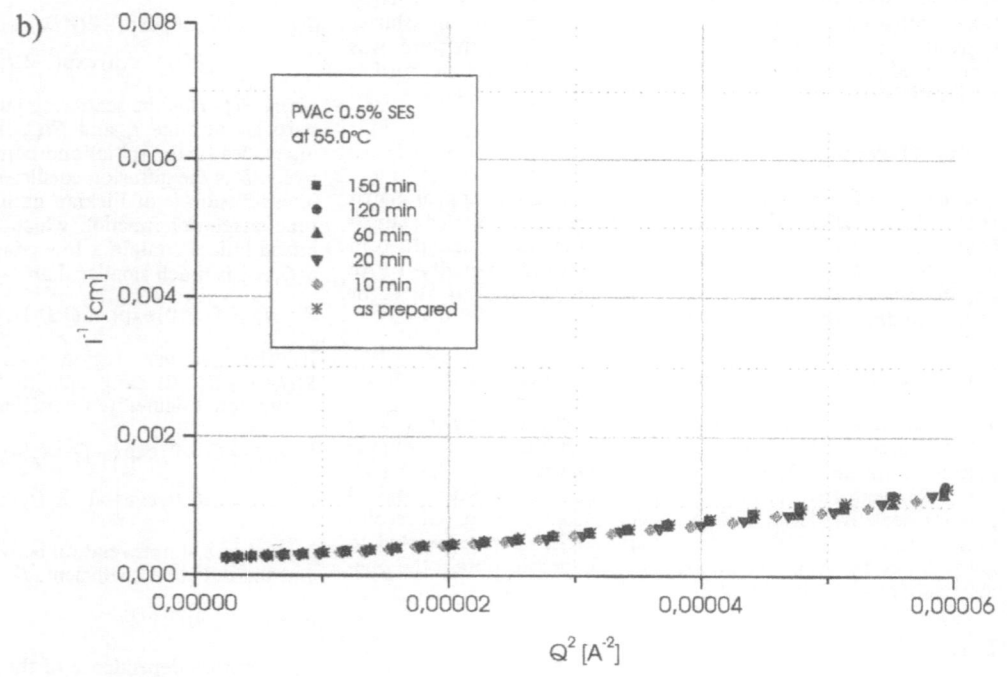

where T_0 is a reference temperature, D_0 is the diffusion coefficient at this temperature and C_1 and C_2 are parameters which depend on the choice of T_0. The parameters were determined by dielectric spectroscopy to be $C_1 = 4.34$ K and $C_2 = 89.0$ K, corresponding to a shift factor of 1.813 at a temperature $T_0 = 60$ °C.

Results and discussion

From SANS experiments of blends of deuterated and protonated latex particles during film formation Zimm

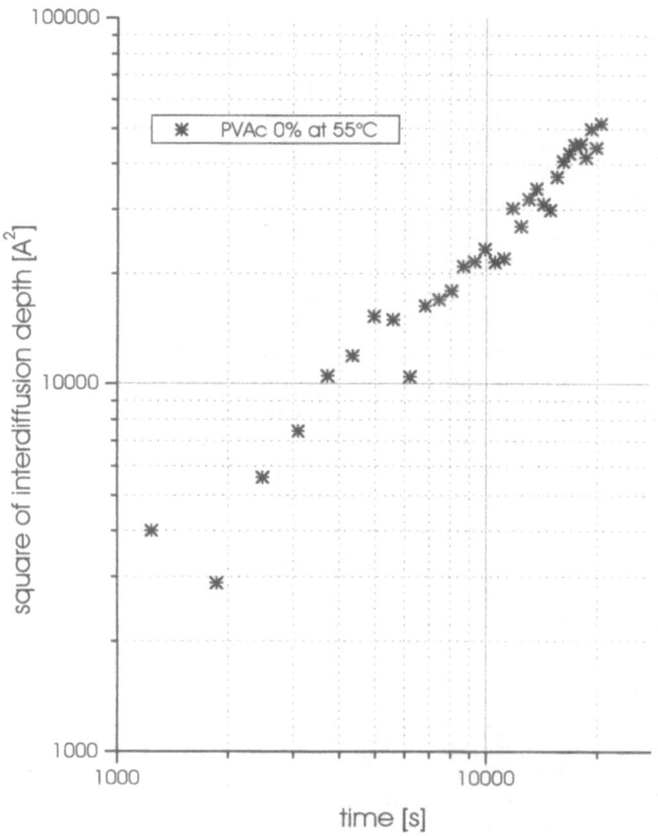

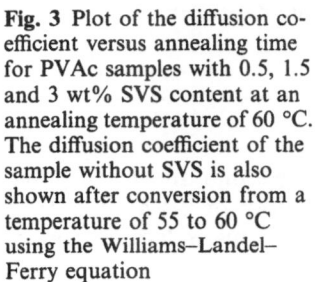

Fig. 2 Log–log plot of the square of the interdiffusion depth versus annealing time of the PVAc sample without SVS measured at an interdiffusion temperature of 55 °C

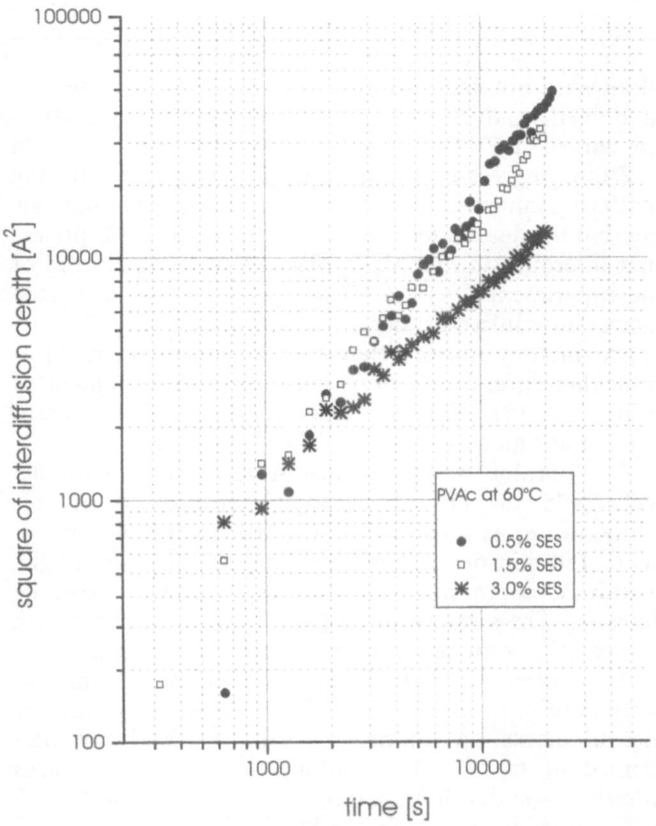

Fig. 4 Log–log plot of the square of the interdiffusion depth versus annealing time t for PVAc samples with 0.5, 1.5 and 3 wt% SVS, respectively, at an interdiffusion temperature of 60 °C

Fig. 3 Plot of the diffusion coefficient versus annealing time for PVAc samples with 0.5, 1.5 and 3 wt% SVS content at an annealing temperature of 60 °C. The diffusion coefficient of the sample without SVS is also shown after conversion from a temperature of 55 to 60 °C using the Williams–Landel–Ferry equation

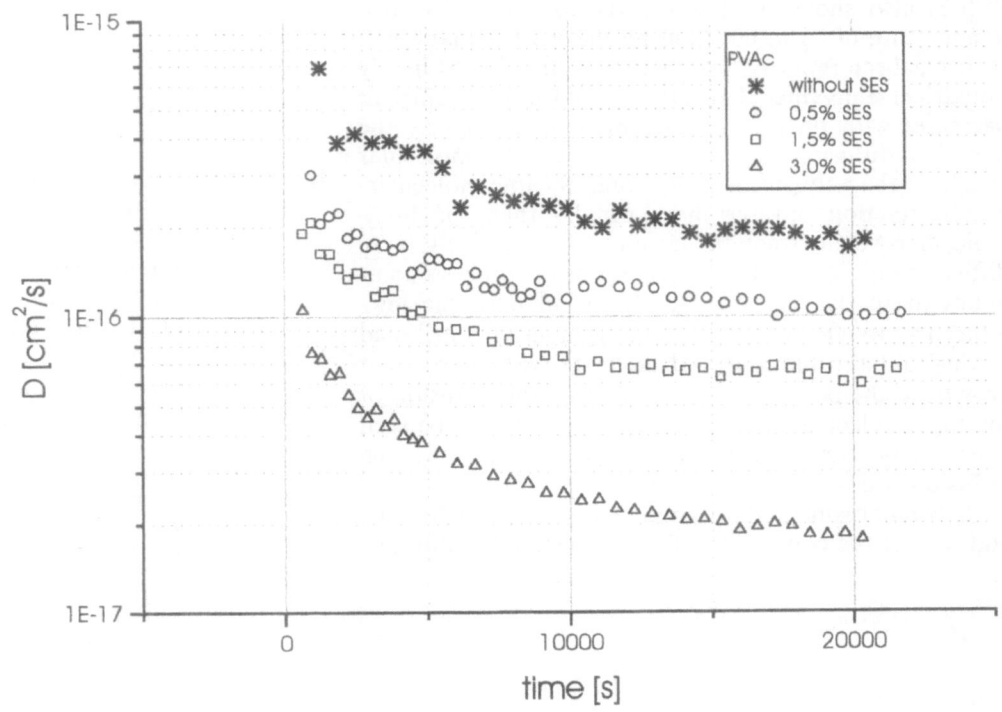

plots were obtained to determine the R_g of the deuterated particles during the annealing process. The scattering curves of PVAc particles without SVS are shown in a Zimm representation in Fig. 1a. The slope of the scattering curves increases with annealing time, indicating that the deuterated particles grow in size. This proves that interdiffusion is taking place at the temperature of the experiment of 55 °C. The square of the calculated mean interdiffusion depth is shown in Fig. 2. The approximately linear behavior indicates Fickian diffusion; therefore, a diffusion coefficient can be calculated using Eq. (8). The diffusion coefficient at different annealing times is plotted in Fig. 4.

Comparable measurements of the sample with 0.5 wt% SVS do not show any interdiffusion at this temperature, as indicated in Fig. 1b. In the time range up to 150 min no significant change in the slope of the scattering curve in the Zimm representation can be detected. The same result holds for all the samples with higher SVS content regardless of the exact amount.

The onset of interdiffusion is observed only at temperatures at or above 60 °C. For this reason further measurements on samples containing SVS were performed at 60 °C. The resulting square of the mean interdiffusion depth for samples with 0.5, 1.5 and 3 wt% SVS, respectively, are shown in Fig. 3.

In all cases the samples exhibit a Fickian-like time dependence. The decrease in the slope of the scattering curves indicates that diffusion is retarded with increasing SVS content. The diffusion coefficients of the three samples with different SVS content at 60 °C are shown in Fig. 4. The diffusion coefficient of the sample without SVS is also shown in Fig. 4 after conversion from a temperature of 55 to 60 °C using the WLF equation.

It has been reported for particles consisting of highly polydisperse polymer chains that the diffusion coefficient decreases with time [3, 8–11]. This is explained by the fact that diffusion depends strongly on the molecular weight of the polymer. At short times the low-molecular-weight fraction diffuses more rapidly than the high-molecular-weight fraction, leading to a relatively large diffusion coefficient. This rapid diffusion process dominates the early stages of interdiffusion until the low-molecular-weight fraction is homogeneously distributed throughout the film [10]. At later stages the apparent growth of the particles is dominated by the diffusion of the high-molecular-weight components. As can be seen from the measurements this leads to a net decrease in the diffusion coefficient.

A comparison of the measurements proves that the addition of the comonomer SVS influences the interdiffusion significantly and confirms the first measurement indicating that even small amounts of SVS retard or suppress diffusion. After 10 000 s, D is of the order 2.3×10^{-16} cm^2/s for the sample without SVS. The addition of only 0.5 wt% SVS almost halves the value of D to 1.2×10^{-16} cm^2/s.

As shown by Rottstegge [1] the copolymer of PVAc and SVS forms a hydrophilic-membrane-like layer at the surface of the particles. Obviously this hydrophilic membrane acts as a barrier which does not completely suppress, but retards, the interdiffusion. Further addition of SVS leads to a successive decrease in the diffusion coefficient. At 3 wt% SVS the coefficient finally decreased to one tenth of the original value ($D = 2.6 \times 10^{-17}$ cm^2/s). Solid-state NMR measurements by Rottstegge [1] showed convincingly that the mobility of the hydrophilic membrane in the dry state changes with SVS content. Whereas a flexible shell is observed at a content of 0.5 wt% SVS, at a significantly higher content a quite rigid shell is found. This effect is in good agreement with our diffusion experiments. The decrease in mobility leads to increasing hindrance of diffusion and explains the retardation of the observed interdiffusion.

The investigations of Kim and Winnik [3, 4] point in the same direction. In these experiments poly (butyl methacrylate) latex particles were coated with a hydrophobic methylacrylate–butylacrylate copolymer. It was found that the diffusion coefficient of these core–shell latices decreased with increasing methacrylic acid concentration in the shell. It was concluded that the T_g of the hydrophilic shell determines the interdiffusion rate. The two effects lead together to increasing acid content, leading to a decrease in the interdiffusion rate.

Conclusion

In summary it can be concluded that the hydrophilic shell formed during the reaction of SVS and VAc copolymer hinders interdiffusion between latex particles, but does not completely suppress it. The addition of only 3 wt% SVS is sufficient to reduce the diffusion coefficient by a factor of 10 with respect to the SVS-free sample. In connection with NMR measurements it can be concluded that the mobility of the hydrophilic layer determines the interdiffusion process.

Acknowledgements The authors acknowledge the help of M. Wilhelm during polymer synthesis. The financial support of BMBF, grant no. FKZ 03N3018A5, and BMBF 03EW5HPZ is also gratefully acknowledged.

References

1. Rottstegge J (1998) PhD thesis Johannes-Gutenberg-Universität, Mainz, Germany
2. Joanicot M, Wong K, Cabane B (1996) Macromolecules 29:4976
3. Kim HB, Winnik MA (1994) Macromolecules 27:1007
4. Kim HB, Winnik MA (1995) Macromolecules 28:2033
5. Fischer JP, Nölken E (1988) Prog Colloid Polym Sci 77:180
6. Summerfield GC, Ullman R (1987) Macromolecules 20:401
7. Summerfield GC, Ullman R (1988) Macromolecules 21:2643
8. Eu MD, Ullman R (1995) Polym Mater Sci Eng 73:10
9. Hahn K, Schuller H, Oberthür R (1988) Colloid Polym Sci 264:1092
10. Hahn K, Schuller H, Oberthür R (1988) Colloid Polym Sci 266:631
11. Wang Y, Winnik MA (1993) J Phys Chem 97:2507
12. Nemoto N, Landry MR, Noh I, Yu H (1984) Polym Commun 25:151

Progr Colloid Polym Sci (2000) 115:134–136
© Springer-Verlag 2000

A. Fernández-Nieves
A. Fernández-Barbero
B. Vincent
F. J. de las Nieves

Deswelling and depletion flocculation of microgel particles under external osmotic pressure

A. Fernández-Nieves
A. Fernández-Barbero
F. J. de la Nieves
Group of Complex Fluids Physics
Department of Applied Physics
University of Almería
04120 Almería, Spain
e-mail: fjnieves@filabres.ualm.es
Tel.: +34-50-215434
Fax: +34-50-215434

B. Vincent
School of Chemistry, University of Bristol
Cantock's Close, Bristol BS8 1TS, UK

Abstract Microgels are novel colloidal gels widely used because of the swelling or deswelling exhibited under specific external conditions. In this work, the effect of added free dextrans on the deswelling of microgel particles has been studied experimentally using photon correlation spectroscopy. Additionally, the influence of the dextran solution over the colloidal stability of the microgel particles has also been studied by turbidimetry. The results show that the extent of deswelling increases with an increase in the dextran concentration in the range where flocculation by depletion is absent. Finally, depletion flocculation is detected.

Key words Microgel · Flocculation · Depletion · Swelling · Dextran

Introduction

The study of microgels is attracting growing interest because of their fast response to external stimuli. The union of gel and colloidal characteristics is the main reason for such extensive attention.

In this work, the osmotic deswelling of microgel particles in the presence of free dextrans and the further flocculation of the colloids are presented, as a clear simultaneous manifestation of the gel and colloid features of a microgel. Dextrans were used as stressing polymers, basically because the osmotic pressure obtained is independent of ionic concentration and temperature [1]. The reaction of the microgel particles to the external osmotic stress was followed by photon correlation spectroscopy (PCS) and turbidimetry. PCS yielded the average diffusion coefficient, and thus the particle mean size, for a given dextran concentration. The turbidimetric technique allowed the determination of the specific turbidity wavelength exponent, that is used for detecting flocculation. The deswelling results are compatible with an osmotic mechanism, which depends on the competition between electrostatics and external stress.

Experimental

Experimental techniques

The viscosity of the dextran solutions was determined with a Schott-Gerate apparatus, using a Ubbelohde capillary in a thermostatic bath (with refrigeration and agitation) assuring constant temperature.

The average hydrodynamic diameter of the microgel particles in the presence of dextrans was determined by PCS (Zetamaster- S, Malvern Instruments). Samples were prepared at a concentration of 5×10^9 cm^{-3}.

Optical density spectra were recorded over the wavelength range 400–650 nm using a Spectronic Genesys 5 spectrophotometer (Milton Roy, USA). The particle concentration for these experiments was set to 5×10^{10} cm^{-3}.

The PCS and turbidity-wavelength measurements of the microgel/free-dextran systems were made after an equilibration time of 90 min. This time was found to be enough for osmotic equilibrium to be reached (similar equilibration times are also found for other microgel/free-polymer systems [2]).

All particle concentrations employed were chosen in the range where multiple scattering is absent. The temperature and salt concentration in all experiments were set to (25.0 ± 0.1) °C and 1 mM (NaCl), respectively.

Microgel particles

The microgel particles consisted of poly(2-vinylpyridine) (2VP) cross-linked with divinylbenzene (0.25 wt%). The initiator used in the synthesis was 2,2′-azobis(2-amidinopropane) dihydrochloride (Wako) [3]. Transmission electron microscopy showed the particles to be spherical and highly monodisperse, with a diameter of (205 ± 8) nm. The microgel size was shown to depend on pH because of the ionization of the 2VP groups. This feature is employed in this work for setting the initial swollen state of the microgel particles to about 1200 nm.

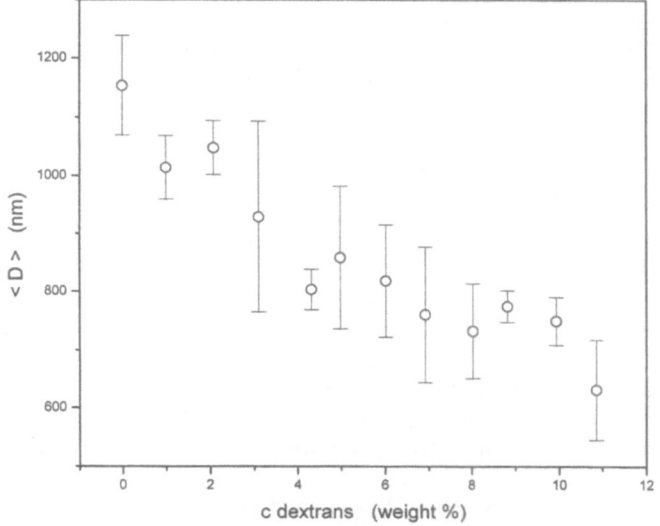

Fig. 1 Deswelling of microgel particles by addition of dextrans

Fig. 2 Wavelength exponent variation with dextran concentration. The specific turbidity is also plotted as a function of the wavelength for different dextran concentrations (*inset*)

Stressing polymer

Dextrans $(C_6H_{10}O_5)_n$ from Fluka were used as stressing polymers. They were derived by enzymatic hydrolysis of polysaccharides and show a common molecular structure with intermolecular hydrogen bonding giving rise to linearity (1.6 linkages) and branching (1.3 and 1.4 linkages) of the molecule. The number-average molar mass, M_n was 70000 g/mol.

Viscosity measurements of dextran solutions are necessary for a proper determination of the microgel hydrodynamic diameter. The results are summarized in a usual second-order polynomial relationship [4].

$$\eta = A + Bc + Cc^2 , \qquad (1)$$

with $A = (0.932 ± 0.024)$cP, $B = (0.119 ± 0.016)$cP and $C = (0.0399 ± 0.0016)$cP; c is the polymer concentration in the range $c \in (0.1, 11)$wt%.

The dextran hydrodynamic diameter was obtained by PCS to be about 25 nm. This value is similar to the end-to-end distance for the polymer, which commonly follows the very simple relation $\langle r^2 \rangle^{\frac{1}{2}} = 0.06 M_n$ [5]. Moreover, the influence of the dextran molecules over the total mean scattered intensity and intensity autocorrelation function was observed to be negligable.

Results and discussion

Osmotic deswelling

The strain of the microgel particles as a function of the bulk dextran concentration is shown in Fig. 1. As can be

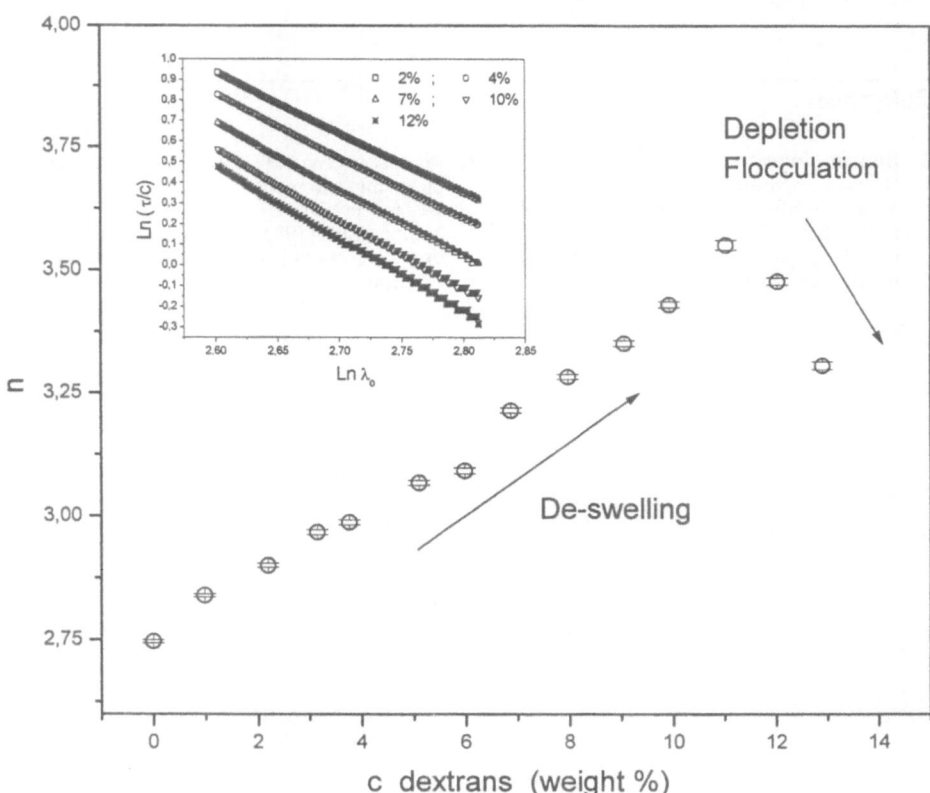

observed, the presence of dextran molecules induces deswelling of the microgel particles. The extent of deswelling becomes more pronounced with increasing polymer concentration. This trend is in accordance with an osmotic deswelling mechanism, consisting of the competition between the electrical osmotic term (due the presence of ionized 2VP groups within the particle), which dominates at high swelling, and the external polymer osmotic pressure. This behaviour is also in agreement with previous studies on nonionic microgel systems [7].

Detection of flocculation

The specific turbidity for nonabsorbing spherical particles is given by the simple formula [6]

$$\frac{\tau}{c} = k\lambda_0^{-n} \ , \tag{2}$$

where τ is the turbidity at concentration c and λ_0 is the vacuum wavelength. k depends on particle size and relative refractive index $(m = \frac{n_{particle}}{n_{medium}})$. For Rayleigh scattering the wavelength exponent, n, reaches a maximum value of 4. When the particle size increases, this approximation fails and n decreases. Therefore, the wavelength exponent is a measure of the particle size. The value of n may then be simply obtained from the slope of a $\ln \frac{\tau}{c}$ versus $\ln \lambda_0$ plot.

The variation of the specific turbidity with wavelength for the microgel particles at different dextran concentrations is shown in Fig. 2 (inset). All curves decrease with wavelength as predicted by Eq. (2), whose slope allows n to be determined. The value of the wavelength exponent is also plotted as a function of the dextran concentration. At low polymer content, the particle is swollen and the scattering is far from the Rayleigh regime. As the dextran concentration increases, the particles deswell (Fig. 1) and n increases. Finally, the exponent falls, indicating an increase in the mean particle size, which could be associated with depletion flocculation, as observed with other nonionic microgel systems [8].

Conclusions

We have observed the deswelling of microgel particles in the presence of free stressing polymer. The experimental trend is explained in terms of competing osmotic forces. Additionally, depletion flocculation has been detected at high free-polymer concentration.

Acknowledgements The authors would like to thank María José García Salinas for help with the viscosity measurements. The financial support under project MAT96-1035-C03-03 and Acción Integrada Hispano-Británica HB 1998-0225 is also greatly acknowledged.

References

1. Bonnet-Gonnet C, Belloni L, Cabane B (1994) Langmuir 10:4012
2. Saunders BR, Vincent B (1997) Colloid Polym Sci 275:9
3. Loxley A, Vincent B (1997) Colloid Polym Sci 275:1108
4. Macosko CW (1994) Rheology. Principles, measurements and applications. VCH, New York
5. Napper DH (1983) Polymeric stabilization of colloidal dispersions. Academic London
6. Heller W, Bhatnagar HL, Nakagaki M (1962) J Chem Phys 36:1163
7. Saunders BR, Vincent B (1996) J Chem Soc Faraday Trans 92:3385
8. Clarke J, Vincent B (1981) J Chem Soc Faraday Trans I 77:1831

Progr Colloid Polym Sci (2000) 115:137–140
© Springer-Verlag 2000

P. Bartlett

The effect of polydispersity on colloidal phase transitions

P. Bartlett
Department of Chemistry
University of Bath
Bath BA2 7AY, UK

Abstract A polydisperse hard-sphere mixture contains particles with a continuous ranges of diameters. It is shown that the observed phase behaviour depends sensitively upon the kinetics of size fractionation. Two limiting situations are considered. In the quenched case, where the size distribution is fixed, the fluid–crystal transition is reentrant at low polydispersities and disappears altogether at high polydispersities. If size partitioning is allowed it is shown that the polydisperse crystal is spinodally unstable to small fluctuations in polydispersity. The direction of the instability suggests that equilibrium corresponds to a set of multiple crystal phases, each containing spheres with a narrow range of diameters.

Key words Colloids · Polydispersity · Hard spheres · Phase transitions

Introduction

A polydisperse mixture of hard spheres with a continuous range of sizes is a simple model of many colloidal suspensions. A well-known example [1] is poly(methyl methacrylate) (PMMA) stabilised by a thin polymeric layer of poly(12-hydroxystearic acid) and dispersed in a good solvent such as decalin. The colloidal particles in this suspension interact via a short-range repulsion, which is well represented by a hard-sphere interaction. While many of the properties of uniform hard spheres have been known for at least 30 years our understanding of polydisperse mixtures is much less developed. Most of the theoretical effort to date has concentrated on exploring the consequences of polydispersity for scattering [2]. In contrast, virtually no theoretical work on polydisperse phase diagrams has been reported (until recently [3–9]) despite the fact that many industrial products are frequently very polydisperse. It is, however, clear that as soon as a suspension is allowed to enjoy a significant degree of polydispersity several interesting new phenomena arise. First, increasing polydispersity can suppress certain phase transitions.

For instance, the fluid–crystal phase transitions reported in dispersions of latex particles of PMMA are only found if the particles have a narrow range of sizes. Pusey reported [1] that while PMMA spheres with a polydispersity, σ, of about 0.075 displayed a fluid–crystal transition similar to that reported for identically sized hard spheres, on increasing σ to about 0.12 no crystallites were found even after several months of observation. Second, those transitions which still remain in polydisperse systems are frequently accompanied by a fractionation of the particle size distribution between coexisting phases [7]. Finally, polydispersity can induce new transitions, not found for monodisperse systems [8].

Despite both the practical importance and the potential richness of polydisperse phase behaviour, the mathematical complexity of treating a mixture with essentially an infinite number of components makes a first-principles determination of a polydisperse phase diagram a formidable task. We examine a simple model for a system of polydisperse hard spheres with the aim of developing some generic insight into the thermodynamics of polydisperse transitions. We calculate the phase behaviour in two limits: the quenched limit, where the size distribution in each phase is fixed, and the corresponding annealed limit, where particles redistribute so as to minimise the total free energy. The physical significance of these two limits is clear if we remember that in a dense colloidal suspension collective diffusion is appreciably faster than self-diffusion [1], so fractionation will be slow. This difference in time scales implies that freezing could occur in two stages in a polydisperse suspension. First, a metastable suspension will relax its density rapidly to equilibrium by growing a crystal without fractionation (quenched behaviour). Then, over much longer times, self-diffusion will occur and the size distributions will relax towards the completely annealed state [9]. We find very different results for the phase behaviour in these two limits. In the quenched situation, the polydisperse crystal is stable at low polydispersities with the fluid–crystal transition only vanishing at polydispersities above a certain terminal level, σ_t. In contrast, in the annealed limit, the polydisperse crystal is thermodynamically unstable and separates into several solid phases, the number of which grows without limit as the polydispersity increases.

The model

The size of each particle in a polydisperse mixture can take any one of a set of essentially continuous values distributed according to a distribution function $x(R)$. The number density of spheres with diameter R is then $\rho x(R)dR$, where ρ is the total number density. Since we do not expect our final results to depend significantly on the exact form of $x(R)$ we choose a specific function for definiteness. The generalised-exponential distribution is widely used to describe suspensions and is defined by

$$x(R) = \frac{(\sigma^{-2})^{\sigma^{-2}}}{\Gamma(\sigma^{-2})} \left(\frac{R}{\bar{R}}\right)^{\sigma^{-2}-1} \exp\left(-\sigma^{-2}\frac{R}{\bar{R}}\right) , \quad (1)$$

where $\bar{R}$ is the mean diameter and σ is the standard deviation in units of $\bar{R}$.

The polydisperse phase equilibrium is calculated using the finite-moment approximation introduced by Sollich and Cates [3] and Warren [4]. The justification for this approximation is the recognition that the excess free energy in a polydisperse hard-sphere mixture depends only on a limited set of moment densities, quantities such as $\phi_n = \rho m_n$, where the nth moment m_n is defined by $\int R^n x(R)dR$, rather than by the explicit form of $x(R)$. The central idea is to treat the ϕ_n as independent thermodynamic density variables. Since the moment variables are simply linear combinations of species densities they acquire many of the properties of conventional particle densities. So, for instance, at equilibrium the moment chemical potentials (defined by analogy to the particle potentials), $\mu_n = \partial f/\partial \phi_n$ $(f = F/V)$, are equal in all coexisting phases. The accurate equation of state (EOS) suggested by Boublik and Mansoori et al. [10] is used for the polydisperse fluid, while the crystal is approximated by a recently introduced EOS [11].

Quenched phase behaviour

In the quenched limit there is no size fractionation. The polydisperse system now behaves essentially as a one-component system with effective properties fixed by the polydispersity. The free energies of the polydisperse fluid and the crystal phases are formally only a function of ρ since m_n are fixed at their initial values. Phase boundaries are located by equating P and $\mu_\rho = \partial f/\partial \rho$. In the monodisperse limit we recover the expected transition from a fluid phase at a volume fraction of $\eta = (\pi/6)\phi_3 = 0.49$ to a crystalline phase at the higher volume fraction $\eta = 0.55$. This is the only transition in the monodisperse limit. Figure 1a shows how the phase boundaries shift with increasing polydispersity. The first effect increasing σ has is to progressively narrow the fluid–crystal coexistence region until it eventually vanishes at the

Fig. 1a, b Quenched phase behaviour for polydisperse hard spheres. **a** Phase boundaries in the polydispersity (σ)–volume fraction (η) plane. The miscibility gap vanishes at the azeotrope, marked by the *filled circle*. **b** The Gibbs-free-energy difference per particle, $\Delta g = g_s - g_f$, as a function of the dimensionless pressure. The *circles* mark the fluid–crystal transitions and the *filled circle* marks the position of the azeotrope

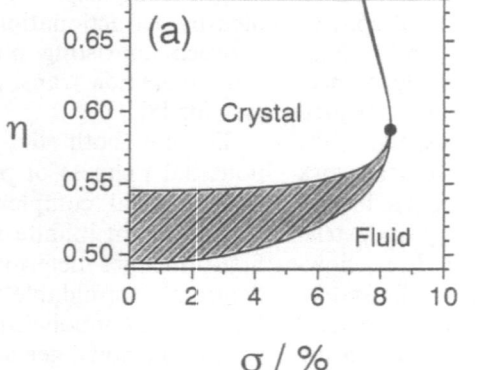

terminal point $\sigma_t = 0.083$ and $\eta_t = 0.59$. At high poly-dispersities, the free-energy difference between the crystal and fluid phases is a nonmonotonic function of the density or equivalently of the pressure (Fig. 1b). This indicates that at high densities there is an additional transition in the polydisperse system from the crystal back to a disordered phase. The range of densities over which the crystal is stable shrinks with increasing polydispersity until, at the terminal polydispersity, it has disappeared completely from the equilibrium phase diagram [5].

Annealed phase behaviour

The phase diagram depicted in Fig. 1a describes quenched freezing where m_n are fixed at their initial values. A natural question is what happens if we relax this constraint and allow particles to redistribute? This is more difficult to answer on two counts: first, because we have to deal with an effective four-component free energy $f(\rho, \phi_1, \phi_2, \phi_3)$ and second, the expression derived by Warren [4] for the reduced entropy, s, is intractable analytically for two or more moments. Here, we do not attempt to derive the equilibrium phase diagram from first principles. Instead, we investigate the stability of the polydisperse crystal phase to small fluctuations in moment densities. We find that a polydisperse crystal is always stable against fluctuations in the total number density but, above a certain density, is unstable against polydispersity fluctuations. This instability suggests that the equilibrium annealed state is one in which the broad initial diameter distribution is split into several narrower fractions. We confirm this hypothesis by comparing the free energies of polydisperse fluid, crystal and multiply fractionated solid phases and find that multiple crystal phases are stable over large regions of parameter space.

The criterion for stability of the polydisperse crystal is the standard one that the matrix of second partial derivatives of f with respect to the moment densities (including $\phi_0 = \rho$ amongst these) should be positive definite. The plane in the moment space where the determinant $|\partial^2 f / \partial \phi_i \partial \phi_j| = 0$ defines the position of the mean-field spinodal. Generally, the determinant is always positive except at high densities and polydispersities. The numerically determined spinodal with the instability region increasing as the degree of polydispersity increases is shown in Fig. 2. The origin of the instability is revealed by the direction in moment space along which the fluctuations diverge as the spinodal plane is crossed. The instability direction is defined by the eigenvector of the matrix $\partial^2 f / \partial \phi_i \partial \phi_j$ whose eigenvalue vanishes at the spinodal. The arrows on the spinodal line in Fig. 2 indicate the direction of the rapidly growing fluctuations, projected into the (σ, η)

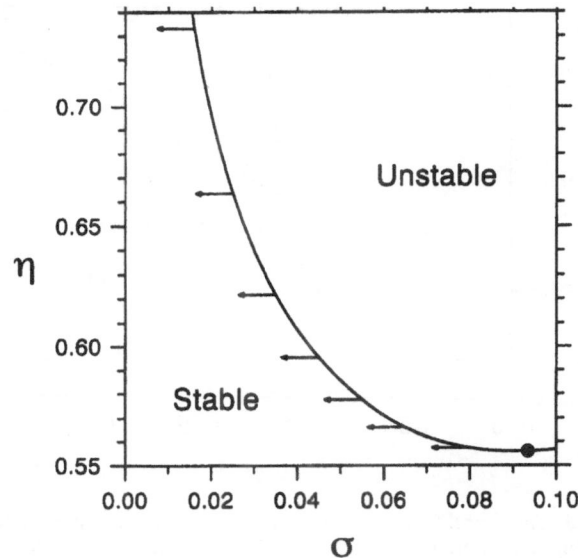

Fig. 2 The annealed spinodal (*solid line*) and critical point (*filled circle*) of the polydisperse hard-sphere crystal in the σ–η plane. The *arrows* indicate the spinodal instability direction

plane. The arrows are almost parallel to the σ-axis so the system is unstable towards a composition fluctuation in which σ and not ρ changes. The physical interpretation of this instability is straightforward. A close-packed crystal of monodisperse hard spheres has a maximum density of $\eta_{cp} \sim 0.74$ at which each sphere contacts its 12 equally sized nearest neighbours. In a polydisperse crystal, by contrast, there is a finite chance that one of the neighbouring spheres will be larger than the mean and these two spheres will then touch at a density $\eta < \eta_{cp}$. Consequently, increased polydispersity lowers the packing efficiency of the crystal. Compressing a polydisperse crystal results in phase separation since at some density the reduction in excess free energy as fractionation occurs will exceed the resulting loss of entropy of mixing.

The physical picture that emerges is that a crystalline lattice can only accomodate spheres with a narrow range of sizes, the width of the sizes being determined by the density. At high densities not all the differently sized spheres can crystallise into a single solid phase. In order to crystallise they must first fractionate before crystallising individually into separate solid phases each containing spheres of a different size. To confirm this picture we compare the free energies at each point in the (σ, η) plane of the polydisperse fluid, the unfractionated crystal (of polydispersity σ) and m coexisting solid phases (each with a polydispersity of σ/m). The resulting stability diagram is shown in Fig. 3. It is clear that there are large regions of parameter space where fractionated crystal phases are stable.

140

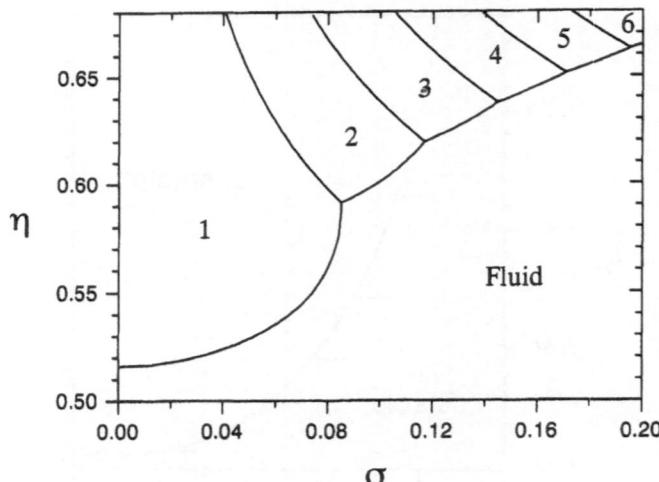

Fig. 3 The stability boundaries of the polydisperse hard-sphere system. The *caption* indicates the number of multiply fractionated crystal phases stable at each η and σ

Conclusions

We have shown that the effect of small levels of polydispersity is to narrow the width of the region of fluid–crystal coexistence. At higher levels of polydispersity, however, the phase behaviour predicted depends sensitively on the time scale for fractionation. If the kinetics of fractionation is slow compared with crystal nucleation and growth, as we conjecture should be the case for dense suspensions, then the fluid–crystal transition is reentrant. Above some density the polydisperse crystal melts back to an amorphous phase and the range of densities over which the polydisperse crystal is stable shrinks rapidly as one approaches a critical level of polydispersity. If there is either no significant separation between the time scales for the relaxation of the total density and the polydispersity or we are interested in the long-time properties then we expect very different behaviour. In this case the polydisperse hard-sphere crystal is thermodynamically unstable with a tendency to separate into multiple crystal phases, each containing spheres of a different size. Finally we comment on the relevance of our predictions to experiment. Although it is not clear yet from the limited experimental data available which (if either) of these two extreme limits is appropriate to the freezing of real colloidal systems our key result is the demonstration that the kinetic behaviour of polydisperse systems could prove to be very rich. Its elucidation remains a challenge to experimentalists.

Acknowledgements It is a pleasure to acknowledge many discussions with Patrick Warren, Wilson Poon, Peter Sollich and Mike Cates.

References

1. Pusey PN (1991) In: Hansen JP, Levesque D, Zinn-Justin J (eds) Liquids, freezing and glass transition. North Holland, Amsterdam, pp 763–942
2. Salgi P, Rajagopalan R (1993) Adv Colloid Interface Sci 43:169
3. Sollich P, Cates ME (1998) Phys Rev Lett 80:1365
4. Warren PB (1998) Phys Rev Lett 80:1369
5. Bartlett P, Warren PB (1999) Phys Rev Lett 82:1979
6. Bartlett P (1998) J Chem Phys 109:10970
7. Evans RML, Fairhurst DJ, Poon WCK (1998) Phys Rev Lett 81:1326
8. van der Kooji FM, Lekkerkerker HNW (2000) Phys Rev Lett 84:781
9. Warren PB (1999) Phys Chem Chem Phys 1:2197
10. Boublik T (1970) J Chem Phys 53:471; Mansoori GA, Carnaham NF, Sterling KE, Leland TW (1971) J Chem Phys 54:1523
11. Bartlett P (1999) Mol Phys 97:685

Progr Colloid Polym Sci (2000) 115:141–145
© Springer-Verlag 2000

C. Pathmamanoharan
N. L. Zuiverloon
A. P. Philipse

Controlled (seeded) growth of monodisperse sterically stabilised magnetic iron colloids

C. Pathmamanoharan · N. L. Zuiverloon
A. P. Philipse (✉)
Van't Hoff Laboratory for Physical
and Colloid Chemistry, Debye Institute
Utrecht University, Padualaan 8
3584 CH Utrecht, The Netherlands
e-mail: a.p.philipse@chem.uu.nl
Tel.: +31-30-2533518
Fax: +31-30-2533870

Abstract Modified polyisobutene and oleic acid appear to be very effective stabilisers for iron colloids formed by thermolysis of iron pentacarbonyl in decalin. The magnetic colloids are fairly monodisperse even for particle radii below 10 nm. The particle size can be increased by seeded growth, and the particle shape can be changed by using a mercaptan stabiliser, which leads to rodlike iron colloids.

Key words Iron ferrofluids · Iron colloids · Magnetic iron nanoparticles

Introduction

To obtain iron ferrofluids [1–3] several methods have been described for the synthesis of zero-valent iron particles [4–13]. Sonolysis of $Fe(CO)_5$ produces amorphous, agglomerated iron particles [9]. Thermal decomposition [4, 5] of $Fe(CO)_5$ generally produces fairly well defined iron colloids, as was also reported earlier in Russian literature (see Ref. 3 and references therein). Such well-defined metal colloids (or "metal nanocrystals") are of interest, for example, to investigate ordered structures of magnetic particles, such as the "super lattices" of cobalt particles in a study by Sun and Murray [14].

Recently, we investigated the thermolysis of dicobalt octacarbonyl in toluene [15], in particular with respect to the use of modified polyisobutene as a stabiliser for cobalt colloids. As a follow-up we report here on polyisobutene (and other stabilisers) for the synthesis of iron nanoparticles with a well-defined size. Metallic iron colloids can be prepared by evaporation of metal iron in a rotating chamber containing modified polyisobutune and alkyl napthalene as solvent [6]. The thermal decomposition of $Fe(CO)_5$ in the presence of a stabiliser, however, is more convenient and the reaction conditions, such as carbonyl and polymer concentration, temperature, and solvent composition, can be easily varied.

Oleic acid is widely used as a stabiliser for iron oxide particles in nonpolar carrier fluids because the kink in the carbon chain prevents the overlap of the chains, thereby stabilising the magnetic particles [16]. We also investigated whether oleic acid is a suitable stabiliser for iron particles, again formed by thermal decomposition of $Fe(CO)_5$. It turns out that both polyisobutene and oleic acid lead to iron particles which are superparamagnetic and which have a narrow size distribution. (Oleic acid coated particles are fairly monodisperse even for particle radii of just a few nanometres).

Some explorative experiments were performed using 3-mercaptopropyltrimethoxysilane (MPS) as a stabilising molecule, since we expected that the mercapto group would adhere strongly to the iron particles. The thermal decomposition of the organometallic precursor $Fe(CO)_5$ in the presence of MPS yields iron particles with a rod shape, which seem not to have been reported earlier.

Experimental

Materials

$Fe(CO)_5$ (97% purity or greater Fluka), decalin (mixture of cis and trans isomers, Merck), oleic acid (Acros), octadecylmercaptan (98% Acros) and MPS (97%, Fluka) were used as received. Modified polyisobutene (code SAP 285) was provided by Shell Research UK. Modification of polyisobutene with polyamine is reported elsewhere [15].

Preparation

The iron nanoparticles were prepared by the thermal decomposition of $Fe(CO)_5$ in dilute solutions containing surface-active agents or polymers. The solvent used was decalin (b.p. 189–191 °C), which allows the decomposition of $Fe(CO)_5$ at elevated temperatures. The thermolysis of $Fe(CO)_5$ was carried out at about 150 °C using an oil bath under an atmosphere of dry nitrogen in a stirred glass flask and the system was left for 24 h. The resulting black suspension of stabilised metal particles in decalin was kept under a nitrogen atmosphere in a Schlenk flask. The details of the apparatus were described earlier [13]. The magnetic fluids were stored (under nitrogen) for long periods as a stable disperse system, without any sign of aggregation or sedimentation of particles due to gravity. The magnetisation of suspensions of iron particles as well as dried particles obeys Langevins law [1] and exhibits no hysteresis.

Experimental techniques

Transmission electron microscopy (TEM) was performed with a Philips CM 10 electron microscope. Specimens for TEM were prepared by dipping carbon-coated copper grids in dilute dispersions. Particle size distributions were calculated using interactive image analysis, always averaging over more than 100 particles. Dynamic light scattering (DLS) was used to determine the diffusion coefficient, from which the hydrodynamic radius was calculated using the Stokes–Einstein relation. Small-angle X-ray scattering (SAXS) experiments were performed using a Kratky camera to obtain plots of scattering intensities, I versus the square of the wave vector, K^2. The slope of linear part of I versus K^2 yields the electronic radius of gyration, R_g. For homogeneous spheres we then obtain the Guinier radius $R_G = (5/3)^{1/2} R_g$ and $D_{SAXS} = 2R_g$ (Tables 1, 2). For particles of low polydispersity, the particle radius is calculated from the minimum in the scattering curves assuming the particles to be homogeneous. Powder diffraction was performed with a Nonius PDS 120 powder diffractometer system. The intensity was measured with a wide angular (2θ) range of 120°. X-ray photoelectron spectra (XPS) were measured using a CLAM-2 vacuum generator system. Magnetisation measurements were performed at room temperature with a MicroMag 2900 alternating gradient magnetometer (Princeton Measurements).

Results and discussion

The thermal decomposition of $Fe(CO)_5$ in the presence of polyisobutene and oleic acid in decalin yields black, stable dispersions of colloidal iron particles. The stabilisers used, the composition, and the dimension of the iron particles produced by thermal decomposition are summarised in Tables 1 and 2. The particle diameters (range 7–30 nm) depend on the type of stabiliser and the concentration of the stabiliser.

A typical TEM micrograph for dispersion Fe2 is shown in Fig. 1. The magnetic fluid Fe2 is superparamagnetic at room temperature and has a saturation

Table 2 Oleic acid stabilisation: amount of reactants and resulting particle sizes. Decalin (60 ml) was used in all preparations

Sample	$Fe(CO)_5$ (ml)	Oleic acid (ml)	Transmission electron microscopy (nm)	D_{SAXS} (nm)
NFe2	5	0.5	10.1 ± 1.0	
NFe3	5	1.0	6.0 ± 0.6	
NFe4	5	1.5	7.3 ± 1.4	
NFe5	5	2.0	5.5 ± 0.3	
Fe8	5	2.0	7.1 ± 0.45	6.0 (7)[a]
NFe6	5	1.5	6.4 ± 0.4	
NFe7	5	2.5	9.3 ± 1.5	
NFe2a5			5.1 ± 0.4	
NFe2a10			5.9 ± 0.5	
NFe2a15			13.5 ± 1.3	
NFe2b5			6.1 ± 0.5	
NFe2b10			10.1 ± 1.5	
NFe2b15			6.6 ± 1.7	
NFe9	10		10.5 ± 1.5	
NFe10	15		11.6 ± 1.2	

[a] Particle size obtained from the first minimum of the scattering curve

Table 1 Amount of reactants and resulting particle sizes

Sample	Modified polyisobutene (g)	Decalin (ml)	$Fe(CO)_5$ (g)	Transmission electron microscopy (nm)	R_H (nm)	D_{SAXS} (nm)
Fe2	2.5	60	5	10.9 ± 1.7	11.4	10.3 (10)[a]
Fe5	1	60	5	11.4 ± 1.7		
Fe6	0.5	60	5	13.5 ± 1.8		
Fe7	0.25[b]	60	5	–		
Fe10	2.5	60	15	14.6 ± 2.7		13.1
Fe13	0.56[c]	60	5	Bidisperse[d]		
Fe15	2.5	60	10	13.9 ± 1.9		
Fe4	1.7	55	5	–		
Fe3	2[e]	50	5	–		
Fe16	2.5	60	2.5	7.9 ± 1.2		
Fe2a5				10.3 ± 1.5		
Fe2a10				11.9 ± 1.7		
Fe2a15				14.4 ± 1.6		12.0

[a] Particle size obtained from the first minimum of the scattering curve
[b] Octadecylmercaptan
[c] + Mercaptosilane (1.5 ml) added
[d] 35.7 ± 6.5 and 9.1 ± 1.5 nm particles

143

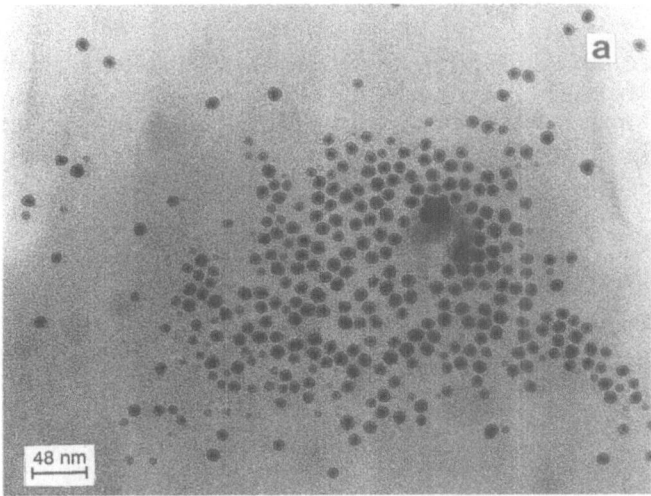

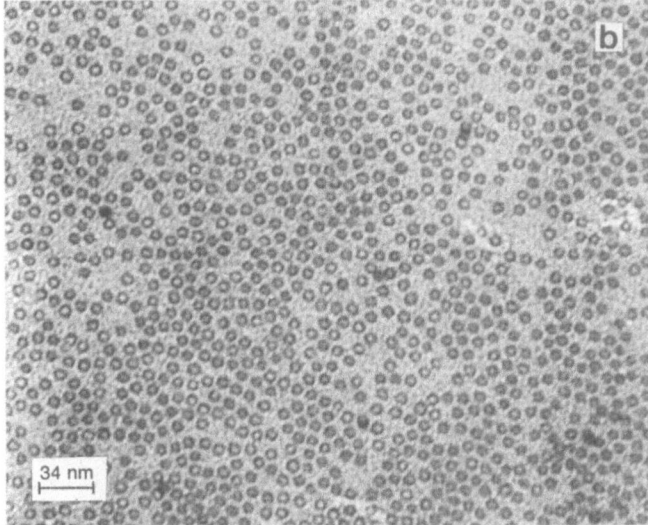

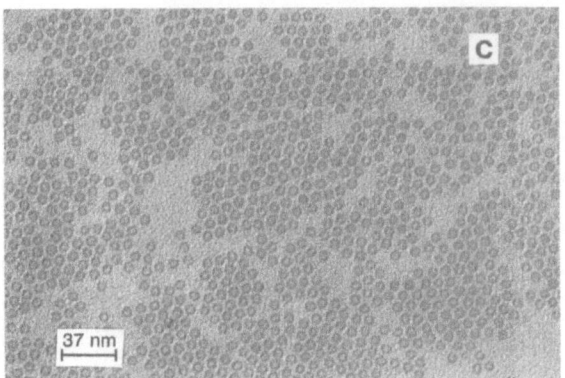

Fig. 1 Transmission electron microscope (*TEM*) micrographs of iron particles formed by thermolysis of iron pentacarbonyl in the presence of polyisobutene (**a**: Fe2) and oleic acid (**b**: Fe8, **c**: NFe5)

magnetisation of 0.055 Am^2/gFe (We used a value of 7.86 g/ml for the density of the iron particles). The saturation magnetisation of bulk Fe is 0.217 Am^2/gFe

[17]. At very low concentration of polyisobutene the particles (Fe7) formed were aggregated and they eventually sedimented at the bottom due to gravity. With smaller amounts of polyisobutene the particle size increases significantly (Fe5 and Fe6). A similar trend was not noticed for particles grafted with oleic acid, although the particle size could be varied by varying the amount of oleic acid (Table 2). The particle dimensions increase with increasing initial concentration of Fe(CO)$_5$ for particles which are grafted with polyisobutene (Table 1, Fe16). The standard deviation in the particle size is in the range 11–16% (Table 1) for polyisobutene-grafted particles and in the case of particles grafted with oleic acid it is in the range 5–20% (Table 2).

X-ray scattering curves of samples Fe2 and Fe8 at very low concentration (0.035 g/ml) are shown in Fig. 2. In Fig. 2 a minimum can be observed which can be attributed to the narrow particle size distribution. In the scattering curves (Fig. 2) higher-order minima are washed out, though for Fe8 such minima could perhaps be expected from the low value of the standard deviation obtained from TEM measurements (6.3%). This difference may be due to the particle shape, which

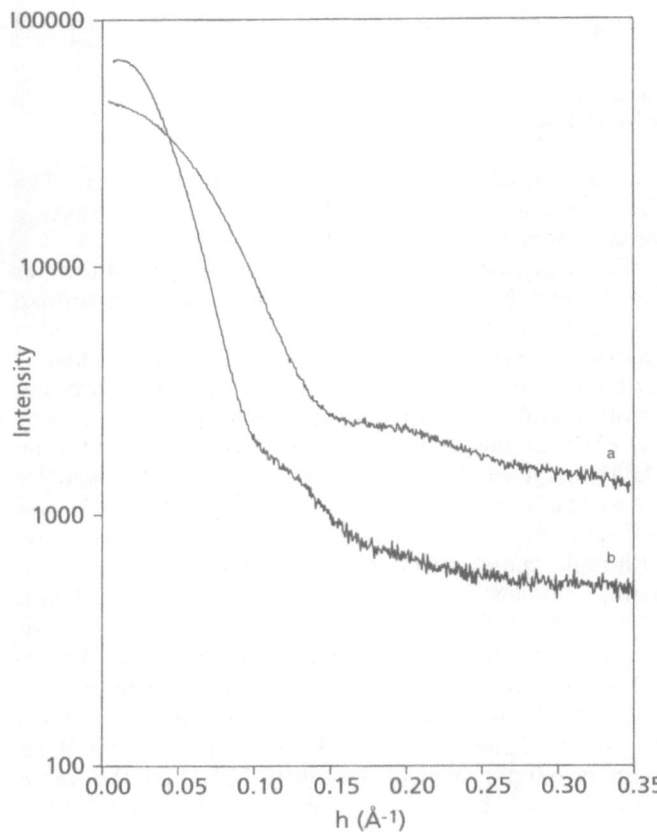

Fig. 2 Scattered intensity (arbitrary units) versus wavevector, *h*, from small-angle X-ray scattering measurements for iron particles stabilised with oleic acid (**a**: Fe8) and polyisobutene (**b**: Fe2)

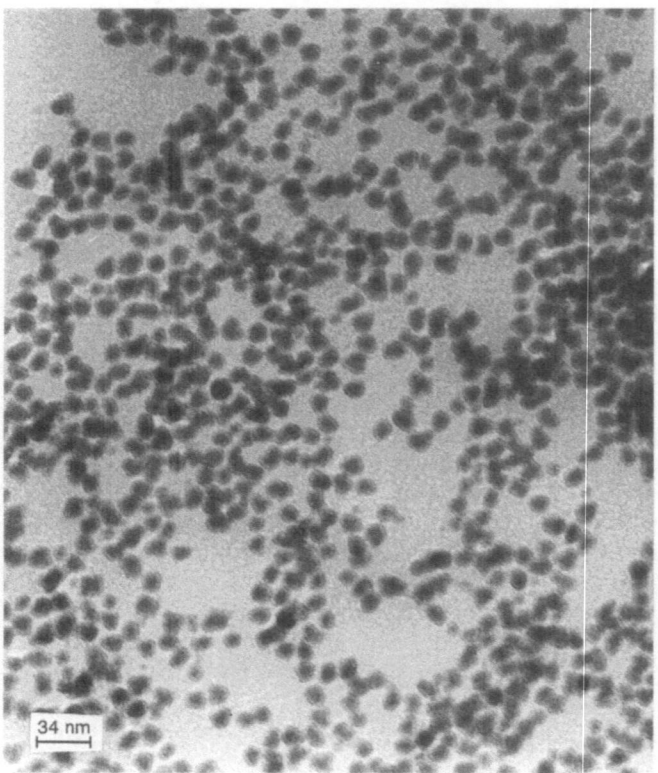

Fig. 3 TEM micrograph of Fe5 iron particles stabilised with polyisobutene

for these small particles is not perfectly spherical. The sizes obtained from the TEM and X-ray scattering measurements agree very well (Tables 1, 2).

Electron micrographs of the samples are shown in Fig. 1a and b. As the particle size increases chainlike structures are formed, a phenomenon which was also reported in Ref. [4]. Figure 3 (Fe5) shows the "necklace" formation on a TEM grid. The Fe2 sample on the foamvar grid was examined by atomic force microscopy (AFM) and magnetic force microscopy (MFM). The MFM image of the particles confirms that the particles are magnetic. The particle sizes obtained from AFM and MFM are 8.9 ± 0.3 nm and 8.7 ± 0.5 nm, respectively. Although these measurements agree well, they are slightly smaller than the size obtained from TEM and SAXS. The diffraction pattern of sample Fe2 is broad and it is difficult to assign it to a particular oxide of iron. When the dispersion was exposed to air the particles were oxidised. The hydrodynamic radius obtained from DLS is 11.4 nm (Table 1). The thickness of the polyisobutene chain is 6.0 nm, which is slightly less than the value found earlier [13]. Unfortunately, light absorption of the particles caused unwanted convection and divergency of the incident laser beam, which to a certain extent was suppressed by using low laser power and by selecting a laser wavelength of 647 nm, far from absorption bands.

Seeded growth

To obtain larger iron particles, the growth of small particles was continued by adding additional $Fe(CO)_5$. After the seeded iron particles had been formed in about 3 h, $Fe(CO)_5$ was added to the reaction mixture in two steps after a time lapse of 3 h. The particle size increases

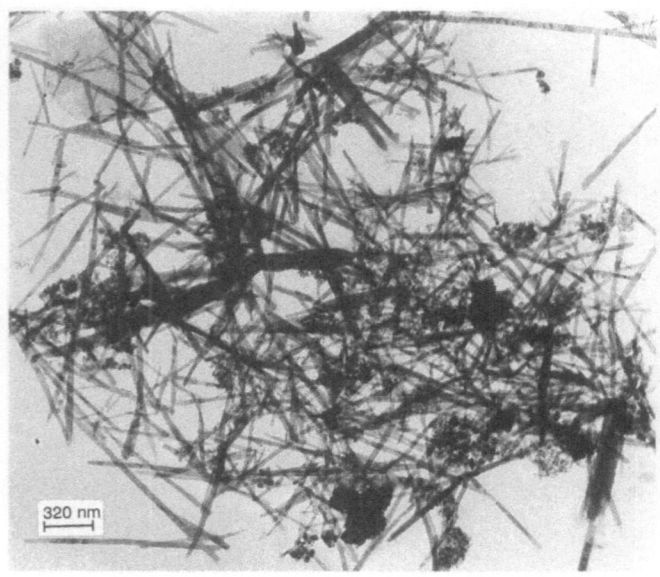

Fig. 4 TEM micrograph of Fe3 iron rods stabilised with mercaptosilane

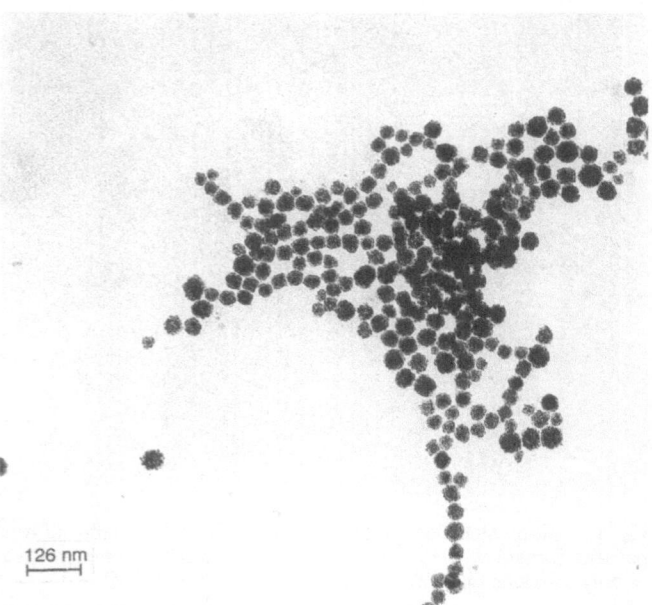

Fig. 5 TEM micrograph of Fe13 iron particles formed by thermolysis of iron pentacarbonyl in the presence of a mixture of polyisobutene and mercaptosilane

from 10.3 to 14.6 nm (Table 1) and the particle size distribution is narrower than that of the seeded particles. The increase in size of the particles agrees with the amount of $Fe(CO)_5$ precursor added. By utilising this method, tailor-made Fe particles can be prepared by controlled addition of the $Fe(CO)_5$ monomer. This procedure for seeded growth by stepwise addition was repeated with oleic acid as the stabiliser. The results are shown in Table 2 (NFe2a5, NFe2a10, NFe2a15). The initial amount of reactants used for NFe2 and NFe2a were the same. The particle size of NFe2a5 is much smaller than that of NFe2 and the final size of the particles after all additions was 13.5 nm (NFe2a15), which is much more than the expected particle size (7.4 nm). The seeded growth reaction was repeated with the same amount of reactants but $Fe(CO)_5$ was added after 24 h instead of 3 h. After the first addition the particle size of NFe2b10 was much larger than the expected value due to the aggregation of particles and after the second addition the particle size decreased because of second nucleation. NFe9 and NFe2b10 have the same final size and NFe2a15 and NFe10 also have nearly the same size. In the case of oleic acid it was not possible to tune the particle size to the required value. The nucleation and growth of particles in the presence of oleic acid as a stabiliser is different from that in the presence of polyisobutene.

Iron needles

The thermolysis of $Fe(CO)_5$ in presence of octadecylmercaptan gave no stable dispersions. With MPS a mixture of needles and spherical particles was obtained (Fig. 4). The particles were magnetic but they sedimented due to gravity and the particles could be separated with a magnet. The particles were washed with toluene 3 times before XPS and powder diffraction measurements. The characteristic iron and sulphur $2p$ peaks were found with a shift from the peak position of pure metal due to the formation of a bond in the XPS spectrum. This confirms that MPS is attached to the iron particles. The diffraction peaks of Fe3 are found to be very sharp, showing the particles to be crystalline and lines characteristic of α-Fe metal (d spacings of 2.03, 1.43, and 1.17 Å) are observed. Magnetisation measurements of Fe3 exhibit hysteresis and this is due to the shape anisotropy. A mixture of MPS and polyisobutene gave a stable ferrofluid dispersion where the particles are bidisperse (Fig. 5). The average sizes of the particles are 35.7 and 9.1 nm (Table 1). It may be that the MPS bridges the small particles, thereby contributing to the formation of aggregates.

Conclusions

Thermolysis of $Fe(CO)_5$ in decalin in the presence of modified polyisobutene or oleic acid yields iron particles which are fairly monodisperse even for particle radii below 10 nm. In particular polyisobutene-stabilised iron colloids are suitable seeds for further growth to tailor the particle size and diminish further the polydispersity. (It is somewhat surprising that the organic shell around the iron core does not obstruct this seeded growth.) The particle aspect ratio can be changed quite drastically by using a mercaptosilane stabiliser. The resulting magnetic iron needles are interesting for future work on the effect of shape anisotropy on magnetic properties. The seeded growth for the spherelike particles is useful, for example, for tailoring magnetic dipole interactions which strongly depend on the particle radius.

Acknowledgements We thank E. Klop of the Akzo Nobel research laboratory for performing the SAXS measurements and C.F.J. Flipse and J. Sturm from the Technical University of Eindhoven for the MFM measurements and for the use of their unpublished results. A. Mens measured and interpreted the XPS spectra. M. Versluis-Helder is acknowledged for performing the powder diffraction measurements and B. Kuipers for the DLS and magnetisation measurements. Mieke Lanen-Vos and Marina Uit de Bulten-Weerensteyn are thanked for assistance in preparing the manuscript. Dr. P.C. Scholten was very helpful by drawing our attention to literature on iron colloids.

References

1. Rosensweig RE (1997) Ferrohydrodynamics. Dover, New York
2. Berkovski B, Bashtovoy V (eds) (1996) Magnetic fluids and applications handbook. Begell House, New York
3. Blums E, Cebers A, Maiorov MM (1997) Magnetic fluids. de Gruyter, New York
4. Griffiths CH, O'Horo MP, Smith TW (1979) J Appl Phys 50:7106
5. Smith TW, Wychick D (1980) J Phys Chem 84:1621
6. Nakatani I, Furubayashi T, Takahashi T, Hanaoka H (1987) J Magn Magn Mater 65:261
7. Lôpez-Quintela MA, Rivas J (1993) J Colloid Interface Sci 158:446
8. de Caro D, Ely TO, Mari A, Chandret B (1996) Chem Mater 8:1987
9. Suslick KS, Choe SB, Cichowlas AA, Grinstaff MW (1991) Nature 353:414
10. Cao X, Koltypin Yu, Kataby G (1995) J Mater Res 10:2952
11. Suslick KS, Fang M, Hyeon T (1996) J Am Chem Soc 118:11960
12. Kataby G, Cojocaru M, Prozorov R, Gedanken A (1999) Langmuir 15:1703
13. Mikhailik OM, Povstugar VI, Mikhailova SS, Lyakhovich AM, Fedorenko OM, Kurbatova GT, Shklovskaya NI, Chuiko AA (1991) Colloids Surf 52:315
14. Sun S, Murray CB (1999) J Appl Phys 85:4325
15. Pathmamanoharan C, Philipse AP (1998) J Colloid Interface Sci 205:340
16. Scholten PC (1978) In: Berkovsky B (ed) Thermomechanics of magnetic fluids. Hemisphere, Washington, DC, pp 43–65
17. Charles S (1996) In: Berkovski B, Bashtovoy V (eds) Magnetic fluids and applications handbook. Begell House, New York, pp 3–13

Progr Colloid Polym Sci (2000) 115:146–150
© Springer-Verlag 2000

Yu. A. Kuznetsov
E. G. Timoshenko

Influence of reptations on conformations of a homopolymer in a Monte Carlo simulation

Yu. A. Kuznetsov · E. G. Timoshenko (✉)
Theory and Computation Group
Department of Chemistry
University College Dublin
Belfield, Dublin 4, Ireland
e-mail: edward.timoshenko@ucd.ie
Tel.: +353-1-7062821
Fax: +353-1-7062127

Abstract We study the role of topological restrictions for the conformational structure of a nonphantom homopolymer chain in a lattice Monte Carlo model. In the athermal regime we find that the standard Metropolis algorithm violates the detailed balance condition if both local monomer and global reptational moves are included; however, if about one reptation is performed per N local moves the balance is recovered and we obtain the Flory exponent value close to $v = 0.588$, where N is the degree of polymerisation. We also find that the structure of the collapsed globule is different at equilibrium from that after the late stage of folding. Namely, due to reptations the end and, even more so, the penultimate monomer groups tend to be buried inside the globule core with other monomers, thus being more exposed to the surface.

Key words Homopolymer ·
Simulation · Conformation ·
Reptation · Nonphantom chain

Introduction

It was recognised early that topological restrictions play an important role in determining the conformation and dynamics of polymer solutions [1]. One of the most important topological restrictions is the integrity of chain links, which implies that different parts of the polymer chain cannot pass through each other. Unfortunately, it is rather difficult to include such restrictions in analytical theories, especially in simple mean–field ones.

Nevertheless, the effects of topological restrictions can be properly studied by computer simulation techniques [2]. One such technique developed by the authors in Ref. [3] is based on a lattice Monte Carlo model. Here, the Monte Carlo updates scheme includes local monomer and reptational chain moves. The former is an attempt to move a randomly chosen monomer to a randomly chosen nearby lattice site. The latter is an attempt to move a randomly chosen end monomer to a randomly chosen lattice site near the other polymer end.

It is important to emphasise that this model does not permit moves that would violate the integrity of links, i.e. the chain is strictly nonphantom. This is ensured automatically due to the particular choice of the connectivity and the excluded-volume parameters.

In this article we study how the probability of performing reptational moves affects the conformations of a single homopolymer chain under good and poor solvent conditions. In particular, we compare the globules obtained by a kinetic process and at true equilibrium. Since the kinetics of polymer collapse proceeds through the formation and growth of locally collapsed clusters along the chain, with their final unification into a single globule, one may expect that after the shape-optimisation stage the homopolymer globule possesses a comparatively simple topological structure [3]. Indeed, a typical coil conformation before the quench possesses a statistically small number of entanglements, or knots, and the folding kinetics adds virtually nothing to that number. Further relaxation of the globule towards the equilibrium requires participation of the chain ends and, thus, can be viewed as an auto reptational stage. This autoreption time can be estimated as $\tau_{\text{rept}} = \tau_0 N^3$ [4], which can yield a time of the order of 10^3 s under usual experimental conditions [5].

The good solvent regime

In this section we shall test the Monte Carlo scheme by determining the value of the swelling exponent, v, [1] in the good solvent regime.

Generally, it seems that to improve the convergence of the system to equilibrium various types of global moves, such as reptations, may be included in addition to local ones. The resulting equilibrium state should not depend on the particular scheme involved. To test this assumption let us study how the probability of performing reptations, P_r, affects the value of the Flory exponent of the coil. The results are presented in Table 1. One can see that the value of v obtained in the scheme without reptations (at the bottom of the left column) is somewhat higher than even the mean-field prediction, $v = 3/5$. As we increase P_r, v starts to decrease. In the limit when only reptations are performed, $P_r = 1$, the swelling exponent value is found to be 0.56. Note that for $P_r = 1/N$ the measured swelling exponent is remarkably close to the most accurate result obtained from renormalisation group theory, $v \approx 0.588$ [1].

We may conclude, therefore, that the scheme with only reptations involved leads to more compact entangled conformations, resulting in an underestimation of the swelling exponent. On the other hand, the scheme with local monomer moves only favours topologically simple conformations as it is rather improbable to create a knot by local movements. If some knots already exist in the initial conformation, the local monomer movements would tend to disentangle them. Indeed, one can imagine that simple "shaking" of an entangled boot strap would more likely disentangle it rather than entangle it more. This weak topological effect reduces the number of entanglements, which leads to a larger radius of gyration and to overestimation of the swelling exponent in the scheme with local moves only.

Such strong dependence on P_r is quite unexpected from the point of view of the standard Monte Carlo paradigm. Of course, the Metropolis check in itself is not sufficient for satisfying the detailed balance condition. One also has to ensure that the phase space of the system is sampled uniformly by attempted Monte Carlo moves [6]. Although, this may be quite simple to ensure for pointlike objects, in our case of a nonphantom chain this is not so. The above observations indicate clearly that the current sampling procedure is not uniform, but biased. The bias is present in both schemes with reptations only and local moves only, but has the opposite effect. We have also seen that if about one reptation is performed per N Monte Carlo steps the topological effects of entanglements and disentanglements balance each other, making the sampling of the phase space essentially uniform and the correct swelling exponent is produced.

However, we should emphasise that this problem of the improper influence of reptations is only present for the Flory coil and it is irrelevant for the ideal coil. Both schemes with local moves only and reptations only would give $v = 1/2$ for the ideal solution. That is why reptation techniques are extremely popular and well justified for studying melts and concentrated solutions, in which reptations may also be the only physically relevant motions. We should also emphasise that if the chain was phantom we would not have this problem either.

Table 1 Values of the mean squared radius of gyration, R_g^2, versus the degree of polymerisation, N, for different simulation procedures. Here Q is the number of statistical measurements, P_r is the probability of making reptations in the scheme, with $1 - P_r$ being the probability of making local monomer moves. The exponents v_1 and v_2 were obtained by a least-squares fit of $\log R_g$ versus $\log N$ in the ranges 100–1000 and 500–1000, respectively

N Q	$P_r = 0$ 80,000	$P_r = 1/N$ 60,000	$P_r = 0.1$ 40,000	$P_r = 1$ 40,000
20	13.14	12.55	12.29	11.50
30	22.21	20.74	19.88	18.44
50	42.16	38.50	35.73	32.63
70	64.89	57.66	52.54	48.45
100	98.17	88.15	77.75	72.54
150	162.2	142.7	122.6	112.9
200	227.2	201.2	169.0	156.8
300	381.2	322.7	261.8	245.9
500	696.5	588.1	460.0	435.9
700	1029	856.1	667.2	637.6
1000	1643	1329	991.5	947.7
v_1	0.608	0.587	0.551	0.559
v_2	0.619	0.588	0.554	0.560

Even though the effect discussed for the Flory coil is clearly of topological origin, it only presents a problem for the simulation procedure and has no implications for real polymers.

Structure of the homopolymer globule

In this section we shall study the structural differences between the two globules of an open homopolymer: one with a relatively simple topological structure corresponding to the late stage of folding and the other with topological entanglements. In practice these simulations were carried out in the following way. Using the lattice Monte Carlo method a large set of homopolymer globules was produced by independent kinetic processes starting from initial coil conformations. Reptational moves were not included during this simulation. To produce a true equilibrium distribution for the globule an additional simulation was applied to the set with the number of reptational moves equal to $P_r = 1/N$.

Let us introduce the mean squared distances along the chain, D_{mn}, and their partially summated combinations, D_k

$$D_{mn} = \left\langle (\mathbf{X}_m - \mathbf{X}_n)^2 \right\rangle, \qquad D_k \equiv \frac{1}{N-k} \sum_{i=0}^{N-1-k} D_{i\,i+k} .$$

(1)

$D_{\hat{k}}$ versus the normalised chain index, $\hat{k}$, for polymers of different degrees of polymerisation is presented in Fig. 1. For very small chain indices D_k does not depend on N and on the particular simulation procedure, which reflects local packing of monomers in the dense globule.

Let us consider first the behaviour of this function for globules prepared without reptations. For small values of the chain index the function is almost linear up to some crossover value that scales as $N^{2/3}$. Then it saturates to some level, which is proportional to the radius of the globule, thus also scaling as $N^{2/3}$. Interestingly enough, for values of the chain index in the vicinity of the chain ends D_k increases once again. This phenomenon actually reflects the mechanism of the polymer collapse during the late coarsening stage. The globule is usually formed by a final unification of two end clusters and, sometimes, a few middle clusters (see e.g. Figs. 4, 6, 10 in Ref. [3]). Thus, the chain ends possess a somewhat higher probability to appear on opposite sides of the globule than the rest of the monomers. This effect in the mean squared distances is fairly weak as the function experiences an increase of about 10% towards the ends.

In fact, this observation is related to the observation that the chain ends are more exposed to the globule surface and it can be better justified by considering the probability of the mth monomer in the chain to appear on the surface of the globule, $P_m^{(\text{surf})}$. This is presented in the left-hand side of Fig. 2. Thus, $P_m^{(\text{surf})}$ for the kinetic simulation is nearly constant, except for a few monomers at the very chain ends. In particular, for the end monomer this probability is about 1.3 times larger than that for a monomer in the centre of the chain.

The effect due to applying reptational moves during later kinetic stages is quite distinguishable in Figs. 1 and 2. For sufficiently large values of the chain indices, $\hat{k} \gtrsim 0.35$–0.4, the function of the mean squared distances obtained from a simulation with reptations (see lines denoted by diamonds in Fig. 1) lies below the appro-

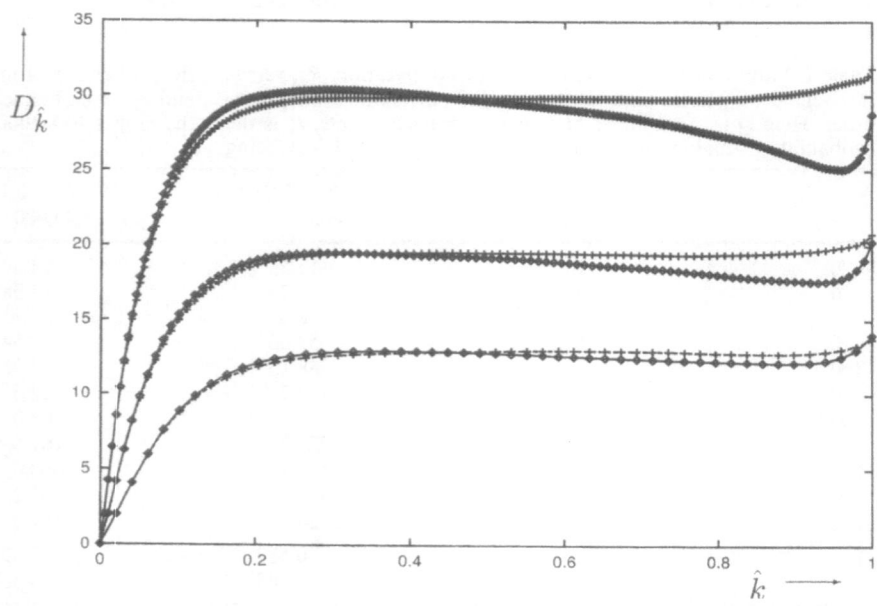

Fig. 1 Plot of the partially summated mean squared distances, $D_{\hat{k}}$, versus the normalised chain index, $\hat{k} = k/(N-1)$, for homopolymer globules of different sizes and different simulation procedures. Pairs of lines correspond to the following values of the degree of polymerisation (from *bottom* to *top*): $N = 50$, $N = 100$ and $N = 200$. Lines denoted by *diamonds* and *pluses* in each pair correspond to the simulation procedures with, $P_r = 1/N$, and without reptations, $P_r = 0$, respectively

Fig. 2 Plot of the probability, $P_m^{(\mathrm{surf})}$, for the mth monomer in the chain to appear on the globule surface versus the monomer index, m, for a homopolymer with a degree of polymerisation $N = 400$, for different simulation procedures. The curves on the *left* and on the *right* correspond to the kinetics (no reptations) and reptational simulation procedures, respectively. Both distributions are symmetric in m and for convenience they are presented only on halves of the interval

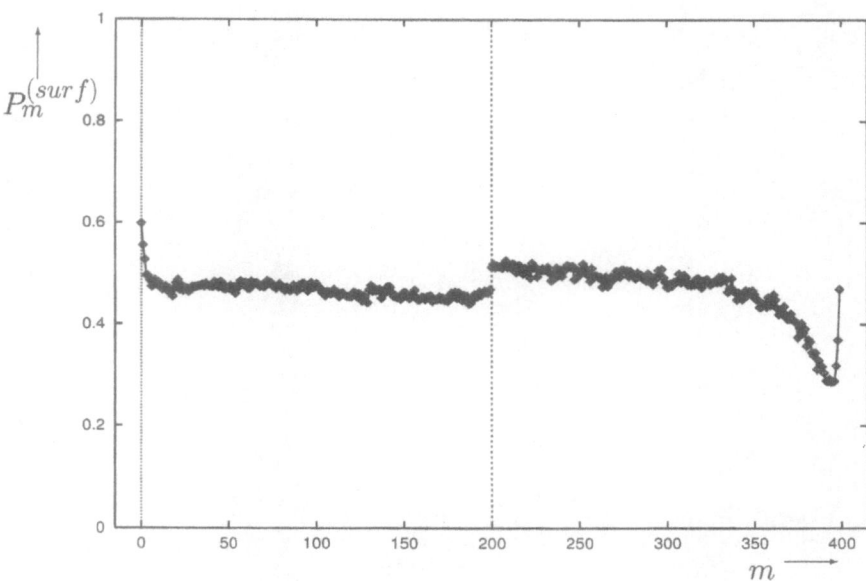

priate curves of the kinetic simulation. The situation is just the reverse for smaller k, with the difference slowly vanishing towards $k = 0$. Thus, the reptational curve for D_k behaves in the following way: it is linear for small k, reaches a maximum at around $\hat{k} = 0.3$ and then slowly decreases, reaching a minimum in the vicinity of the chain ends. At the very end the function increases strongly, still being significantly smaller that for D_k without reptations.

Note that the partially summated mean squared distances are not fully informative due to the role of the end effects in an open chain. Thus, let us consider the mean squared distances as functions of two indices. The ratios of the mean squared distances with and without reptations, $D_{mn}^{(r)}/D_{mn}^{(n)}$, are shown in Fig. 3. Depending on the behaviour of this ratio, the monomers in the chain may be roughly divided into three groups: for monomers in the centre of the chain, $0.1 \lesssim m/N \lesssim 0.9$, the mean squared distances significantly increase due to reptations; for penultimate groups of monomers in the chain $3 \lesssim m$, $N - m - 1 \lesssim 0.1N$, the distances decrease due to reptations, with the effect becoming more pronounced on approaching the ends of the chain, and for a few end monomers the distances decrease but much more weakly. Thus, we can conclude that reptations push the end groups somewhat and the penultimate groups of monomers more so towards the centre of the globule, thus reducing their mean squared distances between each other and monomers from the central group. As the density of the globule does not change here this leads to an increase in the mean squared distances for monomers in the centre of the chain.

Such behaviour of the mean squared distances is quite consistent with the plot of $P_m^{(\mathrm{surf})}$ after the

reptational stage (right-hand side of Fig. 2). This quantity increases for monomers in the central group by about 10% and, instead of being constant, decreases slowly towards the ends of the chain. The drop in the probability is most pronounced for monomers from the penultimate groups. The function rapidly increases for a few monomers at the very ends of the chain, although they still possess a lower probability to be found on the surface than monomers from the central group. This rapid increase in the probability for a few end monomers may be interpreted as the effect of a single end seeking to maximise its entropy. Indeed, the chain ends are freest to explore the surface, whilst the rest of the monomers are more restricted due to connectivity.

Conclusion

In this article we have applied the lattice Monte Carlo model of Ref. [3] to study the role of reptations for conformations of a nonphantom homopolymer chain.

First, we considered the athermal good solvent regime and discovered that the frequency of performing reptations significantly affects the size and even the swelling exponent of the polymer. This problem arises due to a nonuniform sampling of the phase space of the system by attempted Monte Carlo moves when links integrity is strictly insured. For a nonphantom chain it is not clear how to ensure uniform sampling, especially between conformations with different topological numbers. The scheme with reptations only is biased towards entangled conformations, while the scheme with local moves only samples topologically simple states more; however, the balance is recovered when $P_r = 1/N$.

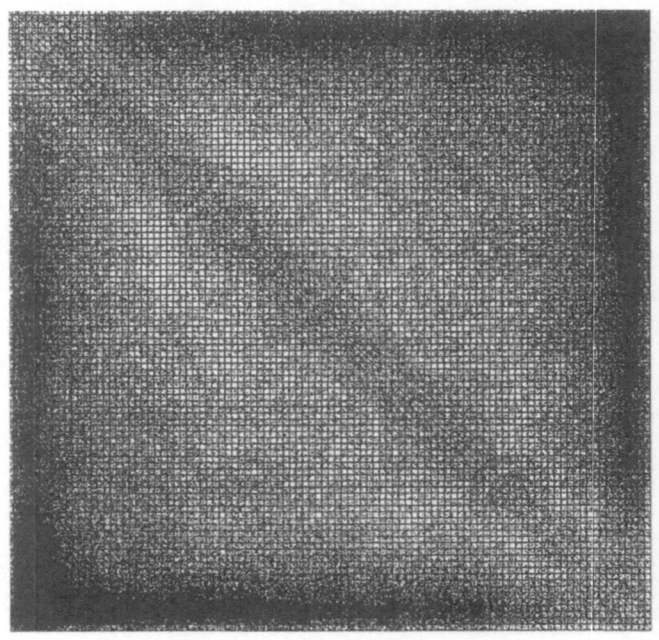

(a)   (b)

Fig. 3a, b Diagrams of the ratios of the mean squared distances with and without reptations, $D_{mn}^{(r)}/D_{mn}^{(n)}$, for different values of the degree of polymerisation. **a** and **b** correspond: $N = 100$ and $N = 200$, respectively. Indices m, n start counting from the *upper left corner*. Each matrix element is denoted by a quadratic cell with a *varying degree of black*; the darkest and the lightest cells correspond to the smallest and to the largest ratios of the mean squared distances, respectively. The intensity of the black colour in the near diagonal elements corresponds approximately to the ratio $D_{mn}^{(r)}/D_{mn}^{(n)} = 1$

Second, we then compared structures of the two globules: one corresponding to late stages of kinetics and the other to the true equilibrium. The globule after the late kinetic stage is characterised by a uniform monomer distribution except for the end monomers, which have a higher probability to be on the surface. Due to reptations the end and, even more so, the penultimate monomer groups are pushed more towards the centre of the globule, with the central monomer group thus being more exposed to the surface. A typical magnitude of this effect in observables expressing the k-dependent properties is about a few dozen percent in the relative change, while the global characteristics, such as the mean energy and radius of gyration, are practically unaffected by this conformational change. The latter is a physically important conclusion for real polymers.

Acknowledgements The authors are grateful to K.A. Dawson, F. Ganazzoli, and especially to A.Yu. Grosberg for interesting discussions. We also acknowledge the support of Enterprise Ireland grants SC/99/186 and IC/1999/01 and the support of the Centre for High Performance Computer Applications, University College Dublin.

References

1. (a) de Gennes PG (1988) Scaling concepts in polymer physics. Cornell University Press, Ithaca; (b) Doi M, Edwards SF (1989) The theory of polymer dynamics. Oxford Science, New York; (c) des Cloizeaux J, Jannink G (1990) Polymers in solution. Clarendon Press, Oxford
2. (a) Binder K (1984) Applications of Monte Carlo method in statistical physics. Springer, Berlin Heidelberg New York; (b) Allen MP, Tildesley DJ (1993) Computer simulation in chemical physics. NATO Advanced Study Institute Series C 397. Kluwer, Dordrecht
3. Kuznetsov YA, Timoshenko EG, Dawson KA (1995) J Chem Phys 103:4807
4. Grosberg AY, Khokhlov AR (1994) Statistical physics of macromolecules. American Institute of Physics, New York
5. Grosberg AY, Kuznetsov DV (1993) Macromolecules 26:4249
6. (a) Parisi G (1986) Les houches. Session XLIII. North–Holland, Amsterdam; (b) Parisi G (1988) Statistical field theory. Addison-Wesley, Reading, Mass

Progr Colloid Polym Sci (2000) 115:151–155
© Springer-Verlag 2000

A. Tsevis
M. Sotiropoulou
P. G. Koutsoukos

Preparation of titania powders in a fluidized-bed reactor

A. Tsevis · M. Sotiropoulou
P. G. Koutsoukos
Department of Chemical Enginnering
University of Patras, 26500 Patras, Greece

A. Tsevis (✉) · M. Sotiropoulou
P. G. Koutsoukos
Institute of Chemical Engineering
and High Temperature Chemical
Processes, FORTH-ICEHT
26500 Patras, Greece
e-mail: ntsevis@iceht.forth.gr
Tel.: +30-61-997579

Abstract Titania powder was prepared in a fluidized-bed crystallizer at 25 °C by mixing titanium isopropoxide in 2-propanol with water at various ratios with two peristaltic pumps at equal flow rates. The column was packed with quartz glass beads of broad size distribution (0.2–0.8 mm). The flow rates of the feed solutions were selected to achieve optimal fluidization of the quartz beads. Various parameters, including the residence time of the reactants in the bed, the excess water with respect to the stoichiometric amounts ($Ti:H_2O$), the presence of foreign ions and the role of the quartz filling beads, were investigated with respect to their importance in the physicochemical and morphological characteristics of the particles formed. The concentration of titanium at the top of the bed was kept constant during the crystallization process after the establishment of a steady state. The difference in concentration from the corresponding value of the feedstock solution was used to calculate the rates of titania formation. The presence of Li^+, Nb^{5+} and W^{6+} in the aqueous medium resulted in the incorporation of these ions in the titania lattice and the reduction of the rates of formation of the particles from the resulting supersaturated solutions.

Key words Titania · precipitation · fluidized bed crystallizer · Doping

Introduction

The synthesis of monosized particles for use as pigments and catalytic substrates may be achieved with batch precipitation techniques [1–6]. Batch processes are not attractive because of high operating costs and the possibility of a variation in the characteristics of the final product. Classifying crystallizers are used for continuous precipitation, but they result in the production of particles of broad particle size distributions due to the extensive residence times [7]. Continuous precipitation of monosized particles may be obtained using narrow ranges of residence times. Doping the TiO_2 supports with altervalent ions of size similar to Ti^{4+} may improve greatly the catalytic activity of the respective catalysts [8–11]. Doping of TiO_2 may be achieved by the introduction of foreign ions in the nucleation and crystal growth steps [12–14].

In the present work, a novel preparation method of TiO_2 particles is presented using a fluidized-bed crystallizer packed with quartz glass beads. The kinetics of formation of TiO_2 in alcoholic solutions as a function of the residence time of the reactants in the bed and the effect of water excess with respect to the total titanium concentration both in the absence and in the presence of Li^+, Nb^{5+} and W^{6+} was investigated. Particle size measurements were carried out in order to estimate the size uniformity of the powders prepared.

Experimental

The experiments were carried out in a double-walled, glass cylindrical-bed crystallizer, thermostated by circulating water. The dimensions of the bed crystallizer were 1.5 cm (inner diameter), 3.8 cm (outer diameter) and 60 cm (length). The reactants were introduced at the bottom of the column from two separated inlets using two peristaltic pumps with adjustable flow rates. The filling material, quartz glass beads with a board size distribution (0.2–0.8 mm), was supported by a glass frit attached in the bottom. The peristaltic pumps were calibrated using 2-propanol. During the precipitation, particles of the nucleated solid were deposited on the surface of the quartz beads. A film of titania was formed on the quartz beads, which served further as seed material for further growth of titania. In order to estimate the contribution of seeded growth to the overall growth rates measured, it was necessary to replace the filling material with fresh material for each experiment.

For the preparation of the titania powder, titanium isopropoxide [$Ti(OC_3H_7)_4$] (Merck above 98% purity) in 2-propanol solvent (high purity) and triply distilled water were used. In cases where precipitation of solid took place in the presence of foreign ions such as W^{6+}, Nb^{5+} and Li^+ ammonium paratungstate, $(NH)_{10}W_{12}O_{41} \cdot 5H_2O$ (Alfa), potassium heptafluoroniobate, K_2NbF_7 (Aldrich), and lithium nitrate, $LiNO_3$ (Merck), respectively, were used for the preparation of the stock solutions. Two alcoholic solutions were prepared: the first was a 0.2 M solution of titanium isopropoxide in 2-propanol, while the second was water in 2-propanol at a stoichiometric ratio of 1:4 or greater. In the experiments done in the presence of foreign ions, the aqueous solutions contained the desired concentrations of the ions under investigation. The reactants were pumped at sufficiently high flow rates to ensure satisfactory fluidization. It should be noted that the fluid coming out of the column did not contain any quartz beads. The hydrolysis reaction resulting in the formation of the titania particles may be represented by the following reactions steps.

$$Ti(OC_3H_7)_4 + 4H_2O \rightarrow Ti(OH)_4(s) + 4C_3H_7OH \ .$$

The solid precipitate condenses next, eliminating water:

$$Ti(OH)_4(s) \rightarrow TiO_2 \cdot xH_2O(solid) + (2-x)H_2O \ .$$

Parameters such as the residence time of the solution in the bed, the excess water with respect to the stoichiometric quantity (TiO_2: H_2O) and the presence of foreign ions are expected to have an effect on the rates of formation and consequently on the morphological characteristics of the particles formed. In all cases the flow rate was adjusted so that the mean residence time was between 6 and 17 min, ensuring laminar flow conditions. The bed was fluidized with pure 2-propanol solvent followed by the continuous injection of the reactants. After a time period equal to the residence time of the solutions in the bed and after the initial stream of the mixed reactants had reached the top of the column samples were withdrawn and filtered through membrane filter (0.22 μm). The filtrate was analyzed for titanium by a spectrophotometric method. The concentration of titanium at the top of the bed was constant during the crystallization process after the establishment of the steady state. The concentration difference between the feedstock and the solution exiting the column was used to calculate the rates of titania formation. The particles produced were centrifuged, the supernatant was decanted and the particles were resuspended in water. The water suspensions were next centrifuged and finally dried. The solids were characterized by powder X-ray diffraction (Phillips PW 1840), scanning electron microscopy (JEOL JSM-5200) and X-ray fluorescence (XRF) spectroscopy (XRF-TN Spectrace). Specific surface areas of the powders were measured by a multiple-point, nitrogen-adsorption Brunauer–Emmett–Teller method (Micromeritics, Gemini 2375). Particle size distribution measurements were done by laser scattering (Malvern, Mastersizer).

Results and discussion

The experimental results showed that the rates measured for fixed residence times were not reproducible when the

initially precipitated titania particles formed an adhering layer on the quartz glass beads used to fill the fluidized bed as confirmed by microscopic examination and XRF analysis of the quartz beads. The Ti/Si ratio increased upon prolonged use of the same column fillings as may be seen in Table 1. Titania particles covering the quartz glass beads acted as seed crystals that accelerated reaction rates even though supersaturation was the same. In order to perform accurate kinetics measurements the column was filled freshly for each experiment.

The effect of parameters including excess water concentration, residence time and the presence of foreign ions on the mass deposition rate, R_G, is presented in Figs. 1–3. The experimental conditions in the present work are summarized in Tables 2 and 3. An increase in the water excess yielded higher rates as may be seen in Fig. 1. This trend may be attributed to the higher supersaturation ratio. The supersaturation ratio, Ω, with respect to the precipitating solid is given by Eq. (1):

$$\Omega = \frac{(Ti^{4+})(OH)^4}{K_s^0} \ , \tag{1}$$

where K_s^0 is the thermodynamic solubility product of anatase at 25 °C and the parentheses denote the activities of the respective ions.

Higher residence times in general resulted in reduced rates of titania formation. As may be seen in Fig. 2 a

Table 1 Ti and Si intensities of X-ray fluorescence spectra

Exp. no. (Quartz sample)	Ti (counts)	Si (counts)	Ti/Si
1	27766	373816	0.074
2	83898	337408	0.248
3	94595	276889	0.341
4	71784	168326	0.426

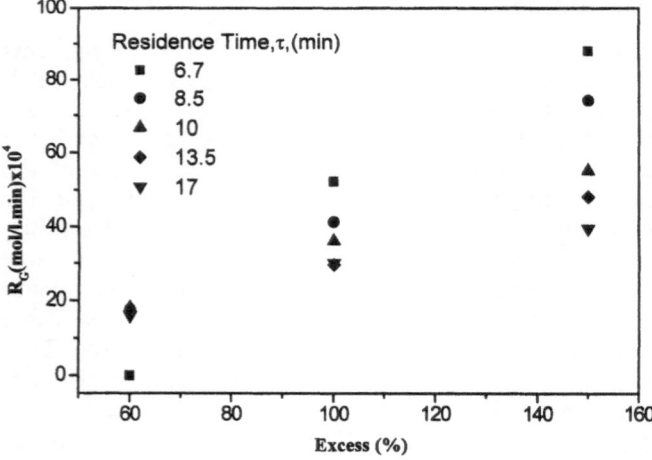

Fig. 1 Rates of titania formation for various water excess values

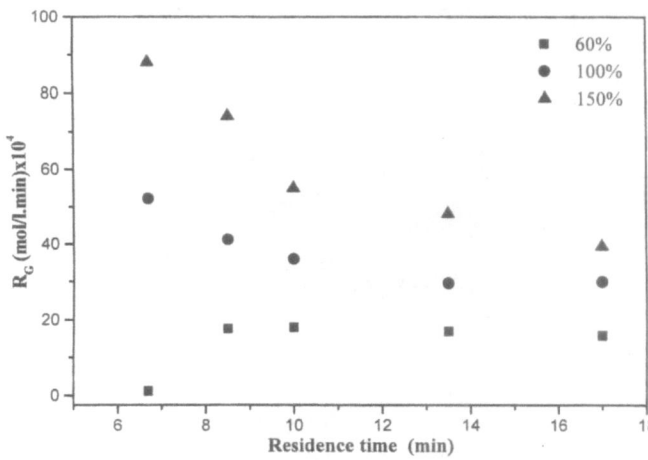

Fig. 2 Variation of the rate of precipitation of titania as a function of the residence time

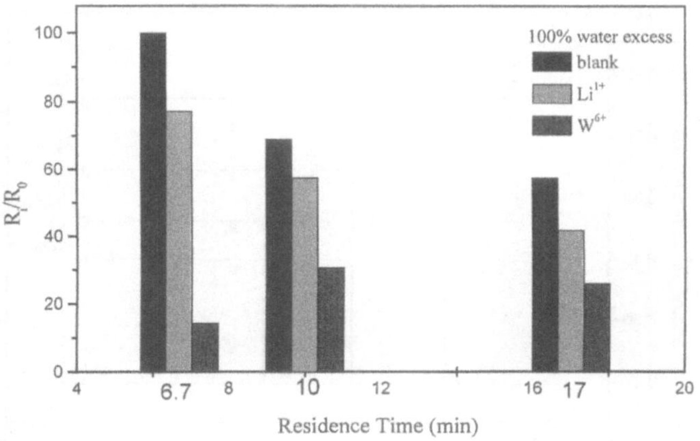

Fig. 3 Relative reduction of the rates of titania precipitation due to the presence of dopant ions

parabolic dependence of the rates of titania formation on the residence time was found. This dependence is similar to the rate–supersaturation dependence of a large number of sparingly soluble salts. It should be noted, however, that at low water excess the rates did not change significantly. For long residence times, the titanium ion concentration at steady state in the bed remained constant. As a result, low values of Ti^{4+} at steady state corresponded to low supersaturations and, therefore, to low mass deposition rates. The effect of residence time on the linearity is shown in Fig. 4. The presence of foreign ions in the aqueous phase during the precipitation resulted in a decrease in the precipitation rate for all water relative excesses tested. The dependence of the precipitation rates on the residence times is shown in Fig. 3.

The powder X-ray diffraction spectra in Fig. 5 showed that the precipitated solid phase was amorphous

titania irrespective of the water excess, the residence time or the presence of foreign ions. The precipitate upon heating at 500 °C for 20 h showed the characteristic peaks of anatase, suggesting the conversion of the initially amorphous precipitate to the crystalline anatase phase. No traces of rutile were detected at this temperature.

The specific surface areas measured for the precipitates formed both in the presence and the absence of foreign ions at different temperatures are shown in Table 4. No differences in the specific surface area value between doped and undoped solids were found. The low values of the specific surface area at 25 °C may be ascribed to the fact that during the course of precipitation solvent was trapped in the pores and could not be removed at the stage of drying. The cleaning procedure of the surface (heating up to 100 °C under vacuum) leads to clean surfaces with a large specific surface area value. This value decreased dramatically when the solid was pretreated thermally. Thermal treatment accelerated

Table 2 Experimental fluidization parameters

τ		u_f (m/h)	Q_f ($\times 10^4$ m³/h)	L (m)[a]	V ($\times 10^5$ m³)[a]
h	min				
0.111	6.7	2.037	3.6	0.22	4.008
0.141	8.5	2.037	3.6	0.28	5.076
0.166	10.0	2.037	3.6	0.33	5.976
0.225	13.5	2.037	3.6	0.45	8.100
0.283	17.0	2.037	3.6	0.57	10.10

[a] Values refer to the fluidization state of the crystallizer

Table 3 Calculations of fluid velocities and porosity of the bed

$Re_f \sim 0.1$					
u_{mf} (m/h)		u_τ (m/h)		ε	
D_p = 0.2 mm	D_p = 0.8 mm	D_p = 0.2 mm	D_p = 0.8 mm	D_p = 0.2 mm	D_p = 0.8 mm
0.22	3.57	61	977	0.48	0.26

154

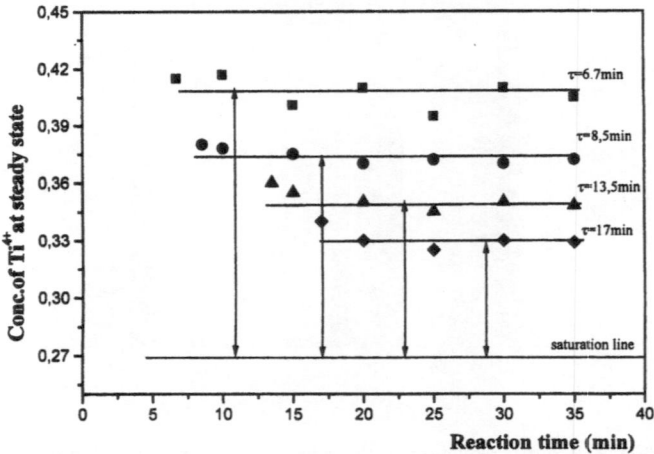

Fig. 4 Concentration versus time profiles of titanium at steady-state operation for various residence times

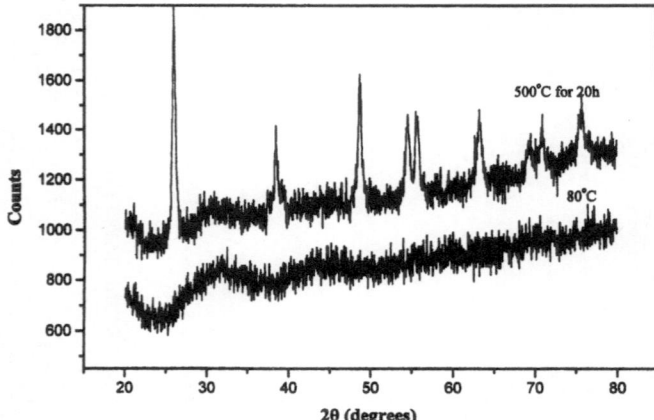

Fig. 5 Powder X-ray diffraction of titania prepared by the fluidized method

the conversion of amorphous to crystalline titania polymorphs. The residence time of the particles in the crystallizer affected the size and the size distribution of the titania particles prepared as may be seen in Table 5. $d_{50\%}$ is the mass media diameter and is taken as the mean diameter of the particles. SD is the geometric standard deviation of the sample and is a measure of the size dispersion. For monosized particles of ceramics and pigments geometric SDs of 1.20 or less are required [15].

As may be seen in Table 5 nanoparticles with a low degree of dispersion were obtained for high residence times; therefore, bed geometries, with high values of L/d and d/d_p should be preferred for the preparation of particles of this order of magnitude. The longer and thinner the bed is in combination with small filler particles the smaller the mean size of the particles obtained. The correlation between the geometric SD and mean particle mass media diameter is given by

$$SD = [(d - d_{50\%})^2]^{1/2}$$

Conclusions

It was shown that titania particles may be synthesized in a quartz glass bead filled-bed crystallizer by the hydro-

lysis of titanium isopropoxide. The bead size affected the size distribution of the crystallites produced and the presence of doping ions resulted in their incorporation into the titania lattice and in a reduction in the titania precipitation rates. In the absence of foreign ions a parabolic dependence of the titania powders formed by precipitation on the driving force was found. This suggested a surface diffusion-controlled mechanism. The particle size of the titania powders depended on the residence times in addition to the bead size. The longer and thinner the bed and the smaller the mean diameter of the filler particles the lower the particle size of the particles obtained.

Table 5 Mean mass diameters, $d_{50\%}$ and standard deviations, SD, for titania prepared under different experimental conditions

τ (min)	Excess (%)	$d_{50\%}$	SD
8.5	60	0.34	0.82
	100	0.35	0.95
	150	–	–
13.5	60	0.12	0.50
	100	0.09	0.52
	150	0.14	0.66
17	60	0.08	0.39
	100	0.09	0.37
	150	0.08	0.40

Table 4 Specific surface areas (m^2/g) of doped and undoped samples of titania preparations

T (°C)	TiO_2 undoped	TiO_2-W^{6+}	TiO_2-Nb^{5+}	TiO_2-Li^+
25	4.52	7.88	6.61	4.35
100	277.05	–	–	–
300[a]	58.83	59.86	56.65	60.69
500[a]	18.75	16.68	22.20	13.67

[a] Samples were subjected to heating for 20 h

References

1. Suyama Y, Kato A (1976) J Am Ceram Soc 56:146
2. Yan M, Rhodes W, Springer L (1982) Am Ceram Soc Bull 61:911
3. Jean JH, Ring TA (1988) Colloids Surf 29:273
4. Jean JH, Ring TA (1986) Langmuir 2:251
5. Ward DA, Ko EI (1995) Ind Eng Chem Res 34:421
6. Brinker CJ, Scherer GW (1990) Sol-gel science: the physics and chemistry of sol-gel processing. Academic, New York
7. Nynlt J (1992) Design of crystallizers. CRC, Boca Raton
8. Akubuiro EC, Verykios XE (1987) J Catal 103:320
9. Akubuiro EC, Ioannides T, Verykios XE (1989) J Catal 116:590
10. Akubuiro EC, Verykios XE, Ioannides T (1989) Appl Catal 46:297
11. Ioannides T, Verykios XE (1994) J Catal 145:479
12. Martin ST, Herrmann H, Chd W, Hoffmann MR (1994) J Chem Soc Faraday Trans 90:3315
13. Gratzel M, Howe RF (1990) J Phys Chem 94:2566
14. Bond GC, Tahir SF (1991) Appl Catal 71:1
15. Ring TA (1984) Chem Eng Sci 39:1731

Progr Colloid Polym Sci (2000) 115:156–160
© Springer-Verlag 2000

T. D. Dimitrova
F. Leal-Calderon

Colloid forces in model food-type emulsions

T. D. Dimitrova · F. Leal-Calderon
Centre de Recherche Paul Pascal
Av. Schweitzer, 33600 Pessac, France

T. D. Dimitrova (✉)
Laboratory of Thermodynamics
and Physicochemical Hydrodynamics
Faculty of Chemistry
Sofia University, 1 James Bouchier Ave.
1164 Sofia, Bulgaria
e-mail: td@ltph.bol.bg
Tel.: +359-2-962 5310
Fax: +359-2-962 5643

Abstract We measured forces between tiny emulsion droplets by using the magnetic chaining technique. Force versus distance profiles were obtained for emulsions stabilized by Tween 20, Tween 20 bovine serum albumin (BSA) mixtures and β-casein. For emulsions stabilized with Tween 20 alone, we observed classical Derjaguin–Landau–Verwey–Overbeek (DLVO) behavior. At high Tween 20 concentrations the force profiles clearly deviate from the double-layer repulsion. An interpretation of the data is proposed combining classical DLVO with stericlike additional effective repulsion due to the micellar condensation at the droplet surface. The systems containing both Tween 20 and BSA qualitatively show the same kind of behavior: the purely electrostatic repulsion is progressively transformed into a steric one in the presence of BSA. The emulsions stabilized by β-casein obey a different type of behavior: in that case the depletion attraction is found to be of great importance. The data are interpreted assuming additivity of DLVO forces and depletion attraction. The threshold flocculation force as a function of the ionic strength was determined for these systems.

Key words Force versus distance isotherms · Micellar condensation · Proteins · Non-Derjaguin–Landau–Verwey–Overbeek interactions

Introduction

Emulsions are one of the most widespread types of colloids. In the present work attention is paid mainly to mimicking food-type emulsions, i.e. protein-stabilized disperse systems. The model experiments could be divided in two typical groups. The first group consists of all the experiments on macroscopic quantities of batch emulsions [1]. The main drawback of this class of methods is the lack of straightforward quantitative theory that can be applied to describe the results. In emulsion systems the overall stability is governed by the properties of the thin liquid film which separates the droplets [2]. The second group of experimental methods consists of all the techniques designed to mimic the interactions between single droplets and surfactant layers, such as a Scheludko-type cell, a drop pressed against a large homophase, etc. [2]. However in the greater number of cases where large protein films are investigated, film thickness measurements are impossible because the protein aggregates [3] in the film make the interface inhomogeneous. On the other hand, the maximum disjoining pressure attainable in these experiments is considerably lower than the capillary pressure in real emulsions. The magnetic chaining technique [4] (MCT) provides the possibility to study the phenomena in conditions very close to those in practical systems. It works with droplets of colloidal size (about 0.2 μm in diameter), i.e. at true capillary pressures and with liquid interfaces. An important feature of the method is the possibility to explore low force levels: the resolution is about 2×10^{-13} N. We applied the technique to model food-type emulsions. In the present work we present the results and propose some explanations.

Experimental

Tween 20, β-casein and NaN_3 were Sigma products. Bovine serum albumin (BSA) was purchased from Acros. The oil used in the emulsion preparation was ferrofluid (Ferrofluidics). It is a 10 vol% dispersion of ferromagnetic (Fe_2O_3) particles in octane. The grains, 10 nm in size, are stabilized by oleic acid and remain dispersed in the range of magnetic fields applied in the experiment (0–500 G). The principle of MCT is described elsewhere [4]. This technique exploits the properties of paramagnetic monodisperse droplets. The magnetic field induces a dipole in each ferrofluid droplet. The interaction between the magnetic dipoles leads to the formation of linear arrays of particles (chains) which are parallel to the external magnetic field. At very low droplet volume fractions ($\phi < 0.1$ vol%) the chains are one droplet thick and the droplets in the chains remain well separated. The repulsive force between the droplets must exactly balance the attractive force between the dipoles induced by the applied magnetic field. Since the dipoles are monodisperse, this force can be calculated exactly. When perfectly aligned particles (at a separation d) are illuminated by incident white light parallel to the chains, the first-order Bragg condition reduces to $d = \lambda/2n$, where n is the refractive index of the suspending medium and λ is the wavelength of the light Bragg-scattered at an angle of 180°. Because the drops are nondeformable one can determine the interfacial separation, $h = d - 2R$, where R is the droplet radius. Hence MCT provides the possibility to measure independently and simultaneously the force between the droplets and the corresponding distance between them.

Results and discussion

Tween 20

Tween 20 is a nonionic surfactant, with a critical micelle concentration (cmc) of 2×10^{-5} M. We examined the force isotherms at different concentrations of the surfactant in the continuous phase and found a strong, long-range repulsion between the drops. A set of curves is presented in Fig. 1. In all cases 3.08×10^{-4} M NaN_3 was present. The pH was 5.6 ± 0.1. For a concentration of 1 cmc we fitted the data with a Derjaguin–Landau–Verwey–Overbeek (DLVO) curve varying as a free parameter only the surface potential, Ψ_0. The inverse Debye length, κ, was calculated from the NaN_3 concentration. The van der Waals attraction was calculated via the Hamaker formula for nondeformed spheres [2], taking $A_H = 4.1 \times 10^{-21}$ J. The expression used for the electrostatic force was [2]

$$F_{el} = 4\pi\varepsilon\varepsilon_0\Psi_0^2 R^2 \left[\frac{\kappa}{h + 2R} + \frac{1}{(h + 2R)^2}\right]\exp(-\kappa h) \ . \quad (1)$$

Hereafter, ε and ε_0 denote the dielectric permeability in solution and in vacuum, respectively. From the fit we found $\Psi_0 = -22$ mV, which is in very good agreement with the experimentally measured value for the ζ potential

of droplets (-17 ± 3 mV). At higher surfactant concentration, one can see a well-pronounced deviation from the classical DLVO profile. This effect is evident for all Tween 20 concentrations above about 50cmc. The curves are roughly linear (in semilog plot) at large distances, but clearly deviate from linearity at short distances. We checked experimentally that no Ostwald ripening or depletion attraction was present. We measured (employing dynamic light scattering) the hydrodynamic radii of our emulsion droplets at different concentrations of Tween 20 in the continuous phase. As shown in Fig. 2, there is a continuous increase in the size of the drops in the interval of surfactant concentrations between 1 and 100 cmc. The diameter of the Tween 20 micelles is 7.3–7.5 nm and the diameter of our ferrofluid droplets at 1 cmc is 176 nm. Hence for concentrations higher than

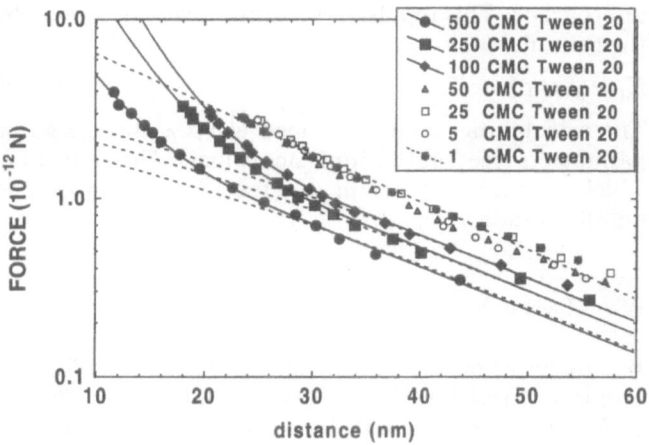

Fig. 1 Force versus distance profiles for emulsions, stabilized with Tween 20 at different surfactant concentrations expressed as multiples of the critical micelle concentration (*cmc*) (*points*). The *dashed lines* are Derjaguin–Landau–Verwey–Overbeek fits and the *solid lines* are the best fits for the highest Tween 20 concentrations (see text for details)

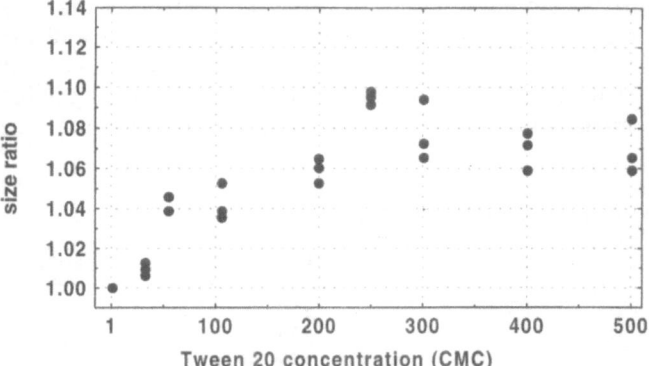

Fig. 2 Relative increase in the mean droplet diameter of the ferrofluid emulsion droplets, stabilized with Tween 20, as a function of the surfactant concentration in the continuous phase

100 cmc the effective size is increased by around two micellar diameters. An explanation for both the force profile evolution and the increase in the hydrodynamic radius at higher Tween 20 concentration may be found in the adsorption of micelles at the surface of the droplets. The micellar layers introduce "roughness" at the surface and, thus, modify the electrostatic repulsion at short distances.

Adopting the hypothesis of micellar condensation at the interface we fitted the observed profiles in the following way. At long distances, we supposed that the operative force is of DLVO-type and from the fit we determined the apparent surface potential. Thus, the values obtained were further used for the calculation of the total force over the whole distance range.

At short distances, we assumed additivity of the DLVO force and the additional stericlike repulsion which arises when the layers of adsorbed micelles are approached. This repulsion was empirically described by the formula

$$F_{st} = f_0 \exp(-h/\lambda) \ . \tag{2}$$

The calculations were performed using λ and f_0 as free parameters (see Table 1 for exact values). The value of λ, which determines the range of the force, is about one micellar radius.

BSA–Tween 20

We determined the force versus distance profiles of droplets covered with mixed layers of Tween 20 and BSA. The ionic strength was 3.08×10^{-4} M, pH 5.8. At constant Tween 20 concentration, when the protein content is increased from 0 to 0.4 wt%, there is a progressive change with respect to classical double-layer repulsion. Typical results are shown in Fig. 3, where the Tween 20 concentration is kept at 5 cmc. The profiles in the presence of the same quantity of BSA at different surfactant content are very similar and do not depend on the Tween 20 concentration. This suggests that the presence of the protein dominates the behavior and the properties of these systems. ζ-potential measurements of the droplets gave values of 1–3 mV, i.e. there is a negligible contribution from the electrostatic repulsion. For that

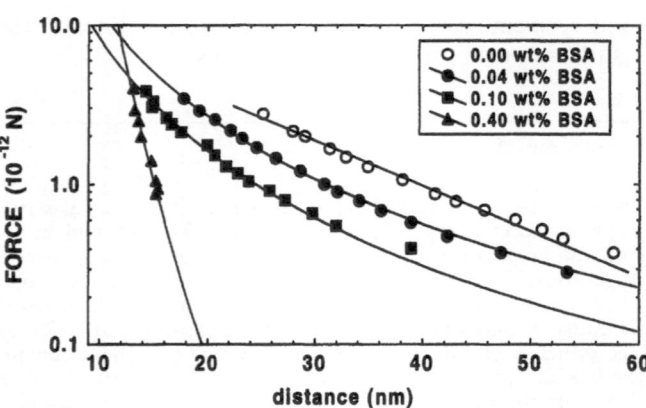

Fig. 3 Force versus distance profiles (*points*) for emulsions stabilized with mixed bovine serum albumin (*BSA*)–Tween 20 adsorption layers. The total concentration of Tween 20 is 5 cmc. The *lines* are the best fits

reason, we fitted the experimental points for mixed Tween 20–BSA systems with a combination of van der Waals attraction and power-law steric repulsion, assuming additivity. The expression used for the steric force was

$$F_s = f_0^{st} (h/h_{min})^{-\alpha} \ , \tag{3}$$

where α accounts for the range of the interaction. The value scaling factor, h_{min}, is the minimum distance measured in each case. The fits were obtained using f_0^{st} and α as free parameters (Table 2).

Note that f_0^{st} and α are effective parameters characterizing the mixed adsorbed layers. The change in α (i.e. the range of the force) with the increase in BSA content can be attributed to some changes in the adsorption and/or structural parameters of the protein–surfactant mixed layer present at the interface. It was proven [5] that a low-molecular-weight surfactant and a globular protein interact in the bulk solution forming a complex. The lack of detailed knowledge concerning the interfacial properties of such complexes makes any further interpretation of the numerical results rather difficult.

β-Casein

β-Casein is a milk protein, which forms micelles or submicelles of various sizes. We investigated the force

Table 1 Fitting parameters for Tween 20 stabilized emulsions

Concentration (criticl micelle concentration)	Ionic strength (M)	Ψ_0 (mV)	f_0 (pN)	λ (nm)
≤50	3.08×10^{-4}	−22.1	–	–
100	3.08×10^{-4}	−17.5 (6[a])	361.3	3.70
250	3.08×10^{-4}	−16.1 (6[a])	140.7	4.12
500	3.08×10^{-4}	−14.3 (5[a])	32.1	4.37

[a] Number of experimental points used for the calculation of the apparent surface potential, Ψ_0, at large distances

Table 2 Fitting parameters for the mixed layers of bovine serum albumin (*BSA*) and Tween 20

| | Tween 20 concentration (critical micelle concentration) | | | | | | | |
| | 5 | | 25 | | 50 | | 250 | |
	f_0^{st}(pN)	α	f_0^{st}(pN)	α	f_0^{st}(pN)	α	f_0^{st}(pN)	α
0.04 wt% BSA	3.61	2.23	3.76	1.73	3.07	2.50	4.46	2.17
0.10 wt% BSA	3.74	2.36	3.21	1.92	3.46	2.44	3.79	2.45
0.40 wt% BSA	3.79	8.97	3.43	3.76	4.89	4.68	–	–

Table 3 Fitting parameters for emulsions stabilized with β-casein

β-casein concentration (wt%)	Ionic strength (M)	Ψ_0 (mV)	Micellar size (nm)
0.10	3×10^{-4}	−26.3	32.2
0.10	1.3×10^{-3}	−43.5	29.2
0.20	3×10^{-4}	−33.9	36.1
0.33	3×10^{-4}	−33.1	24.0
0.50	3×10^{-4}	−48.1	36.3

profiles between β-casein covered droplets at different protein concentrations in the continuous phase (Fig. 4). The ionic strength was 3.08×10^{-4} M, pH 6.2 ± 0.1. The ζ-potential of the droplets was found to be -37 ± 2 mV. None of the profiles measured could be interpreted by DLVO. Instead, we clearly observe that the long-range part of the interaction deviates from linearity (in a semilog plot), which suggests the existence of an additional attractive interaction. The higher the concentration the stronger the deviation from the classical DLVO isotherm. This evolution suggests the occurrence of depletion attraction due to the presence of β-casein micelles. We assumed additivity of the DLVO force and depletion attraction, calculated through the formula [4]

$$F_d = -P_0\pi(R + r + \delta)^2\left[1 - \left\{\frac{R + h/2}{R + r + \delta}\right\}^2\right], \qquad (4)$$

where P_0 is the osmotic pressure exerted by the β-casein micelles and r is the radius of the depleting species. δ is an extra exclusion distance [4] that accounts for the electrostatic repulsion between the β-casein micelles and the droplet surface. For calculating P_0 we adopted the perfect-gas approximation. The surface potential and the size of the depleting species were used as adjustable parameters (Table 3). The calculated surface potential is almost constant and is in agreement with value of the ζ-potential (-37 ± 2 mV). The diameters obtained from the fit do not vary much also. A recent study [6] has revealed that the size of the micelles is found not to vary in the range of protein concentrations between 0.1 and 1 wt%. In Ref. [6] the diameter of the β-casein micelles was found to be 26 ± 1 nm. Our interpretation is also

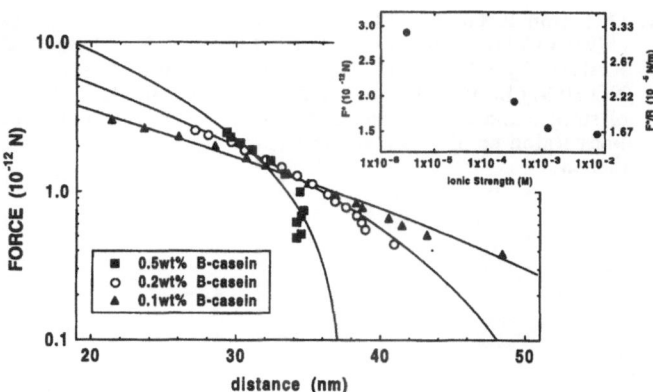

Fig. 4 Force versus distance profiles for emulsions stabilized with β-casein at different protein concentrations (*points*). The *lines* are the best fits. *Inset*: the threshold flocculation force as a function of the ionic strength

consistent with the macroscopic observation of flocculation in emulsions stabilized with sodium caseinate [1].

The depletion effect is well pronounced at all ionic strengths. The increase in the salt content leads to suppression of the electrostatic repulsion, i.e. to larger effective importance of the depletion attraction. We performed a set of experiments, where the β-casein concentration was kept at 0.1 wt%. The data were fitted as described previously. At all salt concentrations we estimated the threshold force, F^*, which induces irreversible flocculation. Indeed, above this threshold force, we observed under the microscope that the droplets remained irreversibly chained even in the absence of a magnetic field. The dependence of F^* on the ionic strength is shown in the inset of Fig. 4. The decrease in the resistance against flocculation is due to the decreased electrostatic repulsion at higher salt concentration.

Concluding remarks

The MCT was applied to food-type emulsions and force versus distance profiles for liquid interfaces covered by proteins were obtained. We investigated the interactions between droplets covered with Tween 20 and found that the presence of a large excess of surfactant modifies the repulsion most probably because micelles condense at the oil/water interface. We studied two proteins – BSA and β-casein. They represent two important classes of proteins – globular and disordered ones, respectively. The presence of protein at the interfaces strongly modifies the force–distance isotherms, in all cases giving rise to specific non-DLVO forces, which are believed to be steric repulsion in the case of BSA and depletion attraction in the case of β-casein. We hope that our results will provide some guidance in understanding the complicated behavior of protein-stabilized emulsions.

Acknowledgements This work is financially supported by Laboratoire Franco-Bulgare. The authors are indebted to Ivan Ivanov and Theodor Gurkov for fruitful discussions.

References

1. Dickinson E, Golding M, Povey MJW (1997) J Colloid Interface Sci 185:515
2. Kralchevsky PA, Danov KD, Denkov ND (1997) In: Birdi KS (ed) Handbook of surface and colloid chemistry. CRC, Boca Raton pp 333–494, and references therein
3. Gurkov TG, Marinova KG, Zdravkov AZ, Oleksiak C, Campbell B (1998) Prog Colloid Polym Sci 110:263
4. Mondain-Monval O, Leal-Calderon F, Bibette J (1996) J Phys II 6:1313
5. Turro NJ, Xue-Gong L, Ananthapadmanabhan KP, Aronson M (1995) Langmuir 11:2525
6. Leclerc E, Calmettes P (1997) Phys Rev Lett 78:150

Progr Colloid Polym Sci (2000) 115:161–165
© Springer-Verlag 2000

FOOD COLLOIDS

L. Ambrosone
A. Ceglie
G. Colafemmina
G. Palazzo

NMR studies of food emulsions: the dispersed-phase self-diffusion coefficient calculated by the least variance method

L. Ambrosone (✉) · A. Ceglie
Consorzio per lo sviluppo dei Sistemi a
Grande Interfase c/o Dept. Food
Technology (DISTAAM)
Università del Molise via De Sanctis
86100 Campobasso, Italy
e-mail: ambroson@hpsrv.umimol.it
Tel.: +39-874-404715
Fax: +39-874-404652

G. Colafemmina · G. Palazzo
Dipartimento di Chimica
Università di Bari, via Orabona 4
70126 Bari, Italy

Abstract A method based on the minimization of the variance is developed for the analysis of the pulsed field gradient NMR echo decays in emulsion systems. The method has been applied successfully to margarines and water-in-olive oil emulsions. Our tests suggest the method might be used in all experimental situations of restricted diffusion in emulsions. For normal and light margarine a different droplet size distribution is observed, the first presenting a bimodal size distribution, while the second has a unimodal one. Also the water-in-olive oil emulsion shows a bimodal distribution.

Key words Polydispersity · Restricted diffusion · Food emulsions

Introduction

To better understand the importance of the droplet size distribution in the study of food emulsions, such as margarine, we discuss briefly some properties of these real-life systems [1–3]. Ordinary margarine is a water-in-oil (w/o) emulsion having fat as a continuous phase. Systems containing 80 wt% fat or more are referred to as "normal margarines", while those containing 60 wt% or less are called "light margarines" or "light spreads". The closed water droplet structure protects the margarine from microbial growth; the fat phase affects the taste (mouthfeel) [2]. The stability of a food emulsion with respect to microbial deterioration is maximized by a closed w/o structure with small radii. In light spreads, the dense emulsion structure may deteriorate the sensory properties.

In heterogeneous systems the diffusion is restricted by the presence of obstacles [4] and pulsed field gradient (PFG) NMR experiments yield information about the confining geometry as experienced by the diffusing particle [5–7]. Recently, there has been a lot of interest in using NMR techniques to obtain information on microgeometry. The reasons for this may be found essentially in four factors:

1. The NMR measurements can be performed on optically opaque sample.
2. The technique is nondestructive, which is very important if one wishes to follow the emulsion over a long time.
3. The results are averaged over the ensemble of droplets.
4. The method is rapid in comparison with some of the others (e.g. optical microscopy).

Experimental

Materials

Two samples of margarine (with 80 wt% and 60 wt% fat) and extravirgin olive oil of a commercial brand were used. The w/o emulsion was prepared by weighing equal amounts of extravirgin olive oil and water and shaking them by hand. The system separates into an aqueous phase and an oil phase. The oil phase constitutes the emulsion where the NMR measurements were performed.

PFG-NMR self-diffusion measurements

The water diffusion behavior in the emulsion samples was measured with the Fourier transform PFG technique by monitoring the -OH peak decay. The measurements were performed in 5-mm NMR tubes on a TESLA BS-587A spectrometer operating at 80 MHz for protons equipped with an Autodif 504 pulsed gradient unit (Stelar). The field gradient strength was calibrated using the dimethyl sulfoxide self-diffusion coefficient [8]. The temperature was fixed at 25.0 ± 0.1 °C with a Stelar VTC9 temperature controller.

Results and discussion

In moderately dilute emulsions, where the interactions between the droplets are negligible and the droplet shape is spherical, it is possible to analyze the self-diffusion data in terms of a collection of isolated droplets. In previous work [9, 10] we showed that under these conditions the echo intensity in a classical PFG experiment is given by

$$I(\delta) = \int_0^\ell E(\delta, R)\Phi(R)\mathrm{d}R \ , \tag{1}$$

where $\ell = \sqrt{2D\Delta}$ (Δ is the experimental time scale) is the diffusion length, $E(\delta, R)$ is the contribution to the echo coming from each isolated sphere, and $\Phi(R)$ is the volume-fraction distribution function defined by

$$\Phi(R)\mathrm{d}R = \frac{R^3 P(R)\mathrm{d}R}{\int_0^\ell R^3 P(R)\mathrm{d}R} \ , \tag{2}$$

$P(R)$ being the droplet fraction with radius ranging from R to $R + \mathrm{d}R$.

Equation (1) is transformed into an equation which is operationally more effective by introducing the dimensionless variable $z = R/\ell$ and the scaled distribution $\Phi(R)\mathrm{d}R = \Psi(z)\mathrm{d}z$.

The echo intensity now reads

$$I(\delta) = \int_0^1 E(\delta, z)\Psi(z)\mathrm{d}z \ . \tag{3}$$

Mathematically speaking Eq. (3) is a Fredholm integral equation of the first kind. The solution of this equation gives the volume-fraction distribution function and can be obtained using our "direct method" [10, 11]; however, whatever the method for solving Eq. (3) the result depends on the diffusion length and then on the choice of the self-diffusion coefficient.

For evaluating the coefficient D directly from experimental data, we showed elsewhere [12] that for $D\delta/R^2 < 1$ Eq. (3) could be expanded as

$$f(\delta) = [1 - I(q)]/\delta^2 = \gamma^2 g^2 \frac{\mu_2}{5} - \frac{1}{3}\gamma^2 g^2 D\delta + \cdots \ , \tag{4}$$

where μ_n is the nth moment of the volume distribution function. For $D\delta/R^2 > 1$, Eq. (3) gives

$$g(\delta) = 1 - I(q, R) = -\frac{166}{7875}\frac{\gamma^2 g^2}{D^2}\mu_6 + \frac{16}{175}\frac{\gamma^2 g^2}{D}\mu_4\delta + \cdots \ , \tag{5}$$

These equations are the basis for the computation of D. Obviously the applicability of these equations is strictly linked to the mean size of the droplets.

For small droplets only Eq. (5) holds [13] and, since the coefficients in Eq. (5) depend simultaneously on D and μ_n, it is impossible to evaluate D.

Then we solve Eq. (3) by means of the direct method, allowing D to assume a spread of values $D^{(1)}, D^{(2)}, D^{(3)}, \ldots, D^{(m)}$. Let $I_i^{(1)}, I_i^{(2)}, \ldots, I_i^{(m)}$ denote the corresponding echo intensities calculated using the volume-fraction distributions obtained with each $D^{(m)}$. The vector which minimizes the quantity

$$\sigma^{(m)} = \frac{\sum_{i=1}^N \left(I_i - I_i^{(m)}\right)^2}{N} \ , \tag{6}$$

represents the mean square deviation and it is used in connection with the direct method to obtain D and $\Phi(R)$. The procedure allows us to obtain D and the volume distribution function. This can be seen in Fig. 1, where σ is drawn as a function of D for normal and light margarine samples. The function $\sigma(D)$ presents a well-marked minimum in both cases, indicating that the procedure is generic and robust, i.e., it should describe the restricted diffusion independently of the distribution's form and remains stable under slight perturbations (errors). Inspecting the figure carefully, one can also see that the water self-diffusion coefficients, $D = (0.75 \pm 0.08) 10^{-9} \mathrm{\ m^2 s^{-1}}$ and $D = (1.48 \pm 0.06) 10^{-9} \mathrm{\ m^2 s^{-1}}$ for normal and light margarine, respectively,

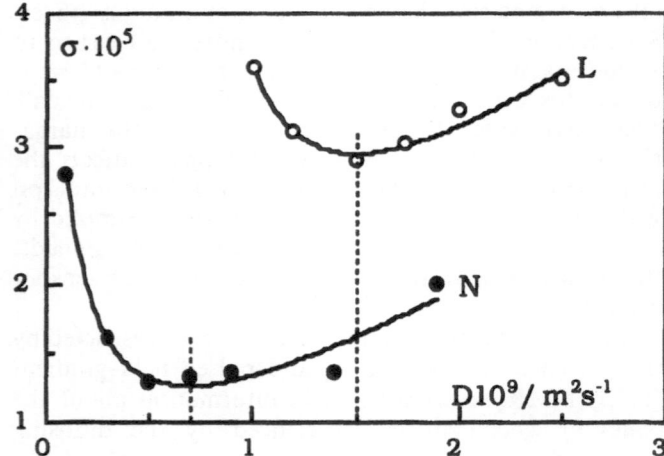

Fig. 1 Variance of the iterative procedure, for the normal (N) and light (L) margarine ($\Delta = 0.220$ s, $g = 0.316$ T m^{-1}). The minimum represents the mean square deviation. *Lines* are a guide for the eye

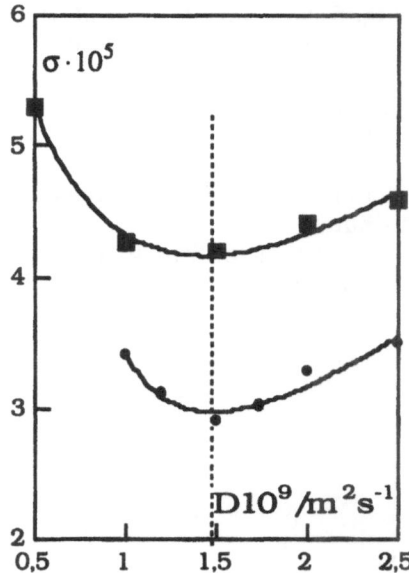

Fig. 2 Variance of the iterative procedure for the light margarine obtained using two different experimental conditions (see text). *Lines* are a guide for the eye

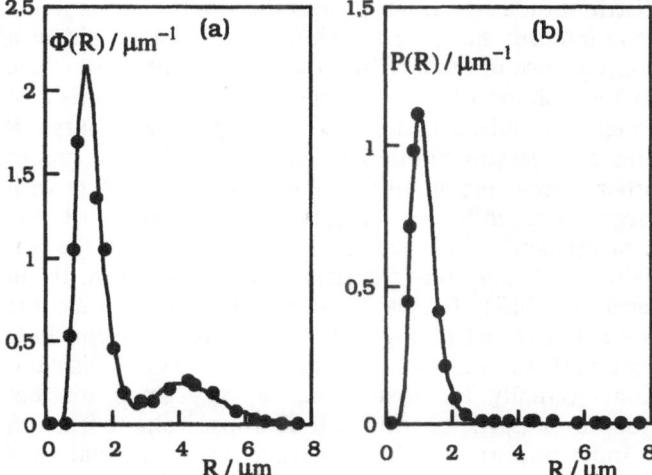

Fig. 3a, b Distribution functions evaluated by the direct method for the normal margarine. **a** Volume-fraction distribution, $\Phi(R)$; **b** number-fraction distribution, $P(R)$. *Lines* are a guide for the eye

are lower than that of pure water ($D = 2.3 \ 10^{-9} \ \mathrm{m^2 s^{-1}}$) for all the systems analyzed.

Presently the only method for determining D in polydispersed systems subjected to restricted diffusion is the polynomial proposed and tested in Refs. [12, 13]. For the light margarine, because of the high water content, the measurements can be performed under different experimental conditions. We chose $\Delta = 0.100$ s and $g = 0.139 \ \mathrm{Tm^{-1}}$ so to have a sufficient number of data points to fit a polynomial to Eq. (4). In Fig. 2 we compare the function $\sigma(D)$ obtained for this case with

that previously analyzed in Fig. 1. As expected, the two minima coincide. The result also suggests that our procedure does not generate artifacts under different conditions. Furthermore, we estimate through Eq. (4) $D = (1.43 \pm 0.4) \ 10^{-9} \ \mathrm{m^2 s^{-1}}$ with a polynomial fit, this being in good agreement with that calculated by minimizing $\sigma(D)$.

$\Phi(R)$ evaluated by the direct method is represented in Figs. 3a and 4a. The normal margarine (Fig. 3a) shows a bimodal distribution, while the light spread shows a single peak (Fig. 4a). Regarding the bimodal distribution, the first peak is located on small radii (about 1 μm) and the second one on larger radii (about 4 μm). If the value of $\Phi(R)$ is averaged, so as to impose a unimodal

Fig. 4a, b Distribution functions evaluated by the direct method for the light margarine. **a** Volume-fraction distribution, $\Phi(R)$; **b** number-fraction distribution, $P(R)$. Lines are a guide for the eye

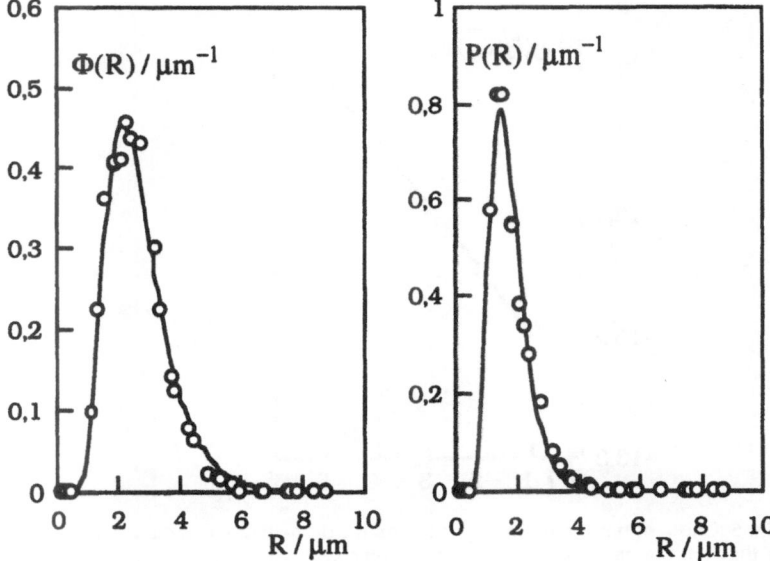

distribution as the other procedures do, the recalculated echo intensity has an error about 30% greater. This is a consequence of the fact that the NMR signal is sensitive to the volume of each droplet (i.e., to the number of water molecules); therefore, the droplets with large R give a large contribution to the decay of the echo. In other words, the small fraction of water droplets with large radii influences greatly the accuracy of the measurements. This fact further demonstrates the advantage of analyzing the experimental data directly in terms of $\Phi(R)$. On calculating $P(R)$ one obtains the result illustrated in Figs. 3b and 4b and can conclude that both the margarines present a unimodal distribution. Actually the system has a very small number fraction of particles with large R whose volume fraction is quite important. This remark is very important for instance, for microbiology since the number of microorganisms in an emulsion is related to the volume fraction of the droplets. There are some reports where the margarine's droplet size has been evaluated [14–16]. Comparing these results with ours, we find that the mean radius of the droplets is in good agreement. Nevertheless, a direct comparison cannot be carried out because the traditional methods for evaluating the droplet size distribution impose a priori the distribution, generally the log–normal one. In our procedure, as described elsewhere [9, 10], there is no hypothesis about the form of the distribution, even though it is well established that many food emulsions are well described by a log–normal distribution [17]. For testing this idea we calculated the first five moments of $P(R)$ for the light spread. Since for this distribution the moments are given by $\mu_n = R_0^n \exp(0.5n^2\sigma^2)$ [9], with $n = 1, 2, 3, \ldots$, a plot of $\ln(\sqrt[n]{\mu_n})$ versus n can be used to check it. The results

reported in Fig. 5 show a good linear fit, thus indicating that the log–normal distribution describes the system quite accurately. In contrast, the data corresponding to normal margarine may be described by a superimposition of two log–normal distributions (the mathematical expression can be found in Ref. [9]). The proposed method is the only one able to discriminate between the n-modal distributions.

It is well known that to disperse a water volume, V, in the form of droplets of radius R in an oil phase an

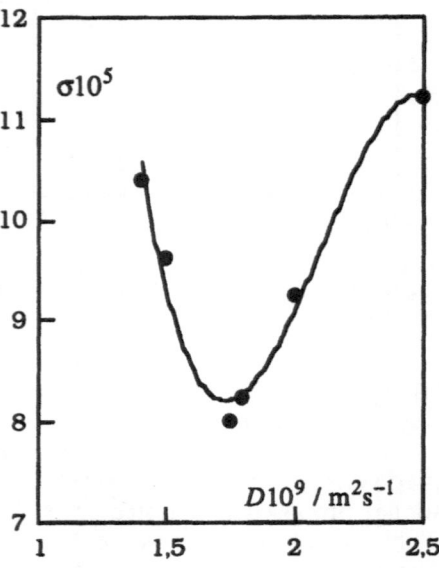

Fig. 6 Variance of the iterative procedure for the water-in-olive oil emulsion. *Lines* are a guide for the eye

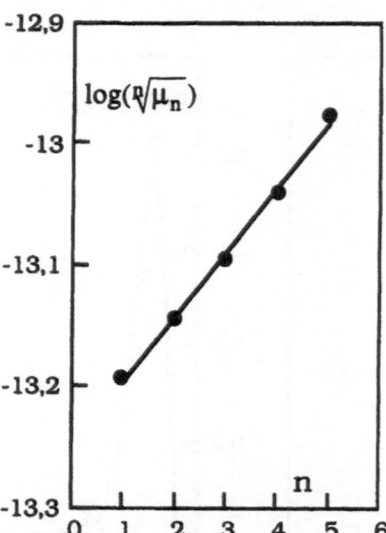

Fig. 5 Check of the log–normal distribution for the experimental data of the light margarine (see text). The *line* represents the best fit

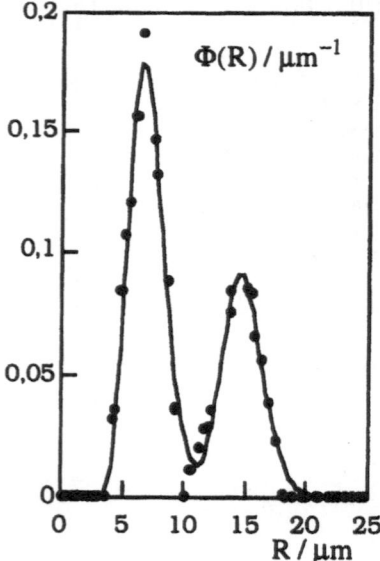

Fig. 7 Bimodal volume-fraction distribution for the water-in-olive oil emulsion. *Lines* are a guide for the eye

energy $E \propto R^{-1}$ is necessary. On the other hand, the small droplets stabilize the emulsion with respect to the separation. Then the evaluated distribution function represents the balance between these two effects. For the normal margarine, where the amount of fat is greater and the chemical energy of dispersion is small, a compromise between the two effects has to be reached. One might say that the first peak of the distribution arises to stabilize the emulsion, while the second makes up for the lack of the dispersion energy. The additives used in light spreads (fat-mimicking agents) exert an emulsifying action, so the dispersion requires less energy compared to that of the normal product. In these occurrences the role of such additives is also to increase the viscosity of the system. On the basis of these observations, for the light margarine case, a unimodal distribution function centered on small radii is expected (Fig. 4b).

Another system where the control of the droplet size is important is olive oil that contains a relatively large amount of water. In edible oils, water droplets play an important role in the oxidation processes as the antioxidant agents are mainly located on the interfacial film; thus, the droplet size determination has technical and economical implications. As shown in Fig. 6, in this case the mean variance also presents a neat minimum located at $D = (1.72 \pm 0.06)10^{-9}$ m^2s^{-1} ($\Delta = 0.220$ s and $g = 0.03$ Tm^{-1}). A low value of g indicates that the droplets have a larger average size. The radius of about 15 μm which is related to the second peak in Fig. 7, is in agreement with this expectation.

Acknowledgement We thank Ministero Universitè Ricerca Scientifica Tecnologica (Italy) Progetti Nazionali Cofinanziati 1998 for financial support.

References

1. McKay RB (1990) Technological application of dispersions, Dekker, New York
2. Dickinson E, Bergensthl B (1997) Food colloids: proteins, lipids and polysaccarides. The Royal Society of Chemistry, Cambridge
3. Malmstern M (ed) (1998) Biopolymers at interfaces. Dekker, New York
4. Kärger J, Pfeifer H, Heink W (1998) Adv Magn Reson 12:1
5. Callaghan PT (1991) Principles of nuclear magnetic resonance microscopy. Clarendon, Oxford
6. Mitra P, Sen Pabitra N (1992) Phys Rev B 143:42
7. Mitra P, Sen Pabitra N, Schwartz LM (1993) Phys Rev B 47:8565
8. Holž M, Mao X, Seiferling D, Sacco A (1996) J Chem Phys 104:669
9. Ambrosone L, Ceglie A, Colafemmina G, Palazzo G (1997) J Chem Phys 107:10756
10. Ambrosone L, Ceglie A, Colafemmina G, Palazzo G (1999) J Chem Phys 110:797
11. Ambrosone L, Ceglie A (1999) Italian Patent Pending FI 99A00004
12. Ambrosone L, Ceglie A, Colafemmina G, Palazzo G (1999) Langmuir 15:6775
13. Ambrosone L, Ceglie A, Colafemmina G, Palazzo G Palazzo G (2000) J Phys Chem B 104:786
14. Van den Enden JC, Waddington D, van Aalst H, Van Kranlingen CG, Packer KJ (1990) J Colloid Interface Sci 140:105
15. Söderman O, Lönnqvist I, Balinov B (1992) In: Sjöblom J (ed) Emulsions – a fundamental and practical approach. Kluwer, Dordrecht, p 239
16. Balinov B, Söderman O, Wärnheim T (1994) J Am Oil Chem Soc 71:513
17. Orr C (1988) In: Becker P (ed) Encyclopedia of emulsion technology, vol 3. Dekker, New York, p 137

Progr Colloid Polym Sci (2000) 115:166–170
© Springer-Verlag 2000

E. O. Arikainen

A diffusing wave spectroscopy study of acidified milk gel under shear

E. O. Arikainen
Department of Pure and Applied Physics
The Queen's University of Belfast
Belfast BT7 1NN, UK
e-mail: k.arikainen@qub.ac.uk
Tel.: +44-1232-273143
Fax: +44-1232-438918

Abstract Diffusing wave spectroscopy (DWS) provides access to information about the internal dynamics of optically dense colloids, which in turn is known to reflect their viscoelastic properties. The rheological properties of optically dense particle gels, such as acidified milk gels (AMG), are of considerable fundamental and applied interest. A cell was specially designed to allow DWS investigations of dense colloids under the condition of shear, in both continuous and oscillatory modes. We demonstrate here the results of the rheo-optical approach to the diagnostics of the condition of an AMG, produced by the addition of glucono-δ-lactone to skimmed milk, under shear. The response of the AMG to shear flow, the structural transitions induced by shear and the effects on the intensity correlation function of the scattered light are discussed qualitatively.

Key words Diffusing wave spectroscopy · Shear · Acidified milk gel · Rheo-optics · Particle gel

Introduction

In particle colloids shear flow can produce a whole variety of effects, for example, shear thickening and thinning, gelation or flocculation, depending on the nature of the interactions involved [1, 2]. The current progress in the rheology of these systems is largely driven by experimental techniques which focus on the microstructure of the systems [3]. Some well-established scattering techniques, for example, static light scattering and small-angle neutron scattering, have been applied in the past to investigate the behaviour of colloidal suspensions under shear [4–7]. Particle colloids have also been investigated by dynamic light scattering for many years; however, the need for dilution or index-matching of the samples precluded its application to systems of high volume fractions (e.g. $\phi > 0.05$) typical of materials of practical importance. The advent of a multiple light scattering technique, called diffusing wave spectroscopy (DWS) [8, 9], that probes the dynamics of the scatterers in optically dense samples opened a whole

range of possibilities. For example, the time-dependent mean-square displacement of the particles in a colloid, the quantity that can be measured by DWS, relates to the bulk viscoelastic properties. This relationship has been demonstrated in a number of chemically and structurally varied systems [10–12].

Acidified milk gels (AMG) are produced by the addition of various acids to milk [13]. They serve as models for coagulated milk products. The main protein component is present in milk in the shape of casein micelles ($d \approx 0.1~\mu m$), which upon acidification coagulate to produce a particle gel [14]. The high volume fraction ($\phi \approx 0.12$) of casein in natural milk means that AMG are optically dense, which makes them an attractive object for DWS studies [12, 15]. Acidification of milk is a complex process involving a reduction in the electrostatic and steric repulsion between the casein micelles and changes to their internal structure [13, 14]. The casein particle aggregates in the AMG are of a fractal nature [16] with a size of the order of a micron. The structure of the AMG network has been established in several electron microscopy studies [18–20]. It consists of the aggregates linked together by strands and sheets with empty spaces between them.

Under applied stress AMG demonstrate non-Newtonian rheological behaviour [17]. In comparison to molecular gels they are rather fragile [14]. Conventional mechanical frequency-dependent rheometric measurements have to be conducted at small values of stress or strain to avoid fracture. Fracture sets an upper limit for the frequency range probed in such measurements as well. The structure and the nature of the mechanical integrity of the AMG suggest that they belong to the class of jammed colloids, the fundamental understanding of which is starting to emerge [21]. In terms of the industrial applications, an understanding of the properties of AMG under the condition of shear flow may lead to improvements in their processing (stirring, flow) and consumer qualities (mouthfeel, spreadability, etc.).

In all rheometric measurements, which involve direct contact between the sample and the probe, the results of such measurements are affected by the probe. The noninvasive nature of rheo-optical measurements [22] allows the on-going processes to be monitored continuously, so rather complex shear/temperature/gelation time/sample history profiles can be observed. Some commercial rheometric devices now include the means of measuring the intensity of the light transmitted through a flowing sample [23]. The approach based on DWS permits investigations of optically dense colloidal systems, particularly that of the diffusion spectrum of the constituent particles on a rather broad timescale. The optically accessible timescale spans the ultrasonic measurements, with frequencies of the order of several megahertz, and mechanical rheometry, where the maximum frequencies are of the order of 10 Hz. The state of the sheared samples can be monitored and quantitatively characterised by recording and analysing the temporal intensity correlation function, the decay of which characterises the diffusion rates. The functional form of the intensity (or field) correlation function depends both on the nature and on the dimension of the system. The difficulties in the interpretation of DWS in flowing systems is that a large class of inhomogeneous flows can be responsible for the same decay law of the temporal correlation function [24]; however, an empirical approach can provide a wealth of information about the condition of a dense colloid under shear flow.

We present here the results of experimental measurements conducted on a rheo-optical experimental setup which allows simultaneous application of shear flow and DWS measurements. The effects of continuous and oscillatory modes of shear in AMG on the intensity correlation function are demonstrated. We attempt to explain them on a qualitative basis.

Experimental

Sample preparation

The samples of AMG used in this study were produced by addition of the acid precursor gluconic-δ-lactone (GDL) to skimmed milk (Dale Farm, UK) with a nominal fat content of 0.05 g, 100 ml at room temperature. The concentration of GDL in the samples was 2 wt%. In the presence of water GDL gradually hydrolyses to regenerate gluconic acid. The time of the gelation process was recorded from the point when the GDL and the milk were added together and mixed by stirring vigorously for 5 min. All the samples were kept at 296 K during gelation and measurements. The acidified milk solution was poured into the cell. The upper plate was lowered until in came into contact with the liquid, which was left to set for 20 h until a rather firm gel formed. For such set unperturbed gels the term "virgin" gels is used throughout this article.

Diffusing wave spectroscopy

The source of light in the DWS setup was an Ar^+ laser (488 nm). The light was delivered to the sample by a single-mode-polarisation-maintaining optical fibre (core size 3.5 μm). A transmission geometry of the DWS experiment with a narrow incident beam [25] was used. The sample was contained in a shear cell (described later). Another single mode optical fibre was positioned on the other side of the sample and, connected to the photomultiplier, it acted as a detector of the scattered light. The polarisation of the scattered light was checked to ensure that it was completely depolarised. The AMG samples were 4–5-mm thick. The single mode optical fibre itself acted as a detector aperture [26]. The signal from the photomultiplier was processed using a Brookhaven BI-9000 AT correlator. The intensity correlation functions [$g^{(2)}(\tau) - 1$] (τ is the delay time) and the transmitted photon counts were recorded.

Shear cell

The cross-section of the shear cell is illustrated schematically in Fig. 1. It consists of two circular, parallel glass plates: a static one at the bottom of the cell and a rotating one (radius 46 mm). The latter is attached to a geared-down motor which can be switched

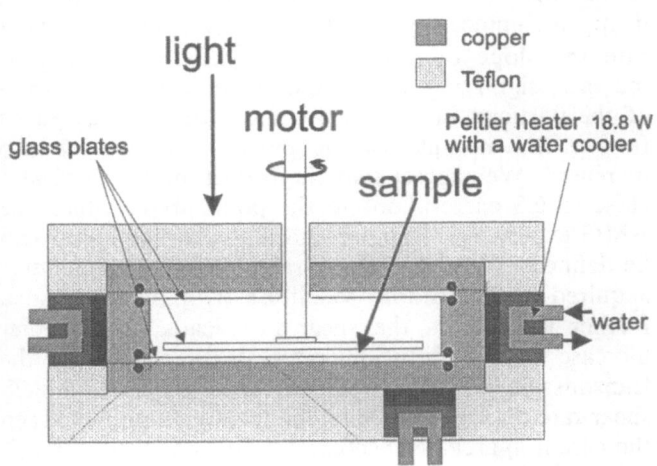

Fig. 1 A schematic illustration of the cross-section of the shear cell

between a continuous rotation mode and an oscillatory one. The speed of the motor was controlled by controlling the applied voltage. The shear rate was calculated as [27]

$$\dot{\gamma} = \Omega r/h \ , \tag{1}$$

where Ω is the angular velocity, r is the distance from the centre of the plate to the point where the incident beam enters the sample and h is the sample thickness. The incident beam of light passes through a window and the scattered light is detected on the other side of the sample through a window on the other side of the cell. The plates are encased in a thermostatic jacket with the temperature maintained by a set of Peltier elements.

Results and discussion

The temporal intensity correlation function $[g^{(2)}(\tau) - 1]$ from the "virgin" AMG and its evolution with increasing shear rate are shown in Fig. 2. The shear rate was increased continuously by increasing the rate of rotation of the upper glass plate. Every time the rate of rotation was increased, 60 s was allowed before the start of the accumulation of the light scattering signal. The amount of transmitted light was recorded as the transmitted photon count averaged over 120 s. The slope of $\ln[g^{(2)}(\tau) - 1]$ (of the fragment of the intensity correlation function at τ ranging between 10^{-4} and 10^{-3} s) and the transmitted photon count measured in the AMG under the continuous and continuously increasing shear are plotted against the shear rate in Fig. 3a and b, respectively. Initially, with the shear rate increasing from 0 to 0.25 the transmission exhibited a slight increase, whereas the slope of $\ln[g^{(2)}(\tau) - 1]$ showed a small decrease. It is possible to explain this effect by the stretching of the gel network in the direction perpendicular to the velocity gradient causing the freezing of some of the degrees of freedom of the casein aggregates. Or, alternatively, the effect points to the "jammed state" [21] in the AMG. At a shear rate value close to 0.3 the amount of transmitted light dropped significantly, by about 9%. At the same shear rate the slope of $\ln[g^{(2)}(\tau) - 1]$ exhibited a dramatic increase, signifying an increase in the rate of diffusion of the particles in the gel. As the shear rate increased further the mobility of the particles in the gel also increased. We suggest that the transition observed at $\dot{\gamma}$ close to 0.3 corresponds to the point above which the AMG experiences fracture and starts to flow, so it can be defined as the yield shear rate. The gel "fragments" acquired more freedom to diffuse with the flow, so a further increase in the shear rate leads to a further increase in the diffusion rates. With regard to the transmission effect, we suggest that at the "critical" shear rate the fractured interconnecting strands between the casein aggregates become free to reorientate (reorientation of the gel strands can be caused by shear flow) and tend to fold in, thus causing an increase in the

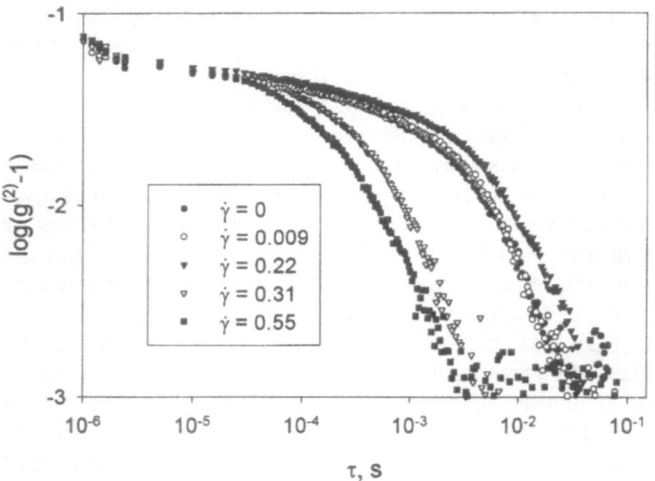

Fig. 2 The effect of the increase in the applied shear rate in the continuous mode on the temporal intensity correlation function

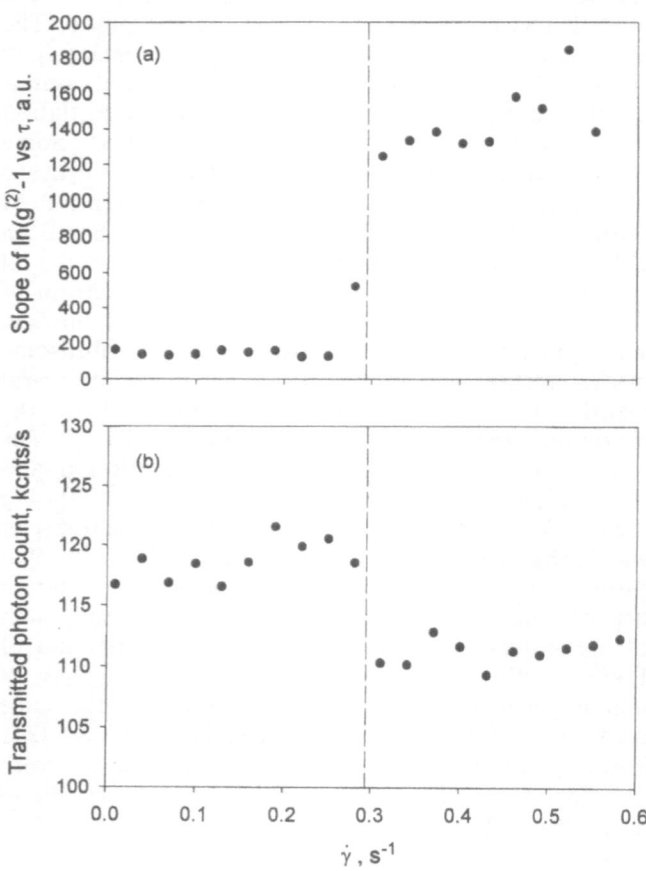

Fig. 3a, b The shear rate dependence of **a** the slope of $\ln[g^{(2)}(\tau) - 1]$ and **b** the transmitted photon count. The *dashed line* indicates the possible position of a structural transition

apparent particle size responsible for the drop in the transmitted photon count.

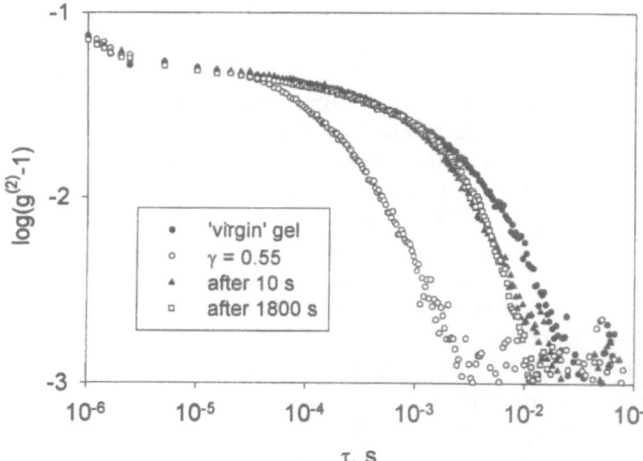

Fig. 4 The temporal intensity correlation function observed in the acidified milk gel (*AMG*) under continuous shear ($\dot{\gamma} = 0.55$) and its recovery after the application of shear was stopped compared to the "virgin" gel

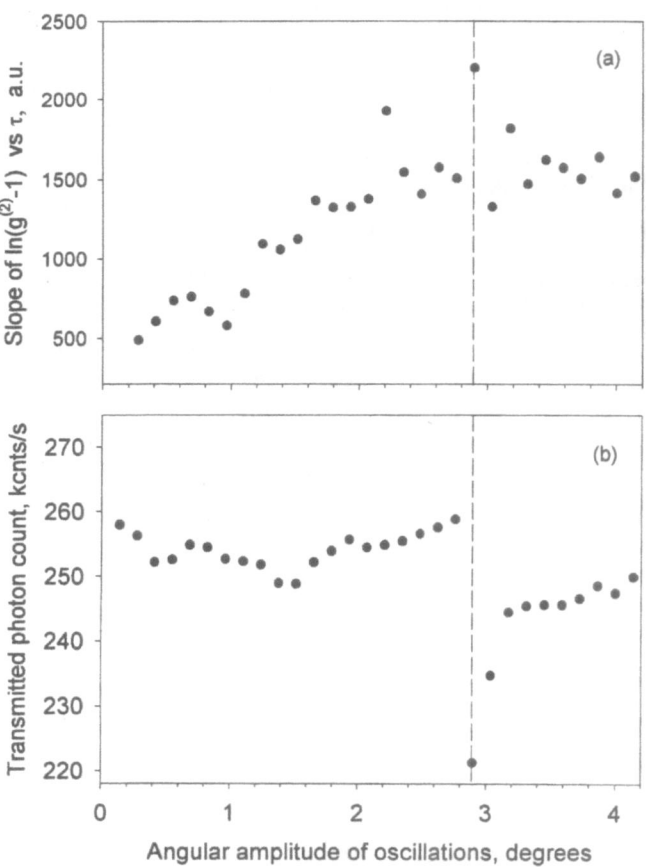

Fig. 5a, b The effect of the angular amplitude of the oscillatory shear on **a** the slope of $\ln[g^{(2)}(\tau) - 1]$ and **b** the transmitted photon count. The *dashed line* indicates the possible position of a structural transition

The process of recovery of the AMG after the previously described shear procedure was monitored. At the point when the shear rate reached a value of 0.55 s^{-1} the motor was stopped. The DWS spectra from the recovering AMG are compared in Fig. 4 to the original spectrum from the unperturbed "virgin" gel. The recovery of the correlation function is rapid. It can be seen that after 10 s the correlation function in the recovering AMG overlaps the original one where τ ranges between 10^{-4} and 10^{-3} s; however, even 1800 s later there is no overlap at τ between approximately 10^{-3} and 10^{-2} s. The longer-time part of the DWS spectrum has been irreversibly affected by the applied shear, at least during the time of observation of the recovery. So after the application of continuous shear, as described earlier, the correlation function in the recovered sample decays faster than in the "virgin" gel. Indeed, the long-time diffusion of the casein aggregates would become faster above the point of gel fracture (the breakup of the links between the particle aggregates) due to the increased probability of the aggregates escaping from the surrounding cages of aggregates. It is reasonable to assume that the extent of "self-healing" in an AMG must depend on the history of the shear it experienced.

An oscillatory shear with a continuously increasing amplitude was applied to an unperturbed gel. The frequency of oscillation was maintained at 1 s^{-1}. A delay of 60 s was allowed between each increase in the amplitude and the start of accumulation of the DWS signal. The slope of $\ln[g^{(2)}(\tau) - 1]$ gradually increases and eventually reaches a plateau (Fig. 5a). This levelling off starts when the transmission drops by about 5% (Fig. 5b). An estimate of the deformation corresponding to the angular amplitude at the point of the observed

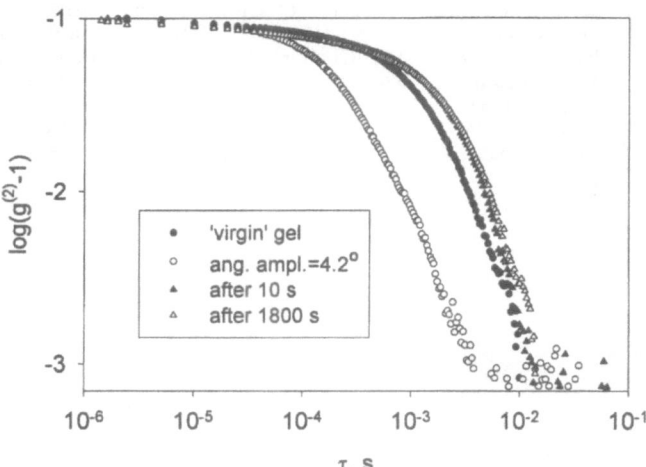

Fig. 6 The temporal intensity correlation function observed in the AMG under oscillatory shear with an angular amplitude of 4.2° and its recovery after the application of shear was stopped compared to the "virgin" gel

"transition" in our experimental geometry gives a value of 0.44. This value of the deformation is close to the ones associated with fracture in AMG determined mechanically [13].

The recovery of the AMG after the application of the oscillatory shear was measured (Fig. 6) and it was very rapid (not longer 10 s). It was possible to see that the recovered gel is "slower" than the "virgin" gel for τ ranging between 10^{-3} and 10^{-2} s. As a result of the application of the oscillatory shear the AMG structure may possibly have rearranged to a more energetically favourable state (the application of (oscillatory) shear is known to induce ordering [28]), which upon the seizure of shear is less likely to be susceptible to rearrangements due to thermal fluctuations.

Conclusions

The effects of the shear flow on the DWS spectrum of the AMG have been demonstrated using a specially designed rheo-optical setup. In the continuous shear mode a sharp increase in the decay rate of the intensity correlation function and a dramatic decrease in the amount of transmitted light were observed. In the oscillatory shear mode a gradual increase in the decay rate of the intensity correlation function and a drop in the amount of transmitted light were observed. We attribute these effects to a structural transition in the AMG, possibly gel fracture. There is an agreement of the trends between our data and the known rheological properties of the AMG. The technique can be effectively used to monitor the recovery/relaxation/"self-healing" of the AMG after the perturbation. The quantitative analysis of the complete DWS intensity correlation functions observed in AMG under shear flow is the subject of our further investigations.

Acknowledgements I am indebted to the late Professor J. C. Earnshaw for the opportunity to work on the project. I am grateful to the International Fund for Ireland for the research grant. Thanks to John Aiken and Alistair Montgomery for the highly skilled technical support.

References

1. van Egmond JW (1998) Curr Opin Colloid Interface Sci 3:385–390
2. Tadros TF (1996) Adv Colloid Interface Sci 68:97–200
3. Goodwin JW, Reynolds PA (1998) Curr Opin Colloid Interface Sci 3:401–407
4. Clarke SM, Ottewill RH, Rennie AR (1995) Adv Colloid Interface Sci 60:95
5. Muzny CD et al (1995) Int J Thermophys 16:337–346
6. Cabane B et al (1997) J Rheol 41:531–547
7. Rennie AR, Clarke SM (1996) Curr Opin Colloid Interface Sci 1:34–38
8. Pine DJ, Weitz DA, Chaikin PM, Herbolzheimer E (1988) Phys Rev Lett 60:1134–1137
9. Maret G (1997) Curr Opin Colloid Interface Sci 2:251–257, and references therein
10. Mason TG et al (1997) J Opt Soc Am A 14:139–149
11. Krall AH, Weitz DA (1998) Phys Rev Lett 80:778–781
12. Arikainen EO et al (1999) Applied Rheology 9:246–253
13. Lucey JA, Singh H (1997) Food Research Int 30:529–542
14. (a) Dickinson E (1992) An introduction to food colloids. Oxford University Press, Oxford, pp 174–199; (b) Dickinson E (1990) Chem Ind 595
15. Dalgleish DG, Horne DS (1991) Milchwissenschaft 46:417–420
16. Walstra P et al (1991) In: Dickenson E (ed) Food polymers, gels and colloids. Royal Society of Chemistry, Cambridge, pp 369–382
17. Lucey JA et al (1997) Int Dairy J 7:381–388
18. Heertje I et al (1985) Food Microstruct 4:267–277
19. Roefs SPFM et al (1990) Colloids Surf 50:141–159
20. Gastaldi E et al (1997) J Food Sci 62:671–675
21. Cates ME et al (1998) Phys Rev Lett 81:1841–1844
22. Wagner NJ (1998) Curr Opin Colloid Interface Sci 3:391–400, and references therein
23. http://www.rheosci.com/oam.htm
24. Bicout D, Maynard R (1993) Physica A 199:387–411
25. Durian DJ (1995) Appl Opt 34:7100–7105
26. Ricka J (1993) Appl Opt 32:2860–2875
27. Hindawi IA, Higgins JS, Weiss RA (1992) Polymer 33:2522–2529
28. Yan YD et al (1994) Physica A 202:68–80

Progr Colloid Polym Sci (2000) 115:171–173
© Springer-Verlag 2000

BIOSYSTEMS

L. Perino-Gallice
E. Bellet-Amalric
A. Braslau
T. Charitat
J. Daillant
G. Fragneto
F. Graner

Adsorbed and free lipid bilayers at the solid–liquid interface viewed by specular and off-specular reflectivity

L. Perino-Gallice · F. Graner
Université Joseph Fourier, Grenoble
France

L. Perino-Gallice (✉) · E. Bellet-Amalric
G. Fragneto
Institut Laue Langevin, LSS Group
BP 156, 38042 Grenoble, France
e-mail: perino@ill.fr
Tel.: +33-4-76207054
Fax: +33-4-76207120

A. Braslau · J. Daillant
Laboratoire SPEC, CEA, Saclay, France

T. Charitat
Institut C. Sadron, Strasbourg, France

Abstract The current biological model membrane constituted of a single bilayer deposited on a substrate has been improved by adding a second bilayer. This one, called "free bilayer", floats 20–30 Å above the first. Carefully optimised X-ray reflectivity measurements have allowed a precise characterisation of the structure out of the plane of the membrane and a good description of the in-plane fluctuations. This work has opened new possibilities for investigating interaction between membrane lipids and soluble proteins, especially for peptides too small to be visible with other techniques.

Key words Reflectometry · Diffuse scattering · Interfaces · Membranes · Fluctuations

The interest of physicists in understanding protein/membrane interactions is growing fast. Simple lipid bilayers deposited on a substrate are often used as biological model membranes [1–3]. The interest in them lies in the fact that they are studied in a bulk aqueous environment and the effect of a small amount of material (e.g. peptides or other small molecules) may be detected by techniques such as neutron or X-ray reflectivity. The limit of this model is the interaction of the bilayer with the solid substrate. We have recently succeeded in depositing two bilayers on flat substrates [4]. The second one, called "free" bilayer, floats 20–30 Å above the first. By interacting less strongly with the substrate, it represents a better model for biological membranes. Double bilayers have been characterised by specular neutron reflectivity [5] and structural information (thickness and composition) has been obtained in the direction normal to the interface [4]. Following these results, a study was started to determine the in-plane structure and fluctuations of the membrane. In the case of a single bilayer a good probe for in-plane imaging is atomic force microscopy (AFM) in noncontact mode. In the case of the free bilayer such measurement is not yet possible. The most promising technique for exploring in-plane structure is off-specular reflectivity [6].

We present here some preliminary results obtained from specular and off-specular reflectivity on single and double bilayers deposited on silicon substrates and studied in bulk water. Since neutron intensities are so far too noisy to extract off-specular data at solid/liquid interfaces, synchrotron radiation was used.

Distearyl phosphatidyl choline lipids were used. They were deposited on 2.5×2.5 cm^2 silicon substrates with the Langmuir–Blodgett and Langmuir–Schaeffer techniques. Details of the deposition procedure are given in Ref. [4]. The specular and off-specular measurements were performed on the BM32 beamline at the ESRF (Grenoble, France).

If we assume that the interface is perfectly smooth, the reflection of the beam on the different interfaces is purely specular. The momentum transfer q, is perpendicular to the plane of the membrane and gives information on the structure in this direction. As a surface is not perfectly smooth the specular signal decreases and a part of the beam is reflected in off-specular directions. The momentum transfer thus has a transverse and an in-plane component. The off-specular signal gives access to the in-plane information: roughness and fluctuations.

The challenge was to measure the signal from X-rays passing through a significant amount of water (27 mm path). X-rays are strongly scattered by water. Nevertheless, the experiment was successful thanks to a very energetic signal of 20 keV. Measurements were made in the incident plane with an incident angle of 1 mrad, smaller than the critical angle for total reflection (1.097 mrad), in order to have maximum signal.

Reflectivity data give information on the interfacial structure via the Fourier transform of the index variation. Data analysis was done by model fitting. The interface was divided into homogeneous layers characterised by constant index, thickness and roughness. The interpretation of the specular data was made in terms of the Born approximation [5], which does not take into account the multiple reflections into a given layer. Since in off-specular measurements an incident angle close to the critical angle is used, this approximation cannot be used and must be replaced by the distorted–wave Born approximation [6]. The characterisation of the single bilayer is essential for understanding the structure of the double bilayer because interpretation of the experimental data is made step by step including the results obtained on simple systems to treat the data of more complicated ones.

We first observed that the diffuse scattering of the substrate alone was very low. The diffusion was mainly due to the adsorbed lipid bilayers. The analysis of the specular signal using a program written by one of us (A.B.) indicates that the single bilayer is strongly adsorbed on the substrate although there are not enough constraints to determine the structure of the bilayer. For the off-specular analysis a mathematical function must be found to model the fluctuations of the membrane (program written by J.D.). We used two kinds of functions. The first one is based on the result obtained by specular reflectivity that the bilayer is strongly adsorbed. Thus, we assumed that the fluctuations could be modelled by a Gaussian function, which is usually used to model the roughness of a solid substrate [7]. By using this model an intensity similar to the experimental one was obtained by adjusting the roughness to a value of the magnitude of the layer thickness. As the specular data indicate that the layer is strongly adsorbed, this conclusion is not convenient. Holes have been observed by AFM in lipid bilayers deposited on the surface of a mica substrate [8]. We, therefore, also modelled the surface by a smooth surface with holes: the fits improved. Nevertheless, the best fit is obtained by an appropriate combination of holes and Gaussian fluctuations (Fig. 1). The holes represent around 20% of the bilayer and have a radius of the order of 70 Å. The in-plane

fluctuations are described by a roughness around 10 Å and a persistence length around 200 Å. The structure is, in fact, much more constrained by the off-specular than by the specular signal. The thickness of the lipid heads is 12.4 Å thick, the contribution from the tails must be distinguished in CH_2 and CH_3 groups with respectively 14.7 Å and 2.6 Å. The parameters obtained from the off-specular fit were introduced in models for the specular signal for both single and double bilayers. For the double bilayer a water layer of 13.5 Å thickness must be added. The results obtained are shown in Fig. 2 and are in good agreement with the experimental points.

Whereas the analysis is still in progress, the first results are promising and the following conclusions can be stated:

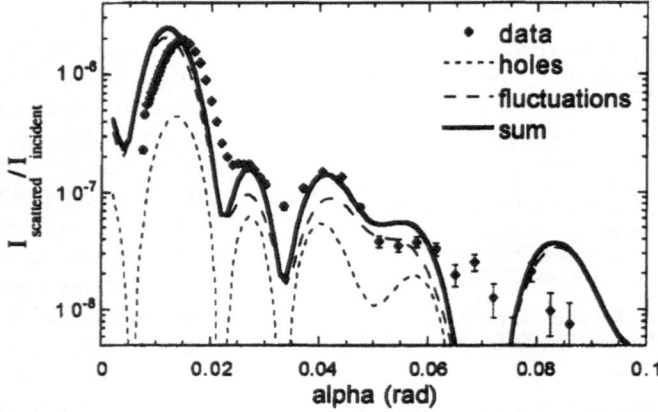

Fig. 1 Experimental (points) and simulated curves (lines) of the off-specular signal as a function of the reflected angle, α, for a distearyl phosphatidyl choline (*DSPC*) bilayer deposited on a silicon substrate at room temperature. (See text for parameters of the fit)

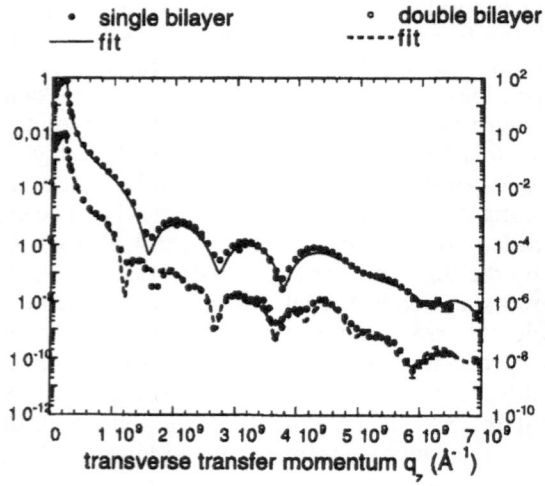

Fig. 2 Experimental (points) and simulated curves (lines) of the specular signal as a function of the transverse momentum transfer, q_z, for a DSPC bilayer and a double bilayer deposited on a substrate

1. As expected, the diffuse scattering of the silicon substrate is negligible due to a very low roughness. The signal is mainly due to the adsorbed lipid bilayers.
2. A very good structural determination has been produced for both single and double bilayers. The off-specular spectra of the single bilayer can be analysed in terms of holes in the structure in addition to Gaussian fluctuations.

This novel technique opens new possibilities for the applications of model membranes. A temperature study is in progress. Its aim is the description of the giant swelling of the system near the gel/liquid main transition. Furthermore, studies on protein/membrane interactions are planned.

References

1. Johnson SJ, Bayerl TM, McDermott DC, Adam GW, Rennie AR (1991) Eur Biophys J 59:289
2. Koening BW, Krueger S, Orts WJ, Majkrzak CF, Berk NF, Silverton JV, Gawrisch K (1996) 12:1343
3. Krueger S, Anker JF, Satija SK, Majkrzak CF, Gurley D, Colombini M (1995) 11:3218
4. Charitat T, Bellet-Amalric E, Fragneto G, Graner F (1999) Eur Phys J B 8:583
5. Sinha SK, Sirota EB, Garoff S (1988) Phys Rev B 38:2297
6. (a) Daillant J, Berlogey O (1992) J Chem Phys 97:5824; (b) Daillant J, Quinn K, Gourier C, Rieutord F (1996) J Chem Soc Faraday Trans 92:505
7. Palasantzas G (1993) Phys Rev B 48:14472
8. Bassereau P, Pincet F (1997) Langmuir 13:7003

Progr Colloid Polym Sci (2000) 115 : 174–180
© Springer-Verlag 2000

H. J. Keh
H. J. Tu

Osmophoresis in concentrated suspensions of spherical vesicles

H. J. Keh (✉) · H. J. Tu
Department of Chemical Engineering
National Taiwan University
Taipei 106-17, Taiwan
e-mail: huan@ccms.ntu.edu.tw
Fax: +886-2-23623040

Abstract The osmophoretic motion of a homogeneous suspension of identical spherical vesicles is considered under conditions of small Peclet and Reynolds numbers. The effects of interaction of the individual vesicles are taken into explicit account by employing a unit cell model which is known to provide good predictions for the sedimentation of monodisperse suspensions of spherical particles. The appropriate equations of conservation of mass and momentum are solved for each cell, in which a spherical vesicle is envisaged to be surrounded by a concentric shell of suspending fluid, and the osmophoretic velocity of the vesicle is calculated for various cases. Analytical expressions of this mean vesicle velocity are obtained in closed form as functions of the volume fraction of the vesicles. Comparisons between the ensemble-averaged osmophoretic velocity of a test vesicle in a dilute suspension and our cell-model results are made.

Key words Osmophoresis · Spherical vesicle · Effects of particle volume fraction · Unit cell model

Introduction

When a vesicle, which is a body of fluid surrounded by a thin semipermeable membrane, is placed in a solution possessing a solute concentration gradient, it advances toward regions of low concentration. This movement of the vesicle is termed osmophoresis [1–3]. In most physically realistic systems, the osmophoretic velocity, $U^{(0)}$, of a spherical vesicle of radius a is related to a uniform solute concentration gradient, ∇c_∞, in an unbounded fluid by the expression [2]

$$U^{(0)} = -\tfrac{1}{2} a L_p RT (1 + \bar{\kappa} + \tfrac{1}{2}\kappa)^{-1} \nabla c_\infty \ , \tag{1}$$

with dimensionless parameters

$$\kappa = \frac{a L_p RT c_0}{D_0} \ , \tag{2}$$

$$\bar{\kappa} = \frac{a L_p RT \bar{c}}{\bar{D}} \ . \tag{3}$$

Here, L_p is the hydraulic coefficient, which is a constant for a given membrane and solvent, $\bar{D}$ and D_0 are the solute diffusion coefficients inside and outside the vesicle, respectively, $\bar{c}$ is the average internal concentration of solute, and c_0 denotes the value of c_∞ at the position of the vesicle center. Typical values in aqueous solutions for the parameters in Eq. (1) are $L_p = 10^{-9}$ m²s/kg, $|\nabla c_\infty| = 10^5$ mol/m⁴ and κ (or $\bar{\kappa}$) = 2.5.

In practical applications of osmophoresis, collections of vesicles are usually encountered, and effects of vesicle interactions will be important. Using a method of reflections, Anderson [3] obtained analytically the migration velocity of two arbitrarily oriented identical spherical vesicles undergoing osmophoresis for the special case of $\kappa = \bar{\kappa} = 0$. He also used this approximate solution for two-vesicle interactions to evaluate the ensemble-averaged osmophoretic velocity in a bounded suspension of identical vesicles to the leading order in the volume fraction of vesicles. Recently, Anderson's analysis has been extended to the general cases of two arbitrary spherical vesicles and of a dilute suspension of spherical vesicles that have a distribution in radius and physical properties [4]. on the other hand, the osmophoretic motion of two arbitrarily oriented spherical vesicles in response to a constant solute concentration gradient was examined by Keh and Yang [5] through an exact representation in spherical bipolar coordinates. Numerical results of correction to Eq. (1) for each vesicle were presented for various values of the size ratio, relative separation, and parameters κ and $\bar{\kappa}$. This numerical solution for two-vesicle interactions was also used to obtain the effect of the volume fraction of vesicles on the mean osmophoretic velocity in a dilute suspension; however, the dependence of the osmophoretic velocity on the vesicle volume fraction for a relatively concentrated suspension has not been investigated yet.

A unit cell model has been employed successfully (and tested against the experimental data) to predict the effect of particle concentration on the mean sedimentation rate in a bounded suspension of identical spherical particles [6–8]. This model involves the concept that an assemblage can be divided into a number of identical cells, one sphere occupying each cell at its center. The boundary-value problem for multiple spheres is thus reduced to the consideration of the behavior of a single sphere and its bounding envelope. Although the Brownian motion of small particles is not included in its analysis, the cell model is of great applicability in relatively concentrated suspensions, where the boundary effect will not be important. In this work, the cell model is used to describe the interactions among osmophoretic spherical vesicles in a monodisperse suspension subjected to a constant solute concentration gradient. The analytical solutions in closed form obtained with this model enable the average osmophoretic velocity to be predicted as functions of the volume fraction of the vesicles for various cases.

Analysis

We consider the steady osmophoretic motion of a uniform three-dimensional distribution of identical spherical vesicles of radius a in a fluid. The uniformly imposed solute concentration gradient, ∇c_∞, equals $-E_\infty \mathbf{e}_z$ and the osmophoretic velocity of the vesicles is $U\mathbf{e}_z$, where $\mathbf{e}_z$ is the unit vector in the positive z direction. As shown in Fig. 1, we employ a unit cell model [6–8] in which each vesicle is surrounded by a concentric spherical shell of suspending fluid having an outer radius b such that the vesicle/cell volume ratio is equal to the vesicle volume fraction, φ, throughout the entire suspension; viz., $\varphi = (a/b)^3$. The origin of the spherical coordinate system (r, θ, ϕ) is set at the center of the vesicle. Our objective is to determine the vesicle velocity, U, in a cell induced by the osmophoretic driving force.

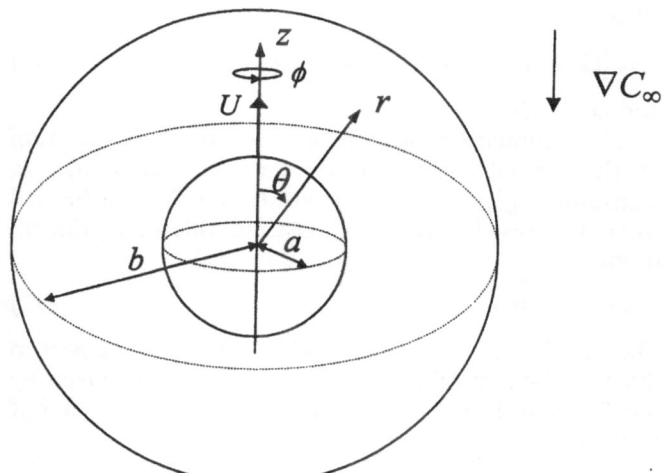

Fig. 1 Geometrical sketch for the osmophoretic motion of a spherical vesicle at the center of a spherical cell

The Peclet number (Ua/D_0) of this axisymmetric problem is assumed to be small. Hence, the equation of continuity governing the concentration distribution, $c_2(r, \theta)$, for the external fluid of constant solute diffusion coefficient is the Laplace equation

$$\nabla^2 c_2 = 0 \quad (a \le r \le b) \ . \tag{4}$$

For the solute concentration field, $c_1(r, \theta)$, inside the vesicle, one has

$$\nabla^2 c_1 = 0 \quad (r \le a) \ . \tag{5}$$

The boundary conditions require that no solute be transferred across the semipermeable membrane of the vesicle. Also, the local solute concentration gradient at the outer (virtual) surface of the cell is parallel to the uniformly applied gradient $-E_\infty \mathbf{e}_z$. Thus [2, 5],

$$r = a: \quad \frac{\partial c_2}{\partial r} = \frac{\kappa}{a}[c_2 - c_0 - (c_1 - \bar{c})] \ , \tag{6}$$

$$\frac{\partial c_1}{\partial r} = \frac{\bar{\kappa}}{a}[c_2 - c_0 - (c_1 - \bar{c})] \ ; \tag{7}$$

$$r = b: \quad \frac{\partial c_2}{\partial r} = -E_\infty \cos\theta \ , \tag{8}$$

where the definition of the parameters κ and $\bar{\kappa}$ is given by Eqs. (2) and (3).

The solution to Eqs. (4)–(8) is

$$c_1 = \bar{c} - 3A\bar{\kappa}E_\infty r\cos\theta \quad (r \le a) \ , \tag{9}$$

$$c_2 = c_0 - A\left[(2 + 2\bar{\kappa} + \kappa) + (1 + \bar{\kappa} - \kappa)\frac{a^3}{r^3}\right]E_\infty r\cos\theta$$

$$(a \le r \le b) \ , \tag{10}$$

where

$$A = [(2 + 2\bar{\kappa} + \kappa) - 2(1 + \bar{\kappa} - \kappa)\varphi]^{-1} \tag{11}$$

and $\varphi = (a/b)^3$.

The boundary condition of the solute concentration at the virtual surface $r = b$ may be taken as the distribution giving rise to the gradient $-E_\infty \mathbf{e}_z$ in the cell when the vesicle does not exist. In this case, Eq. (8) becomes

$$r = b: \quad c_2 = c_0 - E_\infty r \cos\theta \ . \tag{12}$$

The solution of the governing Eqs. (4) and (5) subject to the boundary conditions (Eqs. 6, 7, 12) is also given by the form of Eqs. (9) and (10), but with parameter A defined as

$$A = [(2 + 2\bar{\kappa} + \kappa) + (1 + \bar{\kappa} - \kappa)\varphi]^{-1} \ . \tag{13}$$

For the special case of $\kappa = 1 + \bar{\kappa}$, both Eqs. (11) and (13) give $A = (3\kappa)^{-1}$, and the concentration gradient in the external fluid resulting from Eq. (10) equals the constant imposed value everywhere.

With knowledge of the solution for the solute concentration distribution, we can now proceed to find the flow field in a cell. Due to the low Reynolds number, the fluid motion caused by the osmophoretic migration of the vesicle is governed by the quasisteady fourth-order differential equation for axisymmetric creeping flows:

$$E^4\Psi = E^2(E^2\Psi) = 0 \ , \tag{14}$$

where $\Psi(r, \theta)$ is the Stokes stream function and E^2 is the Stokesian operator. The boundary condition for the fluid velocity at the vesicle surface is [2, 5]

$$r = a: \quad v_r = U \cos\theta + L_pRT[c_2 - c_0 - (c_1 - \bar{c})] \ , \tag{15}$$

$$v_\theta = -U \sin\theta \ , \tag{16}$$

where U is the osmophoretic velocity of the vesicle to be determined. On the outer (virtual) boundary of the cell, the Happel model [6] assumes that the radial velocity and the shear stress are zero; viz.,

$$r = b: \quad v_r = 0 \ , \tag{17}$$

$$\tau_{r\theta} = \eta\left[r\frac{\partial}{\partial r}\left(\frac{v_\theta}{r}\right) + \frac{1}{r}\frac{\partial v_r}{\partial \theta}\right] = 0 \ . \tag{18}$$

A solution to Eq. (14) suitable for satisfying boundary conditions on the spherical surfaces is [9]

$$\Psi = (Cr^{-1} + Dr + Er^2 + Fr^4)\sin^2\theta \ . \tag{19}$$

The constants $C, D, E,$ and F are to be determined from Eqs. (15)–(18) using Eqs. (9) and (10). The procedure is straightforward, with the result (for the flow external to the vesicle)

$$C = 4a^3\omega[U - 3A(-1 + 2\varphi^{1/3})V] \ , \tag{20}$$

$$D = -4a\omega[(3 + \varphi^{5/3})U - 3A(1 + \varphi + 2\varphi^{5/3})V] \ , \tag{21}$$

$$E = 2\omega[(6\varphi^{1/3} - \varphi + 2\varphi^2)U - 3A(2\varphi^{1/3} + \varphi + 4\varphi^2)V] \ , \tag{22}$$

$$F = -2a^{-2}\omega[\varphi^{5/3}U + 3A(\varphi^{5/3} - 2\varphi^2)V] \ , \tag{23}$$

where $V = aL_pRTE_\infty$ (a characteristic migration velocity of the vesicle) and

$$\omega = [4(4 - 6\varphi^{1/3} + \varphi + 3\varphi^{5/3} - 2\varphi^2)]^{-1} \ . \tag{24}$$

The drag force (in the z direction) exerted by the external fluid on the vesicle is [9]

$$F_d = 8\pi\eta D \ . \tag{25}$$

Because the vesicle is freely suspended in the surrounding fluid, the net force exerted by the external fluid on the vesicle must vanish; viz., $D = 0$. With this constraint, Eq. (21) yields the osmophoretic velocity of the vesicle,

$$U = A(1 + \varphi + 2\varphi^{5/3})\left(1 + \frac{1}{3}\varphi^{5/3}\right)^{-1}V \ , \tag{26}$$

where A is given by Eq. (11) or (13).

If the Kuwabara model [7] for the boundary conditions of the fluid flow at the virtual surface of the cell, which assumes that the radial velocity and the vorticity are zero, is used, Eq. (18) is replaced by

$$r = b: \quad (\nabla \times \mathbf{v})_\phi = \frac{\partial v_\theta}{\partial r} + \frac{v_\theta}{r} - \frac{1}{r}\frac{\partial v_r}{\partial \theta} = 0 \ . \tag{27}$$

With this change, the stream function, Ψ, can still be expressed in the form of Eq. (19), and the coefficients $C, D, E,$ and F should be determined by boundary conditions represented by Eqs. (15)–(17) and (27). The result is

$$C = a^3\omega'[(-5 + 2\varphi)U - 3A(5 - 12\varphi^{1/3} + 4\varphi)V] \ , \tag{28}$$

$$D = 15a\omega'[U - A(1 + 2\varphi)V] \ , \tag{29}$$

$$E = -\omega'[(18\varphi^{1/3} - 5\varphi + 2\varphi^2)U - 3A(6\varphi^{1/3} + 5\varphi + 4\varphi^2)V] \ , \tag{30}$$

$$F = 3a^{-2}\omega'[\varphi U - A(\varphi + 2\varphi^2)V] \ , \tag{31}$$

where

$$\omega' = [4(-5 + 9\varphi^{1/3} - 5\varphi + \varphi^2)]^{-1} \ . \tag{32}$$

The fact that there is no drag force exerted on the vesicle requires $D = 0$, and Eq. (29) gives the vesicle velocity as

$$U = A(1 + 2\varphi)V \ . \tag{33}$$

Results and discussion

Due to the differences in the boundary conditions for the solute concentration and fluid velocity distributions at the virtual surface $r = b$ of the unit cell, four cases of the cell model can be defined:

Case I: the boundary conditions at $r = b$ are described by Eqs. (8), (17), and (18).
Case II: the boundary conditions at $r = b$ are described by Eqs. (12), (17), and (18).
Case III: the boundary conditions at $r = b$ are described by Eqs. (8), (17), and (27).
Case IV: the boundary conditions at $r = b$ are described by Eqs. (12), (17), and (27).

The analytical solutions of the concentration and flow fields in the unit cell and the osmophoretic velocity of the vesicle were obtained in the previous section for all four cases.

In case I, the vesicle velocity is given by Eq. (26) with coefficient A defined by Eq. (11). This migration velocity can be expressed as

$$U = U^{(0)}(1 + \varphi + 2\varphi^{5/3})\left(1 - \frac{2 + 2\bar{\kappa} - 2\kappa}{2 + 2\bar{\kappa} + \kappa}\varphi\right)^{-1}$$
$$\times \left(1 + \frac{1}{3}\varphi^{5/3}\right)^{-1} , \tag{34}$$

where

$$U^{(0)} = \frac{1}{2 + 2\bar{\kappa} + \kappa} a L_p R T E_\infty , \tag{35}$$

which is the osmophoretic velocity of the vesicle given by Eq. (1) in the limit $\varphi = 0$. The vesicle velocity, U, can also be written as the following power expansion in φ:

$$U = U^{(0)}[1 + \alpha\varphi + \beta\varphi^{5/3} + O(\varphi^2)] . \tag{36}$$

A comparison between Eqs. (34) and (36) leads to the coefficients α and β,

$$\alpha = \frac{4 + 4\bar{\kappa} - \kappa}{2 + 2\bar{\kappa} + \kappa} , \tag{37}$$

$$\beta = \frac{5}{3} . \tag{38}$$

Note that α is a function of the parameters κ and $\bar{\kappa}$, but that β is a constant not related to these parameters. One can find $\alpha \to 2$ if $\kappa/2 \ll 1 + \bar{\kappa}$ and $\alpha \to -1$ if $\kappa/4 \gg 1 + \bar{\kappa}$ (α is a monotonically decreasing function of κ).

In case II, the osmophoretic velocity of the vesicle can be evaluated by Eq. (26) with A given by Eq. (13) and its expression parallel to Eq. (34) for case I is

$$U = U^{(0)}(1 + \varphi + 2\varphi^{5/3})\left(1 + \frac{1 + \bar{\kappa} - \kappa}{2 + 2\bar{\kappa} + \kappa}\varphi\right)^{-1}$$
$$\times \left(1 + \frac{1}{3}\varphi^{5/3}\right)^{-1} . \tag{39}$$

This vesicle velocity can also be expressed in the expansion form of Eq. (36), with the coefficients α and β given by

$$\alpha = \frac{1 + \bar{\kappa} + 2\kappa}{2 + 2\bar{\kappa} + \kappa} , \tag{40}$$

$$\beta = \frac{5}{3} . \tag{41}$$

Here, β is the same as that in case I; however, in this case, $\alpha = 1/2$ if $2\kappa \ll 1 + \bar{\kappa}$ and $\alpha = 2$ if $\kappa/2 \gg 1 + \bar{\kappa}$ (α is a monotonically increasing function of κ, which is opposite to the corresponding tendency in case I).

In case III, the osmophoretic velocity of the vesicle can be obtained by the substitution of Eq. (11) into Eq. (33), with the result

$$U = U^{(0)}(1 + 2\varphi)\left(1 - \frac{2 + 2\bar{\kappa} - 2\kappa}{2 + 2\bar{\kappa} + \kappa}\varphi\right)^{-1} . \tag{42}$$

When this velocity is expressed in the expansion form of Eq. (36), one has

$$\alpha = \frac{6 + 6\bar{\kappa}}{2 + 2\bar{\kappa} + \kappa} , \tag{43}$$

$$\beta = 0 . \tag{44}$$

Equation (43) indicates $\alpha = 3$ if $\kappa/2 \ll 1 + \bar{\kappa}$ and $\alpha = 0$ if $\kappa/6 \gg 1 + \bar{\kappa}$ (α in this case equals the value of α in case I plus 1).

In case IV, the vesicle velocity is given by Eq. (33) with A defined by Eq. (13) and it can be expressed as

$$U = U^{(0)}(1 + 2\varphi)\left(1 + \frac{1 + \bar{\kappa} - \kappa}{2 + 2\bar{\kappa} + \kappa}\varphi\right)^{-1} . \tag{45}$$

When this formula is written in the form of Eq. (36), one has

$$\alpha = \frac{3 + 3\bar{\kappa} + 3\kappa}{2 + 2\bar{\kappa} + \kappa} , \tag{46}$$

$$\beta = 0 . \tag{47}$$

As in case III, $\beta = 0$, but here, $\alpha = 3/2$ if $\kappa \ll 1 + \bar{\kappa}$ and $\alpha = 3$ if $\kappa/2 \gg 1 + \bar{\kappa}$ (α in this case equals the value of α in case II plus 1).

It can be found from Eqs. (34)–(47) that the mean osmophoretic velocity in a homogeneous suspension of identical spherical vesicles predicted by the cell model is quite sensitive to the boundary conditions specified at the virtual surface of the cell. The boundary condition for the solute concentration at the virtual surface $r = b$ determines the dependence of the normalized vesicle velocity (or mobility), $U/U^{(0)}$, as a function of parameters κ and $\bar{\kappa}$, while the boundary condition for the fluid velocity field at $r = b$ controls the connection of $U/U^{(0)}$ with the remainder part. For given values of κ, $\bar{\kappa}$, and φ, the mean vesicle velocities obtained from cases III and IV (Kuwabara model) are always greater than those obtained from cases I and II (Happel model), respectively. Note that, except for the situation of case I with $\kappa > 4(1 + \bar{\kappa})$, α in Eq. (36) predicted by the cell model has a positive value.

On the basis of the analytical (approximate) solution of the hydrodynamic interaction between pairs of osmophoretic vesicles in a uniformly prescribed solute concentration gradient obtained by the method of reflections correct to $O(r_{12}^{-7})$, where r_{12} is the center-to-center distance between the spherical vesicles, Keh and Tu [4] also derived a formula for the mean osmophoretic velocity in a dilute suspension of vesicles (say, $\varphi < 0.1$) in the expansion form of Eq. (36) with $\beta = 0$ by using the concepts of statistical mechanics. This ensemble-averaged result gives the coefficient α approximately as

$$\alpha = 2 - G + \tfrac{1}{8}[2G^2 + 2(1 - H)G + 5] \ , \tag{48}$$

where

$$G = \frac{1 + \bar{\kappa} - \kappa}{2 + 2\bar{\kappa} + \kappa} \ , \tag{49}$$

$$H = \frac{-5(2 + 2\bar{\kappa} + \kappa)}{6 + 3\bar{\kappa} + 2\kappa} \ . \tag{50}$$

In this statistical model β vanishes, which is similar to that in the Kuwabara cell model (cases III, IV). As examples of limiting situations, Eq. (48) yields $\alpha = 121/48$ if $\bar{\kappa} + \kappa/2 \ll 1$, $\alpha = 3$ if $\kappa \gg 1 + \bar{\kappa}$, and $\alpha = 131/48$ if $\bar{\kappa} \gg 1 + \kappa$.

On the other hand, Keh and Yang [5] used a method of spherical bipolar coordinates to obtain numerical information on the hydrodynamic interactions of two arbitrary spherical vesicles undergoing osmophoresis at all separations between the vesicles. On the basis of this information they also calculated the ensemble-averaged values of α for several values of κ and $\bar{\kappa}$. In these calculations, the effect of the discontinuity of the solute concentration at the surface of each vesicle was inadvertently omitted and, therefore, a term $-3(2 + 2\bar{\kappa} + \kappa)^{-1}$ should be added to their results for α. Compared with these corrected results, the values of α calculated from the approximate formula (Eq. 48) always overestimate its exact results and the error can be significant. Analogous to the general results obtained from the cell model, the statistical model predicts positive values for α.

The normalized osmophoretic velocity in a homogeneous suspension of identical spherical vesicles, $U/U^{(0)}$, as calculated from Eqs. (34), (39), (42), and (45) for the four cases of the cell model, is plotted versus the volume fraction of the vesicles, φ, in Figs. 2–4 for various values

of κ and $\bar{\kappa}$. In all cases, $U/U^{(0)}$ increases monotonically with the increase in φ, with the exception of case I with $\kappa > 4(1 + \bar{\kappa})$, and equals unity in the limit $\varphi = 0$. The calculations are presented up to $\varphi = 0.74$, which corresponds to the maximum attainable volume fraction for a swarm of identical spheres [8]. It is also clear that at volume fractions approaching this, agglomeration due to contacts between vesicles may occur, and the present study does not cover this case.

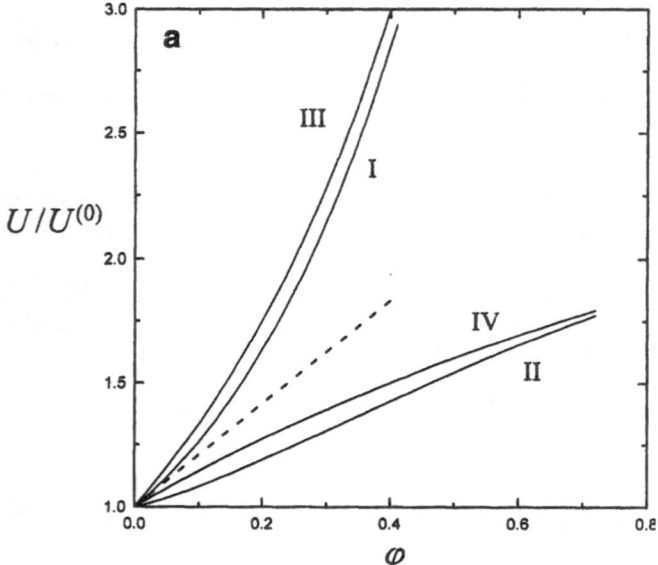

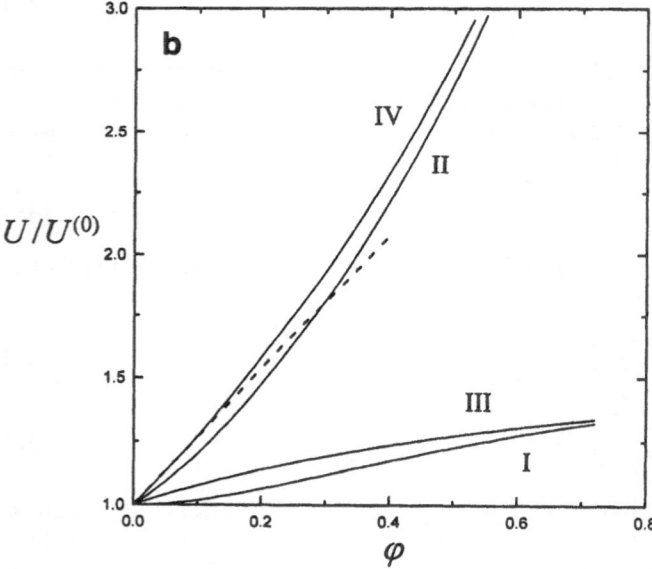

Fig. 2 Plots of the normalized osmophoretic velocity in a monodisperse suspension of spherical vesicles versus the volume fraction of the vesicles: **a** $\kappa = 0$; **b** $\kappa = 5$ and $\bar{\kappa} = 0$. The *solid curves* with labels *I*, *II*, *III*, and *IV* represent the cell-model calculations from Eqs. (34), (39), (42), and (45), respectively, and the *dashed curves* are the statistical-model results calculated from Eq. (36)

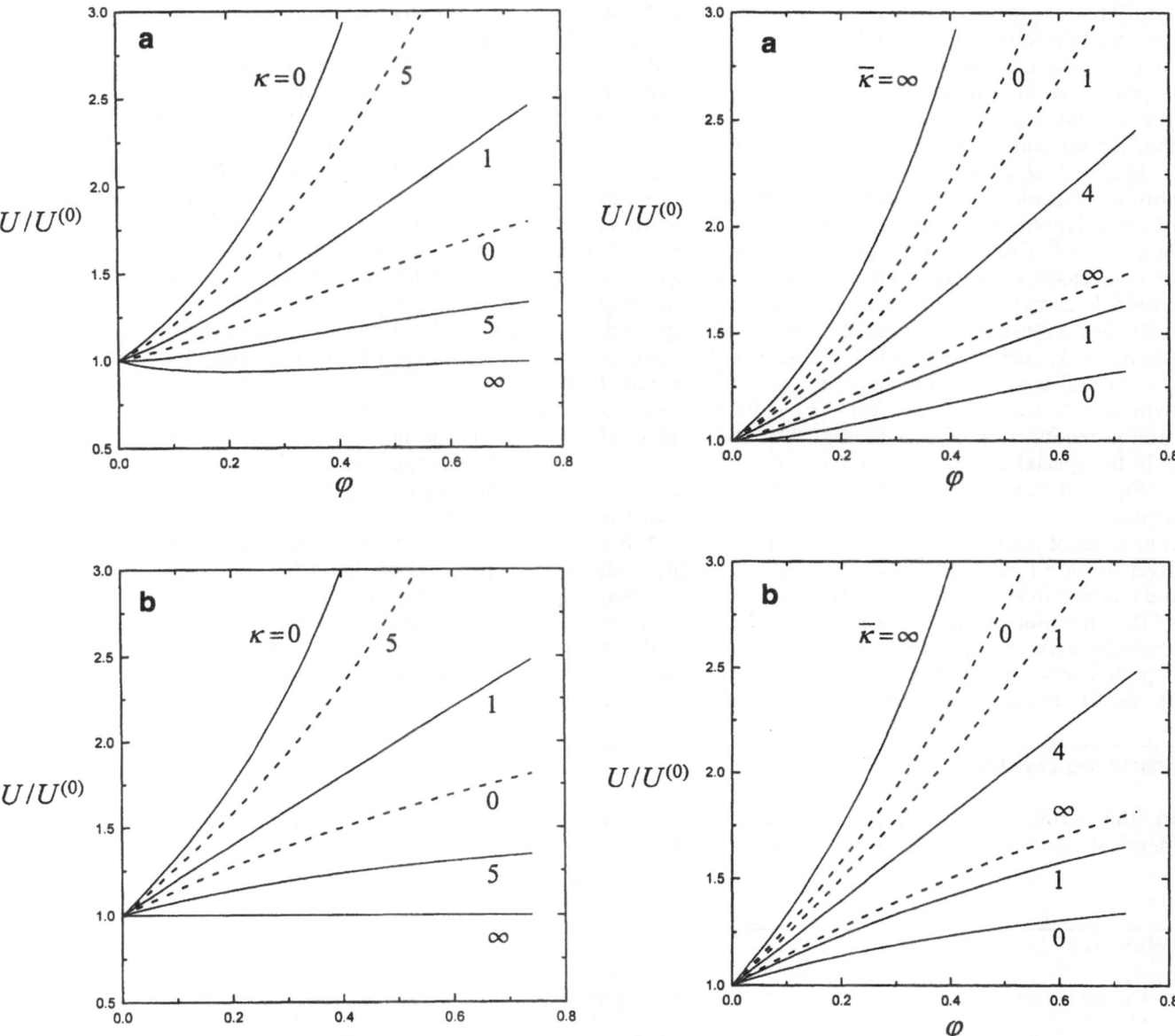

Fig. 3 Plots of the normalized osmophoretic velocity in a monodisperse suspension of spherical vesicles with $\bar{\kappa} = 0$ versus the volume fraction of the vesicles with κ as a parameter: **a** case I (Eq. 34, *solid curves*) and case II (Eq. 39, *dashed curves*); **b** case III (Eq. 45, *solid curves*) and case IV (Eq. 45, *dashed curves*). The *dashed curves* of $\kappa \to \infty$ and $\kappa = 1$ coincide with the solid curves of $\kappa = 0$ and $\kappa = 1$, respectively

Fig. 4 Plots of the normalized osmophoretic velocity in a monodisperse suspension of spherical vesicles with $\kappa = 5$ versus the volume fraction of the vesicles with $\bar{\kappa}$ as a parameter: **a** case I (Eq. 34, *solid curves*) and case II (Eq. 39, *dashed curves*); **b** case III (Eq. 45, *solid curves*) and case IV (Eq. 45, *dashed curves*). The *dashed curves* of $\bar{\kappa} = 4$ coincide with the *solid curves* of $\bar{\kappa} = 4$

The results of $U/U^{(0)}$ as a function of φ for a suspension of identical vesicles with $\kappa = 0$ are shown in Fig. 2a. In this particular situation, all four cases of the cell model predict that $U/U^{(0)}$ is independent of $\bar{\kappa}$. The result obtained from the statistical model with $\kappa = \bar{\kappa} = 0$ (Eq. 36 with $\alpha = 2.08$ and $\beta = 0$, Ref. [5] after correction for α) for dilute suspensions is also shown in this figure to compare with those of the cell model. It can be

seen in Fig. 2a that $U/U^{(0)}$ predicted by cases I and III is greater than that predicted by the statistical model and, in contrast, $U/U^{(0)}$ predicted by cases II and IV is smaller than that predicted by the statistical model. The results for cases I and IV are somewhat close to that for the statistical model in dilute suspensions of vesicles.

The results for $U/U^{(0)}$ as a function of φ for a monodisperse suspension of vesicles with $\kappa > 1 + \bar{\kappa}$ ($\kappa = 5$ and $\bar{\kappa} = 0$) are plotted in Fig. 2b for both cases of the statistical model (Eq. 36 with $\alpha = 2.69$ and $\beta = 0$,

Ref. [5] after correction for α) and the cell model. Now, the osmophoretic velocity predicted by cases I and III is smaller than that predicted by cases II and IV; this is opposite to the situation of $\kappa < 1 + \bar{\kappa}$ illustrated in Fig. 2a. Note that the result for case IV is quite close to that for the statistical model in dilute systems.

In Fig. 3, $U/U^{(0)}$ for a suspension of identical vesicles with $\bar{\kappa} = 0$ is plotted as a function of φ for the four cases of the cell model with κ as a parameter. For a fixed value of φ, $U/U^{(0)}$ decreases monotonically with the increase in κ in cases I and III (opposite to the trend of the ensemble-averaged results), but increases monotonically with the increase in κ in cases II and IV. As expected, when $\kappa = 1$, such that $\kappa = 1 + \bar{\kappa}$, case I is the same as case II and case III is the same as case IV. For case I with $\kappa > 4$, such that $\kappa > 4(1 + \bar{\kappa})$, $U/U^{(0)}$ is not a monotonic function of φ. Also, $U/U^{(0)}$ is independent of φ in the special situation of case III with $\kappa \rightarrow \infty$.

Figure 4 shows plots of $U/U^{(0)}$ for a suspension of identical vesicles with $\kappa = 5$ as a function of φ for the four cases of the cell model with $\bar{\kappa}$ as a parameter. For a given value of φ, $U/U^{(0)}$ increases monotonically with the increase in $\bar{\kappa}$ in cases I and III (opposite to the trend of the ensemble-averaged results), but decreases monotonically with the increase in $\bar{\kappa}$ in cases II and IV. As expected, $\bar{\kappa} = 4$ (such that $\kappa = 1 + \bar{\kappa}$), case I is the same as case II, and case III is the same as case IV.

Concluding remarks

In this article, the osmophoresis of a swarm of identical spherical vesicles suspended uniformly in a fluid with a constant solute concentration gradient has been analyzed using unit cell models with various boundary conditions at the outer (virtual) surface of the cell. On the basis of the assumption of small Peclet and Reynolds numbers, the solute concentration and fluid flow fields in the cell were solved analytically and the vesicle velocities as functions of the volume fraction of the vesicles were obtained in the closed-form expressions (Eqs. 34, 39, 42, 45). Comparisons of the results among the cell models and between the statistical model for dilute suspensions and the cell models have also been provided.

We note that the four cases of the cell model defined in the previous section lead to somewhat different results for the vesicle velocity. Our analysis indicates that the tendency of the dependence of $U/U^{(0)}$ on κ or $\bar{\kappa}$ in cases I and III is not correct, in comparison with the ensemble-averaged results. So, the boundary condition represented by Eq. (8) is not as accurate as the boundary condition represented by Eq. (12), probably due to the fact that the angular component of the solute concentration gradient at the virtual surface of the cell is not specified in Eq. (8). It has been shown that the osmophoretic velocity predicted by case IV of the cell model agrees quite well with that calculated from the statistical model for dilute suspensions of vesicles. Experimental data would be needed to confirm the validity of each case of the cell model at various ranges of κ, $\bar{\kappa}$, and φ.

Acknowledgement This research was supported by the National Science Council of the Republic of China under grant NSC88-2214-E-002-010.

References

1. Gordon LGM (1981) J Phys Chem 85:1753
2. Anderson JL (1983) Phys Fluids 26:2871
3. Anderson JL (1986) Ann N Y Acad Sci 469:166
4. Keh HJ, Tu HJ (2000) Int J Multiphase Flow 26:125
5. Keh HJ, Yang FR (1992) Int J Multiphase Flow 18:593
6. Happel J (1958) AIChE J 4:197
7. Kuwabara S (1959) J Phys Soc Jpn 14:527
8. Levine S, Neale GH (1974) J Colloid Interface Sci 47:520
9. Happel J, Brenner H (1983) Low Reynolds number hydrodynamics. Nijhoff, The Hague

Progr Colloid Polym Sci (2000) 115:181–185
© Springer-Verlag 2000

BIOSYSTEMS

P. Brocca
L. Cantù
M. Corti
E. Del Favero

Thermal fluctuations of small vesicles: observation by dynamic light scattering

P. Brocca · L. Cantù · M. Corti (✉)
E. Del Favero
INFM, Dipartimento di Chimica
e Biochimica Medica
Università di Milano
L.I.T.A., via F. lli Cervi 93
20090 Segrate, Italy
e-mail: mario.corti@unimi.it

Abstract The laser light scattering technique can be used in a nonconventional fashion to study dynamic properties of vesicles which are too small to be observed by microscopy. In fact, in suitable experimental conditions, the correlation function of the scattered light contains a contribution from bilayer fluctuations, besides the usual diffusion one. Characteristic fluctuation times have been determined for single-component phospholipid vesicles of 60 nm radius, prepared by extrusion. The addition of small amounts of a glycolipid (to 2% mole fraction), induces a significative increase in the fluctuation times (of the order of 20%) but still does not affect the diffusive motion, indicating a softening of the membrane. Being so sensitive, this technique is quite promising both for the study of membrane properties in the presence of defects and for applications to biology and pharmacology.

Key words Thermal fluctuations · Small vesicles · Dynamic light scattering

Introduction

The elastic properties of biological membranes are connected to the problem of cell stability and resistance to external influences and to many physiological processes such as shape changes, vesicle endocytosis and specific interactions with transmembrane proteins. Because of their essential role, the elastic properties have been investigated by different techniques, which include the mechanical deformation of the membrane by suction [1, 2] or local compression [3] or the excitation of modes of vibration induced artificially by electrical stress [4, 5] or naturally by thermal effects [6–10]. The experimental observation of the response to perturbations is usually performed by optical video microscopy, which implies that the membranes are arranged in the form of giant vesicles (few microns).

On the other hand, a dynamic scattering technique, namely neutron spin echo, has been profitably used to investigate the elastic properties of the hydrophilic–hydrophobic interface of microemulsion small droplets (size 5–10 nm) [11–13]. This technique is based on some general features of scattering, namely that each different mode of the fluctuation of an interface, characterized by a particular frequency, contributes to the scattered intensity according to different q-dependent functions. Moreover, the zero-order mode, corresponding to the form factor of the average shape of the interface, which is usually the most important contribution to the scattered intensity, has a pronounced minimum in the same q position where the second-order mode, usually the most excitable mode, has a maximum. Then, it is possible to find experimental conditions which allow the observation of both the corresponding characteristic times.

The application of analogous dynamic scattering techniques is promising also in the case of small vesicles (about 100 nm) to assess the characteristic fluctuation times of the closed bilayer, provided that the suitable q range can be attained experimentally. In fact, the extension of the neutron spin echo technique to the case of vesicles has been desired for a long time [14], and recently some experiments with blue laser light have been performed on bicomponent and tricomponent mixed vesicles, stabilized at dimensions above 200 nm by spontaneous curvature effects [15].

Our aim is to reveal the elastic properties of single-component vesicles and their variation as a consequence of the addition of molecules of biological or pharmacological relevance, in the range of dimensions where several experiments on the effects of composition and additives on physico-chemical and biochemical behaviour have already been performed with many different techniques, such as calorimetry or fluorescence spectroscopy [16, 17]. On the other hand this application of the dynamic scattering technique requires very low vesicle polydispersity, both in size and in thickness, in order to be sensitive and to give significative results. For the preparation of single-component vesicles the widely used extrusion technique [18] meets the required conditions if performed on pores of 200 nm and gives rise to vesicles with the desired monodispersities and diameter ($2R \sim 120$ nm). It follows that the q range useful for the observation and separation of both vesicle-translational-diffusion and membrane-fluctuation characteristic times ($qR \sim \pi$) by dynamic light scattering can be approached only by using a laser emitting in the ultraviolet.

Experimental

Vesicle preparation

Dimyristoyl phosphatidyl choline (DMPC) was purchased from Calbiochem-Novabiochem (La Jolla, Calif., USA). Monosialoganglioside (GM1) was extracted and purified according to the method described in Ref. [19] Unilamellar single-component DMPC vesicles were obtained as described by Hope et al. [18]. A small amount of phospholipid (0.5 mg) dissolved in chloroform was dried to a lipid film as thin as possible at the bottom of a tube with constant turning followed by evaporation under vacuum for 2 h. The lipid film was hydrated under a stream of humid nitrogen and dissolved in 500 μl pure water. After a freeze–thaw procedure, each sample was extruded 71 times through two stacked polycarbonate filters of appropriate pore size (200 nm to obtain vesicles of the order of 120 nm in diameter) mounted in an extruder (Avestin, Ottawa, Canada) thermostatted at a temperature above the gel-to-liquid-crystalline transition temperature of the lipid hydrophobic chains, a condition which is required to assure the needed vesicle monodispersity and time stability. The low DMPC transition temperature (23.9 °C) [20] allows the preparation to be carried out close to room temperature. The lipid yield after extrusion was determined as described by Stewart [21]. To perform laser light scattering measurements, samples were diluted to 0.02 mM in a 100 mM NaCl aqueous buffer.

Mixed vesicles, containing small amounts of GM1 up to 2% mole fraction, were obtained after 36 h incubation of single-component DMPC vesicles (after extrusion and dilution) with GM1 micellar solutions with the appropriate concentrations, prepared with the same buffer. Using this procedure, GM1 was localized only in the outer layer of the membrane, reflecting the asymmetric biological situation found in real cells. As different extrusion procedures, although carried out through sieves of identical pore size, led to slightly different vesicle dimensions

(a deviation of $\pm 10\%$ was observed for different preparations) a whole set of samples with increasing GM1/DMPC mole fraction was prepared starting from the same single-component DMPC solution, which was taken as the reference solution of the series. This procedure avoids spurious contributions to the GM1 concentration dependence of the characteristic times arising from the deviation in the dimensions of parent single-component vesicles.

Several solutions of both single-component DMPC and mixed GM1–DMPC vesicles were measured with the dynamic light scattering technique.

Dynamic scattering technique for the measurement of the characteristic fluctuation times of membranes

Both static and dynamic laser light scattering experiments were performed on the same sample. The light scattering apparatus, described in detail elsewhere [21], was improved by an additional argon ion laser source (Coherent) also emitting in the ultraviolet ($\lambda = 363$ nm).

With the dynamic scattering technique the scattered field correlation function, $G_1(t)$, is measured at a given momentum transfer, $q = (4\pi/\lambda)\sin\theta/2$, and

$$G_1(t) \div \left\langle \sum_{i,j} \exp\{iq[r_i(t) - r_j(0)]\} \right\rangle ,$$

where r_l is the position of an infinitesimal volume i at time t.

For a fluctuating spherical vesicle of radius R, Milner and Safran [23] described the shape fluctuations as an overdamped motion (due to the viscosity of the inner and outer liquid). Expanding these fluctuations into spherical harmonics with dimensionless amplitudes u_l and characteristic time constants, τ_l, the correlation function $G_1(q, t)$ can be written as

$$G_1(q,t) = A\left\langle \exp(-Dq^2 r)R^2 \right. \times \left. \left[f_0(qR) + \sum_{l \geq 2} \frac{2l+1}{4\pi} f_1(qR)\langle |u_l|^2 \rangle \exp\left(-\frac{t}{\tau_l}\right) \right] \right\rangle_{R^2} .$$

D is the vesicle diffusion coefficient, A is a normalization factor and $\langle \rangle$ means ensemble average, accounting also for vesicle polydispersity. The expressions for $f_0(qR)$ and $f_1(qR)$ in terms of spherical harmonics can be found in Ref. [23].

The expression for G_1 consists of a sum of exponentials describing the superposition of contributions coming from the translational diffusion of the vesicles, on one hand, (characterized by D) and the fluctuations of the bilayer, on the other hand, (characterized by τ_l). As already stated, each mode contributes to the scattered intensity according to different q-dependent functions, and, among the higher-order f_l, only the second mode term, which accounts for shape fluctuations of elliptical type, is important.

Then, if the scattering sample is a monodisperse vesicle solution, the expression for $G_1(q, t)$ is rather simple, being the sum of two decays

$$G_1(q,t) = a(q)\exp[-\Gamma(q)t] + b(q)\exp[-\Gamma_1(q)t] .$$

The first decay, with amplitude $a(q)$, is due to the pure Brownian diffusion with an inverse time constant depending on q^2, $\Gamma = 1/\tau_{\text{diff}} = Dq^2$. The second one, with amplitude $b(q)$, also accounts for the contribution of the second mode fluctuations with an inverse time constant: $\Gamma_1 = Dq^2 + 1/\tau_2$, which contains a term which does not depend on q^2, as expected for internal motions.

As far as the scattered intensities are concerned, it happens that, for each vesicle dimension, R, two regions can be identified: a very wide q range where $f_0(qR) \gg f_l(qR)$, i.e. $a(q) \gg b(q)$, so usually only the diffusive term can be easily observed, and an interval centered around $qR \cong \pi$ where the $l = 0$ mode goes through a minimum while the $l = 2$ mode has a maximum [23]. Dynamic light scattering measurements performed in the suitable range allow the $l = 2$ mode to be revealed and the corresponding characteristic time, τ_2, to be assessed provided that the stringent requirement of vesicle monodispersity is reasonably met.

It has to be noted that the neutron spin echo technique, which, as already said, was the first dynamic scattering technique to be applied to the study of fluctuating interfaces, allows only the measurement of an effective diffusion coefficient, D_{eff}, coming from the fit of a single-exponential decay $[(y(t) = \exp(-D_{\mathrm{eff}}q^2t)]$ to the first points of the measured correlation function. D_{eff} depends on q and goes through a maximum in the same position where $a(q)$ and $b(q)$ show their minimum and maximum, respectively. Dynamic laser light scattering, instead, allows the separation and independent measurement of the two characteristic inverse time constants appearing in the expression of G_1 and the associated amplitudes.

The measurements were performed as a function of q in the two regimes: one where only the diffusive motion can be seen and the other where fluctuations form an appreciable contribution to the scattered intensity.

It has to be noted that the q range where useful measurements can be made is reasonably extended to allow a good statistical treatment of the data, due to the relative smoothness in the decrease in $a(q)$ and in the simultaneous increase in $b(q)$ as $q = \pi/R$ is approached, giving rise to G_1 functions showing an appreciable double decay starting from $q \sim 0.7\pi/R$.

Results and discussion

The time-dependent part of two field correlation functions performed on the same sample at two different scattering angles, corresponding to different distances from $qR \sim \pi$, is shown Fig. 1 as a function of q^2t.

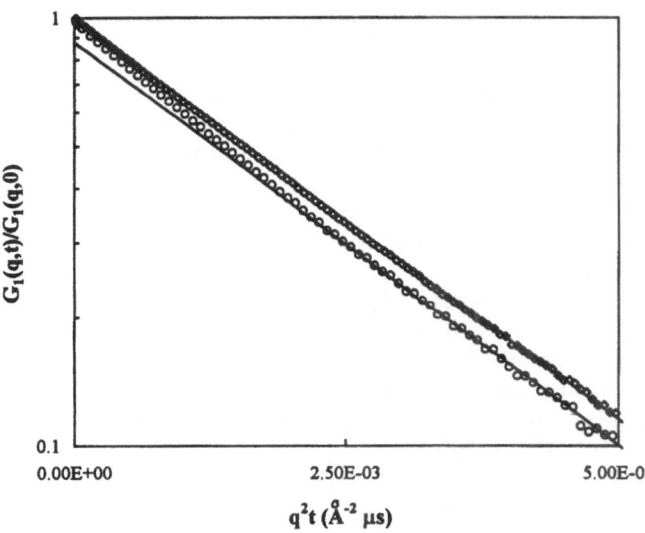

Fig. 1 Time-dependent part of two field correlation functions performed on the same sample (single-component dimyristoyl phosphatidyl choline, *DMPC*, fairly monodisperse vesicles, $R = 66 \pm 4.5$ nm) at two different scattering angles, corresponding to different distances (one further, one closer) from $qR \sim \pi$, reported as a function of q^2t. The *upper curve*, measured at a further angle ($\vartheta = 38°$), contains only the translational diffusion contribution, while the *lower curve*, obtained at a closer angle ($\vartheta = 120°$), contains two contributions: a longer decay being, of course, parallel to that of the upper curve and a shorter decay due to membrane fluctuations

It can be seen that the one obtained at a further angle (upper curve) contains only the translational diffusion contribution, due to the monodisperse vesicles ($R = 66 \pm 4.5$ nm), while at a closer angle (lower curve) two contributions are easily seen, the longer decay being, of course, parallel to that of the upper curve, as expected from the expression of $G_1(q, t)$.

As the scattering angle is varied in an appropriate range, as described before, the shape of the measured $G_1(q, t)$ evolves from single exponential to a double decay: along this path the values of Γ and Γ_1 can be extracted and are reported versus q^2 in Fig. 2 as dots and triangles, respectively. A higher number of dots appears in the figure as also the Γ values obtained in the single-exponential range are reported. It can be easily seen that Γ scales with q^2, as expected in the case of translational motion, irrespective of the range where the data were collected. This should obviously happen, as the diffusing vesicles are the same and it provides a test for good particle monodispersity. Then, of course, the values of D, obtained from $\Gamma = Dq^2$, at $qR \cong \pi$ coincide with the ones at $qR \ll \pi$, where only the diffusional motion is visible and the decay of $G_1(q, t)$ is a single exponential. As far as the values of Γ_1 are concerned, they also lie on a straight line, parallel to the one which fits the Γ values, but with a nonzero value at $q = 0$. This behaviour is, of course, expected, as it perfectly reflects the expression $\Gamma_1 = Dq^2 + 1/\tau_2$. The nonzero intercept is then connected to τ_2 of the second mode fluctuations of the vesicle bilayer, which, for the system of single-component DMPC vesicles with radius 66 nm described in Figs. 1 and 2 is of the order of 100 μs ($99 \pm 3.5 \ \mu s$).

The same procedures for dynamic scattering measurements at different scattering angles and for data treatment were applied to different series of mixed GM1–DMPC vesicles with increasing amounts of GM1

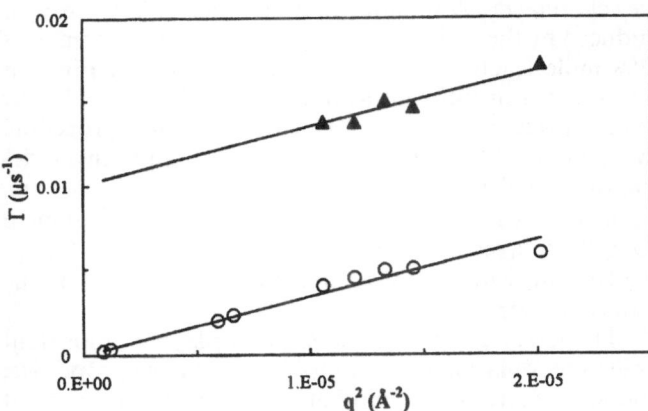

Fig. 2 Values of Γ (*dots*) and Γ_1 (*triangles*) for single-component DMPC vesicles with $R \sim 66$ nm reported versus q^2. The best-fit straight lines going through the two series of points are parallel, as expected, with different intercepts (see text)

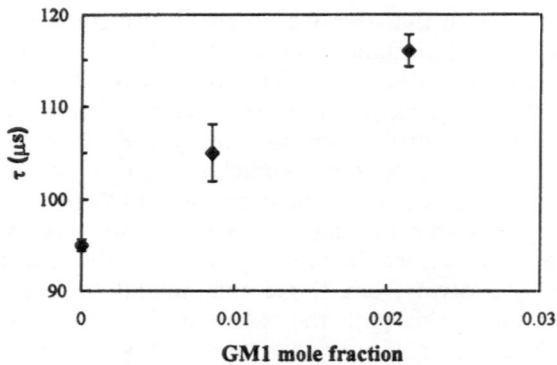

Fig. 3 Characteristic time, τ_2, of the second mode fluctuations of the mixed DMPC-monosialoganglioside (*GM1*) vesicle bilayers as a function of the GM1 content of the outer layer

up to 2% mole fraction, prepared as already described. The comparison of the results to assess for variations and trends was performed within the same series as explained before.

The values of τ_2 for vesicles containing increasing amounts of GM1 are reported in Fig. 3. The parent single-component DMPC vesicles of this concentration series have a radius of 54 ± 2 nm. As already pointed out, the mixed vesicle dimension does not appreciably vary all through the series with respect to the parent solution. This is reflected by the fact the long decay parts of the measured correlation functions, which contain only the diffusive contributions, are seen to be practically unchanged in the series. Instead, the short decay parts, where present, i.e. for the correlation functions measured in the appropriate q range to reveal the fluctuation contribution, exibit a strong variation in slope as the GM1 content is changed, which results in the behaviour of τ_2 shown in Fig. 3. It can be seen that the addition of 0.9% mole fraction is enough to modify appreciably the characteristic fluctuation time of the vesicle bilayer. A modification as large as 25% in τ_2 is induced by the addition of an amount of GM1 as small as 2% mole fraction. The behaviour reported in Fig. 3 is reproduced in different sample series which, as already pointed out, may come out of the extrusion procedure with slightly different dimensions. For example, in a series characterized by a vesicle dimension $R = 65 \pm 4$ nm, τ_2 increases from 99 ± 4 μs for the single-component DMPC bilayer to 125 ± 6 μs for the mixed GM1–DMPC one with 2% GM1 mole fraction inserted into the external layer.

The GM1 ganglioside is seen to play an important role in modulating the membrane mechanical properties, inducing a strong softening of the bilayer even if present in very small amounts. Although much larger than expected, this behaviour is not surprising, but we believe it could easily be a general characteristic feature of gangliosides. In fact some peculiar properties of vesicles

made up of GM3 ganglioside molecules have already been explained in terms of a low bending constant of the bilayer [24, 26], which was reasonably attributed to the mismatch between hydrophilic and hydrophobic lateral hindrances, a mismatch which is even greater for gangliosides other than GM3, including GM1.

As far as the absolute value of the elastic constant is concerned, its derivation from τ_2 requires the use of a model. The theory of fluctuating closed spherical interfaces was developed first for the case of microemulsions, i.e. nearly "geometrical two-dimensional" surfaces [23, 27, 28], for which a connection between fluctuation times and elastic constants was established and a cubic dependence of the characteristic times on the radius was predicted. The theory was then extended to the case of vesicles fluctuating with respect to an average spherical shape, thanks to the presence of excess area. Membranes are treated as continuum media, i.e. their thickness and discrete nature are neglected on the scale of their overall dimension. This is, of course, an excellent approximation in the case of big and giant vesicles, which are then still seen as "geometrical two-dimensional" surfaces. Moreover, the theory developed allows the prediction of relations between both characteristic times and scattering amplitudes associated with the modes of the zeroth and second order [23, 29, 30] or, in other words, relations connecting Γ, Γ_1, $a(q)$ and $b(q)$.

Times and amplitudes were extracted from our experimental data, in the case of both single-component and mixed vesicles, a feature which is offered by the use of dynamic laser light scattering and not by neutron spin echo. The relations among Γ, Γ_1, $a(q)$ and $b(q)$ are astonishingly well followed, within errors of the order of one percent. This would indicate that the model, which is the starting point for the assessment of such relations, developed for microemulsion interfaces and extended to the case of big vesicles, also holds in the case of small vesicles.

Nevertheless, this does not seem really true, if the absolute values of the bending constant of the bilayer are extracted from the measured characteristic times according to the model. In fact, even in the case of single-component small DMPC vesicles, without any contribution coming from spontaneous curvature due to uneven chemical composition of the two sides of the bilayer, a value for the bending constant of the DMPC membrane is calculated which is more than 1 order of magnitude lower than the one appearing in the literature, as measured by other techniques and on a different scale of dimension. We also checked that the effect was not due to osmotic stress, which could play a role [29], with the vesicles being extruded in water and diluted in a buffer: the same discrepancy is, in fact, also observed if water is used as the diluting medium. One fascinating possibility is that the bending characteristics are truly different for big and small vesicles of the same material, i.e. when the membrane thickness is or is not negligible with respect to the overall dimension of the aggregate and its discrete nature with the network of hydrophilic–hydrophobic interactions shows up or not. Should this be the case, it would be of particular biological importance to perform measurements on small vesicles to assess the mechanical properties of the membranes of subcellular organelles, where processes connected to elasticity or discontinuities in elasticity due to the presence of patches often take place, on the same scale.

Conclusions

The technique of dynamic laser light scattering has been proved to be reliable and sensitive for seeing the fluctuating motions of membranes of small phospolipid vesicles and for allowing the extraction of their characteristic times and for determining their variation resulting from the addition of a second component, even in small amounts. Both from a general and a biological point of view, the possibility of also studying the mechanical properties of membranes in the same range of dimension where other features are assessed by widely used techniques, such as calorimetry or fluorescence spectroscopy, could lead to the revelation of significative and illuminating correlations. In particular, the variation of the elastic properties, which is clearly seen by this technique before any calculation involving the use of a model, is of huge biological and pharmacological importance, a relevance which would even be increased by the possibility that the elastic properties of a closed membrane of a given material vary while crossing some threshold regarding the overall dimension with respect to its thickness.

References

1. Evans E, Rawicz W (1997) Phys Rev Lett 79:2379
2. Needham D, Zhelev DV (1996) In: Rosoff M (ed) Vesicles. Dekker, New York, pp 373–444
3. Vinckier A, Semenza G (1998) FEBS Lett 430:12
4. Niggemann G, Kummrow M, Helfrich W (1995) J Phys II 5:413
5. Kummrow M, Helfrich W (1991) Phys Rev E 44:8356
6. Schneider MB, Jenkins JT, Webb WW (1984) J Phys 45:1457
7. Evans E, Rawicz W (1990) Phys Rev Lett 64:2094
8. Méléard P, Mitov MD, Faucon JF, Bothorel P (1990) Europhys Lett 11:355
9. Mutz M, Helfrich W (1990) J Phys 51:991
10. Dobereiner HG, Evans E, Kraus M, Seifert U, Wortis M (1997) Phys Rev E 55:4458
11. Farago B, Richter D, Huang JS, Safran SA, Milner ST (1990) Phys Rev Lett 65:3348
12. Gradzielski M, Langevin D, Farago B (1996) Phys Rev E 53:3900
13. Farago B (1996) Physica B 226:51
14. Van Zanten JH (1996) In: Rosoff M (ed) Vesicles. Dekker, New York, pp 239–295
15. Joannic R, Auvray L, Lasic D (1997) Phys Rev Lett 78:3402
16. Ferraretto A, Pitto M, Palestini P, Masserini M (1997) Biochemistry 36:9232
17. Loughrey HC, Ferraretto A, Cannon AM, Acerbis G, Sudati F, Bottiroli G, Masserini M, Soria MR (1993) FEBS Lett 332:183
18. Hope MJ, Bally MB, Webb G, Cullis PR (1985) Biochim Biophys Acta 812:55
19. Tettamanti G, Bonali F, Marchesini S, Zambotti V (1973) Biochim Biophys Acta 296:160
20. Lewis RN, Mak N, McElhaney RN (1987) Biochemistry 26:6118
21. Stewart JCM (1980) Anal Biochem 104:10
22. Corti M (1985) In: Degiorgio V, Corti M (eds) Physics of amphiphiles: micelles, vesicles and microemulsions. North-Holland, Amsterdam, pp 122–151
23. Milner ST, Safran S (1987) Phys Rev A 36:4371
24. Cantù L, Corti M, Lago P, Musolino M (1991) Photon correlation spectroscopy: multicomponent systems. SPIE 1430:144
25. Cantù L, Corti M, Del Favero E, Raudino A (1994) J Phys 4:1585
26. Cantù L, Corti M, Del Favero E, Dubois M, Zemb T (1998) J Phys Chem B 102:5737
27. Helfrich W (1973) Z Naturforsch 28:693
28. Helfrich W (1990) In: Charvolin J, Joanny JF, Zinn-Justin J (eds) Liquids at interfaces. Elsevier Science, Amsterdam, pp 33
29. Faucon JF, Mitov MD, Méléard P, Bivas I, Bothorel P (1989) J Physique 50:2389
30. Komura S (1996) In: Rosoff M (ed) Vesicles. Dekker, New York, pp 197–236

Progr Colloid Polym Sci (2000) 115:186–191
© Springer-Verlag 2000

BIOSYSTEMS

D. M. McLoughlin
J. J. McManus
A. V. Gorelov
K. A. Dawson

DNA complexes with cationic surfactant in mixed solvents: the influence of excess surfactant and salt

D. M. McLoughlin · J. J. McManus
A. V. Gorelov (✉) · K. A. Dawson
Irish Centre for Colloid Science
and Biomaterials
Department of Chemistry
University College Dublin
Belfield, Dublin 4, Ireland
e-mail: gorelov@pop3.ucd.ie
Tel.: +353-1-7062417
Fax: +353-1-7062127

A. V. Gorelov
Institute of Theoretical
and Experimental Biophysics
Pushchino, Russia

Abstract We have investigated the solubility, thermal stability and structural transitions of DNA complexed with the cationic surfactant dodecyltrimethylammonium bromide (DTAB). The study was done over a broad range of ethanol/water mixtures. The dependence of solubility on surfactant and salt concentration was studied. For low salt concentrations, it was found that DNA–DTA complexes are soluble at an ethanol content higher then 57% v/v; however, addition of excess DTAB increases the ethanol concentration at which the complexes begin to solubilise. At higher ethanol concentrations (about 80% v/v) DNA–DTA complexes were soluble over the entire range of DTAB concentrations investigated.

Also at higher ethanol concentrations it was found that the DNA underwent a helix–coil transition. Circular dichroism data indicated that excess surfactant inhibited the B-to-A transition. In the presence of 5 mM NaBr, the DNA was found to be insoluble in the high-ethanol region, while the surfactant-dependent solubility line was shifted to lower ethanol concentrations. We present the phase diagram describing the solubility and conformational state of the DNA–DTA complex and discuss the influence salt has on its stability and solubility.

Key words DNA–dodecyltrimethylammonium · DNA–surfactant complex A–B transition

Introduction

The occurrence of polyelectrolyte–surfactant mixtures in many technological processes and formulations has resulted in intensive studies of their solution behaviour [1]. While most work in the area has focussed on the complexes formed by synthetic polyelectrolytes, it is recognised that biopolymers such as DNA which can adopt highly ordered structures with a range of architectures may offer special advantages in the development of novel polymer–surfactant complexes with useful properties [2]. The interaction of amphiphiles with DNA has also attracted the interest of researchers due to the potential of DNA–amphiphile complexes in the delivery of genetic materials to cells, in various therapies based on the regulation of gene expression [3]. DNA complexes with synthetic and natural amphiphiles in organic solvents could serve as precursors for the formulation of a well-defined gene delivery system [4]. DNA complexes with surfactant are thus of both basic and applied interest.

The first study of the structure of a DNA–alkyltrimethylammonium complex in ethanol was carried out in 1964 and it was concluded that the DNA existed as a double-stranded compact structure [5], a finding which has been backed up by recent work [6–8]. Two methods of complex preparation were used in these studies. In the first method the required amount of ethanol was simply added to a suspension of the DNA–surfactant complex in a buffer [7]; the second method involved precipitating the complex from solution, drying it and redissolving it in the solvent of interest [9]. However, it is clear that parameters such as the degree of polyelectrolyte neutralisation, ionic strength and concentration of excess surfactant are variable in the preparations. Also the DNA can exist in different solvent-dependent forms, which have distinctive structural parameters. It is the purpose of this work to determine the effect of varying dielectric constant, ionic strength, hydration, excess surfactant concentration and DNA conformation on the structure and solution behaviour of DNA in mixed solvents.

Experimental

Preparation of the DNA–dodecyltrimethylammonium complex

The purification of the reagents and the DNA used in this study is described in a previous publication [10]. The DNA used was in a dilute N-(2-hydroxyethyl)piperazine-N'-ethanesulfonic acid (HEPES) buffer (0.5 mM HEPES, 0.1 mM ethylenediaminetetraacetic acid, EDTA). A 10 mM dodecyltrimethylammonium bromide (DTAB) solution in 0.5 mM HEPES buffer was added dropwise to the DNA solution (concentration 0.75 g/l), with stirring, until a flocculent precipitate appeared; this occurred at a DNA phosphate to DTA ratio of about 0.6. The suspension was then centrifuged at 5000g and as much supernatant as possible was removed. The precipitate was then frozen in liquid nitrogen and freeze-dried.

Solubility study

In the case of the 5 mM Na$^+$ experiment, the precipitated complex was suspended in deionised water and added in aliquot into a 96-well polystyrene microtitre plate, giving a concentration of about 0.012 mg DNA per well. The plate was then frozen in liquid nitrogen and freeze-dried for 1 day. The various solvent/surfactant mixtures were then added to give a DNA concentration of about 0.06 g/l in each well. The wells were sealed with polythene film and the plate was placed on a shaker table at 23 °C in a controlled-temperature incubator and was shaken at low speed for 2 h. Each well was then examined using an inverted microscope and scored soluble or insoluble depending on whether there was any undissolved complex present. The study with a HEPES concentration of 25 μM was done at a DNA concentration of 0.3 g/l.

Circular dichroism spectroscopy

Circular dichroism (CD) spectra were obtained using a Jasco 500 CD spectrometer. The freeze-dried complex was dissolved by gentle shaking in a 3/1 ethanol/water solvent mixture (0.75 ml ethanol, 0.19 ml H$_2$O, 0.06 ml 0.5 mM HEPES, pH 7.5) to a final concentration of about 2 g/l at 4 °C. Measurements of DNA–DTA without excess surfactant in the range 75–67% ethanol were carried out by adding an aliquot of the stock to a 75% ethanol solution. Water was then titrated into the cell, as in the method of Ivanov [11]. All other solutions were made up and run as separate samples. Scan rates were 10 nm/min.

UV measurements

Measurements were carried out on a Pharmacia Ultrospec III (Pharmacia, Uppsala, Sweden) UV/vis spectrometer, with an external microprocessor-controlled Neslab RTE 211 circulating bath. A Pt100 thermal sensor (Radionics, Ireland) was placed in the sample cell and temperature ramps were carried out using the probe

as a reference. Ramping rates were 0.5 °C/min and runs were carried out between 7 and 70 °C. A best fit of the experimental temperature versus absorbance data to a polynomial was obtained. A numerical differentiation procedure was then used to find the maximum of the derivative, which was taken as the melting temperature (T_m).

Results and discussion

The influence of counterions on DNA solubility and conformation

The phase diagram for DNA–DTA in ethanol/water solution as a function of excess surfactant concentration and ethanol/water ratio is shown in Fig. 1a. The buffer concentration was fixed at 25 μM HEPES, 10 μM EDTA. With no excess surfactant present the minimum ethanol/water (v/v) ratio where solubilisation occurred was 57%; as the surfactant concentration was increased to 50 mM there was a sharp increase in this ratio to 78%. As the DTAB concentration was increased still further to 100 mM, there appeared to be no further dependence of solubility on surfactant concentration. To determine whether or not this behaviour was due to differing solubilities of the helix and coil conformation it was decided to characterise the regions where the DNA existed in these forms. This was achieved by running temperature ramp studies of the DNA at fixed surfactant concentrations, while varying the ethanol/water ratio. The ethanol/water ratio where the helix and coil were in equilibrium at 23 °C was found by interpolation of the melting-temperature dependence on the ethanol/water ratio. The line dividing the helix and coil regions is also shown in Fig. 1a. It can be seen that the helix–coil transition is only weakly dependent on surfactant concentration above 5 mM and that the helix–coil line and the solubility line do not coincide; therefore, we can conclude that the solubility behaviour is not due to a conformational change in the DNA. One possible explanation is that in polar solvents such as ethanol, charged polymers will still exhibit polyelectrolyte behaviour [12] and thus addition of excess cationic surfactant will lead to Coulombic screening and precipitation. Preliminary light scattering data for DNA–DTA in 100 and 75% ethanol seemed to indicate that strand separation accompanied denaturation.

The dependence of the solubility phase diagram on Na$^+$ concentration was also studied, and the results for 5 mM Na$^+$ are shown in Fig. 1b. The minimum ethanol concentration where solubilisation occurs has shifted to a lower value, while in the high-ethanol region the complex is insoluble, with the width of the insoluble region decreasing slightly as the concentration of surfactant is increased to 100 mM. A small amount of a concentrated solution of the DNA–DTA complex in 75% ethanol was added to a 95% ethanol/5 mM NaBr

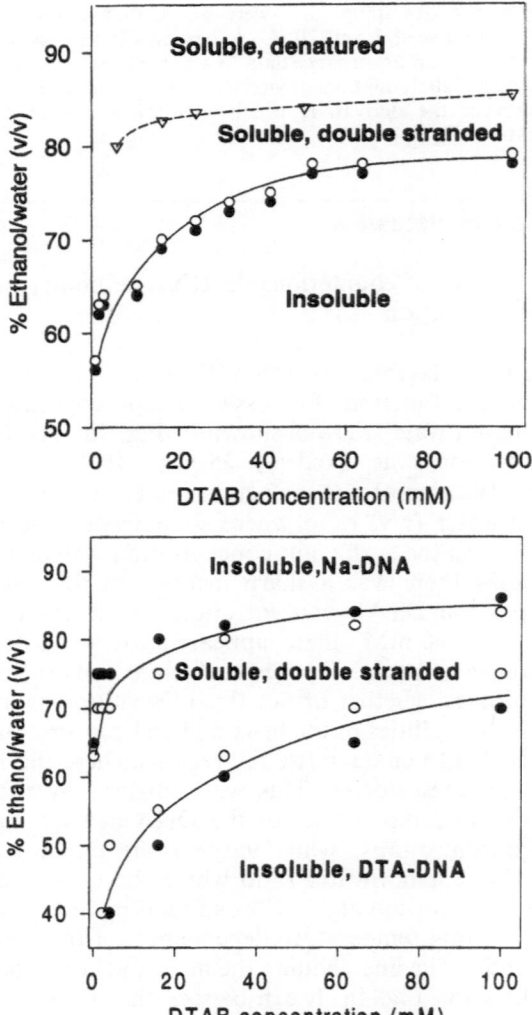

Fig. 1a, b Phase diagram of DNA–dodecyltrimethylammonium (*DTA*) as a function of excess DTA bromide (*DTAB*) concentration and ethanol/water ratio. Open circles represent points where the DNA–DTA complex becomes soluble, *filled circles* represent the onset of insolubility. **a** Buffer concentration 25 μM *N*-(2-hydroxyethyl)piperazine-*N'*-ethanesulfonic acid (*HEPES*), 10 μM ethylenediaminetetraacetic acid (*EDTA*). The *lower line* marks the soluble and insoluble region, while the *upper line* divides the area where DNA exists as a helix from the region where it has a coil conformation. **b** Na⁺ concentration 5 mM. The *lower line* marks the soluble and insoluble regions at lower ethanol/water ratios, while the *upper line* marks the soluble and insoluble regions at high ethanol/water ratios

solution (final DNA concentration 150 mg/l, 0.46 mM) and the resultant precipitate was centrifuged off. It was found that this precipitate was soluble in water. Thus, it appears that DTA^+/Na^+ exchange had occurred.

It has been observed that in solutions containing a mixture of counterions with the same valence the ratio of their bound concentrations is not equal to the ratio of their stoichiometric concentrations in the bulk solution [13]. This effect is due to short-range forces, such as

interaction of charged groups of the polyion, exclusion volume, size effects and changes in hydration number. Manning [14] has proposed a selectivity rule, which can be described by the following equation:

$$\alpha = (\theta_1/\theta_2)/(c_1/c_2) = \exp[-(\delta\mu_1 - \delta\mu_2)/RT] \ ,$$

where α is the selectivity coefficient between the two counterion species, θ_i is the condensed counterion fraction for species i, c_i is the bulk concentration of ionic species i and $\delta\mu_i$ is the free energy of transfer of counterion i from the bulk solution to the condensation layer. The driving force for replacement can possibly be explained by the fact that $\delta\mu$ for Na^+ in ethanolic solution is lower than that for DTA^+, due to its lower solubility; also, Na^+, being smaller, has a closer approach distance and, thus, its repulsive energy of contact will be less. These selectivity effects are especially interesting if the solvent is good for the polyelectrolyte complex with one of the ions, while being poor for the other. This is the case of DNA with DTA^+ and Na^+ as counterions. It has been found for nonstoichiometric synthetic polyelectrolyte–surfactant complexes in organic solvents [12] that when a mixture of low mass counterions and surfactant molecules is present, the polymer adopts a compact conformation, even though the complex is soluble [15]. This has been seen for a number of different polyelectrolytes and solvents [16] and is interpreted as follows: there is an intramolecular aggregation of salt groups which leads to a compact structure. There have been a number of studies reported on the conformation of DNA–surfactant complexes in organic solvents which have claimed the DNA exists as a compact globule [6–8]; however, in these studies steps were not taken to exclude salt from the final solution. It is probable that salt–surfactant exchange could lead to compactisation of the DNA in the manner described previously.

Within the double-stranded soluble region the DNA can exist in a number of defined secondary structures. To determine which of these conformational states were present, and whether any changes occurred on the addition of excess surfactant, CD experiments were performed on the DNA at various ethanol/water ratios and DTAB concentrations. The spectrum of DNA as a function of ethanol concentration with no excess surfactant is presented in Fig. 2a. In ethanolic solution the helix-to-coil transition region is broad (about 20 °C). Because of this, and also because DTAB only weakly stabilises the helix, the temperature at which the CD spectra were run (23 °C) was within the helix–coil transition region and, thus, the DNA was partially denatured. Hence each spectrum is the sum of native and denatured components. Nevertheless, the DNA retains some secondary structure at this temperature, and because the region of ethanol/water ratios where the B/A transition occurs is so narrow (below 8%), we

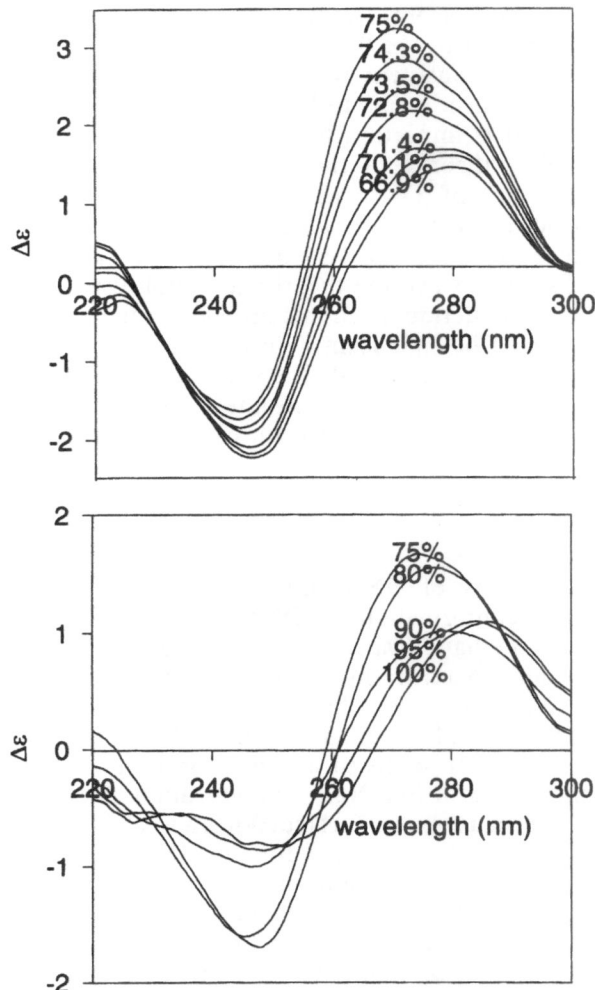

Fig. 2a, b Circular dichroism spectra of DNA–DTA in ethanol/water solution. **a** Without excess DTAB. Buffer concentration 25 μM HEPES, 10 μM EDTA. **b** With 32 mM DTAB. Buffer concentration 25 μM HEPES, 10 μM EDTA. The spectra are labelled with the *volume percentage of ethanol*. The temperature was 23 °C

believe, on the basis of a previous study [10], that the thermal stability of the DNA does not change dramatically over the transition, and it can be assumed that the denatured contribution to the spectrum is essentially constant over the transition region. Therefore, the changes in the CD spectrum are related to conformational changes in the native DNA component.

It can be seen that at 67% ethanol the DNA spectrum exhibits a profile similar to that for Na–DNA in 67% ethanol [11]. A positive band has a peak maximum at 275 nm and a negative band of similar magnitude at 248 nm. The intensities of the two bands are similar, with a band asymmetry ratio of 0.68. Between 67 and 75% ethanol there is a shift of the positive lobe maximum to 270 nm with a concomitant increase in its intensity. The positive-to-negative band ratio increased

to 2. This is similar to data obtained by other workers and appears to indicate that a B-to-A transition occurs in this region [17]; however, the magnitude of the asymmetry change observed was smaller than would be expected for fully native DNA. (Ivanov [11] has observed a positive $\Delta\varepsilon$ value at 270 nm, $\Delta\varepsilon^{270}$, of 9.5 for the A form of DNA, compared to our value of 3.2). It was determined through melting-curve studies that 30% of the sample was native in 75% ethanol, while $\Delta\varepsilon^{270}$ for denatured DNA was found to be 0.39. Assuming 100% of the native sample was in the A form at 75% ethanol, the calculated $\Delta\varepsilon$ is 3.1, which is very close to the observed value and indicates that indeed a B-to-A transition occurred in the native DNA fraction. As the ethanol content is further increased there is a progressive decrease in the positive lobe, along with a shifting of the peak maximum to 258 nm, indicating a transition to the denatured form [18].

The effect of excess surfactant on the B-to-A transition was also studied and the results are presented in Fig. 2b. The spectrum for DNA with 32 mM excess DTAB in 75% ethanol shows a positive lobe maximum of 275 nm, a negative lobe maximum at 245 and a positive-to-negative band ratio of 1. This is different to the situation without excess surfactant, where at 75% ethanol we observed a band asymmetry ratio of 2. Even at 80% ethanol and 32 mM surfactant the positive and negative bands are of similar magnitude. Melting-temperature studies have shown that the DNA was 96% native in 75% ethanol with 32 mM DTAB, and thus this spectrum is not that of denatured DNA. Based on the observation of the band asymmetries it appears that in the presence of excess surfactant at 75 and 80% ethanol the A form is absent and the DNA is in a B-like conformation. As the ethanol content was increased further there was a decrease in the intensity of the positive band, together with a shift in the peak maximum to 285 nm, while the negative lobe also decreased in intensity and shifted to 250 nm. This indicated that there was a progressive denaturation at very high ethanol concentrations. There are several examples in the literature of cases where dehydration does not lead to a B-to-A transition, for example for DNA in methanolic solution [19]. In particular, the alkyltrimethylammonium bromides are known to cause subtle changes in DNA structure on binding [20, 21] and also to inhibit the B-to-A transition in solid films.

The influence of counterions on helix stability

We investigated the influence of salt on the melting temperature in three different ethanol/water mixtures (85, 75 and 67%) and the results are presented in Fig. 3. In all cases there is a linear dependence on the logarithm of Na$^+$ concentration. However, precipitation occurred

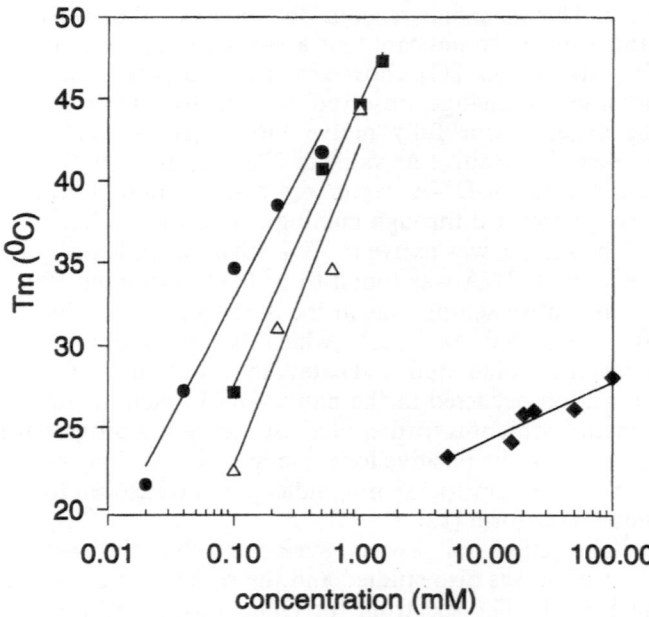

Fig. 3 Melting temperatures of DNA–DTA in various ethanol/water mixtures as a function of NaBr concentration. The percentages of ethanol used were 85 (-△-), 75 (-●-) and 67 (-■-). Also shown is the melting temperature of DNA–DTA as a function of DTAB concentration in 80% ethanol (-◆-). Buffer concentration 25 μM HEPES, 10 μM EDTA. The *lines* are for guiding the eye only

tivity measurements taken as a function of DTAB concentration in the 0–100 mM region did not exhibit any breaks consistent with the formation of micelles. Thus, the effect seems to be related to the structure of DTA^+. It is important to realise that the usual polyelectrolyte theories of the ionic effect on the melting temperature of DNA assume that the predominant influence exerted is through charge and ions are usually treated as point objects; however, this assumption breaks down when the theories are applied to the case of large counterions as in this situation excluded-volume effects may be significant. Work done by Trend et al. [22] has shown that while the melting temperature still varies linearly with the logarithm of ionic strength, the slope of the line is reduced as the cation size is increased.

as the salt concentration was increased, and the range of solubility was narrower as the ethanol content was increased. For a given salt concentration, the values of melting temperature are in the order $^{75\%}T_m > {}^{67\%}T_m > {}^{85\%}T_m$. This is in agreement with previous data, which showed that the stability of the helix is a maximum at 75% [10].

The dependence of melting temperature on excess surfactant was also studied and the results for 80% ethanol are also presented in Fig. 3. It can be seen that, in contrast to the case of excess salt, in the range 5–100 mM there is a much weaker concentration dependence, with T_m only increasing by 5 °C. There are a few possible reasons for this difference. The first is that at high surfactant concentrations micelles are formed, leading to decreased availability of free surfactant ions for stabilising the DNA; however, a plot of conduc-

Conclusions

The solubility of DNA–cationic surfactant complexes and the conformation of DNA in the complex were studied in ethanol/water mixtures as functions of excess surfactant concentration, ionic strength and temperature. We have shown that there is a relatively narrow region in the phase diagram where DNA, in complex with surfactant, exists in a native form and is soluble. In the case of a low NaBr concentration the region is narrow because excess surfactant has a weak influence on the DNA stability and, therefore, the DNA is denatured at high ethanol concentrations. As the concentration of salt is increased the thermal stability of DNA is increased but DNA is no longer soluble at high ethanol concentrations due to exchange between surfactant and sodium cations in the complex. Also we found an indication that in the region where DNA exists in the helical state it can adopt the A or the B form. The present study may shed some light on recent results concerning the coil-to-globule transition of DNA complexed with cationic lipids in organic and mixed solvents [6–8].

Acknowledgements The authors would like to thank Professor Boyd and Mr. Murphy of Queen's University Belfast for use of their CD spectrometer. Also we would like to thank our colleagues in the Irish Centre for Colloid Science and Biomaterials for their support and helpful discussions. J.J.M. acknowledges support from Enterprise Ireland.

References

1. Goddard ED, Ananthapadmanabham KP (eds) (1993) Interactions of surfactants with polymers and proteins. CRC, Boca Raton
2. Ober CK, Wenger G (1997) Adv Mater 9:17–31
3. Flenger PL (June 1997) Sci Am 86–91
4. Reimer DL, Zhang Y, Kong S, Wheeler JJ, Graham RW, Bally MB (1995) Biochemistry 34:12877–12883
5. Sadron G, Beck G, Ebel JP (1964) Biochim Biophys Acta 80:448–455
6. Sergeyev VG, Pyshkina OA, Lezov AV, Melnikov AB, Ryumtsev EI, Zezin AB, Kabanov VA (1999) Langmuir 15:4434–4440
7. Sergeyev VG, Mikhailenko SV, Pyshkina OA, Yaminsky IV, Yoshikawa K (1999) J Am Chem Soc 121:1780–1785
8. Mel'nikov SM, Lindman B (1999) Langmuir 15:1923–1928

9. Ijiro KJ (1992) Chem Soc Chem Commun 18:1339–1341
10. Gorelov AV, McLoughlin DM, Jacquier J-C, Dawson KA (1999) Nuovo Cimento D 20bis:2553–2564
11. Ivanov VI (1992) Methods Enzymol 211:111–126
12. MacKnight WJ, Ponomarenko EA, Tirrell DA (1998) Acc Chem Res 31:781–788
13. Bleam ML, Anderson CF, Record MT (1980) Proc Natl Acad Sci USA 77:3085–3089
14. Manning GS (1984) J Phys Chem 88:6654–6661
15. Bakeev KN, Shu YM, MacKnight WJ, Zezin AB, Kabanov VA (1994) Macromolecules 27:300–302
16. Antonietti M, Forster S, Zisenis M, Conrad J (1995) Macromolecules 28:2270–2275
17. Ivanov VI, Minchenkova LE, Schyolkina AK, Poletayev AI (1973) Biopolymers 12:89–110
18. Girod JC, Johnson WC, Huntingdon SK, Maestre MF (1973) Biochemistry 12:5092–5096
19. Girod JC, Johnson WC (1975) Biochem Biophys Acta 150:35–45
20. Spink CH, Chaires JB (1997) J Am Chem Soc 119:10920–10928
21. Morrissey S, Kudryashov ED, Dawson KA, Buckin VA (1999) Prog Colloid Polym Sci 112:71–75
22. Trend BL, Knoll DA, Ueno M, Evans DF, Bloomfield VA (1990) Biophys J 57:829–834

Progr Colloid Polym Sci (2000) 115:192–195
© Springer-Verlag 2000

A. Pastou
H. Stamatis
A. Xenakis

Microemulsion-based organogels containing lipase: application in the synthesis of esters

A. Pastou · H. Stamatis · A. Xenakis (✉)
Industrial Enzymology Unit
Institute of Biological Research
and Biotechnology
The National Hellenic
Research Foundation
48 Vassileos Constantinou Ave.
11635 Athens, Greece
e-mail: arisx@eie.gr
Tel.: +30-1-7273762
Fax: +30-1-7273758

Abstract Lipases from *Rhizomucor miehei* and *Candida antarctica* have been immobilized in lecithin microemulsion based gels formed with agar and hydroxypropylmethyl cellulose. It was found that both lipases keep its catalytic function after its entrapment in the gels, catalyzing the esterification reaction of 1-propanol with fatty acids in non-polar hydrocarbons at room temperature. Various parameters, which affect the lipase catalytic behavior such as the nature and the concentration of gelling agent, as well as the concentration of alcohol, have been examined. High reaction rates and yields (up to 85%) were obtained with the above microemulsion based gels.

Key words Biocatalysis ·
Microemulsion · Polymer gel ·
Lipase · Organogels

Introduction

Various methods have been reported in the literature for effecting enzyme-catalyzed reactions in apolar media [1–3]. These include

1. Macroheterogeneous biphasic systems, such as liquid–liquid systems composed of a water-immiscible organic solvent and water. These are nearly anhydrous systems where the enzyme is usually suspended as a powder or in an immobilized form adsorbed onto a suitable carrier in organic solvents or gases in a supercritical state.
2. Microheterogeneous systems, such as different types of water-in-oil (w/o) microemulsions or reverse micelles [3].

W/o microemulsions are spontaneously formed, isotropic, thermodynamically stable liquid media with a large interfacial area in which reactants can interact with an enzyme trapped in the dispersed aqueous phase. A major attraction of this procedure is that the enzyme is dispersed at the molecular level. In such a system the enzymes are active as regards the conversion of both hydrophilic and hydrophobic compounds [4, 5].

The major problem, which must be solved for the employment of a microemulsion system in industrial processes, is the recovery of the products and the regeneration of the enzyme. One approach to simplifying the recovery of the product and the enzyme reuse from microemulsion-based media is the use of gelled microemulsion systems [6]. Many w/o microemulsions can be "gelled" by the addition of gelatin above the gelling temperature, yielding a matrix suitable for enzyme immobilization [7, 8]. The preparation of gelled microemulsion systems was first reported in 1986 [7, 8] and involved the gelation of w/o microemulsions based on bis(2-ethylhexyl)sulfosuccinate (AOT) or lecithin into monophasic optically transparent rigid systems by mixing with an aqueous gelatin solution above the gelling temperature. On cooling to room temperature, a transparent gel is formed which has reproducible physical properties. Rees et al. [6] have immobilized lipases in AOT microemulsion-based gelatin gels. These enzyme-containing gelatin-based gels are rigid and stable in various nonpolar organic solvents and may therefore be used for biotransformations in organic media [6, 9–12]. Furthermore, the gel and thus the enzyme may easily be reused several times, with retained activity. Under most conditions the gel matrix fully retains the surfactant, gelatin, water and enzyme components, facilitating the diffusion of nonpolar substrates or products between a contacting nonpolar phase and the gel pellets.

In previous work we investigated the catalytic behavior of the lipase from *Pseudomonas cepacia* immobilized on lecithin microemulsion-based organogels (MGs) formulated with biopolymers such as gelatin, agar and κ-carrageenan [13, 14]. In this work, we report the immobilization of various commercial lipases in lecithin MGs formulated with biopolymers such as agar and hydroxypropyl methylcellulose (HPMC) as well as the application of these lipase-containing gels for the esterification of aliphatic alcohols with fatty acids. Various reaction parameters affecting the activity of immobilized lipases, such as the nature and the concentration of biopolymers and substrates, have been examined.

Experimental

Materials

Lipases *Candida antarctica* and *Rhizomucor miehei* were supplied by Fluka. Lecithin, containing 18–26% phosphatidyl choline, was purchased from Serva and was purified by column chromatography [15]. HPMC, 3600–5500 cP, was obtained from Sigma and malt-extract agar was obtained from Merck.

Preparation of microemulsions and MGs

Lecithin microemulsions were prepared as follows. Appropriate amounts of lipase in 25 mM tris(hydroxymethyl)aminomethane/HCl pH 7.5 were added to 3.8% w/w lecithin in isooctane containing 5% v/v 1-propanol and the final water content was adjusted by the addition of the required amount of buffer.

Lipase-containing MGs were prepared by introducing appropriate amounts of microemulsion containing lipase to a second solution of polymer (HPMC, or agar) in water. In a typical experiment, 1 ml lecithin microemulsion, containing 0.46 mg *C. antarctica* lipase or 0.16 mg *R. miehei* lipase was gelled with 0.5–1.5 g HPMC and 2 ml water at 25 °C, or 0.45–0.85 g agar and 5 ml water at 60 °C. In the case of the agar gels, agar was first solubilized in water at 100 °C. The mixtures were vigorously shaken and stirred until homogeneous (about 5–10 min) and were allowed to cool to room temperature to yield organogels. The gels were stored in a freezer until used.

Lipase-catalyzed reactions

The determination of lipase activity was based on the measurement of the initial rate of synthesis of various fatty acid esters. The gel was cut into approximately 15 pieces and placed into a reaction vial. By adding 10 ml isooctane, which contained various amounts of alcohol and fatty acid, the reaction was initiated. The vial was stirred at 150 rpm at 25 °C. At fixed intervals 50-μl samples were taken from the vial and analyzed by gas chromatography [13].

Results and discussion

Effect of the nature and the concentration of polymer on lipase activity

The aim of our work was to investigate the ability of some natural polymers, such as agar and HPMC, to

form stable MGs that are suitable for enzyme immobilization and application in enzyme-catalyzed reactions in organic solvents. For this purpose, lecithin w/o microemulsions containing two commercial lipases were gelled to pseudosolid gels after mixing with aqueous solutions of natural polymers. The lipase-containing gel was cut into suitable pieces and added to the organic solvent, which contained lipase substrates (1-propanol and fatty acids).

The effect of the biopolymer concentration on the immobilized lipase catalytic activity was determined by comparing the rates of esterification of 1-propanol with lauric acid. Two types of MGs formulated with various concentrations of agar as well as HPMC were used. The effects of the concentration of the two polymers on the catalytic activity of immobilized lipases from *C. antarctica* and *R. miehei* are shown in Tables 1 and 2. The progress of the reaction for the esterification of 1-propanol with lauric acid catalyzed by MGs containing lipase in isooctane is shown in Fig. 1. As can be seen, the lipase activity depends on the nature of the polymers used as well as on their concentration. An increase in the mass fraction of the polymers slightly increased the esterification rate for both polymers used. Moreover, the reaction rate of the esterification was higher in MGs formulated with HMPC than in those formulated with agar. The two lipases retained their catalytic activity in MGs when the mass fraction of the polymers in the gels varied from 0.07 to 0.13 for agar and from 0.16 to 0.35

Table 1 Effect of the mass fraction of agar in lecithin microemulsion-based gels on the esterification rate of 1-propanol (100 mM) with lauric acid (100 mM) catalyzed by lipases from *Rhizomucor miehei* and *Candida antarctica* in isooctane at 25 °C

Mass fraction of agar	Reaction rate (mM min⁻¹)	
	R. miehei lipase	*C. antarctica* lipase
0.07	0.22	0.10
0.09	0.24	0.12
0.11	0.25	0.13
0.13	0.25	0.13

Table 2 Effect of the mass fraction of hydroxypropyl methylcellulose (*HPMC*) in lecithin microemulsion-based gels on the esterification rate of 1-propanol with lauric acid catalyzed by lipases from *R. miehei* and *C. antarctica*. Reaction conditions as described in Table 1

Mass fraction of HPMC	Reaction rate (mM min⁻¹)	
	R. miehei lipase	*C. antarctica* lipase
0.16	0.67	0.15
0.22	0.76	0.17
0.28	0.86	0.17
0.35	0.91	0.18

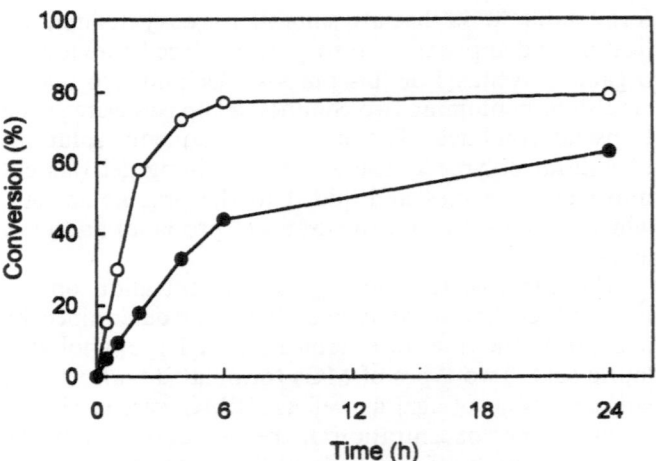

Fig. 1 Typical reaction profiles for the esterification of 1-propanol with lauric acid catalyzed by *Rhizomucor miehei* (○) and *Candida antarctica* (●) lipase in lecithin microemulsion-based gels formulated with hydroxypropyl methylcellulose (*HPMC*). Reaction conditions as described in Table 1. The mass fraction of HPMC was 0.22

for HPMC. It must be noted that MGs prepared in these concentration ranges of polymers were stable and maintained their structural integrity for several days in contact with isooctane. Moreover, the immobilized lipases exhibit good stability, in both gels, for a storage period of 45 days at 4 °C (data not shown). Both lipases catalyzed with high conversion yield (75–85%) the esterification of 1-propanol by various fatty acids with 10–18 carbon atoms in their molecules (data not shown).

Effect of the nature of the organic solvent on lipase activity

The influence of the chain length of the hydrocarbon solvent used on the esterification activity of *C. antarctica* lipase was studied. The effect of the nature of the organic solvent on *C. antarctica* lipase activity immobilized on agar as well as on HPMC organogels is shown in Table 3. As can be seen a higher esterification rate was observed for hydrocarbons with eight carbon atoms such as *n*-octane and isooctane, while the reaction rate dropped slightly for other hydrocarbons. For all solvents used, the final conversion yield at equilibrium was estimated to be about 90% for both MGs. It must be noted that in other studies concerning the activity of immobilized *Chromobacterium viscosum* lipase in AOT MGs formulated with gelatin a decrease in the esterification activity was observed as the alkane chain length increased from six to ten carbon atoms [12]. Similar catalytic behavior was also observed by our group for the catalytic activity of lipase from *Pseudomonas cepacia* immobilized in various lecithin MGs formulated with agar and gelatin [13, 14].

Table 3 Effect of hydrocarbons on the esterification rate of 1-propanol with lauric acid catalyzed by *C. antarctica* lipase in lecithin microemulsion-based gels formulated with agar and HPMC. Reaction conditions as described in Table 1. The mass fraction of the polymers was 0.09 for agar and 0.022 for HPMC

Solvents	Reaction rate (mM min^{-1})	
	Agar gels	HPMC gels
n-Hexane	0.08	0.13
n-Heptane	0.09	0.14
n-Octane	0.10	0.17
Isooctane	0.12	0.17
n-Decane	0.11	0.15
n-Dodecane	0.09	0.12

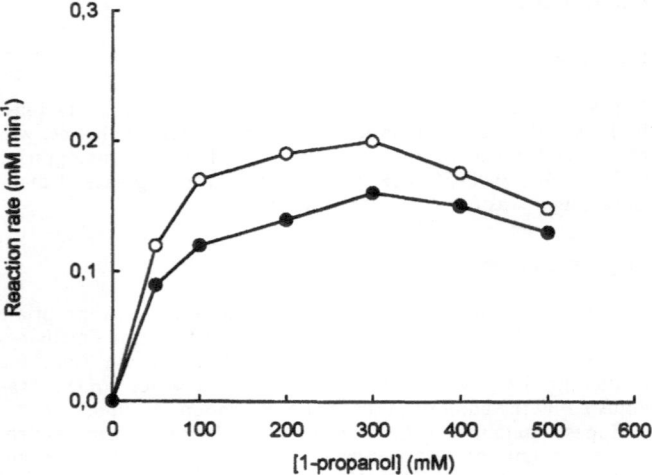

Fig. 2 Effect of the concentration of 1-propanol on the esterification rate with lauric acid catalyzed by *C. antarctica* lipase in lecithin microemulsion-based gels formulated with agar (●) and with HPMC (○). Reaction conditions as described in Table 1

Effect of the alcohol concentration

In order to study the effect of the alcohol concentration on the lipase activity in the MGs, different concentrations of 1-propanol were used in the range 50–500 mM at constant concentration of lauric acid (100 mM). Figure 2 shows the effect of the alcohol concentration on the reaction rate of esterification catalyzed by lipase from *C. antarctica* entrapped in a lecithin microemulsion system or immobilized in MGs formulated with agar or HPMC . As can be seen an inhibitory effect on the lipase catalytic activity was observed in all cases studied at a high concentration of 1-propanol (above 400 mM).

Conclusion

Lipase-containing lecithin MGs formulated with natural gelling agents such as HPMC and agar have considerable potential for their application in biotransformations. The lipases keep their catalytic function after their entrapment in the gels and catalyze various esterification reactions in nonpolar hydrocarbons at room temperature.

Acknowledgements The Greek General Secretariat of Research and Technology and Phaematen S.A., financed this work within the frame of the program 97-IIABE-93.

References

1. Ballesteros A, Bornscheuer U, Capewell A, Combes D, Condoret J-S, Koening K, Kolisis FN, Marty A, Menge U, Scheper T, Stamatis H, Xenakis A (1995) Biocatal Biotransform 13:1
2. Dodrick JS (1989) Enzyme Microb Technol 11:194
3. Khmelnitsky YL, Levashov AV, Klyachko NL, Martinek K (1988) Enzyme Microb Technol 10:710
4. Stamatis H, Xenakis A, Provelegiou M, Kolisis FN (1993) Biotechnol Bioeng 42:103
5. Stamatis H, Xenakis A, Kolisis FN (1999) Biotechnol Adv 17:293
6. Rees GD, Nascimento MG, Jenta TRJ, Robinson BH (1991) Biochim Biophys Acta 1073:493
7. Quellet C, Eicke H-F (1986) Chimia 40:233
8. Haering G, Luisi PL (1986) J Phys Chem 90:5892
9. Backlund S, Eriksson F, Kanerva LT, Rentala M (1995) Colloids Surf B 4:121
10. Nascimento MG, Rezende MC, Vecchia RD (1992) Tetrahedron Lett 33:5891
11. Backlund S, Eriksson F, Hedstrom G, Laine A, Rentala M (1996) Colloid Polym Sci 274:540
12. Jenta TR-J, Batts G, Rees GD, Robinson BH (1997) Biotechnol Bioeng 53:121
13. Stamatis H, Xenakis A (1999) J Mol Catal B 6:399
14. Xenakis A, Stamatis H (1999) Prog Colloid Polym Sci 122:132
15. Avramiotis S, Stamatis H, Kolisis FN, Lianos P, Xenakis A (1996) Langmuir 12:6320

Progr Colloid Polym Sci (2000) 115:196–200
© Springer-Verlag 2000

S. Avramiotis
C. T. Cazianis
A. Xenakis

Membrane spin-probe studies in lecithin and bis(2-ethylhexyl)sulfosuccinate sodium salt water-in-oil microemulsions

S. Avramiotis · C. T. Cazianis
A. Xenakis (✉)
Industrial Enzymology Unit
Institute of Biological Research
and Biotechnology
The National Hellenic
Research Foundation
48 Vassileos Constantinou Ave.
11635 Athens, Greece
e-mail: arisx@eie.gr
Tel.: +30-1-7273762
Fax: +30-1-7273758

Abstract Electron paramagnetic resonance spectroscopy was applied to study the interfacial properties of the surfactant layers in isooctane and in water-in-oil microemulsions. Two systems were studied: one formed with the anionic surfactant (AOT) bis(2-ethylhexyl)sulfosuccinate sodium salt and the other with naturally occurring lecithin. The spectra of the spin probe 5-doxyl stearic acid were monitored in binary mixtures of both surfactants in isooctane to follow the behavior of the continuous phase of the corresponding microemulsions. The probe was quite immobilized as it interfered with the surfactant layers. The differences in mobility observed in the two systems were attributed to the different rigidity of the corresponding surfactant layers. In the case of the microemulsions, the mobility of the probe was altered. In the lecithin systems the formation of reverse micelles induced a curvature of the layer, leading to a less immobilized probe. The mobility of the probe was found to depend on the size of the reverse micelles. In the case of AOT systems, where well-defined water pools are formed, the spectra show the existence of two states of the probe with different mobilities.

Key words Reverse micelles · Surfactant layer · Electron paramagnetic resonance · Probe mobility

Introduction

Water-in-oil (w/o) microemulsions are transparent and thermodynamically stable fine dispersions of water in nonpolar solvents, stabilized by surfactants [1]. These systems, also called reverse micelles, have been used as model systems for a wide range of biological studies. Research studies on enzymes trapped in the aqueous core of these microemulsions have attracted a lot of interest regarding the investigation of the protein functions and conformation in a water-restricted microenvironment [2, 3], leading to potential biotechnological applications in the pharmaceutical or cosmetic industry [4–7]. At the same time the study of surfactant self-association in these organized assemblies as a result of the molecular interactions between hydrophilic as well as lipophilic areas of the constitutents of a microemulsion is also of great interest. Most of the studies in microemulsions have been performed by using a model micellar system based on the extensively studied synthetic surfactant bis-(2-ethylhexyl)sulfosuccinate sodium salt (AOT), that forms spherical reverse micelles. Recently, lecithin, a naturally occuring surfactant, was shown to be a good alternative for forming microemulsions for model studies that simulate biomembranes [8–10]. Lecithin aggregates in nonpolar solvents have been studied and various structures have been reported [11, 12].

Electron paramagnetic resonance (EPR) spectroscopy using spin probes is a technique that can provide important information on a microenvironmental level about conformation, mobility and polarity in biological (e.g. membranes, enzymes) [13, 14] as well as in physicochemical systems (e.g. micelles, microemulsions) [15, 16].

In the present work we have applied EPR spectroscopy to study the spectra of 5-doxyl stearic acid (5-DSA), a membrane spin probe [17], solubilized in lecithin/ isooctane and AOT/isooctane binary systems as well as in microemulsions. This molecule is an amphiphile because of its polar head (-COOH) and its hydrophobic moiety that comprises 18 carbon atoms. When the 5-DSA molecules are solubilized in micellar systems, they are localized at the interfaces intercalating within the surfactant molecules; therefore, an analysis of the EPR spectra can reveal aspects of the mobility of the nitroxide ring located at the membrane as well as the interfacial rigidity fluctuation resulting from the variation of the constituents for both w/o microemulsions.

Experimental

Lecithin, containing 18–26% phosphatidyl choline was purchased from Serva, Heidelberg, Germany. Lecithin was purified as described by Avramiotis et al. [18] and identified by NMR [19, 20] as phosphatidyl choline with an average molecular mass of 800 Da. The purified lecithin was found to contain 0.7–1.5 water molecules per lecithin molecule. The spin-labelled fatty acid 5-DSA was purchased from Sigma (USA). AOT was from Sigma and was used without further purification. Propanol was from Ferak, Berlin, Germany. All chemicals were of the highest available degree of purity. Water of high purity was obtained using a Millipore Milli-Q Plus water-purification system.

Lecithin reverse micelles, with an aqueous core can only be obtained in the presence of cosurfactants such as alcohols with short aliphatic chains [21]. In the present study l-propanol was used in order to stabilize the system. The choice of the compositions of the microemulsions used was based on the phase diagram of the quaternary system lecithin–isooctane–alcohol–water [22]. Lecithin-based microemulsions were prepared as follows. A stock solution of 5% w/w (4.4×10^{-2} M) lecithin in isooctane was prepared and stocked under nitrogen at 4 °C. Microemulsions were formed with the addition of the appropriate amounts of propanol and water. The final propanol concentration ranged from 0.07 to 1.3 M. The total amount of water was adjusted to give the desired value of the molar ratio $w_o = [H_2O]/[lecithin]$. AOT-based microemulsions were prepared by adding water to a stock solution of 0.1 M AOT in isooctane. The total amount of water was similarly adjusted to give the desired value of the molar ratio $w_o = [H_2O]/[AOT]$. A transparent solution was obtained in both cases after gentle shaking for a few seconds. To obtain the desired concentrations of 5-DSA in both microemulsion systems tested, 1 ml of microemulsion was added to a tube into which the appropriate amounts of 5-DSA had previously been deposited. This was done by placing 10 μl of a 7.8×10^{-3} M ethanol solution in the tube and by further evaporating the ethanol. The water content of the initial stock solutions of lecithin/isooctane and AOT/isooctane was periodically checked by Karl Fischer titrations. The amount of water (in general, less than 0.5%) was taken into consideration in the calculation of the total water content.

EPR spectra were recorded at room temperature using a Bruker ER 200 D spectrometer operating at the X-band. Spectra were accumulated and treated using the DAT-200 software for personal computers (University of Lubeck). The samples were contained in an E-248 cell. Typical settings were centerfield, 3471 G; scan width, 100 G; time constant, 0.5 s; scan time, 100 ms; microwave power, 7.5 mW; microwave frequency, 9.76 GHz; modulation amplitude, 1 G.

The rotational correlation time parameter, τ_R, can be used to monitor the dynamics of a spin probe and is calculated using the following relationship [23]:

$$\tau_R = 6 \times 10^{-10} \left[\left(\frac{h_0}{h_{+1}} \right)^{1/2} + \left(\frac{h_0}{h_{-1}} \right)^{1/2} - 2 \right] \Delta H_0 \ . \tag{1}$$

Here h_{+1}, h_0 and h_{-1} are the intensities of the low-field, central and high-field peaks of the EPR spectrum, respectively, and ΔH_0 is the width of the central line. Equation (1) is applicable in the fast motion region, i.e. for rotational correlation times in the range $10^{-11} < \tau_R < 3 \times 10^{-9}$ s [24]. The simulations of the experimental spectra were conducted with a simulation program (PESTWinSim) developed at the National Institute of Environment and Health Sciences, which is available through the Internet (http:/lmb.nie-hs.nih.gov). Using this program four parameters of the simulated spectrum could be controlled independently: the line width, the line shape (percentage of Gaussian or Lorenzian shape), the relative intensity and the hyperfine-coupling constants.

Results and discussion

Lecithin systems

In order to determine the properties of the surfactant interface of w/o microemulsions we applied the EPR spectroscopic technique by using the spin-labeled fatty acid 5-DSA. Figure 1 shows the EPR spectra of 5-DSA recorded in isooctane (Fig. 1, trace a), in solutions of lecithin in isooctane at various concentrations (Fig. 1, traces b–d) and in microemulsions (Fig. 1, traces e, f).

Let us first consider the binary systems. These solutions are used for the preparation of microemulsions and their properties reflect the behavior of the continuous phase of these systems. As can be seen from Fig. 1, trace a, the probe gives an isotropic spectrum in pure isooctane. Important differences appear in the presence of lecithin (Fig. 1, traces b–d). Namely, in the solution with relatively low lecithin concentration, the spectrum of 5-DSA corresponds to a slightly immobilized species, as reflected by the decrease in the high-field line (Fig. 1, trace b). In some cases, depending on the organic solvent, small reverse micelles can form with the polar heads regrouped around the residual 1–2 water molecules bound to the lecithin molecules [12, 25, 26]. The added stearic acid probe is anchored to the lecithin aggregates due to its own amphiphilic character.

The observed differences in the immobilization of the probe spectrum (Fig. 1, traces a–d) are attributed to the position of the nitroxide ring on the fatty acid aliphatic chain [27]. In the 5-DSA molecule, the nitroxide ring is located close to the polar head of the amphiphilic fatty acid (connected to the fifth carbon atom) and, consequently, is closer to the lecithin polar head, thus resulting in restricted mobility. In contrast the nitroxide ring of 12-DSA (another spin-labeled stearic acid probed at the twelfth carbon atom) is much more mobile as it is

198

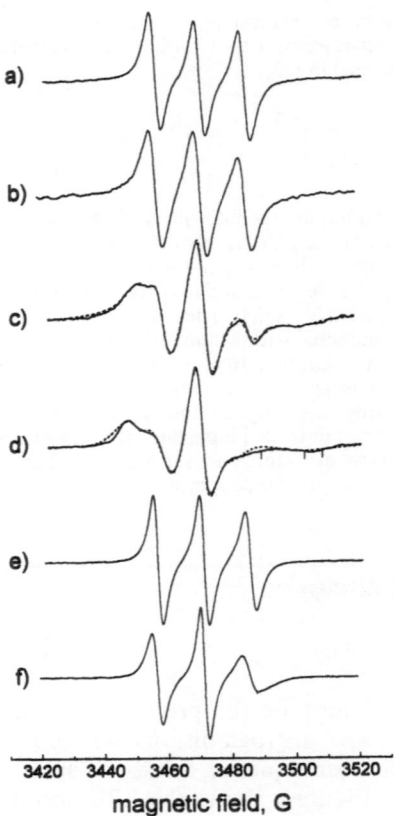

Fig. 1 Electron paramagnetic resonance (EPR) spectra of 5-doxyl stearic acid (*5-DSA*) in isooctane (*a*), lecithin/isooctane, [lecithin] = 1.8×10^{-1} M (*b*), lecithin/isooctane, [lecithin] = 8.8×10^{-4} M, experimental (—), simulated (- - -) (*c*), lecithin/isooctane, [lecithin] = 4.4×10^{-2} M, experimental (—), simulated (- - -) (*d*), lecithin microemulsion, $w_o = 10$, [l-propanol] = 1.3 M, [lecithin] = 4.4×10^{-2} M (*e*), lecithin microemulsion, $w_o = 50$, [l-propanol] = 1.3 M, [lecithin] = 4.4×10^{-2} M (*f*), [5-DSA] = 7.8×10^{-5} M; $T = 25$ °C

not buried in the surfactant layer as the former probe is [28].

By varying the concentration of lecithin, the immobilization of the 5-DSA probe is altered (Fig. 1, traces c, d). The more lecithin is contained, the more immobilized the probe. Comparing the EPR spectra obtained at 1.8×10^{-4} M, (Fig. 1, trace b) and 8.8×10^{-4} M lecithin, (Fig. 1, trace c) concentrations (about 2 and 11 lecithin molecules per 5-DSA molecule, respectively) we observed significant differences in their shape, probably due to some structural changes occurring upon lecithin addition. This could be explained by a possible transition between lecithin monomers and aggregates when the concentration is higher than the critical micelle concentration, which in similar lecithin reverse micelles in benzene has been determined to be about 3 mM [29].

The computer simulations for the spectra of 5-DSA in 8.8×10^{-4} M lecithin solution (Fig. 1, trace c dashed line), showed the presence of two species: one with

$a_N = 13.9$ G (about 57%) and the other, more immobilized, with $a_N = 21.1$ G (about 43%). The computer simulation for the spectra of 5-DSA in 4.4×10^{-2} M lecithin solution (Fig. 1, trace d dashed line) showed the presence of two species both with enhanced immobilization: one with $a_N = 16.2$ G (about 56%) and the other with $a_N = 21.2$ G (about 44%).

The spectra of 5-DSA in lecithin microemulsions at various w_o values at fixed l-propanol content are shown in Fig. 1 (traces e, f). By comparing the spectral shape obtained in these microemulsion systems with the one obtained in the lecithin/isooctane system (Fig. 1, trace d), we can see that the previously discussed immobilization of the probe is almost canceled. This behavior is related to the drastic change of the structure of the lecithin aggregates when water and l-propanol are added to the system to form microemulsions [16]. The increased mobility of the probe is related to the flexibility of the lecithin interface, which is increased, allowing the formation of curved entities with an aqueous core of reverse micellar type.

Depending on the composition of the lecithin microemulsions the spin probes show differences in their mobility. Figure 1 shows the effect of the water content on the spectral characteristics of 5-DSA in lecithin microemulsions with $w_o = 10$, (Fig. 1, trace e) and $w_o = 50$ (Fig. 1, trace f). The calculated rotational correlation time values were $\tau_R = 0.3$ ns and $\tau_R = 1.9$ ns, respectively. It is obvious that as long as w_o increases, τ_R of 5-DSA increases, indicating a decrease in its mobility.

The more interesting point is that the very low mobility of the probe observed in the absence of water and l-propanol is increased in the microemulsions, but when the water content increases the mobility tends to decrease again. We can explain this observation by relating the mobility of the specific probe to the flexibility of the lecithin layers. In the presence of water and alcohol, reverse micelles are formed with the lecithin constituting the interface. At low water content the structure of the reverse micelles is such that the curvature of the lecithin layer is more pronounced than it is in the case of larger reverse micelles formed at high w_o values [30]. The 5-DSA molecule is anchored in the lecithin interface at the external side of the reverse micelles and thus the mobility of the nitroxide ring is affected by the degree of freedom of the hydrophobic chains of lecithin. This can be described as "hedgehog" behavior, i.e. it is an analogy to the hedgehog, which by curving its back can modify the relative position of its spines, thus increasing the space between the edges of the spines. The more water is added to the reverse micelles leading to swollen structures, the less curved is the lecithin layer, thus hindering the mobility of the nitroxide ring.

By adding more l-propanol to the microemulsions the polarity of the water core is altered, rendering the

Table 1 Variation of the rotational correlation time, τ_R, of 5-doxyl stearic acid (*5-DSA*) in lecithin microemulsions with $w_o = 10$ and $w_o = 20$ as a function of 1-propanol concentration: [lecithin] = 4.4×10^{-2} M; [5-DSA] = 7.8×10^{-5} M; $T = 25\ °C$

	[1-propanol] (M)	τ_R (ns)
$w_o = 10$	0.07	2.04
	0.15	1.27
	0.34	0.67
	0.60	0.46
	0.82	0.42
	1.33	0.35
$w_o = 20$	0.28	1.58
	0.34	1.30
	0.42	1.20
	0.60	0.96
	0.82	0.70
	1.33	0.63

dispersed phase less hydrophilic [21] and the curvature of the lecithin layer consequently more pronounced. This can be confirmed by the results shown in Table 1 showing the dependence of the calculated correlation times, τ_R, as a function of the 1-propanol content. As can be seen, the mobility of the nitroxide ring of 5-DSA increases when the 1-propanol concentration increases for the two microemulsions tested at different water contents.

AOT systems

We examined the spectra of 5-DSA in AOT/isooctane solutions and AOT-formulated microemulsions in order to investigate this type of membrane and to compare them to those of the lecithin systems. Figure 2 shows the spectra of 5-DSA in a 0.1 M AOT solution in isooctane (Fig. 2, trace a) and in AOT microemulsions of various water contents (Fig. 2, traces b–e). The addition of AOT to isooctane induces a certain immobilization of the spin probe, as can be seen from the differences in the spectra shown in Fig. 1, trace a and Fig. 2, trace a. This immobilization occurred in a similar way as for the previously described lecithin case, but in a significantly less pronounced manner. Namely, in the lecithin case, the immobilization of 5-DSA is more important, even if the surfactant concentration is smaller than that of AOT. This difference may be attributed to the rigidity of the respective aggregates and the molecular dimensions of the two surfactant molecules, since the hydrophobic chain length of lecithin is at least twice as long compared to that of AOT.

The spectra of 5-DSA recorded in AOT microemulsions with increasing water content as expressed by the w_o values are shown in Fig. 2, traces b–e. On increasing w_o the immobilization increases, whereas for the micro-

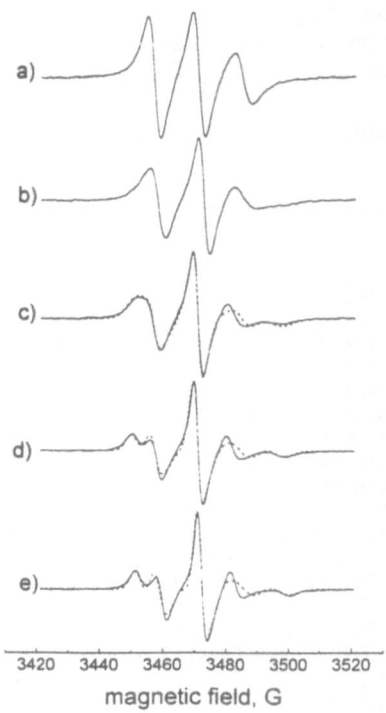

Fig. 2 EPR spectra of 5-DSA in bis (2-ethylhexyl)sulfosuccinate sodium salt (*AOT*) isooctane, [AOT] = 0.1 M (*a*), AOT microemulsion, $w_o = 5$ (*b*), AOT microemulsion, $w_o = 10$, experimental (—), simulated (- - -) (*c*), AOT microemulsion, $w_o = 20$, experimental (—), simulated (- - -) (*d*) and AOT microemulsion, $w_o = 30$, experimental (—), simulated (- - -) (*e*). [5-DSA] = 7.8×10^{-5} M; $T = 25\ °C$

emulsions with w_o higher than 10, a drastic change in the shape of the spectrum occurs. Namely, both the high-field and the low-field lines split into two new ones, indicating the existence of two different states of the spin probe. These two states correspond to two different molecular entities of 5-DSA: a molecular fatty acid and a dissociated one. It is well known that when the water content of these types of microemulsions exceeds the corresponding $w_o = 10$, free water molecules are encountered in the reverse micelles [31]. The presence of well-defined water pools allows the dissociation of the carboxylic group of 5-DSA. The two distinct 5-DSA species have different mobilities, with the dissociated one being buried more in the surfactent membrane and approaching closer to the aqueous microphase.

On analyzing the spectra by means of computer simulations, we obtained the spectral characteristics of each component. The more immobilized species, disso-ciated acid, shows a hyperfine-splitting constant value, a_N, of 20.7–21.6 G for $w_o = 10$–30, while the corre-sponding values of the nondissociated 5-DSA range from 13.9 to 13.4 G. The smaller values of a_N corre-spond to a lower polarity of the environment of the probe, which is consistent with the decreased polarity of the continuous organic phase of the reverse micelles. On the other hand, the dissociated species is closer to the aqueous phase where the polarity is increased. From the areas of the corresponding spectral components, the relative percentages of the two species can be calculated. The percentage of the dissociated 5-DSA was found to vary from 20 to 30% for w_o from 10 to 30. These percentages probably correspond to the dissociation constant of 5-DSA in the specific microenvironment [32].

References

1. Danielson I, Lindman B (1982) Colloids Surf 3:391
2. Luisi PL, Magid L (1986) Crit Rev Biochem 20:409
3. Martinek K, Levashov AV, Klyachko NL, Khmelnitski YL, Berezin YV (1986) Eur J Biochem 155:453
4. Aires-Barros MR, Cabral JMS (1991) Biotechnol Bioeng 38:1302
5. Stamatis H, Xenakis A, Kolisis FN (1993) Biotech Lett 15:471
6. Backlund S, Eriksson F, Hedstrom G, Laine A, Rantala M (1996) Colloid Polym Sci 274:540
7. Protopapa EE, Xenakis A, Avramiotis S, Sekeris CE (1997) Greek Patent 1002706Gr
8. Kahlweit M, Busse G, Faulhaber B (1995) Langmuir 11:1576
9. Shinoda K, Shibata Y, Lindman B (1993) Langmuir 9:1254
10. Avramiotis S, Stamatis H, Kolisis FN, Lianos P, Xenakis A (1996) Langmuir 12:6320
11. Ishii F, Takamoura A, Ishigami Y (1996) Langmuir 11:483
12. Schurtenberger P, Peng Q, Leser ME, Luisi PL (1993) J Colloid Interface Sci 156:43
13. Berliner LJ (ed) (1976) Spin labelling, theory and applications. Academic, New York
14. Marzola P, Forte C, Pinzino C, Veracini CA (1991) FEBS Lett 289:29
15. Cazianis CT, Xenakis A (1989) Prog Colloid Polym Sci 79:214
16. Di Meglio JM, Dvolaitzky M, Taupin C (1985) J Phys Chem 89:871
17. Zana R (ed) (1987) Surfactant solutions. New methods of investigation. Dekker. New York
18. Avramiotis S, Xenakis A, Lianos P (1996) Prog Colloid Polym Sci 100:286
19. Chapman D, Morrison A (1966) J Biol Chem 241:5044
20. Haque R, Tinsley IJ, Schmedding D (1972) J Biol Chem 247:157
21. Shinoda K, Araki M, Sadaghiani A, Khan A, Lindman B (1991) J Phys Chem 95:989
22. Avramiotis S, Papadimitriou V, Cazianis CT, Xenakis A (1998) Colloids Surf A 144:295
23. Kommaredi NS, O'Connor KC, John VT (1994) Biotechnol Bioeng 43:215
24. Smith ICP (ed) (1971) Biological application of electron spin resonance. Wiley-Interscience, New York
25. Sadaghiani AS, Noori A, Khan A (1991) J Surf Sci Technol 7:163
26. Walde P, Giuliani AM, Boicelli A, Luisi PL (1990) Chem Phys Lipids 53:265
27. Costanzo R, De Paoli T, Ihlo JE, Hager AA, Farach HA, Poole CP Jr, Knight JM (1994) Spectrochim Acta Part A 50:203
28. Avramiotis S, Cazianis CT, Xenakis A (1999) Langmuir 15:2375
29. Giomini M, Giuliani AM, Trotta E, Boicelli CA (1989) Chem Phys Lett 158:334
30. Avramiotis S, Bekiari V, Lianos P, Xenakis A (1997) J Colloid Interface Sci 194:326
31. Luisi PL, Straub BE (eds) (1984) Reverse micelles. Plenum, London
32. Sanson A, Ptak M, Rigaud JL, Gary-Bobo CM (1976) Chem Phys Lipids 17:445

Progr Colloid Polym Sci (2000) 115:201–208
© Springer-Verlag 2000

S. Morrissey
E. Craig
V. Buckin

Does the structure of surfactant complexes on the DNA surface depend on the nucleotide sequence?

S. Morrissey · E. Craig · V. Buckin (✉)
Department of Chemistry
University College Dublin
Belfield, Dublin 4, Ireland
e-mail: vitaly.buckin@ucd.ie
Tel.: +353-1-7062436
Fax: +353-1-7062127

Abstract We investigated the binding of surfactant to two double stranded oligonucleotides, each containing a high percentage (76%) of guanine–cytosine or adenine–thymine base pairs. A combination of isothermal titration calorimetry, UV spectroscopy, fluorescence and a high-resolution ultrasonic technique were employed to gain both a structural and a thermodynamic insight into the complex formation. Binding of surfactant to the oligonucleotides was found to occur in two distinct stages. "Micellelike" surfactant aggregates are formed at the first stage of binding. These aggregates have a compressible core and are large enough to accommodate our fluorescence probe. The heat and compressibility effects at the first stage were found to be similar for the two duplexes, indicating that the overall structure and the thermodynamics of the surfactant aggregates are sequence-independent. At the second stage, we observed the formation of large aggregates accompanied by light scattering in the visible region of the spectrum. In contrast, the structure and the thermodynamics of the oligonucleotide–surfactant complexes formed at the second stage are dependent on the nucleotide sequence.

Key words DNA · Surfactant complex · Compressibility · Isothermal titration calorimetry · Oligonucleotide

Introduction

Binding of surfactants to DNA have received a lot of attention in the last few years because of their growing importance in technological and biological applications. [1–3]. In the field of medicine and biotechnology the binding of cationic surfactants to DNA plays an important role in the construction of gene delivery systems [2–4]. Cationic surfactants are also an excellent model system (positively charged ligand with hydrophobic tail) for elucidating the effects of the hydrophobic interactions to the overall binding of various ligands with DNA. This has significance in understanding the major forces involved in more complicated biological processes, such as in the control and regulation of transcription, which depend on protein–DNA interactions [5–7].

Our previous studies of the complexes between dodecyltrimethylammonium bromide (DTAB) and calf thymus DNA (200 ± 70 base pairs) demonstrated at least two stages of binding [8]. At the first stage, the cationic surfactants bind with DNA in dilute solutions, cooperatively forming small aggregates of surfactant on the DNA surface accompanied by a change in the secondary structure of the DNA double helix. The second binding stage results in the formation of large aggregates formed on the surface of the DNA. This fact, shown by the migration of a hydrophobic fluorescent probe into a hydrophobic environment and accompanied by a large change in compressibility close to that of micelles, was interpreted as the formation of micelle-like aggregates of surfactant molecules on the DNA surface with a highly compressible internal hydrophobic core [9].

The major goal of this work is to examine the effects of the nucleotide sequence on the binding of cationic surfactants to DNA. It is directly related to the role of hydrophobic interactions in recognition of DNA sequence by natural and synthetic ligands [10–14]. Studies of the interaction of hydrophobic cations, alkyltrimethylammonium, with DNA revealed that they bind preferentially to adenine-thymine (AT) base pairs but have a general destabilising effect on the "hydrophobic" base-pair stacking interactions due to the increasing nonpolar nature of the solvent [18]. It is logical to expect that the formation of aggregates of surfactant on the DNA surface, accompanied by changes in conformation and hydration of DNA can also be sequence dependent. Recognition by surfactants of the DNA sequence would have useful applications in the biomedical industry, aiding in molecular recognition studies and in the development of drug design.

In order to probe the way sequence influences surfactant binding, we constructed two sets of sequence isomers, one containing a high-percentage (76%) of AT base-pairs and the other containing a high-percentage (76%) of guanine–cytosine (GC) base pairs. The complementary strands of the synthetic 17-mer oligonucleotide sequences

3'd(GAAAATTAAGCAAAAG)5'

(high-percentage AT content)

3'd(CTGGCACCGCACGGTGC)5'

(high-percentage GC content)

were designed to readily promote duplex formation and to minimise any single-strand self-association.

We used a combination of spectroscopic and thermodynamic methods to determine the binding profile, the structure of the surfactant complexes on the oligonucleotide surface and conformational transitions in the duplex that may occur upon binding. Using a fluorescence probe, we detected the existence of hydrophobic regions on the oligonucleotide surface. UV/vis spectroscopy measurements were applied to evaluate any conformational rearrangement of the duplex on the binding of surfactant and to follow the formation of large aggregates. We used isothermal titration calorimetry to characterise the energetics of binding of the surfactant to the oligonucleotides. With a new experimental setup for high-resolution ultrasonic measurements we determined the differential curve of the compressibility changes in the surfactant-to-oligonucleotide binding, which gave us information on the structure of the surfactant aggregates and supplied us with a binding isotherm. A detailed comparison of the differential ultrasonic, fluorescence, calorimetric and absorption (320 nm) curves provided us with direct information on the effects of binding at different relative concentrations of surfactant to oligonucleotide [8]. Such analysis, however, requires high-precision measurements, which were achieved with our computer-controlled automated-titration system. All our measurements were performed in dilute solutions at concentrations of surfactant far below the critical micelle concentration to exclude the presence of free micelles in solution.

Experimental

Materials

The samples of DTAB (Lancaster Synthesis, UK) were recrystallised twice from acetone. They were dried in a vacuum oven at 333 K for 48 h before use. All solutions were prepared by weight using degassed water obtained from a Millipore Super-Q-system. All experiments were conducted in a buffer, which was 5 mM N-(Z-hydroxyethyl)piperazine-N'-ethanesulfonic acid sodium salt (NaHEPES)/5 mM NaCl, pH 7.5. The single-stranded oligonucleotides were synthesised by Oswel DNA and purified by high-performance liquid chromatography. The concentrations of the sequences were determined by UV absorption using the extinction coefficients of each of the sequences calculated theoretically at 298.15 K using the nearest-neighbour method [19]. The strands were dissolved in the buffer and mixed in equimolar amounts to acquire the concentration used in the experiments. The solutions were transferred to a sealed tube, heated to 367 K in a programmable Haake water bath and allowed to cool to 278 K over 24 h. We used UV melting to characterise the helix–coil transitions of our oligonucleotide duplexes. The curves showed sigmoidal melting behaviour with melting temperatures, T_m, of 304 and 340 K for the high-percentage AT and GC duplexes, respectively. All the studies were done at 288.15 K, much below the T_m s. NaHEPES and NaCl (extrapure grade) were purchased from Sigma. The concentration of duplex and single strands in all experiments was 0.1 mM phosphates.

Methods

Ultrasonic measurements

Ultrasonic velocity measurements were made at 7.5 MHz using the resonator technique [20]. Ultrasonic resonator cells with a sample volume of 0.9 cm^3 were thermostated at 288.15 K with a temperature stability of 0.01 K. The new computer-controlled Ultrasonic Scientific frequency synthesiser/precision analyser was used to measure the resonance characteristics of the cells. The reproducibility in the measurements of the ultrasonic velocity was $\pm 10^{-4}$%. The main experimental observable in ultrasound measurements was the concentration increment of the ultrasonic velocity, A, which is determined by the relation

$$A = (u - u_o)/(u_o c \rho_o) , \tag{1}$$

where u and u_o are the ultrasonic velocities in the solution and pure buffer, respectively, c is the molal concentration of the solute in moles per gram and ρ_o is the density of the buffer at 288.15 K in grams per cubic centimeter.

The changes in the values of the apparent molar adiabatic compressibility of the surfactant and oligonucleotide solutions, $\delta K_{\phi s}$, and their changes due to binding of the surfactant with the oligonucleotide were determined from the corresponding experimental values of the concentration increment of the ultrasonic velocity, δA, as follows [8, 21]

$$\delta A = \delta V_\phi - \delta K_{\phi s}/2\beta_o \ , \qquad (2)$$

where β_o is the coefficient of adiabatic compressibility of the buffer at 288.15 K.

The changes in the values of apparent molar volume of surfactant, δV_ϕ, were taken to be 10% of the δA value, which was observed in the previous work on the addition of surfactant to natural DNA [8, 9].

UV spectroscopy measurements

The UV/vis spectroscopic titrations of oligonucleotide solutions by the surfactant were performed in a 1-cm pathlength cell on a Lamda 40 UV/vis spectrophotometer in the wavelength range 200–600 nm. The changes in the absorbance of the oligonucleotides were analysed as a function of the surfactant/phosphate ratio. The corresponding correction on the dilution of the oligonucleotides upon titration was made.

Isothermal titration calorimetry

Enthalpies of binding of DTAB to oligonucleotide samples were measured on the Omega titration (Microcal). In the calorimetry experiment, we measured directly the enthalpy associated with the binding of surfactant with our oligonucleotides. All solutions were carefully degassed before the titrations. The concentrated surfactant solution was injected into the oligonucleotide solutions using a 100-μl syringe. An injection schedule (number of injections, volume of injections and time between injections) was set up using interactive computer software. Successive titrations without the removal of the contents of the calorimeter were required due to the high ratio of binding. Heats of dilution of DTAB were determined by injection of the corresponding volumes of surfactant into buffer alone and were subtracted from the heat of each injection.

Fluorescence measurements

The N-phenyl-1-naphthylamine (Molecular Probe) was used as a hydrophobic fluorescent probe at a concentration of 0.4 μM. The fluorescence excitation and emission spectra of the probe were taken on a Perkin Elmer LS50B spectrofluorometer. The excitation wavelength was 340 nm. The fluorescence intensity of our probe is sensitive to a nonpolar environment [22, 23]. We found a significant increase, approximately 20-fold, in the relative fluorescence of the probe when it migrates into the internal core of micelles. The emission intensity of the probe in oligonucleotide solutions (approximate volume 2.4 cm^3) in the presence of different concentrations of surfactant was analysed at the apparent maximum of the emission peak (about 430 nm). We observed no increase in the fluorescence intensity of the probe in solution with the oligonucleotides alone.

Titration procedure and differential titration curves

All titrations (except isothermal titration calorimetry measurements, which were described earlier) were performed automatically by stepwise injections of small volumes (5–10 μl) of surfactant solutions into the oligonucleotide under constant stirring through small holes in the lids of cells by using a Hamilton Microlab 500 dispenser controlled by a computer. The kinetics of the changes of all experimental parameters after each injection was monitored to ensure that equilibrium was reached. Independent titrations of the surfactant into buffer solution were performed using the same procedure and the values obtained were subtracted from the experimental values of injection of surfactant into the oligonucleotide solution. The dilution of the oligonucleotide solutions in the

course of the titration was 3–6%. In order to carry out a comparative analysis, we calculated the differential titration curves. The derivative of the absorption at 320 nm and the fluorescence intensity versus the ratio of surfactant/phosphate was made to yield the differential absorption and fluorescence curves, respectively (data not shown). Similarly, the differential values of adiabatic compressibility were calculated by taking the derivative of the value of the concentration increment of the ultrasonic velocity of the oligonucleotides.

Results and discussion

Two stages of binding of surfactant to double-stranded oligonucleotides

The experimental titration curves obtained using the spectroscopic techniques are plotted as a function of the surfactant/phosphate ratio in Fig. 1. The enthalpy and the differential of the compressibility curves are presented as a function of the surfactant/phosphate ratio in Fig. 2 for each duplex, with the latter exhibiting a distinct biphasic behaviour, signifying at least two binding stages. The first binding stage occurs at a surfactant/phosphate ratio above 7 and 8 for the high-percentage GC and AT duplexes, respectively, and is accompanied by changes in the fluorescent intensity of the probe, which corresponds exactly to the process of migration of the fluorescent probe into a hydrophobic core, indicating the formation of surfactant aggregates with a hydrophobic core. The second binding stage is observed as a sharp increase in the compressibility at a surfactant/phosphate ratio of 11.1 and 12.9 for the high-percentage AT and GC duplexes, respectively. These changes are accompanied by light scattering measured as the absorption at 320 nm. Both the increase in compressibility and the light scattering indicate the formation of large

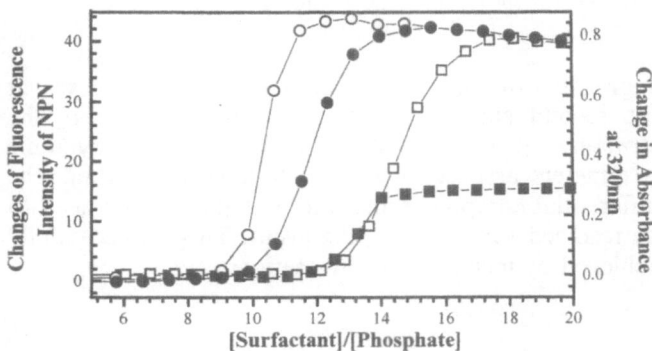

Fig. 1 Titration curves of surfactant binding with the high-percentage adenine–thymine (*AT*) oligonucleotide (*open symbols*) and the high-percentage guanine–cytosine (*GC*) oligonucleotide (*filled symbols*) in 5 mM NaCl, 5 mM NaHEPES pH 7.5 at 288.15 K. The effect of injection on the changes in the relative fluorescence intensity of the probe N-phenyl-1-naphthylamine (*NPN*) (*circles*) and the total changes in the absorbance at 320 nm (*squares*) is also shown

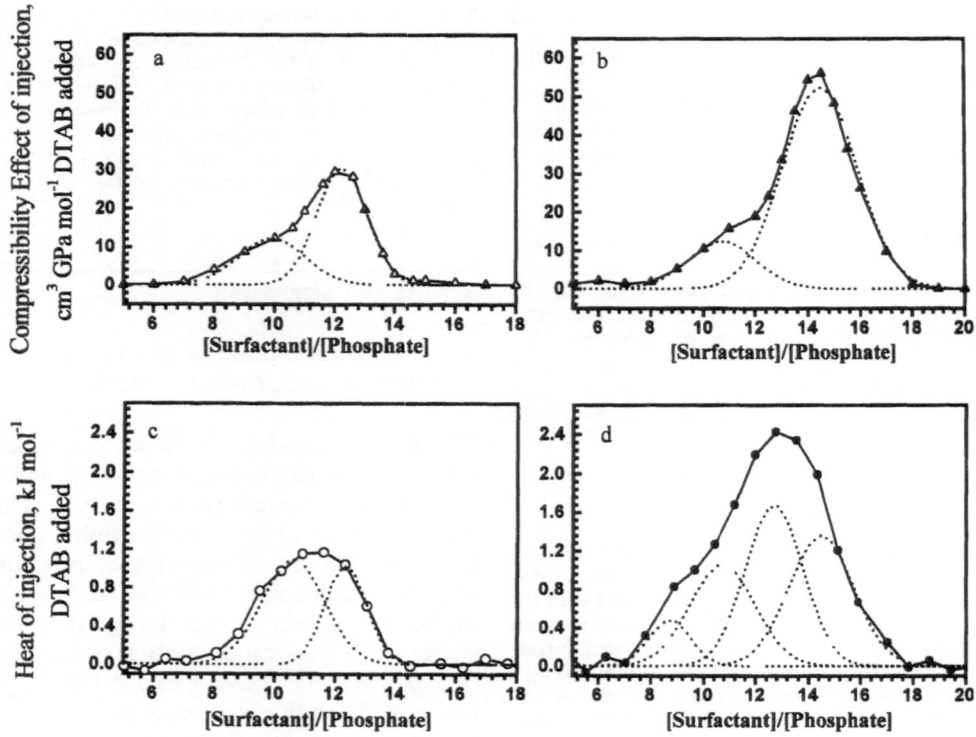

Fig. 2a–d The enthalpy (*circles*) and differential compressibility (*triangles*) titration curves of surfactant (dodecyltrimethylammonium bromide, *DTAB*) binding with the high-percentage AT oligonucleotide (*open symbols*) and the high-percentage GC oligonucleotide (*filled symbols*). The *dashed lines* are the result of the deconvolution procedure. **a** The compressibility effect of injection per mole of surfactant added for the high-percentage AT oligonucleotide and **b** for the high-percentage GC oligonucleotide. **c** The heat of injection per mole of surfactant added for the high-percentage AT oligonucleotide and **d** for the high-percentage GC oligonucleotide. For all curves, the effect of injection of the surfactant into the buffer was subtracted. The *left peak* and the *right peaks* in **a**, **b** and **c** correspond to the first and second stages of binding, respectively. In **d**, the two peaks on the *left* correspond to the first stage of binding and the two peaks on the *right* correspond to the second stage of binding

aggregates of oligonucleotide–surfactant complexes at the second stage. In order to evaluate precisely the structure of the complexes and the thermodynamic parameters associated with each stage of binding, the differential compressibility and enthalpy curves need to be resolved for each binding event. This is successfully achieved by means of deconvolution.

Procedure for the deconvolution of the experimental curves into the first and second stage of binding

As the changes in the absorbance at 320 nm are determined by the light scattering due to the formation of large aggregates at the second stage of binding, the differential of curve of the absorbance at 320 nm (data

not shown) was the basis for the deconvolution procedure.

Compressibility curves

Firstly, the curve of the normalised differential absorption at 320 nm was fitted to a Gaussian function governed by the parameters width, centre and area of the curve. Using the width and centre of the curve as fixed parameters, the differential compressibility curve was then fitted at surfactant/phosphate ratios above 11 for the high-percentage duplex and 13 for the high-percentage GC duplex to a Gaussian function to obtain the compressibility effect corresponding to the second stage of binding (last peaks, Fig. 2a, b). This was subsequently subtracted from the overall compressibility effect and the remaining compressibility effect was fitted to a Gaussian function to yield the compressibility effect at the first stage of binding (first peaks, Fig. 2a, b). The peaks obtained for the first binding stage coincide surprisingly well with the differential fluorescence curves for both oligonucleotides. The maxima of the binding and compressibility effects for both stages are tabulated in Table 1.

Heat effects

The heat effect resulting from the formation of large aggregates at surfactant/phosphate ratios above 11 for the high-percentage AT duplex and above 13 for the

Table 1 Comparison of the effects at the first and second stages of binding for different nucleotide sequences

Experimental values	High-percentage adeninne–thymine duplex		High-percentage guanine–cytosine duplex			
	First stage	Second stage	First stage		Second stage	
			(i)	(ii)	(i)	(ii)
R^a	10.3	12.4	8.5	10.8	12.9	14.5
ΔH^b (kJ mol^{-1} phosphate)	2.8	1.8	1.2	3.3	4.6	4.2
$\Delta K_{\phi S}$ (cm^3 GPa^{-1} mol^{-1} phosphate)	33.7	63.8	–	40.3	–	187.3

[a] R is the surfactant/phosphate ratio at which the maximum of binding occurs
[b] Calculated as the area under the peaks in Fig. 2

high-percentage GC duplex were determined in the same manner as that for the compressibility effect at the second stage of binding (last peaks, Fig. 2c, d) and were then subtracted from the overall enthalpy curves. The remaining heat effect was easily fitted to one Gaussian peak for the high-percentage AT oligonucleotide, in which both the position and broadness of the peak were in good agreement with the compressibility peak at the first stage of binding. However, for the high-percentage GC duplex the remaining heat effect could not be fitted to one peak, which is an indication that other processes take place in the interaction of surfactant with the high-percentage GC duplex. We were able to fit the remaining heat effect with three Gaussian peaks, fixing the position and broadness of one peak (peak 2, Fig. 2d) to the peak corresponding to the compressibility effect at the first stage of binding for the high-percentage GC duplex. The two remaining peaks, peaks 1 and 3 in Fig. 2d, are tabulated as the first stage (i) and second stage (i) of binding, respectively, in Table 1.

First stage of surfactant binding to double-stranded oligonucleotides

The relative changes in the enthalpy, fluorescence and differential compressibility curves are similar for each duplex, indicating that the techniques measured a similar process, i.e. the formation of micelle like aggregates on the surface of the DNA (Figs. 1, 2). The first stage of binding lies between a surfactant/phosphate ratio of 8 and 13 for the high-percentage AT duplex and between 7 and 14 for the high-percentage GC duplex. A small decrease in the absorbance at 260 nm, 2.5 and 4.2% of the initial absorbance at a surfactant/phosphate ratio of 10.3 and 12.1 for the high-percentage AT and GC duplexes, respectively, was observed to occur. It would be unreasonable to assume that the absorbance changes results from the stabilisation of the duplex with increases in the ionic strength as the concentration of surfactant is below 1.3 mM at this stage. Isolated electrostatic binding of the surfactant to the phosphate

groups of DNA is more likely to occur, altering the stability of the duplex. The first peak in the enthalpy curve of the high-percentage GC duplex (Fig. 2d) could be interpreted as the heat required for the some alteration of the high-percentage GC duplex. The heat effect, 0.23 kcal per mole of phosphate, is relatively small especially compared to the other processes (Table 1).

The total compressibility effect for the first stage of binding for the high-percentage AT and GC duplexes are comparable, 33.7 and 40.3 cm^3 GPa^{-1} per mole of phosphate, respectively. Since they are accompanied by the migration of the fluorescence probe into a hydrophobic core, the compressibility effects are considered as the intrinsic compressibility of the micelle core. This compressibility effect is nearly 3 times lower than the compressibility effect for the formation of surfactant aggregates on the surface of natural DNA (200 base pairs), 122.3 cm^3 GPa^{-1} per mole of phosphate, obtained previously [9]. This points to the fact that a third of the amount of surfactant per phosphate is in a micellar environment compared to the micellar-like aggregates on the surface of natural DNA. It was established that the compressibility effect for micelle formation and the formation of surfactant aggregates on the DNA surface is mainly determined by the compressibility of the hydrocarbon-like internal core of the surfactant aggregates and does not depend much on the structure of the micelle [9, 24]. It was therefore possible, from the observed change in the compressibility effect, to estimate the number of surfactants in the micellar core. Indeed, previous studies on the binding of surfactant to DNA revealed that changes in the compressibility of DNA occurred in accordance with the changes in the degree of binding obtained using a surfactant-selective-electrode technique [9]. If the same is true for our system, the ratio of the number of moles of surfactant in the micelles per mole of phosphate should be 0.27 and 0.32 for the high-percentage AT and GC duplexes, respectively, at the first stage of binding. This corresponds to approximately 10 mol of surfactant in a "micellar environment" per duplex, which amounts to the formation of small aggregates of surfactant on the surface of each duplex or the attachment of duplexes on the surface of larger surfactant aggregates. If the aggregate resembled that of a normal micelle, up to five duplexes would be bound to the surface of the micelle since spherical micelles of DTAB are known to contain on average 54 DTAB molecules [25]. The structure of the aggregate may take different forms, resulting in part from a different condensed counterion, which can change the local curvature of the interface of the micelle [26]. The formation of cylindrical-like micelles is one possibility where the duplexes could be arranged more easily, with less steric hindrance, on the surface. In all cases, we cannot exclude the binding of isolated surfactant, which was found to alter the conformation of the helix at low surfactant/phosphate ratios. This binding would not be visible in our compressibility curve because of the absence of the compressibility of the hydrophobic core.

There is a plethora of bonding rearrangements on the complex formation of surfactants to DNA resulting in an overall change in the Gibbs energy of the system. Determining the contributions to this free energy from the enthalpy and entropy helps to characterise the nature of the interactions present. The standard-free-energy changes of the formation of micellar aggregates, ΔG_m^o, per mole of surfactant, is determined through the relation

$$\Delta G_m^o = RT \ln X_f , \qquad (3)$$

where X_f is the mole fraction of free surfactant at each phase transition [27]. The free-surfactant concentration was determined from the amount of surfactant bound per phosphate, obtained previously. Since the total surfactant concentration is much greater than that of the bound surfactant, a precise evaluation of the latter does not significantly affect the overall calculation of the standard free energies. The results for both duplexes are shown in Table 2 and were found to be similar for both nucleotide sequences.

Using the values of the enthalpies of surfactant binding at the first and second stages calculated per mole of phosphate (Table 1) and the number of surfactant molecules forming the micellar aggregates at each stage, obtained from our compressibility curves, we estimated the enthalpy of formation of the micellar aggregates per mole of surfactant in the aggregates, ΔH_m^o. As given in Table 2, these enthalpies are positive and equal to 10.5 and 10.3 kJ mol^{-1} DTAB for the high-percentage AT and GC duplexes, respectively. The values of ΔH_m^o obtained can be considered as the standard heats of formation of the micelles on the oligonucleotide surface [27]. In comparison, the standard enthalpy of micelle formation under the same experimental conditions is much smaller, estimated as 0.8 kJ mol^{-1} DTAB. This large difference in enthalpies arises from a number of factors, including the necessary removal of the condensed sodium ions from the surface of the oligonucleotide and some conformational rearrangements of the duplexes. The different nature of the condensed counterion on the micelle is another contributing factor to the extra energy necessary for the aggregation of the surfactants on the oligonucleotide surface compared to micelle formation. Condensed counterions (chlorides) on the surface of the free micelle reduce the repulsion between the head groups, contributing favourably to the heat of micellisation. In contrast micelle-like aggregates formed in our oligonucleotide solutions are stabilised by electrostatic interactions, with the negative phosphate groups unable to orient themselves around the surface of the aggregate in the same manner as the condensed counterions. The lack of such favourable interactions can contribute to the heat required for the binding process.

Positive values of ΔH_m^o demonstrate that the binding of surfactant to our duplexes is entropy-driven. The entropy changes required to overwhelm the endothermic heat effects (determined by combining ΔG_m^o with ΔH_m^o) are large and typical for hydrophobic interactions when accompanied by the observed enthalpy changes [27]. Other conformational features such as the bending of our duplexes may be responsible for

Table 2 Comparison of the thermodynamic parameters at the first and second stage of binding for different nucleotide sequences

Thermodynamic parameters	High-percentage adenine–thymine duplex		High-percentage guanine–cytosine duplex	
	First stage	Second stage	First stage	Second stage
$\Delta H_m^{o\,a}$	10.5	3.6	10.3	6.0
$\Delta G_m^{o\,a}$	−26.3	0.4	−26.1	0.6[c]
$\Delta S_m^{o\,b}$	127.7	−14.1	126.3	−22.6

[a] Given in kilojoules per mole of surfactant in a micellar environment. It was calculated from the enthalpy per mole of phosphate given in Table 1 and the amount of surfactant in the micellar aggregate. The amount of surfactant in the micellar aggregate was calculated from the compressibility effects of surfactant binding, suggesting that the molar compressibility effect of the formation of the surfactant hydrocarbon internal core of the surfactant aggregate is 128 cm^3 GPa^{-1} per mole of surfactant, which was found previously in free micelles [9] and in aggregates of surfactant on the surface of 200-base-pair DNA duplexes [8]
[b] Given in joules per Kelvin per mole of surfactant in a micellar environment
[c] Calculated as the difference in the Gibbs energy between the first and second stage

the observed differences between the duplexes at the second stage of binding. The dominant contribution to the entropy effect originates from the release of water molecules from the hydration shell of the alkyl groups in the transfer from the aqueous environment to the hydrocarbon-like interior of the micellelike aggregates. Other contributions include the favourable release of condensed counterions from the oligonucleotide surface, while the reduced mobility of the duplexes acts against complex formation.

Second stage of surfactant binding
to double-stranded oligonucleotides

The surfactant/phosphate ratios considered as the second stage of binding range from 10.8 to 14 for the high-percentage AT duplex and from 10 to 18 for the high-percentage GC duplex. Considerable differences in the experimental parameters between the duplexes are observed at the second stage of binding, resulting in the formation of aggregates between the oligonucleotide–surfactant complexes. The high-percentage GC oligonucleotide–surfactant complex exhibits over twice the absorbance at 320 nm compared to the high-percentage AT oligonucleotide–surfactant complex (Fig. 1), indicating that larger complexes are formed for the high-percentage GC duplex. This is also reflected in the changes in the compressibility effect (Table 1) corresponding to 3 times the amount of surfactant in a micellar environment per phosphate for the high-percentage AT duplex (0.50) compared to the high-percentage GC duplex (1.51). The extra amount of surfactant bound may be explained by the extra heat effect observed for the high-percentage GC duplex between the surfactant/phosphate ratios of 10 and 15.5 (Fig. 2d).

The observed differences in the binding with the high-percentage AT and GC duplexes can be explained in two ways. Firstly, the inherent structural features of the particular sequences may influence the binding. The grooves (major and minor) of DNA differ significantly in their hydration, electrostatic potential, hydrogen bonding, groove width and steric effects depending on the nucleotide sequence [28]. Secondly, conformational changes may occur and these could differ depending on both the composition and the sequence of the DNA. It is well known, for instance, that the stability of B-DNA is affected by an increase in salt concentration in systems of GC-rich sequences resulting in Z-DNA structures [28]. Oligomers containing a run of alternating GC sequences can exhibit salt-induced B–Z transitions costing an average 1 kJ per mole of phosphate [29], which can be influenced by drugs, including intercalators [30, 31] and minor-groove binding ligands [32].

Conclusion

Binding of surfactant to high-percentage AT and GC duplexes occurs in two distinct stages. At both stages, the binding is entropy-driven. Micelle-like aggregates, with a compressible hydrophobic core, are formed at the first stage with similar thermodynamic and compressibility effects for both duplexes, indicating that the binding of surfactant is independent of the DNA sequence. Oligonucleotide–DTAB complexes are found to aggregate at the second stage, which is observed by light scattering in the visible region of the spectrum and a further increase in positive enthalpy and compressibility. However, the structure and thermodynamics of the aggregate differ between the duplexes, demonstrating that the nucleotide sequence influences the formation of oligonucleotide–surfactant complexes at the second stage.

Acknowledgements We acknowledge support from the Forbait Basic Research Grant SC/97/529. We would also like to acknowledge E. Kudryashov for his help. We gratefully acknowledge the support of Ultrasonic Scientific Ltd who provided the equipment for high resolution ultrasonic measurements.

References

1. Goddard ED (1993) In: Goddard ED, Ananthapadmanabhan KP (eds) Interactions of surfactants with polymers and proteins. CRC, Boca Raton, pp 396–411
2. Felgner A, Ringold GM (1989) Nature 337:387–388
3. Behr JP (1994) Bioconjugate Chem 5:382–389
4. Fendler JH (1982) In: Fendler JH (ed) Membrane mimetic chemistry. Wiley, New York, pp 6–47
5. Chatterjee R, Chattoraj DK (1979) Biopolymers 18:147–166
6. Maulik S, Chattoraj DK, Moulik SP (1998) Colloids Surf B 11:57–65
7. Werner MH, Gronenborn AM, Clore GM (1996) Science 271:778–784
8. Morrisey S, Kudryashov ED, Dawson KA, Buckin VA (1999) Prog Colloid Polym Sci 112:71–75
9. Kudryashov E, Kapustina T, Morrissey S, Buckin V, Dawson K (1998) J Colloid Interface Sci 203:59–68
10. Neide S (1997) Biopolymers 44:105–121
11. Dervan PB (1986) Science 232:464–471
12. Kopka ML, Larson TA (1992) In: Propst CL, Perin TJ (eds) Nucleic acid targeted drug design. Dekker, New York, pp 304–374
13. Zimmer C, Wahnert U (1986) Prog Biophys Mol Biol 47:31–112
14. Berman PB (1994) Curr Biol 4:345–350
15. Trauger JW, Baird EE, Dervan PB (1996) Nature 382:559–561
16. Gottesfeld JM, Neely L, Trauger JW, Baird EE, Dervan PB (1997) Nature 387:202–205
17. White S, Szewczyk JW, Turner JM, Baird EE, Dervan PB (1998) Nature 391:468–471
18. Bhattacharya S, Mandal SS (1997) Biochim Biophys Acta 1323:29–44

19. Richards EG (1975) In: Fasman GD (ed) Handbook of biochemistry and molecular biology: nucleic acids 3rd ed., CRC Press, Cleveland, OH, Volume 1, pp 597

20. Buckin V, Smyth C (1999) Seminars in food analysis 4:89–104

21. Owen BB, Simons HL (1957) J Phys Chem 61:479–482

22. Grieser F, Drummoligonucleotide CJ (1988) J Phys Chem 92:5580–5593

23. Brito RMM, Vaz WLC (1986) Anal Chem 152:250–255

24. Kudryashov E, Kapustina T, Morrissey S, Buckin V, Dawson K (1998) J Colloid Interface Sci 203:59–68

25. Lianos P, Zana R (1981) J Colloid Interface Sci 84:100

26. Ilekti P, Martin T, Cabane B, Piculell L (1999) J Phys Chem B 103:9831–9840

27. Tanford C (ed) (1973) The hydrophobic effect: formation of micelles and biological membranes. Wiley, New York

28. Blackburn GM, Gait MJ (eds) (1996) Nucleic acids in chemistry and biology, 2nd edn. Oxford University Press, New York

29. Sheardy RD, Levine N, Marotta S, Suh D, Chaires JB (1994) Biochemistry 33:1385–1391

30. Pohl FM, Jovin TM, Baehr W, Holbrook JJ (1972) Proc Natl Acad Sci USA 69:3805–3809

31. Mirau PA, Kearns DR (1983) Nucleic Acids Res 11:1931–1941

32. Zimmer C, Marck C, Guschlbauer W (1983) FEBS Lett 154:156–160

Progr Colloid Polym Sci (2000) 115:209–213
© Springer-Verlag 2000

BIOSYSTEMS

O. Cavalleri
L. De Michieli
C. Natale
M. Novi
R. Rolandi
S. Thea
A. Gliozzi

Forces between carboxyl and amide groups measured by atomic force microscopy

O. Cavalleri (✉) · L. De Michieli
R. Rolandi · A. Gliozzi
Department of Physics
University of Genoa, via Dodecaneso 33
I-16146 Genoa, Italy
e-mail: cavalleri@fisica.unige.it
Tel.: +39-10-3536309
Fax: +39-10-314218

M. Novi · S. Thea
Department of Chemistry and Industrial
Chemistry, University of Genoa
Genoa, Italy

O. Cavalleri · L. De Michieli · C. Natale
R. Rolandi · A. Gliozzi
INFM, Research Unit of Genoa
Genoa, Italy

Abstract The atomic force microscope (AFM) has been used to measure the interaction force between a hydrophilic (COOH-functionalised) AFM tip and an oligopeptide (Cys-Gly-Ala-Ala-Ala-Ala amide) self-assembled film as a function of their separation distance. These measurements produce force–distance curves characterised by two relative minima in the regions in which the tip is strongly attracted by the film surface. We have performed control experiments to prove the relationship of one of the two minima with the formation of hydrogen bonds between tip carboxyl groups and oligopeptide terminal amide groups. Under conditions less favourable to hydrogen-bond formation, curves with only one minimum are recorded. Other experiments performed by using the same functionalised tips and self-assembled alkanethiol films confirm that the less deep minimum is related to hydrogen bonding between the tip and the sample.

Key words Atomic force microscopy · Force measurements · Force–distance curves · Hydrogen bonding · Self-assembled film

Introduction

Since the early years after its development [1], atomic force microscopy (AFM) has been employed not only to inspect sample topography, but also to probe local forces at high resolution as a function of tip–sample separation. Because the AFM probe is very small – the typical curvature radius of the tip is 40 nm – the force due to the interaction of a relatively small number of molecules is measured.

The force as a function of tip–sample separation (force–distance curve) provides information on the nanomechanical properties of hard and soft samples such as graphite [2], gold [2–4], bone [5], living cells [6, 7], gelatin [8] and other polymeric materials [9, 10].

Surface forces such as those due to electrostatic [11–13] and van der Waals interactions [2, 14] have been identified on the basis of their distance dependence. Furthermore since self-assembly techniques provide methods for functionalising the tip, adding chemical sensitivity to the system, the forces between ligands and receptors [15–18], antigens and antibodies [19–21] and complementary DNA strands [15, 22] have been measured.

Notwithstanding the fast development of this field, the way the physical and chemical properties of the interacting surfaces affect the curve shape has not yet been understood in detail. Especially, for chemically active surfaces made of biological molecules a general explicative picture of the measured forces cannot yet be made. The main difficulty is that the interfacial phenomena between biological macromolecules are very complicated and can occur on many different levels of molecular complexity.

Having in mind that a correct physical interpretation of the forces measured by AFM is based on the knowledge of the structure of both the interacting molecular systems and the surrounding environment we chose to study simple and well-defined model systems based on the buildings blocks of proteins.

In this article we address the effect of hydrogen bonding between tip and sample on the shape of the force–distance curve and report the results of AFM force measurements performed with functionalised tips and self-assembled monolayers. The tip was gold-coated and functionalised with a mercaptoacid self-assembled monolayer. The system investigated was a hexapeptide film deposited by self-assembly on crystalline gold. The oligopeptide had a cysteine and an alanine amide as end groups. Cysteine provides the thiol group which binds gold. We also carried out control measurements on self-assembled films of two different thiols, one having a terminal carboxyl and the other having a terminal methyl.

Experimental

The oligopeptide (Cys-Gly-Ala-Ala-Ala-Ala amide, more than 95% pure) was synthesised for us by Tib-Molbiol (Italy). Ethyl alcohol (99.9% pure), 1-decanethiol (98% pure or greater) and urea (analytical reagent) were purchased from Fluka and were used without further purification. 11-Mercaptoundecanoic acid was prepared from the corresponding bromide (Aldrich, 99% pure) according to a method described in the literature for similar compounds [23]. The product, after purification by flash chromatography on silica gel (eluant: ether:petroleum ether 1:1), was characterised by ^{1}H NMR spectroscopy. Water was Milli-Q grade. For self-assembly, 1 mM solutions of the previous compounds were employed. The oligopeptide was dissolved either in a 9:1 mixture of water and ethyl alcohol or in pure water, while 1-decanethiol and 11-mercaptoundecanoic acid were dissolved in ethyl alcohol.

The gold substrates were prepared by vacuum evaporation of 120–150 nm gold (99.99% purity, Corradi, Milan, Italy) onto cleaved mica sheets (Lot Oriel Italia, Milan, Italy). The evaporations were carried out at a base pressure of 2×10^{-6} mbar, at a substrate temperature of 600 K, and were followed by 1–2 h annealing in vacuum at the same temperature. Before use, the gold films were flame-annealed in a butane flame to red glowing, quenched in ethanol (p.a., Fluka) and dried in a stream of nitrogen. The gold surfaces prepared in this way have atomically flat (111) terraces, a few hundred nanometres in size.

The samples for the self-assembly experiments were transferred into the suitable solution immediately after quenching in order to minimise air exposure. The samples were kept in the solution overnight at room temperature and after extraction they were thoroughly rinsed with solvent.

For the AFM measurements a Dimension 3000 equipped with a "G" scanner head (92.8-μm scan range) and controlled by a Nanoscope III (Digital Instruments, Santa Barbara, Calif., USA) was used. Force–distance measurements were made using microcantilevers from Digital Instruments ("V"-shaped, length = 193 μm, width = 20 μm, spring constant = 0.06 N/m), gold-coated and modified by Bioforce Laboratory with a carboxyl (hydrophilic) surface.

Results and discussion

The force acting on the tip was measured as a function of the sample position. This measure produced a "force–distance curve" that was obtained by allowing the tip to approach the sample along the vertical axis (z-axis) and the resulting cantilever deflection, Δs_c, and piezo displacement were recorded. The force acting on the cantilever is provided by Hooke's law,

$$F = -k_c \Delta s_c , \qquad (1)$$

where k_c is the elastic constant of the cantilever.

The piezo displacement, ΔZ, which is the parameter directly controlled during the measurement, is related to the tip–sample distance, D, according to

$$Z_0 - \Delta Z = D + \Delta s_c + \Delta s_s , \qquad (2)$$

where Δs_s is the sample deformation and Z_0 the position of the sample surface with respect to the cantilever rest position before the approach. The force was recorded while approaching and withdrawing the tip and is reported as a function of the piezo displacement on a scale whose origin was chosen arbitrarily. Each curve has three distinct regions: the zero line, where the cantilever remains in its resting position because the tip/sample distance is too large for any interaction to occur, the noncontact region, where noncontact forces appear and the contact region, where, in the absence of sample (and tip) deformation, the piezo displacement and the cantilever deflection are equal. The noncontact forces can be repulsive and attractive. The attractive forces can cause a tip jump onto the sample in the approach phase and a jump off the sample in the withdrawal phase.

A typical force–distance curve obtained at pH 2 by using a tip, functionalised with carboxyl groups, on an oligopeptide film deposited on a gold substrate is shown in Fig. 1. The jump-to-contact region, in the approaching curve, and the jump-off-contact region, in the retracting curve, have two relative minima. Usually, when nonspecific interactions occur, these parts of the curves have only one minimum. In our case, while the tip approaches the film, it encounters a first attractive force causing the first jump. This force is followed by a repulsive force that acts for a few nanometres and, after this one, by another attractive force causing the second jump, which is usually larger than the first one. Subsequently the tip undergoes a repulsive force roughly proportional to the piezo displacement, indicating that the tip is in contact with the film surface. The retracting curve indicates similar behaviour, suggesting the presence of two kinds of attractive forces. The big hysteresis in the jump-off-contact region is a common feature of the retracting curves and is likely to be due to the fact that the number of interactions increases after tip and sample have been in contact. In fact, while the tip approaches the sample, at small tip–sample separations, only the few molecules at the outermost tip interact with the surface molecules. However, once the tip encounters the sample, the contact area of the touching tip is sufficiently large for additional molecules to interact.

The reproducibility of the force–distance curves is very high. Similar curves were obtained in different

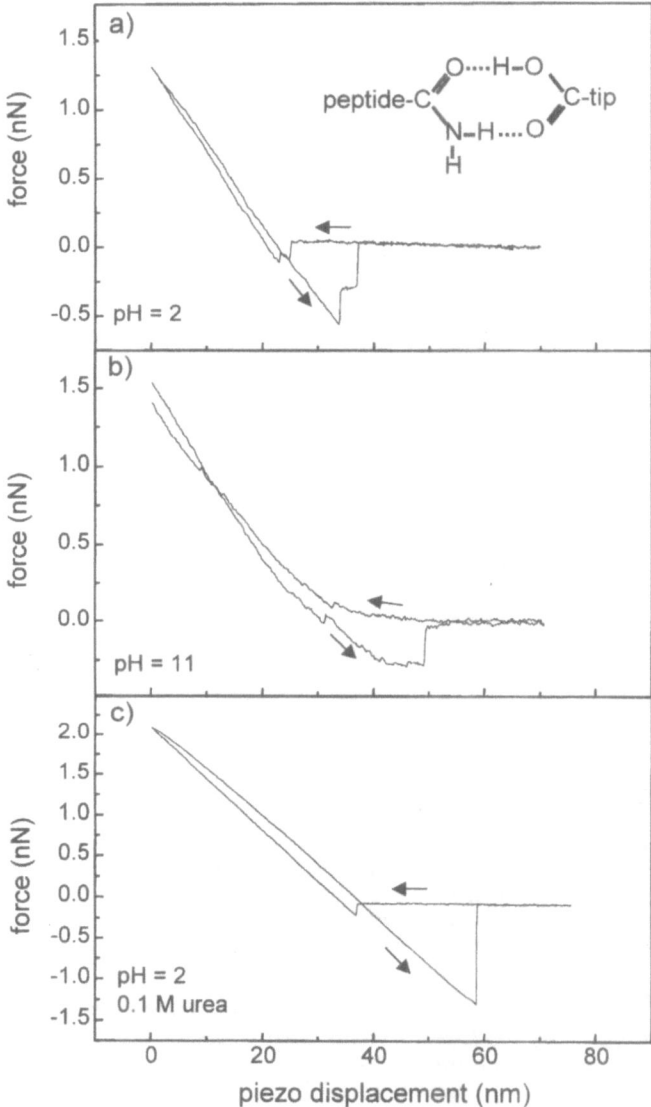

Fig. 1 Force–distance curves acquired with a COOH-functionalised tip on a Cys-Gly-Ala-Ala-Ala-Ala amide covered gold surface: at **a** pH 2 **b** pH 11 and **c** pH 2 in the presence of 0.1 M urea . The origin of the piezo displacement scale was chosen arbitrarily. Raw data were converted into force values using the nominal spring constant of the cantilever (0.06 N/m). The inset in **a** is a sketch of the two-hydrogen-bond interaction between the carboxyl and amide groups

regions of the same sample and with different samples. In a few cases the first minimum, which is always present in the jump-off-contact region, cannot be detected in the jump-to-contact region.

The shape of these curves is similar to those reported by Boland and Ratner [22], who measured the interaction forces between complementary DNA nucleotide bases. These authors attribute the first step to attractive forces due to hydrogen bonding between tip and sample.

The conditions of the experiment reported in Fig. 1a are favourable to the formation of hydrogen bonds between the molecules on the tip surface and those on the sample surface. In particular, since at pH 2 the carboxyl groups on the tip as well as the terminal amide groups of the oligopeptide are not ionised, two hydrogen bonds for each pair of interacting groups can be formed (Fig. 1a, inset).

We assumed as a working hypothesis that the presence of the two minima in the force–distance curve indicates hydrogen bonding and we carried out control experiments to prove this hypothesis.

As a first control we changed the pH. One of the force–distance curves recorded at pH 11 is reported in Fig. 1b. The only difference with respect to the experiment reported in Fig. 1a was the pH value. The approaching curve shows a repulsive force which gradually increases when tip-sample separation decreases. The following jump-to-contact region has only one small minimum. The jump-off-contact region in the retracting curve has perhaps two minima but these are not so clearly distinguished as in Fig. 1a. It is likely that at this pH the carboxyl groups are mostly ionised and charge the tip surface negatively. Hydrogen-bond interactions between carboxyl functions and amide groups, which are not ionised under these conditions, will be considerably modified. The changed conditions could justify the absence of the two minima in the jump-to-contact region. At present we cannot give an unambiguous interpretation for the shape of the jump-off-contact region.

The gradual repulsive force is a double-layer force caused by the surface charge density on the tip surface. The double-layer force between a spherical tip and a flat sample, under the conditions that the surface potential is low (below 25 mV) and the distance between the tip and the sample is longer than the Debye length, is described by the following equation introduced by Butt [11]:

$$F_{\text{d.l.}} = \frac{2\pi R \lambda_{\text{D}}}{\varepsilon_1 \varepsilon_0} \left[(\sigma_{\text{T}}^2 + \sigma_{\text{S}}^2) \exp\left(\frac{-2D}{\lambda_{\text{D}}}\right) + 2\sigma_{\text{T}}\sigma_{\text{S}} \exp\left(\frac{-D}{\lambda_{\text{D}}}\right) \right],$$

(3)

where D is the distance between the surfaces, σ_{T} and σ_{S} are the surface charge densities of tip and sample, respectively, ε_1 is the dielectric constant of the liquid, R is the tip radius and λ_{D} is the Debye length, given by

$$\lambda_{\text{D}} = \left(\sum_i \frac{\rho_i e^2 z_i^2}{\varepsilon_1 \varepsilon_0 k T} \right)^{-\frac{1}{2}},$$

(4)

where ρ_i is the concentration of the ith electrolyte, $z_i e$ its charge, T the temperature and k the Boltzmann constant. Even if the conditions under which Eq. (3) holds are not fulfilled, Butt [11] showed that this equation can provide the order of magnitude of the repulsive double-layer force, describing the exponential behaviour of the repulsive force as a function of the tip–sample position.

In our particular case there are surface charges only on the tip; therefore, the second term vanishes and only the first one, which is an exponential with a decay length equal to half the Debye length, remains.

To evaluate the exponential decay length, the experimental force–distance curves were corrected in order to take into account the cantilever deflection and were then fitted with exponential functions. The fits provide decay lengths that agree nicely with the expected values. In the particular case of the curve reported in Fig. 1b the fit provides a decay length of (5.2 ± 0.6) nm, while the calculated Debye length is 9.6 nm.

The relationship between the presence of two minima and hydrogen bonding is confirmed by experiments in which we inhibited the hydrogen-bond formation between tip and sample by adding urea to a solution at pH 2. We used urea since it is known to form hydrogen bonds with both carboxyl and amide groups, thus hindering the formation of hydrogen bonds between tip and sample. A force–distance curve recorded at pH 2 in the presence of 0.1 M urea in the solution is reported in Fig. 1c. The approaching and retracting curves have only one minimum according to the suppression of hydrogen bonding between tip and sample. The jump-to-contact and jump-off-contact are of the same order of magnitude as those observed in Fig. 1a, thus indicating that the deepest minimum is related to the presence of the nonspecific attractive forces already observed in the interaction of organic surfaces [24, 25]. It has been previously reported that at small separation distances the tip is pulled towards the sample surface by long-range electrostatic forces predicted by Derjaguin–Landau–Verwey–Overbeek theory [26]. Also in this case hysteresis is observed when the tip pulls away from the surface.

To check if the two-minima behaviour is a peculiar feature of the oligopeptide films we obtained force–distance curves by using the same kind of tips, functionalised with carboxyl groups, and self-assembled thiol monolayers. One of the curves, obtained with a 11-mercaptoundecanoic acid film at pH 2, is reported in Fig. 2a. The minima are smaller than those observed with the oligopeptide film; nevertheless, two minima are clearly visible in the jump-to-contact and jump-off-contact regions.

Further proof that the two minima are related to hydrogen bonding between tip and sample comes from the fact that the first minimum is absent when the surfaces expose chemical groups that do not form hydrogen bonds.

The force–distance curve obtained for a 1-decanethiol film is reported in Fig. 2b. The conditions are the same as in the experiment of Fig. 2a. The sample surface is made of methyl groups, which cannot form hydrogen

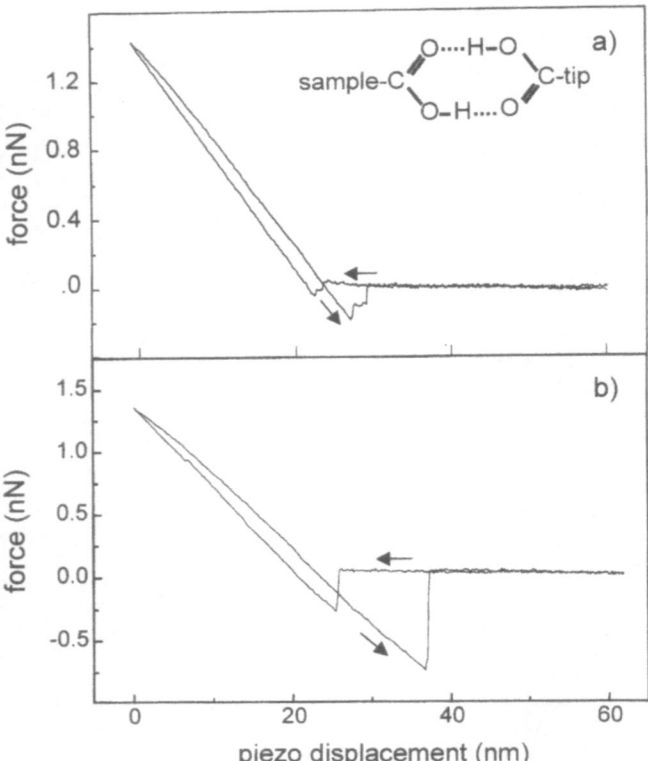

Fig. 2 Force–distance curves acquired at pH 2 with a COOH-functionalised tip **a** on a 11-mercaptoundecanoic acid covered gold surface and **b** on a 1-decanethiol covered gold surface. The origin of the piezo displacement scale was chosen arbitrarily. Raw data were converted into force values using the nominal spring constant of the cantilever (0.06 N/m). The inset in **a** is a sketch of the two-hydrogen-bond interaction between two carboxyl groups

bonds with the carboxyl groups on the tip, and both the jump-to-contact and the jump-off-contact regions have only one minimum.

Conclusions

A series of control experiments shows a correlation between the presence of two minima in the attractive regions of force–distance curves and hydrogen bonding between tip and sample. Force–distance curves with two minima in the jump-to-contact and jump-off-contact regions were obtained by using tips functionalised with carboxyl groups and samples exposing either amide or carboxyl groups at a pH value favourable for hydrogen-bond formation. The two minima were not observed in the approaching curves at high pH, when carboxyl groups were mostly ionised, and in the presence of urea, which hinders hydrogen-bond formation between tip and sample. Furthermore, force–distance curves with only one minimum were recorded on films exposing methyl groups, which cannot give rise to hydrogen bonds.

Acknowledgements We would like to thank C. Dell'Erba for useful discussions and suggestions. This work has been supported by grants from ASI contract ARS-98-174, MURST (Cofinanziamento 1997) and the CNR projects "Mesoscopic and microscopic physico-chemical properties of ordered lipoprotein films" and "Biotecnologie".

References

1. Binnig GC, Quate F, Gerber C (1986) Phys Rev Lett 56:930
2. Burnham NA, Colton RJ (1989) J Vac Sci Technol A 7:2906
3. Landman U, Luedtke WD, Burnham NA, Colton RJ (1990) Science 248:454
4. Agrait N, Rubio G, Vieira S (1995) Phys Rev Lett 74:3995
5. Tao NJ, Lindsay NM, Lees S (1992) Biophys J 63:1165
6. Hoh JH, Schoeneberger C-A (1994) Biophys J 107:1105
7. Radmacher M, Fritz M, Kacher CM, Cleveland JP, Hansma PK (1996) Biophys J 70:556
8. Radmacher M, Fritz M, Hansma PK (1995) Biophys J 69:264
9. Mate CM (1992) Phys Rev Lett 68:3323
10. Châtellier X, Senden TJ, Joanny J-F, Di Meglio J-M (1998) Europhys Lett 41:303
11. Butt H-J (1991) Biophys J 60:1438
12. Weisenhorn AL, Maivald P, Butt H-J, Hansma P-K (1992) Phys Rev B 45:11226
13. Ishino T, Hieda H, Tanaka K, Gemma H (1994) Jpn J Appl Phys 33:4718
14. Burnham NA, Dominguez DD, Mowery RL, Colton RJ (1990) Phys Rev Lett 64:1931
15. Lee GU, Chrisey LA, Colton RJ (1994) Science 266:771
16. Florin EL, Moy VT, Gaub HE (1994) Science 264:415
17. Moy VT, Florin E-L, Gaub HE (1994) Science 266:257
18. Lo Y-S, Huefner ND, Chan WS, Stevens F, Harris JM, Beebe TP (1999) Langmuir 15:1373
19. Dammer U, Hegner M, Anselmetti D, Wagner P, Dreier M, Huber W, Guntherodt HJ (1996) Biophys J 70:2437
20. Hinterdorfer P, Baumgartner W, Gruber HJ, Schilcher K, Schindler H (1996) Proc Natl Acad Sci USA 93:3477
21. Allen S, Chen X, Davies J, Davies MC, Dawkes AC, Edwards JC, Roberts CJ, Sefton J, Tendler SJB, Williams PM (1997) Biochemistry 36:7457
22. Boland T, Ratner BD (1995) Proc Natl Acad Sci USA 92:5297
23. Zheng T, Bukart M, Richardson D (1999) Tetrahedron Lett 40:603
24. Sarid D (1991) Scanning force microscopy. Oxford University Press, New York
25. Weisenhorn AL, Hansma PK, Albrecht TR, Quate CF (1989) Appl Phys Lett 54:2651
26. Derjaguin B, Landau L (1941) Acta Physiochim 14:633

Progr Colloid Polym Sci (2000) 115:214–221
© Springer-Verlag 2000

J. Eastoe
A. M. Downer
A. Paul
D. C. Steytler
E. Rumsey

Adsorption of fluoro surfactants at air-water and water-carbon dioxide interfaces

J. Eastoe (✉) · A. M. Downer · A. Paul
School of Chemistry
University of Bristol
Bristol BS8 1TS, UK
e-mail: julian.eastoe@bristol.ac.uk
Tel.: +44-117-9289180
Fax: +44-117-9250612

D. C. Steytler (✉) · E. Rumsey
School of Chemical Sciences
University of East Anglia
Norwich NR4 7TJ, UK
e-mail: d.steytler@uea.ac.uk
Tel.: +44-1603-592033
Fax: +44-1603-25985

Abstract Aqueous phase behaviour, and water-in-carbon dioxide micro-emulsion formation, were studied with two model fluorinated anionic surfactants, sodium bis(1H,1H-per-fluoropentyl)-2-sulfosuccinate (di-CF4) and sodium bis(1H,1H,5H-octafluoropentyl)-2-sulfosuccinate (di-HCF4). The properties are compared to the common, branched-chain, hydrocarbon surf-actant Aerosol-OT (AOT). For the fluoro surfactants, surface excesses at the air-solution interface, mea-sured by tensiometry and neutron reflection, agreed well, and the γ-ln(activity) curves were consistent with a pre-factor of 2 in the Gibbs equation. Both di-CF4 and di-HCF4 were effective at stabilising water-in-carbon dioxide (w/c) microemul-sions. Pressure-temperature studies indicated that di-CF4 is the most

effective, since it gave microemul-sions at lower pressures, e.g. for $w = 20$ ([water]/[surf]) at 30 °C the w/c phase was stable down to 120 bar. High-pressure small-angle neutron scattering showed the mi-croemulsions have a simple spherical droplet structure. In water-CO_2 mixtures these surfactants behave in a similar fashion to AOT in con-ventional hydrocarbon systems. Therefore, the fluoro-succinates represent a well-characterised model series of surfactants, which may be useful for understanding the effects of surfactant chemistry on w/c microemulsion phase properties.

Key words Fluorocarbon surfactants · Adsorption · Microemulsions · Carbon dioxide

Introduction

Carbon dioxide is cheap, non-toxic, highly volatile, chemically inert, non-flammable and potentially recy-clable. However, owing to weak intermolecular inter-actions, CO_2 is generally a poor solvent, especially for polar solutes, and this limits its applications. Mixtures of CO_2 and water have a vast potential as "green" alternatives to petrochemical solvents, and water-in-carbon dioxide (w/c) microemulsions show much prom-ise for this. Unfortunately, the solubility of water in CO_2 is low, typically ~0.2%, and recently a significant research effort has focused on enhancing this level, e.g. [1–19]. Making the analogy with water-oil (w/o) emul-sions, surfactants could be used to enhance miscibility.

After extensive solubility studies with numerous, readily available, surfactants it was found that very few dissolve in CO_2 to any appreciable extent [1]. Nonetheless, it was found that fluorinated amphiphiles had a greater solubility, and in recent years a number of groups have demonstrated w/c microemulsion formation with spec-ialised fluoro surfactants [2–18]. These were either custom made, for example the di-chain hybrid surfactant $(C_7H_{15})(C_7F_{15})CHSO_4^-Na^+$ (H7-F7) [3, 4], or commer-cially available technical products like the ammonium carboxylate perfluoropolyether $CF_3O(CF_2CF(CF_3)O)_3$ $CF_2COO^-NH_4^+$ (PFPE) [5–9, 11–16]. Lately, molecular simulation methods have been used to model reverse micelle formation in CO_2 [9]. The use of these w/c systems as reaction media has been demonstrated for enzymatic [10], organic [11, 12] and inorganic [13, 14] reactions. Furthermore, the important issue of controlling the pH in the water pools has recently been addressed [15]. This PFPE surfactant has also been show to stabilise w/c emulsions, where pressure (density) can be used to invert the system from c/w to w/c [16]. In a related area, DeSimone's group (e.g. [17]) have pioneered research into CO_2 soluble fluoropolymers, block co-polymers and dendrimers. These ideas have been rapidly commercialised into a new dry-cleaning technology, which has the potential to replace the traditional perchloroethylene-based process (see http://www.micell.com). More recently it has been shown that functionalised polydimethylsiloxanes are CO_2 soluble [19], suggesting they could be used instead of high-cost fluorinated polymers. Over a similar period, Poliakoff et al. (e.g. [20–22]) have demonstrated various applications of CO_2 as a novel reaction medium.

In terms of surfactants for w/c microemulsions, the issue of how chemical structure controls stability remains largely unresolved, and this is due to the lack of suitable compounds. It is well known that for w/o systems an efficient, and versatile, surfactant is Aerosol-OT [sodium bis(2-ethylhexyl)sulfosuccinate, AOT]. This consists of two branched hydrocarbon (ethylhexyl) chains, linked via a succinate residue, and the molecular shape is believed to be partly responsible for its action. Here, it is demonstrated how fluorinated analogues of AOT are effective at stabilising w/c microemulsions. The focus is on two custom-synthesised anionics, sodium bis(1H,1H-perfluoropentyl)-2-sulfosuccinate (di-CF4) and sodium bis(1H,1H,5H-octafluoropentyl)-2-sulfosuccinate (di-HCF4). Preliminary studies of the di-HCF4 alone are discussed elsewhere [18], but now it is possible to compare the behaviour of the two analogues. The molecular structures of di-CF4 and di-HCF4 are shown in Fig. 1.

Since both have bulky, CO_2-compatible fluorocarbon chains and a water-soluble sulfonate group, a tendency to form reversed curvature structures at water-CO_2 interfaces is expected, in a similar fashion to AOT in water-oil mixtures. The reasons for studying di-CF4 and di-HCF4 are twofold. Firstly, for di-HCF4 the precursor alcohol 1H,1H,5H-octafluoropentanol is around 10 times cheaper than that of the more fluorinated di-CF4. Clearly, if these surfactants were to be used in any commercial product, cost will be a prime consideration. Secondly, the differences at the hydrophobic chain tips are that a terminal fluorine atom in di-CF4 is substituted for hydrogen in the di-HCF4. This has the effect of introducing a large permanent dipole moment into the chain; for example, in the related molecule CF_3CF_2H it is 1.54 D [23]. Therefore, with the economics of surfactant production in mind, it is of interest to examine how this affects the surfactant properties, and especially the efficiency of w/c microemulsion formation.

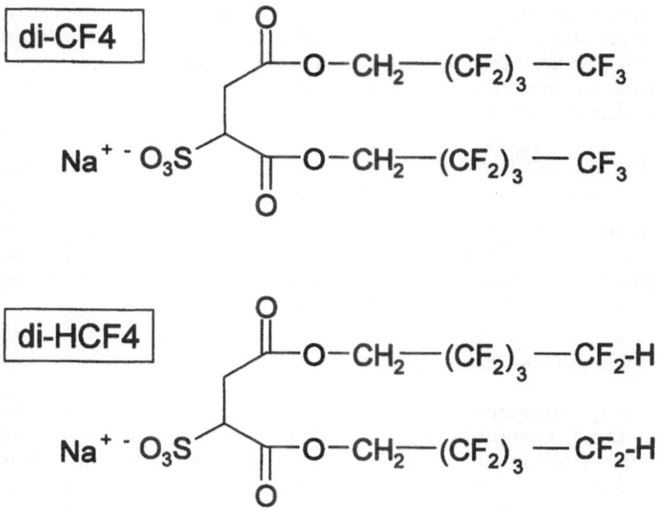

Fig. 1 Chemical formulae of the fluorinated surfactants

The first part of the paper describes the properties of aqueous solutions around the cmc, especially adsorption, studied by tensiometry and neutron reflection (NR). Next, the phase stability of w/c microemulsions as a function of temperature T, pressure P and composition w ([water]/[surf]) is covered. Finally, structural studies of the w/c systems by high-pressure small-angle neutron scattering (SANS) are described.

Materials and methods

Chemicals

The preparation and purification of di-CF4 and di-HCF4 has been described previously [24, 25]. As a final purification step, aqueous solutions, made up below the cmc, were foam fractionated. H_2O of resistivity 18.2 MΩ cm was taken from either a R0100HP Purite water system or a Millipore Milli-Q Plus system and was used throughout. D_2O (99.9% D-atom, Fluorochem) was distilled prior to use and CO_2 (BOC) was used as received.

Surface tension and NR

Tensiometric measurements were carried out using the drop-volume (DV) method (Lauda TVT1). As explained elsewhere [25–27], a trace level of background EDTA (99.5% tetrasodium salt hydrate, Sigma) was included in the solutions to chelate any polyvalent cation contaminants. The surfactant:EDTA ratios and temperatures were 25:1 and 60:1 and 30 °C and 25 °C for di-CF4 and di-HCF4, respectively (thermostatting to ±0.1 °C by Grant LTD6G bath). The slightly higher temperature for di-CF4 was to avoid phase separation at high multiples of the cmc.

NR measurements were performed on the CRISP and SURF reflectometers at ISIS, Rutherford Appleton Laboratories, Didcot, UK (see http://www.isis.ac.uk), using the standard set-up for a free liquid surface [28]. Measurements were made at selected concentrations in null reflecting water (NRW, 8.0 mol% D_2O in H_2O), at the same temperatures as the tensiometry. A full account of the theory of NR can be found in the literature, e.g. [28]; a summary relevant to the present work follows. The specular reflection from a surfactant film adsorbed at the surface of NRW can be modelled in terms of a single, uniform layer using the optical matrix method [29]. The fitted thickness, τ, and scattering length density, ρ, derived from the model lead directly to an expression for the area per molecule, A, which is given by:

$$A = \frac{1}{\Gamma N_a} = \frac{\Sigma b_i}{\rho \tau} \qquad (1)$$

In the above expression, Σb_i is the sum of nuclear scattering lengths over a single molecule. It is useful to compare the absolute adsorption measured by tensiometry and NR in terms of the surface excess Γ, and N_a is the Avogadro number.

Water-in-CO₂ systems and SANS

The P-T microemulsion phase stability was determined visually in a stirred, high-pressure optical cell described elsewhere [30]. The microemulsion composition is defined in terms of the surfactant concentration and the water-to-surfactant molar ratio w. Experiments were carried out using the LOQ time-of-flight instrument at ISIS, as described previously [18, 30]. SANS determines the scattering cross-section $I(Q)$ (cm^{-1}) as a function of momentum transfer Q (Å^{-1}) $= (4\pi/\lambda)\sin(\theta/2)$; λ is the incident neutron wavelength (2.2 → 10 Å) and θ the scattering angle ($< 7°$) [31]. In addition to the usual transmission, empty cell and solvent background corrections, the pressure-induced sample volume changes and the true cell path-length were taken into account, as described in [18]. Under these T-P conditions the solubility of water in CO₂ is negligible (0.13%) as compared with the 5% or so of D₂O present, and the w values were not corrected for this small effect. For liquid CO₂ the neutron scattering length density may be taken as $\rho_{CO_2} = 2.2 \times 10^{10}$ cm^{-2}, varying by 10% over the pressures studied. For the surfactants, ρ_{surf} is about 2.0×10^{10} cm^{-2}, and since $\rho_{D_2O} = 6.4 \times 10^{10}$ cm^{-2} the scattering comes principally from the contrast step at the D₂O interface. Therefore, the water radius R_c can be determined from the scattering pattern.

Data analysis

The $I(Q)$ data were analysed using the FISH fitting programme [32], and from a number of different possible models it was found that a Schultz distribution of spherical particles gave the best fits and most physically reasonable parameters. For samples with di-HCF4, which were close to the low-pressure phase boundary, it was necessary to introduce an attractive structure factor. This scattering law is:

$$I(Q) = \phi(\rho_{D_2O} - \rho_{CO_2})^2 \Sigma_i [V_i P(Q, R_i) X(R_i)] S(Q, \zeta) \qquad (2)$$

The parameters ϕ, R and V are the particle volume fraction, radius and volume. The spherical form factor is $P(Q)$, and $X(R_i)$ is the Schultz function [33], which is characterised by an average radius R^{av} and RMS deviation $\sigma = R^{av}/(Z+1)^{1/2}$, where Z is a width parameter. $S(Q)$ is the Ornstein-Zernicke structure factor, which describes a decaying particle distribution with ζ a correlation length, so that:

$$S(Q, \zeta) = 1 + \left[\frac{S(0)}{1 + (Q\zeta)^2} \right] \qquad (3)$$

$S(0)$ is related to the strength of interactions via the isothermal compressibility [33]; however, here it acts as an effective parameter. Since the sample composition and scattering length densities are all known, the parameters R_c^{av}, σ/R_c^{av}, ζ and $S(0)$ were adjusted. For the first three of these the uncertainties may be taken as ± 1 Å, ± 0.02 and ± 10 Å, respectively.

Results and discussion

Aqueous systems: surface tension and NR

Surface tension measurements were used to check the surface chemical purity of the surfactants, and Fig. 2a shows the γ-ln(activity) curves. Activity coefficients were obtained from the Debye-Hückel limiting law. Both plots show clean breaks at the cmc with no indication of

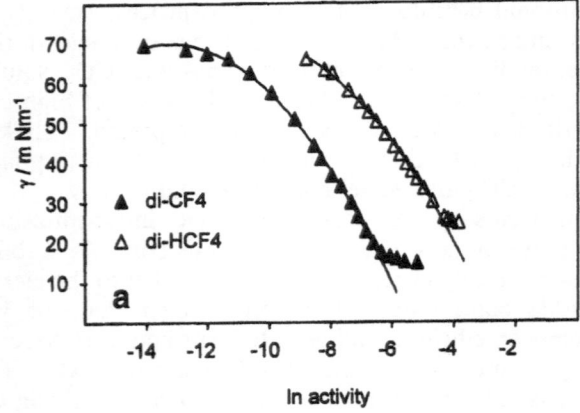

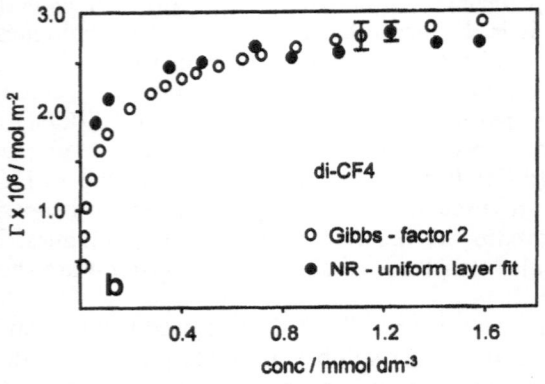

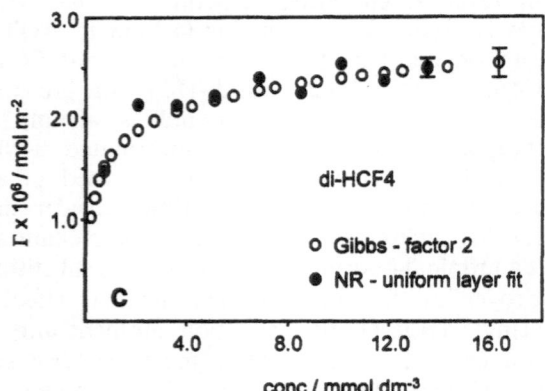

Fig. 2 a Surface tensions of surfactant solutions. The lines are polynomial fits to the pre-cmc data. **b, c** Surface excesses at the air-solution interface measured by two methods

a minimum. The cmc's were also determined conducti-metrically, and the values agreed well with those from tensiometry. Since the data were obtained by the (dynamic) DV method, these are true equilibrium tensions. With DV tensiometry it is possible to follow the time dependence of the tension over effective surface ages of up to 15 min. With many different surfactants, and for bulk concentrations less than about 1 mmol dm^{-3}, it is found that the tension decays (gently) with time. Therefore, in order to attain equilibrium it is necessary to follow the decay until a stable reading is achieved. Also shown on Fig. 2a are polynomial fits to the pre-cmc data, and the equations of these lines were used to generate gradients, and hence surface excesses Γ using the Gibbs equation:

$$\Gamma = -\frac{1}{2RT}\frac{d\gamma}{d\ln a} \qquad (4)$$

The symbols in the above expression have their usual meanings. This approach assumes electrical neutrality of the interface through a pre-factor of 2. The validity of Eq. (4) is tested by NR measurements, which are discussed below. Parameters derived from the surface tension measurements are listed in Table 1.

It is known that the presence of trace quantities of polyvalent cations can affect adsorption measurements with anionic surfactants (e.g. [25–27]), and it is impor-tant that these contaminants are eliminated. When present at a low, constant ratio to the surfactant, EDTA has been shown to be effective [25–27]. It has been established that EDTA is not itself surface active [25–27]; therefore once a M^{2+} ion is complexed, it is effectively removed from the interface. Therefore, extreme care, as described fully elsewhere [25, 27], was taken with the tension and NR experiments. It has recently been noted that D_2O, which is used in NR, can also be a source of ionic contamination [27], and so it was distilled prior to use.

Looking at the tension values at the cmc (Table 1), for the terminal-H compound γ_{cmc} is approximately 9 mN m^{-1} higher that for the F-terminal analogue. The H atom at the external surface of the monolayer is expected to increase the surface free energy with respect to the fully fluorinated surface. An additional effect is the H-CF$_2$- dipolar repulsion, which should also increase the tensions. A similar observation has been made with H(CF$_2$)$_7$COOH and the perfluorinated analogue: the limiting tensions were 21.8 and 15.2 mN m^{-1}, respectively [34]. Air-water tensions of related nonionic surfactants, with either CF$_3$- or H-CF$_2$- chain tips, have also been studied [35]. The switch to H-CF$_2$- chains increased the cmc by a factor of 3.3, whilst γ_{cmc} increased by about 18 mN m^{-1}, which is a more dramatic effect than seen here.

The NR data (not shown) as a function of concen-tration, made up in NRW, were analysed by the single-layer model to yield the effective film thicknesses τ and scattering length densities ρ (examples given in Table 1). These fitted values were used to obtain the molecular areas and surface excesses Γ with Eq. (1). Figure 2b and c shows the Γ-concentration plots for di-CF4 and di-HCF4 determined by DV tensiometry and NR. Clearly, NR is a direct method, which essentially "counts" molecules in the film, whilst tensiometry is indirect and the interpretation of γ-ln a data always involves as-sumptions in terms of an adsorption isotherm. As can be seen, for both surfactants there is good agreement. The results indicate that, for these fluoro-sulfosuccinates, Eq. (4) is valid, namely there is electrical neutrality of the interface and a 1:1 cation:anion ratio in the film. This is a significant result since there has been an on-going debate concerning the validity of the Gibbs equation with ionic surfactants, e.g. [36]. However, some discrepancies in the molecular areas can be seen (Table 1). Firstly, consider the uncertainties: for tensio-metric measurements it is ± 3 Å^2, and for NR, Γ can be measured to ± 3–5%. Therefore, at the cmc of di-CF4 the differences between A_{cmc} by tensiometry and NR are outside the errors. Given the care taken to achieve surface chemical purity, this is difficult to explain. Note that the tensions show no minimum at the cmc, nor any long-time decay of γ below the cmc, at least outside of the dynamic region, i.e. >60 s, and this tends to rule out any organic/surface active impurities. A key indication of this is the value of Γ, measured by NR, at 2× cmc, which was identical to that at the cmc, as one would expect for a pure surfactant. (If surface-active impurities

Table 1 Parameters derived from surface tension and neutron reflection (NR) measurements. The Gibbs equation, Eq. (4), was used to obtain the molecular areas at the cmc. The NR data were analysed with Eq. (1). The NR experiments with di-CF4 were at the cmc, but with di-HCF4 the concentration was 13.4 mmol dm^{-3} ≈ 0.84× cmc. For AOT the NR data were taken from [38–40]

	Surface tension				Neutron reflection		
	T (°C)	cmc (mmol dm^{-3})	γ_{cmc} (mN m^{-1})	$A_{cmc} \pm 3$ (Å^2)	$A_{cmc} \pm 2$ (Å^2)	$\tau_{cmc} \pm 1.5$ (Å)	$\rho_{cmc} \times 10^{-6}$ (Å^{-2})
di-CF4	30	1.57	17.7	56.0	62.7	21.6	1.59
di-HCF4	25	16.0	26.8	65.0	65.8	19.0	1.56
AOT	25	2.56	30.8	77.0	78.0	18.0	2.80

were present they would adsorb strongly below the cmc, but above would be dissolved in micelles, thereby altering the surface composition and the adsorbed amount.)

The cross-sectional area of a fluorocarbon chain is approximately 28 $Å^2$ [37], and so these monolayers appear to be quite densely packed. For di-CF4, A_{cmc} is 56 ± 3 $Å^2$ (surface tension) and repeat measurements at different EDTA concentrations showed this value to be reproducible, whereas NR give a slightly larger A_{cmc} of 62.7 $Å^2$. Even at this larger area the relative packing efficiency of chains is high: for di-CF4, 89% of A_{cmc} can be accounted for by the fluorocarbon chains alone. Together, the data show that the fluoro-succinates are strongly adsorbed, with the terminal-H surfactant di-HCF4 having a slightly lower surface activity.

These molecular areas can be compared with that for the branched hydrocarbon AOT (Table 1). Tensiometric experiments on AOT were carried out in our laboratory, and results from neutron reflection experiments by Thomas and co-workers [38–40] are also given.

Water-in-CO_2 microemulsions:
phase stability and SANS

Figure 3 shows examples of P-T phase stability maps for di-CF4 and di-HCF4 w/c microemulsions. Owing to CO_2-hydrate formation, temperatures below 10 °C were not investigated. At high pressures, single-phase (1ϕ), transparent microemulsions formed. On lowering the pressure a critical pressure P_c is reached, below which the samples turn rapidly opaque (2ϕ). The phase transitions are entirely reversible. If the sample is not stirred a macroscopic separation occurs, and a dense surfactant-water phase appears to separate from a transparent CO_2 phase. Similar phase instability is also observed for AOT-stabilised w/o systems with low-density alkanes, such as ethane and propane [30, 41]. For both CO_2 and alkane-continuous systems the phase changes are thought to be driven by a loss of compatibility between the surfactant chains and the solvent (Hildebrand solubility parameter).

For di-CF4 at $w = 10$, the phase transition pressures are quite low, for example at 20 °C the microemulsion is stable at close to bottle pressure for CO_2. This value represents the lowest reported P_c, indicating that di-CF4 is one of the most efficient compounds in terms of w/c formation. Comparison can be made with another surfactant, the commercial product PFPE: for $w = 20$ at 30 °C the P_c is 120 bar for di-CF4, whilst for PFPE it is 178 bar [15]. Changing to di-HCF4 causes a reduction in the phase stability; for example, with $w = 20$ at 30 °C the critical pressure increases to around 200 bar. Clearly, the hydrophobic chain structure, and especially the extent of fluorination, are important factors for micro-

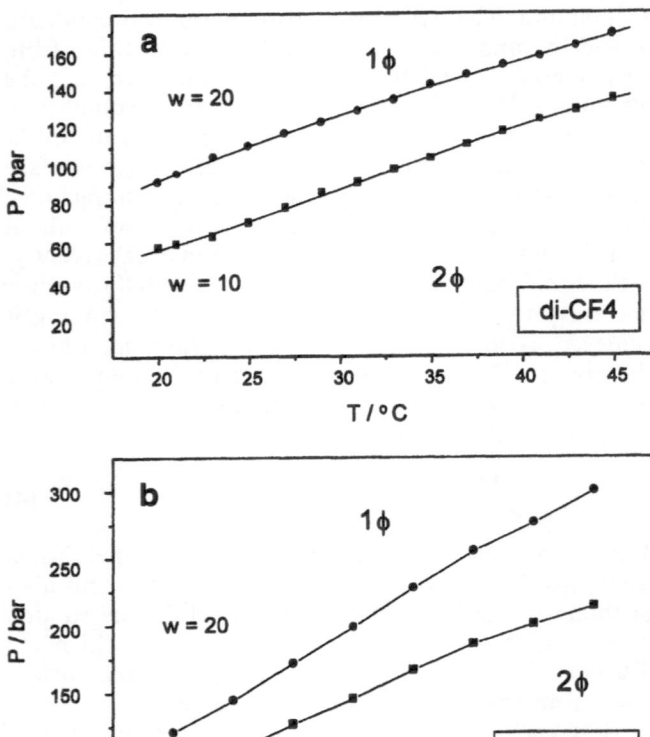

Fig. 3a, b Phase behaviour of water-in-CO_2 (w/c) microemulsions. [surf] = 0.05 mol dm^{-3}

emulsion stability. The phase transition pressure is also reduced with electrolyte solutions (NaCl). This is consistent with more efficient packing in the film, induced by screening of the repulsive headgroup interactions. Alternatively, it could be that a proportion of the surfactant is present as free monomer, either in the water pools or the CO_2 (unlikely because it is an ionic compound), and the salt causes a stronger partitioning to the interface. These effects highlight the importance of working with well-defined, pure compounds, and also justify the efforts taken in synthesis, purification and careful characterisation of the a-w adsorption behaviour, prior to studies in CO_2.

When used for SANS experiments the pressure cell is limited to around 600 bar, and so the measurements were all carried out at 15 °C. A pressure of 500 bar was chosen in order to make direct comparisons between the two surfactants (see Fig. 4 and Table 2). Figure 4 shows scattering curves from single-phase microemulsions as a function of w, and least squares fits to a model of interacting polydisperse spheres, as described in the Materials and methods section. The fitted parameters are given in Table 2.

The results give clear evidence for the presence of water droplets. For all the samples the absolute intensity scaling of the fitted $P(Q)$ was typically within 5% of that

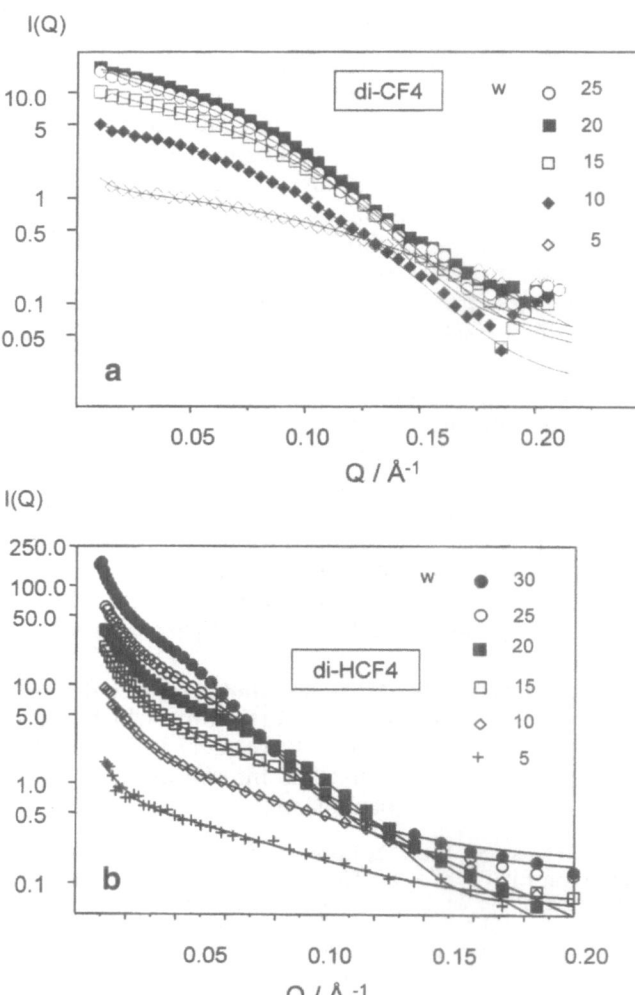

Fig. 4a, b High-pressure SANS from D_2O droplets in w/c microemulsions at 15 °C and $P = 500$ bar. [surf] = 0.10 mol dm^{-3}

Table 2 Parameters fitted to SANS data. [surf] = 0.10 mol dm^{-3}, $T = 15$ °C and $P = 500$ bar. * indicates that the correlation length ζ was fixed. The polydispersity σ/R_c^{av} was 0.20, except for $w = 5$ systems where it was 0.4

| w | di-CF4 | di-HCF4 | | |
	R_c^{av} (Å)	R_c^{av} (Å)	$S(0)$	ζ (Å)
5	13.6	12.0	10	1000*
10	20.0	13.0	82	150
15	21.1	18.6	52	155
20	23.0	26.4	154	376
25	–	30.9	636	1000*
30	–	35.8	400	1000*

expected based on the sample composition, suggesting the model is physically reasonable. Certain repeat experiments showed and the effects of pressure, temperature and composition to be reproducible. In the 2ϕ region, below P_c, the SANS signal disappeared, but with a stirred system the scattering was recovered on returning to the single-phase microemulsion by increasing the pressure.

With di-CF4 at 500 bar the microemulsions are far from P_c (see Fig. 3), and well within the single-phase region. There is little evidence of critical-type scattering, and the SANS is characteristic of the pure form factor. Therefore, $S(Q)$ was not included in the modelling. On the other hand, even at these relatively high pressures, with di-HCF4 the samples are closer to P_c, and enhanced scattering is evident at low Q. As w increases the change in the $S(Q)$ parameters is consistent with enhanced droplet clustering on approach to P_c. It has been shown elsewhere [18] that dilution of the droplets reduces $S(Q)$, whereas the form factor hardly changes. However, it is inappropriate to read too much into the structure factors since it has been included as a necessary step, to take into account the effects of clustering. For analogous AOT-stabilised w/o phases with low-density alkanes like propane, the issues of $S(Q)$ and transition pressures have been covered elsewhere [30, 41].

Figure 5 shows a plot of R_c versus w for the two surfactants. A linear dependence is to be expected with spherical droplets, since $R_c = (3V_{D_2O}w)/A_h$ with V_{D_2O} the molecular volume of water and A_h the effective area per headgroup at the interface.

With di-HCF4 it was possible to study a wider range of w values, and the fitted line shown gave $A_h = 87 \pm 5$ Å^2. Furthermore, the intercept is consistent with the size of the polar core of a dry reversed micelle. This result compares favourably with that for AOT, which was found to have $A_h = 72$ Å^2 in a range

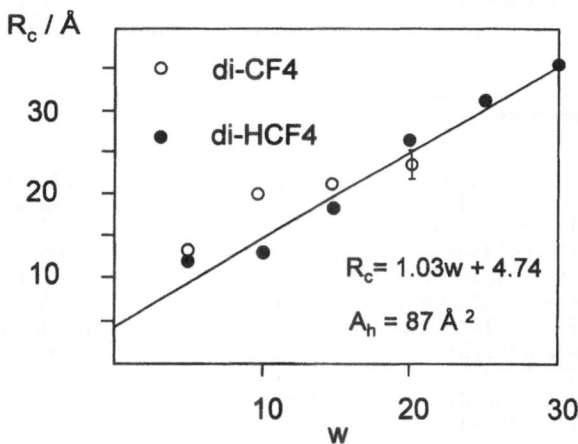

Fig. 5 Dependence of droplet radius, as determined by SANS, on water content

of n-alkanes from propane to decane [30]. For di-CF4 the range of accessible w values is more limited, and the linear dependence less convincing, especially at high w values. This may be due to a phase transition that is difficult to detect visually in the cell, namely a maximum w corresponding to an emulsion failure, or Winsor II-type instability. Since the radii do not increase much above $w = 20$, this may be the maximum swelling possible with this surfactant. However, it can be seen that both surfactants can be used to form simple droplet microemulsions.

Conclusions

Two fluorinated anionic surfactants, di-CF4 and di-HCF4, have been synthesised and characterised. The main aim was to investigate the effects of surfactant chemistry on structure and stability of w/c microemulsions.

To provide a foundation for studies at the water-CO_2 interface, check for chemical purity and fully characterise the compounds, their dilute aqueous phase behaviour was studied first of all. Once appropriate measures were taken to purify the compounds to a sufficiently high level, the surface excesses measured by both tensiometry and NR agreed well. Furthermore, adsorption studies were consistent with a pre-factor of 2 in the Gibbs equation, which is to be expected for a pure 1:1 electrolyte surfactant in the absence of any surface dissociation. The areas per molecule at the cmc were consistent with close-packed fluorocarbon layers, and about 90% of the A_{cmc} can be accounted for by the chains alone. In summary, the F-terminated surfactant is more strongly adsorbed, has a lower cmc and has a lower limiting surface tension than the H-tipped analogue. Although there is an effect on A_{cmc}, a full partial structure factor analysis, which is reported elsewhere [25], indicates the chain-tip dipole of the terminal-H surfactants does not affect the layer structure normal to the interface, e.g. the extent of water penetration into the layer. As found by Pitt et al. [42], the chemical structure of the chain tip has a strong effect on limiting surface tension.

Both of these fluoro surfactants are effective at stabilising w/c microemulsions. Visual studies of the phase behaviour show that di-CF4 is the most efficient compound, since it gives microemulsions at significantly lower pressures. With this surfactant, microemulsions may be formed close to bottle pressure of a normal CO_2 cylinder, and this low pressure may have advantages in terms of associated plant costs in any future practical application. However, SANS experiments indicate that the maximum water uptake is slightly lower than for di-HCF4 (this transition is difficult to detect in the view cell). Added electrolyte has an important effect on the phase diagram, and this highlights the need for working with pure surfactants if meaningful results are to be obtained from such structure-performance studies. Even in water-CO_2 mixtures these fluoro surfactants behave in a straightforward way, rather akin to the structurally related AOT in low-density hydrocarbons such as propane [30, 41]. The SANS data were consistent with a low concentration of free monomer in CO_2, and an effective head group area of around 90 $Å^2$, which is some 20 $Å^2$ greater than for AOT in hydrocarbon-water phases [30].

In conclusion, the fluoro-succinates represent a well-characterised model series of surfactants which can be employed in studying the links between surfactant structure and stability of these interesting, and potentially useful, w/c microemulsions.

Acknowledgements The CO_2 work was funded by a joint research grant from EPSRC (GR/L05532 and GR/L25653). A.P. and E.R. acknowledge the support of EPSRC in terms of studentships. Kodak and EPSRC are thanked for providing a CASE studentship to A.M.D. We also thank CLRC for allocation of beam time at ISIS and a grant towards consumables and travel. Drs. Jeff Penfold and Richard Heenan are thanked for assistance with neutron scattering experiments. Sandrine Nave (Bristol) carried out the tensiometric study of Aerosol-OT.

References

1. Consani KA, Smith RD (1990) J Supercrit Fluids 3:51
2. Beckman EJ, Hoefling TA, Enick RM (1991) J Phys Chem 95:7127
3. Harrison K, Goveas J, Johnston KP, O'Rear EA (1994) Langmuir 10:3536
4. Eastoe J, Steytler DC, Bayazit Z, Martel S, Heenan RK (1996) Langmuir 12:1423
5. Johnston KP, Harrison KL, Clarke MJ, Howdle SM, Heitz MP, Bright FV, Carlier C, Randolph TW (1996) Science 217:624
6. Clark MJ, Harrison KL, Johnston KP, Howdle, SM (1997) J Am Chem Soc 119:6399
7. Zielinski RG, Kline SR, Kaler EW, Rosov N (1997) Langmuir 13:3934
8. Heitz MP, Carlier C, deGrazia J, Harrison KL, Johnston KP, Randolph TW, Bright FV (1997) J Phys Chem B 101:6707
9. Salaniwal S, Cui ST, Cummings PT, Cochran HD (1999) Langmuir 15:5188
10. Holmes JD, Steytler DC, Rees GD, Robinson BH (1998) Langmuir 14:6371
11. Jacobson GB, Lee CT, Johnston KP (1999) J Org Chem 64:1201
12. Jacobson GB, Lee CT, daRocha SRP, Johnston KP (1999) J Org Chem 64:1207
13. Ji M, Chen X, Wai CM, Fulton JL (1999) J Am Chem Soc 121:2631
14. Holmes JD, Bhargava PA, Korgel BA, Johnston KP (1999) Langmuir 15:6613
15. Holmes JD, Ziegler KJ, Audriani M, Lee CT, Bhargava PA, Steytler DC, Johnston KP (1998) J Phys Chem B 103:5703

16. Lee CT, Psathas PA, Johnston KP, deGrazia J, Randolph TW (1999) Langmuir 15:6781
17. DeSimone JM, McClain J, Betts DE, Canelas DA, Samulski ET, Londono JD, Chochran HD, Wignall GD, Chillura-Martino D, Triolo R (1996) Science 274:2049
18. Eastoe J, Cazelles BMH, Steytler DC, Holmes JD, Pitt AR, Wear TJ, Heenan RK (1997) Langmuir 13:6890
19. Fink R, Hancu D, Valentine R, Beckman EJ (1999) J Phys Chem B 103:6441
20. Jawwad AD, Poliakoff M (1999) Chem Rev 99:495
21. Poliakoff M, George MW, Howdle SM (1996) In: Van Eldik R (ed) Chemistry under extreme and non-classical conditions, chap 5. Spektrum, Heidelberg
22. Poliakoff M, Meehan NJ, Ross SK (1999) Chem Ind (London) 19:750
23. Buckley GS, Rodgers AS (1983) J Phys Chem 87:126
24. Yoshino N, Komine N, Suzuki J-I, Arima Y, Hirai H (1991) Bull Chem Soc Jpn 64:3262

25. Downer A, Eastoe J, Pitt AR, Simister EA, Penfold J (1999) Langmuir 15:7591
26. An SW, Lu JR, Thomas RK (1996) Langmuir 12:2446
27. Downer A, Eastoe J, Pitt AR, Penfold J, Heenan RK (1999) Colloids Surf A 156:33
28. Penfold J, Thomas RK (1990) J Phys Conden Matt 2:1369
29. Lekner J (1987) Theory of reflection. Nijhoff, Dordrecht
30. Eastoe J, Robinson BH, Young WK, Steytler DC (1990) J Chem Soc Faraday Trans 86:2883
31. King SM, Heenan RK (1996) The LOQ instrument handbook. Rutherford Appleton Laboratory Report RAL-TR-96-036, CCLRC, Didcot, UK
32. Heenan RK (1989) FISH data analysis program. Rutherford Appleton Laboratory Report RAL-89-129, CCLRC, Didcot, UK
33. Kotlarchyk M, Chen S-H, Huang JS, Kim MW (1984) Phys Rev A 29:2054

34. Hudlicky M, Pavlath AE (1995) Chemistry of organic fluorine compounds 2, chap 6. American Chemical Society, Washington
35. Achilefu SA, Selve C, Stebe M-J, Ravey J-C, Delpuech J-J (1994) Langmuir 10:2131
36. Bae S, Haage K, Wantke K, Motschmann H (1999) J Phys Chem B 103:1045
37. Mengyang L, Acero A, Huang Z, Rice S (1994) Nature 367:151
38. Li ZX, Lu JR, Thomas RK (1997) Langmuir 13:3681
39. Li ZX, Lu JR, Thomas RK, Penfold J (1995) Prog Colloid Polym Sci 98:243
40. Li ZX, Lu JR, Thomas RK, Penfold J (1997) J Phys Chem B 101:1615
41. Kaler EW, Bilman JF, Fulton JL, Smith RD (1991) J Phys Chem 95:458
42. Pitt AR, Morley SD, Burbidge NJ, Quickenden EL (1996) Colloids Surf A 114: 321

Progr Colloid Polym Sci (2000) 115:222–226
© Springer-Verlag 2000

M. Ferrari
L. Liggieri
R. Miller
F. Ravera

Adsorption of n-alkyl polyoxyethylene glycol ethers at liquid–vapour and liquid–liquid interfaces

L. Liggieri · M. Ferrari (✉) · F. Ravera
CNR, Istituto di Chimica Fisica Applicata
dei Materiali (ICFAM)
via De Marini 6, 16149 Genova, Italy
e-mail: ferrari@icfam.ge.cnr.it
Tel.: +39-010-6475723
Fax: +39-010-6475700

R. Miller
Max-Planck-Institut für Kolloid- und
Grenzflächenforschung
Max-Planck-Campus, Haus 2
Am Mühlenberg 2
14476 Golm/Potsdam, Germany

Abstract n-Alkyl polyoxyethylene glycol ethers (C_iEO_j's) show peculiar adsorption properties at liquid interfaces. In fact, the partial hydrophobic character of the ethylene groups in the hydrophilic chain allows different orientations of adsorbed molecules. As a consequence, the usual adsorption models hardly predict the experimental observations, which calls for a more complex description. To this aim, a study on the adsorption properties of some C_iEO_j's is presented. The equilibrium data are interpreted according to a model assuming the partitioning, depending on the surface coverage, of the adsorbed molecules between two orientation states. Correspondingly, dynamic experiments are interpreted by a model considering both the transport of surfactant in the bulk phase by diffusion and the process of reorientation at the interface. These models provide a more accurate description of the experimental results for C_iEO_j's at the water–air interface. For adsorption dynamics at the liquid–liquid interface, the transfer of surfactant between the two phases must be explicitly considered.

Key words Non-ionic surfactants · Adsorption dynamics · Surface isotherms · Molecular orientation · Partitioning

Introduction

Among non-ionic surfactants, n-alkyl polyoxyethylene glycol ethers (C_iEO_j) are one of the most studied class of surface active substances, owing to their involvement in many industrial and technological fields, being as well of interest for their lower toxicity and environmental impact. In addition, synthetic pathways allow the chains to be easily "tuned" in order to obtain specific properties.

The characterisation of the adsorption of some C_iEO_j's is the aim of this work. In particular, a theoretical and experimental study to increase the knowledge of surface properties of these surfactants both at liquid–air and liquid–liquid interfaces is performed. In the interpretation of the data, the Langmuir adsorption-type isotherm has been often used [1]. However, this model is not always appropriate to describe the equilibrium and the dynamics behaviour, especially at lower concentrations. Thus Frumkin-like models [2] or a mixed adsorption approach [3–5] for the equilibrium and the dynamics, respectively, have been also used.

A model which improves the description of the properties of this class of surfactants takes into consideration two different states for the adsorbed molecules at the interface [6–8]. These states are characterised by two different partial molar areas which could be due to different orientations of the molecules with respect to the surface owing to the features of the hydrophilic chain. Moreover, in the liquid–liquid case the transfer across the interface is important [9] and it has to be taken into account in the surface tension data interpretation both for the equilibrium and the dynamic adsorption.

In the current work the adsorption is studied using two different tensiometric techniques: the pendent/emerging drop and the maximum bubble pressure methods.

Theoretical background

The data interpretation is performed on the basis of the theoretical model proposed firstly by Fainerman et al. [6], where two states are considered in the adsorption layer characterised by two molar surface areas, ω_1 and ω_2. The corresponding γ-c isotherm depends on four parameters: the two surface areas ω_1 and ω_2 and other two parameters b_1 and b_2 related to the surface activities of the two states:

$$c = \frac{1 - \exp\left(-\frac{(\gamma_0 - \gamma)\omega}{RT}\right)}{b_2\left[\left(\frac{\omega_1}{\omega_2}\right)^{\alpha}\exp\left(-\frac{(\gamma_0 - \gamma)\omega_1}{RT}\right) + \exp\left(-\frac{(\gamma_0 - \gamma)\omega_2}{RT}\right)\right]} \quad (1)$$

where γ and γ_0 are the surface tension and the surface tension of the pure system, respectively; ω is the average molar surface area, weighted by the adsorption Γ_1 and Γ_2, corresponding to the two states:

$$\omega = \frac{\omega_1\Gamma_1 + \omega_2\Gamma_2}{\Gamma_1 + \Gamma_2} \quad (2)$$

and α is a parameter which can alternatively be used instead of b_1, defined by:

$$b_2 = b_1\left(\frac{\omega_2}{\omega_1}\right)^{\alpha} \quad (3)$$

For $\alpha = 0$ the surface activity is independent of the molar surface area, and for $\alpha > 0$ the molecules with the larger surface area are more surface active.

The physical reason for the two surface molar areas can be the different possible orientations of the adsorbed molecules with respect to the surface. Under this hypothesis the state with the larger surface area corresponds to the flat orientation along the surface and prevails at low surface pressure, while the other state corresponds to the normally oriented adsorbed molecules and prevails at high surface pressure.

It is worth noting that the Langmuir adsorption model is obtained, as a particular case, when $\omega_1 = \omega_2$, and $b_1 = b_2$ (or $\alpha = 0$), which means that only one state is possible in the adsorbed layer.

This two-state isotherm, besides being used to interpret the equilibrium surface tension data, can be used in the framework of diffusion-controlled adsorption, improving in several cases the description of the dynamic experimental data [10, 11]. Moreover, a theoretical approach to describe the evolution of the surface tension during the adsorption process has also been developed [8] on the basis of this thermodynamic model, considering three dynamic processes, i.e. the adsorption-desorption exchange between the surface and the bulk, the orientation of the adsorbed molecules and the diffusion process in the bulk.

Materials and methods

The surfactants used were n-decyl polyoxyethylene glycol ethers, $C_{10}EO_j$, with $j = 4$, 5, 8, and n-dodecyl polyoxyethylene glycol ethers, $C_{12}EO_j$, with $j = 5$, 8. These surfactants are characterised by a linear alkyl chain. They were supplied by Nikko (Japan) at high purity grade with a monodisperse polyoxyethylene chain length.

Water was produced by a Millipore-MilliQ system fed with distilled water. The surface tension of pure water was checked to be greater than 72.5 mN/m, at 20 °C. The oil was hexane Uvasol grade, used after purification on an alumina column. All the measurements were performed at $T = 20$ °C. The dynamic and the equilibrium surface tension data were obtained according to the pendant drop method, using ASTRA (automatic surface tension realtime acquisition), a computer-assisted drop-shape analysis apparatus, which allows the evolution of the surface tension to be monitored in a time range from a few seconds to hours. In particular, the equilibrium surface tension data have been evaluated from dynamic surface tension signals long enough to warrant the attainment of the adsorption equilibrium.

For the water–air systems, for adsorption process characterised by a shorter time (10^{-3}–10 s) the dynamic surface tension was measured by means of a dynamic maximum bubble pressure (DMBP) tensiometer (GammaLab, Germany).

Results and discussion

In the experimental studies of adsorption, in particular for very low concentrations (below 10^{-8} mol/cm^3), some care has to be used to avoid or take into account depletion effects. For example, if the surface tension of an emerging bubble inside the surfactant solution is measured, the adsorption onto the solid walls of the cell can affect the concentration. To avoid this problem, a pre-equilibration of the cell with the solution is necessary.

A similar problem can be met when a pendant drop is used. In this case the depletion of the solution inside the drop is due to the adsorption at the water–air interfaces. In order to know the actual surfactant concentration inside the drop, an iterative procedure is applied as explained previously [8].

When liquid–liquid systems are considered, besides these, some other problems concerning the solubility of the surfactant in both phases have to be considered. For equilibrium measurements the actual surfactant concentration in water has to be known and the partition coefficient values are necessary. For the systems used in this work, the partition coefficients have been evaluated with the technique already described [9], and they are reported in Table 1. For dynamic interfacial tension measurements the initial conditions with respect to the

Table 1 Partition coefficient in water–hexane systems at $T = 20$ °C

	k_p (c_h/c_w)	Conc. range (mol/cm^3 × 10^{-8})
$C_{10}E_4$	38 ± 2	0.4–7
$C_{10}E_5$	13.9 ± 0.8	2.5–15
$C_{10}E_8$	0.82 ± 0.04	4–40
$C_{12}E_5$	21 ± 2	2–4

distribution of the surfactant are very important and the transfer in the second liquid phase can be considered an ulterior controlling process of the adsorption dynamics.

Results for water–air systems

As shown in Fig. 1, the two-states model accurately describes the equilibrium properties of $C_{10}EO_j$ adsorption at the water–air interface, with best fit values of the parameters reported in Table 2. In Figs. 2–4 the dynamic surface tension data, for $C_{10}EO_4$, $C_{10}EO_5$ and $C_{10}EO_8$, respectively, are plotted, with the theoretical curves generated by the diffusion-controlled model using the two-state adsorption isotherms with the parameters found from the equilibrium results (Table 2). The diffusion coefficients have been extrapolated from PFGSE-NMR measurements for C_8EO_4 by assuming the diffusion coefficient of C_iEO_j's to scale as $M^{-1/2}$, where M is the molecular weight.

The diffusion-orientation controlled adsorption model, described previously [8], does not improve the accuracy of the description, indicating that the rate of the transformation process between states 1 and 2, or the orientation process, is much higher than the diffusion one.

By applying the two-state model for the interpretation of the equilibrium data of the $C_{12}EO_i$ aqueous solutions (Fig. 1), it has been only possible to find the best fit values for ω_2 and b_2, while the quantities ω_1 and α remain undetermined. Thus, these data seem to be described by the particular case of a Langmuir isotherm (only one determined state), with $1/\Gamma^\infty = \omega_2$ and $a = 1/2b_2$. However, the diffusion-controlled adsorption together with the Langmuir isotherm are not adequate to describe the dynamics of adsorption. Therefore, state 1 possibly exists, but can be appreciated only at very low concentrations.

Table 2 Isotherm parameters for water–air systems at $T = 20$ °C

	ω_1 (cm^2/mol × 10^9)	ω_2 (cm^2/mol × 10^9)	b_2 (cm^3/mol × 10^8)	α
$C_{10}E_4$	6.7	2.6	1.4	2.2
$C_{10}E_5$	7.0	2.6	1.1	2.6
$C_{10}E_8$	9.9	4.0	2.9	3.3
$C_{12}E_5$	–	3.1	11	–
$C_{12}E_8$	–	4.0	17	–

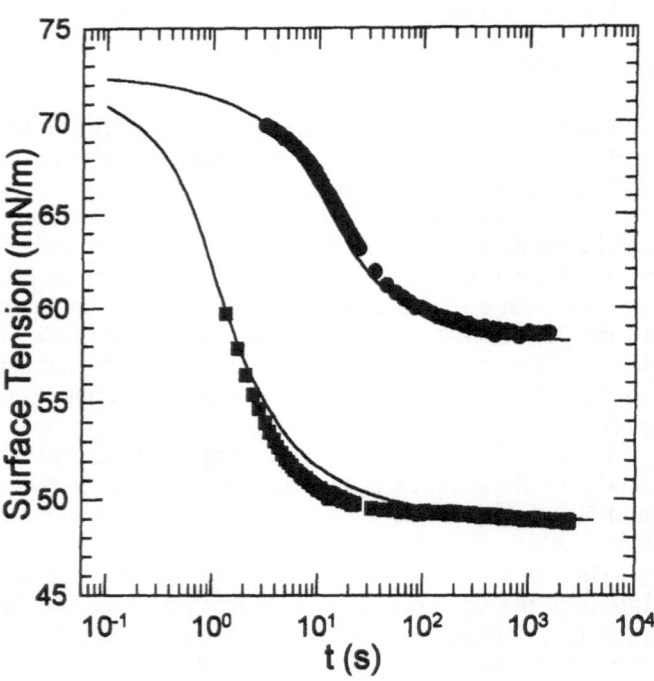

Fig. 1 Surface tension versus surfactant concentration for $C_{10}EO_4$ (●), $C_{10}EO_5$ (■), $C_{10}EO_8$ (▲), $C_{12}EO_8$ (○), and $C_{12}EO_5$ (□), at the water–air interface, with best fit isotherms (*continuous lines*) by the theoretical Eq. (2)

Fig. 2 Dynamic surface tension of $C_{10}EO_4$ at the water–air interface during adsorption for initial concentrations of 2 (●) and 6 (■) mol/cm^3. Theoretical curves are from the diffusion-controlled adsorption model ($D = 4.6 \times 10^{-6}$ cm^2/s) with the two-state isotherm (*solid lines*)

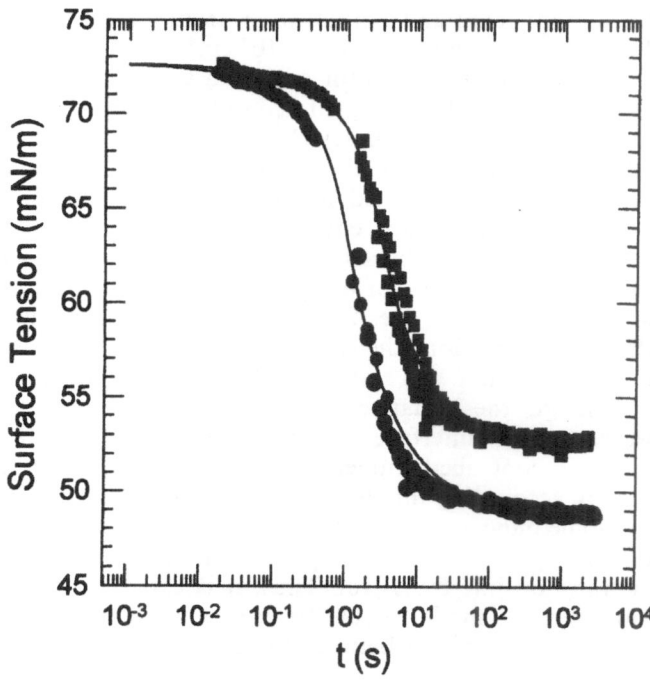

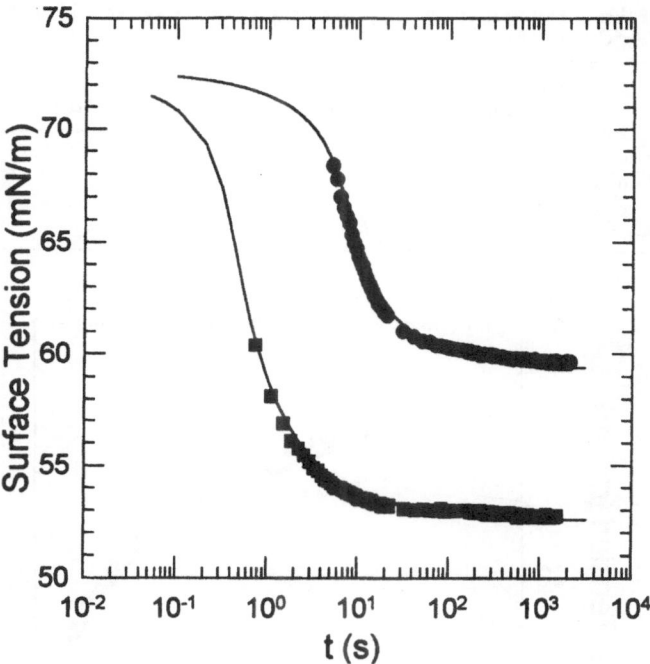

Fig. 3 Dynamic surface tension during the adsorption of $C_{10}EO_5$ at the water–air interface for initial concentrations of 6 (■) and 10 (●) ×10^{-8} mol/cm^3. Theoretical curves are from the diffusion-controlled adsorption model ($D = 4.3 \times 10^{-6}$ cm^2/s) with the two-state isotherm (*solid lines*). Short-time data have been obtained by the dynamic maximum bubble pressure technique

Fig. 4 Dynamic surface tension during the adsorption of $C_{10}EO_8$ at the water-air interface for initial concentrations of 2 (●) and 8 (■) ×10^{-8} mol/cm^3. Theoretical curves are from the diffusion-controlled adsorption model ($D = 3.7 \times 10^{-6}$ cm^2/s) with the two-state isotherm (*solid lines*)

Results for water–hexane systems

Figure 5 shows the equilibrium interfacial tension versus surfactant concentration in water, for C_iEO_j in water–hexane. The concentration has been evaluated for each system by taking into account the partition coefficient value. The theoretical curves are the two-state adsorption isotherms with best fit values of the parameters reported in Table 3.

As in the case of water–air, the equilibrium results for $C_{12}EO_5$ at the water–hexane interface seem to follow a Langmuir isotherm, but this single-state model is not good for the description of the dynamic data. The adsorption dynamics at water–hexane have been investigated with two different initial conditions: pre-equilibrated phases and non-equilibrated phases, i.e. with the surfactant initially present only in water. As shown in Fig. 6, the transfer and the diffusion in the second phase strongly affect the adsorption dynamics.

All these results seem to be in agreement with the hypothesis considering the orientation process to be responsible for the two states in the adsorbed layer. In fact, by observing the equilibrium results, both for water–air and water–hexane, b_2 increases with the length of the alkyl chain, and ω_1 increases by increasing the length of the oxyethylene and the alkyl chain. Moreover, ω_1 and ω_2 are in general of the order of magnitude of

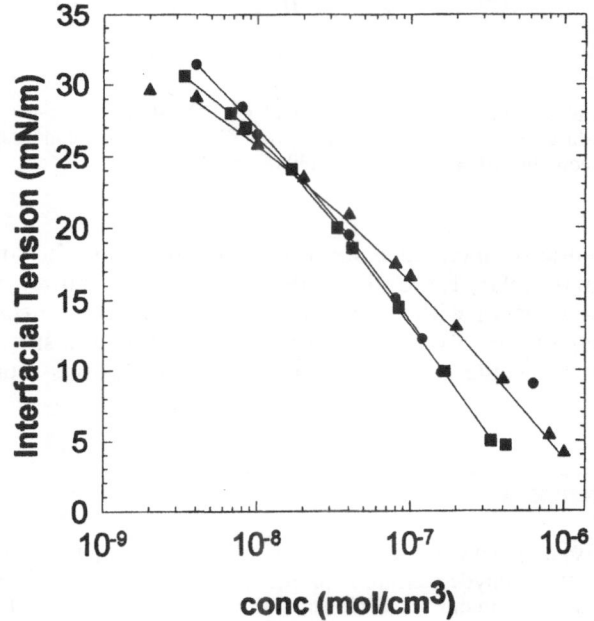

Fig. 5 Equilibrium interfacial tension versus surfactant concentration of $C_{10}EO_4$ (●), $C_{10}EO_5$ (■), $C_{10}EO_8$ (▲) and $C_{12}EO_5$ (○) at the water–hexane interface, and the two-state isotherm best fit (*solid lines*)

the molar areas occupied by the molecules lying along and normal to the surface, respectively, and they are

Table 3 Isotherm parameters for water-hexane systems at $T = 20\,°C$

	ω_1 (cm^2/mol $\times 10^9$)	ω_2 (cm^2/mol $\times 10^9$)	b_2 (cm^3/mol $\times 10^8$)	α
$C_{10}E_4$	5.8	3.0	7.3	5.1
$C_{10}E_5$	9.4	3.6	27.0	5.6
$C_{10}E_8$	8.7	4.1	30.1	7.1
$C_{12}E_5$	–	4.6	33.8	–

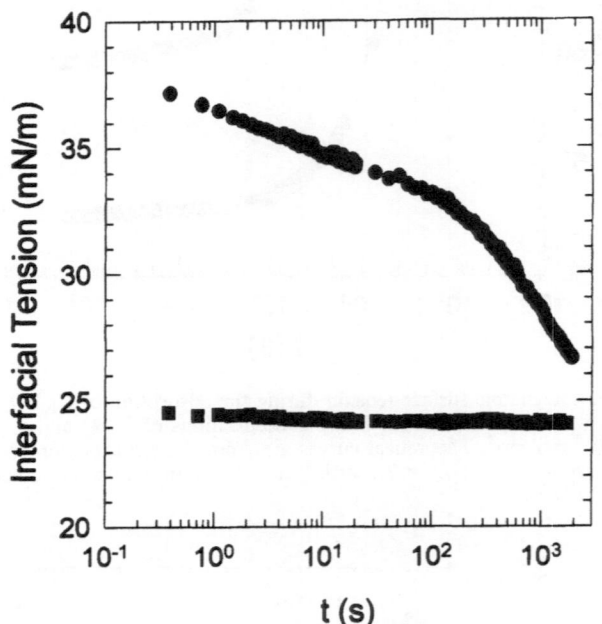

Fig. 6 Dynamic interfacial tension of $C_{10}EO_5$ at the water-hexane interface ($c = 4 \times 10^{-8}$ mol/cm^3) with different initial conditions: pre-equilibrated (●) and non-equilibrated (■)

essentially unchanged for water–air and water–hexane. In particular, for $C_{10}EO_8$, the values of ω_1 and ω_2 are in agreement with the maximum and minimum surface areas obtained by a computer simulation of the surfactant molecule corresponding to the orientations along the surface ($\omega_=$) and normal ($\omega_\perp$), from which $\omega_= = 1.1 \times 10^{10}$ cm^2 and $\omega_\perp = 4.1 \times 10^9$ cm^2.

Different orientations for adsorbed C_iEO_j molecules can be explained by taking into consideration the partial hydrophobic (or lipophilic) character of the ethylene groups in the hydrophilic chain, which allows the molecules to be oriented along the surface direction as long as the surface coverage is small. This latter consideration could be also an explanation for the fact that the found values of parameter α for water–hexane are about twice the ones for water–air. This means that the orientation along the surface is promoted in the case of water–hexane with respect to the water–air case [8].

Though the dynamic behaviour of C_iEO_j at the water–hexane interface is described neither by the diffusion-controlled model with a two-state isotherm nor by accounting for the molecules' orientation kinetics, some indications exist about the orientation and the diffusion processes with comparable characteristic times at water-oil interfaces [10]. Thus, it is expected that a more realistic model should be used to account for the transfer of surfactant between the two finite phases.

Conclusions

The introduction of a model considering two surface molar areas (ω_1 and ω_2) for the adsorbed molecules has improved the description of the adsorption equilibrium properties of this class of surfactants. Moreover, the adsorption dynamics in liquid–air systems is well represented by a diffusion-controlled model together with the two-state isotherms. However, this does not occur for the adsorption dynamics in water–hexane systems.

In liquid-liquid systems the partitioning properties of the surfactant affects the dynamic behaviour, and has to be taken into account in a more complex theoretical model. The role of the chains in the adsorption process also needs further investigation, allowing for a physical explanation accounting both for orientation and conformation changes of the hydrophilic tail at the interface.

References

1. Van Os NM, Haak JR, Rupert LAM (1993) Physico-chemical properties of selected anionic, cationic and nonionic surfactants. Elsevier, Amsterdam, p 250
2. Chang H-C, Hsu C-T, Lin S-Y (1998) Langmuir 14:2476
3. Eastoe J, Dalton J, Rogueda PGA, Crooks ER, Pitt AR, Simister EA (1997) J Colloid Interface Sci 188:423
4. Lin S-Y, Tsay R-Y, Lin L-W, Chen S-I (1996) Langmuir 12:6530
5. Eastoe J, Dalton J, Rogueda PGA (1998) Langmuir 14:979
6. Fainerman VB, Miller R, Wüstneck R, Makievski AV (1996) J Phys Chem 100:7669
7. Ferrari M, Liggieri L, Ravera F (1998) J Phys Chem B 102:10521
8. Liggieri L, Ferrari M, Massa A, Ravera F (1999) Colloid Surf A 156:455
9. Ravera F, Ferrari M, Liggieri L, Miller R, Passerone A (1997) Langmuir 13:4817
10. Miller R, Aksenenko EV, Liggieri L, Ravera F, Ferrari M, Fainerman VB (1999) Langmuir 15:1328
11. Aksenenko EV, Makievski A, Miller R, Fainerman VB (1998) Colloid Surf A 143:311

Progr Colloid Polym Sci (2000) 115:227–232
© Springer-Verlag 2000

SOFT MATTER INTERFACES

K.-D. Wantke
H. Fruhner
K. Lunkenheimer

The influence of molecular exchanges on surface dilational properties of surfactant solutions

K.-D. Wantke (✉) · H. Fruhner
K. Lunkenheimer
Max-Planck-Institut für Kolloid- und
Grenzflächenforschung
14424 Potsdam/Golm, Germany
e-mail: wantke@mpikg-golm.mpg.de
Tel.: +49-331-5679233
Fax: +49-331-5679202

Abstract Molecular exchange processes between fluid surfaces and the adjacent bulk phases were investigated in the frequency range $1 \text{ Hz} \leq f \leq 500 \text{ Hz}$ using a new version of the oscillating bubble method. The measuring data result in a complex dilational modulus of the fluid surface which depends on elastic, viscous and transfer properties of the surface. Three models of the surface exchange mechanism are considered to explain the frequency behavior of the modulus. It was possible to approximate all chosen measurements by one of these models. The parameters of the models could be determined by a fit-procedure. Using our measurements we have demonstrated the influence of an intrinsic surface dilational viscosity for special solutions which is independent of bulk diffusion effects. This can be interpreted as a consequence of the molecular exchange dynamics near the surface in a non-equilibrium state. An appropriate theoretical model allows estimation of the molecular exchange rate and the dissipative loss.

Key words Surface rheology ·
Surface dilational viscosity ·
Surface elasticity · Molecular
kinetics · Surfactant

Introduction

Surface rheological properties of surfactant solutions have influence on many technological processes such as flotation, foamability, and coating processes. The understanding of these processes requires well-defined and reliable measurements and suitable rheological models, in particular models of surfaces and interfaces. A fluid surface is a highly compressible phase due to the molecular exchange processes between the surface and the adjacent subsurface. At a pure water surface these exchange processes are very fast and their influence on the surface rheological properties is negligible. On the contrary, surface active substances exhibit a strong delay of the molecular motion, and an expansion or compression of the surface of a surfactant solution causes a disturbance of the thermodynamic equilibrium. The theoretical aspects of this mechanism are widely investigated, whereas, additional, reliable experiments are required to verify the theory.

Many devices use deformations in the frequency range <1 Hz. These concern normally only equilibrium states at the surface combined with bulk diffusion processes. Therefore, for investigations of non-equilibrium states at the surface, experiments with a higher deformation rate must be realized.

Materials and methods

The established methods for the investigation of dilational properties of fluid surfaces use longitudinal and transversal waves, as well as motions of barriers, bubbles, and drops [1–4]. A common problem of all these methods is the separation of the bulk influence. This procedure becomes simple and reliable if the involved bulk flux possesses a simple form. In particular, oscillating bubbles and drops can fulfill such conditions.

Our new version of the oscillating bubble method allows determination of the complex dilational modulus $\varepsilon(f, c)$ of a fluid surface in a simple and reliable manner. The principle of this method is shown in Fig. 1. Within a closed chamber a small hemispheric bubble is formed at the tip of a capillary. A piezo-

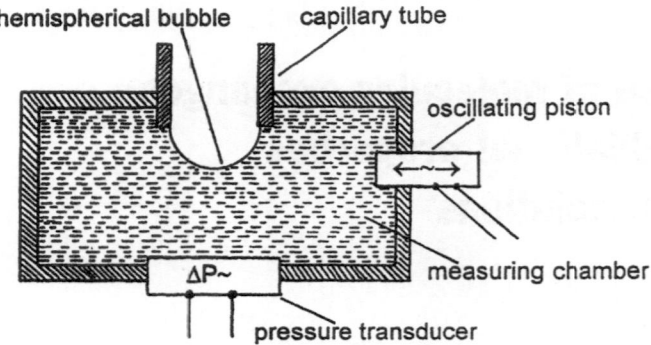

hemispherical bubble   capillary tube

oscillating piston

measuring chamber

ΔP~

pressure transducer

Fig. 1 Schematic diagram of the oscillating bubble measuring device

electric driver generates oscillations of the bubble volume which cause sinusoidal changes in the surface area, the radius, and the pressure within the chamber. These pressure oscillations are monitored by a sensitive pressure transducer. The three-phase contact line must be stabilized by a sharp edge at the tip of the capillary and by special surface properties.

Using this system one obtains information about the pumped volume, the change in area, the change in pressure, and the phase angles between these oscillations. With these measurements we can easily determine the complex surface dilational modulus in the frequency range 1 Hz < f < 500 Hz. It is defined by the equation:

$$\varepsilon(f,c) = \Delta\gamma/\Delta\ln A \qquad (1)$$

where $\Delta\gamma$ represents the change in surface tension, $\Delta\ln A = \Delta A/A$ is the relative change in surface area, f the frequency, and c the bulk concentration. A simple function describes the relation between $\Delta\gamma$ and the corresponding changes in measured Δp [5, 6].

Results and discussion

The modulus $\varepsilon(f,c)$ includes elastic, viscous, and kinetic properties. Using an appropriate model the separation of the influence of the different processes on $\varepsilon(f,c)$ should be possible. Therefore, the knowledge of $\varepsilon(f,c)$ in a broad frequency and concentration range represents an appropriate basis for detailed studies of dynamic processes at a fluid surface. It can be used to verify or modify the established theoretical models. We have tested three different theoretical models using our experimental results.

The diffusion-controlled model

In the standard model the effective surface is reduced to a monolayer composed of surfactant and water molecules. In addition, a small sublayer with bulk properties (assumed thickness $d \approx 20$ nm) is introduced for theoretical reasons. If the deformation of the surface is not too fast, a permanent equilibrium state between the monolayer and the adjacent sublayer can be assumed. Therefore, the sinusoidal homogeneous dilation $\Delta A/A = |\Delta A/A| \exp(i\omega t)$ causes only an oscillating

bulk diffusion process with the amplitude Δc which influences the surface concentration Γ. The thickness d of the sublayer must be small in comparison to the wavelength λ of the diffusion wave $[\lambda(500\,Hz) \approx 3500$ nm]. Then, $\varepsilon(f,c)$ becomes the form of the Lucassen/van den Tempel modulus, which reads [7–9]:

$$\varepsilon = \varepsilon_m \frac{1 + \xi + i\zeta}{1 + 2\xi + 2\xi^2}, \quad \xi = \sqrt{\frac{\omega_0}{\omega}} . \qquad (2)$$

The behavior of this complex modulus $[\varepsilon(f,c) = |\varepsilon(f,c)| \exp(i\phi(f,c))]$ is characterized by the limiting cases: $|\varepsilon(f,c)| \to 0$, $\phi(c,f) \to 45°$ for low frequencies $(f \to 0)$ and $|\varepsilon(f,c)| \to \varepsilon_m = $ const., $\phi(c,f) \to 0$ for high frequencies $(f \to \infty)$. The curves relative ascent depends on the parameter:

$$\omega_0 = \omega_m = D(dc/d\Gamma)^2 . \qquad (3)$$

This value determines the molecular exchange between monolayer and sublayer. It describes the ratio of the changes in sublayer concentration to the change in monolayer concentration according to:

$$\Delta c = (dc/d\Gamma)\Delta\Gamma = (\sqrt{\omega_m/D})\Delta\Gamma$$
$$(D = \text{bulk diffusion coefficient}) . \qquad (4)$$

The experimental verification of the standard model using the oscillating bubble equipment applied revealed the following results:

1. The data of all low concentrated surfactant solutions ($c \ll$ CMC or $c \ll$ solubility limit) can be approximated by Eq. (2).
2. The data of certain solutions (e.g. fatty acids, cf. Fig. 2) show the same behavior also for higher concentrations.
3. The parameters ε_m, ω_m resulting from the fit-procedure of these data are significantly smaller than the corresponding theoretical values ε_0, ω_0 (calculated from the surface equation of state data [6]).
4. The amplitude of the bulk diffusion process is, according to Eq. (4), also smaller than its theoretical value and, therefore the molecular exchange between bulk and surface becomes smaller. This is a hint at an influence of the sublayer on the dynamic surface tension.

The mixed kinetic diffusion-controlled model

The description of the frequency behavior of the modulus $\varepsilon(f,c)$ of many surfactant solutions near the CMC requires a modification of the standard model. The simplest way is the introduction of a kinetic effect. This means a non-equilibrium state between monolayer and sublayer is caused by a delaying molecular exchange. Then, in Eq. (2) the real function ζ must be replaced by the complex function $\zeta = h(f,c)\zeta'$ with

$$h(f,c) = k/(k + (1+i)\sqrt{\omega D/2}) \ . \tag{5}$$

The influence of the kinetic effect can be neglected if the exchange constant k is large enough and therefore $h(c,f)$ becomes 1. However, in some cases the fit-procedure of $\varepsilon(f,c)$ leads to k values <5 cm s^{-1}, causing a non-equilibrium state at the surface in the mentioned frequency range. Then, the limits of $\varepsilon(f,c)$ are: $|\varepsilon(f,c)| \to 0$ and $\phi(c,f) \to 90°$ for $f \to 1$ and $k \ll 1$ cm s^{-1}, as well as $|\varepsilon(f,c)| \to \varepsilon_m$ and $\phi(c,f) \to 0°$ for $f \to \infty$.

The amount $|\varepsilon(f,c)|$ is only a little larger than in the standard case and therefore a phase angle $\phi(c,f) > 45°$ for low frequencies ($f < 30$ Hz) is the main reason for the discussed modification of the model. Suprisingly, such a behavior is related with an increasing phase angle in the higher frequency range ($f > 100$ Hz, cf. Fig. 3). This means the phase angle passes a minimum between 50 Hz and 400 Hz. In addition, the amount $|\varepsilon(f,c)|$ also exhibits unexpected behavior (cf. Fig. 4). After an ascent it passes a predicted level range, followed by a further increase for higher frequencies. Such behavior can be approximated by the following equation:

$$\varepsilon = \varepsilon_m \frac{1 + \zeta' + i\zeta'}{1 + 2\zeta' + 2\zeta'^2} + i\omega\kappa, \quad \zeta' = h(f,c)\sqrt{\frac{\omega_0}{\omega}} \tag{6}$$

where κ represents the intrinsic surface dilational viscosity. Its introduction is a formal step. However, an acceptable approximation of a complex function requires a realistic model, and in many cases the approximation of $\varepsilon(f,c)$ is only possible by Eq. (6). In addition, evidence was given that a solution revealing a measurable intrinsic surface viscosity is able to stabilize foams [10]. This is a hint at a dissipative process at the surface.

For the physical interpretation of the intrinsic viscosity we consider the molecular dynamics at the surface. The relaxation processes within the monolayer and within a small adjacent sublayer are too fast for the explanation of this effect. However, in the mentioned non-equilibrium state between monolayer and sublayer, the energy and momentum balances of molecular exchange processes are not zero and these may cause viscous losses. Therefore, the intrinsic surface viscosity can be interpreted as a consequence of the delaying molecular exchange between monolayer and sublayer. Such a model leads to the discussed frequency behavior of $\varepsilon(f,c)$ with a phase minimum in the medium frequency range. The fit-procedure of the measured data yields normally to reliable values of ε_m, ω_m, and κ. However, the calculation of the molecular exchange constant k and of the loss energy results in significant values for $k < 5$ cm s^{-1} only.

The general viscoelastic model

The above-discussed models describe the surface dilational properties of many solutions of soluble surfactants satisfactorily. However, the phase angles of some higher concentrated solutions decrease for $f \to 0$, although the amplitude shows normal behavior. Examples are n-decyl- and n-nonyldimethylphosphine oxide (cf. Figs. 5 and 6). This effect increases with decreasing chain length and increasing CMC, respectively. The continuity equation requires the assumption of a higher concentrated sublayer for the explanation of such behavior. The qualitative description of the molecular

Fig. 2 Surface dilational modulus of various concentrated decanoic acid solutions (phase angle inserted as numbers)

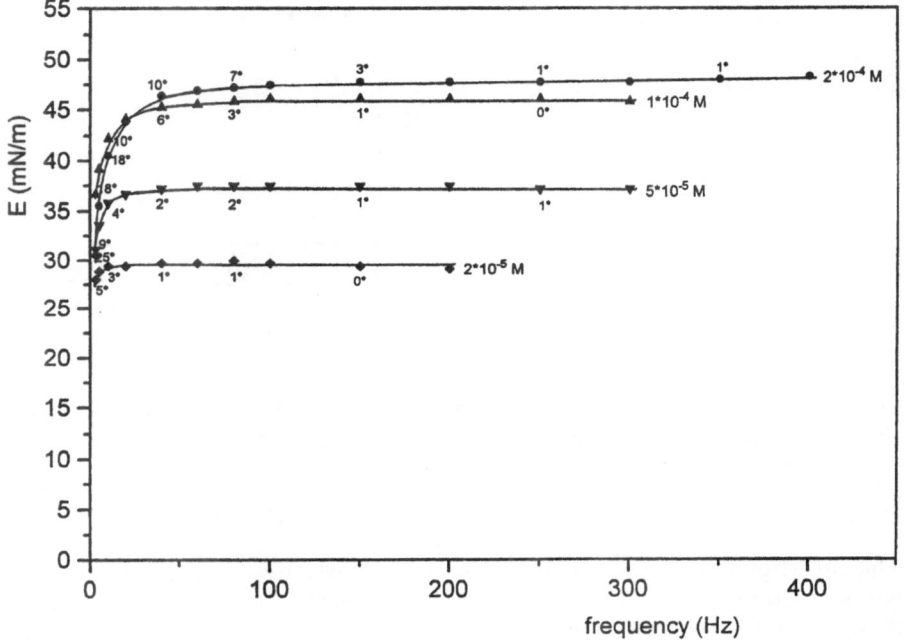

Fig. 3 Phase angle of the surface dilational modulus of various concentrated n-undecyl-dimethylphosphine oxide solutions

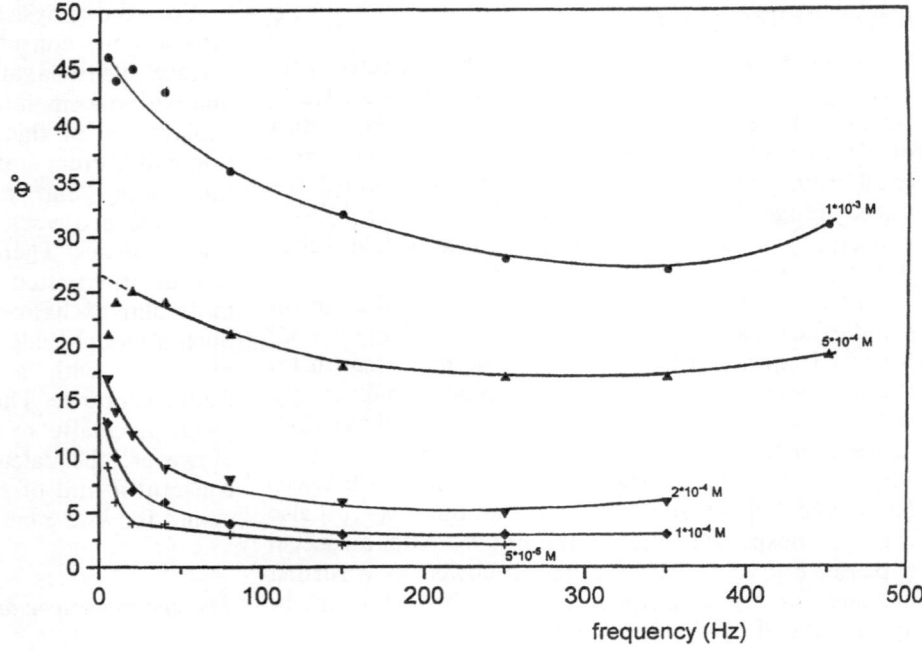

Fig. 4 Amount of the surface dilational modulus of various concentrated n-undecyldimethyl-phosphine oxide solutions

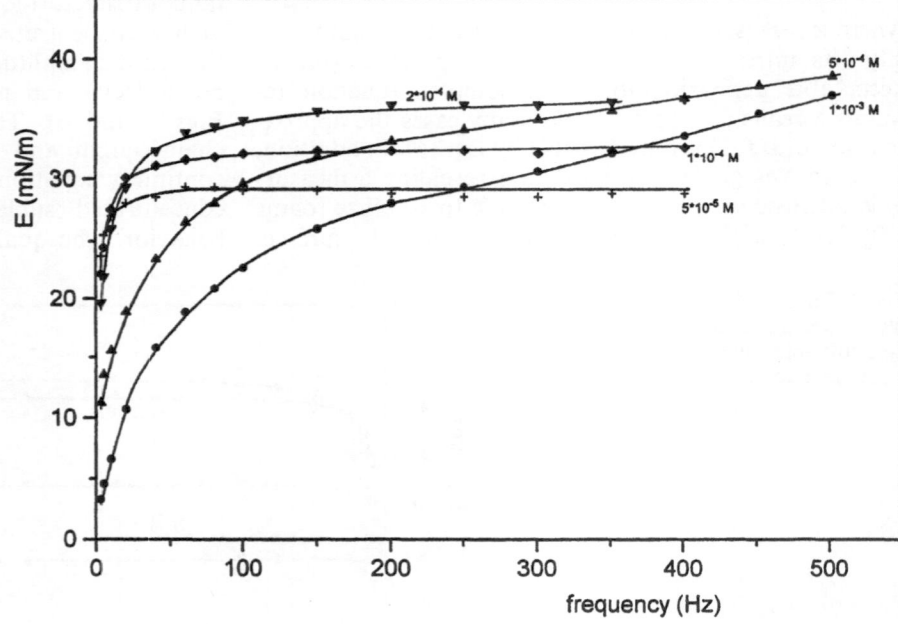

dynamics and its quantitative determination are not yet clear and one has to take into account the general storage and loss modulus:

$$\varepsilon(f,c) = G'(f,c) + iG''(f,c) . \tag{7}$$

The investigation of special mixtures, e.g. polyelectrolyte/surfactant mixtures, leads to similar problems. In these cases the determination of the surface dilational

modulus is also the prerequisite for understanding the mechanism.

Conclusions

The dynamic behavior of a complex fluid system is to a high degree determined by the rheological properties of its internal and external surfaces or interfaces. Reliable

Fig. 5 Phase angle of the surface dilational modulus of various concentrated n-nonyldimethylphosphine oxide solutions

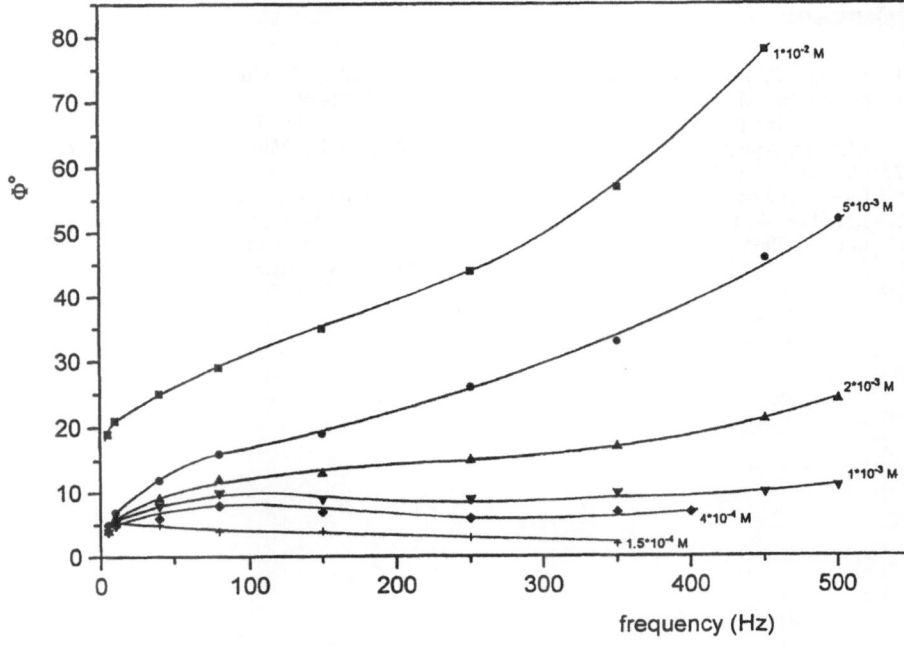

Fig. 6 Amount of the surface dilational modulus of various concentrated n-nonyldimethyl-phosphine oxide solutions

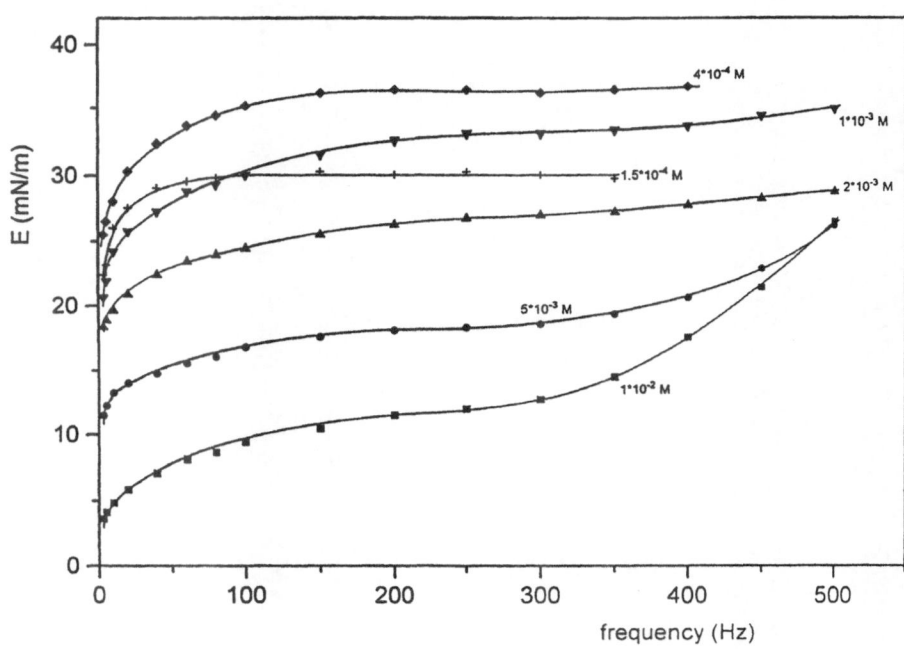

measurements of these properties are the precondition for the investigation of complex fluid systems. Using a new version of the oscillating bubble method we studied the surface dilational properties of various surfactant solutions. On the basis of these results we have verified different rheological models and specified their parameters. It was possible to demonstrate the influence of an intrinsic surface dilational viscosity that special solutions exhibit. This effect can be interpreted as a consequence of the molecular exchange processes near the surface under non-equilibrium conditions. However, not all results can be described by known models. Hence, further investigations are required to explain these experiments.

References

1. Dukhin S, Kretzschmar G, Miller R (1995) In: Möbius D, Miller R (eds) Studies in interface science, vol 1. Elsevier, Amsterdam, p 1
2. Noskov BA (1996) Adv Colloid Interface Sci 69:63
3. Bonfillon A, Langevin D (1994) Langmuir 11:2965
4. Earnshaw JC, Hughes CJ (1991) Langmuir 7:2419
5. Wantke KD, Fruhner H (1998) In: Möbius D, Miller R (eds) Studies in interface science, vol 6. Elsevier, Amsterdam, pp 327–365
6. Wantke KD, Fruhner H, Fang J, Lunkenheimer K (1998) J Colloid Interface Sci 208:34
7. Lucassen J, Hansen RS (1967) J Colloid Interface Sci 23:319
8. Lucassen J, van den Tempel M (1972) Chem Eng Sci 27:1283
9. Lucassen J, van den Tempel M (1972) J Colloid Interface Sci 41: 491
10. Fruhner H, Wantke KD, Lunkenheimer K (2000) Colloids Surf A 162:193

Progr Colloid Polym Sci (2000) 115 : 233–237
© Springer-Verlag 2000

S. Siegel
M. Kindermann
M. Regenbrecht
D. Vollhardt
G. von Kiedrowski

Molecular recognition of a dissolved carboxylate by amidinium monolayers at the air-water interface

S. Siegel · M. Regenbrecht
D. Vollhardt (✉)
Max-Planck-Institut für Kolloid- und
Grenzflächenforschung
14424 Potsdam/Golm, Germany

M. Kindermann · G. von Kiedrowski
Ruhr-Universität Bochum
Organische Chemie 1
44780 Bochum, Germany

Abstract The specific properties of a molecular recognition system based on the acid-base interactions of amidinium and carboxylate components are studied. The cationic heptadecylbenzamidinium chloride spread as a Langmuir monolayer is the "host" component and sodium phenylacetate dissolved in the aqueous subphase is the "guest" component. The interaction of the heptadecylbenzamidinium chloride with the dissolved carboxylate causes drastic changes in the surface properties. The specific features of the surface film of the amidinium-carboxylate complex are characterised by coupling the results of the surface pressure-area (π-A) isotherms with the Brewster angle microscopy and atomic force microscopy studies. In the plateau region of the π-A isotherms of the amidinium-carboxylate monolayer, round domains consisting of numerous filigree branches are formed. At higher resolution, strings of molecular thickness are observed which with further compression increasingly overgrow with a second molecular layer.

Key words Molecular recognition · Langmuir monolayers · Amidinium-carboxylate complex · Brewster angle microscopy · Atomic force microscopy

Introduction

An important goal of supramolecular chemistry is to find methods to control the assembly of molecules into larger structures [1, 2]. One approach is to use the air-solution interface to regulate the assembly process by incorporating strong directional interactions for the manufacture of thin films [3]. Based on the principle of molecular recognition, the specific composition of two various molecular components should result in systems with new properties and functions [4, 5]. There are interesting studies on the formation of two-dimensional highly organised structures by molecular-specific recognition on the complementary of hydrogen bonding [6, 7].

In this work we focus on the construction of surface films using acid-base interactions to control the assembly. In the simple molecular recognition model studied, a stable "host" monolayer is spread on the aqueous surface, while the "guest" component is dissolved in the aqueous subphase.

Materials and methods

The cationic heptadecylbenzamidinium chloride spread as a Langmuir monolayer is the "host" component. Sodium phenylacetate, used as "guest" component, was dissolved in the aqueous subphase.

The synthesis of p-heptadecylbenzamidinium chloride commenced with a Wittig reaction of 4-formylbenzonitrile with the appropriate long-chain triphenylphosphonium bromide to give the styrene derivative which was hydrogenated (10% Pd/C, H_2, MeOH) to the corresponding saturated compound. An efficient way was found to convert, in a single step, aromatic nitriles to amidines by reaction with methylchloroaluminium amide [8]. Prior to use, the aluminium reagent was prepared by addition of NH_4Cl to commercially available trimethylaluminium.

The long-chain heptadecylbenzamidinium chloride, dissolved in chloroform/ethanol (4:1), was spread on both a clean water surface

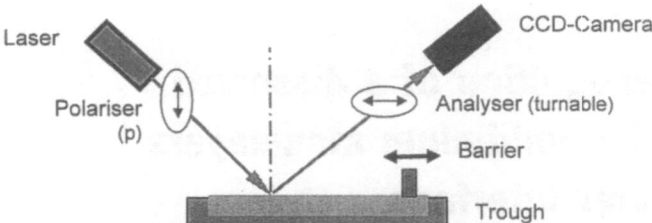

Fig. 1 Schematic presentation of the set-up of the Langmuir film balance and the Brewster angle microscope

and on water containing 1 mmol of the soluble component. The experimental set-up is described in detail elsewhere [9]. All studies were performed at 20 °C using a thermostated Langmuir film balance. The surface pressure π was measured with a Wilhelmy plate of filter paper. The isotherms were recorded at a compression speed of 5 Å^2 per molecule per min.

The imaging of the surface layer was performed with a Brewster angle microscope (NFT, Göttingen, Germany) mounted on the film balance. A schematic diagram of an experimental set-up with a BAM 2 Brewster angle microscope is shown in Fig. 1. A green laser is used so that images with a resolution of about 3 μm can be produced. A special scanning technique permits sharp images, although the angle of view is not perpendicular. For atomic force microscopy (AFM) studies after the recognition process, the Langmuir films were deposited onto Si wafers using the Langmuir-Blodgett (LB) technique because a direct observation at the aqueous surface is impossible. The investigations were performed using a NanoScope III (Digital Instruments, Calif.).

Results and discussion

Figure 2 shows the surface pressure-area (π-A) isotherms of heptadecylbenzamidinium chloride monolayer

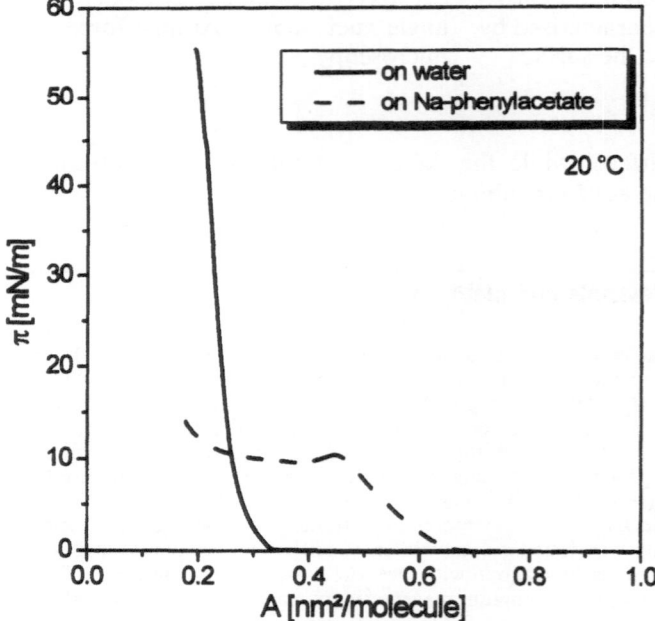

Fig. 2 π-A isotherms of heptadecylbenzamidinium chloride on water and on a 1 mM aqueous solution of sodium phenylacetate

on clean water and on a 1 mM aqueous solution of sodium phenylacetate. As can be seen, heptadecylbenzamidinium chloride forms stable monolayers on water. The absence of a plateau region in the isotherm indicates that these monolayers are already spreading in a state of two-phase coexistence between a non-

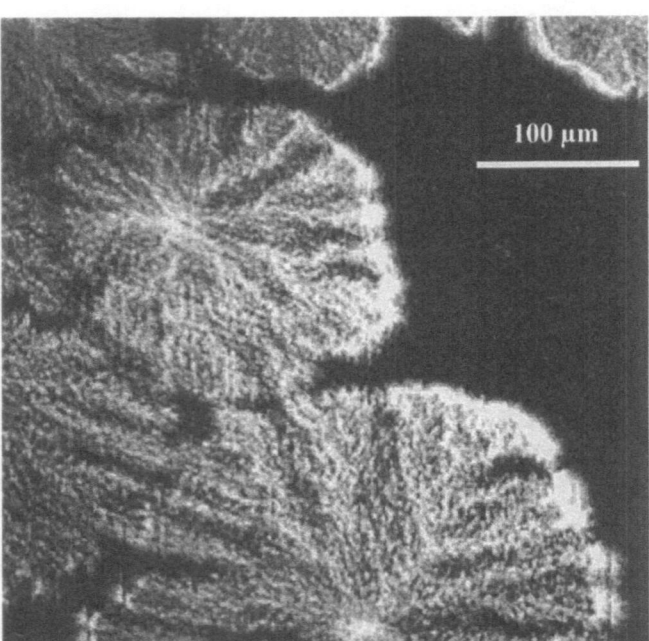

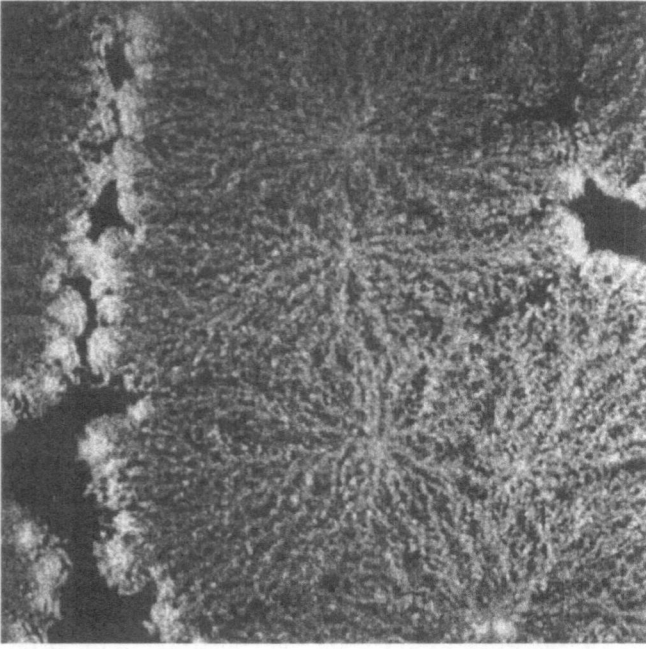

Fig. 3 BAM images of a heptadecylbenzamidinium chloride monolayer on a 1 mM aqueous solution of sodium phenylacetate. The images are taken within the plateau region

Fig. 4 AFM images of a heptadecylbenzamidinium chloride
layer transferred onto a silicon
wafer. The topography of the
domains from Fig. 3 is visible in
two different scales

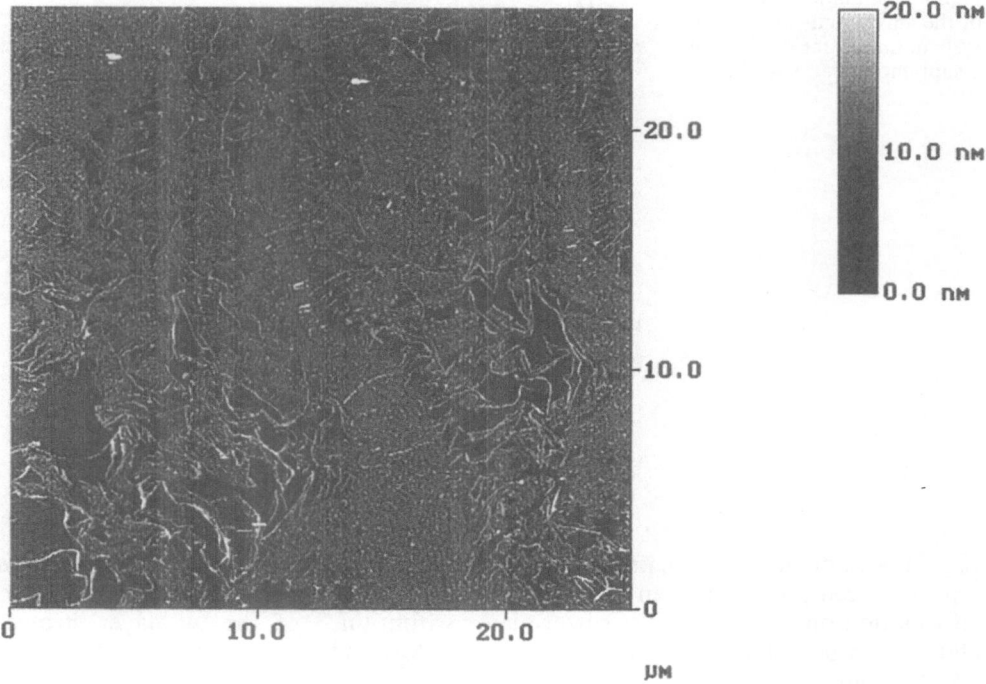

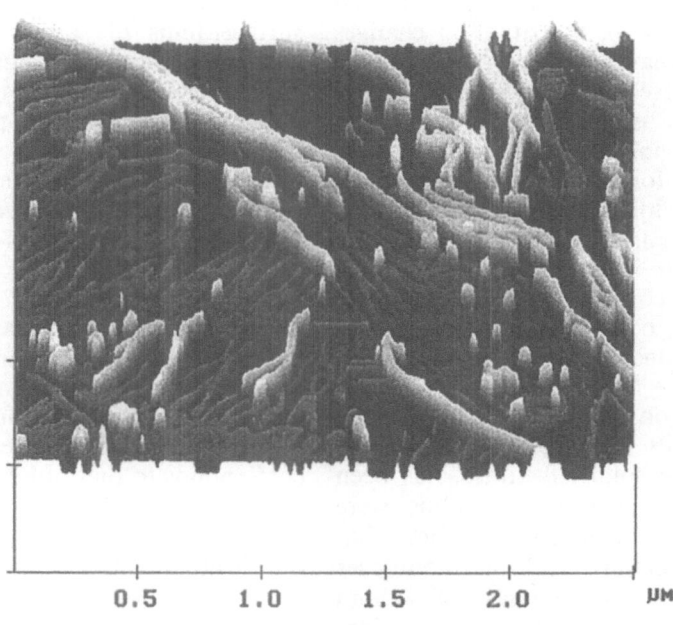

textured condensed state and the surrounding gaseous
state. This is supported by Brewster angle microscopy
(BAM) studies, which are not shown here. At all
pressures for $\pi > 0$ mN/m, a homogeneously reflecting
condensed phase is formed.

The behaviour of the heptadecylbenzamidinium chloride monolayer on the solution of the dissolved "guest"
component sodium phenylacetate is completely different.
This concerns not only the π-A isotherms, but also the
area per molecule at tightest packing is much larger (see

Fig. 5 Possibilities interaction of the amidinium-carboxylate system: **a** supramolecular pair; **b** supramolecular assembly

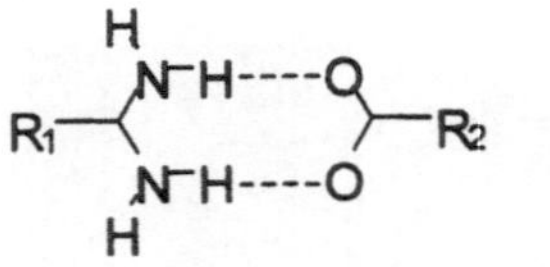

Fig. 2). At a definite surface pressure of about 10 mN/m, a pronounced plateau is seen in the isotherm. The best information on the state of the surface layer within this plateau can be obtained by a coupling of the BAM and AFM studies.

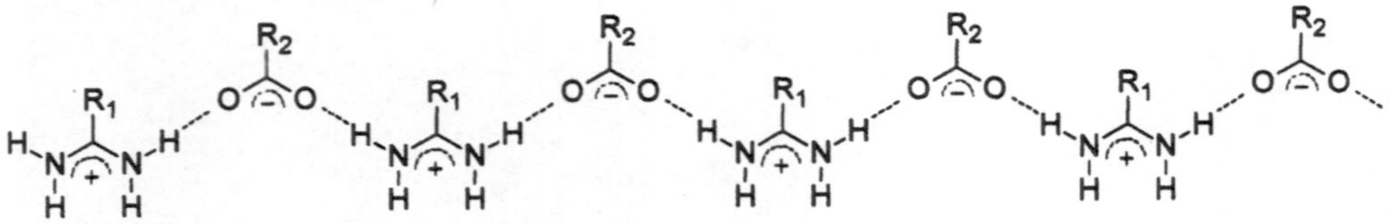

The BAM images (Fig. 3) are taken within this pressure plateau region of the π-A isotherm. Whereas the Langmuir monolayer of heptadecylbenzamidinium chloride on pure water forms a homogeneous reflecting condensed monolayer at $\pi > 0$, the situation changes drastically when the monolayer can interact with the sodium phenylacetate dissolved in the aqueous subphase. After the kink point in the isotherm, regularly shaped domains having a pronounced texture are formed. Figure 3 shows a snapshot for an advanced state of the domain growth. The round domains consist of numerous filigree branches having small gaps between them. The domains grow from a centre in all directions to a large size and finally they make contact with each other. As can be clearly seen (Fig. 3, below), the domains merge into a system with many branches, but the initial centres remain visible.

Additional information on the domain texture, such as thickness, textural details at higher resolution, and stability after transfer onto solid substrates, has been obtained by AFM studies. The surface films were prepared in the same way as for the BAM studies. That means the heptadecylbenzamidinium chloride monolayer was compressed on the aqueous subphase of sodium phenylacetate to molecular areas of the plateau region. Then the surface film was transferred onto a silicon wafer.

Figure 4 shows that the filigree texture of the domain strings remain preserved in spite of the transfer procedure. The strings can become dense (Fig. 4, above), but also some free gaps remain visible. More information provides the topography of the surface film on a larger scale (Fig. 4, below). Strings of molecular thickness are partly overgrown by a second molecular layer. The ratio of the two-layer strings increases within the plateau region of the π-A isotherm with compression to smaller areas.

The results of the π-A isotherms and the BAM and AFM measurements show that owing to molecular recognition of the dissolved sodium phenylacetate by the heptadecylbenzamidinium chloride monolayers, surface films of specific properties are formed. The condensed phase domains consist of well-defined substructures with a defined thickness of 1 or 2 layers.

Figure 5 shows two possibilities for the kinds of interactions which can be expected between the amidinium and carboxylate components. Although the above studies provide unequivocal evidence for molecular recognition based on acid-base interactions, they cannot clarify the bonding between both components. However, the favoured formation of bilayers overgrowing the filigree strings in the plateau region of the isotherms indicates interdigitated amidinium-carboxylate interaction. Further experimental techniques, such as grazing incidence X-ray diffraction (GIXD) or IR spectroscopy, must be used to clarify the question whether the bonding corresponds to that of interdigitated films.

Conclusions

The combination of π-A isotherms and BAM and AFM studies can provide detailed information on molecular recognition models based on acid-base interactions. The spread amphiphilic compound heptadecylbenzamidinium chloride forms stable non-textured condensed monolayers on water. The monolayer properties change drastically if the aqueous subphase contains sodium phenylacetate. Already the π-A

isotherms indicate remarkable changes in the properties of the heptadecylbenzamidinium chloride monolayer. At a pronounced plateau pressure, condensed phase domains with a specific texture start to grow. In the present case, round domains consisting of numerous filigree branches merge into a system with many branches after growth to large size. The molecules are ordered and tilted.

The topography of these surface films, visualised in larger scale by AFM, reveals strings of molecular thickness increasingly overgrown by a second molecular layer with further compression. The specific features of the studied molecular recognition system demonstrate that surface films can be constructed based on the acid-base interactions of amidinium and carboxylate components.

References

1. Lehn J-M (1990) Angew Chem 102:1347
2. Desiraju GR (1995) Angew Chem 107:2541
3. Kusmenko I, Buller R, Bouwman WG, Kjaer K, Als-Nielsen J, Lahav M, Leiserowitz L (1996) Science 274:2046
4. Zerkowski JA, Mathias JP, Whitesides GM (1994) J Am Chem Soc 116:4305
5. Zerkowski JA, MacDonald JC, Seto CT, Wierda DA, Whitesides GM (1994) J Am Chem Soc 116:2382
6. Kimizuka N, Kawasaki T, Kunitake T (1993) J Am Chem Soc 115:4387
7. Weck M, Fink R, Ringsdorf H (1997) Langmuir 13:3515
8. Moss RA, Ma W, Merrer DC, Xue S (1995) Tetrahedron Lett 36:8761
9. Vollhardt D (1996) Adv Colloid Interface Sci 64:143

Progr Colloid Polym Sci (2000) 115:238–242
© Springer-Verlag 2000

Formation of rigid nanodiscs: edge formation and molecular separation

M. Dubois
L. Belloni
Th. Zemb
B. Demé
Th. Gulik-Krzywicki

M. Dubois · L. Belloni · Th. Zemb (✉)
Service de Chimie Moléculaire (CEA/
Saclay), 91191 Gif sur Yvette, France
e-mail: zemb@nanga.saclay.cea.fr
Tel.: +33-01-69086328
Fax: +33-01-69086640

B. Demé
Institut Laue-Langevin
38042 Grenoble, France

Th. Gulik-Krzywicki
Centre de Génétique Moléculaire
CNRS, 91190 Gif sur Yvette, France

Abstract We show that the control of size in catanionic nanodiscs relies on the balance between charge excess favoring the disc edge and frozen ion pair formation favoring rigid faces. In the absence of salt, average diameters in the range of 60 nm to 3 μm have been obtained by varying the molar ratio x (cationic/anionic + cationic) from $x = 0.39$ to 0.46. The cost in entropy of mixing associated with partial molecular separation inside each crystallized nanodisc is lower than the work associated with osmotic pressure in the region of the phase diagram where catanionic nanodiscs are produced.

Key words Catanionic surfactant · Nanodics · Osmotic pressure

Introduction

Catanionic mixtures of two amphiphilic molecules (myristic acid and cetyltrimethylammonium hydroxyde), with H^+ and OH^- as counter-ions, form swollen lamellar phases. The bilayers composing these lamellar phases are made of amphiphilic ion pairs. These infinite bilayers fragment upon dilution and form rigid nanodiscs in the absence of salt [1]. The large in-plane persistence length, exceeding one micrometer, is due to the sandwich structure of the nanodiscs: a two-dimensional alternated charge lattice is formed on each face. The core of the nanodiscs is made of frozen hydrocarbon chains [1]. Hence, the Young's modulus of this new type of colloidal object exceeds 10^8 Pa [2].

The formation of nanodiscs is reversible with temperature [1], since above the ion pair melting temperature (composition dependent, but of the order of 30 °C [3]), closed vesicles are formed. The high-temperature thermodynamically stable state is a dispersion of small unilamellar vesicles. This dispersion is similar to the equilibrium microstructure which has been described in the presence of excess salt when anionic and cationic components are mixed with their counter-ions [4, 5, 6]. At concentrations of around 1% by weight, a two-phase regime is observed by electron microscopy where large crystallites of a lamellar phase are in equilibrium with a dispersions of discs (Fig. 1).

Statistics obtained after explicit counting of disc sizes in the electron microscopy pictures have demonstrated a large variation in disc size induced by only a small variation in composition [2]. The situation is similar to the ternary system which has been described as producing "bicelles" by Benedek and co-workers [7]. These authors used a combination of short-chain and long-chain neutral phospholipids. "Bicelles" are uncharged flat rigid discs, with curved edges [7]. In their classical work, Benedek and co-workers have shown that the available volume of short-chain phospholipids combined with the volume of long-chain phospholipids quantitatively explains the observed bicelle diameter, even in cases where nanodisc formation is not reversible and is only a metastable state in the ternary phase diagram.

The aim of the study presented here is to evaluate the charge separation needed to form disc edges and faces. This allows an evaluation of the molecular composition of both edges and faces. From the composition of these two

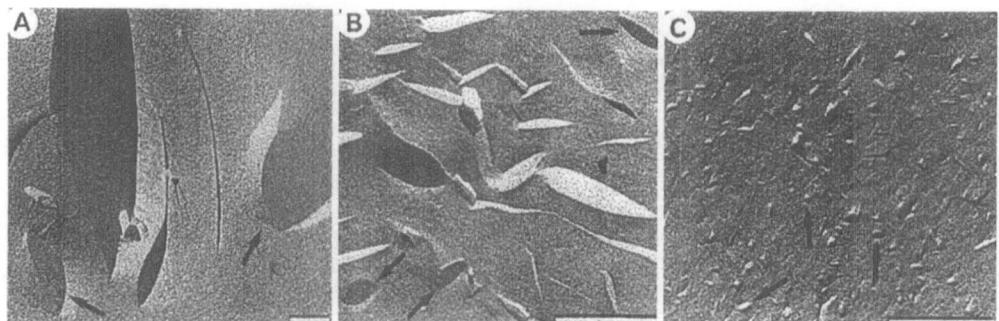

Fig. 1 Images obtained after cryofracture of nanodisc dispersions of different sizes. The initial weight content of the catanionic surfactant in solution is of the order of 10 g/l. With a molar ratio $x = 0.45$ on the basic side of the phase triangle, one obtains an average disc size of 2–3 μm (**A**); for $x = 0.43$ the average nanodisc size is 250 nm (**B**). In the case of a larger excess of cationic component ($x = 0.39$) the average nanodisc size is 30 nm (**C**). The *bar* represents 500 nm

parts of the discs, the cost in entropy can be compared to osmotic pressure due to electrostatic repulsions.

Composition of edge and faces

A schematic view of a nanodisc of thickness $2t = 43$ Å (measured) and radius R is shown in Fig. 2. The two faces (total surfaces S_f) contains N_f surfactant molecules of composition x_f, each occupying the surface $\sigma_f = 25$ Å^2 (area per headgroup for frozen chains involved in ion pairs): $S_f = 2\pi R^2 = N_f \sigma_f$. The edge (surface S_e) contains N_e surfactant molecules of composition x_e with a area per chain $\sigma_e \approx 50$ Å^2 (twice the previous value for molecules in excess): $S_e = 2\pi^2 Rt = N_e \sigma_e$. The total number of

molecules per disc is $N = N_f + N_e$ of global composition $x = (N_f x_f + N_e x_e)/N$. The disc size is a function of the three compositions, $R(x_f, x_e, x)$. Assuming fixed values for the composition x_f, x_e (independant of the total composition), the best fit of the experimental data $R(x)$ are obtained with $x_f = 0.454$ and $x_e = 0.18$ (Fig. 3). Table 1 gives the different parameters describing the disc for each measured size.

The faces are close to equimolar equilibrium, which ensures that the faces are less charged than the edge, a physical situation opposite to the case of clay particles, where faces are more hydrophilic than the edge. The composition of the edge is consistent with the molar ratio found in the neighboring phase of giant cylindrical micelles [8]. Indeed, the disc edge shape is close to a long wormlike micelle cut in half. We find, by fitting, a disc edge composition (0.18) close to the ratio where the single phase of large wormlike micelles has the lowest free energy of formation. The molar ratio where cylindrical micelles are observed is between 0 and 0.3 for myristic acid. Therefore, it is consistent to find a disc edge composition of 0.18, which is in the middle of the L_1 phase region.

Fig. 2 Schematic view of a nanodisc: thickness of bilayer is $2t$, disc radius is R. Edge composition with a higher molar ratio of the excess cationic component is the proposed explanation for disc size control via composition

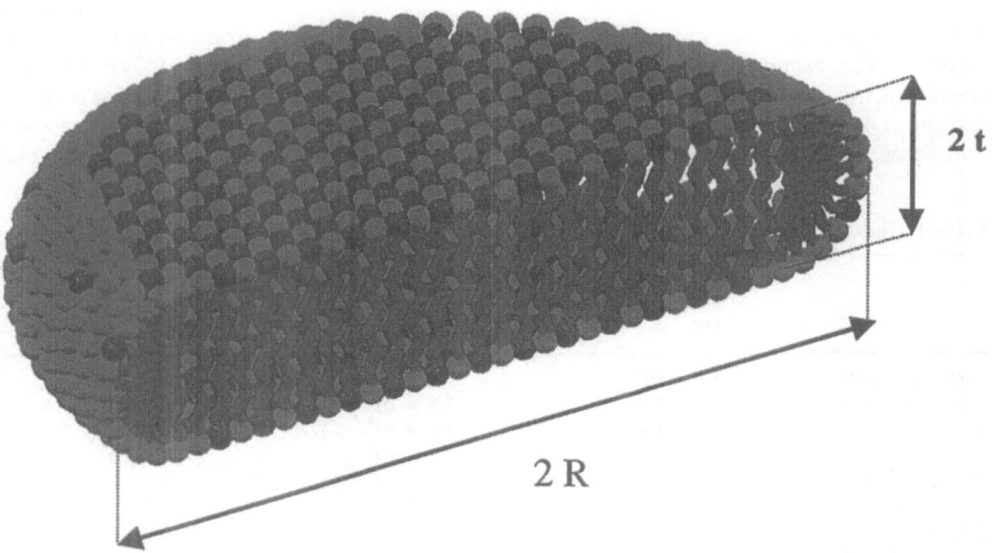

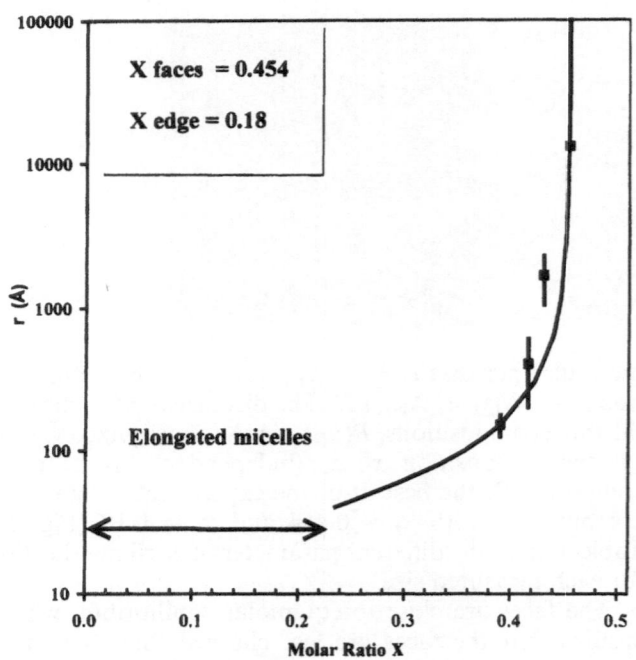

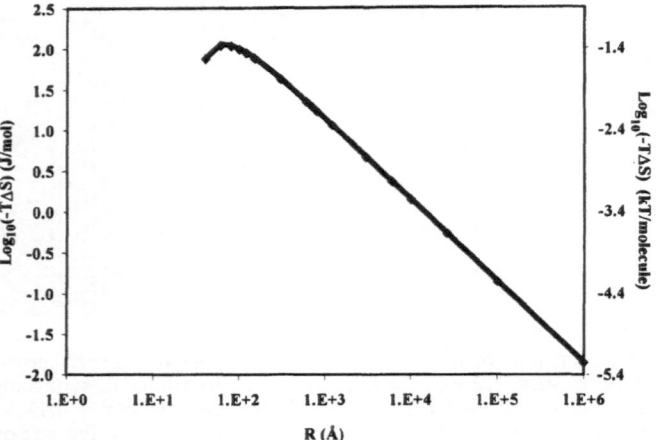

Fig. 3 Observed equilibrium radii of nanodiscs compared to the evolution expected (*line*) if the composition of edges and faces are 0.18 and 0.454, respectively. The region where micelles are found instead of nanodiscs is indicated

Fig. 4 Free energy cost of molecular separation $T\Delta S$ needed to form edges and faces of different molar ratio, as a function of nanodisc diameter R

of magnitude of the free energy cost of molecular separation is less than 0.1 kJ/mol.

How does this figure compare to the main interaction energy in these solutions in the absence of salt? We consider now the case where discs are found in equilibrium with a partially swollen charged catanionic lamellar phase. Using the two control variables, i.e. surface charge controlled by molar ratio and weight fraction controlling the interlayer distance, the molar ratio phase diagram can be remapped in terms of water layer thickness between bilayers and area per charge. This is shown in Fig. 5. On the same diagram are drawn isobar lines of constant osmotic pressure, calculated assuming parallel lamellae or parallel cylinders geometries. In order to allow comparison, the molecular volume is identical in both geometries while the area per head group is twice as large in cylinders than in plans. In terms of free energy associated with $\Pi\Delta V_{m}$, where V_{m} is the molar volume, the interaction term is predominant over charge separation effects, as can be seen by comparing Figs. 4 and 5.

From the comparison of the location of the discs and lamellar phase in the phase diagram and the ab initio

Neglecting charge effects, the cost in free energy associated with the entropy of mixing per molecule is derived from:

$$
-\frac{\Delta S}{k} = \frac{N_e}{N}[x_e \ln x_e + (1 - x_e)\ln(1 - x_e)]
$$
$$
+ \frac{N_f}{N}[x_f \ln x_f + (1 - x_f)\ln(1 - x_f)]
$$
$$
- [x \ln x + (1 - x)\ln(1 - x)] \qquad (1)
$$

This quantity can be calculated from the values of Table 1. The result is shown in Fig. 4, where the free energy of mixing is plotted versus the molar ratio of the sample, i.e. the macroscopic value, averaged over edge and faces. In the range where nanodiscs form, the order

Table 1 Characterization of nanodiscs with a molar ratio of 0.454 on the faces and 0.18 on the edge (see text for the notation)

	Disc radius (Å)			
	150	400	1650	13000
Surf. faces (Å^2)	1.0×10^5	9.0×10^5	1.7×10^7	1.1×10^9
Surf. edge (Å^2)	6.4×10^4	1.7×10^5	7.0×10^5	5.5×10^6
N_f	4.1×10^3	3.6×10^4	6.7×10^5	4.2×10^7
N_e	1.3×10^3	3.4×10^3	1.4×10^4	1.1×10^5
$N = N_f + N_e$	5.4×10^3	3.9×10^4	6.8×10^5	4.2×10^7
X calculated	0.390	0.430	0.448	0.453
X measured	0.390	0.416	0.430	0.454

Fig. 5 Using total surfactant volume fraction and surface per charge (controlled by the composition) as variables, this phase diagram superposes the location of the three states found for nanodiscs: U is the unbound dispersion of interacting nanodiscs; L_β is the lamellar phase with frozen chains; L_1 is a solution of wormlike charged micelles. Isobar lines of constant osmotic pressure (labeled in Pa) are drawn on the same diagram for flat bilayers (*thick line*) and for parallel cylinders of same molar volume and double area per headgroup (*dashed lines*)

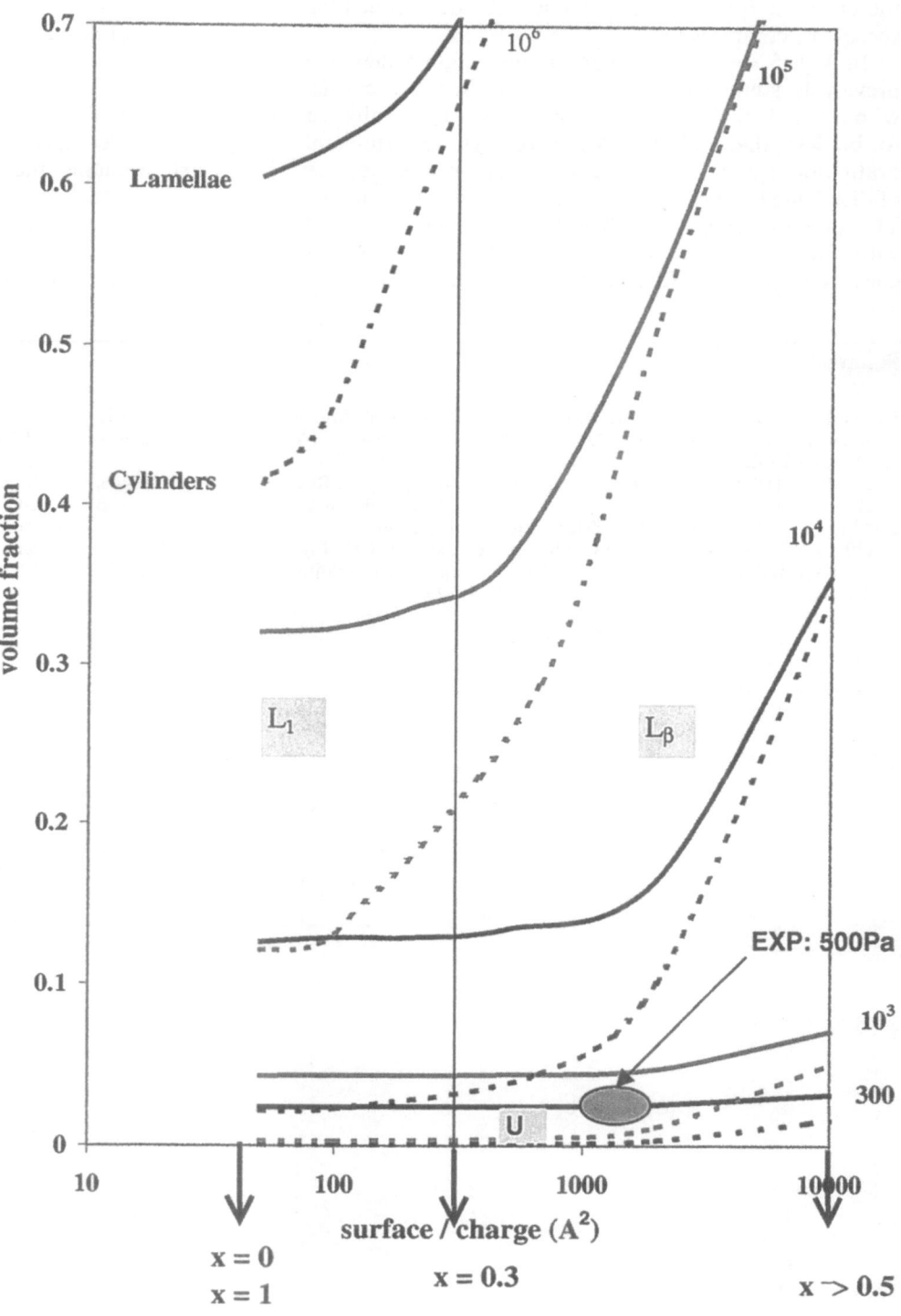

evaluations of the free energy, one expects that the bilayers fragment into discs when the electrostatic term of the free energy corresponds to a pressure of the order of 500 Pa. At a mole fraction $x = 0.45$ and a global concentration of 1%, the sample spontaneously microphase separates into stacks of lamellae as shown in Fig. 1. The osmotic pressure of this sample, as measured by a conventional osmometer, is between 400 and 600 Pa. The relatively low precision is due to partial permeability of the membrane to the catanionic mixture;

therefore the osmotic pressure has to be determined via successive comparisons.

In Fig. 5 we can see that in the region where the previously studied nanodiscs of controlled size coexist with a lamellar phase, the osmotic pressure is evaluated to be less than 300 Pa. Moreover, an experimental evaluation using a membrane osmometer of the pressure of disc-lamellar phase coexistence region has been made. The pressure range is 500 ± 200 Pa, well correlated with the theoretical expectation in the region where coexistence is found (dashed region in Fig. 5).

From the known area per charge and residual ionic strength of solution, we derive an estimation of the surface potential of the faces of 200 mV using a mole fraction for the faces of $x = 0.45$. From the molar ratio of the faces, which constitutes the major part of the disc volume, we expect a new type of packing of the hydrocarbon chains and therefore that the position of the large angle peak corresponding to frozen chains is in unconventional position and shape, since ion pairing should induce shifts towards large q or a superstructure at lower q. Investigations of this point are in progress.

References

1. Zemb Th, Dubois M, Demé B, Gulik-Krzywicki Th (1999) Science 283:816
2. Dubois M, Gulik-Krzywicki Th, Demé B, Zemb Th (1998) CR Acad Sci Paris Ser II C 567–575
3. Martinet AC, Dubois M, Zemb Th (1998) Rapport CEA-R-5820 (available at CEA-DIST, Saclay, F91191 Gif/Yvette)
4. Kaler EW, Murthy AK, Rodriguez BE, Zasadzinski JAN (1989) Science 245:1371–1374
5. Marques EF, Regev O, Khan A, Miguel MG, Lindman B (1998) J Phys Chem B 102:6746
6. Sanders CR, Prosser RS (1998) Structure with folding and design 6:1227–1234
7. Mazer NA, Benedek GB, Carey MC (1980) Biochemistry 19:601–615
8. Dubois M, Gulik-Krzywicki Th, Demé B, Zemb Th (1999) In: Manne S, Warr GG (eds) Supramolecular structures in confined geometrics. (ACS symposium series 736) American Chemical Society, Washington, pp 86–101

Progr Colloid Polym Sci (2000) 115:243–248
© Springer-Verlag 2000

METHODS

Y. Chevalier
Z. Elbhiri
J.-M. Chovelon
N. Jaffrezic-Renault

Chemically grafted field-effect transistors for the recognition of ionic species in aqueous solutions

Y. Chevalier (✉) · Z. Elbhiri
Laboratoire des Matériaux Organiques à
Propriétés Spécifiques, UMR 5041 CNRS
Université de Savoie, BP 24
69390 Vernaison, France
e-mail: yves.chevalier@lmops.cnrs.fr
Tel.: +33-4-78022271
Fax: +33-4-78027187

Z. Elbhiri · J.-M. Chovelon
N. Jaffrezic-Renault
Ingénierie et Fonctionnalisation des
Surfaces, IFoS, École Centrale de Lyon
BP 163, 69131 Écully, France

Abstract The direct chemical grafting of complexing agents onto the silica grid of field-effect transistors (FETs) was realized, allowing the selective complexation of ions at the surface of the FETs. The chemical recognition of ions at the surface provides sensitivity to the presence of such ions. Electrochemical devices known as ion-sensitive field-effect transistors (ISFETs) for the quantitative analysis of ions were elaborated and their electrochemical behaviour was studied. Thus, the grafting of the crown ether benzo-15-crown-5 onto the silica grid brought about sensitivity to potassium ions that the bare silica had not. In the same way, the grafting of phosphonate groups led to ISFET devices sensitive to calcium ions. The electrical response was related to the surface potential at the silica-water interface and could be rationalized by means of a modified site-binding model.

Key words Electrochemical potential · Adsorption · Chemical sensors · Ion sensitive field-effect transistors

Introduction

The in situ detection of ionic species is a key in various domains such as the control of industrial processes, the analysis of aqueous wastes or the in situ monitoring of physiological events involving ions in medicine [1]. Robust devices having a fast response time and a low drift are required; their size has also to be reduced for medicinal applications. The technology of conventional ion-selective electrodes (ISEs) allows the fabrication of microelectrodes which can be placed at the end of a catheter for in situ measurements of physiologically relevant ions such as sodium, potassium or calcium. Conventional ISEs are fragile and their miniaturization is expensive; very small devices cannot be realized. Ion-sensitive devices derived from the technology of field-effect transistors (FETs) introduced by Bergveld [2, 3] may allow real progress since their size may reach some μm^2 and because the microelectronics technology allows the collective fabrication of robust solid-state sensors at low cost [4]. Ion-sensitive field-effect transistors (ISFETs) are becoming a reliable alternative to ISEs and pH meters equipped with an ISFET as a H^+-sensitive element are currently on the market [5].

As a schematic principle of an ISFET, the FET measures the electric field created when ions adsorb from the solution to the surface of the device. An ISFET is then sensitive to ions which are able to adsorb at its surface. Such ions are called "potential-determining ions" in electrochemistry.

According to this principle, pH-sensitive ISFETs could be realized since the surface potential of most insulators currently available (SiO_2, Si_3N_4, Al_2O_3, Ta_2O_5) depends on the concentration of H^+. Alkaline and alkaline-earth ions are not able to bind at the silica surface. The detection of ions different from H^+ then requires a modification of the surface of the silica insulator where surface sites for the binding of ions are created. The first attempts at the sensitization of ISFET to ions other than H^+ consisted in the immobilization of ion-complexing species (ionophores) into a thin polymer film deposited at the FET's surface [6–8]. This technique, derived from the ISE's membrane technology, suffers the same drawbacks as for ISEs: the need for a plastifier when PVC is used, the poor adhesion of the polymer film to the silica surface, and the slow leaking of the plastifier and the ionophore into the aqueous solutions during the utilization [9]. This latter phenomenon is always present even when the ionophore has a low solubility in water. The consequences of these phenomena are drifts occurring in the course of measurements, a continuous loss of sensitivity and the need for frequent calibration of the devices.

In order to overcome most of these problems, an alternative to the immobilization into polymer coatings has been developed where the complexing molecules were chemically grafted onto the silica surface [9, 10]. ISFETs sensitive to potassium and to calcium ions could be elaborated in this way [11–13].

We are presenting in this paper the elaboration of such ISFET devices by means of chemical grafting onto the silica grid of the FETs, and investigations into the physical chemistry involved in the detection of ionic species by ISFETs.

Principle of an ISFET

An ISFET is realized by a modification of a conventional MOSFET (metal-oxide-semiconductor field-effect transistor) [14]. In a MOSFET (Fig. 1), two spots of a silicon chip doped by n-type charge carriers (the source and the drain) are separated by a region of p-type silicon called the channel. This structure with a p-n interface is blocked. A significant current I_D between the source and the drain can only be obtained if the channel is depleted from the positive charge carriers by a positive potential

at the metal grid exceeding the threshold potential V_T. In this regime of inversion where $V_G \gg V_T$, the intrinsic n-type charge carriers are the majority in the channel and the current is given by:

$$I_D = \mu_e \frac{W}{L} C_i \left(V_G - V_T - \frac{V_D}{2} \right) V_D \qquad (1)$$

where μ_e = electron mobility, W = channel width, L = channel length and C_i = SiO$_2$ insulator capacity per unit area. At constant V_D, the potential of the grid V_G controls the current I_D. In an ISFET (Fig. 1), the ionic solution is put in place of the metal grid. The role of the potential V_G is then played by the surface potential of the silica insulator with respect to a reference electrode. If the potential of the reference electrode ensures the inversion regime conditions, the I_D current at constant V_D is a measurement of the potential at the silica-solution interface. The value of this potential is set by the potential-determining ions of silica. The ISFET is then a measurement device of the concentration of potential-determining ions of the insulator surface.

V_T is related to the potentials at the insulator-solution and the semiconductor-insulator (Si-SiO$_2$) interfaces. The potential drop at the Si-SiO$_2$ interface determines the electronic state of the channel, and thus the value of I_D. In the detection mode called "source-follower" where V_D and I_D are maintained constant, the Si-SiO$_2$ interface does not vary and any variation of V_G is related to potential variations at the SiO$_2$-solution interface only. Thus, an ISFET measures the potential ψ_0 of the SiO$_2$-solution interface [15, 16] where the recognition of ions takes place:

$$V_T = E_{ref} + \psi_0 + \chi$$
$$+ \text{constant potentials in the Si-SiO}_2 \text{ transducer}$$

$$(2)$$

Fig. 1 Schematic principles of the MOSFET and ISFET

Field-Effect Transitor, MOSFET

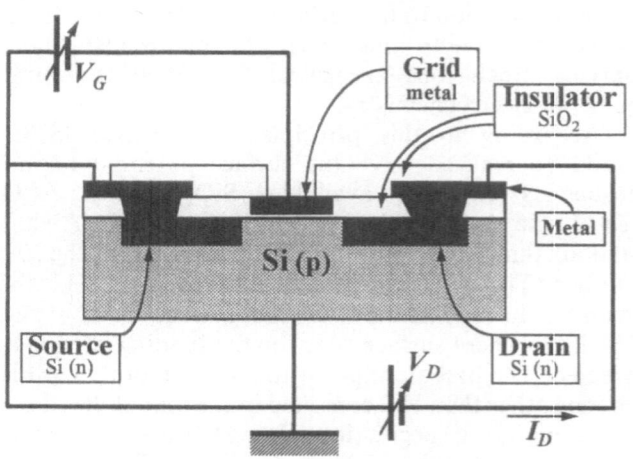

Ion-Sentive Field-Effect Transitor, ISFET

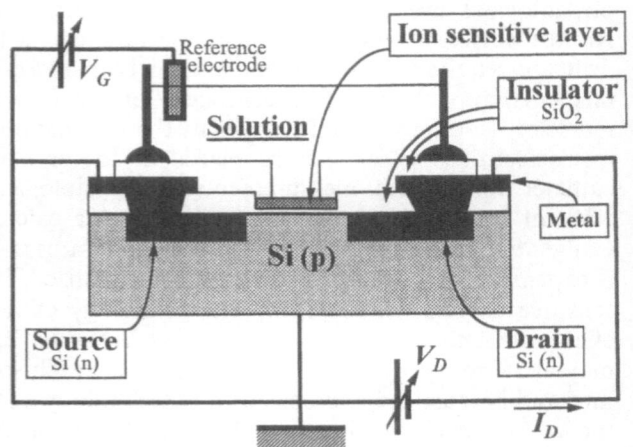

The reference electrode potential E_{ref} and the potential drop χ due to surface dipoles are eliminated by means of measurements in a differential mode. The difference of the electrical signals coming from a grafted and a non-grafted ISFET (bare silica) dipped in the same solution is measured. This mode of measurement eliminates most of the non-specific responses (to pH variations, temperature changes, etc.) and considerably reduces the drifts [17]. The measured voltage simply reads: $\Delta V_G = \Delta \psi_0$.

Potassium-sensitive ISFETs

The crown ether benzo-15-crown-5 ($B_{15}C_5$) was selected because of its known complexing ability towards K^+ and Na^+ ions. The grafting of $B_{15}C_5$ onto silica was carried out by means of a silane intermediate. Silylated crown ethers were synthesized for that purpose. The choice of the silane, which should be reactive towards the silanol (Si-OH) groups of the silica surface, is important and was investigated. A polycondensation side reaction takes place in the presence of residual water with trifunctional silanes (trimethoxysilanes, trichlorosilanes), which are widely used in other applications. A thick and hairy, ill-defined layer results from the polycondensation at the surface. In order to ensure a monolayer formation at the silica surface, monofunctional silanes were used in the present study, but their lesser reactivity as compared

to trifunctional ones had to be compensated by the choice of very reactive groups: chlorosilane or dimethylaminosilane for the grafting onto silica [18], and hydroxysilane for the grafting onto aminated silica [19]. The synthesized multifunctional molecules containing the $B_{15}C_5$ moiety and the various silane groups are shown in Fig. 2 together with their corresponding grafting reactions [12]. The efficiency of the grafting reactions could be controlled on silica powders (fumed silica) of large specific area for which several analytical methods were available (elemental analysis, IR and CP-MAS NMR spectroscopies), but the very small area of the silica grid of the ISFETs (10000 μm^2) did not allow such analyses. The transfer of grafting reaction processes from silica powders to the ISFET surface is still an open problem because the surfaces of fumed silica and of the thermal silica of ISFETs (obtained by oxidation of silicon) are quite different.

The electrical responses of ISFETs were measured at pH 10 and in a concentrated background electrolyte ($[N(CH_3)_4^+ \ Cl^-] = 0.5$ M). This electrolyte was chosen instead of the more usual NaCl or NaNO$_3$ because the $B_{15}C_5$ crown ether can bind to Na$^+$ ions.

The responses of ISFETs are often discussed with respect to the Nernst law by analogy with conventional ISEs. The responses of ISFETs as a function of log([K$^+$]) indeed resemble closely that of ISEs (Fig. 2): they are quasi-linear for concentrations of K$^+$ higher

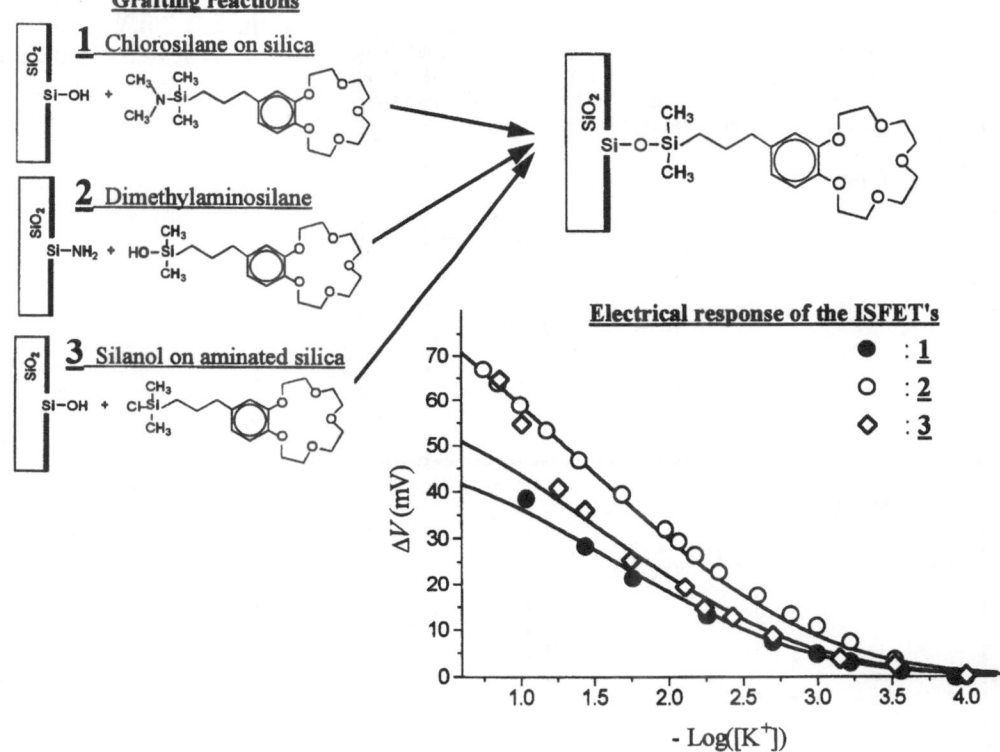

Fig. 2 *Left*: grafting reactions of the silylated $B_{15}C_5$ *Bottom right*: electrical responses of the ISFETs grafted according to reactions **1**, **2** and **3** in 0.5 M $N(CH_3)_4^+Cl^-$ background electrolyte at pH 10. The *solid lines* were calculated according to the site-binding model

than a threshold which is defined as the detection limit ($\sim 10^{-3}$ M in the present case), but the slopes are always lower than the 59 mV/p(K$^+$) of the Nernst law. The sensitivity of ISFETs depends on the type of grafting reaction in the order: dimethylaminosilane **2** > chlorosilane **1** (for the grafting reactions carried out onto silica), following the reactivity of the silanes [18]. The obvious advantage of using highly reactive silanes such as dimethylaminosilane **2** is a higher grafting density, which gives a better sensitivity. The larger grafting density of **2** could not be measured at the surface of the ISFETs but was easily measured on fumed silica powders by means of carbon elemental analysis; it is also well documented in the literature [18]. The grafting densities on ISFETs as estimated according to the site-binding model (presented in the third section, solid lines in Fig. 2) were also in the same order. The hydroxysilane **3** fell between **1** and **2**, but it could not be directly compared since the grafting reaction was carried out on aminated silica instead of silica.

Calcium-sensitive ISFETs

Ca^{2+}-selective ISFETs could be elaborated according to the same route as for K$^+$, but the phosphonate complexing group was selected instead of B$_{15}$C$_5$. The technology and measurement mode were the same. It was not possible to prepare a multifunctional molecule bearing both the phosphonate and a reactive silane group because of their chemical incompatibility. Cyclization and polycondensation reactions take place because the nucleophilic phosphonate group attacks the silane at the silicon atom. Two alternatives are proposed: the grafting of the phosphonate group by means of a silane intermediate may be performed in several reaction steps, or the phosphonate group has to be protected in the multifunctional molecule. The deprotection reaction should be as smooth as possible. Two grafting schemes pertaining to each path were checked.

In the multistep process (Fig. 3A), silica was grafted with the commercial 3-chloropropylchlorodimethylsilane in a first step; the second step was a chlorine to iodine exchange reaction, and sodium dimethylaminopropylphosphonate was reacted with the grafted silica in a third step [11]. The same reactions carried out on fumed silica could be monitored at the different steps by means of elemental analysis, IR and ^{13}C and ^{29}Si CP-MAS NMR [13]. The analyses have shown that the grafting of the chlorosilane in the first step was successful: the grafting densities of **G1** were found in the 3 μmol/m^2 range. However, about half of the grafts were lost during the second step and less than 10% of the original grafting density remained after the third step. This considerable loss of grafting density as further reactions were carried out in refluxing methanol containing ionic species was ascribed to the cleavage of the grafts at the Si-O-Si linkage. As a consequence, the ISFETs grafted by means of this process showed a very low electrical response to Ca^{2+} ions (Fig. 3). The response as a function of log([Ca^{2+}]) was not linear; the upper bound of the sensitivity, taken as the tangent at high Ca^{2+} concentration, was 7 mV/p(Ca^{2+}) [11].

Since repeated chemical reactions at the silica surface have to be avoided, a ready-to-graft molecule was synthesized where the phosphonate group was protected

Fig. 3 *Left*: chemical reactions of the multistep and one-step grafting processes. *Right*: electrical responses of the Ca^{2+}-sensitive ISFETs at pH 10 and in 0.5 M KCl background electrolyte. The *solid lines* were calculated according to the site-binding model

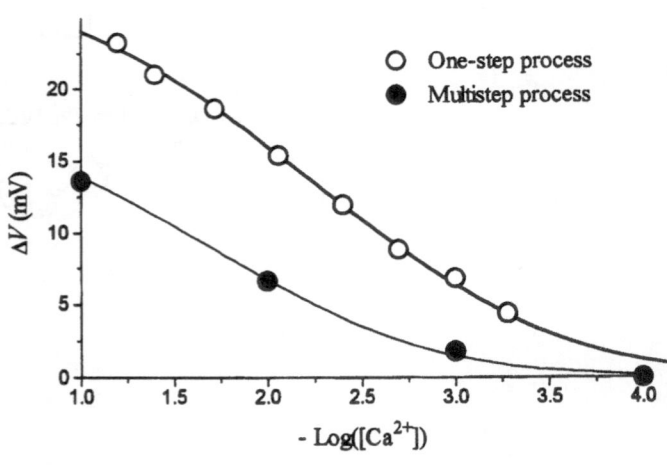

A: Multistep process

B: One-step process

as an ester function. The smooth deprotection of a phosphonic ester was performed with bromotrimethyl-silane (2 h at 50 °C); the resulting trimethylsilyl ester cleaved very rapidly when the ISFET was dipped into water at room temperature, yielding the corresponding phosphonic acid [20]. Thus, diethyl [3-(chlorodimethyl-silyl)propane]phosphonate was synthesized, together with the corresponding cyclic silyl ester which formed during the synthesis (Fig. 3B) [13]. The grafting of the cyclic silyl ester onto silica by means of ring opening by the silanol yields the same grafted group after deprotection. The electrical responses of the ISFETs were measured in differential mode at pH 10 in 0.5 M KCl as a background electrolyte. The ISFETs grafted according to the one-step process showed a larger electrical response than for the multistep process (Fig. 3). The sensitivity was 10 mV/p(Ca^{2+}), well below the 30 mV/p(Ca^{2+}) of the Nernst law for divalent ions. In contrast to the ISFETs grafted by means of the multistep process, the response was linear over a large concentration domain (10^{-3}–10^{-1} M); the detection limit was around 10^{-4} M.

The one-step grafting process clearly appears superior, but its practical application was more difficult. A multifunctional ready-to-graft molecule has to be synthesized and a suitable protective group has to be selected. On the one hand, the protective group should be compatible with the synthesis scheme and the grafting reaction onto silica. On the other hand, the deprotection reaction should be smooth since it should keep the molecules grafted. Moreover, the one-step process allowed us to graft a simpler chemical structure, while a supplementary functional group was added to the graft at each step in the multistep process. Thus, as the result of the last reaction step, a quaternary ammonium group was present in the graft; it was not useful for calcium detection but was sensitive to anions, so that the ISFET's response depended on the type of anion associated with the Ca^{2+} ions or present in the background electrolyte.

The electrical response of ISFETs and the site-binding model

In the differential mode, the ISFET gave directly the variation of the surface potential at the SiO_2-solution interface caused by the adsorption of potential-determining ions ($\Delta V_G = \Delta \psi_0$). The measurements presented in this paper were performed in the presence of a concentrated background electrolyte, which ensured a constant and high ionic strength. In this condition, the electrical potential decreased sharply away from the SiO_2 surface. For a 0.5 M 1–1 electrolyte, the Debye length

$$\kappa^{-1} = \left(\frac{e^2 N_A \times 10^3}{\varepsilon_0 \varepsilon k T} \sum C_i z_i^2 \right)^{-1/2} = 4 \text{ Å} \tag{3}$$

was smaller than the size of a hydrated ion. The contribution of the diffuse layer to the potential was thus negligible ($\psi_D \ll \psi_0$); the measured potential drop took place in the Stern layer. If the capacity C_S of the Stern layer is assumed constant, the potential is related to the surface charge density σ_0 ($\psi_0 = \sigma_0/C_S$) and the measured potential difference gives the variation of surface charge density brought about by the adsorption of ions: $\Delta \psi_0 = \Delta \sigma_0/C_S$ [21].

The ions M bind to the grafted complexing molecules S according to a complexation equilibrium taking place at the surface:

$$K = \frac{[\text{S-M}]}{[\text{S}] [\text{M}]_S} = \frac{\theta}{(1 - \theta)[\text{M}]_S} \tag{4}$$

where [S-M], [S] and θ denote the concentrations of occupied and free surface sites and the fraction of occupied sites. The total concentration of surface sites is the grafting density, $N_S = [\text{S-M}] + [\text{S}]$. The concentration of ions at the surface $[\text{M}]_S$ is given by Boltzmann statistics:

$$[\text{M}]_S = [\text{M}] \exp(-ze\psi_0/kT) \tag{5}$$

The variation of surface charge density caused by the complexation of z-valent ions at the surface is $\Delta \sigma_0 = ze\theta N_S$. This model introduced by Yates, Levine and Healy [22, 23] was successful because the Stern capacity of oxides was really constant within experimental accuracy [24]. This led to the approximate equation valid at the isoelectric point of silica [17]:

$$\log(K) + \log[\text{M}] = \frac{ze \, \Delta \psi_0}{2.3kT} - \log\left(\frac{ze}{\Delta \psi_0} \frac{N_S}{C_S} \right) \tag{6}$$

which can be fitted to the experimental data with K and N_S/C_S as free parameters. The grafting density N_S can be estimated, assuming C_S takes the value of the Stern capacity of silica ($C_S = 20 \ \mu\text{F/cm}^2$) [24].

This model could fit most of the experimental data. However, a validation of the model was not possible. It should be noticed that the same complexation constant $K = 10^{2.1}$ was found for the complexation of K^+ by grafted $B_{15}C_5$, whatever the grafting density. For the ISFET grafted according to reaction 3, the fit was very poor at high concentrations of K^+. Presumably, the surface chemistry was more complex because of the heterogeneity of surface sites of aminated silica. $K = 10^{2.7}$ was found for the complexation of Ca^{2+} by the phosphonate group. These values of K were in agreement with the complexation equilibrium constants pertaining to bulk solutions. For the multistep grafting process of the phosphonate group, the electrical response was so weak that the sensitivity to the K and N_S/C_S parameters was poor in looking for the best fit of the model to the experimental data (Table 1). It was nevertheless not possible to obtain a good fit with the

Table 1 Parameters giving the best fit of the site-binding model to the experimental data

	log(K)	Grafting density N_S (μmol/m^2)
K$^+$ detection with B$_{15}$C$_5$		
Chlorosilane on silica (**1**)	2.1	0.10
Dimethylaminosilane on silica (**2**)	2.1	0.22
Hydroxysilane on aminated silica (**3**)	2.1	0.13
Ca^{2+} detection with phosphonate groups		
One-step grafting	2.7	0.03
Multistep grafting	$\leq$2.0	$\leq$0.01

same value of K as for the one-step process. This is an indication that the repeated chemical modifications of the surface have created different surface sites which can bind Ca^{2+} ions.

ISFETs could work in solutions of low ionic strength as well, but the situation would be more complex. The potential of the diffuse layer ψ_D cannot be neglected at low ionic strength, and the relationship between the ionic concentration and V_G depends on many parameters which are difficult to assess. As a consequence, the use of ISFETs for in-line measurements or for in vivo medicinal applications, where the ionic strength and pH cannot be modified, will require their prior calibration in the conditions of utilization.

Conclusions

ISFETs sensitive to ions other than H$^+$ could be elaborated by means of the direct chemical grafting of complexing molecules on the silica grid. Using a differential measurement mode in solutions of high ionic strength, the measured potential is directly related to the surface density of bound ions. This simple situation allows for systematic studies on the physical chemistry at the interface.

This simplicity is only apparent, however. The present paper was focused on one aspect of a multidisciplinary research work involving chemistry, electrochemistry of the solutions and of the semiconductors, and a large part of the technology of microsystems.

Concerning the transducer-solution interface, the requirements for an efficient detection of ions are a high grafting density and a large complexation constant of the grafted molecules towards the target ions. The choice of the chemical species and grafting reaction are then crucial. The use of ready-to-graft complexing molecules yields higher grafting densities and more reproducible results than a multistep grafting process, but their synthesis may appear difficult because such molecules are necessarily multifunctional. The chemical stability of the grafts is also important for continuous utilization conditions. In particular, ester functions have to be avoided because of their slow hydrolysis in water. This is the reason why phosphonate groups have been preferred to the more readily available organophosphate esters.

References

1. Bailey PL (1979) Ion-selective Electrode Rev 1:81
2. Bergveld P (1970) IEEE Trans Biomed Eng 17:70
3. Bergveld P (1972) IEEE Trans Biomed Eng 19:342
4. Cléchet P, Jaffrezic-Renault N, Martelet C (1994) J Phys IV 4:C1–283
5. Orion pHuture; Leeds & Northrup Durafet
6. Moss SD, Janata J, Johnson CC (1975) Anal Chem 47:2238
7. Moss SD, Johnson CC, Janata J (1978) IEEE Trans Biomed Eng 25:49
8. McBride PT, Janata J, Comte PA, Moss SD, Johnson CC (1978) Anal Chim Acta 101:239
9. Cléchet P, Jaffrezic-Renault N, Martelet C (1992) In: Yamauchi S (ed) Chemical sensor technology, vol 4. Kodansha, Tokyo, pp 205–225
10. Bataillard P, Cléchet P, Jaffrezic-Renault N, Martelet C, Morel D, Serpinet J (1986) J Electrochem Soc 133:1759
11. Jaffrezic-Renault N, Chovelon J-M, Perrot H, Le Perchec P, Chevalier Y (1991) Sensors Actuators B5:67
12. Elbhiri Z, Chovelon J-M, Jaffrezic-Renault N, Chevalier Y (1999) Sensors Actuators B58:491
13. Elbhiri Z, Chevalier Y, Chovelon J-M, Jaffrezic-Renault N (2000) Talanta (in press)
14. Bergveld P, deRooij NF, Zemel JN (1978) Nature 273:438
15. Schenck JF (1977) J Colloid Interface Sci 61:569
16. Cichos C, Geidel Th (1983) Colloid Polym Sci 261:947
17. Perrot H, Jaffrezic-Renault N, de Rooij NF, van den Vlekkert HH (1989) Sensors Actuators 20:293
18. Lork KD, Unger KK, Kinkel JN (1986) J Chromatogr 352:199
19. Rocher V, Poyard S, Jaffrezic-Renault N, Ajoux C, Lemiti M, Sibai A (1994) Sensors Actuators B18–19:342
20. McKenna CE, Higa MT, Cheung NH, McKenna MC (1977) Tetrahedron Lett 155
21. Bousse L, de Rooij NF, Bergveld P (1983) IEEE Trans Electron Devices 30:1263
22. Levine S, Smith AL (1971) Discuss Faraday Soc 52:290
23. Yates DE, Levine S, Healy TW (1974) J Chem Soc Faraday Trans 170:1807
24. Westall J, Hohl H (1980) Adv Colloid Interface Sci 12:265

Progr Colloid Polym Sci (2000) 115:249–254
© Springer-Verlag 2000

METHODS

L. Davoust
J.-L. Achard
A. Cartellier

Detection of waves at an interface by way of an optical fibre

L. Davoust (✉) · J.-L. Achard
A. Cartellier
LEGI, Domaine Universitaire, BP 53
38041 Grenoble, France
e-mail: laurent.davoust@hmg.inpg.fr
Fax: +33-4-76825271

Abstract Coalescence of two interfaces in two-phase flows (e.g. bubble flows, sprays) is strongly conditioned by the unavoidable presence of natural surfactants. These surfactants, through various mechanisms like the Marangoni effect, sorption barriers at the involved interfaces, and diffusion-convection within the adjacent phases, yield a specific interfacial rheology characterized by a given level of elasticity and viscosity. The aim of this paper is to present a wave damping experimental set-up which gives insight into the rheology of a contaminated air–water interface. Electromechanically driven cylindrical capillary waves are detected with the help of an interferometric technique based on the use of an optical fibre; this technique allows measurement of both the amplitude and the vertical velocity of the capillary waves so-created. Contamination is achieved by spreading a non-ionic fatty acid surfactant over the air–water interface.

Key words Interface · Fluid · Capillary waves · Optical fibre · Surfactant

Introduction

Coalescence of two fluid–fluid interfaces is currently known as a process dependent on a balance between two competitive times [1]: (1) a typical "interaction time" between the two interfaces interacting with each other, and which is nothing but the consequence of the local hydrodynamical conditions (e.g. shear, turbulent fluctuations), and (2) a "draining time" during which the interstitial film localized between the two interfaces disappears for coalescence to occur (Fig. 1). Draining of the interstitial film can be dramatically inhibited by a non-homogeneous distribution of the surfactant along the interfaces and the subsequent gradients of the interfacial concentration in surfactant yield a Marangoni force which competes against the motor pressure gradient in the interstitial film. At first sight, the presence of a surfactant might be merely thought of as a damping mechanism of the coalescence process. Actually, when two droplets or bubbles are approach-

ing, a dimple grows locally at the two involved interfaces before coalescence occurs. However, this deformation ability of the interfaces depends on the nature of the available surfactant and also on the typical frequency of the interfacial deformations. Moreover, the hydrodynamic fluctuations in the surrounding continuous phases are able to yield interfacial deformations and are often characterized by a quite wide spectrum of frequencies (0.1–1000 Hz at least). Subsequently, one cannot clearly state whether surfactants decrease (or not) the coalescence efficiency. Then, the rheology of contaminated gas–liquid or liquid–liquid interfaces shows itself as an unavoidable step to overcome when investigating coalescence of bubbles or droplets. As a surfactant experiences various kinetics such as sorption barriers, diffusion, or convection, the subsequent rheology of a contaminated interface is currently investigated in the light of the so-called Gibbs elasticity. Wave damping experiments prove themselves to stand as a relevant tool to investigate interfacial rheology. The aim of this paper

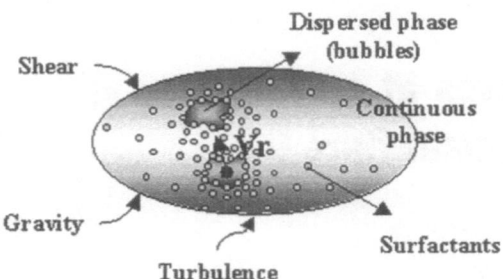

Fig. 1 Encounter of two bubbles in a contaminated media

is to present an optical technique devoted to the detection of cylindrical capillary waves at an air–water interface.

Materials and methods

Experimental set-up

An axisymmetric float, made of PTFE, is submitted to a periodic motion: a sinusoidal electric current flows in the cylindrical coil attached to the inner centre of the float; the whole is sliding vertically within the annular space managed between the two poles of an axisymmetric permanent magnet (Fig. 2). The travelling cylindrical capillary waves so excited are damped along the radial direction in a cylindrical tank whose diameter is large enough for any reflection of the waves at the outer wall to be avoided. The tank is filled with distilled water; the bath so-obtained is covered with a hexanoic acid monolayer. The amplitude of the waves is measured with the help of an optical fibre, whereas the measurement method is based on interferometry (Fig. 3). Interestingly, whatever the optical fibre involved in our measurements (monomodal or polymodal fibres), the measured amplitudes were found to be equal: this result suggests that the measurement technique was quite robust.

Optical detection of the waves

A IR laser beam is inserted inside an optical fibre whose cleaved end, aligned with gravity, is accurately positioned a few micrometers above the air–water interface. As shown by Bohanon et al. [2], the laser beam here referred to as E is refracted at the end. The

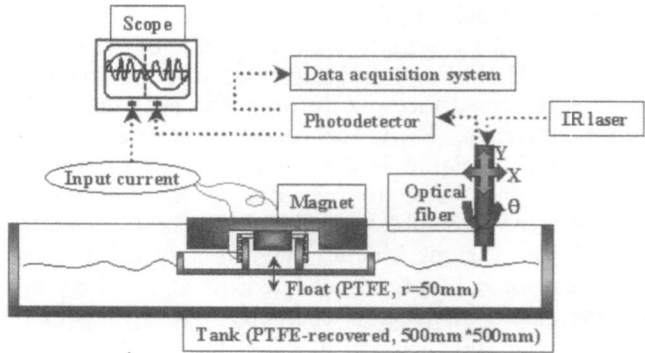

Fig. 2 Experimental set-up as sketched

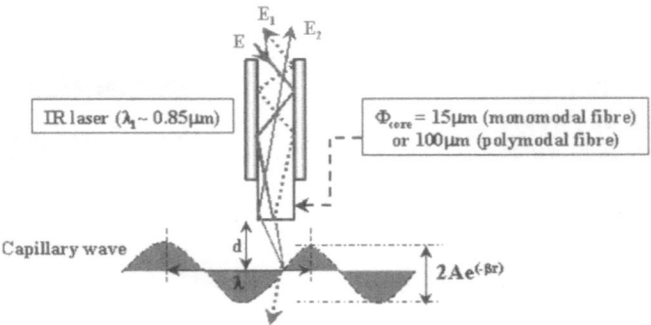

Fig. 3 Optical detection of the waves

beam E_1 denotes the reflected part of E whereas the transmitted part goes out the fibre and is subsequently refracted at the air-water interface (Fig. 2). After this second refraction, the reflected part of the transmitted beam, referred to as E_2, is assumed to enter again inside the optical fibre. The difference in the optical paths between E_1 and E_2 depends on the vertical displacement of the interface. As demonstrated hereafter, the capillary waves, responsible for a sinusoidal time-dependence of the displacement, yield a frequency-modulated fringe pattern. Raising and lowering the interface yield the same fringe patterns, so that the measured optical signal exhibits a fundamental frequency twice as high as one of the capillary waves.

Analytics on the optical signal

Every capillary wave of whatever nature (Cartesian, circular, standing or travelling) has an amplitude ζ which can always be locally approximated by that of a harmonic function depending on the radius:

$$\zeta = A(r) \exp(-\beta r) \sin(2\pi r/\lambda + \omega t) \tag{1}$$

where $A(r)$, β, ω, λ and t denote respectively the amplitude of the wave at the radius r, its spatial damping coefficient, its frequency, its wavelength and time. The difference in the optical paths between E_1 and E_2 is therefore responsible for the phase shift:

$$\phi(t) = 2\pi \frac{(2(d - \zeta))}{\lambda_1} \tag{2}$$

where λ_1 and d stand for respectively the IR laser wavelength ($\lambda_1 \approx 0.85\ \mu m$ in the fibre) and the distance between the end of the optical fibre and the interface at rest. The optical signals E_1 and E_2 are subsequently written as the complex expressions:

$$E_1 = A_1 \exp\left(-2\pi I \frac{tc}{\lambda_1}\right) \quad \text{and}$$

$$E_2 = A_2 \exp\left(-4\pi I \frac{(d - \zeta)}{\lambda_1}\right) \exp\left(-2\pi I \frac{tc}{\lambda_1}\right) \tag{3}$$

in which the variable c denotes the speed of light. At the output of the photo-receptor, the measured voltage for the optical signal is a linear combination of the light intensity $E \times E^*$ where E is the sum of the two former signals E_1 and E_2, and E^* is the conjugate complex field of E:

$$\text{voltage} = K\left(A_1^2 + A_2^2 + 2A_1A_2\right.$$
$$\left. \times \cos\left[\left(\frac{4\pi}{\lambda_1}\right)\left(d - A(r)\exp(-\beta r)\sin\left(\omega t + 2\pi \frac{r}{\lambda}\right)\right)\right]\right) \tag{4}$$

How to obtain the wave amplitude?

As soon as the amplitude of the wave increases [$A(r) = \{2\lambda, 4\lambda\}$, see e.g. Fig. 4], we find that it is somewhat awkward to define an unequivocal criterion allowing us to count the number of fringes and to correlate it to the requested wave amplitude. As a matter of fact, it is convenient first to normalize the former voltage (new bounds between -1 and $+1$, Fig. 4) and to differentiate it: the envelope of the derived signal is found to stand for the vertical velocity of the interface (Fig. 5):

$$\frac{\partial(\text{normalized voltage})}{\partial t}$$
$$= V_z \frac{4\pi}{\lambda_1} \sin\left(\left(\frac{4\pi}{\lambda_1}\right)\left(d - A(r) \times \exp\left(-\beta r\right)\sin\left(\omega t + 2\pi\frac{r}{\lambda}\right)\right)\right) \quad (5)$$

From the experimental viewpoint, the measured voltages are first filtered by way of two analog filters (one high-pass filter of cut-frequency $f_c = 0.1$ Hz and one low-pass filter of cut-frequency $f_c = 2 \times 10^4$ Hz). Then, before any data processing, subsequent voltages are numerically processed with the help of wavelet decomposition: the retained approximated signal is calculated at the level 3 of a decomposition in a wavelet basis composed of Daubauchies wavelets (of kind 5). Straight afterwards, the signal is submitted to the (numerical) normalization procedure detailed above and is differentiated in order to obtain both the vertical velocity and the amplitude of the capillary waves.

How to obtain the wavelength

The input electric voltage in the coil and the excited capillary wave at the bottom of the float exhibit the same phase and spectral content. The wavelength was measured making use of this point of interest: at each radial displacement of the optical fibre, both the electric voltage in the coil (which stands as a time schedule) and the optical signal were stored (Fig. 6). Between two successive synchronisations of the two signals, the radial displacement of the optical fibre was clearly identified as the wavelength of the capillary waves.

Fig. 4 Superimposed theoretical curves of the normalized optical signal and of the sinusoidal interfacial deformation ($\lambda = 0.01$ m, $\lambda_1 = 0.85$ μm, $d = 10$ μm, $r = 0.05$ m, $\beta = 10$ m^{-1} and $\omega = 10 \times 2\pi$). Left: $A(r) = 2\lambda_1$; right: $A(r) = 4\lambda_1$

Results

The interfacial contamination is achieved by depositing some hexanoic acid over the considered air–water interface. Petroleum ether was used as the spreading solvent. The water bath was acidified (0.005 M HCl) in order to prevent the involved (carboxylic) hexanoic acid from any otherwise natural dissociation. The dispersion equation for the case of cylindrical capillary waves has already been established in the literature with the following assumptions: (1) excitation is assumed to be locally imposed at the remote centre of the liquid bath, whereas (2) the Gibbs elasticity is assumed purely dilatational. Clearly, one does not expect any interfacial shear to be provided by such cylindrical waves. As we found the dispersion relation exhibited in the paper by Jiang et al. [3] to be inexact, the following dispersion relation (involved in our study) is calculated from the analytical developments by Bock [4]:

$$k^3/\omega^2 \left[2\eta^2\omega^2\left(k^2 + m^2\right) + k\varepsilon(k - m)\left(\rho g + \sigma k^2\right) + m\omega^2\left(\rho\varepsilon - 4k\eta^2\right)\right]$$
$$+ I\left(k\eta/\omega\right)\left[k\left(k^2 - m^2\right)\left(\rho g + \sigma k^2\right) + \rho\omega^2\left(k^2 + m^2\right)\right] = 0$$

$$(6)$$

with $(I^2 = -1)$, $k = \left(2\pi/\lambda + I\beta\right)$, $m^2 = \left(k^2 - I\omega\rho/\mu\right)$ and $\varepsilon = \varepsilon_d + I\omega\eta_d$. The symbols m, k, ε_d, and η_d denote, respectively, the vertical and the radial wave numbers, and the dilatational elasticity and viscosity. For the case of the cylindrical waves, far enough from the float ($|kr| \gg 1$) the coefficient $A(r)$ involved in the writing of the interface amplitude can be developed as $A(r) \sim 1/\sqrt{kr}$ (asymptotic development of a zeroth-order Hankel function). This last approximation is currently practised by Bock [4] and also Jiang et al. [3]. As a consequence and not surprisingly, the former dispersion relation is the same as the one obtained by

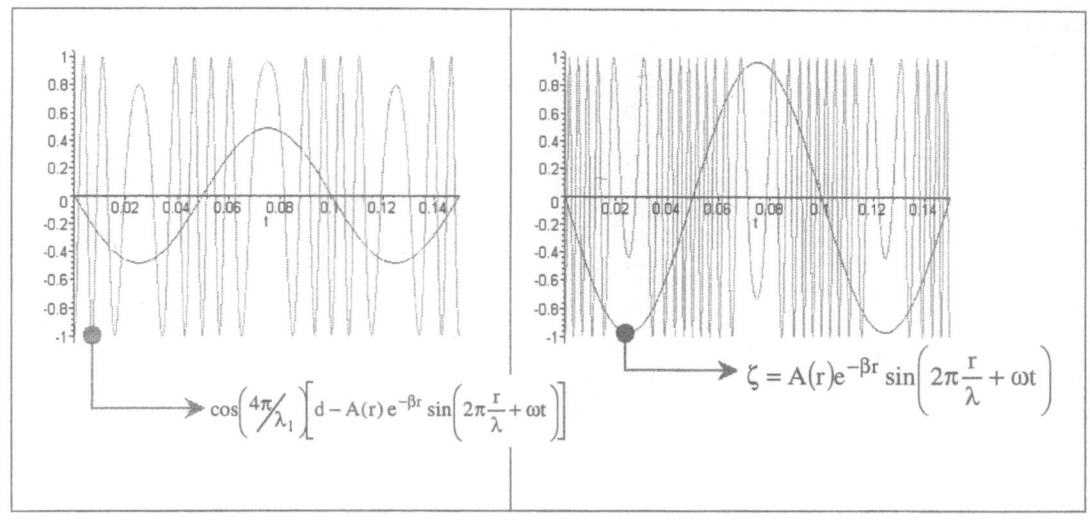

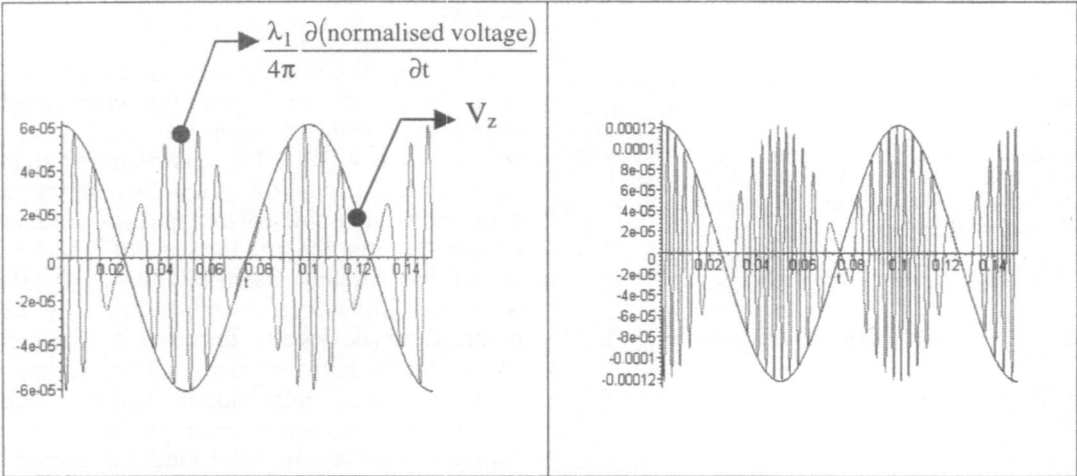

Fig. 5 Derivation of the vertical velocity as an envelope (same parameters: $\lambda = 0.01$ m, $\lambda_1 = 0.85$ μm, $d = 10$ μm, $r = 0.05$ m, $\beta = 10$ m^{-1} and $\omega = 10 \times 2\pi$). Left: $A(r) = 2\lambda_1$; right: $A(r) = 4\lambda_1$

Lucassen and Lucassen [5] in the case of Cartesian 2-D capillary waves; in fact, far enough from the moving float ($|kr| \gg 1$), the cylindrical capillary waves can be seen locally as quasi-plane.

In this paper, both the dilatational elasticity ε_d and the dilatational viscosity η_d are directly calculated from the former dispersion relation and from the experimental values of ω, λ, and β (Fig. 7, top); our strategy differs from the widespread identification procedure chosen by many authors in the literature. Rather than identifying ε_d, η_d, and the interfacial tension σ from the dispersion relation with the help of a minimization technique (least-square methods or conjugate-gradients methods), here we make use of the adsorption isotherm

for the hexanoic acid. Our measurements of σ, based on bubble tensiometry, are displayed in Fig. 7 (bottom, upper corner on the right-hand side). Thus, the (complex) dispersion equation separates into two arithmetic equations which are solved in order to provide the dilatational elasticity and the dilatational viscosity. We do not need any assumption about the value of the dilatational viscosity. Jiang et al. [3] disregarded the dilatational viscosity; for the particular case of hexanoic acid (small carbon chain: C_6) examined in this paper, the dilatational viscosity is effectively demonstrated to be much smaller than the dilatational elasticity.

The whole of our experimental results are consistent with the available literature (e.g. see [6–9]). Our curves of the damping coefficient versus the acid concentration present non-monotonic variations in qualitative agreement with the curves obtained by Lucassen and Hansen [6] at higher frequency (200 Hz) for a set of fatty acids

Fig. 6 The two analog signals received from the photoreceptor (after a polymodal fibre) and from the coil attached to the float

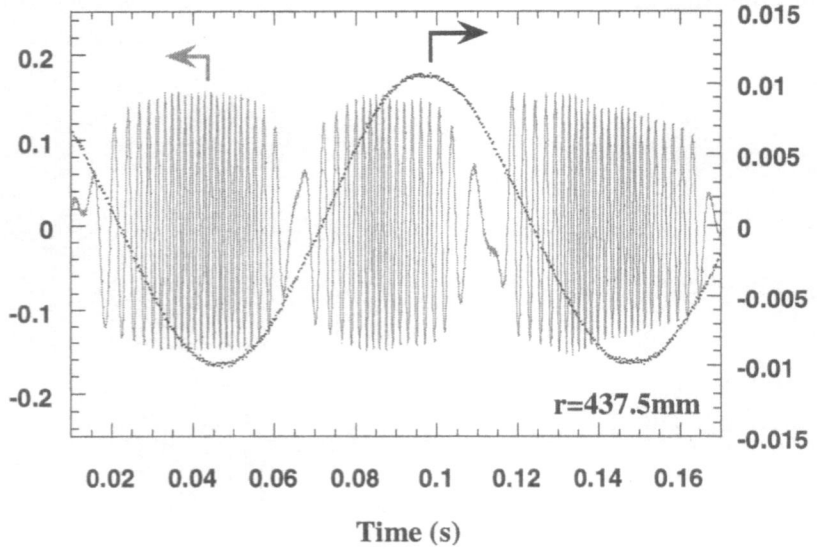

Fig. 7 Top: damping coefficient curves. Bottom: the Gibbs elasticity curves and the adsorption isotherm for hexanoic acid

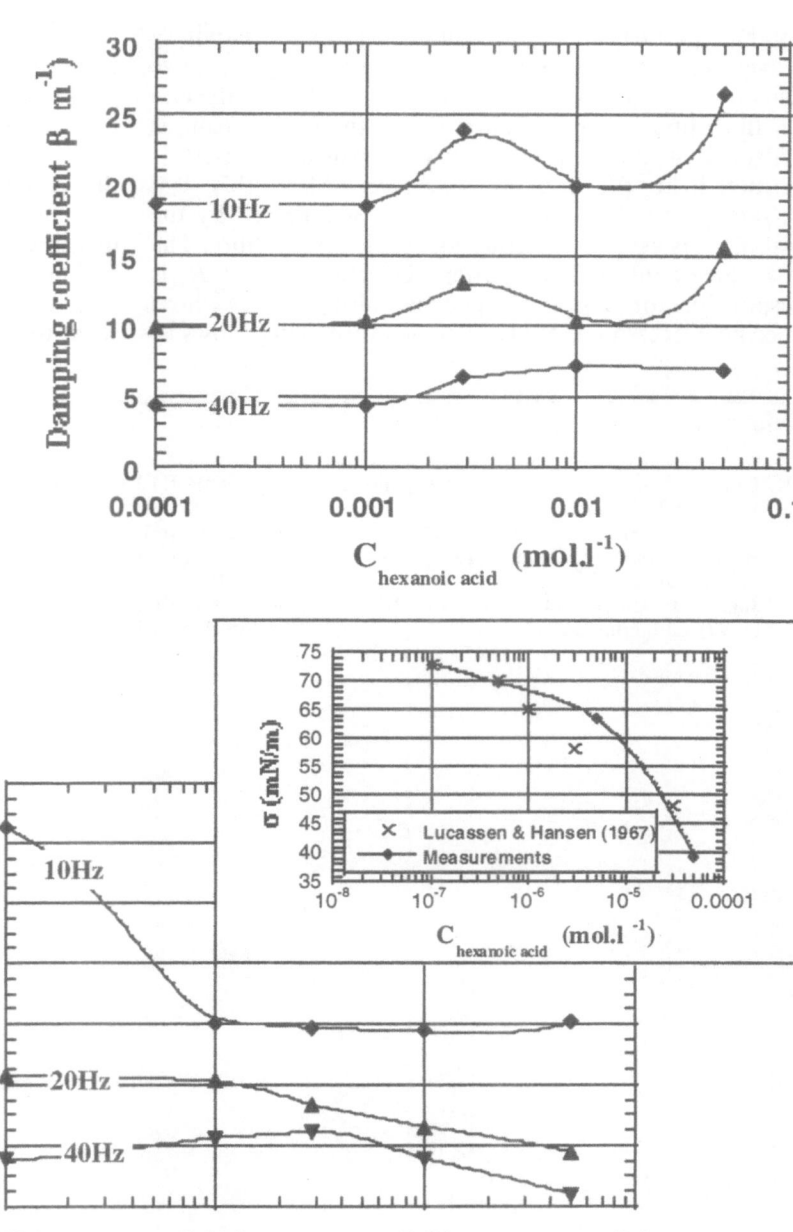

(including ours). As our frequencies cover a low range (10–40 Hz), the soluble hexanoic acid was free to experience various kinetics as sorption and also diffusion within the bulk. As a consequence of this solubility, the dilatational elasticity ε_d is found to be much lower than the level of elasticity ε_d which characterizes an octanoic monolayer [9] and even more a decanoic monolayer [7]. The β curves (Fig. 7, top) exhibit a smooth peak whose magnitude decreases as the frequency of the capillary waves increases. This phenom-

enon is correlated with the simultaneous change observed for the slope of the ε_d curves ($C_{\text{hexanoic acid}} \approx 2 \times 10^{-3} \text{ mol} \cdot \text{l}^{-1}$).

Conclusions

This paper presents a wave damping apparatus equipped with a non-protruding optical fibre allowing us to measure accurately the motion of an air–water interface weakly perturbed by travelling cylindrical capillary waves. The optical fibre is located vertically just above the interface. As an input IR laser is refracted at the end of the optical fibre, a part E_1 of the laser beam is reflected at the end whereas the transmitted part is refracted at the moving interface. After this last refraction, the laser beam E_2, which is reflected by the interface, is assumed to enter again the optical fibre. The subsequent interference pattern between E_1 and E_2 is responsible for an optical signal exhibiting fringes whose specific processing is analytically justified. This allows us to measure unambiguously the amplitude of the interfacial deformation.

Some experimental results are also obtained when the interface is contaminated with a monolayer of fatty acid surfactant. A dispersion relation calculated from the theoretical developments by Bock [4] allows the dilatational rheology of the interface to be calculated from the measurements of: (1) the frequency, (2) the wavelength, and (3) the spatial damping coefficient of the waves. Our results, established for several excitation frequencies, seem to be in qualitative agreement with the literature.

References

1. Chester AK (1991) Trans Inst Chem Eng A, 69:259–270
2. Bohanon TM, Mikrut JM, Abraham BM, Ketterson JB, Dutta P (1991) Rev Sci Instrum 62:2959–2961
3. Jiang Q, Chiew YC, Valentini JE (1992) Langmuir 8:2747–2752
4. Bock EJ (1991) J Colloid Interface Sci 147:422–432
5. Lucassen-Reynders EH, Lucassen J (1969) Adv Colloid Interface Sci 2:347–395
6. Lucassen J, Hansen RS (1967) J Colloid Interface Sci 23:319–328
7. Maru HC, Wasan DT (1978) Chem Eng Sci 34:1295–1307
8. Christov CI, Ting L, Wasan DT (1982) J Colloid Interface Sci 85:363–374
9. Ting L, Wasan DT, Miyano K, Xu SQ (1984) J Colloid Interface Sci 102:248–259

Progr Colloid Polym Sci (2000) 115:255–258
© Springer-Verlag 2000

H. Ruf

Effects of experimental errors in dynamic light scattering data on the results from regularized inversions

H. Ruf
Max-Planck-Institut für Biophysik
Kennedyallee 70
60596 Frankfurt am Main, Germany
e-mail: ruf@mpibp-frankfurt.mpg.de
Tel.: +49-69-6303347
Fax: +49-69-6303346

Abstract Three of the selection methods used with regularized data inversion were tested with dynamic light scattering data from a binary mixture of polystyrene latex particles. The three methods were the F-test (the standard method of CONTIN), the stability plot (one of the methods used with ORT) and the L-curve method. Two experimental situations differing in the scattered intensity were studied. The first was that of low mean count rates, where the random photon detection noise was dominant; the second was that of high mean count rates, where the non-random intensity fluctuation noise prevailed. The F-test proved to be very sensitive for non-random statistical and systematic errors, and then yielded too weakly regularized solutions associated with complex and poorly reproducible size distributions. The stability plot and L-curve methods, on the other hand, proved to be largely insensitive to these kinds of errors in the data and yielded reliable results, as concluded from a comparison with the size characteristics of the particles determined with electron microscopy.

Key words CONTIN · ORT · F-test · Stability plot · L-curve

Introduction

The various mathematical methods devised to overcome the ill-posed problem of the inversion of a Fredholm integral equation of the first kind, associated with the determination of a continuous size distribution from dynamic light scattering data (DLS), concentrate on finding the simplest size distribution whose autocorrelation function is consistent with the data. Regularization techniques like CONTIN [1] or ORT [2] produce simplified distributions by means of a regularizor, which is included as an additional constraint into the minimum condition of the fit. Accordingly, a sum of the variance and the regularizor, called the objective function $F_{ob}(\alpha)$

$$F_{ob}(\alpha) = \|\sqrt{M_\varepsilon}(y - Ax)\|^2 + \alpha^2 \|r - Rx\|^2 = \min \quad (1)$$

is minimized. The term $\|y - Ax\|$ is the residual norm or mean deviation; its square is the variance; $\|r - Rx\|$ is the solution seminorm; y denotes the experimental data of the field autocorrelation function and x the solution of the fit; A is a matrix involving the kernel of the Fredholm integral equation of the first kind; M_ε is the covariance matrix of the residuals, which can be used to weight the data; α is the regularization parameter, which controls the degree to which the regularizor contributes to the minimum condition of the fit.

The regularizor is a property of the size distribution suited to impose smoothness. In the case of CONTIN it is the integral over the second derivative of the distribution or the overall curvature. In the case of ORT it is built up from the first derivatives of the coefficients of the spline functions used to approximate the distribution in this method. Regularization means that the variance of the fitted autocorrelation function and a fraction α of the overall curvature of the distribution associated with this are minimized together. A simplification of the size distribution, however, is always accompanied by an increase in variance, and hence a compromise has to be searched for, where the distribution is simplified but where the variance is not much different from that of the least-squares solution.

In a recent study using data from polystyrene latex beads and lipoprotein particles [3], we tested three methods for the determination of such a compromise solution. The first was Fisher's F-test, the standard method of CONTIN, which is utilized to test the variance of each regularized solution on whether this is significantly different from that of the least-squares solution. The F-test provides a confidence interval rather than a single solution. Solutions with a PROB1 value greater than 0.9 are to be rejected because of a too large variance. Solutions with a PROB1 value smaller than about 0.2, whose variance is nearly that of the least-squares solution, on the other hand, are suspected of having too complex size distributions. Provencher [1] recommended that solutions with PROB1 values near to or equal to 0.5 should be selected. The second method was the stability plot used with the regularization method ORT, where the logarithm of the solution norm is plotted versus the logarithm of the regularization parameter. The optimally regularized solution is taken from a point of inflection of a plateau region of this curve, where the additional condition must be fulfilled that the mean deviation is nearly equal to that of the least-squares solution. The third method was the L-curves method [4, 5], which was originally suggested by Lawson and Hanson [6]. This method combines the dependencies of variance and regularizor on the regularization parameter in a single plot. The resulting curve is L-shaped. The point of maximum curvature in the corner of the L-curve represents a good compromise between achieving a small residual norm and keeping the solution seminorm reasonably small [4].

The results presented in this paper are from an extension of this study to a binary mixture of polystyrene latex particles with a mean size ratio of about 1:4.

Materials and methods

The polystyrene latex beads LB1 and LB3 were purchased from Sigma. The number-weighted mean radii determined from electron microscopy (EM) were 27 nm for LB1 [3] and 155 nm for LB3. Preparation of the sample for DLS measurements was described elsewhere [7]. The two experimental set-ups used for the DLS measurements are described in [8]. In the case of the second set-up, the detection unit used here consisted of a few mode fiber (OZ SMF-1064-FC, LINOS Photonics, Göttingen, Germany) and an avalanche diode (SPCM-AQ-131-FC, EG&G, Canada). A photo multiplier (Thorn EMI 9863B) was used with the first set-up, with which the measurements at low count rates were carried out. The measurements at high count rates were performed with the second set-up using two Brookhaven 9000 AT correlators to cover the delay time range from 10 μs to 3.84 ms in steps of 10 μs. All measurements were carried out in the batch mode, where the duration of a single measurement was one minute. Each individual measurement was checked for outliers by comparing the contents of the first data channels. A series of 960 of such measurements was

Table 1 Experimental conditions of the dynamic light scattering measurements. R denotes the mean count rate in kcps, N_p the number of data channels and $\Delta\tau$ the sampling time in μs. B is the baseline of the photon count autocorrelation function obtained by summing the raw data of the corresponding individual measurements. Values for number of samples, N, baseline, B, and total number of counts, A_{tot}, are the averages of four measurements rounded to the first decimal place. The coherence factor β was obtained from cumulant analysis. It is equal to the intercept of the corresponding normalized photon count autocorrelation function minus one

R	N_p	$\Delta\tau$	N	B	β	A_{tot}
50	136	15	9.6×10^8	4.7×10^8	0.66	6.7×10^8
820	512	10	1.4×10^9	9.5×10^{10}	0.40	1.1×10^{10}

performed for each scattering intensity. The data sets finally evaluated were constructed from 240 of these measurements. The characteristic parameters of the data are summarized in Table 1. CONTIN analysis was performed with 99 grid points for the distribution and without using weights. In the case of the L-curve method, the modification discussed in [3] was applied and the next more weakly regularized solution was taken. This will be denoted as the L-curve solution. For data analysis the following general scheme was applied. If the first analysis with an appropriate size range did not provide any indication of considerable normalization errors due to an error in the experimental baseline of the measured photon count autocorrelation function, this was taken as the final one. Otherwise, baseline errors were determined according to the procedure given in [3], and data were corrected as described elsewhere [9]. Since the stability plot gave practically the same results as the L-curve method when a plateau region with a clear inflection point could be identified, it will not be mentioned specifically.

Results and discussion

The size distributions obtained from the measurements with the low count rate are shown in Fig. 1. The residuals (Fig. 1c) indicate that the noise is largely random. The size distributions determined from the original data with the F-test (Fig. 1a) showed in two cases very narrow peaks, and the regularization strengths of these solutions were nearly one order of magnitude smaller than those of the other two solutions showing broader peaks. These narrow peaks hinted at relative baseline errors with a positive sign, and consistently values of 1.4×10^{-3} and 1.8×10^{-3} were obtained. After correction of these two data sets, the regularization strengths of the F-test and L-curve solutions were nearly the same. The size distributions of the L-curve solutions are depicted in Fig. 1b.

When the mean scattering intensity is increased, the non-random intensity fluctuation noise becomes dominant (Fig. 2c). The solutions determined with the F-test from the four data sets were two to three orders of magnitude more weakly regularized than the solutions determined with the L-curve method. Accordingly, the size distributions of the F-test solutions show very

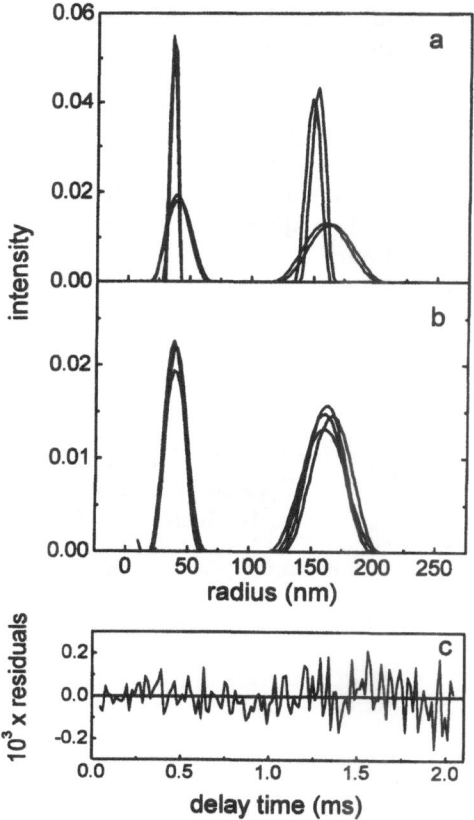

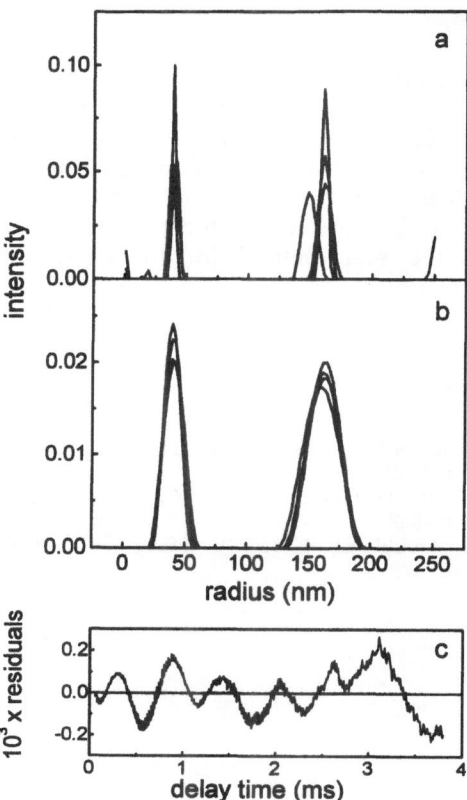

Fig. 1a–c Results from CONTIN analyses of the measurements on a binary mixture of polystyrene latex beads LB1 and LB3 with a mean count rate of 80 kcps. **a** Intensity-weighted size distributions obtained with the F-test from the original data (the ordinate scale is chosen such that the area under each peak is equal to one). **b** Size distributions obtained with the L-curve method, where the two data sets yielding extremely narrow peaks in **a** were corrected for normalization errors [9]. **c** Residuals of the field autocorrelation function of the F-test solution of one of the data sets showing the narrow peaks in **a**

Fig. 2a–c Results from CONTIN analyses of the measurements on a binary mixture of polystyrene latex beads LB1 and LB3 with a mean count rate of 820 kcps. **a** Intensity-weighted size distributions obtained with the F-test from the original data. **b** Size distributions obtained with the L-curve method. **c** Residuals of the field autocorrelation function of the F-test solution of one of the data sets

narrow peaks (Fig. 2a). The size distributions of the L-curve solutions show broader peaks and, as before, a much better agreement. In the case of non-random noise, the narrow peaks of the F-test solutions are not good indicators for normalization errors and the residuals gave no direct evidence for the existence of such errors in the data, which was supported by the good reproducibility of the L-curve results.

ORT analysis of the different data sets yielded practically the same size distributions. The distributions obtained with the L-curve method from CONTIN analysis not only showed a better reproducibility but also represented more reasonable results. DLS measurements on the individual LB1 and LB3 particles with relatively low mean count rates and of longer durations yielded intensity-weighted mean radii, r_I, of 39 nm and 158 nm, and standard deviations over the corresponding

mean radius, $\Delta r/r_I$, of 0.16 and 0.05, which, except for the width of the LB3 distribution, agreed very well with the corresponding values determined from electron microscopy (EM) [3, 7]. The relative width of the LB3 beads determined from EM was only 0.02. The mean sizes of 40 nm (LB1), 162 nm and 161 nm (LB3) and the widths of the LB1 beads of 0.19 and 0.16 determined from the mixture (Figs. 1b and 2b) compare well with these numbers. Only the relative widths of 0.09 and 0.08 of the LB3 peaks are considerably larger. It is generally difficult to determine the correct width of such a nearly monodisperse sample from DLS measurements with a size distribution algorithm from data of this statistical accuracy, which was of the order of 8×10^{-5} in [7] and of the order of 2×10^{-4} in this study. This was concluded from a comparison of the field autocorrelation function of a truly monodisperse sample of this size with one calculated by assuming a Gaussian distribution and using the width from EM. The maximum differences of the two functions is only about 3×10^{-5}.

Conclusions

The results presented here confirm our previous findings [3, 8]. The F-test is only useful when the data contain randomly distributed noise and are free from normalization errors. The stability plot and the L-curve method are much less affected. Since inflection points are not always clearly expressed in the stability plots, the L-curve is the best method for the determination of appropriately regularized solutions. Here, the difference in the mean count rate or the content of non-random noise has practically no effect on the L-curve results, which is probably due to the fact that the data are of relatively high statistical accuracy. Finally, a polydispersity equal to or greater than that of the LB1 beads can also be determined with reasonable accuracy from such a mixture when applying a measurement duration of about four hours.

References

1. Provencher SW (1982) Comput Phys Commun 27:213; 27:229
2. Schnablegger H, Glatter O (1991) Appl Optics 30:4889
3. Ruf H, Gould BJ, Haase W (2000) Langmuir 16:471
4. Hansen C (1992) SIAM Rev 34:561
5. Hansen C, O'Leary DP (1993) SIAM J Sci Comput 14:1487
6. Lawson CL, Hanson RJ (1974) Solving least squares problems. Prentice-Hall, Englewood Cliffs, NJ
7. Ruf H, Haase W, Wang WQ, Grell E, Gärtner P, Michel H, Dufour JP (1993) Prog Colloid Polym Sci 93:159
8. Ruf H, Gould BJ (1998) Eur Biophys J 28:1
9. Ruf H (1989) Biophys J 56:67

Progr Colloid Polym Sci (2000) 115:259–264
© Springer-Verlag 2000

METHODS

A. Meister
G. Förster
A. Blume

Energy calculations related
to aliphatic chain-packing modes

A. Meister · G. Förster · A. Blume (✉)
Martin-Luther-University
Halle-Wittenberg
Institute of Physical Chemistry
Mühlpforte 1, 06108 Halle, Germany
e-mail:blume@chemie.uni-halle.de

Abstract X-ray measurements reveal that the orthorhombic gel phase of long-chain lipids shows a large variation in structural properties. These can be classified by systematic lattice energy calculations. Typical values are reported for the space group *Pbnm*. The experimental X-ray data of two samples (dihexadecylphosphatidylethanolamine, n-docosylphosphocholine) are used to obtain energy data for different indexing combinations in each phase. The most probable polymorphism was selected by applying a continuous energy criterion. It allows a discussion of packing frustration or freedom of the respective lipid in the bilayer or monolayer structure on heating.

Key words Long-chain compounds · Hydrocarbon chain-packing modes · Orthorhombic subcell · Van der Waals energy · Contour plot

Introduction

The chemical and physical properties of substances like paraffins, polyethylene, fatty acids, fats, lipids, soaps, or detergents in the absence or presence of water are strongly determined by their long aliphatic hydrocarbon chains [1]. Chain-packing modes in the crystalline state are usually analyzed using the concept of subcells within the main cells, the latter describing the three-dimensional packing of the molecules. In the case of mesomorphic subgel and gel phases in bilayers or in ordered phases of monolayers, the packing of the aliphatic chains is often the only structural information available using X-ray scattering techniques.

The conformation of the hydrocarbon chain $(-CH_2-)_n$ in crystals was described by Müller [2] as a zigzag plane (all-*t*). Bunn [3] calculated the electron-density distribution and the atomic positions of normal long-chain paraffins in an orthorhombic lattice (*Pnam*) from X-ray powder patterns. The development of the subcell concept by Vand [4] represents the basis of any further description of hydrocarbon chain-packing modes. After the pioneering work of Kitaigorodski [5] on packing principles for long-chain molecules, a first crystallographic overview of subcell modes was given by Segerman [6] followed by Abrahamsson et al. [7]. Based on single-crystal analysis they classified the packing in orthorhombic, monoclinic, and triclinic subcells, collected characteristic data, and also suggested hybrid subcells. Meanwhile, many experimental data exist which are not in close agreement with the reported typical values. To analyze changes in the X-ray powder diffraction patterns of simple and complex lipids, Maulik et al. [8] systematically varied the dimensions of hydrocarbon subcells and examined the effect of chain rotation at fixed lattice sites on the scattering behavior.

Considering the attractive forces between long saturated chains (van der Waals-London dispersion forces), Salem [9] calculated the total dispersion energy for two linear chains as a function of chain distance. Segerman [6] applied Salem's calculations to describe the various hydrocarbon chain-packing modes of long-chain compounds determined by single-crystal analysis. However, his attempt to relate the frequency of occurrence of the subcells to the van der Waals energy failed. Recently, Kuzmenko et al. [10] compared packing characteristics of alkyl chains in Langmuir monolayers by means of lattice energy calculations. At progressively increasing values of the dihedral angle ϕ between neighboring backbone planes of the chains, the cell dimensions were varied in order to minimize the lattice energy. The calculations revealed three minima for either a- or b-glide symmetry, corresponding to different packing modes. Apart from the common herringbone packing, Kuzmenko et al. [10] localized for the first time a so-called pseudo-herringbone arrangement.

In most cases, a clear differentiation of packing modes for long-chain compounds suffers from a lack of crystals suitable for single-crystal analysis. Although the measurement of X-ray powder diffraction patterns in the wide-angle range is very sensitive, the indexing of the reflections is based on trial and error, which prevents an unambiguous calculation of the lattice parameters. Therefore a distinction between the two common subcells $O\perp$ and $O'\perp$, for instance, is impossible without further information [11]. Tenchov et al. [12], who investigated dihexadecylphosphatidylethanolamine (DHPE) in a temperature range from -15 to 60 °C, observed a so-called "Y" transition and indexed the subcell reflections on the basis of $O'\perp$ (*Pbnm*) to bring the data in agreement with the additional density measurements.

The aim of this work is to demonstrate that lattice energy calculations are a powerful tool to differentiate between possible chain-packing modes. For this purpose, temperature-dependent X-ray powder diffraction measurements were carried out on two different lipids. Lattice energy simulations were performed to analyze the experimental data. A comparison with ideal packing modes gives an insight into the respective phase behavior.

Materials and methods

DHPE was purchased from Fluka Chemie (Buchs, Switzerland) and n-docosylphosphocholine (C22-PC) was synthesized by Heiser and Dobner [13]. Samples with 50 wt% water were prepared by dispersing a weighed amount of lipid into the required amount of water. The dispersions were homogenized by cycling 10 times between 75 °C and 0 °C and vortexed at these temperatures for 2 min.

X-ray studies were performed with a powder diffraction equipment in transmission technique (Stoe, Darmstadt) using $CuK_{\alpha1}$ radiation. On a stage, the monochromator and a $2\theta = 40°$ position sensitive detector was arranged, and a computer controlled temperature attachment (N_2 gas cooling) allowed 1 K step measurements between $-40 °C < T < 100 °C$ within 15 min. Data were processed with the program Origin (Microcal). The different peak parameters were obtained with a least squares fitting procedure for the whole pattern (correlation coefficient > 0.995). Peak positions were estimated with an accuracy of $\Delta s = \pm 5 \times 10^{-4}$ nm^{-1}. After indexing, the calculated lattice constants have typical error estimates of ± 0.001 nm. The calculated energy is as accurate as the breakoff criterion used in the software (0.01 kJ/CH$_2$). Designer (Micrografx) was used to handle the subcell data and Cerius^2 (Molecular Simulations) for the van der Waals energy calculations.

A new notation was used (Lβ_Y, Y = A...G) to differentiate between different gel phases. It is based on the van der Waals energy contour plot of the space group *Pbnm* and involves all possible orthorhombic packing modes [14].

Results and discussion

X-ray scattering

Two examples are given to demonstrate that the evaluation of the temperature dependence in the form of a contour plot is a prerequisite for the investigation of chain packing modes.

For comparison with recent literature data [12], a DHPE water suspension (50 wt%) was reinvestigated (Fig. 1). The X-ray contour plot in the range of the fingerprint reflections implies that in the temperature range between -30 °C (frozen excess water) and 10 °C only one orthorhombic gel phase exists, with a pronounced temperature dependence of the packing parameters. The strong fingerprint peaks merge in a characteristic "Y" shape. Usually, for a phase transition, a jump in the position of the reflections is observed together with a symmetry change and a two-phase coexistence. The temperature range for the high-temperature hexatic phase Lβ_B is surprisingly large and extends from 12 °C to the melting of the chains. Following the interpretation of Tenchov et al. [12] that the chain packing is of the type $O'\perp$ (*Pbnm*), the indexing of the fingerprint reflections is shown in Fig. 1 {(110), (200)}. In the recorded diffraction range, reflections with higher Miller indices can be observed {(020), (310)}, which also show a "Y"-shaped temperature dependence. The increase of the reciprocal spacing (020) is very distinct and can be used as an indicator for the phase Lβ_E. In the hexatic phase Lβ_B the $\sqrt{3}$ peak of the fingerprint reflection (Fig. 1, $s \approx 4.1$ nm^{-1}) becomes better visible the higher is the temperature.

The contour plot for the second example (C22-PC) shows that C22-PC has a pronounced gel phase polymorphism (Fig. 2). The reciprocal spacings in the wide-angle region for the low-temperature Lβ_C phase decrease with temperature. This temperature dependence is in contrast to the increase in the reciprocal spacings for the (110) reflection for the Lβ_E phase of DHPE (Fig. 1). Above the transition at -6 °C in C22-PC, the temperature dependence changes and becomes similar to that of the phase Lβ_E of DHPE. The intensity ratio of the two fingerprint reflections as well as their linewidths change at the transition Lβ_C-Lβ_E. At the transition from Lβ_E to Lβ_A and Lβ_F, the (200) reflection disappears. The observed shift of the fingerprint reflections to smaller reciprocal spacings corresponds to a shift in the lattice constants which is typical for the pseudo-herringbone packing [10]. Starting at the transition Lβ_E-Lβ_A, the peak positions as a function of temperature form a "needle"-like pattern. The hexatic phase Lβ_F, observed for C22-PC below chain melting, is different to the one observed for DHPE. The indexing given in Fig. 2 is relevant to the gel phase Lβ_C with the

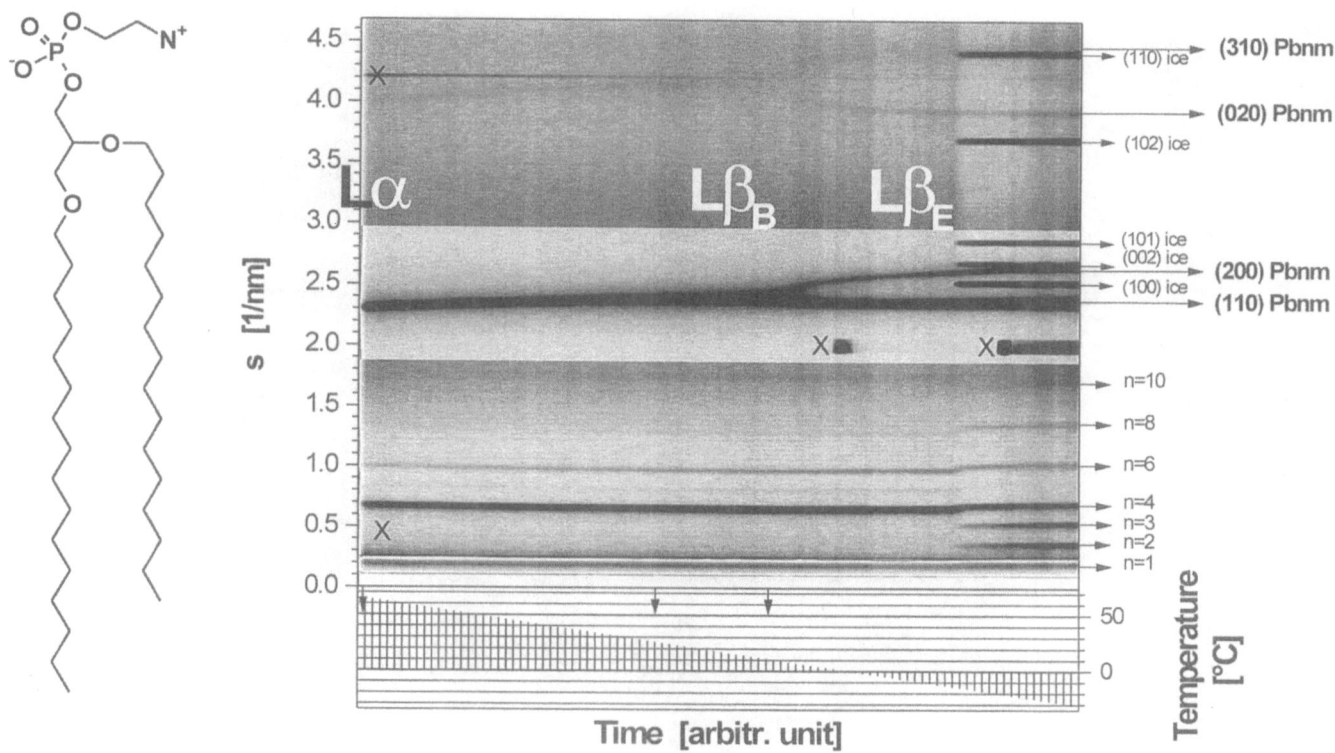

Fig. 1 Contour plot of the X-ray diffraction pattern of DHPE in the small- and wide-angle regions. The indices on the right are related to (1) the orders of the repeat distance, (2) the chain-packing mode $O'\perp$ (*Pbnm*), and (3) the ice reflections. Phase transitions are marked on the bottom by *arrows*, where the temperature is plotted versus the time. The phases are differentiated by a new notation. X marks artifacts in the pattern of unknown origin

assumption of an $O\perp$ subcell (*Pbnm*) and has to be changed for each of the other gel phases. For the first time it could be demonstrated that a crossing of the positions of the fingerprint reflections takes place at the transition $L\beta_E$-$L\beta_A$. It is an indication for the appearance of the pseudo-herringbone packing of the chains and is responsible for the "needle"-like shape in the contour plot (Fig. 2).

The presentation of the two contour plots clearly demonstrates that the common orthorhombic gel phase shows more variations in its appearance than was theoretically predicted until now.

Energy calculations

The indexing of the two fingerprint reflections in X-ray powder patterns is based on trial and error, and produces four orthorhombic lattices with inclined backbone planes (*Pbnm*). In order to determine the appropriate packing mode, static lattice energy calculations were carried out for all different packing modes using

the Lennard-Jones interaction potential of the DREIDING force field of the Cerius^2 software. By changing the dihedral angles of the chain fragments, the van der Waals energy of a given subcell was minimized. Systematic variations of the lattice constants a and b give a set of energy data which were plotted as a van der Waals energy contour map projected into the reciprocal lattice plane [14]. This map shows three local energy minima corresponding to the pseudo-herringbone arrangement of $O\perp$ ($L\beta_A$), the herringbone arrangements of $O\perp$ ($L\beta_C$) and $O'\perp$ ($L\beta_E$), and a latent minimum for the inverse pseudo-herringbone arrangement of $O'\perp$ ($L\beta_G$) (Fig. 3). Their energy values are given in Table 1 together with those of selected points with singular geometry and saddle points.

A comparison of these theoretically predicted energies and the energies of the most probable chain packing deduced from the experimental data shows that none of the measured values crosses the minima in the energy contour map. The energy increases more or less continuously (Table 1), and the s-values are localized on lines which are in close agreement to the literature [15, 16]. At a transition, the selection of the appropriate next phase was based on the criterion of a continuous increase in energy.

For DHPE, the temperature dependence of the fingerprint reflections is in agreement with the behavior of paraffins in the crystalline state (rotator phase [17]). Both the behavior of DHPE in the $L\beta_E$ phase and the

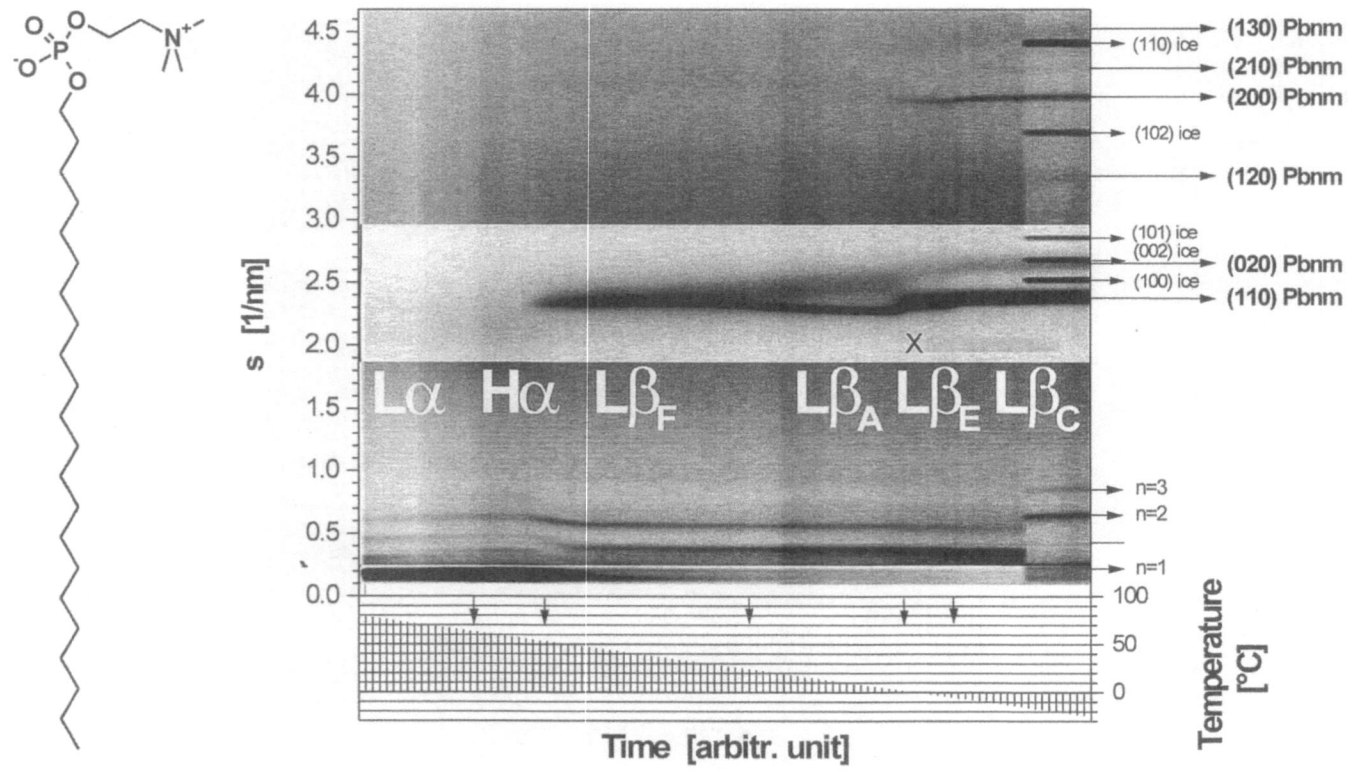

Fig. 2 Contour plot of the X-ray scattering of C22-PC in the small- and wide-angle regions. The indices on the right are related to (1) the orders of the repeat distance, (2) the chain-packing mode O⊥ (*Pbnm*) and (3) the ice reflections. Phase transitions are marked on the bottom by *arrows*, where the temperature is plotted versus the time. The phases are differentiated by a new notation. *X* marks artifacts in the pattern of unknown origin

pronounced temperature dependence of the reflections for the hexatic phase Lβ_B reflect the stabilizing influence of the headgroups. This, on the other hand, acts as a frustration with respect to the chain-packing modes. The structure of the Lβ_E phase was characterized by additional density measurements by Tenchov et al. [12]. Our calculations show that it does not have the lowest possible energy. The cooling to −30 °C was not able to transform this gel phase into an Lβ_C phase. In the predicted models for the hexatic phase the Lβ_B phase is energetically favored. Also, different models are discussed in the literature which consider dynamic contributions [18–20].

For C22-PC, several phase transitions are observed and one is forced to use a different indexing of the experimental scattering data. The resulting lattice energies could be sorted using the continuous energy criterion (Table 1). The chosen polymorphism is the most probable one, including the appearance of the pseudo-herringbone packing in Lβ_A. In the hexatic phase, the interpretation of the pattern as arising from an Lβ_F phase is based on the scattering in the wide-angle region. The rich polymorphism of C22-PC in comparison to DHPE seems to be due to the molecular shape (single-chain compound) and the freedom which is given by the full interdigitation of the molecules within the lamellar structure.

Fig. 3 Selected chain-packing modes in the space group *Pbnm*. *A* Lβ_A: O⊥, pseudo-herringbone, energy minimum; *B* Lβ_B: O⊥, hexatic, saddle point; *C* Lβ_C: O⊥, herringbone, energy minimum; *D* Lβ_D: squared, saddle point; *E* Lβ_E: O′⊥, herringbone, energy minimum; *F* Lβ_F: O′⊥, hexatic, transient state; *G* Lβ_G: O′⊥, inverse pseudo-herringbone, latent energy minimum (symmetry break)

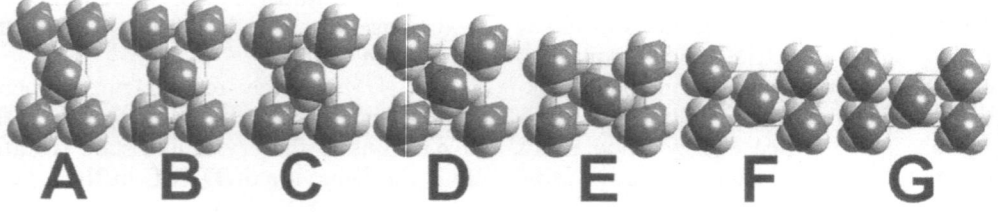

Table 1 Lattice energies of selected points in the van der Waals energy contour map, and of experimental data of DHPE and C22-PC

Temp. (°C)	Phase	a (nm)	b (nm)	Energy (kJ/CH$_2$ group)
Van der Waals energy contour map: local minima, saddle points, and points with singular geometry [14]				
	Lβ_A	0.431	0.851	−8.46
	Lβ_B	0.462	0.800	−8.21
	Lβ_C	0.508	0.727	−8.69
	Lβ_D	0.605	0.625	−7.91
	Lβ_E	0.733	0.506	−8.45
	Lβ_F	0.784	0.465	−8.37
	Lβ_G	0.816	0.459	−8.59
DHPE				
−30	Lβ_E	0.741	0.507	−8.39
	(Lβ_C)[a]	(0.507)	(0.741)	(−8.65)
5	Lβ_E	0.781	0.492	−8.12
	(Lβ_C)	(0.492)	(0.781)	(−8.33)
18	Lβ_B	0.474	0.821	−8.09
	(Lβ_F)	(0.821)	(0.474)	(−8.26)
50	Lβ_B	0.485	0.840	−7.74
	(Lβ_F)	(0.840)	(0.485)	(−7.84)
C22-PC				
−25	Lβ_C	0.500	0.741	−8.67
	(Lβ_E)	(0.741)	(0.500)	(−8.34)
−5	Lβ_C	0.503	0.749	−8.62
	(Lβ_E)	(0.749)	(0.503)	(−8.34)
−2	Lβ_E	0.757	0.503	−8.28
	(Lβ_C)	(0.503)	(0.757)	(−8.51)
2	Lβ_E	0.766	0.504	−8.20
	(Lβ_C)	(0.504)	(0.766)	(−8.42)
10	Lβ_A	0.448	0.863	−8.25
	(Lβ_G)	(0.863)	(0.448)	(−8.21)
25	Lβ_A	0.468	0.850	−8.02
	(Lβ_G)	(0.850)	(0.468)	(−8.21)
30	Lβ_F	0.825	0.491	−7.83
	(Lβ_B)	(0.491)	(0.825)	(−7.80)
50	Lβ_F	0.829	0.489	−7.86
	(Lβ_B)	(0.489)	(0.829)	(−7.80)

[a] Data in parentheses are less favored for energetic reasons, because of indexing arguments, or owing to additional experimental data

Conclusions

Temperature-dependent X-ray measurements reveal a large variability in the chain-packing modes in the orthorhombic gel phase. For the space group *Pbnm*, all possible arrangements were classified by means of static lattice energy calculations. Energy calculations were combined with experimental data to establish, in a trial and error process, the indexing of the X-ray powder patterns.

The progress and advantage of the experimental method presented here is based on the following steps: (1) use of the reciprocal space to values of at least 4.5 nm^{-1} to collect the characteristic scattering necessary for indexing, (2) study of the temperature dependence of the scattering behavior in a wide temperature range, (3) use of small temperature steps to characterize all phases and their temperature dependence.

The comparison between two long-chain lipids (DHPE, C22-PC) indicates that not all systems have the freedom for changing their chain packing within an overall layer structure. It should be the aim of further studies to find out a phase sequence for the different chain-packing modes which is applicable to all long-chain compounds.

Acknowledgements We thank U. Heiser and B. Dobner in the group of Prof. P. Nuhn at the Institute of Pharmaceutical Chemistry, Martin Luther University Halle/Wittenberg for the syntheses of C22-PC and the Deutsche Forschungsgemeinschaft (SFB 197) and the federal state of Saxony-Anhalt for financial support.

References

1. Small DM (1968) J Am Oil Chem Soc 45:108
2. Müller A (1928) Proc Roy Soc A 120:437
3. Bunn CW (1939) Trans Faraday Soc 35:482
4. Vand V (1951) Acta Crystallogr 4:104
5. Kitaigorodski AI (1961) Organic chemical crystallography. Consultants Bureau, New York
6. Segerman E (1965) Acta Crystallogr 19:789
7. Abrahamsson S, Dahlén B, Löfgren H, Pascher I (1978) Prog Chem Fats Other Lipids 16:125
8. Maulik PR, Ruocco MJ, Shipley GG (1990) Chem Phys Lipids 56:123
9. Salem L (1962) J Chem Phys 37:2100
10. Kuzmenko I, Kaganer VM, Leiserowitz L (1998) Langmuir 14:3882
11. Förster G, Reihs T (1997) Pharmazie 52:697
12. Tenchov B, Koynova R, Rappolt M, Rapp G (1999) Biochim Biophys Acta 1417:183
13. Heiser U, Dobner B (1997) unpublished results
14. Förster G, Meister A, Blume A (1999) in preparation
15. Peterson IR, Brezezinski V, Kenn RM, Steitz R (1992) Langmuir 8:2995
16. Moenke-Wedler T, Förster G, Brezesinski G, Steitz R, Peterson IR (1993) Langmuir 9:2133
17. Doucet J, Denicolo I, Craievich A (1981) J Chem Phys 75:1523
18. McClure DW (1968) J Chem Phys 49:1830
19. Yamamoto T (1985) J Chem Phys 82:3790
20. Ungar G, Masic N (1985) J Phys Chem 89:1036

Progr Colloid Polym Sci (2000) 115:265–269
© Springer-Verlag 2000

METHODS

S. Holzheu
H. Hoffmann

Influence of non-ionic adsorbing substances on the anomalous Kerr effect of hectorite dispersions

S. Holzheu · H. Hoffmann (✉)
Department of Physical Chemistry
University of Bayreuth, Germany

Abstract There is a long tradition in investigating colloidal systems by transient electric birefringence. One interesting phenomenon is the observation of an anomalous signal in the semi-dilute concentration regime of charged colloidal dispersions. The occurrence of the anomalous Kerr effect is experimentally well documented but only poorly understood theoretically. Some years ago we investigated the influence of the block copolymer F127 on the anomalous Kerr effect in dispersions of the synthetic clay mineral saponite. The adsorption of F127 led to a disappearing of the anomaly. In this study we expanded the investigation to other systems. We carried out experiments with the synthetic clay mineral hectorite and three different adsorbing substances: the non-ionic surfactant $C_{12}EO_9$, the block copolymer F127 and the polymer polyvinylpyrrolidone. There seems to be a general rule that the adsorption of non-ionic substances reduces the anomaly. Additional measurements of electric conductivity, surface tension and electrophoretic mobility are presented to clarify this general observation.

Key words Anomalous Kerr effect · Synthetic clay mineral · Adsorption · Hectorite

Introduction

In 1875, Kerr [1] observed that optical isotropic materials become birefringent when exposed to an electric field. This was termed the Kerr effect and is attributed to the orientation of single molecules or whole molecular assemblies in the electric field. Especially particles of colloidal dimensions are easy to align. Therefore the so-called electric birefringence has developed as a powerful characterisation technique for colloidal dispersions [2].

One popular experimental version of electric birefringence is transient electric birefringence (TEB). In TEB the orientation of the particles is achieved by applying a rectangular electric pulse. The resulting signal is normally particularly simple. With the onset of the pulse the birefringence rises monotonously and reaches a steady-state value (Fig. 1, left). After the pulse offset the birefringence decays and becomes zero again. The time constants of the buildup and decay are directly related to

the rotational diffusion constant of the particle. However, in some systems the TEB signal is more complicated [3]: the birefringence increases initially but then passes through a maximum and decreases (Fig. 1, right). Signals like this are usually referred to as the anomalous Kerr effect or just called the "anomaly". They can be described as an superimposition of two effects of opposite sign. If the amplitude of the second dominates, the curve passes through zero. The initial increase of the birefringence is attributed to an orientation of the particles with their longitudinal axis in the direction of the electric field. Hence the opposite sign of the second effect must be due to a perpendicular orientation of the particles. First attempts to explain the anomaly were based on an assumption of a permanent dipole perpendicular to the particle [4]. A more recent theory postulates a dipole moment originating from ion fluctuations [5]. These models work very well in reproducing the measured TEB signals, but it is difficult to predict the

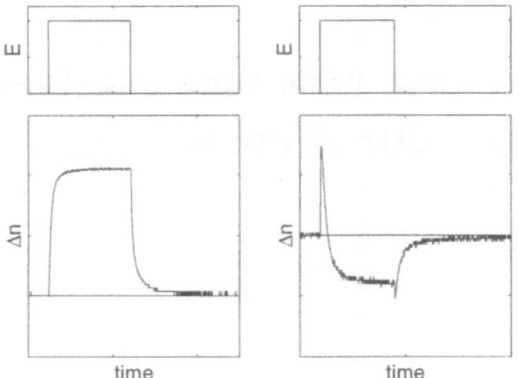

Fig. 1 Normal and anomalous transient electric birefringence (TEB) signals

influence of different parameters on the anomaly on the basis of these models. Therefore there is still a need for more systematic experimental data.

Some years ago we investigated the influence of the block copolymer F127 on the anomaly of the synthetic clay mineral saponite [6]. In this study we will present further measurements with other non-ionic adsorbing substances and another clay mineral.

Materials and methods

Materials

The synthetic clay mineral hectorite was obtained as a gift from Clariant, Germany. $C_{12}EO_9$ was prepared in a reaction of dodecanol with ethylene oxide in the mole ratio of 1:9. Pluronics F127 was received from BASF/Wyandotte and has a molecular weight of 12.500 g/mol. The EO content is 70% w/w. This corresponds to an average composition of $EO_{97}PO_{69}EO_{97}$. The molecular weight of the polymer polyvinylpyrrolidone (PVP) (K60, Fluka, Germany) is 160.000 g/mol. The chemicals were used without further purification. All samples were prepared in double distilled water.

Transient electric birefringence

TEB measurements were performed with a self-built instrument. The samples were filled into a quartz cell and thermostated at 25 °C. High-voltage rectangular pulses were generated by a high power pulse generator (Cober model 606). A He/Ne laser was used as a light source. All data were collected in the linear detection mode and converted into birefringence.

Other methods

For an independent characterisation of the samples, additional measurements were performed. Zeta potential was recorded using a Zetasizer 3000 (Malvern Instruments) and the Smoluchowski approximation. Surface tension measurements were carried out with a Lauda tensiometer TE1C based on the method of du Noüy. Electric conductivity was automatically measured in a titration experiment with a WTW LF2000 [7]. During all measurements the samples were thermostated at 25 °C.

Results

The behaviour of hectorite dispersions in the electric field is comparable to saponite dispersions. At very dilute concentration the TEB signal is normal and has a positive steady-state birefringence. With increased concentration a second effect with a negative sign develops and overcomes the first one. When approaching the sol-gel transition (about 7 g/l) the time constants increase and it is no longer possible to reach the stationary birefringence. After the sol-gel transition the anomaly has disappeared. For our study of the influence of non-ionic adsorbing substances on the anomalous Kerr effect of hectorite, we kept the hectorite concentration constant at 2 g/l. At this concentration the anomaly has already developed but the influence of the sol-gel transition can be excluded.

Figure 2 shows the influence of the block copolymer F127 on the anomaly of hectorite. Without the block copolymer the steady-state birefringence of the dispersion is negative. With enhanced F127 concentration the amplitude of the first effect increases and that of the second decreases. At 0.8 g/l F127 the steady-state birefingence is already slightly positive. At 2 g/l the second effect is no longer detectable and the signal is totally normal. However, the increase of the remaining first effect still goes on up to a concentration of about 3 g/l. Beyond that concentration the signal does not change very much and probably the maximum amount of adsorption is reached.

The influence of F127 on the anomaly of hectorite is consistent with former measurements with the clay mineral saponite, where we also found a disappearing of the anomaly with increasing block copolymer concentration. To clarify the question if it might be a general observation that non-ionic adsorbing substances make the anomaly disappear, we extended our examinations to a non-ionic surfactant and a polymer.

The influence of $C_{12}EO_9$ on the anomaly of hectorite is shown in Fig. 3. The effect of $C_{12}EO_9$ is similar to F127. At a concentration of 1.3 mM (0.76 g/l) $C_{12}EO_9$ the steady-state birefingence is positive. Up to a concentration of about 7 mM the amplitude of the first effect increases slightly. However, the second effect does not disappear totally. Even at very high concentrations a clear hump is visible in the TEB signal.

TEB signals with the polymer PVP are presented Fig. 4. Here the situation is more complicated as PVP leads to a flocculation of the hectorite particles. At high concentrations of PVP a redispersion can be observed. TEB measurements are restricted to one-phase samples. Therefore only two TEB signals are plotted in Fig. 4: one before flocculation and one after redispersion. The comparably low concentration of 0.2 g/l PVP is sufficient to yield a positive steady-state birefringence. After the redispersion at a concentration of 5 g/l PVP the anomaly totally disappeared.

Fig. 2 TEB signals of hectorite dispersions (2 g/l) with increasing amounts of block copolymer F127

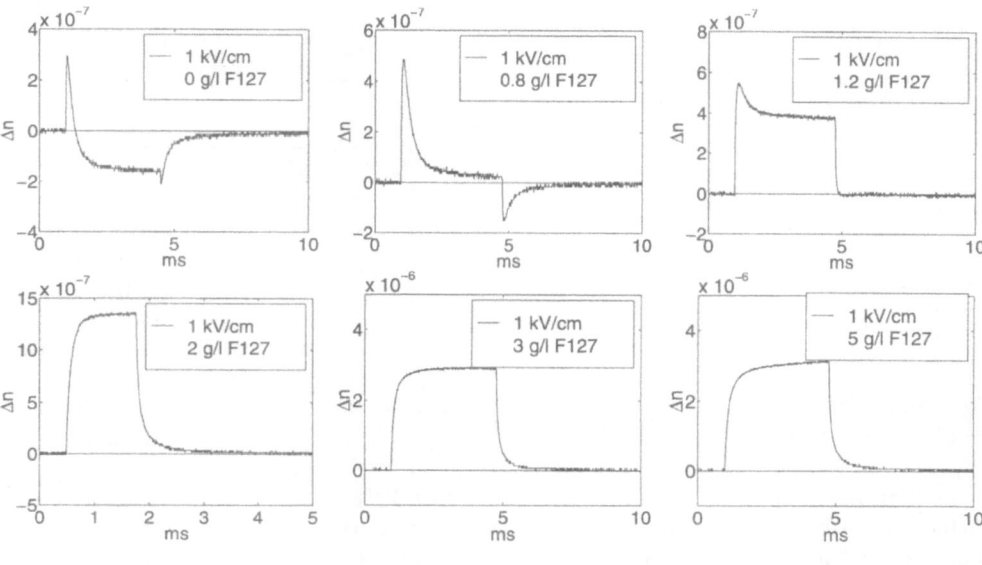

Fig. 3 TEB signals of hectorite dispersions (2 g/l) with increasing $C_{12}EO_9$ concentration

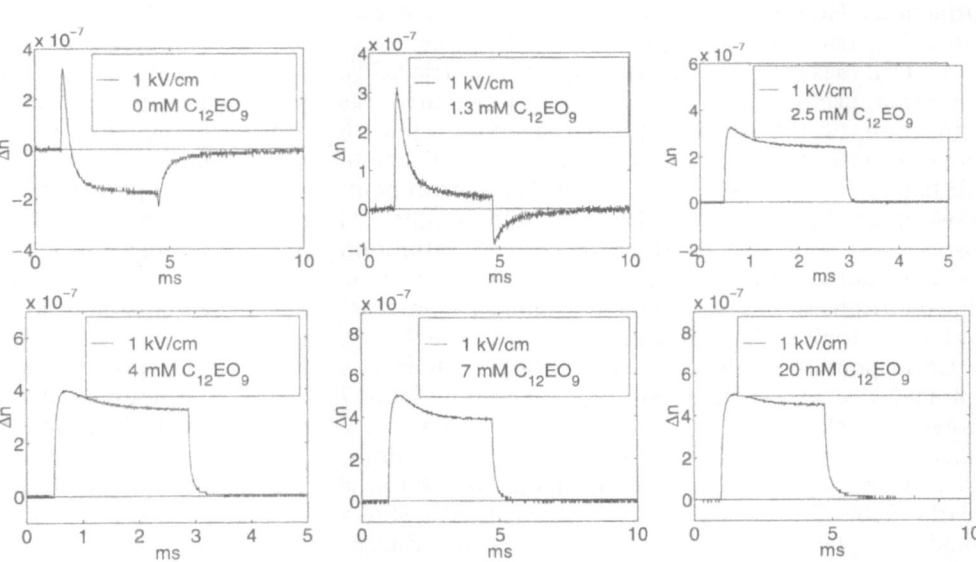

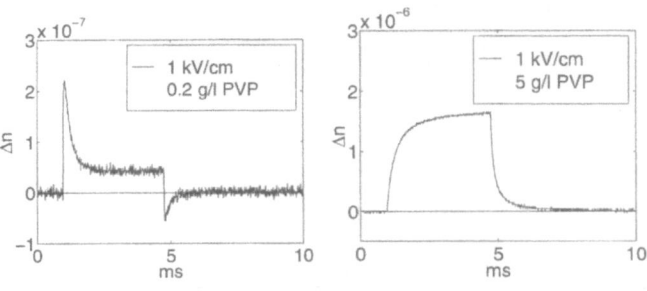

Fig. 4 TEB signals of hectorite dispersions (2 g/l) with increasing polyvinylpyrrolidone (PVP) concentration

To make sure that the observations are really an effect of adsorption of the non-ionic molecules, we performed surface tension measurements. Figure 5 shows the surface tension of $C_{12}EO_9$ with and without 2 g/l hectorite. The surface tension is related to the activity of the surfactant molecule. In the presense of hectorite the curve is shifted towards higher concentration. This means that on adding hectorite a given activity of the free surfactant is reached at higher concentrations. The difference between the concentration with and without hectorite yields the adsorbed amount of surfactant. The curve with hectorite shows two breakpoints. The first at about 0.1 mM can be related to the critical adsorption density at which admicell formation starts [8]. After the second breakpoint at about 6 mM the surface tension stays constant and saturation is reached.

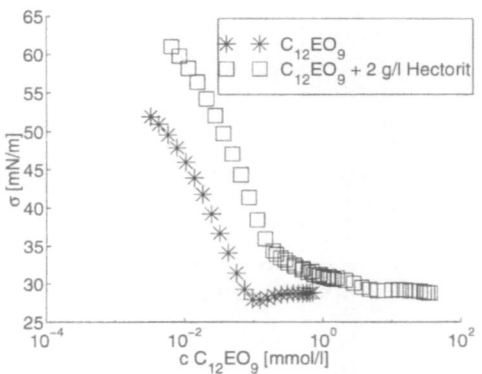

Fig. 5 Surface tension as a function of the $C_{12}EO_9$ concentration in the presence and in the absence of hectorite

Similar results can be obtained for F127. For PVP, surface tension measurements are not very sensitive as the surface activity of PVP is less than those of surfactants. However, we conclude from the flocculation and redispersion that adsorption does take place.

As TEB measurements are dealing with electric fields, it seemed promising to determine zeta potentials as a function of the surfactant concentration. Figure 6 shows the measurements for F127 and $C_{12}EO_9$. For both substances we observe two regions. The breakpoint between the fast and slow decrease of zeta potential corresponds well to the above-mentioned saturation concentrations. However, it is interesting to note that the decrease in zeta potential is more pronounced for the non-ionic surfactant than for the block copolymer.

Corresponding to the decrease in zeta potentials, one can find a decrease in electric conductivity. Figure 7 shows the relative variation of the conductivity as a function of the F127 and $C_{12}EO_9$ concentrations, respectively. Again the saturation of the adsorption is clearly visible. The drop in conductivity of about 8 $\mu S/cm$ could be explained by an additional counterion condensation of about 0.6 mM, equivalent to 11% of the total cation exchange capacity in the system. The absolute conductivity of a 2 g/l hectorite dispersion of 220 $\mu S/cm$

is dominated by background electrolyte originating from some impurities always present in synthetic clay minerals [9].

Discussion

Our experiments clearly indicate that the added non-ionic substances indeed adsorb on the hectorite particles. Consequently, the disappearing or – as in the case of $C_{12}EO_9$ – reduction of the anomaly must be due to this adsorption. Taking into account our former results [6, 10], we think that this is a general observation.

In spite of this simple experimental result we still have problems to understand the observation on a microscopic scale. A theory should be capable of explaining the reduction of the anomaly due to the adsorption as well as the difference between, for instance, $C_{12}EO_9$ and F127. As mentioned in the Introduction, most recent explanations for anomlous TEB signals are based on the ion-fluctuation theory [5, 11]. However, the authors point out that the description is phenomenological in the sense that the values of the parameters that enter are arbitrary and that it is difficult to relate them to microscopic properties. For this reason we want to pursue an empirical approach looking for microscopic parameters which could be correlated with the reduction and the disappearance of the anomaly.

Conductivity and zeta potential measurements indicate increased counterion condensation, but even at the saturation adsorption the hectorite particles show a negative zeta potential and hence a significant negative charge. It is not possible to relate the disappearing of the anomaly with a certain zeta potential: at 2 g/l F127 at a zeta potential of −40 mV the anomaly is not detectable, whereas the addition of 20 mM $C_{12}EO_9$ drops the zeta potential to −25 mV but does not make the anomaly disappear totally.

Also the maximum amount of adsorption does not vary very much between F127 and $C_{12}EO_9$. A difference will arise from the adsorption layer thickness and

Fig. 6 Zeta potential of hectorite particles as a function of F127 and $C_{12}EO_9$ concentration; the hectorite concentration is 2 g/l

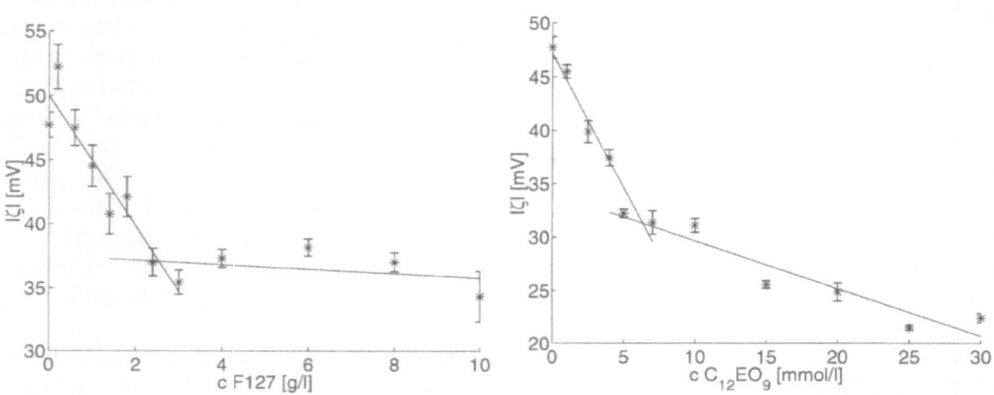

Fig. 7 Relative conductivity of hectorite as a function of F127 and $C_{12}EO_9$ concentration; the hectorite concentration is 2 g/l

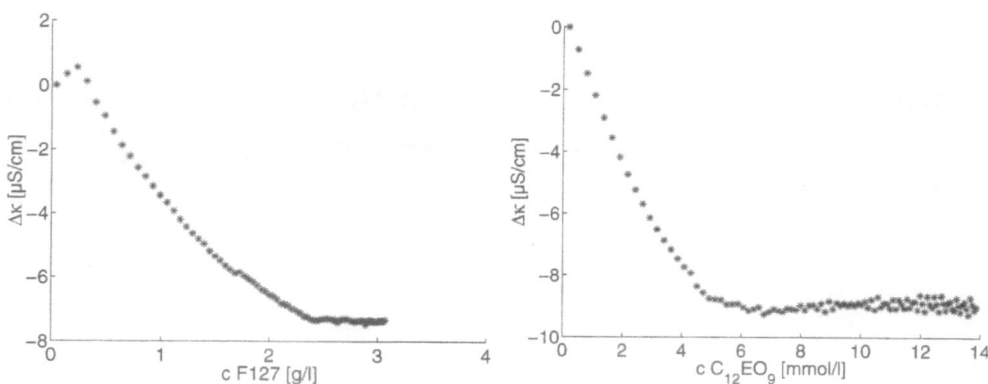

dynamics. For the non-ionic surfactant the adsorption layer thickness is restricted to the double length of the surfactant (about 4 nm). Adsorption layers formed by F127, in contrast, are known to have a thickness of more than 10 nm, which is mainly a result of EO chains penetrating into the solution. PVP at least should form an adsorption layer which resembles a random coil configuration.

Exchange kinetics is expected to decrease in the sequence $C_{12}EO_9$, F127, PVP. The larger the molecules, the more bonds can be formed between the hectorit surface and the adsorbing molecule. Consequently, the adsorption enthalpy becomes large compared to the thermal energy and the exchange kinetics slow down. Hence adsorption layer thickness and dynamics can be correlated to the degree of anomaly reduction. Of course, this does not give an explanation but these parameters could lead us one step further in the understanding of the anomaly and it may be worthwhile to look at them more carefully.

Acknowledgements We want to thank T. Behrends from the Department of hydrology for his assistance in zeta potential measurements.

References

1. Kerr J (1875) Philos Mag 50:337
2. O'Konski CT (1973) Molecular electro-optics. Dekker, New York
3. Hoffmann H, Krämer U (1990) In: Bloor DM, Wyn-Jones E (eds) The structure, dynamics and equlibrium properties of colloidal systems. Kluwer, Dordrecht, pp 385–396
4. Thureston GB, Bowling DI (1969) J Colloid Interface Sci 30:34
5. Yamaoka K, Sasai R (1999) J Colloid Interface Sci 209:408
6. Hecht E, Hoffmann H (1998) Tenside Surf Deterg 35:185
7. Hornfeck U (1998) Thesis, University of Bayreuth
8. Harwell JH, Hoskins JC, Schechter RS, Wade WH (1985) Langmuir 1:251
9. Tompson DW, Butterworth JT (1991) J Colloid Interface Sci 151:236
10. Yamaguchi Y, Hoffmann H (1997) Colloid Surf 121:67
11. Szabo A, Haleem M, Eden D (1986) J Chem Phys 85:7472

Progr Colloid Polym Sci (2000) 115:270–274
© Springer-Verlag 2000

METHODS

C. Urban
S. Romer
F. Scheffold
P. Schurtenberger

Structure, dynamics and interactions in concentrated colloidal suspensions and gels

C. Urban · S. Romer · F. Scheffold
P. Schurtenberger (✉)
Department of Physics
University of Fribourg
1700 Fribourg, Switzerland
e-mail: peter.schurtenberger@unifr.ch
Tel.: +41-26-3009115
Fax: +41-26-3009747

S. Romer
Polymer Institute, ETH Zürich
8092 Zürich, Switzerland

Abstract The application of static and dynamic light scattering to many colloidal systems of practical interest has often be considered too complicated owing to strong multiple scattering. There are two new approaches to overcome this problem. One of them aims at suppressing contributions from multiple scattering using novel cross-correlation schemes. While this relies on the suppression of multiple scattering, the so-called diffusing wave spectroscopy (DWS) works in the limit of very strong multiple scattering. DWS can be used for the characterization of dynamic and static properties of colloidal systems on a large range of time and length scales ranging from a few Ångstroms to hundreds of nanometers. We demonstrate that a wealth of information can be obtained from these methods on the structure, dynamics, interaction effects, stability, aggregation and sol-gel transition in colloidal dispersions.

Key words Cross correlation experiments · Diffusing wave spectroscopy · Static structure factor · Colloidal gels · Sol-gel transition

Introduction

Dynamic light scattering (DLS) is one of the most popular experimental techniques in the study of colloidal suspensions, and it has played a major role in these recent developments, where it has helped to gain invaluable data on the dynamic properties of colloidal systems. However, its application to many systems of industrial relevance has often be considered to be too complicated owing to the very strong multiple scattering in undiluted solutions. There are two recent solutions to this problem, which we applied in our study.

A very interesting approach aims at sufficiently suppressing contributions from multiple scattering from the measured photon correlation data [1]. The general idea is to isolate singly scattered light by performing two scattering experiments simultaneously on the same scattering volume and to cross-correlate the signals obtained. If both experiments then share the same scattering vector q, but use different scattering geome-

tries, only singly scattered light will produce correlated intensity fluctuations on both detectors. A particularly interesting scheme is the so-called 3D cross-correlation experiment in which initial and final wave-vector pairs are rotated by some angle about the common scattering vector $q_1 = q_2$. Several research groups [2–5] have recently demonstrated the feasibility of such an experiment and clearly shown that dynamic and static light can be used to successfully characterize extremely turbid suspensions. This opens up a completely new field of suspension characterization using DLS experiments combined with novel cross-correlation schemes in numerous areas of soft condensed matter research and technology.

The interaction between charged colloidal particles is of fundamental importance for the phase behavior of colloidal suspensions, as well as for their mechanical properties and, most importantly for applications in the paint and food industry, their stability against aggregation induced by van der Waals forces or depletion attraction. However, 50 years after the advent of DLVO theory the interactions between charged colloids are still under debate. A survey of the existing data on structure factors $S(q)$ in deionized suspensions reveals, for example, a characteristic discrepancy between theoretical predictions and experimental data. While integral equations are capable of reproducing the peak position and the height of the minima and maxima in $S(q)$ quite accurately if polydispersity is properly taken into account, experiments on deionized suspensions with light scattering show an enhancement of the low-angle scattering which by far exceeds the values one expects from a purely repulsive interaction, even if one accounts for the incoherent scattering due to polydispersity in particle size [6]. This enhanced forward scattering may originate from multiple scattering or may reflect an increased compressibility due to attractive long-range interactions, and finding a conclusive answer to this question remains one of the most challenging open problems in colloid physics. In order to exclude multiple scattering as a source for this discrepancy we have therefore performed a systematic study of $S(q)$ for latex suspensions using a 3D cross-correlation spectrometer in order to suppress contributions from multiple scattering.

A second approach to overcome problems caused by multiple scattering is based on a completely different principle. While DLS relies on the suppression of multiple scattering, the so-called diffusing wave spectroscopy (DWS) works in the limit of very strong multiple scattering, where a diffusion model can be used in order to describe the propagation of the light across the sample [7]. Using such a diffusion approximation, one can then determine the distribution of scattering paths and calculate the temporal autocorrelation of the intensity fluctuations analogous to DLS. It is thus still possible to study the dynamics of a colloidal suspension by measuring the intensity fluctuations of the diffusing light observed either in transmission or backscattering geometry. While DWS does not yield explicit information on the q-dependence of the so-called dynamic structure factor $S(q,t)$, it is capable in providing unique information on particle motion on very short time scales. DWS can probe particle motion on very short length scales, and it has been demonstrated, for example, that it can measure motions of particles of the order of 1 μm in diameter on length scales of less than 1 nm [7]. It is in fact possible to quantitatively determine the average mean-square displacement of the scattering particles $\langle \Delta r^2(t) \rangle$ from the measured intensity autocorrelation function over a very broad time scale of 10^{-6} s $\leq t \leq 10^4$ s. This allows us to obtain very important information on the dynamics of particle suspensions and gels that can, for example, directly be compared with the results from computer simulations.

The structural and dynamic properties of concentrated and strongly interacting colloidal suspensions have been the subject of intense experimental and theoretical investigations. The interest in these systems has been based on the use of colloids as ideal model fluids in soft condensed matter physics as well as from their considerable technological importance. However, despite this enormous effort made, there is still a considerable lack of knowledge in particular on the dynamics of concentrated particle suspensions. An especially interesting area is particle aggregation and gelation, and a significant number of experimental studies and computer simulations have been published during the past years [8]. However, the majority of studies have concentrated on dilute suspensions, and there is still a lack of knowledge on the sol-gel transition and on the dynamics of concentrated colloidal gels. In the second half of this article we demonstrate how DWS can be used to obtain quantitative information on such systems.

Materials and methods

Static structure factor measurements were done with suspensions of sulfate latex spheres (diameter: 110 ± 10 nm, IDC Corporation) with volume fractions in the range from 0.05% to 1% by volume. A mixture of ethanol and water with a volume ratio of 70% ethanol and 30% water was used in order to prevent crystallization of the fully deionized samples in the investigated concentration range. Alcohol changes both the Debye length and the association-dissociation equilibrium of the surface charge. The suspensions were filtered using a 0.8 μm membrane filter (Versapor acrodisc, Gelman) into the quartz sample cells (rectangular cells, 10×10 mm, and cylindrical cells with an inner diameter of 8 mm, Hellma). In order to obtain fully deionized samples, the suspensions were kept atleast 10 days in contact with equal volumes of mixed bed ion exchanger resin (Dowex) prior to a measurement.

The static light scattering experiments were performed using the 3D cross-correlation instrument described previously [4]. Combined static and dynamic light scattering experiments were made in order to suppress multiple scattering and obtain the full static structure factor as described in detail elsewhere [4, 5]. Highly diluted aqueous suspensions of the latex particles were also characterized by means of dynamic and static light scattering with a commercial goniometer system (ALV/DLS/SLS-5000F single mode fiber compact goniometer system). Furthermore, we determined the particle size distribution with transmission electron microscopy.

Aggregation and the sol-gel transition in concentrated suspensions were studied using monodisperse polystyrene spheres (sulfate latex particles, diameter 298 nm, 6% standard deviation, kindly synthesized and provided by Frank Horn, University of Freiburg, Germany) at a volume fraction of 20% in a buoyancy-matching mixture of H_2O and D_2O. The coagulation of the electrostatically stabilized dispersion is induced by the urease-catalyzed hydrolysis of urea [9]. The urea represents 10% of the solvent volume. The urease concentration (125 Units for 1 mL of H_2O and D_2O) is chosen such that the aggregation at a temperature of 20 °C occurs slowly over a period of days to ensure complete gelation. Similar results have also been collected with 44 Units/mL of urease. The gels thus obtained appear homogeneous and their features are fully reproducible.

The DWS set-up consists of solid state laser (diode pumped Nd: YVO4 solid state laser, 2 W at 532 nm, "Verdi" from Coherent). The beam is expanded to a diameter of about 7 mm and illuminates the sample as a uniform planar source. The light, passing through a rectangular cuvette (path length 1 mm, Hellma), is then collected in transmission geometry with a single mode fiber after a polarizer, which is perpendicular to the incident beam polarization. This ensures that only multiple scattered light will be detected. Because of possible after-pulsing effects of the detector, the signal is split and fed into two photomultipliers (Hamamatsu). A digital correlator (ALV-5000E, ALV) performs finally a pseudo-cross-correlation measurement.

Results and discussion

Static structure factor measurements

We have performed a systematic study of the static structure factor as a function of particle volume fraction with our 3D instrument, which allows us to determine the amount of multiple scattering and subsequently correct for it. Particular care was given to the precise determination of the sample polydispersity and a careful preparation in which all possible precautions have been made to prevent aggregation and remove all existing large particles. Examples of experimentally determined $S(q)$ together with the corresponding theoretical calculations using the polydisperse Rogers-Young closure [6] are shown in Fig. 1. The experimental data shown in Fig. 1 are in excellent agreement with the theoretical calculations. At low concentrations we do not yet reach

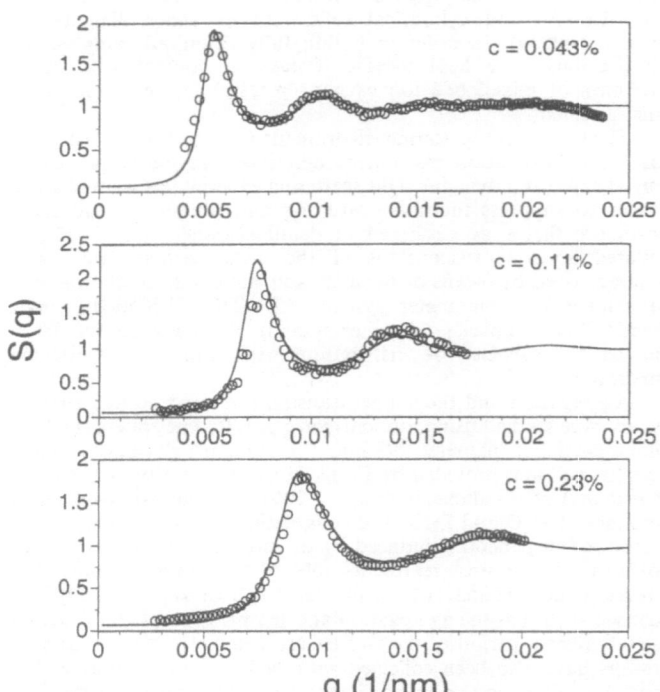

Fig. 1 Static structure factors measured with de-ionized latex suspensions (particle radius $r = 59$ nm) at different concentrations. The data points have been corrected using the 3D instrument as described elsewhere [5]. Solid lines are polydisperse Rogers-Young calculations

the asymptotic limit $S(0)$ required for an unambiguous test of the theoretical predictions. Under these conditions we are also limited by the very strong influence of trace amounts of dust particles, which cause enormous intensity fluctuations at scattering angles $\theta < 16°$. It is clear that a systematic test of the current theories requires an additional effort in sample preparation as well as an improved measurement protocol with a closed filtration/ion exchange loop and possibly a software dust filter. However, it is important to point out that, even under these conditions, multiple scattering cannot be neglected, and that the use of a classical static light scattering instrument would result in significantly too high values of the intensity at low values of q.

Our experiments clearly demonstrate that we can successfully measure $S(q)$ up to very high volume fractions and thus can now obtain important information on particle interaction effects in colloidal suspensions that previously were not accessible. This not only allows us to re-investigate the unsatisfactory situation with respect to the complete and quantitative understanding of colloid interactions and their interpretation via analogies to the theory of simple liquids. It also permits us to use light scattering methods in order to perform in situ investigations of structure, interaction and stability in a large variety of industrially relevant suspensions.

Sol-gel transition and gel dynamics

The coagulation of charge-stabilized concentrated latex dispersions with volume fractions typically in the range of $0.1 \leq \Phi \leq 0.4$ is induced through an increase of the ionic strength. In order to achieve a homogeneous variation of the ions throughout the sample without any gradient, we have chosen the urease-catalyzed hydrolysis of urea [9]. The onset of aggregation, the subsequent sol-gel transition and evolution of the gel is then followed in situ by means of DWS. By treating the transport of light in dense suspensions as a random walk, DWS extends the traditional light scattering techniques to the strongly multiple scattering regime and allows the study of very turbid suspensions [7]. As in all DLS experiments, the motion of the particles is then probed by monitoring the temporal fluctuations of the scattered light and their correlation function $g(\tau)$, where τ are the correlator lag-times. Figure 2A shows several typical snapshots out of a sequence of correlation functions $g(\tau) - 1$ during destabilization.

For ergodic systems, the mean square displacement of the correlated Brownian particles can be successfully modeled by means of an averaged short-time diffusion coefficient. This leads to a correlation function well approximated by a single exponential decay [7]. In Fig. 2B we present the correlation function of the stable suspension together with the corresponding fit to

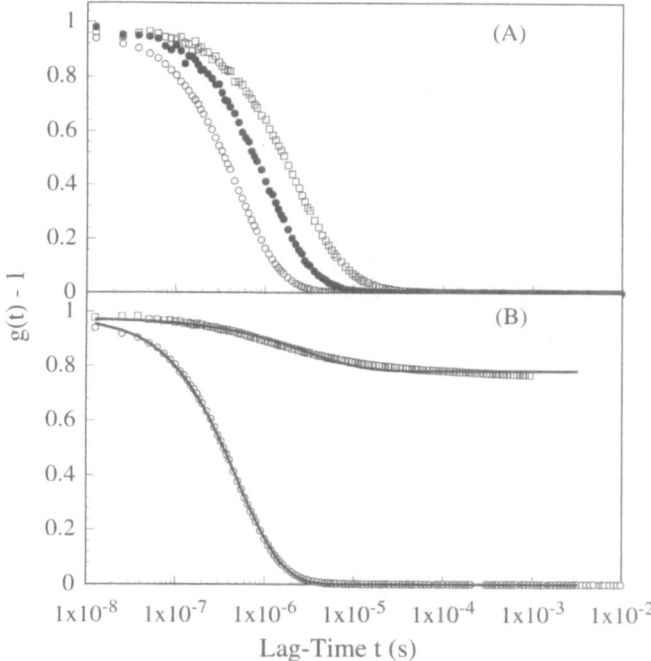

Fig. 2 A, B Autocorrelation functions $g(\tau)-1$ during aggregation of a destabilized latex suspension. **A** Stable suspension (*open circles*) and evolution for 10 and 25 min after destabilization. **B** Fits of the autocorrelation functions (*solid lines*) for the stable system described by a suspension of interacting Brownian particles and for the gel described by a stretched exponential decay to a finite plateau

shown in detail elsewhere [12], we overcome the nonergodicity limitation of light scattering in concentrated systems by using a combination of two cuvettes. The first cuvette contains the gelling sample which can be either ergodic or nonergodic, whereas the second one is a turbid ergodic system that leads to an additional decay of the correlated signal at long delay times. We thus obtain the true gel contribution in a wide lag-time interval from $10 \text{ ns} \leq \tau \leq 1 \text{ ms}$.

We can also compare our findings for concentrated particle gels with those obtained for dilute fractal gels using DLS. In a recent article, Krall and Weitz [13] have developed a simple theoretical model for such systems. The contributions of internal elastic modes lead to the arrested decay observed in the experimentally measured correlation functions. The dynamics of the trapped particles can be described by a stretched exponential decay to a finite plateau. We have extended this approach to DWS in order to study dense systems [12]. From the fit of the data for the terminal stage we see that the proposed gel model also describes the arrested dynamics quite well (Fig. 2B). However, it is clear that additional theoretical work will be required for a better understanding of the dynamics of concentrated colloidal gels, as some of the underlying assumptions in the model by Krall and Weitz may not be applicable in our case.

Conclusions

This work demonstrates clearly that colloidal systems of high turbidity can quantitatively be investigated by using either cross-correlation techniques or DWS. In cross-correlation experiments, a maximum of information can be recovered owing to the strong suppression of multiple scattered light, which allows for a measurement of the full q-dependence of the static and dynamic structure factor. An important result is the clear demonstration of the influence of multiple scattering when working with quite dilute systems. Significant amounts of multiple scattered light also can be present under these conditions, for example at scattering angles where the particle form factor or the static structure factor have a minimum, and strongly influence a quantitative analysis of the light scattering data. With the 3D instrument, such influences of multiple scattering can be avoided easily. While the application of the 3D cross-correlation technique still requires the presence of sufficient singly scattered photons, DWS allows measurements with highly turbid systems. We have in particular be able to show that we can follow the aggregation, sol-gel transition and ageing of the gel and obtain information on the (local) particle dynamics under these conditions from time-resolved DWS experiments.

a single exponential decay. Figure 2B clearly demonstrates that the initial stable system can be described very well as a suspension of interacting Brownian particles. However, at later stages in the aggregation process the clusters fill the entire sample volume and gelation occurs. Because of the high volume fraction the network is very stiff, and as a consequence the trapped particles can only execute limited, very slow motions about their fixed averaged positions. After about 10 days the gel changes only a very little and we take that time as the terminal state (Fig. 2B). The correlation function now exhibits a distinctly different decay, which is described by a stretched exponential with an arrested decay leading to a plateau.

However, at later stages we are confronted with an additional difficulty due to the nonergodicity of these systems. In solid-like media, such as gels, the scatterers are only able to make limited Brownian excursions. As a consequence, the time-averaged intensity correlation function of the scattered light generally measured in a DLS experiment is different from the ensemble-averaged correlation function. For dilute, nonergodic samples, different approaches have been proposed in order to properly average the signal [10, 11]. The extension to turbid systems is, however, not always practicable. As

References

1. Schätzel K (1991) J Mod Opt 38:1849–1865
2. Overbeck E, Sinn C, Palberg T (1997) Prog Colloid Polym Sci 104:117–120
3. Aberle LB, Hulstede P, Wiegand S, Schroer W, Staude W (1998) Appl Opt 37:6511
4. Urban C, Schurtenberger P (1998) J Colloid Interface Sci 207:150–158
5. Urban C (1999) Dissertation, No 13067, ETH Zürich, Zürich, Switzerland
6. Gisler T, Schulz SF, Borkovec M, Sticher H, Schurtenberger P, D'Aguanno B, Klein R (1994) J Chem Phys 101:9924–9936
7. Weitz DA, Zhu JX, Durian DJ, Pine DJ (1992) In: Chen S-H, Huang JS, Tartaglia P (eds) Structure and dynamics of strongly interacting colloids and supramolecular aggregates in solution. Kluwer, Dordrecht, pp 731–748
8. Brinker CJ, Scherer GW (1990) Sol-gel science: the physics and chemistry of sol-gel processing. Academic Press, San Diego
9. Graule TJ, Baader FH, Gauckler LJ (1994) J Mater Educ 16:243–267
10. Pusey PN, van Megen W (1989) Physica A 157:705–741
11. Xue J-Z, Pine DJ, Milner ST, Wu X-I, Chaikin PM (1992) Phys Rev A 46:6550–6563
12. Romer S, Scheffold F, Schurtenberger P (1999) Submitted to Phys Rev Lett
13. Krall AH, Weitz DA (1998) Phys Rev Lett 80:778–781

Progr Colloid Polym Sci (2000) 115:275–281
© Springer-Verlag 2000

J. O. Uhomoibhi
K. A. Dawson

Light scattering studies of the long time dynamics of foam

J. O. Uhomoibhi (✉)
Irish Centre for Colloid Science
and Biomaterials[1]
Department of Pure and Applied Physics
The Queen's University of Belfast
Belfast BT7 1NN
Northern Ireland, UK
e-mail: j.uhomoibhi@qub.ac.uk
Tel.: +44-1232-273110
Fax: +44-1232-438918

K. A. Dawson
Department of Chemistry
University College Dublin, Belfield
Dublin 4, Ireland

[1] Established at the Queen's University of
Belfast and University College Dublin

Abstract Foams are exemplars of non-equilibrium systems and are of fundamental scientific interest. They are naturally evolving and highly multiply light scattering. Information about the structure and dynamics of foam are largely inaccessible to traditional experimental measurements such as surface observation with microscopes (TEM, SEM, AFM, etc.), electrical conductivity or external pressure. The strong scattering of light from foams gives them their familiar white colour appearance, which precludes direct visualisation. We have exploited the strong multiple scattering of light inherent in foam to quantitatively probe its structural dynamics over long times (hours). Our foam sample comprised commercially available Gillette shaving cream contained in slab cells of varying thickness (1–20 mm). The 488 nm line of a CW argon ion laser was used. The results are in agreement with simulations. Multiple laser light scattering is a direct and non-invasive technique of potential industrial use in process monitoring for studies of opaque bulk materials.

Key words Dynamics · Foam · Waves · Diffusion · Scattering

Introduction

Foams are random dispersions of gas bubbles separated by thin, liquid films. By the term "bubble" we mean a spherical or near-spherical thin envelope of liquid together with another phase, which it encloses and which is usually a gas. When foam structure is destroyed, a homogeneous aqueous solution is left behind [1]. Foams have a myriad of important uses, from smothering fires, containing explosions and trapping toxic materials to dying garments, enhancing oil recovery and shaving [2]. They are also of broad scientific interest for their ability to efficiently fill space with random packing of bubbles and for the coarsening of this disordered structure with age [3]. Owing to their low density and unique rheological properties, foams have wide ranging applications. They behave elastically like a solid at low shear stress but flow like a fluid at higher shear stress. Foams are intrinsically unstable and tend to coarsen over time. The stability and temporal evolution forms the most important aspects of all foams. This is due to the fact that coarsening of foam is an example of a large class of non-equilibrium processes for which the dynamics are poorly understood [4]. Until now, most studies have been devoted to the understanding of the morphology of two-dimensional foams, which are much simpler and easier to understand than three-dimensional foams (the object of the present paper). The dynamics of the evolution of a two-dimensional foam structure is now well known [5]. Foams evolve in several ways. The liquid between bubbles may drain due to gravity or adjacent bubbles may coalesce if the liquid film becomes too thin and ruptures. Coarsening of foams can result from diffusion of gas from smaller bubbles to larger bubbles owing to pressure differences between bubbles of different sizes. This serves to decrease the total interfacial surface area with time. The coarsening process can be described in terms of a "statistically self-similar" size distribution, where the shape of the asymptotic distribution is independent of time when scaled by a single time-dependent length that grows as a power law [4].

In this paper we report on light scattering studies of the long time dynamics of three-dimensional foams. The techniques employed include diffusing wave spectroscopy (DWS) measurements in both transmission and backscatter modes and direct measurement of transmitted intensity through foam samples of varying thickness at different times (foam age). In contrast to the case of two-dimensional foams, the scaling behaviour and internal structure for three-dimensional foams is difficult to observe because the strong multiple scattering of light, which gives foams their white appearance, precludes them from direct visualisation. In recent times, following the development of DWS and other light scattering techniques, studies of optically thick media exhibiting a very high degree of multiple scattering are being undertaken. Here the diffusive nature of the transport of light in strongly scattering media is exploited to relate the temporal fluctuations of multiply scattered light to the motion of the scatterers [6]. Our light scattering programme of studies of foams and other opaque disordered materials is shown in Fig. 1.

An important consequence of multiple scattering is that DWS probes particle or other motions over length scales much shorter than the wavelength of light in the scattering medium [7].

In the multiple scattering regime, the characteristic time dependence is determined by the cumulative effect of any scattering events and hence by particles or bubble motion over length scales much less than the wavelength. This then implies that the characteristic time scales are much faster and the corresponding characteristic length scales are much shorter than for conventional dynamic light scattering (DLS), used on systems that exhibit relatively weak multiple scattering. Our present work focuses on understanding the changes in the dynamics of foams as it evolves into the long time

regime. We utilised a commercial shaving cream, Gillette Foamy Regular, which provides convenient and reproducible samples.

Theory

The goal of light scattering studies including DWS is to describe the propagation of light in a multiply scattering medium in terms of the diffusion approximation. The phase correlations of the scattered waves within the medium are ignored and only the scattered intensities are considered [8]. The path followed by an individual photon is described as a random walk. The validity of the diffusion approximation in a given system depends on the length scales over which the transport of light is described. For length scales less than the mean free path, l, of the coherent field, the wave equation for the electric field is used. On the other hand, for those longer than l, the phase correlation is considered only for limited specialised cases [6]. The use of the full wave equation is essential in cases of extremely strong scattering where it is predicted that light localisation will occur [8]. The transport of light can be described in terms of the intensity or energy density. For scatterers which are small compared to the incident radiation wavelength, λ, the scattering is anisotropic. The transport of the light energy density $U(r, t)$ can be described by the diffusion equation:

$$\frac{\partial}{\partial t} U(r, t) = D_l \nabla^2 U(r, t) \tag{1}$$

where D_l is the diffusion coefficient of the light. The average number of scattering events, n_0, required to randomise the direction of light propagation is greater than one. the length scale over which this randomisation occurs is the transport mean free path l^*, defined by:

$$l^* = n_0 l \tag{2}$$

where l is the distance over which the direction of light propagation is randomised by a single scattering event. So for $n > 1$ the diffusion equation is valid only over length scales longer than l^* and the diffusion coefficient of the light is:

$$D_l = cl^*/3 \tag{3}$$

where c is the speed with which light travels in the medium. When the sample dimensions and the distance that photons travel through the sample are greater than the transport mean free path, l^*, then the theory of DWS based solely on the diffusion approximations will provide an accurate description of temporal fluctuations of the scattered light. This approach considers the diffusive transport of individual photons directly in its description of the fluctuations in the scattered electric field. The path of each photon is determined by random, multiple scattering from a sequence of parti-

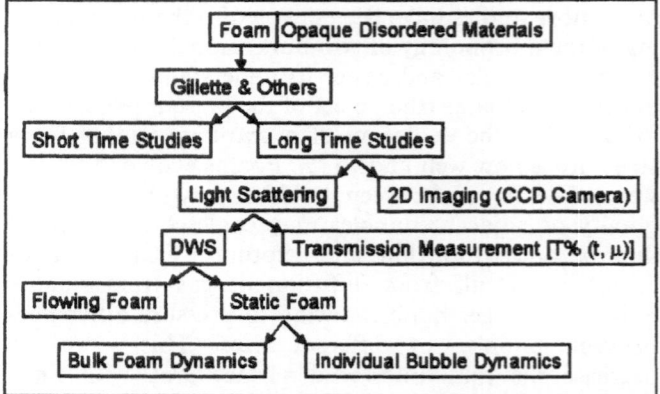

Fig. 1 Programme of studies of foam and opaque disordered materials using laser light scattering techniques

cles or other objects. The fluctuation in the phase of the electric field due to the motion of the scatterers is calculated for each light path. The contributions of all paths, approximately weighted within the photon diffusion approximation, are then summed to obtain the temporal autocorrelation function. The approach was originally developed to analyse temporal fluctuations in backscattering [9] and was later extended to transmission geometries.

Intensity and transmission for multiply scattered light

The light scattered by all the bubbles in our foam interferes and forms a random speckle pattern. If the bubbles move, the speckles fluctuate and hence the intensity varies with time. The variation of intensity in the speckle patterns reflect the motion of the bubbles [10]. For square-low detectors, like the one in our case, their $i(t)$, the instantaneous current output, is proportional to the square of the incident electric field, $E(t)$. That is, $i(t)$ is proportional to $|E(t)|^2$, which is in turn proportional to the intensity of the light.

The correlator computes the time autocorrelation function of the photomultiplier output and

$$\langle i(t)i(t)\rangle = B\langle |E(0)|^2|E(t)|^2\rangle \qquad (4)$$

where B is proportionally constant [11]. The scattered intensity autocorrelation function is given by

$$I_2(t) = \langle |E_s(0)|^2|E_s(t)|^2\rangle \qquad (5)$$

Since the scattered light, $E(t)$, in Eq. (4) is equal to the scattered field, $E_s(t)$, then $\langle I(0)I(t)\rangle$ is proportional to $I_2(t)$.

All the bubbles lying within the scattering volume at a given instant contribute light to the scattered field. The scattered field can be regarded as a superposition of fields from each of the scatterers so that

$$E_s = \sum E_s^{(n)} \qquad (6)$$

where $E_s^{(n)}$ is the scattered field from the nth bubble.

The static transmission, T, of light through a sample of thickness L is a simple optical quantity to measure. The diffusion of light is characterised by the transport mean free path, l^*, defined as the average distance a photon travels before its direction is randomised [8].

If an absorption length $l_a \gg l^*$ is included, the solution of the diffusion equation, Eq. (1), is given by

$$T \approx \frac{(5l^*/3L)\beta}{[1+(4l^*/3L)]\sinh\beta} \qquad (7)$$

where $\beta = \sqrt{3L^2/l^*l_a}$. The value of l^* is defined by the diffusion coefficient of light in the medium [4]. However, for large L and no absorption ($L \gg l^*$ and $\beta \ll 1$), Eq. (7) reduces to

$$T \approx \frac{5l^*}{3L} \qquad (8)$$

Correlation function in transmission and backscatter

The autocorrelation function for light scattering can be calculated by solving the Laplace transform of the diffusion equation. In this case, the source of diffusing intensity is taken to be a distance $z = z_0$ inside the illuminated face, where it is expected that $z_0 \approx l^*$. The solution of the diffusion equation, considering a point source on axis with the detector, is given using the Laplace transform $g(\tau)$ [13]:

$$g(\tau) \propto \int_{(L/l^*)\sqrt{(6\tau/\tau_0)}}^{\infty} \left[A(s)\sinh(s) + e^{-s(l-z_0/L)} \right] ds \qquad (9)$$

where

$$A(s) = \frac{(\varepsilon z - 1)\left[\varepsilon s e^{sz_0/L} + (\sinh(s)+\cosh(s))e^{-s(l-z_0/L)}\right]}{(\sinh(s)+\varepsilon s\cosh(s))^2 - (\varepsilon s)^2} \qquad (10)$$

and all other symbols retains their usual meaning. This time scale reflects the diffusive nature of the transport of the light. Since $l^*/L \ll 1$, the typical length scale over which bubble motion is probed by DWS in transmission is much smaller than the wavelength. Therefore, the decay of the autocorrelation function is due to the cumulative effect of many scattering events, so that the contribution of individual bubbles to the total decay is small. In the limit of large L and no absorption, an expression for $g(\tau)$ in transmission is given by [2]

$$g_T(\tau) \approx \frac{\sqrt{6\Gamma_1\tau}}{\sinh(\sqrt{6\Gamma_1\tau})} \qquad (11)$$

with $\Gamma_1 = (L/l^*)^2(1/\tau_0)$, where Γ_1, the first cumulant, is defined by the small τ behaviour, $-\ln[g_T(\tau)] \approx \Gamma_1\tau$. Γ_1 is evaluated numerically. The correlation function in backscattering has been calculated [6] and is given by

$$g_B(\tau) \approx \exp\left[-\gamma(6\tau/\tau_0)^{1/2}\right] \qquad 12$$

where $\gamma = \langle Z_0\rangle/l^* + 2/3$. The value of γ reflects the relative contribution of short paths to the backscattered light. These paths involve few scattering events and hence decay more slowly. γ depends on both polarization and on anisotropy of scattering as characterised by l^*/l [6]. A larger contribution of short paths leads to a slower decay of $g_B(\tau)$.

Materials and methods

For the measurements of the transmission and correlation function, the foam was sealed into rectangular glass (Hellma) standard cells of thickness L from 0.1 to 1.0 cm. We used Gillette shaving cream as our sample. this is the type of foam that is made up of bubbles of gas, suspended in a fluid and stabilised by surfactant molecules at the interfaces [2, 4]. The shaving cream is white due to the very strong multiple scattering of light from the bubble interfaces. It is highly reproducible with very low absorption of light. Its dynamics arise from coarsening which is attributed to two distinct mechanisms. In the first case, two bubbles approach and coalesce, forming a single layer bubble. In the second case, which is analogous to Ostwald ripening, coarsening arises from the slight solubility of the gas in the fluid. This introduces force due to internal pressure difference to drive the gas from the smaller bubbles to the larger ones. In both cases, the net result is the disappearance of the smaller bubbles so that the average bubble size increases over time [7]. The volume fraction of our foam sample is of the order of ~92% of space while still remaining relatively spherical because there is a distribution of bubble sizes at any given time in the life of the foam. The experimental set-up is as in an ordinary dynamic light scattering (see Fig. 2).

The foam sample was illuminated with a plane wave of argon ion laser light with wavelength 488 nm. A pinhole and photomultiplier detector arrangements were set up to observe a few speckles of the scattered light in both the forward and backscatter directions. Different behaviour is expected in DWS for forward and backscattering correlation functions. The correlation function is exponential in time in forward scattering, and in the square root of time in backscattering. For the intensity or correlation function in transmission, the experimental set-up was arranged as in Fig. 2.

For the transmission case, the laser beam is incident on the sample on the one side, scattered light being collected directly opposite the input point on the other. On the other hand, the correlation function of the backscattered light was measured by illuminating one side of the scattering volume and collecting the scattered light from the same side. The angle of scattering was set at 30°. In both cases the photomultiplier output pulses were processed by an amplifier and discriminator and the train of photon-detected pulses was analysed using the Brookhaven Instruments BI-9000AT digital correlator. The correlator computes the temporal intensity correlation function by multiplying the intensity at time t by the intensity at time $(t + \tau)$, where τ is the delay time, and then averages the product. Hence, the intensity correlation function that is measured is $g(\tau) = \langle I(t)I(t = \tau)\rangle$.

Results and discussion

Multiple light scattering has been exploited in this work to develop quantitative probes of the long time dynamics of foam by approximating the transport of light as a diffusive process. This is a completely different approach compared with the earlier assumption that foam can be modelled as periodic layers of liquid and gas. In this section we present the light scattering results for transmission of light through the foam and the correlation functions of the scattered light using the forward and backscatter geometries. It has been emphasized that l^* is important because it is determined by the foam structure and directly reflects the average bubble size [4]. This implies that the foam structure may be parametrised by a single scale length, that is the bubble diameter. A small value of l^* implies a long optical path in the medium, and consequently more scattering. This is to be expected for an optically dense material such as foam. It was suggested that l^* can be determined independently by measuring the static transmission, so that the transmission provides a measuring probe to noninvasively determine the bubble size and its time evolution.

The first measured quantity was the static intensity of light transmitted through a thickness, L, of foam, which is the simplest optical property of the system. The thickness dependence is shown for foams of different ages in Fig. 3. From this, we make the following observations. First, the transmitted intensity, I, increases with the foam age as the average bubble size grows. We note that light transmitted through the foam slab was completely depolarised. The second point is the diffusive nature of light, which is characterised by the transport mean free path, l^*. In line with what has been suggested [4], l^* was found to be determined by the foam structure and reflected the average bubble size. The value of l^* is defined by the diffusion coefficient of light, $cl^*/3$, where c is the speed with which the light travels through the

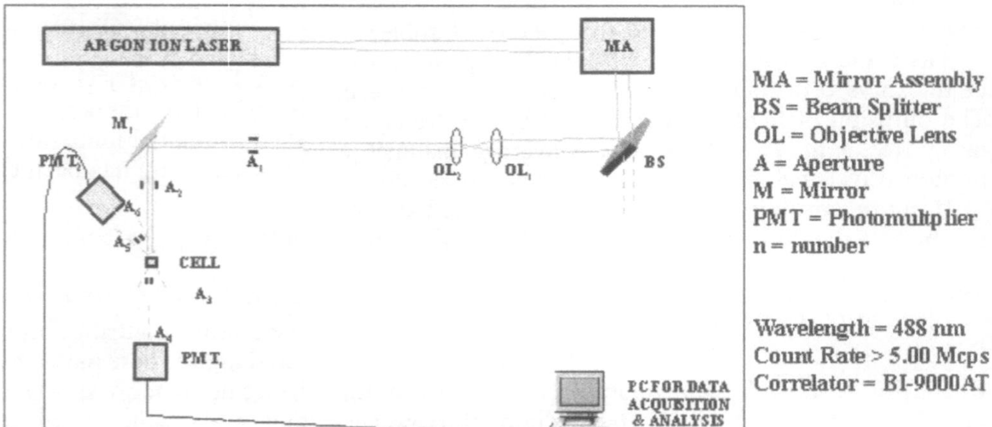

Fig. 2 Experimental arrangement for transmission and backscatter DWS measurements

MA = Mirror Assembly
BS = Beam Splitter
OL = Objective Lens
A = Aperture
M = Mirror
PMT = Photomultiplier
n = number

Wavelength = 488 nm
Count Rate > 5.00 Mcps
Correlator = BI-9000AT

effective background medium. Light is absorbed in the system, as well as scattered. To help understand the effect of l^* on the transmission, we recall Eq. (7).

Here it becomes clear that, at large l^*, the first term in the equation, $(5l^*/3L)/[1 + (4l^*/3L)]$, dominates, whereas for small l^* it is entirely the second term, $\beta/\sinh\beta$, which affects the transmission. Our data seem to relate more to the small l^* curves. This may be related to the approximate nature of the boundary condition used. It may also be due to geometry effects. The time evolution of the transmitted intensity from our multiple light scattering experiment with foam samples of varying thickness was found to be similar to other observations already made [4].

This illustrates the scaling behaviour in the coarsening foam and suggests that the transmitted intensity, and hence l^*, do directly reflect the foam structure and bubble size. Bubbles do not merge in foam. They change their size as gas diffuses from the smaller to the larger ones. Analysis of the transmission data shows that the transmitted intensity increases exponentially with the average bubble diameter. This confirmed the argument that the transmitted intensity is smaller at low values of the transport mean free path.

The time variation of intensity results in a broadening of the spectrum of scattered light. The optical spectrum is mathematically related to the autocorrelation function of the scattered electric field by Fourier transformation. This measures how quickly the field loses all similarity to its value at a given instant. Theoretically, if the correlation function of the spectrum is measured it is possible to infer the other [10]. The fundamental correlation function, known as the first-order autocorrelation function, $g_T(\tau)$, is defined by the product of the fields at time, t, and a later time, $t + \tau$, $g_T(\tau) \approx$

$\langle E_s(\tau)E_s^*(t + \tau)\rangle$, where the angle brackets indicate averaging the product for many values of t.

The normalised electric-field correlation function for the transmitted light is investigated here. In Fig. 4, typical data collected at different foam ages in a sample of 0.5 cm thickness show the evolution of the correlation function. The shape of correlation function, $g_T(\tau)$, is as expected nearly exponential in τ with a first cumulant defined by the small τ behaviour, $\ln[g_T(\tau)] \approx \Gamma_1\tau$. Over long time, Γ_1 displays scaling behaviour for foams at specific age (see Fig. 4A).

In contrast to transmission, in backscattering there is no definite characteristic path length set by the sample thickness [6]. Paths of all lengths contribute in backscattering and the correlation function therefore consists of contributions from all orders of multiple scattering. There is consequently a significantly broader distribution of time scales in the decay. The longer paths consist

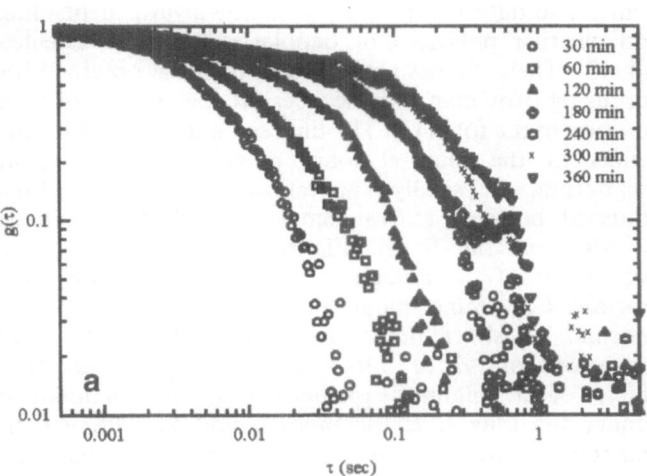

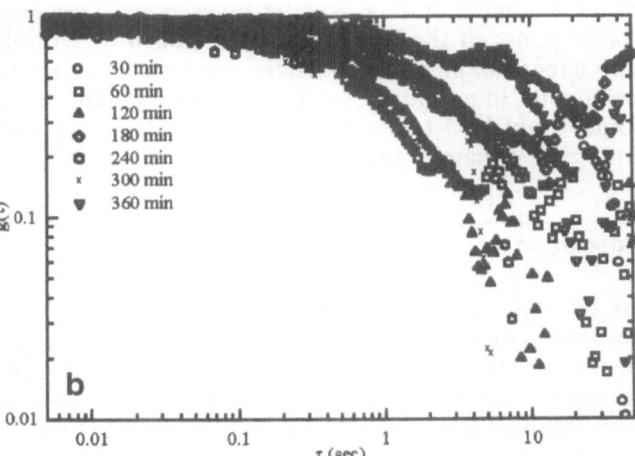

Fig. 4 Normalised field autocorrelation functions for light **a** transmitted through and **b** backscattered from a 0.5 cm thick cell. Data were collected for 5 min for the different foam ages up to 360 min as indicated on the graph

Fig. 3 Light intensity transmission as a function of foam thickness at foam ages between 30 and 360 min for wavelength $\lambda = 488.0$ nm. An inverse relationship is exhibited between foam age and foam thickness. More light is transmitted through the sample with increasing foam age

of a larger number of scattering events and thus decay more rapidly. On the other hand, the shorter paths consist of a smaller number of scattering events decaying more slowly.

A set of correlation functions measured in backscattering at different foam ages from a slab thickness of 0.5 cm at 30° scattering angle are shown in Fig. 4B. Here the normalised correlation function is plotted as a function of time after subtracting the background. On further analysis and within the precision of experiment, the data shown in Fig. 3 were found to be nearly linear when plotted against the square root of the delay time. This suggests that, in the limit of large L and no adsorption, the data can be simply described using $g_B(\tau) \approx \exp(-2\sqrt{6}\tau/\tau_0)$, where τ_0 is the time constant.

The behaviour of the correlation functions in Fig. 4 seems to be a general occurrence. For young foams, and hence small bubbles, there is some curvature for data plotted logarithmically versus the square root of the time. The data curve upward or downward, depending on whether polarised or depolarised light is detected [4, 12]. These shapes are identical to those obtained for diffusing Brownian particles, despite the absence of such motion in the foam [2]. The time constant, τ_0, reflects the effect of the internal foam dynamics on the light scattering. Physically, τ_0 reflects the average time interval between rearrangement events at any single location in the foam. Figure 5 shows the scaling behaviour of the dynamics of foam in the long time regime. For young foams, the first cumulant Γ_1, under normal or short time conditions with experimental durations of ~2 or 5 min, gives rise to values of a factor of ten magnitude greater than the results obtained under the long time dynamics condition with experimental durations >30 min. This outcome may be attributed to the method of averaging utilised by DWS in the treatment of data. At later times as the foam ages, the response of the foam is still very much discernible. An understanding of the underlying reasons for this behaviour, in naturally evolving complex systems of this kind, could lead to the development of techniques for long time dynamics process monitoring in industries.

Conclusion

In general, time elements enter into all rheological responses of matter and, in this case, foams. This is because it takes time for a sample to comply with stress (external and internal). The basis of rheology rests on two idealised cases: relaxation time at infinity (elastic) and retardation time at zero (flow). Very slow or long time experiments reveal mechanisms that are not discovered in normal or short time experiments. In all, these cases show components that are finite and measurable, provided sufficient time is given.

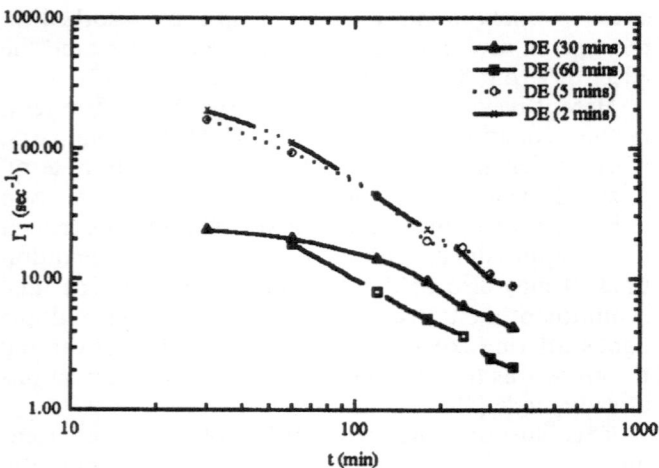

Fig. 5 Variation of the first cumulant Γ, (s^{-1}) with foam age t (min) for different experimental durations (DE) from normal (2 and 5 min) and long time (30 and 60 min) dynamics measurements

The aim of the set of experiments reported here was primarily to investigate the long time dynamic behaviour of a commercial foam, which provides a convenient and reproducible experimental sample. Secondly, it was also to demonstrate the usefulness of light scattering techniques in studying foam dynamics by exploiting the strong multiple scattering that is typical in the sample through DWS and transmission measurements. The extraction of meaningful information from experimental data is intrinsically linked to the experimental procedure.

On one hand, the static intensity of light transmitted through several different thickness of foam was measured and the data points have a reasonable agreement with the theory. The diffusive nature of light characterised by the transport mean free path has been investigated. The transmitted intensity values of l^* are comparable to those already reported.

The coalescence of bubbles via film rupture has never been observed; foam coarsening results solely from gas diffusion between bubbles. One of the most important results obtained is the scaling behaviour of foam. Correlation techniques have been used to study the dynamic evolution of fluctuations caused by the rearrangement process of bubbles in foam. The shape of the correlation function $g_T(\tau)$ in transmission is nearly exponential in τ and the first cumulant Γ_1 of $g_T(\tau)$ reflects the rearrangement events. Evolution of Γ_1 with time shows a power-law relation illustrating a scaling behaviour. Moreover, the time constant τ_0 of the correlation function in backscattering again exhibits scaling behaviour and the backscattering correlation function, $g_B(\tau)$, was found to be nearly exponential in $\sqrt{\tau}$. A physical interpretation of τ_0 requires the identification of the nature of the internal dynamics and development of a model for the resultant temporal fluctuations in the scattered light. The shapes of $g_B(\tau)$ are identical to those obtained for diffusing Brownian particles, despite the absence of such motion in the foam. The rate of rearrangement events can be determined from temporal fluctuations of scattered light and it has a scaling behaviour, which decreases as a power-law in time. The rearrangement events may be due to changes in packing conditions.

The data presented here provide clear evidence of scaling behaviour in the coarsening of a three-dimensional foam. However, the physical mechanisms that lead to the rearrangement events require further investigation. Finally, the present results point to the fact that laser light scattering is an effective probe of long time dynamics of self-evolving, non-equilibrium systems such as foam.

References

1. Sebba F (1987) Foams and biliquid foams aphrons. Wiley, New York
2. Durian DJ, Weitz DA, Pine DJ (1991) Science, 252:686
3. Tritton DJ (1988) Physical fluid dynamics, 2nd edn. Clarendon Press, Oxford
4. Durian DJ, Weitz DA, Pine DJ (1991) Phys Rev A 44:R7902
5. Weaire D, Rivier N (1984) Contemp Phys 25:59
6. Pine DJ, Weitz DA, Zhu JX, Herbolzheimer E (1990) J Phys, France 51:2101
7. Weitz DA, Zhu JX, Durian DJ, Gang H, Pine DJ (1993) Phys Scr T49:601–621
8. Ishimaru A (1978) Wave propagation and scattering in random media, vols 1 and 2. Academic Press, London
9. Pine DJ, Weitz DA, Maret G, Wolf PE, Chaikin PM, Helbolzheimer E (1990) In: Sheng P (ed) Classical localisation of waves. World Scientific, Singapore, pp 313–372
10. Steer MW, Picton JM, Earnshaw JC (1985) In: Callow JA, Woolhouse HW (eds) Advances in botanical research, vol 2. Academic Press, London, pp 1–69
11. Berne BJ, Pecora R (1976) Dynamic light scattering with applications to chemistry, biology and physics. Wiley, New York
12. Mackintosh FC, John S (1989) Phys Rev B 40:2383
13. Carslaw HS, Jaeger JC (1990) Conduction of heat in solids, 2nd edn. Clarendon Press, Oxford

Progr Colloid Polym Sci (2000) 115:282–286
© Springer-Verlag 2000

METHODS

C. Dwyer
P. Allen
V. Buckin

Temperature dependence of the ultrasonic parameters of bovine muscle: effects of muscle anisotropy

C. Dwyer · V. Buckin (✉)
Department of Chemistry
University College Dublin
Belfield, Dublin 4, Ireland

P. Allen
National Food Centre
Dunsinea, Castleknock
Dublin 15, Ireland
e-mail: vitaly.buckin@ucd.ie
Tel.: +353-1-7062371
Fax: +353-1-7062127

Abstract Ultrasonic velocity and attenuation in muscle were measured as a function of temperature using a high-resolution ultrasonic resonator technique. The effect of anisotropy on these ultrasonic parameters over the temperature ramp was also investigated. High resolution of our measurements enabled us to obtain the derivative of ultrasonic velocity as a function of temperature. The ultrasonic parameters in both muscles, with fibres of parallel and perpendicular orientation to the direction of propagation of ultrasonic waves, underwent profound changes with temperature. Muscles with fibres of parallel orientation showed several marked peaks in the velocity differential plots in the temperature interval between 280 and 315 K. The perpendicular samples showed significantly less temperature dependence. Attenuation increased in the parallel sample between 290 and 295 K and decreased between 300 and 310 K, however, these changes were absent in the sample with fibres of perpendicular orientation. Some of these transitions are well correlated with known phase transitions in muscle lipids and proteins. Overall, our results demonstrate that the resonator technique can be used to determine the structural changes in muscles with temperature.

Key words Ultrasound · Ultrasonic velocity · Ultrasonic attenuation · Muscle · Temperature transitions

Introduction

Muscle is an important biological tissue that also has commercial value as a food and a food-processing ingredient. The structure of muscle undergoes many physical and chemical changes with temperature, which are of extreme importance in its biological function and for subsequent processing in the food industry. Owing to the optical non-transparency of biological tissues, many of the methods traditionally used to analyse structural changes in biological systems cannot be easily applied to muscle. Therefore, for both research purposes and industrial applications, there is a need for new techniques that can rapidly, cheaply and non-destructively analyse these temperature-related transitions and determine their effect on muscle structure.

One of the new techniques with significant potential in tissue studies is ultrasonic spectroscopy, which is based on the measurements of parameters of low-intensity ultrasonic waves propagating through materials. An advantage of ultrasonic techniques is the ability of ultrasonic waves to propagate through optically non-transparent materials, and therefore they can be used in muscle studies. Ultrasonography has been used for many years as a diagnostic medium to detect abnormalities in body tissues, such as muscle [1–3], and also to determine muscle composition and muscle/fat depth in animals [4, 5]. Measurements of ultrasonic velocity and attenuation have been carried out in food systems and these parameters have been shown to be sensitive to composition, microstructure and phase transitions in different foods [6–8]. Most of these measurements were performed using the so-called pulse technique [9, 10]. In the pulse technique, the ultrasonic velocity, u, is determined through direct or indirect measurements of the time of propagation through the food sample of the transmitted ultrasonic pulse and ultrasonic attenuation, α, through the change of amplitude of the pulse. The pulse technique has been used previously in studies of the transmission of ultrasound in mammalian tissues [11–14], to determine the composition of muscle [15, 16], to measure the back fat thickness of animals [17, 18] and in ultrasonic spectral analysis of muscle sensory attributes [19]. In spite of its wide use, the pulse technique has some limitations. For high-resolution measurements, a relatively large volume of measuring sample is required. The large sample volume makes it difficult to control the temperature of the sample, which is essential in ultrasonic velocity measurements and for the analysis of temperature transitions.

A new high-resolution ultrasonic resonator technique was developed in the last decade [20, 21]. In this technique, the ultrasonic velocity, u, and ultrasonic attenuation, α, are determined through measurements of the frequency of the acoustical resonance and from the energy losses in the resonance, respectively. The resonator technique allows us to perform the measurements with higher resolution and it does not require a large volume-measuring cell as in the pulse technique, enabling ease of temperature control. It has been successfully applied previously in the analysis of biopolymers in dilute solutions, where high resolution is a key factor [22–24]. In the present work, we used the resonator technique to analyse temperature-related transitions in skeletal muscle. We measured ultrasonic velocity and ultrasonic attenuation in muscle samples as a function of temperature and the changes in these parameters correlated with the known phase transitions that occur in muscle with temperature. Anisotropy (muscle fibre direction) was shown to have an important effect on both velocity and attenuation.

Materials and methods

The ultrasonic resonator technique

Our measurements were carried out in a specially designed resonator cell. The resonator cell consists of two main components: the resonator chamber, where the acoustical resonance is formed, and the piezotransducers, which excite and detect ultrasonic vibrations. One piezotransducer excites the ultrasonic wave travelling in the direction of the second and this wave is reflected from the second transducer back to the first and reflected again to the second transducer. At the frequencies corresponding to the whole number of half-wavelength between the piezotransducers, a resonance occurs resulting in an increase in the amplitude of the signal at the second piezotransducer and the change in the phase (Fig. 1A).

The experimental set-up is shown in Fig. 1B. Lithium niobate piezoelectric transducers, which had a resonance frequency of

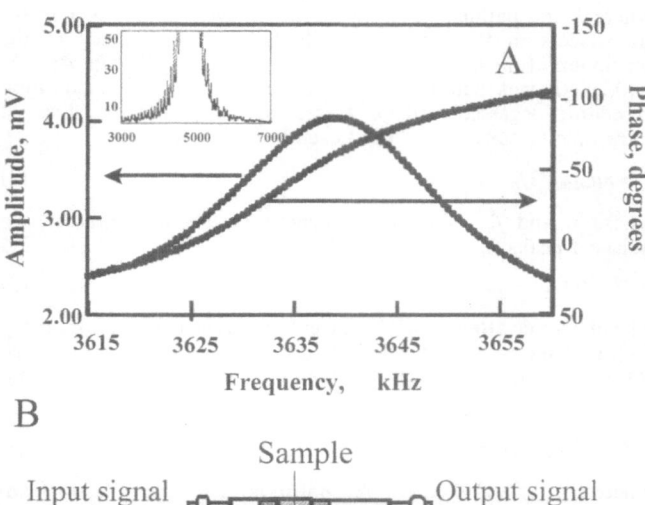

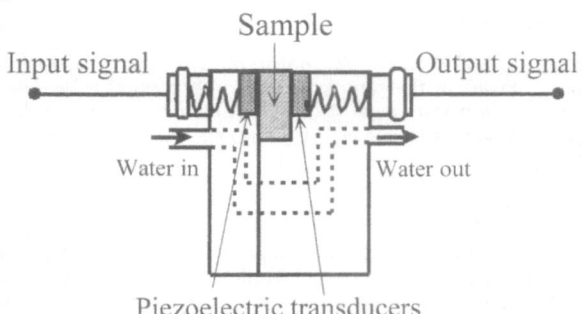

Fig. 1 A The amplitude-frequency and phase-frequency characteristics of the resonator shown in **B** filled with muscle at 278 K. The amplitude of the input signal was 4 V. **B** Cross section of the stainless steel resonator cell used for ultrasonic muscle measurements. 5 MHz piezotransducers were made of lithium niobate. *Insert* Resonance peaks for the resonator shown in **B**, filled with muscle at 278 K

5 MHz, were used. The measurements were carried out at frequencies between 3.6 and 4.0 MHz. Temperature control is an essential element in these measurements, as ultrasonic velocity is highly temperature-dependent. The cell temperature was controlled using a Haake F8 water bath circulator, which has temperature stability better than 10^{-2} K, and monitored using a specially calibrated temperature probe connected to a PC-controlled Ultrasonic Scientific frequency synthesiser/analyser. The measurements were automatically carried out using the PC-controlled Ultrasonic Scientific frequency synthesiser/analyser.

Degassed MilliQ water was used as the calibration standard for our resonator cell [25]. The cell was filled with degassed MilliQ water and heated from 278 to 353 K over 5 h, the digital controller of the Haake F8 being used to programme the temperature ramp. Muscle samples were then filled into the cell and subjected to identical temperature ramp measurements.

Strip-loin steak from bovine muscle *Longissimus dorsi* was purchased locally and stored at 278 K. The water and fat contents of the muscle sample were 72.85% and 3.5%, respectively. The muscle samples were prepared by cutting strips of muscle using a double scalpel instrument. The strips were placed into the sample cell. The muscle fibres were aligned either parallel or perpendicular to the direction of propagation of the ultrasound waves.

In these measurements the ultrasonic velocity, u, was calculated using the equation [21]:

$$u = 2lf_n/n \tag{1}$$

where l, the pathlength, is the distance between the two piezo-transducers in the resonator cell, f_n is the frequency at the maximum of the nth resonance and n is the the wave number. The pathlength was determined by measuring f_n of the calibration standard, degassed MilliQ water, and the known values of ultrasonic velocity for water at each temperature [25]:

$$l = nu_{water}/2f_n \qquad (2)$$

where f_n and f_{n-1} are the frequencies at the maximum of two adjacent peaks:

$$n = f_n/(f_{n+1} - f_n) \qquad (3)$$

The ultrasonic attenuation coefficient, α, was calculated from the Q factor of the resonator, which is determined by the total energy losses in the resonator cell and has two contributions, Q_{sample} and Q_{cell}:

$$1/Q = 1/Q_{sample} + 1/Q_{cell} \qquad (4)$$

where Q_{sample} and Q_{cell} are the contributions to the Q factor from the energy losses in the sample and the cell, respectively. The Q factor for the nth resonance, Q_n, was evaluated from the slope of the phase, φ, versus frequency at the maximum of the resonance, $(\partial\varphi/\partial f)_{f=f_n}$:

$$Q_n = f_n(\partial\varphi/\partial f)_{f=f_n}/2 \qquad (5)$$

and the value of Q_{cell} was determined from calibrations with standard liquids, such as degassed MilliQ water. The contribution of Q_{cell} to the total Q factor of the resonator was small compared with the contribution from the attenuation of our muscles. The attenuation coefficient was calculated using the equation [21]:

$$\alpha = \pi f_n(1/Q_n - 1/Q_{cell})/u \qquad (6)$$

Results

Temperature dependence of ultrasonic parameters of muscle

Ultrasonic velocity

The temperature dependence of the ultrasonic velocity, u, in muscle is shown in the insert in Fig. 2. Ultrasonic velocity was found to change with temperature for muscles of both parallel and perpendicular (to the direction of propagation of the ultrasonic wave) fibre alignments. The high resolution of our ultrasonic measurements allowed us to calculate the differential of ultrasonic velocity with respect to temperature, $\partial u/\partial T$, for parallel- and perpendicular-aligned muscles and is shown in Fig. 2. Ultrasonic velocity values for perpendicular-aligned muscles were found to be, in general, greater than velocity values for parallel-aligned muscles. The difference in velocity values for parallel- and perpendicular-aligned muscles increased with temperature. Muscles with parallel fibre orientation exhibited marked changes at 285 K, between 290 and 295 K and at 310 K. These changes were not apparent in the perpendicular muscle samples at 295 K.

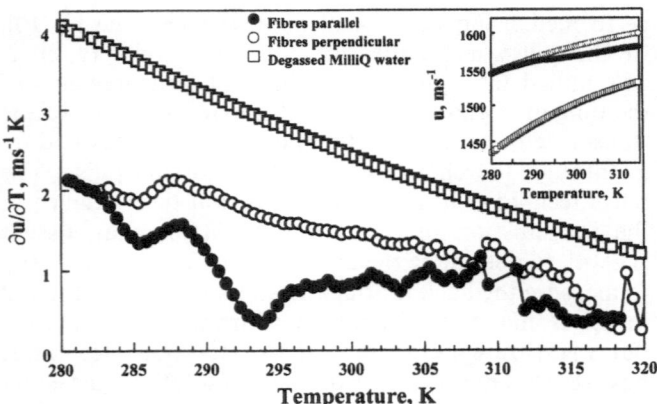

Fig. 2 The differential of ultrasonic velocity with respect to temperature for degassed MilliQ water and muscles with parallel and perpendicular alignment to the direction of propagation of ultrasound waves. *Insert* Temperature dependence of the ultrasonic velocity for the muscles with parallel and perpendicular orientation to the direction of propagation of ultrasonic waves

Ultrasonic attenuation coefficient

The ultrasonic attenuation coefficient in muscle was also profoundly affected by temperature and this is shown in Fig. 3. In the muscles with parallel and perpendicular fibre orientation, attenuation was at a minimum at 285 K. The value of attenuation coefficient increased dramatically in muscle with parallel fibre orientation between 290 and 295 K, but not in the perpendicular sample. In this temperature range the values of the attenuation coefficient for the parallel sample were up to three times higher than the values for the perpendicular sample.

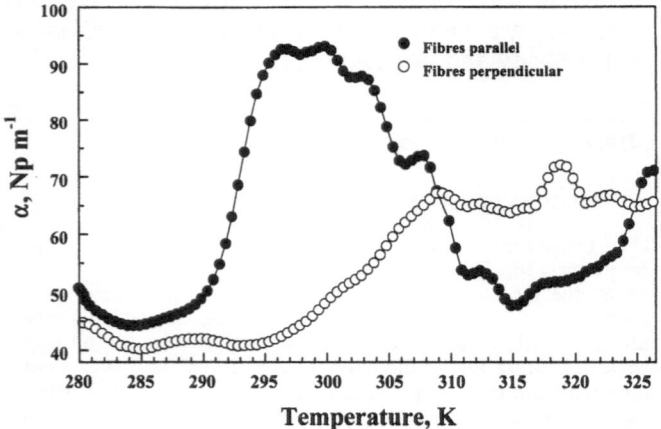

Fig. 3 Temperature dependence of ultrasonic attenuation coefficient for muscles with parallel or perpendicular orientation to the direction of propagation of ultrasound waves. The attenuation coefficient values for degassed MilliQ water were found to be less than 1 neper/m at all temperatures

Discussion

Temperature dependence of ultrasonic parameters of muscle

We found five temperature intervals in which the ultrasonic parameters of muscle change with temperature significantly.

The first change in the ultrasonic parameters of muscle was at around 285 K. The differential of ultrasonic velocity plot and the attenuation curve for both parallel- and perpendicular-aligned muscles have a peak at this temperature. The peak in the attenuation plot for the parallel sample at 285 K was slightly masked by the following large peak at higher temperature. Lipids in muscle undergo melting transitions with temperature [26]. Oleic acid, which accounts for over 50% of the fatty acids present in beef muscle tissue [27], has a melting point of 287 K. The corresponding change in the ultrasonic parameters for muscles at this temperature can be attributed to this melting point.

At about 295 K the derivative of ultrasonic velocity plot and the attenuation curve for parallel muscle samples have a peak. This did not appear in the plots for the perpendicular muscle. The reason for this change in ultrasonic parameters is unclear.

In the temperature range between 297 and 309 K, ultrasonic attenuation decreases in the parallel muscle and increases in the perpendicular muscle. In the same temperature interval, the derivative of ultrasonic velocity versus temperature in the parallel muscle approaches the level for the perpendicular one. This indicates a significant rearrangement in the state and/or the structural parameters of muscles. Various processes can contribute to the observed transition and further investigation is needed for comprehensive physicochemical interpretation of this result.

At 311 K there was a small change observed in the ultrasonic parameters of both parallel and perpendicular muscles. It is interesting to note that the body temperature of beef animals is known to be around this temperature [28].

Between 313 and 330 K we observed several peaks on the temperature dependence of ultrasonic parameters of muscles (Figs. 2 and 3). Myofibrillar proteins, of which the muscle fibres are predominately composed, undergo denaturation with temperature, which commences at around 313 K and proceeds until coagulation is complete at 338 K [29]. The change in the ultrasonic parameters of muscle around these temperatures indicates that the resonator technique is capable of detecting denaturation of these proteins.

Above 343 K there was a lot of scattering in the ultrasonic parameters measured. This could be attributed to collagen denaturation at about 331 K, which

results in the inherent loss of structure of the muscle as the muscle fibres and connective tissue network undergo a shrinking effect [30] and which could have resulted in the muscle losing true contact with the piezotransducers.

The dependence of ultrasonic parameters on muscle anisotropy

Ultrasonic velocity and attenuation coefficient values for muscles with parallel and perpendicular orientation to the direction of propagation of ultrasonic waves were found to be very different. This is in good agreement with the previous publications [31–33], where ultrasonic parameters of muscle with parallel and perpendicular alignment to the direction of propagation of the ultrasonic waves were measured at one temperature. Our high-resolution measurements allow us, for the first time, to compare the differences in the temperature profiles of the ultrasonic parameters of muscles with parallel and perpendicular alignment. Muscles with perpendicular orientation were found to have higher velocity values than parallel muscles over the entire temperature range. Attenuation values were lower for muscles with perpendicular orientation between 278 and 308 K but higher between 309 and 330 K. Earlier in this paper, several temperature-related changes in the ultrasonic parameters of muscles were described. Some of these changes observed in parallel muscles were absent in perpendicular muscles.

In ultrasonic measurements, muscle undergoes compressions and extensions. The elasticity of muscle is an important contributor to ultrasonic velocity and the energy losses in these deformations determine the ultrasonic attenuation. In the muscles with fibres aligned parallel to the direction of ultrasonic wave propagation, compressions and extensions are along the axes of the muscle fibres, while in the muscles with fibres aligned perpendicular to the direction of propagation, the compressions and extensions are transverse to the muscle fibre axes. We could expect that the elasticity and energy losses of muscles along the axes of fibres are more sensitive to their structural arrangement compared with perpendicular orientation. Therefore, any changes in the structural arrangements of fibres should be more "visible" in the measurements of muscles with fibres aligned parallel to the direction of propagation of ultrasonic waves.

Conclusion

The ultrasonic resonator technique was found to be capable of analysing the temperature transitions in muscles. Some of the observed transitions have a simple physicochemical interpretation, while the origin of the other transitions needs further investigation. Muscle anisotropy was shown to have a strong effect on the temperature profile of ultrasonic velocity and attenuation coefficient values. Overall, our results demonstrate that the resonator technique is a rapid, non-expensive and non-destructive method, which can characterise the structural parameters of muscle and the structural changes that occur with temperature.

Acknowledgements This work was supported by Grant 97/R&D/T/178 from the Department of Agriculture, Food and Forestry and the research Walsh Fellowship from Teagasc. We are grateful to Ultrasonic Scientific Ltd for their generous contribution of their equipment for our high-resolution ultrasonic measurements.

References

1. Banerjee A, Kohl T, Silverman NH (1995) Am J Cardiol 76:1284
2. Heckmatte J, Rodrillo E, Doherty M, Willson K, Leeman S (1989) J Child Neurol 4 (suppl):5101
3. Bedi DG, John SD, Scwischuk LE (1988) J Clin Ultrasound 26:345
4. Boyd JS, Omran SN, Aycliffe TR (1988) Vet Rec 123:8
5. Brethour JR (1994) J Anim Sci 72:1425
6. Alliston JC, Kempster AJ, Owen MG, Ellis M (1982) Anim Prod 35:165
7. Smith DE, Wittinger SA (1986) J Food Proc Pres 10:227
8. McClements DJ, Povey DJ, Dickinson E (1993) Ultrasonics 31:433
9. McClements DJ (1997) Crit Rev Food Sci Nutr 37:1
10. Javanaud C (1988) Ultrasonics 26:117
11. Ludwig GD (1950) J Acoust Soc Am 22:862
12. Goss SA, Frizzell LA, Dunn F (1979) Ultrasound Med Biol 5:181–186
13. Goss SA, Johnston RL, Dunn F (1980) J Acoust Soc Am 68:93
14. Goldman DE, Hueter TF (1956) J Acoust Soc Am 28:35
15. Gresham JD, McPeake SR, Bernard JK, Henderson HH (1992) J Anim Sci 70:631
16. Faulkner DB, Parrett DF, McKeith FK, Berger LL (1990) J Anim Sci 68:3, 604
17. Smith MT, Oltjen JW, Dolezal HG, Gill DR, Behrens RD (1992) J Anim Sci 70:29
18. McLaren DG, McKeith FM, Novakofski J (1989) J Anim Sci 67:1657
19. Park B, Whittaker AD, Miller RK, Hale DS (1994) J Food Sci 59:697
20. Eggers F, Kaatze U (1996) Meas Sci Technol 7:1
21. Buckin V, Smyth C (1999) Semin Food Anal 4:113
22. Buckin V, Kudryashov E, Morrissey S, Kapustina T, Dawson K (1998) Prog Colloid Polym Sci 110:214
23. Kudryashov E, Kapustina T, Morrissey S, Buckin V, Dawson K (1998) J Colloid Interface Sci 203:59
24. Morrissey S, Kudryashov E, Dawson KA, Buckin VA (1999) Prog Colloid Polym Sci 112:71
25. Del Grosso VA, Mader CW (1972) J Acoust Soc Am 52:1442
26. Skala D, Bastic L, Remberg G, Jovanovic J (1989) Proceedings of the international congress of meat science and technology, Copenhagen, Denmark, vol 11, p 573
27. Kinney Sweeten M, Cross HR, Smith GC, Smith SB (1990) J Food Sci 55:43, 118
28. Jones SB, Carroll RJ, Cavanaugh JR (1977) J Food Sci 42:125
29. Cheng CS, Parrish FC (1976) J Food Sci 41:1449
30. Palka K, Daun H (1999) Meat Sci 51:237
31. Goss SA, Johnston RL, Dunn F (1978) J Acoust Soc Am 64:430
32. Johnston RL, Goss SA, Maynard V, Brady JK, Frizzell LA, O'Brien WD, Dunn F (1979) In: Linzer M (ed) Ultrasonic tissue characterization, vol 11. (National Bureau of Standards, spec. publ. 525) US Government Printing Office, Washington, pp 19–27
33. Bowen T, Connor WG, Nasoni RL, Pifer AE, Sholes RR (1979) In: Linzer M (ed) Ultrasonic Tissue Characterization vol 11. (National Bureau of Standards, spec. publ. 525) US Government Printing Office, Washington, pp 19–27

Progr Colloid Polym Sci (2000) 115: 287–294
© Springer-Verlag 2000

METHODS

E. Kudryashov
C. Smyth
G. Duffy
V. Buckin

Ultrasonic high-resolution longitudinal and shear wave measurements in food colloids: monitoring of gelation processes and detection of pathogens

E. Kudryashov · C. Smyth · V. Buckin (✉)
Department of Chemistry
University College Dublin
Belfield, Dublin 4, Ireland
e-mail: vitaly.buckin@ucd.ie
Tel.: + 353-1-7062436
Fax: + 353-1-7062127

G. Duffy
The National Food Centre
Teagasc, Castleknock
Dublin 15, Ireland

Abstract In the present paper we describe the applications of the ultrasonic high-resolution longitudinal and shear wave measurements for food and bio-colloids. In the first example, both ultrasonic methods were used for the monitoring of the acidified milk gelation induced by glucono-δ-lactone (formation of gel network in yoghurt). Ultrasonic measurements demonstrated a high sensitivity to pre-gelation and gelation processes during the formation of acid milk gels. The hydration of colloidal calcium phosphate released into serum and the swelling of casein in micelles at pH 5.6–5.0 are suggested as the main contributors to the ultrasonic velocity and attenuation changes during the pre-gelation. The increase in shear loss modulus of acidified milk at pH 5.0–4.85 can be explained by the aggregation of the casein micelles into clusters. Subsequent reformation of these clusters into a gel network at pH 4.85–4.6 is observed as a sharp rise in the storage moduli of acid milk gels and an increase in the ultrasonic velocity. The second example is the application of the ultrasonic shear wave measurements for the detection of *Salmonella* in liquids. The antigen-antibody binding monitored by impedance measurements of a quartz crystal at 5, 15 and 25 MHz results in both the decrease in resonant frequency and an increase in the imaginary part of the quartz impedance. The analysis of the data indicates that the bacteria cells on the sensor surface do not exhibit pure mass-load behaviour, and the viscoelastic properties of the interfacial layer must be taken into account for quantitative analysis. Overall, our ultrasonic measurements demonstrate their high potential as non-destructive methods of analysis of complex foods and bio-colloids.

Key words Ultrasonic waves · Acid milk gelation · Viscoelastic moduli · Piezosensor · Bacteria detection

Introduction

The development of different technological and academic research concerned with complex colloids, such as food and bio-colloids, has emphasised the need for sensitive and highly efficient methods of analysis. Techniques such as different kinds of optical spectroscopy and rheology methods, which play an important role in many analytical applications, often fail either because of optically opaque samples (spectroscopy) or because of their fracture behaviour during testing (e.g. rheology). Therefore, the development and evaluation of new experimental techniques is required.

High-resolution ultrasonic spectroscopy is one of the potential techniques for the analysis of complex foods and bio-colloids. In this technique, two kinds of ultrasonic waves (of megahertz frequency range) can be generated by the vibrating surface of a piezotransducer into an analysed medium: (1) longitudinal, where the particle oscillations are in the direction of the wave,

(2) shear, where these oscillations are perpendicular to the direction of the wave [1]. The shear and volume viscoelastic moduli of the medium determines the propagation of the longitudinal wave, while the shear wave is solely determined by the shear moduli. A major advantage of ultrasonic techniques is the ability of the ultrasonic waves to propagate through optically non-transparent materials. The low amplitude of mechanical displacements in ultrasonic waves makes them ideal for non-destructive characterization of complex colloids.

Longitudinal waves dominated the ultrasonic analysis of liquids and colloids in the last two decades. The applications of ultrasonic measurements for the study of milk (milk fat, milk proteins) [2, 3], monitoring of rennet and acidified milk coagulation [4, 5] were previously reported. The majority of these studies were based on ultrasonic attenuation and velocity measurements using pulse ultrasound spectrometry. Although this method provides a reasonable resolution in measurements of attenuation, it has a low resolution in ultrasonic velocity measurements and requires a large volume of samples for analysis. The development of a new high-resolution resonator technique [6] has allowed for the expansion of the applications of this technique for complex colloids. The modern ultrasonic resonator technique provides a combination of high-resolution, automatic control with easy-to-use facilities [6]. Applications of the technique for the precise determination of the heat coagulation temperature of calcium fortified milk were recently presented [7].

The shear wave propagation measurements were used for analysis of various systems [8]; however, their application in liquid media and weak gels was limited. This is particularly due to the high attenuation of shear waves in these systems. An alternative experimental approach has been developed, which is based on the thickness-shear mode resonator technique [9–11]. In this technique a piezoelectric crystal resonator launches ultrasonic shear waves into the substrate attached to the crystal surface. The mechanical resonance of the crystal is dependent on the crystal parameters, and viscoelastic properties of the attached medium. This can be represented in terms of a mechanical equivalent diagram, which includes mass, elastic and damping (friction) elements (Fig. 1). A simple physical interpretation of it is a mass on a spring with a friction force. Surface loading of the crystal gives an extra contribution to all these elements. The mechanical vibrations of the quartz crystal are coupled to its electrical properties via the piezoelectric effect and therefore can be evaluated from the measurement of the electrical impedance of the quartz crystal. This method was successfully applied to different colloid systems [9, 12].

The expansion of this method resulted in the impressive development of new (relatively cheap) piezo-

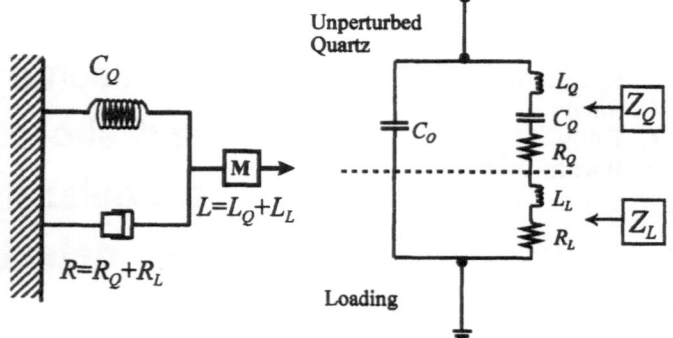

Fig. 1 The mechanical model (*left*) and equivalent circuit (*right*) for a shear wave quartz crystal under liquid loading. The motional arm values (R_Q, L_Q, C_Q) correspond to the unloaded crystal, and the values of R_L, L_L characterise the effect of media in direct contact with one quartz electrode

electrical biosensors for the detection of a wide range of species from small molecules to bacteria cells [13–17]. The majority of these sensors were based on the measurement of the changes of the crystal resonant frequency, Δf_s (e.g. quartz crystal microbalance) as a result of changes of surface mass loading, caused by the specific adsorption of molecules onto a specially modified surface. The value of Δf_s is given by the modified Sauerbrey's equation (see [9] and refs therein):

$$\Delta f_s \approx -\gamma 2 f_s^2 / N[\rho_s + (\rho_L \eta_L / 4\pi f_s)^{1/2}] \tag{1}$$

where γ is constant for a given resonance, N is the number of resonance harmonics, and $\rho_L \eta_L$ is the product of density × viscosity of the layer in contact with the resonator. A value of ρ_s corresponds to the surface density of the solid layer attached to the crystal surface.

The quartz crystal microbalance demonstrated a high sensitivity similar to immunoassays [15] (i.e. ELISA [18]). However, the response of the piezocrystal to bacteria was found complex, and depends on viscoelastic properties of the attached layer, $\rho_L \eta_L$ [16]. This leads to a complicated relationship between the resonant frequency of the piezocrystal and the mass of attached bacterial cells [Eq. (1)]. Therefore a more complete analysis of the resonator parameters (e.g. impedance analysis) must be applied for quantitative detection of biomaterials.

In this paper we describe the application of ultrasonic longitudinal and shear wave techniques for food and bio-colloids. Firstly, we report the application of both ultrasonic methods for the monitoring of the acid milk gel formation, which is the main constituent of various dairy products, including yoghurts, soft cheeses and caseinates. The velocity and attenuation of longitudinal ultrasonic wave measurements were used to analyse the pre-gelation and gelation processes in acidified milk (colloidal calcium phosphate solubilisation, casein micelle aggregation, gel formation). The evolution of the viscoelastic moduli of acidified milk gel was monitored using ultrasonic shear wave measurements.

Secondly, we describe the design of the ultrasonic piezoelectric biosensor for *Salmonella* and its application for the detection of bacteria cells, as well as factors important for the quantitative applications of the sensor.

Materials and methods

Materials

Low-heat pasteurised skim milk (Dale Farm, Ballymena) and the acid precursor glucono-δ-lactone (GDL, Sigma) were used to produce acid milk gels.

Protein A (Sigma) and rabbit anti-*Salmonella* (Biogenesis) were used for the construction of the sensitive layer. Dulbecco's phosphate buffered saline (DPBS) is composed of 137 mM NaCl, 2.7 mM KCl, 8 mM Na_2HPO_4 and 1.5 mM KH_2PO_4, pH 7.42. Heat inactivated *Salmonella* antigen (stock solution of 10^{10} cells ml^{-1}) was supplied by National Food Centre (Teagasc).

Acidification of milk

The samples of skim milk (about 20 ml) were degassed and then the GDL was added (giving a final concentration of GDL of 2% w/w), at which time we started to record the gelation time. The samples were kept under constant stirring for 5 min before being quickly loaded into the testing cells of the ultrasonic device. The kinetics of acidification were monitored by the measurement of pH (using an Orion 710A pH meter with a Gelplas combination pH electrode) and the ultrasonic technique simultaneously for a period of 16 h at 20 °C.

Ultrasonic piezosensor for Salmonella

A piezoelectric sensor for *Salmonella* has been constructed by immobilizing the anti-*Salmonella* antibody on the crystal electrode of 5 MHz AT-cut polished quartz crystal. Firstly, 3 μl of Protein A (1 mg/ml in a 1:1 solution of DPBS and 0.1 M sodium acetate, pH 5.5) were spread onto a cleaned and dried sensor electrode, and allowed to dry. The crystal was washed several times with DPBS and distilled water. Secondly, 5 μl of rabbit anti-*Salmonella* (Biogenesis) in DPBS was spread onto the crystal coated with protein A. Then the crystal was washed and dried again, and exposed to 400 μL of DPBS, before addition of the bacteria suspension. A drop of *Salmonella* antigen solution either from stock solution or after its dilution in DPBS was added to the cell to obtain the final concentration of the bacteria in the sensor cell. The antigen-antibody binding was monitored by impedance measurements (at 5, 15 and 25 MHz) until it had reached saturation. Then the sensor was washed out with DPBS, and the next portion of bacteria was added. Finally, the crystal was washed and allowed to dry. All subsequent steps of sensor assembly, as well as antigen-antibody binding reaction, were investigated using impedance analysis of the quartz piezosensor.

Methods

Longitudinal acoustical wave measurements

The velocity (u) and attenuation coefficient (α) of longitudinal ultrasonic wave propagation in acidified milk was measured by a high-resolution resonator method. The details of cell construction and measurement procedures are described elsewhere [6, 7]. All measurements were done differentially with two identical resonator cells of 1 cm^3 volume, at a frequency of about 7.2 MHz. The resonance frequencies and half bandwidth of selected resonances in the reference and measuring cells were automatically measured by a PC-controlled Ultrasonic Scientific frequency synthesiser/analyser. The resolution of our measurements was about 10^{-5}% for ultrasonic velocity and better than 0.1% for ultrasonic attenuation.

The measurements were carried out by the following procedure. First the reference cell was filled with non-acidified skim milk and the measuring cell was filled with acidified skim milk. The resonant frequencies, f_n (n is the harmonic number), and bandwidth of the resonant peaks at the −3 dB level in the reference and measuring cells, Δf_n, were measured continuously over 16 h.

The changes of the ultrasonic velocity relative to the corresponding values for skim milk ($\delta u/u$) and the changes in the ultrasonic attenuation coefficient (α) of the acidified milk are calculated from the following equations [6,7]:

$$\delta u/u = \delta f_n/f_n \tag{2}$$

$$\alpha = \pi f_n(1/Q_n - 1/Q_{cell})/u \tag{3}$$

$$Q_n = f_n/\Delta f_n \tag{3a}$$

where δf_n is the change in f_n relative to the corresponding values for non-acidified skim milk and Q_n is the Q factor of the resonance determined by the total energy losses in the resonator cell. There are two contributions to Q_n, one from the sample and the other from the cell. The value of Q_{cell} is obtained from calibration of the cell with standard liquids.

Shear wave measurements

The experimental set-up consisting of the quartz crystal resonator and a temperature-controlled quartz crystal holder was described previously [19]. The AT-cut, 5 MHz quartz crystals (Roditi International, UK, and International Crystal Manufacturing, USA) were used in this work. The working diameter of the gold electrodes on both sides of the quartz plate was 6 mm. The quartz crystal was placed between two Viton O-rings, and a glass tube was fixed onto the crystal in order to form a liquid sample container. One side of the crystal was immersed in the liquid, while the other face was kept in air. The volume of the glass liquid cell was approximately 1 cm^3. Gold foil electrodes were attached to both sides of the crystal, providing electrical connection with the HP 4194A Impedance Analyser (Hewlett Packard, Japan) via very short shielded cables.

Impedance analysis involving the measurement of current at a known applied voltage was carried out near frequencies of 5, 15 and 25 MHz. A programme based on HP VEE software was designed to measure the magnitude, $|Z|$, and phase angle of impedance, θ (Z-θ spectra), as well as conductance, G, and susceptance, B (G-B spectra), and to collect the characteristic parameters continuously.

Quantitative analysis of data

The Butterworth-van Dyke equivalent circuit (Fig. 1) simulates the electrical characteristics of the shear resonator over a range of frequencies close to the resonance region [9, 11]. According to the model, the electrical impedance of the loaded quartz resonator can be described as the motional impedance of the resonator, Z, in parallel with the static capacitance, C_0. Two components of the motional impedance, the motional impedance of the unperturbed resonator, Z_Q, and the motional impedance created by the surface load, Z_L, have real and imaginary parts, which are related to the parameters of the equivalent circuit by: $Re(Z_Q) = R_Q$,

$\mathrm{Re}(Z_L) = R_L, \quad \mathrm{Im}(Z_Q) = 2\pi f L_Q - 1/(2\pi f C_Q), \quad \mathrm{Im}(Z_L) = X_L = 2\pi f L_L.$ The parameters C_0, C_Q, and L_Q depend only on the quartz physical properties and dimensions relative to the electrode area [9, 11]. The inductance, L_L, corresponds to the effect of the surface loading on the inertia mass of the crystal, and the resistance represents the dissipation of electrical energy due to the damping of quartz vibrations in the attached medium.

The values of R_L and X_L were determined by measuring the frequency dependence of the electrical impedance of the quartz near the frequencies of 5, 15 and 25 MHz, which is then fitted using the Butterworth-van Dyke equivalent circuit. They were used for calculations of the storage, G', and the loss, G'', moduli of milk and acid milk gel according to the following equations [1, 11]:

$$G' = (R_L^2 - X_L^2)/(\rho A^2)$$
$$G'' = 2(R_L X_L)/(\rho A^2) \qquad (4)$$

where ρ is the density of the medium, A is a constant determined by the physical properties and the geometry of the quartz resonator, and the resonant frequency at which the measurements are performed. The coefficient A was determined by calibration of our quartz with liquids (water-glycerol mixtures) of known viscosity.

Results and discussion

Ultrasonic measurements of the kinetics of acidified milk gelation

Shear wave measurements

Figure 2 illustrates the evolution of the storage, G', and the loss, G'', moduli of milk in the process of formation

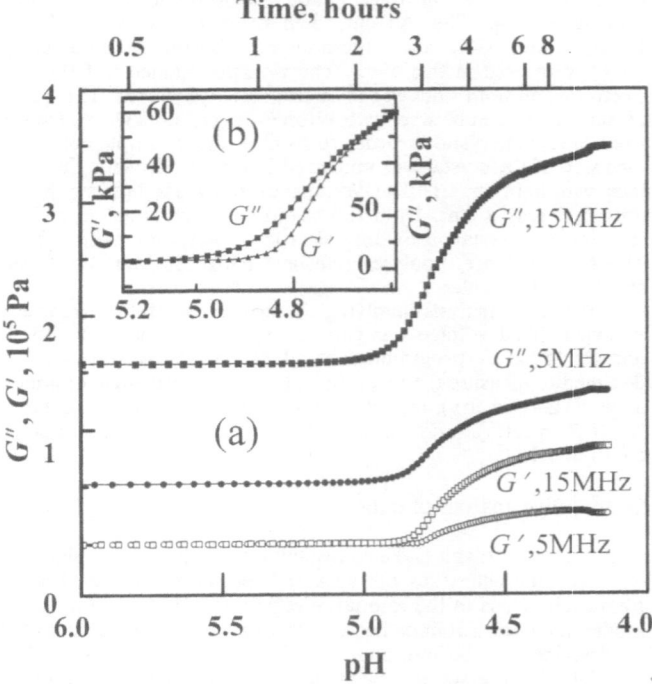

Fig. 2 a The evolution in the high-frequency viscoelastic moduli of milk (G', G'') during acidification of milk by GDL. **b** The changes in viscoelastic moduli around the sol-gel transition point are enlarged

of the acidified milk gel. Three stages of the milk acidification can be distinguished clearly.

The first stage. At a pH higher than 5.1, no changes in rheological parameters were observed. The loss modulus remains the same as those for skim milk. The value of the skim milk viscosity was calculated as $\eta = G''/2\pi f$. A value of 1.7 ± 0.1 mPa s was obtained. This was found to be the same for all measured frequencies (5, 15 and 25 MHz), and is in good agreement with the value of 1.68 mPa s determined previously at low frequencies [20].

The second stage. In the pH range between 5.1 and 4.85 the loss modulus, G'', rises with pH (Fig. 2). This observation is in agreement with previous rheological studies. Low-frequency dynamic viscosity measurements in acidified milk indicated a growth in the viscosity of milk starting at approximately pH 5.1 [19, 21]. The increase in the loss modulus of the milk in this pH range, 5.1–4.8, was suggested to be due to the formation of large aggregates of casein particles, which precluded the formation of the gel network [21, 22]. According to electron microscopy studies, starting from pH 5.2 the casein micelles lose their integrity. At the same pH, particles start growing in size, as has been seen by turbidity and light scattering studies at a pH of about 5.1 [22].

The storage modulus remains at a very low value, at all measured frequencies, which indicates the absence of gel network formation in acidified milk in this pH range.

The third stage of acid milk gel formation starts at pH 4.8. The storage modulus, G', starts rising quickly with decrease of pH, which indicates the formation of the three-dimensional gel network of casein particles in milk. Previously this sol-gel transition was observed by both a sharp increase in stiffness and a decrease in loss tangent (G''/G') of the acid milk gel at pH 4.8–4.7 [21]. At pH below 4.2 the high-frequency rheological parameters show a tendency to level off. In contrast, according to our and previously reported dynamic rheology measurements of the acid milk gel, the storage modulus, G', rises with time even when pH had reached its saturated value of 4.0 [19, 23]. This indicated the long-range rearrangements in the gel structure (observed at low frequency), which exceeds the structural scale of the high-frequency measurements (~1 μm) [19].

Longitudinal wave measurements

Ultrasonic velocity in dispersed systems is determined by the density and viscoelastic moduli of the individual component phases, the size of disperse particles and their volume fraction [24]. The overall attenuation in a colloidal dispersion is composed of absorption and scattering losses [25]. The absolute value of the ultrasonic attenuation and the relative changes of ultrasonic

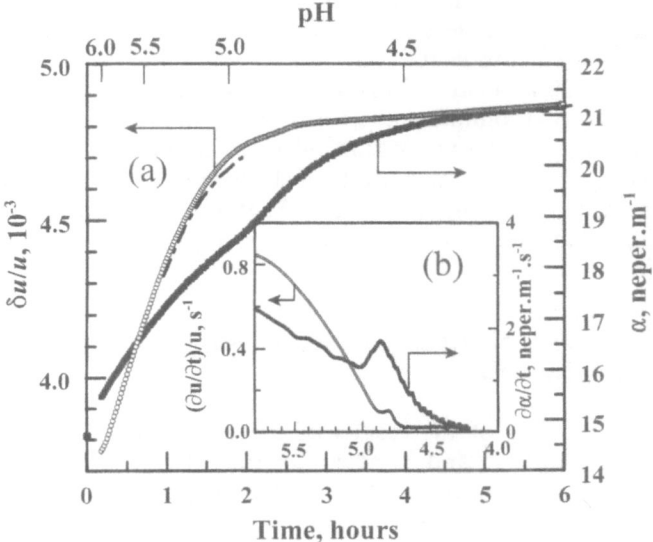

Fig. 3 a Ultrasonic attenuation in acidified skim milk and relative velocity, $\delta u/u$ (which is the relative change in ultrasonic velocity difference between the acidified skim milk and the non-acidified skim milk), plotted against acidification time. The contribution of colloidal calcium phosphate solubilization [calculated using Eq. (5)] to the changes of ultrasonic velocity is shown by the *broken line*. **b** Time differential of ultrasonic attenuation ($\partial\alpha/\partial t$) and relative velocity ($\partial u/\partial t)/u$ plotted against the pH of milk

velocity in acidified skim milk are shown in Fig. 3a. Both ultrasonic attenuation and relative velocity in skim milk increased gradually over several hours until a plateau was reached after 5 h of acidification. The pre-gelation and gelation stages of acid milk formation can be separated if one considers the rate of the changes in ultrasonic parameters. The calculated differentials of both attenuation and velocity for acidified milk were plotted as a function of the pH of milk, and they are shown in Fig. 3b.

Pre-gelation processes. As can be seen from Fig. 3a, the main changes in ultrasonic parameters occur in the pre-gelation stage between pH 5.6 and 5. Pre-gelation processes such as the release of colloidal calcium phosphate from micelles and partial liberation of casein protein result in the modification of the chemical composition of casein micelles. These processes cause the swelling of casein micelles and the aggregation of casein micelles. The apparent/partial effects of the changes in the micelles physicochemical properties on the ultrasonic parameters are discussed below.

The effect of casein liberation from the micelles into serum during acidification seems to be small. This conclusion follows from our analysis of casein hydration within micelles at pH 6.7 (initial pH of milk). According to our results, the specific apparent volume and adiabatic compressibility of total casein proteins inside

micelles are nearly the same as the corresponding values for α-casein in solution, indicating a similar level of hydration of proteins in both states.

A significant contribution to the increase of the ultrasonic velocity is expected owing to the dissociation of colloidal minerals from the casein micelle. It is well known that multivalent ions such as calcium and magnesium are highly hydrated in water solutions, and these ions have a significant effect on ultrasonic velocity in aqueous solution, because the compressibility of water in the hydration shell is different (generally less) from bulk water [26]. Therefore the release of colloidal calcium phosphate into serum during acidification is supposed to have a significant increase on the ultrasonic velocity. In our calculations we have used data on the dissociation of micellar calcium and inorganic phosphate in GDL acidified skim milk reported by Gastaldi et al. [21] to calculate the relative change in ultrasonic velocity, $\delta(u/u)_{hydr}$, according to the equation:

$$\delta(u/u)_{hydr} = \delta(u/u)(Ca^{2+})_{hydr} + \delta(u/u)(H_2PO_4^-)_{hydr}$$

(5)

The values of $\delta(u/u)(Ca^{2+})_{hydr}$ and $\delta(u/u)(H_2PO_4^-)_{hydr}$ were calculated from the apparent molar volume and apparent adiabatic compressibility for the hydrated calcium and phosphate ions [27].

There is quantitative consistence between the experimental and theoretical dependencies of relative ultrasonic velocity changes over the pH range 5.6–5.0 (Fig. 3a). This supposes that the dissociation of colloidal calcium phosphate into serum makes an important contribution to the change of ultrasonic velocity in acidified skim milk.

An increase in particle size can have a significant effect on both ultrasonic velocity and ultrasonic attenuation. We calculated the contributions of the aggregation and the increase in particle size into the ultrasonic parameters in skim milk using the Truell-Waterman theory ([25] and refs. therein). The effect of aggregation or change in casein particle size on ultrasonic parameters of skim milk during acidification is small. The changes in ultrasonic attenuation and relative ultrasonic velocity corresponding to the increase of particle size (from 0.1 to 1 μm) resulting from the aggregation are only 10% of the total changes in ultrasonic parameters during milk acidification. Next we considered the effect of water transfer from the milk serum to the casein micelle. We calculated the effect of the change in voluminosity and water content of casein micelles on both ultrasonic parameters. It was found that the increase in the volume of the casein particles results in negligible changes in the ultrasonic velocity in comparison with the experimental value for milk acidification. We would expect this, as there is little change in the overall density and compressibility of the system. On the other hand, the ultrasonic attenuation was found to be highly sensitive to changes in the voluminosity and water content of the casein micelle. The total changes in ultrasonic attenuation during pre-gelation can be explained by a 25% rise of casein volume fraction compared to its initial value of 0.1.

Gelation process. The sharp peaks in both differentials of ultrasonic velocity and attenuation around pH 4.8 may correspond to the point of gelation in acidified milk gel, which is in agreement with our shear wave measurements. From the deconvolution of the curves, we established the contribution from gelation to relative ultrasonic velocity as 1×10^{-5} and attenuation as 0.27 neper m^{-1}, which consists approximately of 1% and 4% of the total changes of the corresponding ultrasonic parameters during acidification of skim milk.

The contribution of the changes in elasticity and viscosity of acid milk gel into ultrasonic parameters can be estimated from the magnitudes of the changes in ultrasonic velocity and attenuation upon the sol-gel transition using the following expressions [1]:

$$\Delta(\alpha)_G \cong \frac{2\pi^2 f^2}{\rho u^3} \left(\Delta\eta_v + \frac{4}{3}\Delta\eta_s \right)$$

$$2\rho u \Delta(u)_G \cong \Delta K' + \frac{4}{3}\Delta G' = \Delta M' \quad (6)$$

Here $\Delta(u)_G$ and $\Delta(\alpha)_G$ are the changes in ultrasonic velocity and attenuation upon sol-gel transition (between pH 5.0 and 4.6). The values of $\Delta M'$, $\Delta K'$, $\Delta G'$ and η_v, η_s correspond to the changes in real parts of complex, bulk and shear moduli, respectively, η_v and η_s are the changes in the volume and shear viscosities upon gelation.

By using the above Eqs. (6) and the values of the changes in ultrasonic velocity and attenuation (integral area of seconds peaks in Fig. 3b), we have obtained the value of the changes in viscoelastic parameters of the acidified milk during the sol-gel transition. The change in complex elastic modulii ($\Delta K' + \frac{4}{3}\Delta G'$) is estimated to be about 100 kPa at the frequency of 7.2 MHz used in longitudinal wave measurements. This value is of the

same order as the value of the shear storage modulus of acid milk gel determined from ultrasonic shear wave measurements ($G' \cong 27$ kPa at 5 MHz, and $G' \cong 130$ kPa at 25 MHz). The change in effective viscosity (volume and shear) of acidified skim milk during gelation was estimated to be about 3.9 mPa. In comparison, the magnitude of shear viscosity changes in acidified skim milk is about 2.5 mPa determined by the ultrasonic shear wave technique.

Impedance analysis of ultrasonic piezosensor for *Salmonella*

A summary of the results of the impedance measurements (at 5, 15 and 25 MHz) for the subsequent steps of the piezosensor assembly as well as the antigen-binding reaction is presented in Table 1. As can be seen, the deposition of both protein A and antibody are accompanied by the positive values of X_L, and by zero values of R_L, which indicate the formation of solid-like layers on the crystal surface. The magnitude of the changes of resonant frequencies corresponds to the protein monolayer thickness. This is confirmed by our analysis of the surface profile of the deposed layers using atomic force microscopy.

The observed changes in the resonant frequency, Δf_r, and imaginary part of impedance, X_L, after inoculation of *Salmonella* antigen into the sensor (giving a final concentration of 10^6 cell per ml) are shown in Fig. 4. The resonant frequency of the sensor drops continuously over 3 h after the addition of the bacteria, indicating the adsorption of the *Salmonella* antigen on the antibody-modified sensor surface. These changes in resonant frequency correlate well with the changes in the value of X_L. Since the rise in the value of X_L corresponds to the attachment of an additional mass to the sensor (Fig. 1), the adsorbed bacteria appear to be strongly bound to the crystal surface. The decrease in the value of R_L (Table 1) can also be explained by the lowering of the viscosity of the interfacial antibody-liquid layer (due to the restriction of antibody molecule movements on the surface) upon bacteria binding.

Table 1 The changes in resonant frequency, (Δf_r), real (R_L) and imaginary (X_L) components of the motional part impedance of the quartz crystal at 5, 15 and 25 MHz after the deposition of sensitive layers on the quartz crystal surface and adsorption of *Salmonella* cells

Adsorption processes	$-\Delta f_r$ (Hz)			X_L (Ω)			R_L (Ω)		
Frequency (MHz)	5	15	25	5	15	25	5	15	25
Protein A	9.8	30	55	7	21	37	0.7	−0.2	2
Anti-*Salmonella* antibody	50	156	263	36	106	179	1.7	0.8	−3.4
10^6 cells ml^{-1}, DPBS	35	7	7	18	4	7	−20	2	3
10^9 cells ml^{-1}, DPBS	58	40	110	30	47	70	−8	15	19
10^9 cells ml^{-1}, air	60	190	320	44	127	217	−1	2	0.6

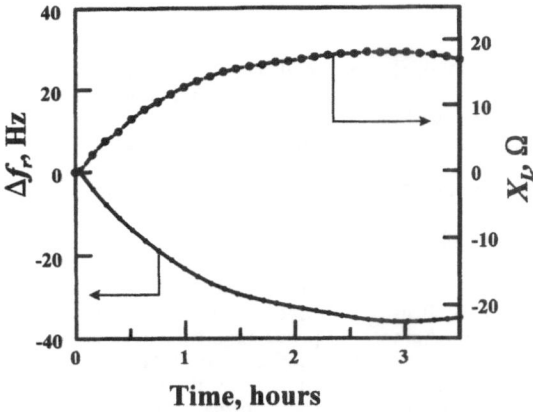

Fig. 4 Typical curves for changes in the resonant frequency f_r and the imaginary part of the impedance X_L of a crystal, observed after inoculation of a *Salmonella* antigen (10^6 cells ml^{-1}) into the sensor

The antigen-antibody binding reaction reaches saturation in about 3 h for bacteria concentration of 10^6 cells ml^{-1} and several minutes for a concentration of 10^9 cells ml^{-1}. The final magnitude of the resonant frequency changes is about 100 Hz. In order to interpret this value using Eq. (1), first we assume that the bacteria layer behaves as a rigid elastic mass. For cell density of about 1 g cm^{-3} and mean rod-like *Salmonella* cell (1.1 × 3.5 μm) of volume 3.3 μm^3 [18], the calculated amount of mass in a cell monolayer on the 29 mm^2 piezoelectrically active surface was about 25 μg. This mass corresponds to a frequency shift of 4 kHz during cell attachment and is substantially bigger than observed experimental values. This result can be explained by poor coverage of our sensor by *Salmonella* cells (about 1%). However, similar changes in frequency were found previously for different bacteria, even though there was complete coverage of the sensor surface [14, 16]. Therefore further studies must be done to explain this phenomenon.

To explain these findings we should consider that the size of the *Salmonella* cell is several times larger than the effective thickness of the sensitive layer (decay length of shear wave in water is about 0.3 μm, at 5 MHz). Therefore it can be assumed that only a small portion of the adhered bacterial cell, in particular that in contact with the sensor surface, contribute to the changes of the mechanical properties of the sensitive layer (mechanical impedance or its electrical equivalent, Z_L). Thus the viscoelastic properties of this layer may be determined mainly by the surrounding viscous liquid. A similar observation was reported previously by Rodahl et al. [17]. The fact that the contributions of the attached bacterial layers to impedance (R_L and X_L) and the resonant frequency changes in the liquid media are smaller (in particularly at higher frequencies) compared to those obtained for dried layers (Table 1) supports these finding.

Conclusions

Ultrasonic longitudinal and shear wave measurements demonstrate a high sensitivity to pre-gelation and gelation processes during the formation of acid milk gels. The hydration of colloidal calcium phosphate released into serum and the swelling of casein in micelles at pH 5.6–5.0 are suggested as the main contributors to the ultrasonic velocity and attenuation changes during the pre-gelation process. The increase in the attenuation of the longitudinal wave and loss moduli of acidified milk indicates the aggregation of the casein micelles into clusters. Subsequent reformation of these clusters into a continuous gel network at pH 4.8–4.6 is observed as a sharp rise in the storage moduli of acidified milk gel and an additional increase in ultrasonic velocity and attenuation.

The ability of the ultrasonic shear wave biosensor to monitor bacteria in liquids was demonstrated. However, impedance measurements of the antigen-binding reaction indicates that the attachment of the bacteria to the sensor does not exhibit pure mass-load behaviour, and the viscoelastic properties of the interfacial layer must be taken into account.

Overall our ultrasonic measurements show a high potential as non-destructive methods of analysis of complex foods and bio-colloids.

Acknowledgements This work was supported by grant 97/R&D/T178 from the Department of Agriculture Food and Forestry of Ireland and EU grant FAIR-CT97-3096. We are grateful to Mr. James Mahon, an undergraduate student, for his help in ultrasonic sensor construction. We gratefully acknowledge the support of this work by Ultrasonic Scientific Ltd, who provided the equipment for high resolution longitudinal ultrasonic measurements.

References

1. Herzfeld KF, Litovitz TA (1959) Absorption and dispersion of ultrasonic waves. Academic Press, New York
2. Miles CA, Shore D, Langley KR (1990) Ultrasonics 28:394
3. Griffin WG, Griffin MCA (1990) J Acoust Soc Am 87:2541
4. Gunasekaran S, Ay C (1994) Trans ASEA 37:857
5. Benguigi L, Emery J, Durand D, Busnel JP (1994) Lait 74:197
6. Buckin V, Smyth C (1999) Semin Food Anal 4:89
7. Smyth C, Dawson K, Buckin V (1999) Prog Colloid Polym Sci 112:221
8. Ferry JD (1980) Viscoelastic properties of polymers. Wiley, New York
9. Buttry DA, Ward MD (1992) Chem Rev 92:1355
10. Johannsmann D (1999) Macromol Chem Phys 200:501
11. Bandey HL, Martin SJ, Cernosek RW, Hillman AR (1999) Anal Chem 71:2205
12. Martin SJ (1997) Faraday Discuss Chem Soc 107:463
13. Muramatsu H, Kajiwara K, Tamiya E, Karube I (1986) Anal Chim Acta 188:257
14. Gryte DM, Ward MD, Hu WS (1993) Biotechnol Prog 9:105
15. Bovenizer JS, Jacobs MB, O'Sillivan CK, Guilbault GG (1998) Anal Lett 31: 1287
16. Tessier L, Schmitt N, Watier H, Brumas V, Patat F (1997) Anal Chim Acta 347:207
17. Rodahl M, Höök F, Fredriksson C, Keller C, Krozer A, Brzezinski P, Voinova M, Kasemo B (1997) Faraday Discuss Chem Soc 107:229
18. Prescot LM, Harley JP, Klein DA (1996) Microbiology. WCB, Dubuque
19. Kudryashov ED, Hunt N, Arikainen EO, Buckin VA (2000) J Dairy Sci, (to be published)
20. Walstra P, Jenness R (1984) Dairy chemistry and physics. Wiley, New York
21. Gastaldi E, Lagaude A, Tarodo de la Fuente B (1996) J Food Sci 61:59
22. Banon S, Hardy J (1992) J Dairy Sci 75:935
23. Lucey JA, Singh H (1998) Food Res Int 30:529
24. Povey MJW (1998) In: Povey MJW, Mason WP (eds) Ultrasound in food processing. Blackie, London, pp 30–65
25. McClements DJ (1998) In: Povey MJW, Mason WP (eds) Ultrasound in food processing. Blackie, London, pp 85–103
26. Sarvazyan AP (1991) Annu Rev Biophys Chem 20:321
27. Millero FJ, Ward GK, Chetirkin PV (1977) J Acoust Soc Am 61:1492

Progr Colloid Polym Sci (2000) 115:295–299
© Springer-Verlag 2000

METHODS

C. Neto
D. Berti
G. D. Aloisi
P. Baglioni
K. Larsson

In situ study of soft matter with atomic force microscopy and light scattering

C. Neto · D. Berti · G. D. Aloisi
P. Baglioni (✉)
Department of Chemistry and CSGI
University of Florence
via Gino Capponi 9
50121 Florence, Italy
e-mail: baglioni@apple.csgi.unifi.it
Tel.: +39-55-2757567
Fax: +39-55-240865

K. Larsson
Camurus Lipid Research Foundation
Ideon, Gamma 1, Solvesgatan 41
223 70 Lund, Sweden

Abstract We present the results of a morphological and structural study of a stable colloidal dispersion of two liquid crystal phases of cubic and hexagonal structure. The study was made with an atomic force microscope operated in situ and with dynamic and static light scattering. The results obtained with the three techniques are in good agreement with each other and with previous studies. In situ imaging makes it possible to directly observe possible structural modifications of the particle's morphology as a consequence of in situ addition of polymers, enzymes, salts, etc., to the dispersion. Light scattering can be consequently used to confirm and strengthen these results.

Key words Cubic liquid crystals · Hexagonal liquid crystals · Atomic force microscopy · Dynamic light scattering · Static light scattering

Introduction

Lyotropic liquid crystals occur abundantly in nature, being ubiquitous in living systems; they are normally made of an amphiphile and water. The structure of the mesophase depends on the water content: in the cubic phase the bilayers are bent to form spherical units, with the polar head on the surface, and the spherical units form a body-centered cubic arrangement, water taking up the space between the units. The hexagonal phase (H_I) consists of cylinders of indefinite length of lipid molecules, which expose their polar groups to the surface of water with the core hydrocarbons in a disordered liquid-like conformation. The cylinders are ordered in a hexagonal array in water.

Colloidal dispersions of these two amphiphilic phases can be prepared [1]. The submicron particles that constitute the dispersion are called cubosomes and hexosomes, and have been already characterized by means of cryo-TEM (transmission electron microscopy) and X-ray diffraction [2, 3]. Some atomic force microscopy (AFM) images of these particles have been presented in a previous paper [4].

Cubosomes' similarity with liposomes suggests their possible biological relevance. Cubic phases are unique in their ability to accommodate proteins: the molecules are located in the water channel systems and there they retain their native structure.

As shown by Larsson and Lindbolm [5], a very hydrophobic wheat fraction, gliadin, can be dispersed in monoolein and a bicontinuous cubic phase formed on addition of water, with the protein in the lipid regions of the cubic phase. Other examples have been reviewed [4–6]. Immobilized enzymes offer many advantages over enzymes in solution, including strongly increased stability, higher activity and specificity, broader temperature and pH ranges, and fewer interferences.

The main amphiphilic constituent of cubosomes and hexosomes is glycerol monooleate (GMO), which is an uncharged, biocompatible lipid (e.g. present in sunflower oil). Landh [3, 7] found that if a bicontinuous cubic GMO-water phase is mechanically fragmented in the presence of a dispersing and stabilizing agent (an amphiphilic block copolymer), a stable dispersion can be formed. A non-ionic triblock polymer, called Polaxamer 407 $(PEO_{98}-PPO_{67}-PEO_{98})$, was used for this purpose. In mixtures of GMO and GTO (glycerol trioleate) a reversed hexagonal phase is formed.

Materials and methods

The GMO was a distilled monoglyceride (RYLO MG 90, Danisco Ingredients, Brabrand, Denmark) with the following fatty acid composition: oleic acid 92%, linoleic acid 6%, saturated acids 2%. The GTO was a high monosaturated sunflower oil (Trisun 80 SVO, Eastlake, Ohio). Polaxamer 407 was obtained from BASF (Germany). The dispersion procedure has been described in a previous work [1].

The cubic phase composition is 97% wt of water and 3% of a composition containing an amphiphilic part made of GMO (94%) and Polaxamer 407 (6%).

The hexagonal phase composition is 97% wt of water and 3% of the following composition: 94% GMO/GTO (88/12 wt ratio respectively) and 6% Polaxamer 407.

The sample preparation for the AFM experiments was described in a previous paper [4]. The particles have a polar surface and they can be adsorbed on the mica surface, which is negatively charged, and then imaged with an AFM liquid scanner under water. The sample was scanned in non-contact with an Explorer 2000 Topometrix.

The experimental acquisition of images in liquid and on such soft samples requires long equilibration times before the sample can be scanned steadily and with low tip pressure on the sample. The electrostatic interaction of the particles with the surface is not so strong as to make the imaging procedure straightforward. Slow scan rates and large scan sizes are generally more delicate to the soft particles. Images were taken with online filtering and subsequently processed by flattening to remove the background slope.

Static and dynamic light scattering (SLS and DLS) measurements were performed with a Brookhaven apparatus consisting of a computer-controlled and stepping motor-driven variable-angle detector system. The light source was the doubled frequency of a Coherent Innova diode pumped Nd-YAG laser ($\lambda = 532.0$ nm), linearly polarized in the vertical direction, whose power has been attenuated in order to avoid sample heating. The laser long-term power stability was $\pm 0.5\%$. The signal was detected by an EMI 9863B/350 photomultiplier, and for DLS analyzed by a digital autocorrelator (BI 9KAT).

Thermostated (30 °C) decahydronaphthalene was used as index matching liquid in order to avoid interference from reflection at the sample cell-air interface.

SLS experiments were performed at 30 different angles ($25° \leq \theta \leq 150°$). For each angle, 25 individual measurements were collected and averaged. Each series of measurements was repeated at least three times. The data were then corrected for background (cell and solvent) scattering and plotted as $I(Q)$, where $Q = (4\pi n/\lambda_0) \sin(\theta/2)$ is the modulus of the scattering vector. Our experimental setup covered a Q range from 7×10^{-4} to 3×10^{-3} Å^{-1}.

Results

Cubosomes (Figs. 1 and 2) and hexosomes (Figs. 3 and 4) were imaged with a non-contact atomic force microscope in aqueous solution. In non-contact mode, the AFM tip is oscillated at its resonant frequency: as the probe gets closer to the sample surface (a few nm or less from it), the interaction forces between the tip and the sample change both the oscillation amplitude and phase of the probe tip. These changes can be detected and used

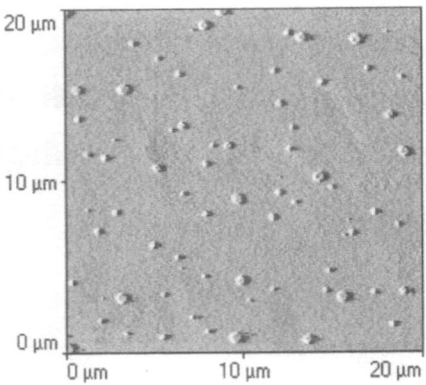

Fig. 1 In situ AFM image (non-contact mode) of the dispersion of cubosomes adsorbed on mica. Image dimensions are 20×20 μm^2. The cubosomes in this image appear to have a side of dimensions varying between 500 and 220 nm and a height around 9 nm. Small objects visible in this and following figures are probably agglomerates of the degraded crystal phase

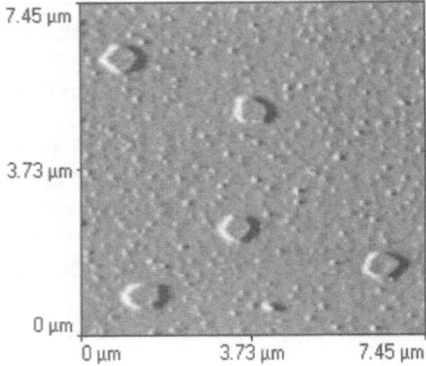

Fig. 2 In situ AFM image (non-contact mode) of the dispersion of cubosomes adsorbed on mica. Image dimensions are 7.45×7.45 μm^2. The cubosomes in this image appear to have a side of dimensions varying between 550 and 400 nm and a height of about 9 nm

by a feedback loop to control the tracking of the probe over the surface. The distance between the scanning tip and the sample surface being so small, we cannot exclude some physical contact between the probe and the particle surface, possibly because of mechanical vibrations and environmental noise. The actual contact sometimes may cause limited damage to the particles (especially for the hexosomes, see Fig. 4) and the tip may drag them in the scanning direction. However, both in non-contact mode and in contact-mode, AFM certainly remains the less perturbative in situ imaging technique available.

In AFM images, cubosomes appear as faceted cubic particles, oriented with one side parallel to the mica surface, probably as a consequence of the drag experienced by the particles during the scanning of the tip. The angles at the corners of the squares (in projection)

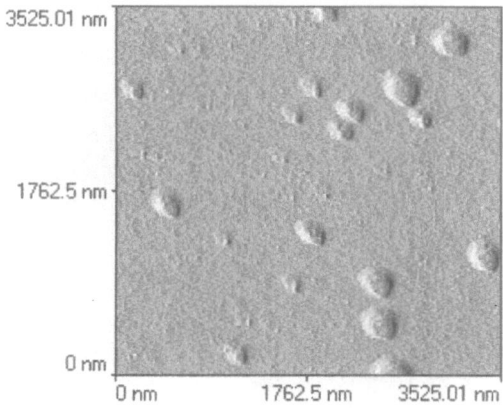

Fig. 3 In situ AFM image (non-contact mode) of the dispersion of hexosomes adsorbed on mica. Image dimensions are 3525 × 3525 nm². The hexosomes in this image appear to have a side of dimensions varying between 280 and 230 nm and a height around 20 nm

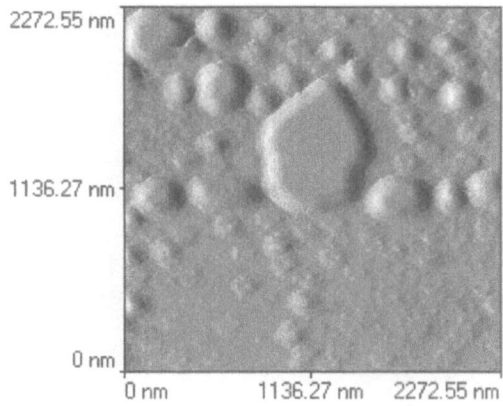

Fig. 4 In situ AFM image (non-contact mode) of the dispersion of hexosomes adsorbed on mica. Image dimensions are 2272 × 2272 nm². The hexosome in this image appears to be damaged since only two of the expected sides are visualized intact. Its dimensions are about 400 × 400 × 25 nm

sometimes do not appear clearly as 90°. This appearance was also observed in previously published cryo-TEM images [2], where vesicular structures are often found at the corners of the faceted particles, as a consequence of the difficulty in terminating the structure at a point where three orthogonal planes meet. The observed cubosomes show a narrow size distribution with sides of about 300 nm.

A similar size distribution appears from the SLS and DLS experiments made on the same samples, even though on average the particles in AFM appear slightly bigger because the resulting image is a convolution of the tip shape with the particle shape. DLS experiments were performed at an angle of 90°. A cooperative diffusion coefficient, $\langle D \rangle_z = \langle \Gamma \rangle / Q^2$, and a normalized

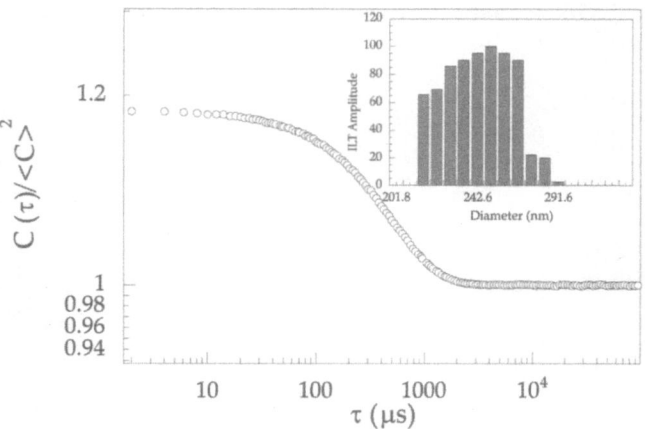

Fig. 5 Autocorrelation function of the intensity of the light scattered at 90° by a dispersion of cubosomes in water. A cumulant fit of this curve yields a hydrodynamic radius of 127.9 nm and a distribution width index of 0.129. An inverse Laplace transform performed by means of the CONTIN algorithm supports the fairly narrow size distribution of the particles, as shown in the *inset*

second moment, $\mu_2 / \langle \Gamma \rangle^2$ (with $\langle \Gamma \rangle$ the initial decay rate of the intensity autocorrelation functions), were obtained by means of a second-order cumulant analysis [8]. From $\langle D \rangle_z$ a hydrodynamic radius can be calculated according to the Stokes-Einstein relation.

The autocorrelation functions were also Laplace inverted by CONTIN [9] in terms of a continuous distribution of exponential decay times. The best solutions were chosen with the smoothing constraint (probability to reject) of 0.5.

Figure 5 shows the autocorrelation function of the intensity of the light scattered by a solution of cubosomes (diluted 1:250 from the original concentration). A second-order cumulant analysis yields a hydrodynamic radius of 127.9 nm and a polydispersity index of 12.9%, thus pointing out a fairly monodisperse population of particles. The inset shows the CONTIN distribution of the exponential decay rates, which strengthens the previous observation.

From static light scattering measurements the root-mean-square radius of gyration was evaluated from a simple Guinier approximation [10], that is

$$I(Q) \propto \exp\left(-\frac{R_g^2 Q^2}{3}\right) \tag{1}$$

This equation is generally applicable in the small-Q region, that is $QR_g < 1$. Figure 6 shows a Guinier plot, $\ln I(Q)$ versus Q^2, which yields a straight line in the low Q region. The slope of this curve gives the radius of gyration of the particles that in this case turns out to be 121.8 ± 1.4 nm.

Strictly speaking, cubosomes cannot be assumed as spherical particles of uniform refractive index; neverthe-

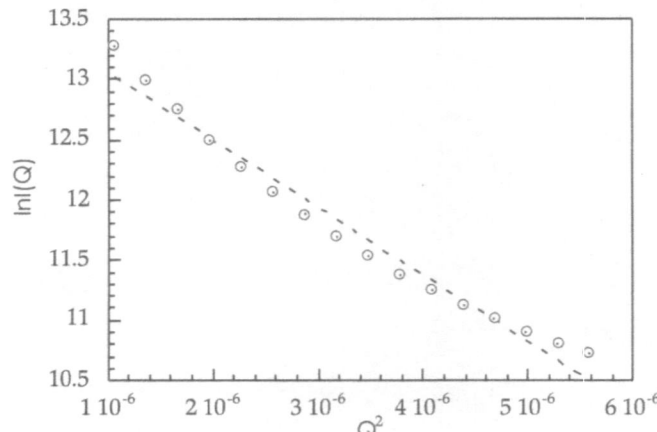

Fig. 6 Guinier plot of the intensity of the light scattered by a dispersion of cubosomes in water. The slope of the linear fit allows the evaluation of the radius of gyration of the aggregate, which in this case is 121.8 ± 1.4 nm

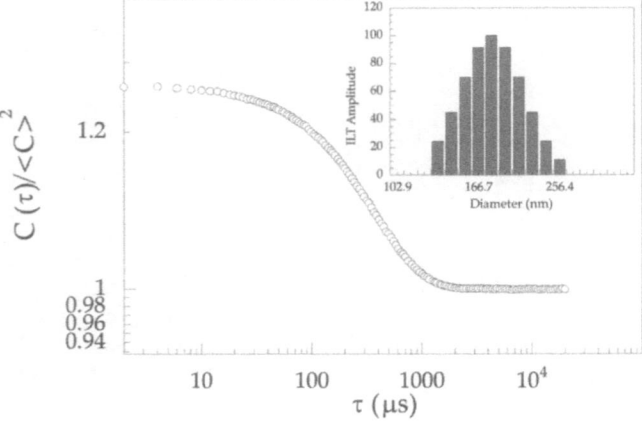

Fig. 7 Autocorrelation function of the intensity of the light scattered at 90° by a dispersion of hexosomes in water. The hydrodynamic radius obtained by means of a cumulant analysis is around 94 Å, while the polydispersity index is 11.2%. An inverse Laplace transform performed by means of the CONTIN algorithm supports the fairly narrow size distribution of the particles, as shown in the *inset*

less we can evaluate the radius of a spherical particle possessing that particular radius of gyration, and then compare this value with the data obtained by AFM. Since $R_g = (3/5)^{1/2}$ for a sphere, one obtains for the cubosomes a geometric radius of 157.2 nm, which shows a fair agreement with observed side imaged by AFM.

The captured images of the hexosomes show faceted hexagonal particles. The number of the sides of the visualized hexagonal structures is not always equal to six as expected (see Fig. 4). This is probably a consequence both of the orientation of the hexosome with respect to the mica surface and of the friction exerted by the scanning of the tip (as mentioned before) on the sample. Anyway, the same portion of the sample could be scanned repeatedly without changing the image.

The hexagonal geometry is clearly distinguishable, especially from the nearly 120° inner angle. The main portion of the observed hexosomes was found having sides 200 nm long, with a mean lateral dimension shorter than the cubosomes.

The medium apparent height of the cubosomes and hexosomes on the surface of the mica is about 11 nm, definitely below the expected height, which should be close to the lateral dimension. This is probably due to two factors. First, the oscillating tip may sink into the soft particles and the AC detection method of the non-contact AFM mode, which is sensitive to force gradients in a direction perpendicular to the sample surface, may underestimate their height over the mica surface. Moreover, the adsorption phenomenon of the particles on the mica surface certainly has a flattening effect.

Figure 7 shows the intensity autocorrelation function for a dispersion of hexosomes (diluted 1:400 from the original concentration). The second-order cumulant analysis gives a hydrodynamic radius of 94 nm. The

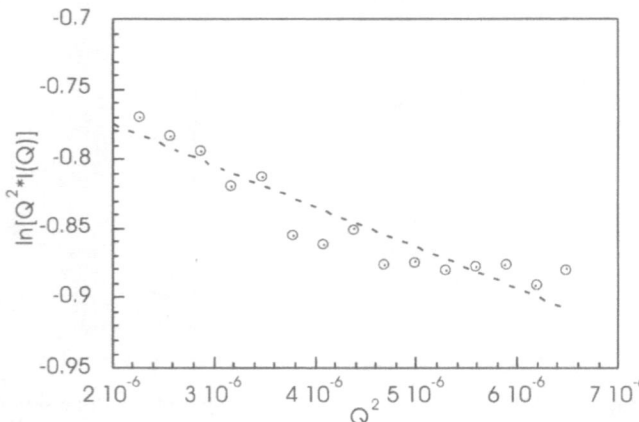

Fig. 8 Plot of $\ln[Q^2 \times I(Q)]$ versus Q^2 for the intensity of the light scattered by a dispersion of hexosomes. Assuming a discoidal shape for the aggregate, the fit of the linear portion of the curve yields the thickness of the disk, which is 59.7 nm

Laplace inversion of the autocorrelation function performed by means of the CONTIN routine (shown in the inset) confirms the narrow particle size distribution already pointed out by the second-order moment that indicates a polydispersity index of 11.2%.

To further demonstrate that the flattening effect is likely to be an AFM artifact, we have tried to interpret the SLS applying a data treatment for disklike particles. Guinier analysis is a model independent method that allows an evaluation of the radius of gyration of the aggregate, but does not provide structural details, such as the shape of the scattering object. This means that particles of different shape may exhibit the same radius of gyration. However, some extensions of the Guinier have been derived for particular shapes, such as cylinders and flat particles.

For flat particles it can be shown that

$$I(Q)Q^2 \propto T^2 \exp(-R_t^2 Q^2) \tag{2}$$

where T is the thickness of the flat particle and $R_t = T/\sqrt{12}$. Once the thickness has been obtained, the radial dimension can be determined by the Guinier plot, since for a disklike object

$$R_g^2 = (R^2 + T^2)/4 \tag{3}$$

where R is the disk long radius. A plot of the SLS intensities according to Eq. (2) shown in Fig. 8 yields R_t from the slope of the intermediate Q region; thus a thickness of 59.7 nm can be evaluated.

From a Guinier plot we can determine the radius of gyration of the whole object, which is 80.0 ± 1.3 nm. The geometric radius for the disc would be 148.4 nm according to Eq. (3). The SLS picture, with the initial assumptions for the shape derived from an AFM hypothesis of flat disks, clearly suggests that the thickness observed by means of the microscopic technique is highly underestimated. Light scattering indicates that both hexosomes and cubosomes are globular particles, similar to spheres from a hydrodynamic point of view.

Conclusions

AFM, DLS and SLS have been applied to the in situ investigation of the size and structure of submicron particles of very soft matter, cubosomes and hexosomes. Structural and morphological information about these particles has been obtained. Our results point out that the observations provided by these complementary experimental methods should be critically compared to highlight possible pitfalls and artifacts and to provide a global picture of the system. In fact AFM measurements provide a good valuation of the lateral dimensions and of the shape of the particles, but not as good an estimate of their height on the substrate. Operating in situ is a determinant requisite when working with soft matter and the results of this study are very encouraging towards the investigation of the many possible biological applications of these liquid crystalline phases. Cubic phases of GMO, a biocompatible material, can be investigated as drug-delivering devices for controlled-release applications with high stability and versatility. It is clear that non-invasive techniques such as AFM and light scattering will give a strong aid to future studies in this field.

Acknowledgements Thanks are due to MURST, CNR and CSGI for financial support.

References

1. Gustafsson J, Ljusberg-Wahren H, Almgren M, Larsson K (1997) Langmuir 13:6964–6971
2. Gustafsson J, Ljusberg-Wahren H, Almgren M, Larsson K (1996) Langmuir 12:4611–4613
3. Landh T (1994) J Phys Chem 98:8453–8467
4. Neto C, Aloisi G, Baglioni P, Larsson K (1999) J Phys Chem 103:3896–3899
5. Larsson K, Lindbolm G (1982) J Disp Sci Techol 3:61
6. Ericsson B, Larsson K, Fontell K (1983) Biochim Biophys Acta 729:23
7. Landh T (1992) Thesis, University of Lund
8. Koppel DE (1972) J Chem Phys 57:4814–4820
9. Provencher SW (1982) Comput Phys Commun 27:213–218
10. Porod G (1982) In: Glatter O, Kratky O (eds) In: Small angle X-ray scattering. Academic Press, London pp 17–51

Progr Colloid Polym Sci (2000) 115:300–306
© Springer-Verlag 2000

RHEOLOGY

R. Biehl
T. Palberg

Assessment of shear-induced structures by real space and Fourier microscopy

R. Biehl · T. Palberg (✉)
Institut für Physik der Universität Mainz
Staudinger Weg 7, 55099 Mainz
Germany
e-mail: thomas.palberg@uni-mainz.de

Abstract We report preliminary measurements of the shear-induced sliding layer structure in an aqueous suspension of highly charged polystyrene spheres. Particle interaction was controlled by advanced conditioning procedures to result in fluid or body–centred cubic equilibrium structures. Shear was applied in an optical plate–plate shear cell of variable slit width. Fourier microscopy yielded complementary information to real space analysis. The accessible range of scattering vectors was $(3.5 \leq k \leq 7.2)$ μm^{-1}. We checked the experimental performance by recording the form factor of a non-interacting suspension and structure factors of less dilute suspensions in dependence on electrolyte concentration c. For deionized suspensions of particle number densities of $n = 0.34$ μm^{-3} we observed crystallisation. Shearing this sample we obtained two-dimensional structure factors which are compatible with sliding layer formation. The observed quantitative discrepancies to the theoretical model calculations are discussed.

Key words Colloidal suspensions · Shear-induced structure · Microscopy · Light scattering

Introduction

Colloidal suspensions in many ways may be considered as scaled-up models of atomic condensed matter [1]. Both scattering techniques and real space microscopic approaches have widely been used to study phase behaviour [2], phase transition kinetics [3], crystal morphologies [4], twinning and twin annealing [5–7], glass formation [8] or behaviour in restricted geometry [9]. Also for experiments on shear-induced states, optical techniques proved valuable in exploring non-equilibrium structures [10–12]. The comparison of time and ensemble averaged scattering data with exemplary real space observation yielded valuable insights into both analogies and colloid specific differences of soft condensed matter solids as compared to their atomic counterparts. In particular, for model systems of monodisperse spheres with analytically tractable interaction, such investiga-

tions are backed by a large body of theoretical and simulation work [13, 14].

In this contribution we shall demonstrate that both experimental approaches may under certain conditions be used on the very same ensemble of particles. This applies both to equilibrium states but, even more importantly also to stationary non-equilibrium states. To avoid problems of multiple scattering or background scattering, we here used an optical plate–plate shear cell of variable but small gap width $d = 10$–2000 μm. For the same reason we decided to work with dilute suspensions with typical particle densities of $n \leq 1$ μm^{-1} only. At such densities we obtain a good resolution of particle positions and in addition the first Bragg reflections fall within the angular range of the collecting optics. Thus we were able to use our microscope both in the conventional high-resolution real space mode and in the Fourier mode.

In what follows we briefly describe particles, particle conditioning and phase behaviour before we give a more detailed description of the optical techniques used. Special emphasis is put on the optical and mechanical performance of our shear cell. Examples of both real space and Fourier space microscopy are provided. Taking the example of shear-induced layer structures, we conclude with a discussion of the instrumental performance, its possibilities and limitations.

Materials and methods

We used polystyrene spheres of nominal diameter $2a = 301$ nm (IDC, Portland, Ore., USA). These we carefully characterized to obtain a static light scattering radius of $a_{SLS} = (155.5 \pm 1.2)$ nm [15]. The titratable number of surface charges was $N = 2.3 \times 10^4$ and the effective charge from conductivity was $Z^* = 1980$ [16, 17]. All sample conditioning follows procedures recently described in more detail [16]. We only briefly note that the actual measuring cell is integrated into a closed tubing system. During conditioning the suspension is peristalticly pumped through a set of devices, allowing precise adjustment and in situ control of the interaction parameters, particle number density n and concentration of added electrolyte c. We further note that the concentration of residual impurities in general is negligibly small ($c_B < 10^{-7}$ mol l^{-1}) and that the contamination with airborne carbonate can be kept below 5×10^{-7} mol l^{-1} h^{-1} [16]. At typical conditions of $n \approx 1$ μm^{-3} and $c = 5 \times 10^{-6}$ mol l^{-1}, residual uncertainties are on the order of 1% and 5%, respectively.

Samples were observed from below using an inverted microscope (Leica DM IRB, Leitz, Wetzlar, Germany) mounted on a vibration-isolated table. The instrument provides sufficient space for mounting the optical shear cell. It allows for a flexible change of illumination from a tungsten halogen lamp to a $\lambda_0 = 635$ nm laser diode (Laser Grafics, Klein Ostheim, Germany). In both cases, illumination is with parallel light. A central beam stop of 1.3 mm was applied to eliminate zeroth-order diffracted light from the detection path. Note that this is similar to dark field or ultramicroscopic illumination, where owing to wide angle illumination the zeroth-order light passes outside the objective. Accordingly, our images are constructed from diffracted light of order $O \geq 1$. Since the back focal plane of the objective is not directly accessible, the beam stop was adjusted directly in front of the objective lens. In the Fourier mode an additional Bertrand lens is inserted in the light path. This leads to a real image of the back focal plane of the objective on the CCD chip. The central beam stop covers angles of $\Theta \leq 15.1°$. The aperture of the objective (PL Fluotar, Leitz) collects light under angles $\Theta \leq 44.4°$. Thus the accessible range of scattering vectors $k = (4\pi n/\lambda_0) \sin(\Theta/2)$ is 3.5 $\mu m^{-1} \leq k \leq 7.2$ μm^{-1}. For high-resolution real-space observations we used a long-distance, dry, $63 \times$ objective able to cover practically the whole range of plate-plate distances adjustable in the shear cell (maximum working distance ca. 1800 μm). The front focal plane (i.e. the objective in the z-direction) can be adjusted over a range of 100 μm with an accuracy of 10 nm. This is achieved with a piezoelectric drive (PI, Göttingen, Germany). We note that all optical components except the Bertrand lens were chosen to be phase corrected. Resulting images are digitally recorded with a CCD camera (Jai, Copenhagen, Denmark) and stored in a computer.

Several requirements had to be fulfilled in the construction of the plate-plate shear cell. The plate-plate distance should be variable to study influences of the cell walls on order formation, and the shear rate should be variable over a large range to study a great variety of mechanical non-equilibrium conditions. To have reproducible shear conditions the cell was constructed to allow for adjustments of several degrees of freedom. A sketch of the instrument is shown in Fig. 1. It is mounted on a sturdy manual x-y stage to chose the observation volume. A massive brass ring with Plexiglas fitting takes the lower quartz plate of 70 mm diameter and thickness 1 mm. The brass ring can be tumbled by a set of three micrometer screws to obtain the lower plate parallel to the x-y stage and thus to the optical axis. On top of the brass ring the rotation platform for the upper plate is mounted. Here a stainless steel ring rotates in an outer brass part. The upper plate of height 18 mm and a lower diameter of 32 mm (top of the sheared volume) is fixed below this ring. Both quartz plates are polished to better than $\lambda/4$ planarity (Hellma, Müllheim, Germany). The lower plate can be adjusted by a second set of micrometer screws to obtain its lower surface perpendicular to the rotation axis. Its distance and co-planarity to the lower plate is independently adjusted by a third set of micrometer screws tumbling the whole upper part of the shear cell against the lower.

A cock-wheel is fixed on top of the stainless steel ring and rotated with a toothed belt drive. Rotation velocities can be varied from 0.025 to 0.5 rpm. At a plate separation of 60 μm and a radial distance of $r = 10$ mm off the centre this corresponds to shear rates of 0.45 Hz and 9 Hz, respectively. Shear rates were checked to be constant in time within 2%.

The interaction parameters for sheared suspensions should be precisely adjusted and kept stable inside the shear cell over several hours. To this end the Plexiglas fitting contains bores to rinse the cell with suspension and connect it to the preparation circuit. A delicate point is the sealing against air at the gap between the lower and upper plates. Coverage with oil films or lamellae were observed to lead to incomplete sealing or wearing of the lamellae. Good results were achieved with a free-gliding Viton O-ring sealing.

Fig. 1 Sketch of the optical plate–plate shear cell

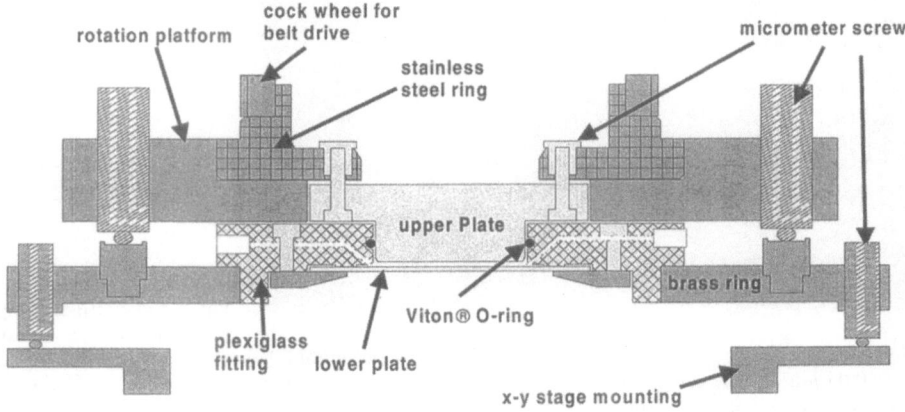

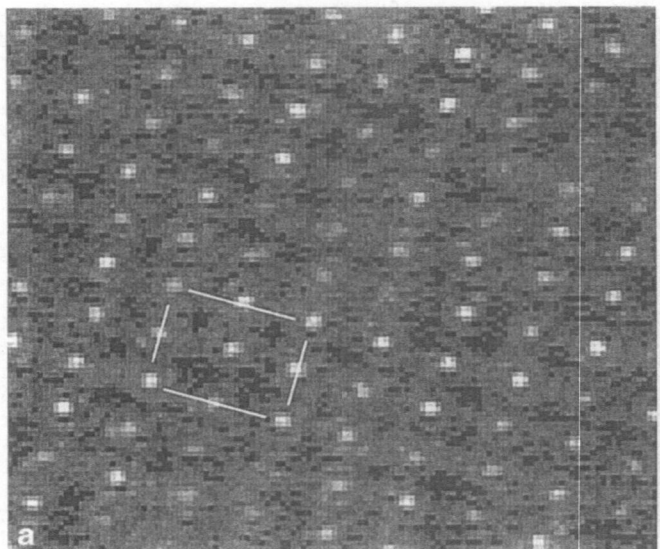

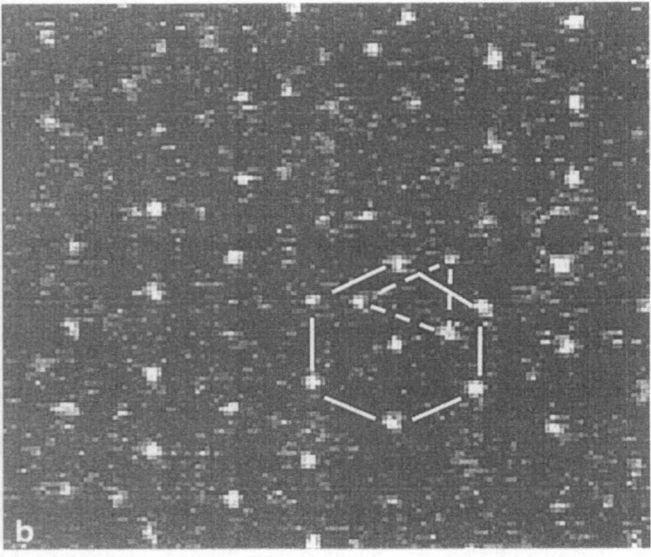

Fig. 2a,b Real space images of equilibrium and non-equilibrium structures as observed in a suspension of $n = 0.35 \ \mu m^{-3}$ and deionized conditions. **a** System at rest; bcc lattice with the $\{110\}$ plane parallel to the focal plane; the lattice constant is $a = 1.8 \ \mu m$. **b** System under shear; $d\gamma/dt = 1$ Hz. Planes of hexagonal order are sliding over each other parallel to the focal plane. Note that both images show two layers of the corresponding lattices. See text for details

Results

Instrumental performance

Using real space microscopy we first checked the suspension to display the expected structural behaviour. At mechanical equilibrium, dilute, salty systems were found to stay in a disordered state. With increasing particle number density n and decreasing electrolyte concentration c a pronounced fluid order evolves. Under deionized conditions (no added salt) the system shows a fluid-solid phase transition located approximately at $n \approx 0.08 \ \mu m^{-1}$. With increasing salt concentration the phase boundary shifts to higher n-values and a coexis-

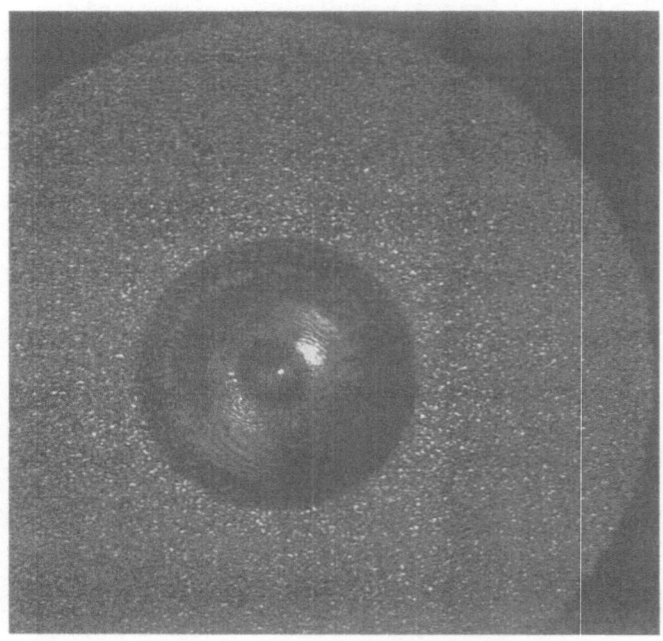

Fig. 3 Fourier image of a dilute salty suspension ($c = 0.7 \ \mu mol \ l^{-1}$, $n = 0.05 \ \mu m^{-3}$). The central dark part is the central beam stop, the outer dark part originates from the lens aperture

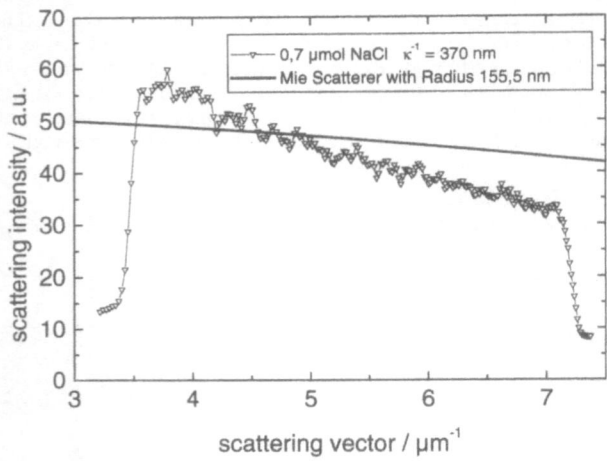

Fig. 4 Qualitative comparison of the angular averaged scattered intensity of a dilute salty suspension ($c = 0.7 \ \mu mol \ l^{-1}$, $n = 0.05 \ \mu m^{-3}$) (same as in Fig. 3) to a calculated Mie form factor using $a_{SLS} = 155.5$ nm. Note that in both cases arbitrary units are used. For details see text

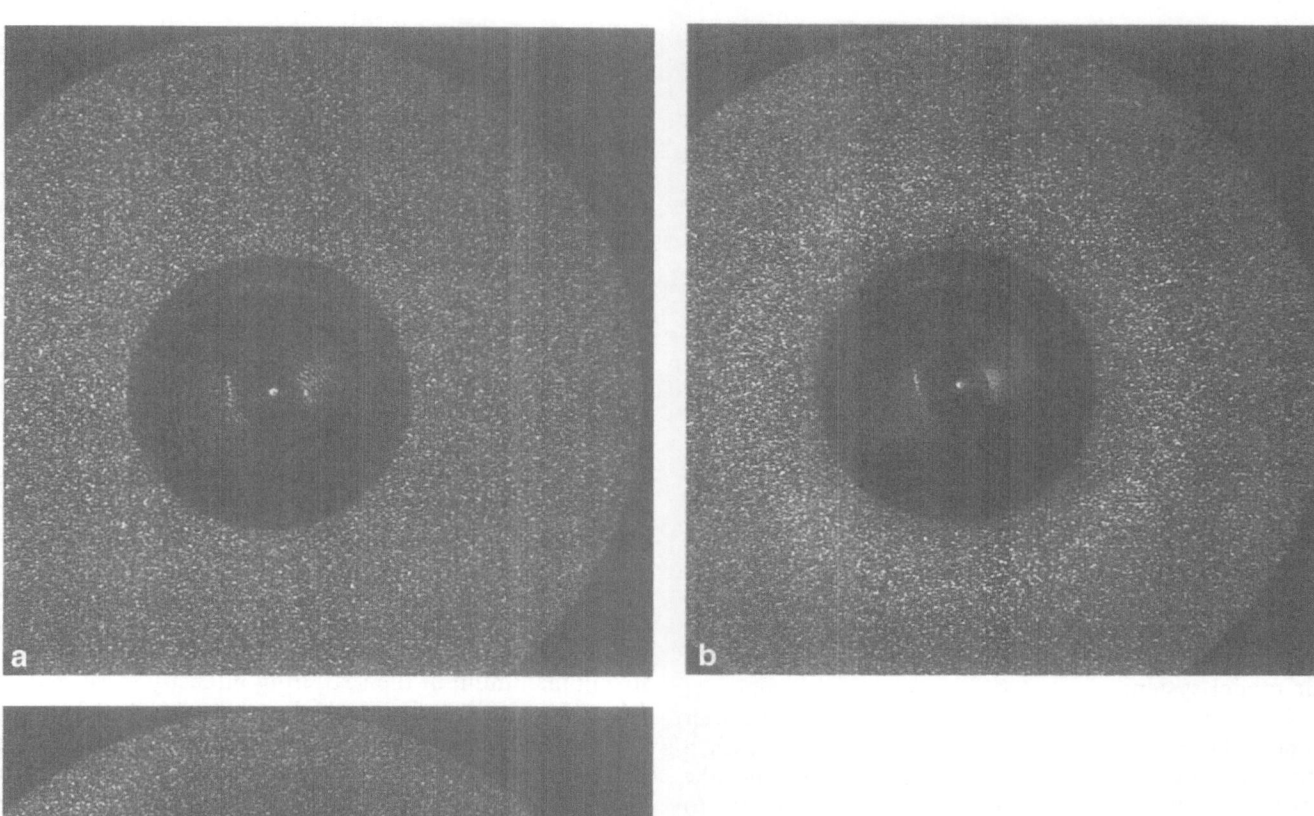

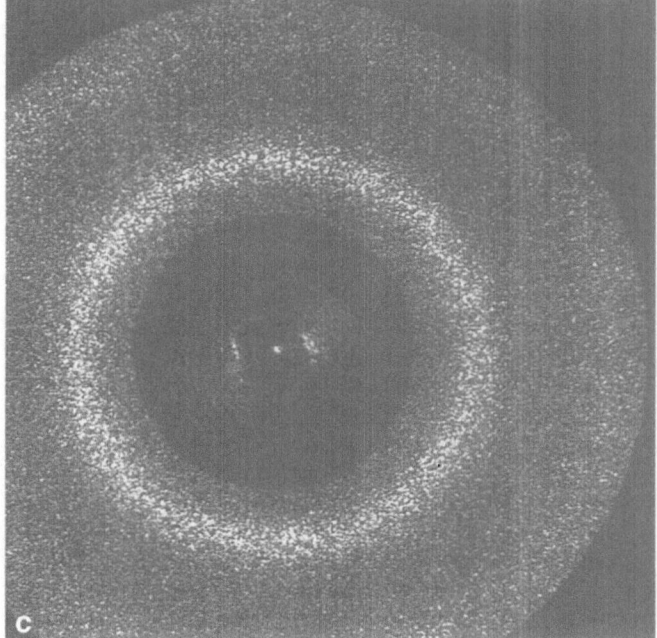

Fig. 5a–c Fourier images of a suspension with $n = 0.35$ μm^{-3} and different concentrations of added NaCl. **a** $c = 39$ μmol l^{-1}; **b** $c = 2.6$ μmol l^{-1}; **c** $c = 0.34$ μmol l^{-1}. The corresponding Debye screening lengths are 50, 170 and 350 nm, respectively

tence region is observable. The crystal structure was observed to be bcc. This is shown in Fig. 2a for a suspension at deionized conditions and $n = 0.35$ μm^{-3}. This corresponds to a volume fraction $\Phi = n(4/3)a_{SLS}^3 = 0.0048$ and a body-centred cubic (bcc) lattice constant of 1.8 μm. The lines are guides to the eye, indicating a portion of the imaged {110} plane. Note a second plane is visible in this image shifted by half a lattice constant. If subjected to shear, stationary non-equilibrium struc-

tures may be formed. At low shear, previous work indicates the existence of planes with hexagonal in-plane order, sliding above each other [10, 11]. Figure 2b shows such layers of nearly perfect hexagonal order, observed for the same suspension at a shear rate of 1 Hz. Note that in this snapshot the layers are shifted by an odd fraction of the lattice constant. At high shear rates an isotropic shear melt may be obtained (not shown). Thus we observe the expected

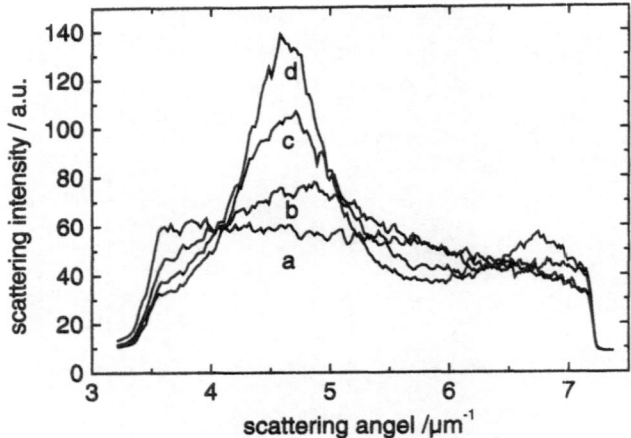

Fig. 6 Angular averaged scattered intensity of a suspension with $n = 0.35\ \mu m^{-3}$ and different concentrations of added NaCl. Curve a: $c = 39\ \mu mol\ l^{-1}$; curve b: $c = 5.1\ \mu mol\ l^{-1}$; curve c: $c = 2.6\ \mu mol\ l^{-1}$; curve d: $c = 0.23\ \mu mol\ l^{-1}$

equilibrium and non-equilibrium phase behaviour for our model system.

We then checked the performance in the Fourier mode, where we observe the back focal plane with a Bertrand lens. After subtracting any background the scattered intensity I_S should be proportional to $I_S \propto I_0\, nb_0^2 P(k)S(k)$, where I_0 is the illuminating intensity, b_0^2 the zero wave vector single particle scattering length squared, $P(k)$ the particle form factor and $S(k)$ the structure factor [18]. Figure 3 shows the Fourier image (scattering pattern) of a diluted, salty suspension. At this low particle density and high concentration of added salt, no structure formation is expected [$S(k) = 1$]. We observe a slight but smooth decrease of scattered intensity.

In Fig. 4 the angular averaged intensities are shown plotted versus the corresponding scattering vectors k. The latter were calculated using the radial position of a pixel in relative units (i.e. 0 at the centre and 1 at $r_{aperture}$, the position of aperture):

$$k(w) = \frac{4\pi n_w}{\lambda} \sin\left(\frac{1}{2}\arcsin\left(\frac{1}{n_w}\sin\left(\arctan\left(w\frac{r_{aperture}}{f_{objective}}\right)\right)\right)\right)$$

(1)

Here $f_{objective}$ denotes the focal length of the objective and n_w is the refractive index of water. At large $k > 7.4\ \mu m^{-1}$ a darkcount background of about 10 units is visible. The background at small k is slightly larger owing to parasitic stray light. (Note that the inner beam stop is not completely black in this image.) The scattered intensity shows some statistical scatter on the order of some 10 units. The statistical accuracy could be further enhanced by averaging over several images. Apart from this scatter, the signal decreases monoto-

nously over the accessible range of scattering vectors. We compare our data to a form factor $P(k)$ calculated using Mie theory for monodisperse spheres of radius 155.5 nm. The observed decrease obviously is stronger than expected theoretically. As seen by the small background at small k, stray light is not likely to be responsible for this effect. Rather we suspect a smooth apparatus function to be present which possibly originates from phase shifts between central and off-centre beams passing the Bertrand lens, which in this instrumental version was not phase corrected. A correction for this effect could in principle be performed by monitoring the suspected apparatus function for a true Rayleigh scatterer for which also $P(k) = 1$. At the moment we refrain from further quantitative evaluation.

As a further qualitative check we performed measurements on suspensions of decreasing salt concentration. Figure 5a–c shows Fourier images for a particle density of $0.35\ \mu m^{-3}$ corresponding to an average interparticle spacing of $d_{NN} = 1/n^{-1/3} = 1.4\ \mu m$. A pronounced ring pattern evolves upon decreasing c. This is also seen in the angular averaged data of Fig. 6. A first strong maximum in the scattering intensity is located at $4.6 \pm 0.1\ \mu m^{-1}$ and a second one at $7.0 \pm 0.3\ \mu m^{-1}$. The main maximum location corresponds to an average particle distance of $1.35\ \mu m$, in good agreement with the selected particle density. We attribute this pattern to the formation of a pronounced short-range order in the suspension.

Sliding layer formation

Finally we recorded the Fourier images of crystalline suspensions subjected to shear. Figure 7a and b shows the patterns for a low shear rate of 0.5 Hz and a higher rate of 5 Hz. Similar patterns of twofold symmetry perpendicular to the applied shear direction have been observed before in many systems and interpreted as due to the formation of hexagonal layers sliding over each other [10, 12, 19]. In particular, the set of inner reflections are well compatible with the ones calculated by Loose and Ackerson [11] for plane layers moving in a zig-zag fashion. Our outer reflections, however, are only qualitatively in accordance with these predictions. Most remarkably, the intensities of the second and third order seem to be exchanged compared to Fig. 5b in [11]. In Fig. 7 the intensity of the third order in fact is larger than that of the second-order reflections. This may be compared to the image recorded after cessation of shear. There the layers are expected to register, giving rise to a sixfold symmetric scattering pattern. This indeed is observed, but again the intensities are interchanged between the second and third order.

We note that this effect cannot be due to our apparatus function, since that was found to monoto-

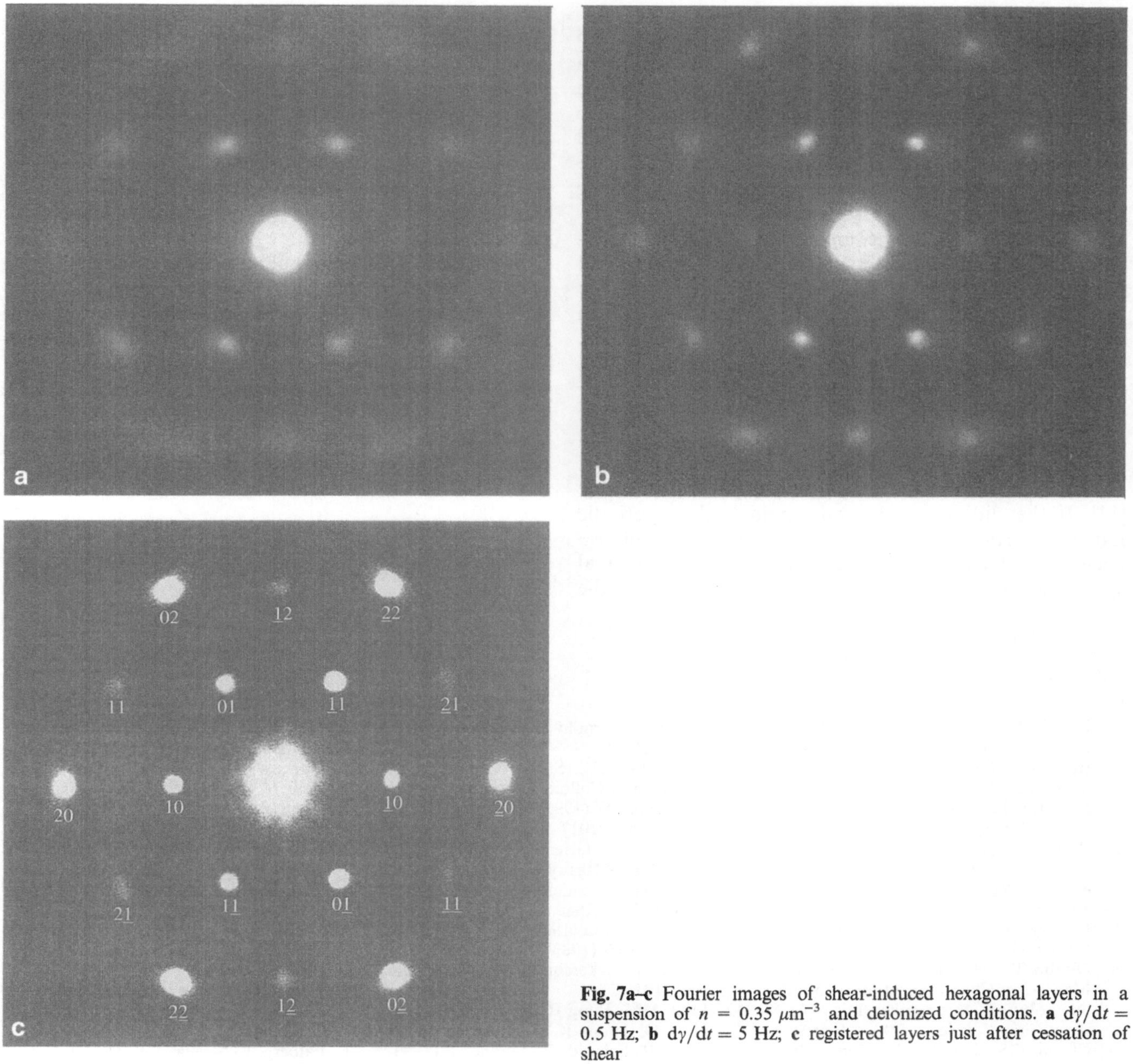

Fig. 7a–c Fourier images of shear-induced hexagonal layers in a suspension of $n = 0.35\ \mu m^{-3}$ and deionized conditions. **a** $d\gamma/dt = 0.5$ Hz; **b** $d\gamma/dt = 5$ Hz; **c** registered layers just after cessation of shear

nously decrease. A possible physical explanation could be a violation of the boundary conditions assumed in the calculation, e.g. the condition of bulk samples with an infinite number of layers or the condition of plane layers. In our case the plate-plate distance was 60 μm, corresponding to some 30 layers. Investigations with varied gap width should solve this point. Further, in some large area scans in the real space mode we could identify a considerable amount of particles being "out of plane". Distortions in z-directions are also known from other thin cell preparations at low layer number [9, 20]. They have been interpreted in terms of optimized accommodation of particles in the restricted geometry given. At present we cannot solve this interesting question and refrain from further speculation.

Conclusions

A newly constructed shear cell in plate-plate geometry with a constant shear rate for improved structural investigations of the shear mechanism was presented. The cell is designed for the investigation of dilute crystalline suspensions over a wide range of shear rates. The cell can be combined with advanced conditioning procedures to work either under complete deionized conditions or with a controlled addition of salt. To enhance the possibilities of optical access to equilibrium and non-equilibrium structures, we used both real space and Fourier microscopy. We presented data on the equilibrium bcc phase and several metastable hexagonal layer phases under and after shear. To our surprise we observed significant deviations from predicted scattering patterns and discussed possible origins.

The elucidation of the underlying physical reasons can and will be the next step of our research. To this end we may use further capabilities of our instrument. For the data presented the exposure time was 1 ms. This in principle should allow for two additional experiments. First, the trajectories of the particles in real space may be monitored with high accuracy. Second, the Fourier images could also be checked for periodic components. In particular for layers sliding over each other in a continuous fashion, a periodic change in intensity of second- and third-order reflections should be observable [11]. If the sliding in addition would lead to periodic lattice distortions, changes in the peak position will also occur [21]. Height calibrated large area scans in real space, would further allow us to characterize the

distribution and amplitude of layer position fluctuations in the z-direction. In such height-dependent measurements we exploit the accuracy of our plate surfaces and the possibilities of piezoelectric positioning of the objective. Finally, the dependence on the gap width may be checked using the adjustment of the upper plate. Assuming a similar particle density as in the present experiment, the numbers of planes could be varied between 6 and 600 layers.

All these experiments could deepen our understanding of the sliding mechanism. However, also the metastable states after shear are accessible and can be investigated with regard to their relaxation into equilibrium. Neutron scattering studies on concentrated samples relaxing into stacking faulted fcc structures have already been reported. The symmetry breaking transition to the bcc state and the formation of lateral twin patterns, however, is still an unsolved problem.

Acknowledgements We have profited from many stimulating discussions with Th. Preis and Ch. Dux. Financial support of the Deutsche Forschungsgemeinschaft (DFG Pa459/8-1) is gratefully acknowledged.

References

1. Pusey PN (1991) In: Hansen JP, Levesque D, Zinn-Justin J (eds) Liquids, freezing and glass transition (51st summer school in theoretical physics, Les Houches, France, 1989). Elsevier, Amsterdam, pp 763–942
2. Bartlett P, van Megen W (1994) In: Mehta A. (ed) Granular matter. Springer, New York, Berlin Heidelberg, pp 195–257
3. Palberg T (1999) J Phys Condens Matter 11:R323
4. Okubo T (1994) In: Schmitz KS (ed) Macro-Ion Characterization: From dilute solutions to complex fluids (ACS symposium series 548) American Chemical Society, Washington, p 364–380
5. Elliot MS, Bristol BTF, Poon WCK (1997) Physica A 235:216
6. Dux C, Versmold H (1997) Phys Rev Lett 78:1811
7. Maaroufi MR, Stipp A, Palberg T (1998) Prog Colloid Polym Sci 108:83
8. van Megen W (1995) Transport Theory Stat Phys 24:1017
9. Crocker JC, Grier DG (1998) MRS Bull 23: 24; Murray C (1998) MRS Bull 23:33; van Blaaderen A (1998) MRS Bull 23:39; Asher SA, Holtz J, Weismann J, Pan G (1998) MRS Bull 23:44
10. Ackerson BJ (1983) Physica A 128:221
11. Loose W, Ackerson BJ (1994) J Chem Phys 101:7211
12. Laun HM, Bung R, Hess S, Loose W, Hess O, Hahn K, Hädicke E, Hingmann R, Schmidt F, Lindner P (1993) J Rheol 36:1057
13. Buttler S, Harrowell P (1995) Phys Rev E 52:6424
14. Stevens MJ, Robbins MO (1993) Phys Rev E 48:3778
15. Garbow N, Müller J, Schätzel K, Palberg T (1997) Physica A 235:291
16. Evers M, Garbow N, Hessinger D, Palberg T (1998) Phys Rev E 57:6774
17. Hessinger D, Evers M, Palberg T (1999) Phys Rev E (in press)
18. D'Aguanno B, Klein R (1991) J Chem Soc Faraday Trans 87:379
19. Dux Ch, Musa S, Reus V, Versmold H, Schwahn D, Lindner P (1998) P J Chem Phys 109:2556
20. Neser S, Bechinger C, Leiderer P, Palberg T (1997) Phys Rev Lett 79:2348
21. Palberg T, Streicher K (1994) Nature 367:51–54

Progr Colloid Polym Sci (2000) 115:307–314
© Springer-Verlag 2000

M. Evers
T. Palberg
N. Dingenouts
M. Ballauff
H. Richter
T. Schimmel

Vitrification in restricted geometry: dry films of colloidal particles

M. Evers · T. Palberg (✉)
Institut für Physik der Universität Mainz
Staudinger Weg 7, 55099 Mainz
Germany
e-mail: thomas.palberg@uni-mainz.de

N. Dingenouts · M. Ballauff
Polymer-Institut der Universität Karlsruhe
Kaiserstraße 12, 76128 Karlsruhe
Germany

H. Richter · T. Schimmel
Institut für angewandte Physik der
Universität Karlsruhe, Engesserstraße 7
76131 Karlsruhe, Germany

Abstract We prepared dry films of colloidal particles from surfactant co-stabilized polystyrene and poly(methyl methacrylate) latex spheres of moderate polydispersity. The film thickness was varied from monolayer to some 20 μm. The average structure and the influence of polydispersity was analyzed by small-angle X-ray scattering. An analysis of local microscopic structures was performed by force microscopy on the least polydisperse sample. Irrespective of the thickness, we observe the formation of only short-ranged order which is rather pronounced but changes its qualitative character in the vicinity of the substrate. The observations suggest a bulk structure of small, defect-rich and stacking faulted clusters of local face-centred cubic symmetry and the co-presence of small compact voids. Closer to the substrate we find clear evidence of a tendency to form layers parallel to the substrate. This is interpreted as induction of orientational order by the presence of the restricting wall. While the local environs are often well ordered, the range of crystalline or in-layer order is restricted to only a few dozen particle diameters. This behaviour differs significantly from that observed in other systems and may be due to particularly small adsorbate-adsorbate, adsorbate-substrate and capillary forces.

Key words Hard spheres · Restricted geometry · Vitrification · Layering · Colloids

Introduction

Bulk colloidal suspensions of spherical particles may in many ways be considered a mesoscopic model of atomic substances. If the interaction between the particles is sufficiently strong, short- and long-range ordered states are formed which have been termed colloidal fluids, crystals and glasses. Owing to the colloid specific length and time scales, their structure and dynamics are conveniently accessible by comparably simple, yet powerful optical methods like scattering techniques or microscopy. At the same time, suspended spherical particles often exhibit an analytically tractable interaction potential and thus may be treated by various theoretical approaches. This has lead to a considerable interest in their properties

and in particular their phase behaviour, solidification kinetics and vitrification have been studied in great detail [1–6].

Most investigations so far were on bulk suspensions. Recently, a growing interest also appeared in their behaviour in restricted geometries. Most of these studies focused on the suspended state and the formation of long-range order therein. This often was motivated by the unusual structural, optical and mechanical properties observed which promise interesting applications e.g. as optical devices [7]. Dried films, on the other hand, have been studied extensively to optimize the performance of paints, coatings, inks, etc. [8]. Again, a long-range order is often desired, as it provides a dense packing of spheres [9]. From the colloid physics point of view, this first step of film formation, the assembly of particles into a more or less compact object (followed by particle deformation and polymer inter-diffusion), is the most challenging. It can directly be compared to the order formation, crystallization or vitrification in the freely suspended state with [7, 10–14] or without [1–6] the presence of restricting geometries.

While in many cases the model character of the systems with respect to atomic condensed matter is quite pronounced, there are also specific differences originating from the soft condensed matter nature of colloidal systems [15]. Colloidal particles may be tailored with a diversity of rather well-defined interactions ranging from sterically stabilized hard spheres to long-range repulsive charged spheres, and further to attractive interactions, e.g. introduced by depletion forces. They are further easily influenced by the presence of external fields ranging from gravity or shear to electric or magnetic fields. This gives rise to an extremely rich phase behaviour and a diversity of self-organized structures [15]. Unavoidable polydispersity, on the other side, may lead to significant deviations from ideally expected structures.

We here study systems of dry colloidal spheres which upon drying develop a pronounced short-range order [9]. It does, however, neither form crystals and nor show a pronounced layering phenomenon in multi- and mono-layer samples. Both the latter have been observed in many previous studies on other dried systems [16–18], but also in suspension near container walls [7]. The vitreous state observed thus was quite unexpected. We therefore conducted experiments to explore two counteracting influences on the order formation process and report our preliminary results in this contribution. Polydispersity on the one hand will tend to decrease order formation, while the presence of a flat substrate should actually increase order formation and in particular for mono- and few layer films should lead to a well-developed layering phenomenon.

After a brief introduction of materials used and techniques we shall present the results taken with small-angle X-ray scattering (SAXS) on powdered thick film samples. We then shall show the results of scanning force microscopy (SFM) on the least polydisperse sample dried in films of various thickness d. The nature of the bulk short-range order and the characteristic changes in the vicinity of the surface are characterized and possible reasons discussed.

Materials and methods

Sample preparation and thick film SAXS measurements

We synthesized three samples of polystyrene (PS) and poly(methyl methacrylate) (PMMA) latex spheres by standard emulsion polymerization using potassium peroxodisulfate as starter and sodium dodecylsulfate (SDS) as surfactant [19]. Chemicals used were of analytical grade and styrene was purified by distillation in vacuo. We obtained spherical particles of moderate to pronounced polydispersity. These were PMMA 71/12 of diameter $2a = 71$ nm and polydispersity $\sigma = 12\%$, PS 69/11 with $2a = 69$ nm and $\sigma = 11\%$, and PS 106/4.8 with $2a = 106$ nm and $\sigma = 4.8\%$. All three were purified by prolonged dialysis against Milli-Q water. Dried thick film samples ($d = 15$–4000 μm) were obtained by spreading a small amount (2–4 ml) of suspension on a petri-dish and left to evaporate over a period of several days at ambient temperature. For comparison, some samples were covered by a beaker to enhance humidity. Further samples were dried in a cylindrical bore of roughly 6 mm diameter and 10 mm depth or on flat glass substrates, either confined by a Teflon ring of 15 mm diameter or without. No perceptible difference was observed in the scattering experiments for these thick film formation techniques. Note that $d > 15$ μm corresponds to a bulk situation and that an influence of the substrate is expected to be operative for samples containing only a few layers of particles.

For the SAXS measurements the thick film samples were powdered to obtain isotropic specimens. The sample holder of approximately 2 mm height and 15 mm width had thin aluminium foil windows in the back and front. The depth in direction of the beam was 1.5 mm and the total sample volume was 43.7 mm^3. It was filled with weighted amounts of dry powder. Using the densities of PS ($\rho = 1.054$ g cm^{-3}) and PMMA ($\rho = 1.19$ g cm^{-3}) leads to an overall particle volume fraction of some 40%. This value is much lower than the packing fraction obtained for the spheres within the densely packed grains. The difference is due to the packing of the grains.

All SAXS measurements were performed using a modified Kratky camera with a position-sensitive counter. The range of scattering vectors covered $0.03 \leq k \leq 4$ nm^{-1}. A detailed description of this instrument and data treatment used has recently appeared [20]. Details of measurements on latices may further be found elsewhere [21, 22]. Special care was taken to properly account for the change in contrast between particles, surfactant and surrounding medium when going from the wet to the dry state. This was comprehensively discussed previously [9, 22]. There also the details of the data treatment and of the scattering theory used are given. In addition to the procedure known for monodisperse samples [23], we explicitly accounted for the influence of polydispersity on both particle form factor [21] and structure factor $S(k)$. Following the theoretical treatment of D'Aguanno and Klein [24], we discuss this issue in terms of a measurable structure factor $S_M(k) = I(k)/nI_0(k)$. It results from a comparison of intensities $I(k)$ measured at the finite particle number density n in the dried state and $I_0(k)$ measured in the dilute suspended sample at vanishing n.

Thin film preparation and SFM measurements

The least polydisperse sample PS 106/4.8 was used to study the formation of order in thin films of less than 40 particle diameter thickness. During thin film formation the conditioning of the sample (starting packing fraction, salt content, surfactant, etc.) as well as the preparation of the substrate were observed to have a pronounced influence on the range and homogeneity of order. Further, the drying velocity was found to play a major role. This is in contrast to our thick film samples but well known in the literature [8, 16, 17].

Drying of a drop on an untreated substrate generally leads to a situation where the combined action of few pinning centres, Coulomb repulsion in the wet state and lateral capillary forces is clearly visible. There subtle changes of preparation parameters exert a pronounced influence on the resulting structure. In particular, the area density of pinning centres may be varied between a completely sticky surface (e.g. if the substrate is covered with a polyelectrolyte of opposite charge) to a completely unsticky surface (e.g. using the two step on oil technique of Lazarov et al. [25]. In the first case, open structures result either from random settling of particles or from Coulomb repulsion in the wet state. In the second case, compact rafts of particles are observed with practically no influence of Coulomb repulsion. The oil is evaporated after complete water evaporation and the dry raft then settles upon the substrate.

We here worked with concentrations between $\Phi = 0.001$, leading to the formation of monolayers, and $\Phi = 0.1$, leading to the formation of multilayers. All samples had initially been deionized, but drying was performed in contact with air. We note that not in all cases were flat, homogeneous films formed. At elevated concentration the drop dried from the outer rim to form a coffee ring-like structure. If enclosed in a ring, often the formation of concentric ripples was observed. This was known from literature [18], but was rather pronounced in the case of PS 106/4.8.

Characterization of the resulting films was performed mainly by scanning force microscopy (SFM) using a commercial instrument (NanoScope IIIa, Digital Instruments, Mannheim, Germany) in the tapping mode. Cantilevers of length $125\ \mu m$ and tip radius 10 nm were used. Figures 5 and 6 were taken with a home-built SFM, details of which are given elsewhere [26]. For selection of scanning areas and for determination of film thickness the SFM is mounted on an inverted optical microscope (DM IRB, Leitz, Stuttgart, Germany). A principle difficulty in the determination of absolute layer thickness arises from the fact that a single scan usually does not cover both areas of substrate and intact multilayer surface. Thus owing to the restricted scan range of $100 \times 100\ \mu m^2$, only relative height differences can be accessed. We here resorted to optical microscopy and focused on the substrate and upper layer particles with the focus height controlled by a piezoelectric drive (PI, Göttingen, Germany). This allowed an estimate of layer thickness with an uncertainty of roughly two layers. We note that for films of strongly varying height the uncertainty is increased considerably.

Results and discussion

Residual order in vitreous thick films

Figures 1–3 show the results of SAXS measurements for the three samples, PS 106/4.8, PS 69/11 and PMMA 71/

12 [9]. In all three cases a pronounced first peak is observed, while the following maxima are much less pronounced and much lower in height. Thus in all cases no significant long-range order had been formed. The peak height decreases from 8.5 to 3.5 to 2.05 from Fig. 1 to Fig. 3, indicating a decrease in the degree of ordering. The comparison between the two PS samples indicates a disordering effect of polydispersity. The PMMA sample, however, has a significantly lower height of the first peak and at the same time pronounced higher-order peaks than the comparably polydisperse PS 69/11, indicating the presence of further factors.

To obtain a more quantitative description we fitted the three $S(k)$ with theoretical curves using the Percus-Yevick-Vrij (PYV) theory for polydisperse hard spheres [27, 28]. This is an extension of the well-known

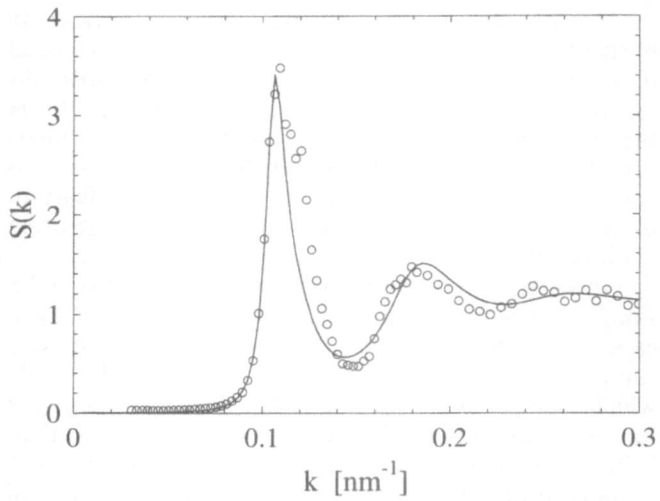

Fig. 2 Structure factor of PS 69/11

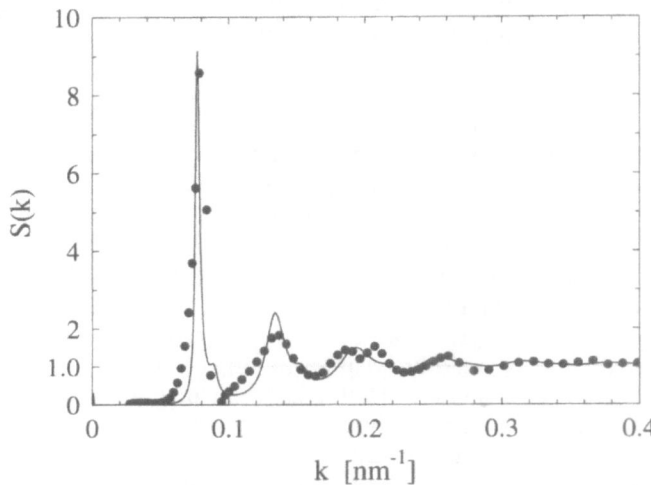

Fig. 1 Structure factor of PS 106/4.8

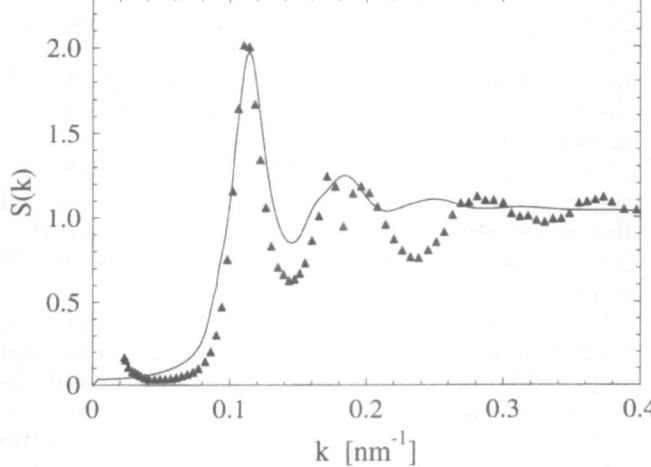

Fig. 3 Structure factor of PMMA 71/12

Percus-Yevick (PY) theory for monodisperse hard spheres [29], which has been tested against computer simulations. As discussed elsewhere [9], we consider PYV to be a suitable theory to compare to our scattering data in order to obtain at least a semi-quantitative analysis of the average structure.

The solid lines in Figs. 1–3 show the best fits obtainable. Since the particle radius and the polydispersity had been determined before, the only free fitting parameter was the packing fraction of the spheres in the powder grains. We obtained values of 0.66, 0.62 and 0.59 for the three samples. The first value is slightly above the other two, but slightly below the value of random close packing of monodisperse spheres. Note that a strong polydispersity may increase this value. Note, on the other hand, that the decrease of packing fraction is correlated with the decrease of order as qualitatively inferred from the peak heights.

The fits capture the form of experimental $S(k)$. In particular, the height and position of the first and second maxima are reproduced near quantitatively, with the exception of the second maximum of PS 106/4.8. There the experimental second maximum appears much broader and less high than seen for PYV. From this comparison we conclude that all samples show features of amorphous solids. In particular, no long-range order is observed.

However, a significant deviation is present in both PS samples. It is located at the main peak, where PYV is expected to be rather accurate. Both samples show a shoulder-like feature on the high k side of the peak, which is not present in the PMMA data. This does not interfere with our exclusion of long-range order. Rather, it indicates the possibility of a short-range order, which is more pronounced than a random close packing of spheres.

We further performed comparisons to simple model calculations of $S(k)$ and were able to exclude perfect face-centred cubic (fcc) crystals of strong but purely optical polydispersity and perfect fcc nanocrystals [9]. Both would have lead to pronounced higher-order maxima. However, taking the packing fractions obtained from the PYV fits and indexing the first maximum (111), the position of the shoulder was found to be corresponding to the (200) fcc reflection. This was suggestive of a residual local fcc order in our overall amorphous solids. We thus tried further structure identification with SFM for the least polydisperse PS sample, PS 106/4.8.

In Fig. 4 we show the cleaved surface corresponding to a vertical cut through the cylindrical bore preparation sample. No long-range order is visible except the presence of a preferred orientation of particle rows. At smaller scale, however, clusters with faulted fcc structure may be identified. These are non-compact, rather irregular in shape and sometimes heavily distorted

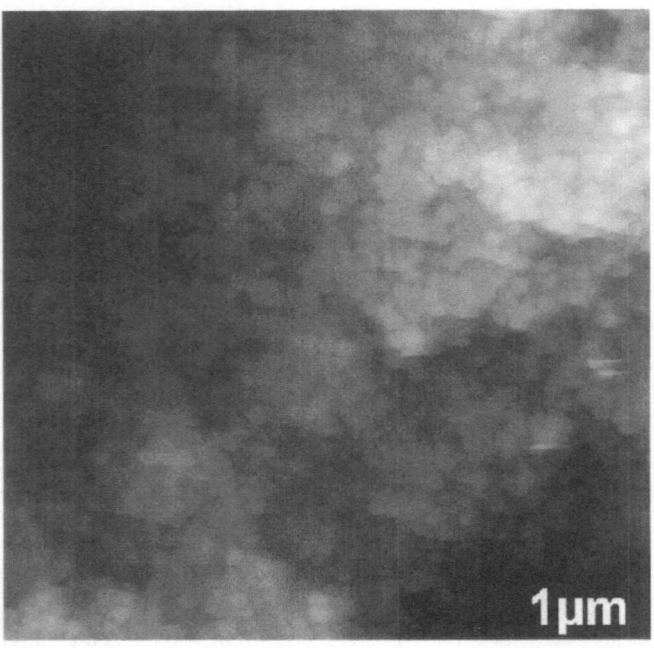

Fig. 4 Cleaved bore preparation, $5 \times 5\ \mu m^2$ scan of the cleave plane, $2\ \mu m$ Z-range

(bent). The number of particles in such clusters was estimated from counting several clusters and assuming an overall spherical shape. We note that the major difficulty in doing so was not the pure statistics of having only few clusters. The problem rather was to decide whether a particle or cluster segment belonged to the parent or had to be considered disconnected. The average cluster size thus was determined with about one order of magnitude uncertainty to be on the order of 10^2 particles per cluster. This surely is in the nanocrystalline domain. In addition to clusters, small, rather compact voids are visible. Their average size was some dozens of particles. We can, however, not prove that they had been there already in the split state. It may as well be that loose bits and pieces of the brittle material simply fell out of the sample upon cleaving.

Further, we studied cleavage surfaces of thick film splinters. The general impression was very similar to that observed in the bore sample. In a few cases, however, we were able to access small terraces of better defined in-plane order. We recall that the cleavage surfaces are not parallel to the substrate. Two examples taken close to the former film bottom and some 10 μm above are shown in Figs. 5 and 6. A pronounced triangular ordering is readily visible in Fig. 5 with a height variation of some 5–15 nm between neighbouring particle tops. At second sight one may trace the same order also in the picture taken somewhat away from the film bottom. The height variation of neighbouring sphere tops, however, amounts to some 80 nm (e.g. particle

bulk structure may be heavily biased by the individual selection. An additional difficulty in our case arises from the lack of quantitative measures of disorder. Thus only cautious suggestions for the bulk structure should be made. Here we propose the following. The largest structures in the thick film samples are of crystalline short-range order. They are nanocrystal-sized clusters with a high concentration of defects of various kinds and a highly irregular, non-compact surface. Most prominent are conventional stacking faults and disclinations, but layer corrugation seems to be another important feature. Further, small voids seem to be present. The restriction of well-defined fcc environments to only few lattice constants may be considered responsible for the presence of the (200) shoulder in the SAXS data and the absence of higher-order peaks.

Layering close to the substrate and in thin films

The proceeding section has shown by scattering techniques that for the three dry film systems under study the bulk structure is amorphous. For the least polydisperse sample the residual short-range order was further characterized to be crystal-like. We now shall investigate the changes occurring due to the influence of the geometric restriction by the substrate. The latter could already be sensed from Figs. 5 and 6, where an increase of layer corrugation was observable with increasing distance from the substrate. We now shall focus on surfaces parallel to the substrate. Again PS 106/4.8 is investigated.

In particular we study a preparation in a ring on untreated quartz substrate from a suspension of $\Phi \approx 0.05$. In this case we observed the formation of a coffee-ring structure, i.e. a radial increase of film thickness towards the outer rim and a maximum thickness of about 20 layers. The height variation is not monotonous; rather, we observe the appearance of concentric ripples with multilayers of various height. In the sample centre the coverage is low, as can be seen in Fig. 7. Particles appear to prefer grouping, but no close-packed order has evolved. In Fig. 8 (some mm off-centre) the density is enhanced, but still no triangular lattice formation is observed. Rather, groups are formed with compact voids in between. The first ripple appears 2 mm off-centre with a radial extension of about 0.3 mm. Further out, larger terraces are observed. Scans across two extended multilayer regions are shown in Figs. 9 and 10. They correspond to approximately the third and the seventh layer. They were located on the ripple of Fig. 8 and in an extended multilayer region some 5 mm off-centre, respectively. In the first case the local particle environs have obtained a higher degree of regular coordination than in the monolayer. A horizontal grain boundary is visible in the image centre. While the layer surface seems

Fig. 5 PS 106/4.8. Cleaved thick film, $1 \times 1 \ \mu m^2$ scan, 60 nm Z-range, former film bottom

Fig. 6 PS 106/4.8. Cleaved thick film, $1 \times 1 \ \mu m^2$ scan, 75 nm Z-range, some 10 μm above former film bottom

rows in the upper middle). Thus in Fig. 6 we observe an increased roughness, albeit with a smaller typical length scale than before.

One has to bear in mind that SFM only gives examples of local structures and conclusions for the

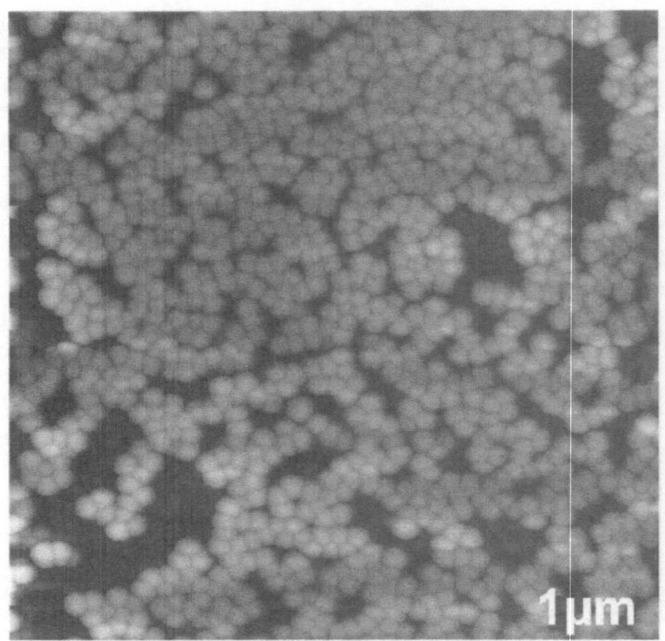

Fig. 7 PS 106/4.8. Dried sample 5 × 5 µm² scan, 200 nm Z-range

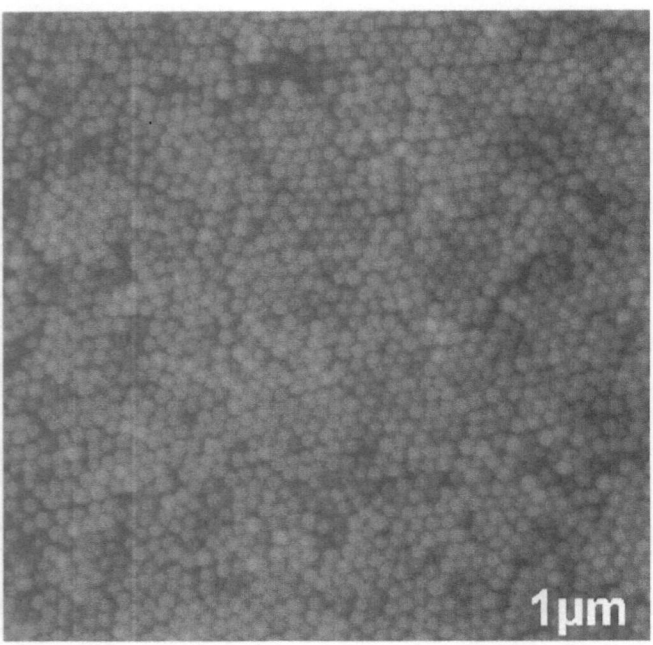

Fig. 9 PS 106/4.8. Dried, highly concentrated sample, more than 20 layers, 5 × 5 µm² scan, 200 nm Z-range

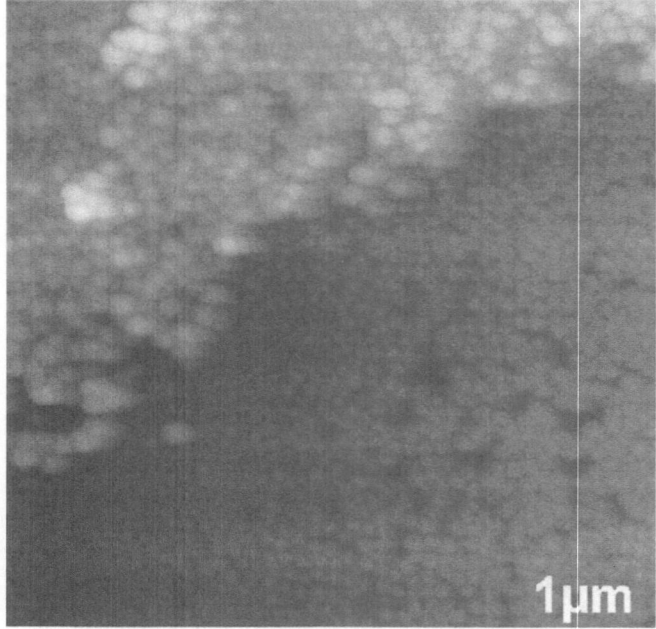

Fig. 8 PS 106/4.8. Dried sample, rim of a dune, substrate is visible, 5 × 5 µm² scan, 300 nm Z-range

rather compact and flat, there are some "holes" present, too. The height variation (except the holes) is on the order of some 10–20 nm and its typical length scale comprises several particle diameters. In the second

example the roughness had increased and its corrugation length decreased. Note that now triangular, square and rhombic ordering are observed to coexist, indicating a loss of orientational order induced by the substrate.

At heights of some 10 layers and larger, a different situation is observed for isolated but extended regions. These are located mainly in the outer terrace-like multilayer regions. As shown in Fig. 11, there the surface appearance seems to be very similar to that in Fig. 4. Strong height variations, locally crystalline ordered clusters, various kinds of defects and void-like structures are encountered. We note that such structures coexist with more regularly ordered regions.

Our SFM investigations thus have shown that the bulk disorder is not retained as the surface is approached. Instead, the formation of layers or layer-like sheets of particles is observed. The presence of a flat substrate induces an orientational order in our films. This phenomenon as such is well known from both suspended [7, 8, 10–12, 14] and dried suspensions [8, 16–18]. Orientational order may further be influenced by lateral constraints [13]. In all these cases, however, the samples under study would readily crystallize. Their densest packed plane is oriented parallel to the substrate, in order to maximize packing fraction or minimize particle interactions. This phenomenon has also been interpreted in terms of wetting behaviour. In computer simulations it was found that a hard-sphere crystal equilibrium in coexistence with its melt would wet a hard wall [30]. This was recently confirmed experimentally

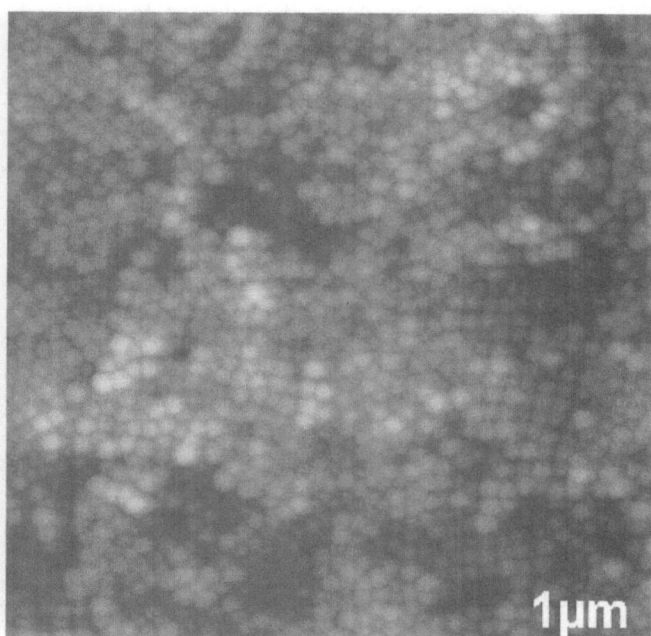

Fig. 10 PS 106/4.8. Dried, highly concentrated sample, some 7 layers, 5 × 5 μm^2 scan, 200 nm Z-range

Fig. 11 Dried sample, rim of a dune, 3–5 layers, 5 × 5 μm^2 scan, 200 nm Z-range

[31]. Charged spheres have been known to wet a charged wall for quite some time and the phenomenon was comprehensively investigated. Here wetting may lead to a layered structure which extends several layers into the fluid. It further may significantly influence the crystallization process via heterogeneous nucleation [1], while in slit geometries it seems to be a necessary pre-requisite for crystal formation [7].

In the present work, layering was observed in a dried film and further for a polydisperse sample which shows an amorphous bulk state. We conclude that layering seems to be a general feature of colloidal suspensions near a flat substrate and not to be restricted to crystallizing systems.

Origin of short-range order and weak layering

In the present study the influence of the substrate on orientational ordering is clearly visible and changes in the character of the disorder are conceivable. However, both the bulk structure and the mono- and multilayer structure apparently are of short-range order only. The global order is not as pronounced as previously observed for other particles of similar polydispersity. Layers are observed to be non-compact and the in-plane order may be crystal-like but nearly always only short ranged. A pronounced corrugation is observed for larger film thickness as well as the presence of different crystallographic orientations. Combining these observations with

those on the thick film samples, it seems that this particular latex shows a tendency to form only pronounced short-range order.

This immediately raises the question of the underlying cause. The comparison of PS samples of different polydispersity showed a decrease of residual order. The PMMA particles of similar polydispersity had qualitatively different scattering patterns. At present we are therefore not sure whether the differences from other dry film preparations are due to the polydispersity of the sample alone.

To identify possible further origins of the short-rangedness of order, it is instructive to recall that also the results of the monolayer regions are quite different to those known from literature. In particular, no crack formation is observed, indicating a very low concentration of pinning centres [16]. This is expected for vanishingly small particle-substrate interaction only and usually can be achieved only by a two step on oil preparation. There, however, compact rafts are observed in most cases [25], owing to a dominance of capillary forces [8, 17, 18]. Here compact areas are observed, too, but are of small size and intersected by compact voids of smooth outer boundary. We thus may speculate that also these are very small and do not suffice to herd the particles together to form large compact rafts. On the other hand, attractive inter-particle forces seem to be very small, too. These would result in non-compact lateral structures. In the limiting case of sticky particles on an unsticky substrate, one would even expect fractal-like structures to form. Here the average local coordination number is about 3.9 for the central region, which is significantly lower than that of a compact crystalline region (6) but larger than that expected for two-dimensional diffusion-limited aggregation.

We thus suggest that the smallness of inter-particle, particle-substrate and capillary forces may be one further significant reason for the short-rangedness of the observed residual order and thus a condition strongly supporting vitrification. This may also explain why vitrification in other samples was rare to observe up to now.

We finally speculate that the absence of significant forces may be due to the co-stabilization with SDS. It is known that the presence of surfactant may have a strong influence on the structural properties of dry films [8]. As SDS is a short, stiff molecule, it may sterically stabilize the dry particles against van der Waals coagulation. Further, it is not chemisorbed to the particle surface, but a small amount is always in solution. During the last stage of evaporation, this may be sufficient to cover the substrate also with a monolayer of surfactant. On the other hand, the SDS concentration at the water-air surface would be enhanced as well, and thus the surface tension would be decreased. While this at present is purely speculative, it would be consistent with our observations and preliminary data. To confirm this picture, further experiments are clearly needed. A systematic comparison between samples of comparable polydispersity but different stabilization is under way.

Acknowledgements It is a pleasure to thank E. Bartsch and M. Stamm for many fruitful discussions. Financial support was given to the Mainz group by the Sonderforschungsbereich 262 and by the MWFZ Mainz. This work was further supported by the Bundesministerium für Forschung und Technologie, by the AIF and by the Deutsche Forschungsgemeinschaft. This is gratefully acknowledged.

References

1. Palberg T (1999) J Phys Condens Matter 11:R323
2. Bartlett P, van Megen W (1994) In: Mehta A (ed) Granular matter. Springer, New York Berlin Heidelberg, pp 195–257
3. Bartsch E (1998) Curr Opin Colloid Interface Sci 3:577
4. van Megen W (1995) Transp Theory Stat Phys 24:1017
5. Lai SK, Ma WJ, van Megen W, Snook IK (1997) Phys Rev E 56:766
6. Cummins HZ, Li G, Hwang YH, Shen GQ, Du WM, Hernendez J, Toa NJ (1997) Z Phys B 103:501
7. Crocker JC, Grier DG (1998) MRS Bull 23:24; Murray C (1998) MRS Bull 23:33; van Blaaderen A (1998) MRS Bull 23:39; Asher SA, Holtz J, Weismann J, Pan G (1998) MRS Bull 23:44
8. Winnik A (1997) Curr Opin Colloid Interface Sci 2:192
9. Dingenouts N, Ballauff M (1999) Langmuir 15:3283
10. van Blaaderen A, Witzius P (1995) Science 270:1177
11. Neser S, Leiderer P, Palberg T (1997) Prog Colloid Polym Sci 102:194
12. Neser S, Bechinger C, Leiderer P, Palberg T (1997) Phys Rev Lett 79:2348
13. Bubeck R, Neser S, Bechinger C, Leiderer P (1998) Prog Colloid Polym Sci 110:41
14. Zahn K, Lenke R, Maret G (1999) Phys Rev Lett 82:2721
15. Witten TA (1999) Rev Mod Phys 71:367
16. Skjeltorp AT, Meakin P (1988) Nature 335:424
17. Kralchevsky PA, Paunov VN, Denkov ND, Nagayama K (1994) J Colloid Interface Sci 167:47; Velev OD, Denkov ND, Paunov VN, Kralchevsky PA, Nagayama K (1994) J Colloid Interface Sci 167:66
18. Dushkin CD, Lazarov GS, Kotsev SN, Yoshimura H, Nagayama K (1999) Colloid Polym Sci 277:914
19. Weiss A, Pötschke D, Ballauff M (1996) Acta Polym 47:333
20. Dingenouts N, Ballauff M (1998) Acta Polym 49:178
21. Dingenouts N, Bolze J, Pötschke D, Ballauff M (1998) Adv Polym Sci 144:1
22. Dingenouts N, Ballauff M (1998) Macromolecules 31:7423
23. Glatter O, Kratky O (eds) (1982) Small angle X-ray scattering. Academic Press, New York
24. D'Aguanno B, Klein R (1991) J Chem Soc Faraday Trans 87:379
25. Lazarov GS, Denkov ND, Velev OD, Kralchevsky PA, Nagayama K (1994) J Chem Soc Faraday Trans 90:2077
26. Richter H (1999) Diploma work, University of Karlsruhe
27. Vrij A (1979) J Chem Phys 71:3267
28. van Beurten P, Vrij A (1981) J Chem Phys 74:2744
29. Hansen JP, McDonald IR (1986) Theory of simple liquids. Academic Press, London
30. Courtemanche DJ, van Swol F (1992) Phys Rev Lett 69:2078
31. Heymann A, Stipp A, Sinn C, Palberg T (1998) J Colloid Interface Sci 207:119

Progr Colloid Polym Sci (2000) 115:315–319
© Springer-Verlag 2000

B. M. van der Horst
H. C. Langelaan
A. D. Gotsis

Rheological studies of water-in-oil-in-water double emulsions

B. M. van der Horst
H. C. Langelaan (✉)
Agrotechnological Research Institute
Postbus 17, 6700 AA Wageningen
Netherlands
e-mail: h.c.langelaan@ato.wag-ur.nl
Tel.: +31-317-475231
Fax: +31-317-475347

A. D. Gotsis
Delft University of Technology
Department of Polymer Technology
Julianalaan 136, 2628 BL Delft
Netherlands

Abstract In this study the rheological behaviour of water-in-oil-in-water ($W_1/O/W_2$) multiple emulsions was investigated. Different emulsions varying in volumetric ratio of the components were prepared and characterised using small-amplitude oscillatory rheological measurements. The microstructure of the multiple emulsions was examined using laser diffraction. The results indicate that the microstructure of the multiple emulsions plays a dominant role in its rheological behaviour. $W_1/O/W_2$ multiple emulsions are found to be more viscoelastic than simple W_1/O emulsions of comparable total volumetric composition. The rheological response of $W_1/O/W_2$ emulsions depends highly on the amount of internal phase (W_1/O) but also on the amount of the internal water, W_1. An attempt was made to fit the results into the Krieger-Dougherty and Palierne models, taking into account the contributions of the individual components. It seems, however, that the amount of internal interfacial area is the determining factor for the rheology of stabilised multiple emulsions.

Key words Water-in-oil-in-water · Multiple emulsions · Rheological behaviour · Interfacial area

Introduction

Double emulsions can be considered as emulsions within emulsions. In the case of a water-in-oil-in-water ($W_1/O/W_2$) emulsion, the oil (O) droplets have smaller water (W_1) droplets dispersed within them and the oily phase itself (W_1/O) is dispersed as droplets in the continuous water phase (W_2). Multiple emulsions are ideal systems for applications in the cosmetics, pharmaceutical and food industries [1]. The rheology during the different processes that occur in the emulsion (such as creaming, coalescence, Ostwald ripening, etc.) and the composition of the multiple emulsion affect their stability, spreadability, organoleptical response, etc. In the present paper the rheological properties of double emulsions were investigated using dynamic mechanical measurements.

$W_1/O/W_2$ emulsions were prepared by a two-step emulsification procedure. Commercial sunflower oil was supplied by a local supermarket. A polyglycerol polyricinoleate lipophilic emulsifier, Grinsted PGPR 90 (Danisco Ingredients), was used to stabilise the internal, W_1/O, emulsion and a polyoxyethylene sorbitan monolaurate hydrophilic emulsifier, Tween 20 (Sigma Holland), was used to stabilise the outer, $(W_1/O)/W_2$, emulsion.

The composition of the double emulsions was varied in two different ways. For the first set of samples we prepared several single (W_1/O) emulsions with composition from 10% to 80% W_1 in oil (O), and used them as the internal dispersed phase for the double emulsion, keeping the ratio of the internal phase (W_1/O) to external water (W_2) constant and equal to 50:50. In the second set of samples the composition of the internal W_1/O phase was always 50% water in oil, while the ratio of this phase over the external water (W_2) changed from 1:9 to 9:1.

Two models were tried here to describe the rheology of the emulsions. The Krieger-Dougherty [2] model [Eq. (1)] describes the relative viscosity, η_r, of suspensions of rigid particles but it is also used for the description of emulsions with highly stabilised interfaces [3]:

$$\eta_r = \left[1 - \frac{\phi}{\phi_m}\right]^{-[\eta]\phi_m} \tag{1}$$

Here ϕ is the volume fraction of the dispersed phase, ϕ_m is the maximum packing fraction and $[\eta]$ is an intrinsic viscosity (with a value of 2.5 for hard spheres [2]). The maximum packing fraction depends on the shape and the polydispersity of the particles and it is usually estimated from the viscosity data. In the case of the present polydisperse emulsions, the volume fraction of the dispersed phase can easily exceed the maximum packing concentration for monodisperse sized spheres. The Krieger-Dougherty model is restricted to Newtonian systems and cannot account for elasticity.

The Palierne model [4] [Eq. (2)] may be a better model for emulsions that use emulsifiers, since it predicts viscoelasticity even when the components are non-elastic by introducing extra elasticity due to the deformation of the interface:

$$G^*(\omega) = G_m^*(\omega)\left(\frac{1 + \frac{3}{2}\phi H}{1 - \phi H}\right) \tag{2}$$

Here $G^*(\omega)$ and $G_m^*(\omega)$ are the complex moduli of the emulsion and the matrix, respectively, at frequency ω, ϕ is the volume fraction of the inclusions and H is a function of the complex moduli of the components, the interfacial tension and the radii of the inclusions [4].

Results and discussion

Figure 1 shows the curves of the loss modulus, G'', versus the frequency, ω, for several single emulsions. The elasticity modulus, G', gives a similar graph. When the volume fraction of W_1 increases, both the viscosity and the elasticity of the emulsion increase. The relative loss modulus data (G''_{emul}/G''_{oil}) of the single emulsions are shown in Fig. 2 as a function of the volume fraction of W_1. The data can be fitted well by the Krieger-Dougherty model [Eq. (2)] and ϕ_m is evaluated to be 0.85. These results imply that the water droplets in the single emulsion are polydisperse and behave like hard spheres because the interface/emulsifier layer cannot be easily deformed. The Palierne model cannot fit these data.

Figure 3 shows the G' data for the second set of the double emulsions (i.e. when the phase W_1/O of all these emulsions has a ratio of W_1 to O equal to 1:1), together with the predictions of Palierne's model. Both G' and G'' increase when the volume fraction of the internal single emulsion (W_1/O) increases and they exceed the moduli of the single 50:50 emulsion. On the other hand, in the first set of emulsions (i.e. when the double emulsions were made from the same amount of W_1/O phases containing different W_1 to O ratios) the increase of the amount of W_1 results in a decrease of the moduli of the double emulsion. This is in contradiction to the increase of the moduli with the amount of W_1 in the single emulsions. All these data do not fit in Palierne's or Krieger-Dougherty models, even for low volume fractions of W_1.

The dependence of the moduli on the composition of the internal phase (% W_1 in the W_1/O phase) is shown in Fig. 4 for the double emulsions of the first set. The dependence of the moduli on the amount of (W_1/O) in W_2 is shown in Fig. 5 for the emulsions of the second set. In the former case the elasticity and loss moduli decrease with the volume fraction of W_1, while they increase with

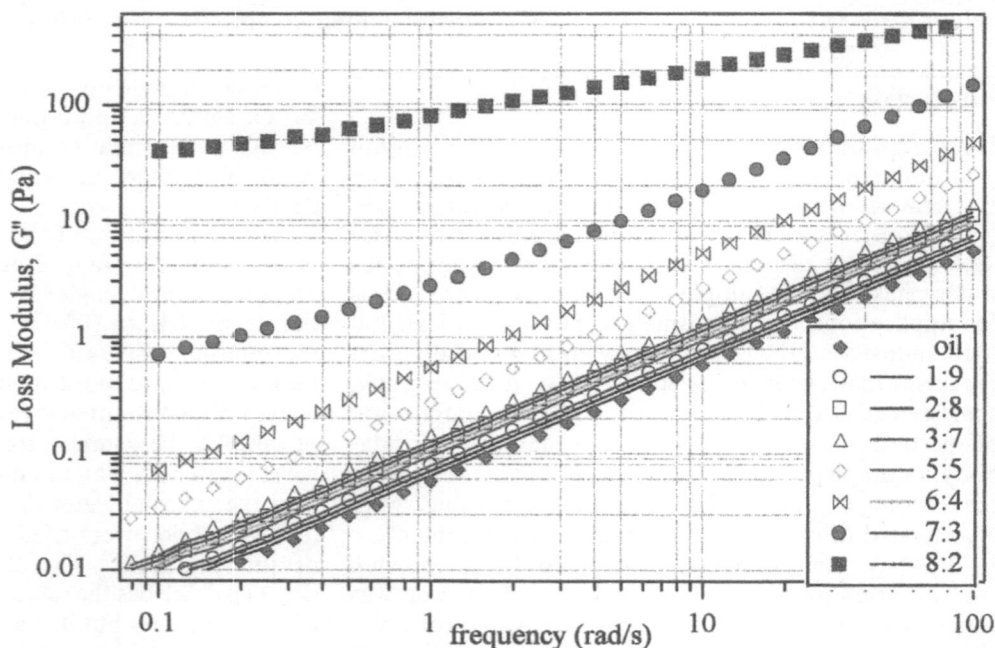

Fig. 1 G'' (Pa) versus ω (rad/s) of the single emulsions with W_1:O ratios of 1:9 to 8:2

Fig. 2 Relative loss modulus (G''_{em}/G''_{oil}) versus the volume fraction of the internal water phase (single emulsions), fitted by the Krieger-Dougherty relation; $\phi_{max} = 0.85$ (polydisperse)

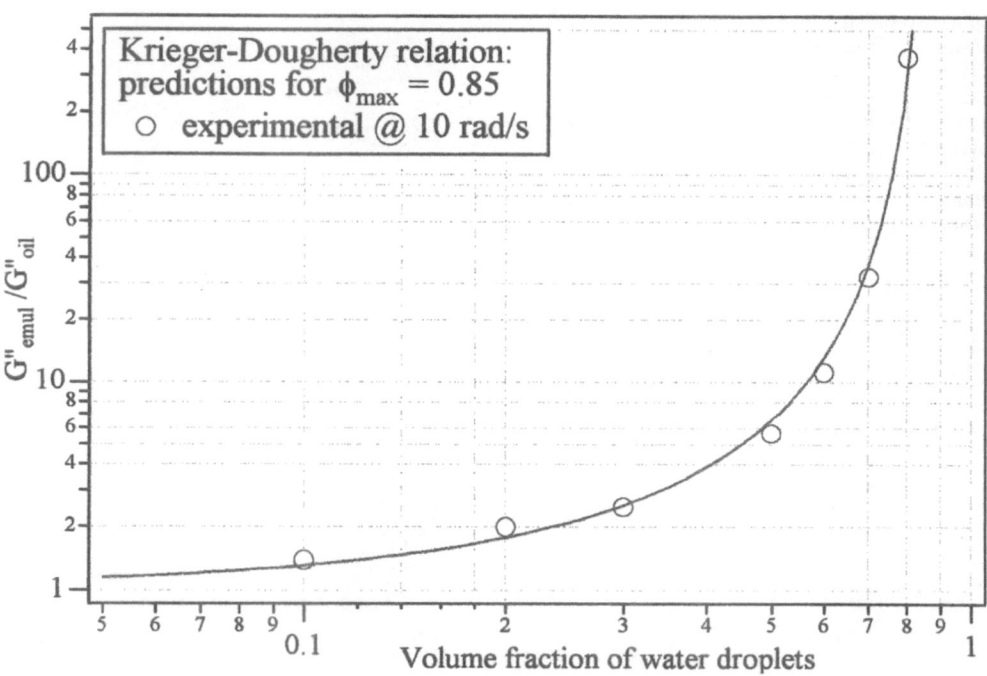

Fig. 3 Elasticity modulus (Pa) versus frequency (rad/s) of the double emulsions and the predictions of the model of Palierne. Internal phase composition is 50:50 in all emulsions and the ratio of the internal/external phase varies from (1:1):18 to (9:9):2

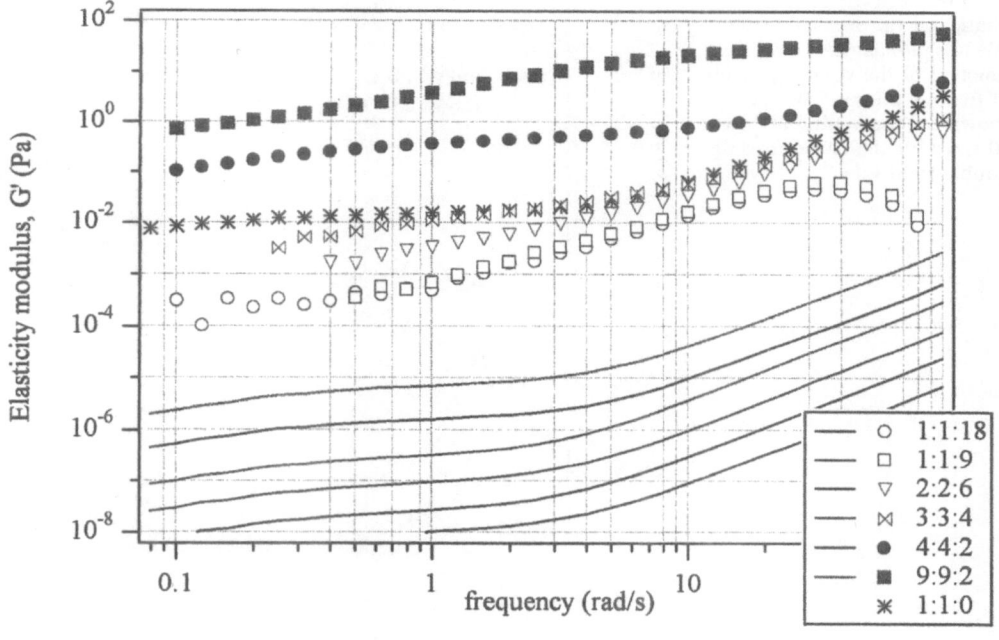

the volume fraction of (W$_1$/O) in the latter case. These figures also show the interfacial area per volume ($S_V = 3\phi_{disp}/R_{drop}$) between the (W$_1$/O) phase and W$_2$. S_V also decreases with the amount of W$_1$ in W$_1$/O and increases with the amount of (W$_1$/O) in W$_2$.

The double emulsions in Fig. 4 contain a constant amount of an increasingly stiffer dispersed phase (the single emulsion becomes stiffer as the amount of W$_1$

increases), while their moduli decrease. One possible explanation for this apparent contradiction is that the moduli are influenced mostly by the change of the (highly stabilised and, therefore stiff) interface, while the stiffness of the bulk of the dispersed phase has a secondary effect. This seems to be the case also in Fig. 5. The inner phase here consists always of the same single emulsion. The disagreement of the measurements with the predictions

Fig. 4 Moduli at 1 and 10 rad/s (*left-hand side axis*) and interfacial area (per volume) between the droplets of the inner oily phase and the external water (W$_2$) matrix (*right-hand side axis*) as a function of the volume fraction of water W$_1$ in the oily phase

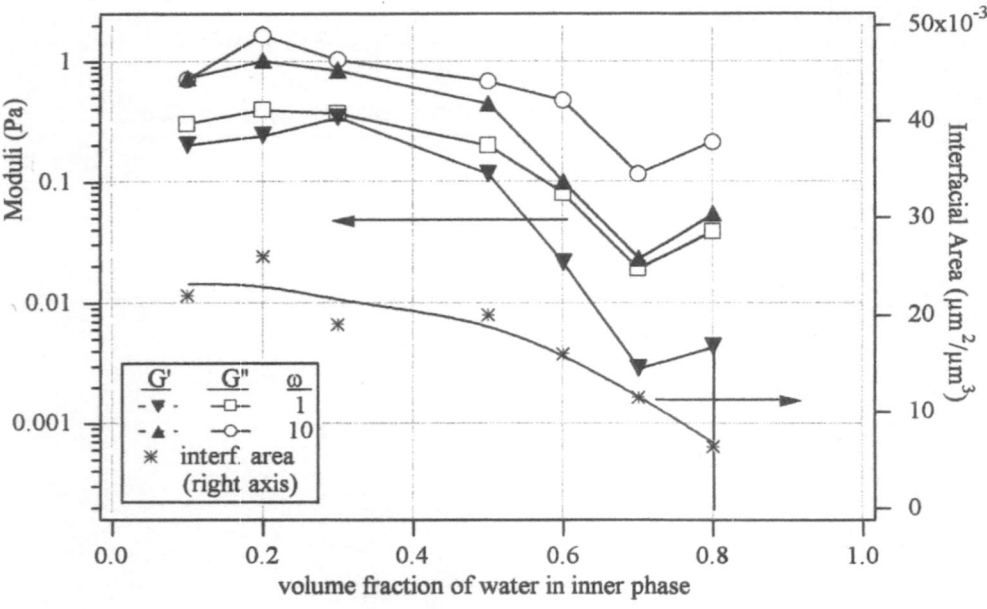

Fig. 5 Interfacial area (per volume) between the internal oily phase and the external water (W$_2$) (*right-hand side axis*) as a function of the volume fraction of the oily phase in W$_2$. Also shown are the moduli at 1 and 10 rad/s for the corresponding double emulsions

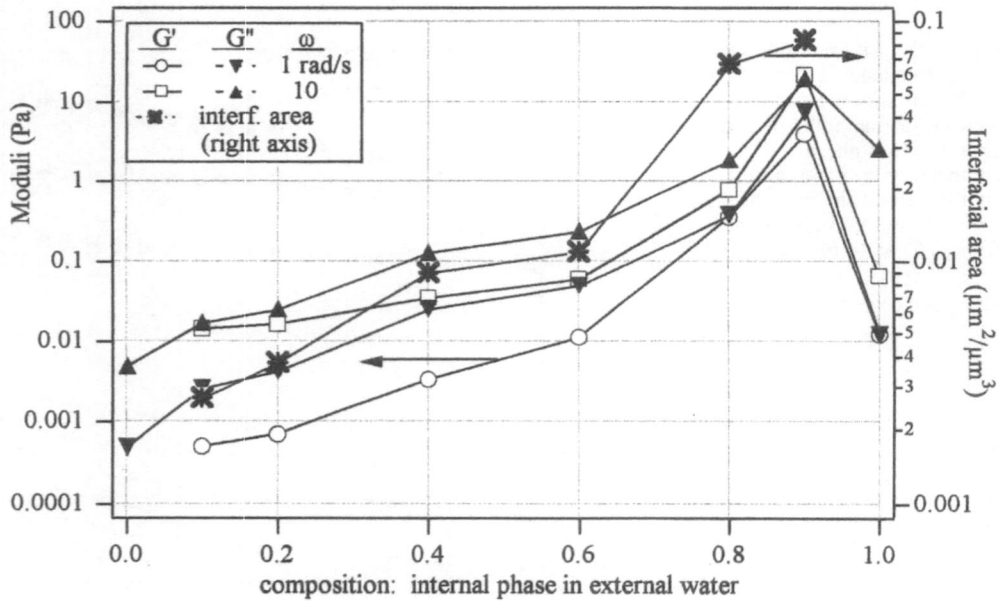

of the models and the fact that the moduli of the double emulsions are higher than the moduli of their components that the increase of the moduli in Fig. 5 cannot be attributed to the relative increase of the amount of highly elastoviscous dispersed phase. Also in this case the higher moduli are due to the increase of the interfacial area.

The bulk stiffness/viscosity of the inner phase plays, however, another role. The interfacial area, S_V, decreases in Fig. 4 as a result of the increased average size of the composite W$_1$/O droplets. The higher stiffness of this phase at high volume fractions of W$_1$ hinders a finer dispersion and results in larger droplet sizes [5] and lower values for S_V. On the other hand, in the emulsions in Fig. 5 the droplet sizes do not depend so much on the composition, since the viscosity of their bulk remains the same. Coalescence may increase somewhat these sizes but the major effect at higher volume fractions in this case is the presence of more composite droplets of the same size, which results in more interfacial area and higher moduli. Above the maximum packing capacity for monodisperse spheres, both the area and the moduli accelerate as the polydispersity increases or the droplets lose their spherical shape.

Conclusions

The storage and loss moduli of the single emulsions increase with the increase of the volume fraction of the water droplets. Because the emulsified water droplets behave like hard spheres and the emulsifier layer is stiff the viscosity of these emulsions can be fitted well with the Krieger-Dougherty model. They also show elasticity. The moduli and the interfacial area of the multiple emulsions increase non-linearly when the amount of external water decreases and the composition inside the internal phase remains constant. Double emulsions with variable composition of the internal phase have moduli and interfacial area that decrease with the increase of the internal water phase. In the heavily stabilised double emulsions the interfacial surface is the most important parameter that determines their viscoelasticity: the more interfacial area there is the higher the moduli. On the other hand, the composition of the W_1/O phase affects its viscosity, which, in turn, determines the droplet size in the double emulsion, therefore the interfacial area and finally the moduli.

References

1. Garti N, Bisperink C (1998) Colloid Interface Sci 3:657
2. Krieger IM, Dougherty TJ (1959) Trans Soc Rheol 3:137
3. Tadros ThF (1994) Colloids Surf A 91:39
4. Palierne JF (1990) Rheol Acta 29:204
5. Taylor GI (1932) Proc R Soc A 138:41; Grace (1982) Chem Eng Commun 14:225

Progr Colloid Polym Sci (2000) 115 : 320–324
© Springer-Verlag 2000

RHEOLOGY

N. Hunt
V. Buckin

Temperature dependence and the effects of heat treatment on the rheological properties of various butters

N. Hunt · V. Buckin (✉)
Department of Chemistry
University College Dublin
Belfield, Dublin 4, Ireland
e-mail: vitaly.buckin@ucd.ie
Tel.: +353-01-7062371
Fax: +353-01-7062127

Abstract Rheological characteristics of butter are the key physical properties for its consumer acceptability. They are determined by a three-dimensional network of fat crystals, which can be affected by composition, mechanical treatment and the temperature history of the butter. It is well known that butters possessing desirable rheological qualities, which are made utilising novel industrial technologies, lose these properties on being exposed to temperature treatment. In the present paper, we have quantitatively established the dependence of rheological parameters of various butters as a function of temperature. In addition we examined the effects of heat treatment on butter rheology. It was shown that there is good relation between yield stress curves and the percentage of solid fat in a butter and that the relative change of the yield stress value, which is 4.5 to 5.5 times between 278 and 288 K (5–15 °C), is approximately the same for all butters. This result exhibits that we can predict the yield stress of a butter over a range of temperatures, from a value measured at just one temperature.

Key words Butter · Yield stress · Microfixing · Soft butter fractions · Cream tempered

Introduction

Butter possesses flow properties that are between those of an ideal solid and liquid. When an applied stress is below a critical value, the yield stress [1], butter undergoes deformation like a solid, with a definite recovery being observable on removal of the stress. If this yield stress is exceeded the butter flows like a liquid; thus butter is described as a viscoelastic material. Rheological characterisation of foods, such as butter, is important for product development, finished product quality control and ingredients [2].

Butter is an emulsion of water droplets in a semisolid matrix of milk fat. It has an average composition of approximately 80% milk fat, 16% water, 1.2% milk proteins and optional salt of less than 2% [3]. The fat matrix, consisting of a three-dimensional network of fat crystals, held together by primary (non-reversible) and secondary (reversible) bonds, surrounded by a liquid fat medium, is primarily responsible for the texture of butter [4]. The complex fatty acid composition of milk fat [5] is shown by its melting behaviour. 233 K (−40 °C) and is complete at approximately 313 K (40 °C) and it is this fatty acid composition and the interactions of the fatty acids with each other within the matrix that controls the rheological properties.

Overall the rheological properties of butter are determined by several factors, such as the ratio of solid to liquid fat, size and shape of fat network [6, 7, 8]. Methods used industrially to improve the rheological properties of butter can be divided into three main areas: (1) chemical treatments are those where soft fractions are incorporated into butter, whether by the process of fractionation [9], where low melting fractions are isolated from milk fat and added to the butter, or by controlling the feed that the herd take [10];

(2) temperature treatments, where cream is subjected to heating and cooling regimes referred to as rebodying and tempering [11]; and (3) mechanical treatments, where the butter is subjected to a stress or shear so as to enhance the rheological properties, such as homogenising cream, which has the effect of giving smaller crystals and thus a less hard network structure, and microfixing where the butter crystal network structure is partially broken, thus allowing the butter to flow more readily.

In the present paper we studied the dependence of the rheological parameters of butter on temperature in a quantitative manner and analysed various butters of different composition, made using different techniques by yield stress experiments. We also investigated the effects of heat treatment on the rheological characteristics of these butters.

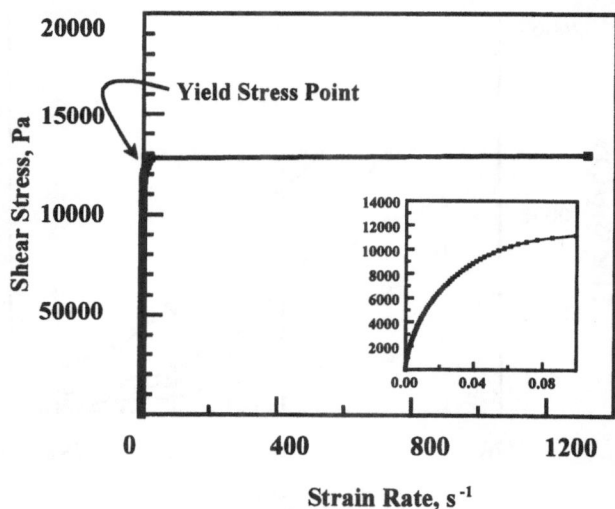

Fig. 1 Typical yield stress graph of a butter at 280.15 K

Materials and methods

Butter

All butters are commercially available and were purchased from supermarkets. The butters analysed were: traditionally manufactured natural butter, butter with 30% added soft fractions and 15% added soft fractions, cream tempered butter, and microfixed cream tempered butter.

Butter temperature treatment

Triplicate butter samples, maintained at approximately 278 K, were extracted (40–50 mm depth) from foil packs or tubs using a modified 10 cm³ syringe (Luer Loc, Becton-Dickinson) and were sealed with a rubber bung and parafilm (American Can Company). A syringe was then placed from the fridge (maintained at 278 K) into a water bath (Haake C25, with F8 controller) set to 298 K, leaving the butter sample there for 40 min. After this time the butter-filled syringe was replaced back in the fridge, and left overnight for about 21 h. This treatment will be referred to as treatment "a". Treatment "b" refers to the butter sample undergoing the whole of procedure "a" twice. One syringe of each type of butter contained a non-temperature-treated sample.

Rheological measurements

Yield stress measurements

The yield stress of the non-temperature-treated butter samples were compared as a function of temperature, and were subsequently compared to their "a" and "b" temperature-treated counterparts. Measurements were carried out between 278 and 292 K, at 2 K intervals. From 278 to 292 K, butter changes from being hard to soft, and above 293 K it is very soft and rheological measurements are of little interest [12].

For the yield stress measurements a rheometer, used in all the measurements (Dynamic Stress/Strain Rheometer, SR 2000, Rheometric Scientific), was set in continuous motion and the stress built up over a set time until the yield value was reached and then a little more with the onset of flow. The stress was set to reach 18,000 Pa over 33 min (100 points per decade). A cone and plate geometry

was used, the diameter of the cone being 16 mm and the gap being set at 0.8 mm. Controlled temperature around the peltier plate and cone tool was maintained by construction of an insulated thick aluminium block and reflective insulated covering.

The cone tool was used to apply a shear stress to the sample of butter, which was on the lower plate. This shear stress on the sample produced a resulting strain rate, or rate of deformation. Shear stress was plotted as a function of strain rate. Figure 1 shows a typical example of a yield stress curve. As is seen when flowing ensues, as a result of the crystal network being broken a shoulder appears in the plot of stress against strain rate. The extrapolation of this "flow" region of the curve to the y-axis gives us the yield stress point.

The butter samples were loaded onto the lower peltier temperature-controlled plate by pushing out 1 mm thickness of butter from the syringe and cutting it using a tight cheese wire. The first 1 mm was discarded, and the next 1 mm thick sample was loaded on to the lower fixture.

Results and discussion

Non-heat-treated butters

From Fig. 2 we observe that the chemically modified butters have the lowest yield values, or flow the easiest, at all temperatures as compared to the mechanically and temperature technologically modified butters. The other three butters have similar composition but show different rheological properties. Butters exposed to heat (cream tempering) and mechanical treatment have lower yield stress values compared to untreated butters, thus showing that there is a rheological benefit gained from these treatments. From this graph one can distinguish the different methods used to produce the butters and their subsequent effects on the rheological properties.

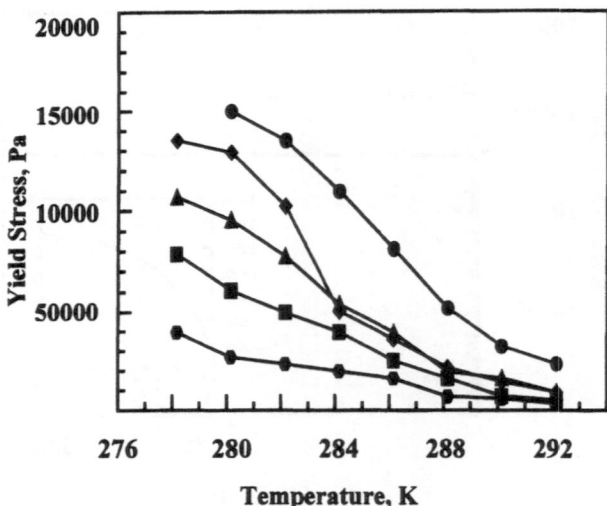

Fig. 2 Comparison of the yield stress values, as a function of temperature, of traditionally manufactured butter with butters made using rheologically enhancing techniques: ● – Natural Butter; ◆ – Cream Tempered; ▲ – Cream Tempered and Microfixed; ■ – 15% Added Soft Fractions; ● – 30% added Soft Fractions

Correlation of yield stress measurements with solid fat content

The yield stress graphs resemble closely the solid fat content (SFC) graphs which have been frequently published in the past [see for example 15, 16]. The amount of the fat portion of butter which is solid, the SFC, is one of the major factors affecting the rheological properties of butter. Our results on the chemically modified butters generally agree with this. The rigidity of the network is mainly controlled by the proportion of fat that is solid. The two main factors which influence the amount of solid fat are temperature and composition. A butter composed of a large amount of unsaturated glycerides, for example, will have a low solid fat content and thus will be soft. Also the combination in triglycerides and the arrangement of fatty acid residues in the triglycerides have an effect on the SFC and in turn the melting curve of butter. Over the temperature range we studied, owing to the presence of many glycerides with close melting points, the variation of SFC with temperature drops almost linearly and generally drops from an initial value of about 75% to 25% [13].

These observations have also been correlated in the past with previous differential scanning calorimetry (DSC) [17] measurements interpreted in terms of the existence of three predominant melting species, with a low-melting species around 281 K, a middle-melting species between 281 and 293 K and a high-melting species above 293 K, indicating that the majority of solid fat has melted above 293 K [13].

SFC has been shown to have a good correlation with rheological properties at 278 and 295 K [14]. Our yield stress curves exhibit qualitatively similar behaviour to SFC curves, showing overall that there is a good relationship between our yield stress curves and the percentage of solid fat. The relative change of the yield stress value, within the temperature range 277 to 288 K, is 4.5 to 5.5 times, and is nearly the same for all butters (within the limits of our resolution). This observation can be used for the prediction of the yield stress over a range of temperatures from the value measured at just one temperature. The change of SFC value over the same temperature range is smaller than that of the yield stress by about three times. This indicates the complexity of the relationship between SFC values and the rheological parameters of butter. It also indicates that other factors such as the crystal network structure are extremely important in determining both the absolute values and the temperature dependence of the rheological parameters of butter. Therefore, SFC graphs cannot always help to predict the rheological properties of butter. An example of this is a traditionally manufactured butter compared with a microfixed sample of this butter will have completely different rheological characteristics, due to the different crystal network they possess even though their SFC is the same.

Effect of heat treating butter and the changes in the yield stress value

The heat treatment increases the yield stress of butters made with temperature and mechanical treatments and reduces the yield stress of natural butter. The curves essentially all fall into the same region. The butters made by chemical modification demonstrate an increase in the yield stress. This can be explained by an involvement of some elements of mechanical and/or heat treatment during its manufacture.

We see in Fig. 3 that traditionally manufactured natural butter does not show the classic problem of hardening after our heat treatment. Instead, its rheological characteristics actually improve, whereas this is the opposite for all butters which were made using non-traditional techniques. When butter is cooled after melting, the fat matrix reforms but in a different manner than the original non-heated butter; thus the history of the butters becomes erased and they are not as distinguishable from each other as before. On cooling the different polymorphic forms of fatty acid crystals {α (hexagonal), β' (orthorhombic) and β (triclinic), these polymorphic forms have different melting points which increase in the order $\alpha < \beta' < \beta$ [18]} may be present in different amounts than were present originally, thus altering the matrix structure. After heat treatment it is

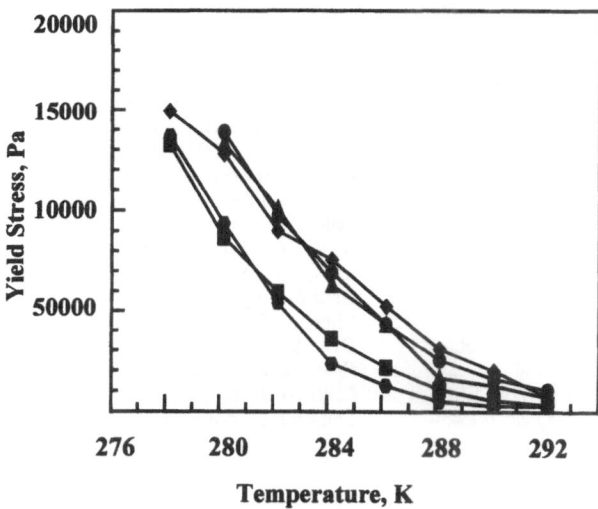

Fig. 3 Comparison of the yield stress values, as a function of temperature, of traditionally manufactured butter with butters made using rheologically enhancing techniques, after temperature treatment "a": ● – Natural Butter; ◆ – Cream Tempered; ▲ – Cream Tempered and Microfixed; ■ – 15% Added Soft Fractions; ◗ – 30% added Soft Fractions)

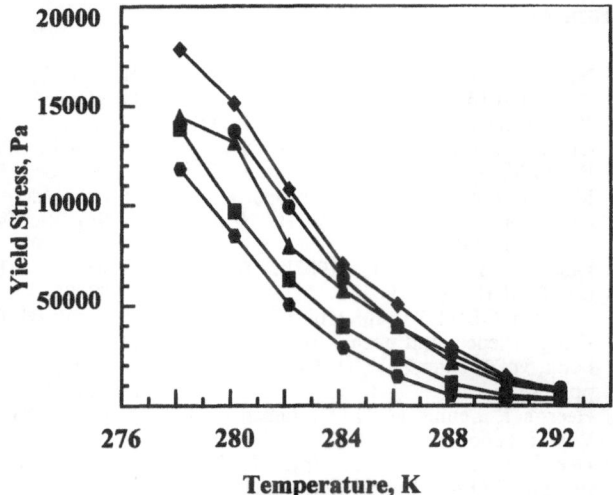

Fig. 4 Comparison of the yield stress values, as a function of temperature, of traditionally manufactured butter with butters made using rheologically enhancing techniques, after temperature treatment "b": ● – Natural Butter; ◆ – Cream Tempered; ▲ – Cream Tempered and Microfixed; ■ – 15% Added Soft Fractions; ◗ – 30% added Soft Fractions)

quite apparent that composition is the major factor controlling rheological properties.

Effects of a second heat treatment of butter and changes in the yield stress value

It can be seen in Fig. 4 that there is little difference between the curves in this graph and those in Fig. 3. The original structure has been completely destroyed and the crystals in the matrix exist in a completely different arrangement from those in Fig. 2, thus implying that relatively little further rearrangement of the fat crystal network occurs. It is obvious that there is some form of physical/chemical limit to crystal reorganisation which can be brought about in butter by heat treatment.

Conclusion

We have successfully shown that the rheological parameters of various non heat-treated butters can be quantitatively analysed as a function of temperature so as to distinguish the different methods used for their manufacture. The butter which demonstrates the best

rheological properties at low temperature was that containing 30% added soft fat fractions, a chemically modified butter, with the least rheologically attractive being traditionally manufactured natural butter.

We have found that there is good relation between yield stress curves and the percentage of solid fat for a butter and that the relative change of the yield stress value, which is 4.5 to 5.5 times between 278 and 288 K (5–15 °C), is approximately the same for all butters. This result exhibits that we can predict the yield stress of a butter over a range of temperatures, from a value measured at just one temperature.

We have shown that the rheological parameters of butter, such as yield stress value, can totally change after heat treatment.

We demonstrated quite clearly that temperature treatment could increase or decrease the good rheological characteristics of butter depending on the composition and manufacturing history. However after the second heat treatment no effect is observed as previous history has been destroyed along with the original matrix structure.

Acknowledgement We wish to acknowledge Enterprise Ireland for their help in funding this project (Grant no.'s HE/98/231 and HE/98/232) and also Glanbia plc.

References

1. Nguyen QD, Boger DV (1992) Annu Rev Fluid Mech 24:47–88
2. Velez-Ruiz JF, Canovas GVB (1997) Crit Rev Food Sci Nutr 37:311–359
3. Rosenthal I (1991) In: Rosenthal I Milk and dairy products: properties and processing, chap 4. VCH, New York, pp 143
4. DeMan JM, Beers AM (1987) J Texture Stud 18:303–318
5. Christen GL (1993) In: Hui YH (ed) Dairy science and technology handbook, vol 1, chap 2. VCH, New York, pp 83–155
6. Heertje I, Leunis M (1997) Lebensm Wiss Technol 30:141–146
7. Heertje I, Leunis M, van Zeyl WJM, Berends E (1987) Food Microstructure 6:1–8
8. Bornaz S, Fanni J, Parmentier M (1994) J Am Oil Chem Soc 71:1373–1380
9. Kaylegian KE, Lindsay RC (1992) J Dairy Sci 75:3307–3317
10. Fearon AM, Mayne CS, Marsden S (1996) J Sci Food Agric 72:273–282
11. Schaffer B, Szakaly S, Lorinczy D, Belagyi J (1999) Milchwissenschaft 54:82–85
12. Prentice JH (1992) In: Hui YH (ed) Dairy rheology: a concise guide, chap 7. VCH, New York, pp 67–84
13. Shukla A, Bhaskar AR, Rizvi SSH, Mulvaney SJ (1994) J Dairy Sci 77:45–54
14. Taylor MW, Norris R (1977) NZ J Dairy Sci Technol 12:166–170
15. Lane R (1998) In: Early R (ed) The technology of dairy products, chap 5. Blackie, London, pp 158–197
16. Schaap JE (1993) In: Protein and fat globule modifications by heat treatment, homogenisation and other technological means for high quality dairy products. IDF special issue 9303 International Dairy Federation, Brussels, Belgium, pp 408–412
17. Sherbon JW, Dolby RM (1972) J Dairy Res 39:319–324
18. Walstra P (1987) In: Blanshard JMV, Lillford P (eds) Food structure and behaviour, chap 5. Academic Press, London, pp 67–85

Progr Colloid Polym Sci (2000) 115:325–328
© Springer-Verlag 2000

Acoustic spectroscopy of aerogel precursors

C. Sinn

C. Sinn*
Institut für Physik der
Johannes-Gutenberg-Universität
Staudingerweg 7
55099 Mainz, Germany

Present address:
* Institut für Angewandte Physik der
Universität Bern, Sidlerstrasse 5
3012 Bern, Switzerland
Tel.: +41-31-6318926
Fax: +41-31-6313765

Abstract We investigate the acoustical properties of silica gels, which are precursors in the aerogel production process. These gels exhibit a strong "ringing gel" behavior, that is they emit a characteristic sound if one knocks against the container. We study this sound emission with a very simple spectroscopic technique and observe resonances which are characteristic for natural frequencies of a cylindrical body. From a fit of the experimental frequency positions to calculated values, we determine a sound velocity of $c_T = 4$ m/s for a gel sample with porosity $\phi = 97.5\%$. This low sound velocity can only be interpreted as the transverse sound mode predicted by Biot's theory for sound propagation in porous media.

Key words Acoustic properties · Porous materials · Silica gels · Aerogels

Introduction

Some years ago, Oetter and Hoffmann [1] discovered a strange behavior of a class of gels they were investigating. Their gels emitted a characteristic sound upon hitting the container. Oetter and Hoffmann termed this a "ringing gel" or "humming gel" behavior and established this property in a class of gels consisting of ternary phases of a hydrocarbon, a surfactant, and water. The structure of these phases has been investigated subsequently in detail [2], although the physical explanation of the effect remained rather indeterminate. A similar observation had been reported earlier by Bacri et al. [3], but without mentioning an audible sound emission of their polymer gels.

Recently, we observed the same phenomenon to occur for the precursors of the aerogel process that consist of a highly porous silica network filled with an alkanol. These precursors are frequently called the "wet gel" or alcogel. If the alkanol is removed supercritically, one obtains the aerogel as a rigid, lightweight, and transparent body, where the silica frame contributes only approximately 5% to the total volume.

Having this porous structure in mind, we were tempted to interpret the "ringing gel" effect of the alcogels in terms of Biot's theory for sound propagation in porous media [4], developed originally for the description of sound propagation in submarine sediments. This theory has been fruitfully applied to explain the elastic properties of gels [5]. A review by Johnson [6] of the theory and its relation to published experimental results is available.

In view of Biot's theory, we suggest the following explanation for the "ringing gel" behavior. Upon knocking against the container, one excites a shear wave in the gel, opposite to the more familiar compressional waves. A compressional wave is present, too, but has a sound velocity higher by some orders of magnitude. The gel within the container forms a resonant body for the shear wave. These resonances are responsible for the audible sound that can be perceived if one hits the container. This suggestion has been tested experimentally by simple acoustic spectroscopy, which will be introduced in this contribution. We will further report the first quantitative results of these experiments.

In what follows, we will start with a crude outline of the theory involved, followed by a description of our experiments and a concluding outlook.

Sound propagation in porous media

Biot's theory describes phenomenologically the propagation of acoustic waves in a porous, fluid-filled, macroscopically homogeneous and isotropic body. Each volume element of this body experiences an average displacement of the fluid and the solid part, respectively. The respective equations of motion are coupled by viscous, as well as inertial, forces. The macroscopic material properties that enter are the bulk moduli of the fluid (K_f) and the solid (K_s) phase, the bulk modulus K_b of the skeletal frame, the shear modulus G, and the porosity ϕ. For details, we refer the reader to Johnson [6], the notation of which we will follow as closely as possible.

General hydrodynamics predicts a crossover between two distinct regimes that is governed by the viscous skin depth, $\delta = (2\eta/\rho_f\omega)^{1/2}$ [7], with η the viscosity and ρ_f the mass density of the fluid phase; ω is the frequency of a disturbance. For porous media this crossover separates a high-frequency from a low-frequency regime, the crossover frequency $\omega_c = 2\eta/\rho_f a^2$ being related to the average pore size a of the medium. Porous materials exhibit pore sizes of the order of 100 nm and less. Therefore, we are essentially always in the low-frequency regime, indicating that the viscous skin depth is much larger than the characteristic pore size. Furthermore, for gels the skeletal frame is generally much more deformable than the pore fluid. This "gel limit" is identified therefore by K_b, $G \ll K_f$.

Applying these assumptions, the theory predicts the following sound modes in the porous medium.

First, there is a fast compressional mode, which is very much like the sound mode in pure liquids. The sound velocity of this longitudinal mode is given by:

$$c_{FL} = c_0 \left[1 + \frac{\xi_1 K_b + \xi_2 G}{2K_f} \right] \tag{1}$$

where $c_0^2 = K_m/\rho_m$ is Wood's result [8] for the sound velocity in a composite medium with $K_b = G = 0$, such as a dispersion of air bubbles in water, or a colloidal suspension. The bulk modulus of the composite medium can be calculated assuming two different elastic media in parallel, yielding for our case $K_m^{-1} = \phi/K_f + (1 - \phi)/K_s$. The total density is $\rho_m = \phi\rho_f + (1 - \phi)\rho_s$. The ξ_i in Eq. (1) represent the deviations from Wood's formula owing to the finite stiffness of the frame: $\xi_1 = \phi^2(K_mK_f)(K_f^{-1} - K_s^{-1})^2$ and $\xi_2 = \frac{4}{3}K_f/K_m$. The sound attenuation can be calculated correspondingly, but is of little interest in the present context. The compressional mode treated so far can be visualized as a cooperative in-phase motion of fluid and skeletal frame.

Second, there exists a slow compressional mode, corresponding to the out-of-phase motion of fluid and skeleton. The sound velocity of this wave is purely imaginary:

$$c_{SL}^2 = -i\omega \frac{k(K_b + \frac{4}{3}G)}{\eta} \tag{2}$$

corresponding to an overdamped sound mode; k is the permeability defined through Darcy's law $Q = -(kA/\eta)\nabla p$, which relates the volume flow rate Q through a sample area A due to an applied hydrostatic pressure gradient ∇p. Equation (2) can be used to calculate elastic constants of gels from linewidth measurements by dynamic light scattering, as first performed by Tanaka et al. [9].

Third, there is only one trivial shear mode, corresponding to the in-phase movement of fluid and skeleton. The sound velocity of this transversal wave is given by:

$$c_T = \left(\frac{G}{\rho_m} \right)^{1/2} \tag{3}$$

Accordingly, from theory one expects two propagating sound modes in porous media with different sound velocities, which are determined by Eqs. (1) and (3).

Notice that from the "gel limit" $G \ll K_f$, as mentioned above, one can already estimate that the longitudinal sound velocity c_{FL} should be some orders of magnitude higher than the transversal one, c_T.

Materials and methods

Materials

The gels under investigation are prepared by base-catalyzed hydrolysis of tetramethoxysilane (TMOS) in the presence of a large amount of methanol. We use an aqueous solution of 0.1 mol/dm^3 ammonia as hydrolyzing agent. The molar ratio of water to TMOS is chosen to 4:1, which is two times the stoichiometric amount. For a gel sample with porosity of nominal $\phi = 97.5\%$, we use 82.6 cm^3 methanol. Half of this amount is mixed with 12.5 cm^3 TMOS; the remaining amount is mixed with 6.1 cm^3 of the aqueous ammonia solution. The aqueous solution is added dropwise to the stirred TMOS solution at room temperature. After some minutes of further stirring, the liquid is filled into cylindrical glass vials with plastic snap-in lids. All reagents are of grade "purum" and were used without further treatment. TMOS is stored under Ar atmosphere.

Gelation occurs for this composition within approximately 12 h. The composition has been chosen to yield the highest porous material within acceptable gelation times.

The gels are known to exhibit an aging behavior. Accordingly, the samples under study were stored approximately one week at room temperature before use.

Experimental setup

For the acoustic spectroscopy, we use a small condensor microphone capsule (Conrad Electronic), which is equipped with an amplifier and a load resistor within the metal housing. Power is provided by an external 3 V battery. The data sheet shows a linear response over almost the whole frequency range of 20–18000 Hz. The microphone output is AC coupled to the high impedance input of a digital oscilloscope (LeCroy 9314). The oscilloscope performs a fast Fourier transformation (FFT) of 5000 data points of the signal trace. The Nyquist frequency was 2.5 kHz, yielding a frequency step of 1 Hz. The FFT was calculated using a Hamming window, which results in an only slightly poorer frequency resolution of 1.4 Hz. Thirty FFT sweeps are averaged to yield the final spectrogram.

The experimental procedure is as follows. The glass vial is brought into contact with the condensor microphone, the vial being hold by hand at its plastic lid. The glass is hit with a metallic screwdriver to excite the sound modes of the gel. The oscilloscope is pre-triggered onto this signal with a delay of 0.05 s, the time-base being 0.1 s. These settings assure that the acoustic signal is dominated by the sound modes of the gel, the excitement pulse contributing only insignificantly. This statement was checked by hitting the glass vial again during the recording period.

In order to eliminate further resonance frequencies stemming from the glass container, we divide the data by those obtained from a similar vial filled with methanol to the corresponding height. We note that at low frequencies only contributions from the mains frequency and the oscilloscope cooling fan were noticeable, the latter one with comparably small amplitude.

Data evaluation

The result of the procedure as described is shown in Fig. 1 and Fig. 2 as a thick solid curve. Many individual resonances can be observed with varying strengths. We here concentrate on the frequency position only.

The thin vertical lines represent the expected resonance frequencies for a cylindrical sample. These can be calculated from the wave equation in cylindrical coordinates r, θ, z, using a separation ansatz for these coordinates. The well-known solution leads to natural frequencies with wave vectors:

$$k^2 = \frac{\omega^2}{c^2} = k_{r,\theta}^2 + k_z^2 \tag{4}$$

which can be specified applying the appropriate boundary conditions. For acoustically hard walls, which require a wave node on the surface of the cylinder of height H and radius R, we obtain:

$$f = \frac{c}{2}\left[\left(\frac{u_{l,m}}{\pi R}\right)^2 + \left(\frac{n}{2H}\right)^2\right]^{1/2} \tag{5}$$

where $\omega = 2\pi f$ has been used; $n = 2, 4, 6 \ldots u_{l,m}$ is the m-th root of the Bessel function of order l, which can be found, for example, in Abramowitz and Stegun [10].

In our case, the glass walls, to which the gel properly sticks, can be well approximated by hard walls. The gel-air surface, however, may be regarded as acoustically soft. In this case, we would have a node at the bottom of the glass vial and a crest at the surface. Accordingly, the lowest mode in z-direction exhibited a quarter wavelength instead of a half. The corresponding boundary conditions would lead to Eq. (5) with $n = 1, 3, 5 \ldots$

Our experimental results, however, show that we need both contributions, i.e., we observe all modes with $n = 1, 2, 3, 4, 5, 6 \ldots$ We interpret this observation in terms of a reflection coefficient (of intensity) that does not equal 1. Indeed, using a sound velocity of 4 m/s we obtain a reflection coefficient of approximately 0.6. We note, however, that shear waves do not propagate in isotropic bodies like fluids, implying a reflection coefficient of 1. This apparent contradiction cannot be resolved at present.

Results and discussion

Figure 1 shows exemplarily the results for a gel sample with nominal porosity $\phi = 97.5\%$. The thick curve is the experimental result, whereas the vertical thin lines

represent the calculations of the natural frequencies as described above. For clarity, only resonances with $l = 0$ are shown. The inner diameter of the glass vials, $2R = 27$ mm, and the height of the gel within the vial, $H = 27$ mm, have been determined independently. The sound velocity is then chosen such that the first calculated resonance fits the first experimental peak. A subsequent fine-tuning of c_T and the sample dimensions is performed for an optimal fit of all the resonances. We note that the gel meniscus and the vial bottom are slightly curved and the vial dimensions are not specified very accurately. We observe a very close agreement between experimental and calculated frequency positions, though the peak at $f = 175$ Hz is not present in the calculation. Note that the lowest resonance for $l = 1$ is expected at $f = 196$ Hz. From Fig. 1, we determine a sound velocity of $c_T = 4.1$ m/s.

Figure 2 shows the experimental result for the same gel, but with different sample dimensions; the glass vial has been simply filled with a smaller amount of liquid. The fitted sound velocity is $c_T = 4.2$ m/s, in close agreement with the previous one. However, it is clearly visible that the agreement between experimental data and calculated resonance positions is not as good as before. We regard this as an estimate for possible systematic sources of error in the determination of the sound velocity.

It remains to show that the determined sound velocity is due to the shear mode as claimed above. To do so, we have to determine the shear modulus and the density of our samples as a function of porosity independently.

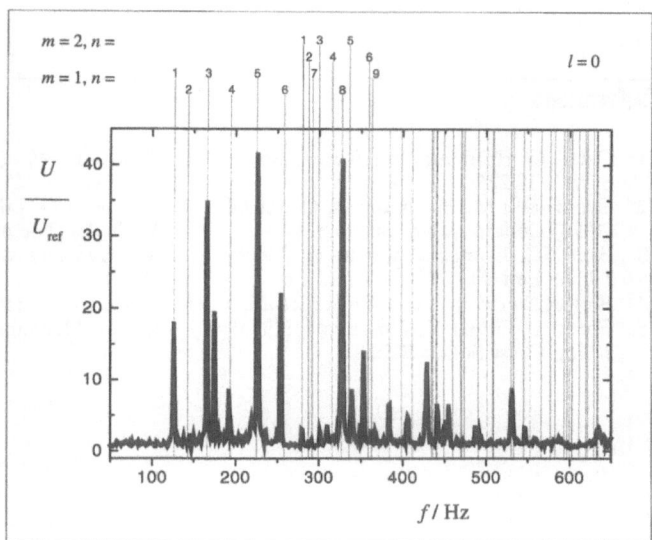

Fig. 1 Normalized sound amplitude as a function of frequency for a typical gel sample, $\phi = 97.5\%$. The sample dimensions, as obtained from the fit, are $H = 27$ mm, $2R = 26$ mm. The sound velocity is $c_T = 4.1$ m/s. Indices $l = 0$, m, n are indicated for a few modes

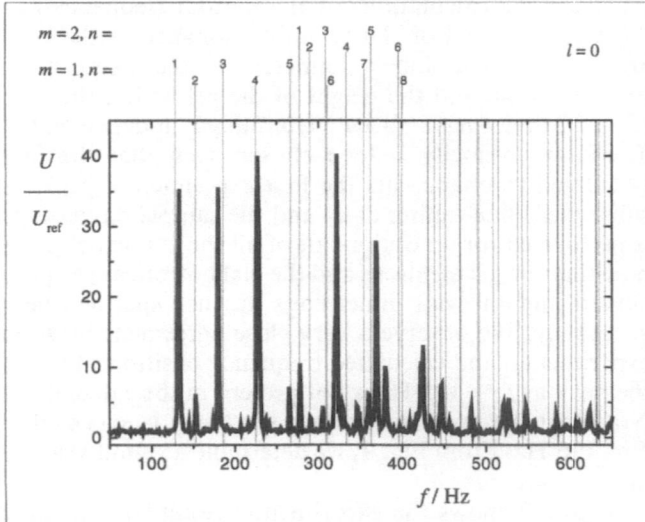

Fig. 2 Normalized sound amplitude as a function of frequency of the same gel as in Fig. 1, but with different sample dimensions: $H = 22$ mm, $2R = 27$ mm. The sound velocity is $c_T = 4.2$ m/s. Indices $l = 0$, m, n are indicated for a few modes

These data can then be used to prove that Eq. (3) quantitatively describes the sound velocity as measured with the method introduced here. This will be demonstrated in a future publication. For this contribution, we restrict ourselves to numerical evidence.

The value of c_0 can be calculated according to Eq. (1) using $\phi = 0.975$, $\rho_f = 800$ kg/m^3, $\rho_s = 2000$ kg/m^3, $K_s = 30$ GPa, and $K_f = 0.71$ GPa. The moduli have been calculated from experimental sound velocity data

and the density of the bulk materials [5]. With these data we obtain $c_0 \approx 940$ m/s, a value which is clearly incompatible with the result of our measurements. A quick calculation shows that the ξ_i do not alter this statement.

On the other hand, if we use Eq. (3) for the calculation of the shear modulus of our sample, we obtain $G = 14$ kPa. This value seems to be reasonable in view of the data of Forest et al. [5], where the porosity was significantly lower ($\phi \approx 90\%$) than in the present study. In addition, they report a longitudinal sound velocity of the order of magnitude of c_0 as calculated above.

These calculations are a clear evidence for the statement that the "ringing gel" effect is connected to shear waves propagating in the porous material. In addition, Biot's theory is obviously applicable to the systems under study, though a rigorous proof requires the quantitative prediction of the sound velocity from macroscopic data, as mentioned above.

The results presented in this contribution allow a deeper insight into the nature of the "ringing gel" effect, which is an immediately noticeable and striking property of these soft materials. With some restrictions, one might even think of assembling a jelly xylophone from these materials, as the porosity of the samples can be varied easily.

Acknowledgements I am indebted to thank Andreas Emmerling (University of Würzburg, Germany) for providing me with the recipe for the preparation of the gels. I would like to further mention the fruitful and pleasant collaboration with Guido Bucher and Ivo Flammer at the University of Bern. I gratefully acknowledge stimulating discussions with Jaro Rička and Vitali Buckin. This work has been supported by a research scholarship of the Deutsche Forschungsgemeinschaft (Si713/1-1) and the Swiss National Foundation.

References

1. Oetter G, Hoffmann H (1989) Colloids Surf A 38:225
2. Gradzielski M, Hoffmann H, Oetter G (1990) Colloid Polym Sci 268:167
3. Bacri JC, Dumas J, Levelut A (1979) J Phys Lett 40:L231
4. Biot MA (1962) J Appl Phys 33:1482
5. Forest L, Gibiat V, Woignier T (1998) J Non-Cryst Solids 225:287
6. Johnson DL (1982) J Chem Phys 77:1531
7. Landau LD, Lifshitz EM (1991) Lehrbuch der Theoretischen Physik, Band VI: Hydrodynamik. Akademie-Verlag, Berlin
8. Wood AW (1941) A Textbook of sound. Macmillan, New York
9. Tanaka T, Hocker LO, Benedek GB (1973) J Chem Phys 59:5151
10. Abramowitz M, Stegun IA (1972) Handbook of mathematical functions. Dover, New York

Progr Colloid Polym Sci (2000) 115 : 329–333
© Springer-Verlag 2000

N. M. Kovalchuk
V. I. Kovalchuk
D. Vollhardt

Direct numerical simulation of the dynamic behaviour of a system with a surfactant droplet under the free water surface: theoretical consideration of the auto-oscillation of surface tension

N. M. Kovalchuk
Institute for Problems of Material Science
Kiev, Ukraine

V. I. Kovalchuk
Institute of Biocolloid Chemistry
Kiev, Ukraine

D. Vollhardt (✉)
Max-Planck-Institute of Colloids
and Surfaces, 14424
Potsdam/Golm, Germany

Abstract Numerical simulation of the dynamic behaviour of a system containing a surfactant droplet under a free air-water surface is performed on the basis of a set of fluid mechanics equations, taking into account of the dynamics of the adsorption on the interface and the Marangoni effect. The behaviour of the system is different in two time intervals. During a long induction period the convective transfer of surfactant is negligible, the surface tension remains nearly constant and the system parameters change rather slowly. After overcoming a threshold, the system becomes unstable and it results in a jump in the convection velocity, the adsorption on the surface and the surface tension.

Key words Surface tension · Auto-oscillation · Numerical simulation · Dynamic behaviour, Marangoni instability

Introduction

It is well known that a liquid layer with a free surface can exhibit various dissipative structures when a gradient of temperature or concentration is imposed [1]. One of the mechanisms leading to the formation of these structures is related to the action of the surface tension gradient. The theoretical investigation of this problem was started by Pearson [2] and Scriven and Sternling [3]. They examined the behaviour of a thin liquid layer, heated from below, on the assumption of a constant temperature gradient. Using a small disturbance analysis they determined the condition for the onset of instability. Many other researchers have later considered analogous systems under different conditions [4–6]. Systems were also comprehensively theoretically investigated where not a normal temperature gradient but rather a gradient along the surface is imposed [7]. Oscillatory instabilities produced by heat from a temperature-controlled hot wire which was situated below the air-liquid interface were reported in an interesting experimental paper [8].

New experiments have recently shown the existence of an interesting effect related to the surface-tension-driven instability. Oscillation of the surface tension is initiated when a surfactant droplet is placed in bulk water under the interface [9]. A qualitative explanation of the latter phenomenon considers two different stages in the development of the instability [9]. The first stage is a slow stage where convection is small and does not practically disturb the diffusion flux of the surfactant. In the second fast stage the convection develops rapidly and accelerates the surfactant transfer to the surface. In the present paper we focus on theoretical evidence for the described mechanism of the surface tension oscillations by direct numerical simulation of the hydrodynamic processes in this system.

Materials and methods

The phenomenon of auto-oscillation of surface tension was obtained for diethyl phthalate (DEP) as the surface active substance. The experiments were performed in a tensiometric set-up with a measuring cell presented schematically in Fig. 1. A cylindrical measuring cell filled by pure water was placed in a thermostated glass vessel. A glass capillary and a platinum Wilhelmy plate were inserted in the measuring cell through the openings of a covering plate. The capillary was submerged in the

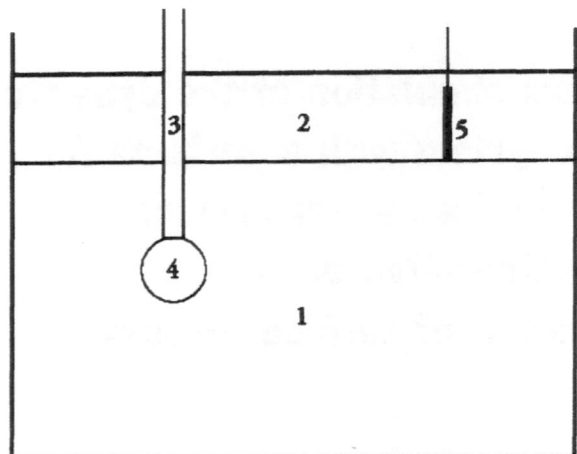

Fig. 1 Measuring cell for the study of the auto-oscillations of surface tension: *1* water, *2* gaseous phase, *3* capillary, *4* surfactant droplet, *5* Wilhelmy plate

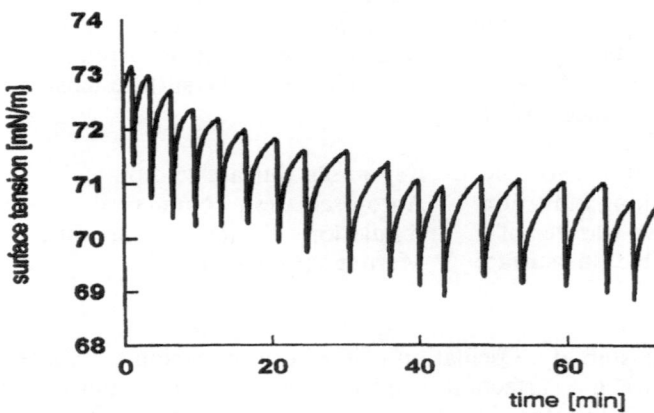

Fig. 2 Experimental data of auto-oscillations of surface tension

water. A droplet of DEP was carefully formed at the tip of the capillary under the water surface. The measurement of the surface tension was started immediately. During a certain rather long time interval t (depending on the immersion depth of the capillary , $t \geq 20$ min), the surface tension did not change remarkably. Not until after this induction time were the auto-oscillations of the surface tension observed (see Fig. 2). The period of this oscillation is about 5 min; the amplitude is about 1.5–2 mN/m. The oscillation has an asymmetric shape; a sharp decrease of the surface tension is followed by a gradual increase (Fig. 3). The oscillations can be observed for more than 8 h.

A qualitative analysis of the surface tension oscillation mechanism was performed previously [9]. To understand the behaviour of such systems in more detail, the processes of diffusion and convection in a semi-infinite liquid volume with free upper surface and with a small surfactant droplet in the bulk is numerically simulated in this paper. The behaviour of the system under consideration is described by a set of non-steady-state Navier-Stokes equations, the continuity equation and the convective diffusion equation. In the cylindrical coordinate system with the Z coordinate directed downwards, the equations written in terms of the vorticity and the stream function are

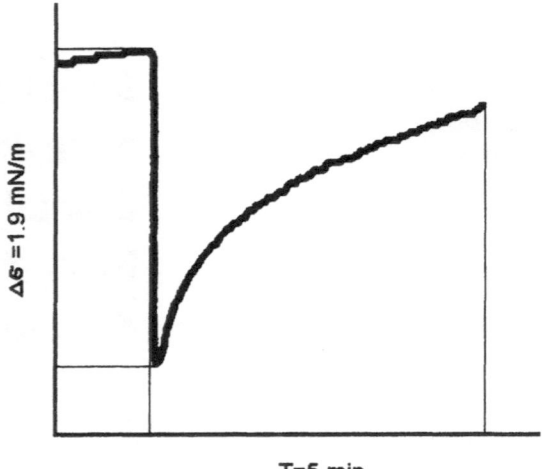

Fig. 3 Shape of a separate oscillation (experimental data)

$$\frac{\partial \omega}{\partial t} + v_r \frac{\partial \omega}{\partial r} + v_z \frac{\partial \omega}{\partial z} - \frac{v_r}{r}\omega = v\left(\Delta\omega - \frac{\omega}{r^2}\right) \tag{1}$$

$$\Delta\Psi - \frac{2}{r}\frac{\partial\Psi}{\partial r} = \omega r \tag{2}$$

$$\frac{\partial c}{\partial t} + v_r \frac{\partial c}{\partial r} + v_z \frac{\partial c}{\partial z} = D\Delta c \tag{3}$$

$$v_r = \frac{1}{r}\frac{\partial\Psi}{\partial z}; \quad v_z = -\frac{1}{r}\frac{\partial\Psi}{\partial r} \tag{4}$$

where v_r and v_z are the velocity components, t is time, v is the cinematic viscosity of the liquid, c is the concentration of the surfactant, D is the diffusion coefficient of the surfactant in the liquid, $\omega = \frac{\partial v_r}{\partial z} - \frac{\partial v_z}{\partial r}$ is the vorticity, and ψ is the stream function

$$\Delta = \frac{1}{r}\frac{\partial}{\partial r}r\frac{\partial}{\partial r} + \frac{\partial^2}{\partial z^2}$$

Initially, the liquid is motionless. In the first stage, the distribution of the surfactant dissolved from the droplet takes place in the system only by diffusion. During the process the surfactant reaches the initially clean surface and adsorbs there. Of course, the surfactant adsorbs more in regions having a relative smaller distance from the droplet. Thus there will be a surfactant concentration gradient on the surface and a corresponding gradient of the surface tension. This surface tension gradient produces a convective motion in the liquid. The viscose stress balance on the free liquid surface is given by the relation

$$\frac{\partial v_r}{\partial z} = -\frac{1}{\mu}\frac{\partial\sigma}{\partial r} = -\frac{1}{\mu}\frac{d\sigma}{d\Gamma}\frac{\partial\Gamma}{\partial r} \tag{5}$$

where μ is the dynamic viscosity of the liquid, σ is the surface tension, and Γ is the surface concentration. If the surfactant concentration near the surface is small, proportionality can be assumed between the surface concentration and bulk concentration:

$$\Gamma = \alpha c \tag{6}$$

where α is Henry's constant.

The quantity of the absorbed surfactant changes with time according to

$$\frac{\partial \Gamma}{\partial t} + v_r \frac{\partial \Gamma}{\partial r} + \Gamma \left(\frac{\partial v_r}{\partial r} + \frac{v_r}{r} \right) - D_s \frac{1}{r} \frac{\partial}{\partial r} r \frac{\partial \Gamma}{\partial r} - D \frac{\partial c}{\partial z} = 0 \qquad (7)$$

where D_s is the surface diffusion coefficient.

Equations (1)–(3) are solved by using a suitable set of initial and boundary conditions. Initially the liquid is supposed motionless and the surfactant concentration is set zero everywhere in the bulk, with the exception of the liquid-surfactant droplet interface. For the liquid-surfactant droplet interface the concentration is set constant and equal to the solubility of the surfactant in the liquid. The boundary condition for the vorticity at $z = 0$ is obtained from Eqs. (2) and (5) assuming $v_z = 0$ at the free liquid surface:

$$\omega = -\frac{1}{\mu} \frac{d\sigma}{d\Gamma} \frac{\partial \Gamma}{\partial r} \qquad (8)$$

The surfactant concentration on the free liquid surface is given by Eqs. (6) and (7). The other boundary conditions for the vorticity and the stream function are the same as usually used [10]. It was also used the condition that the droplet surface is motionless.

For the system under consideration the mathematical simulation is carried out on an irregular grid using 31×31 mesh points. In the region of interest the maximum density of the grid lines is near the z-axis in the z interval from 0 to the surface of the droplet. The one-step explicit computational method for the parabolic Eqs. (1) and (3) was applied for the simulation of this system. Evaluations made according to [10] show that in the present case the maximum time step for this procedure should be roughly 0.001 s. To avoid instability in the numerical scheme the unilateral difference "against the stream" was used.

The elliptic Eq. (2) was solved by the method of successive approaches. The values from the preceding time step were used as initial approach. The iteration process is performed until the needed precision is reached. The relative inaccuracy for the stream function was used as a precision criterion.

The parameter corresponding with the substance properties of DEP (solubility in water $c_0 = 6.7 \times 10^{-6}$ mol/cm^3, Henry constant $\alpha = 6.9 \times 10^{-4}$ cm) were chosen for the simulation process. The diffusion coefficients in the volume D and in the surface D_s were both set to 5×10^{-6} cm^2/s. The values of the droplet radius r_0 and the immersion depth h are chosen as $r_0 = 1$ mm and $h = 1$ cm.

Results and discussion

As a result of the computational procedure, concentration and stream function values as functions of time at the grid points were obtained. Based on these data, the distributions of surface velocity, adsorption and surface tension were calculated. In agreement with the experiments, an induction time interval of approximately 45 min was obtained where the surface tension remains constant. At the beginning of this time interval, only diffusion flux takes place in the system. The stream function and vorticity remain zero. The surfactant accumulates gradually in the surface owing to diffusion, and then, slow convective motion develops according to the Marangoni effect. During the induction time, surface velocity and adsorption are growing close to exponentially. The adsorption change with time during a certain interval of the induction period is presented for $r = 0.5$ cm in Fig. 4 (curve 2). In [9] the approximate expression for adsorption change was

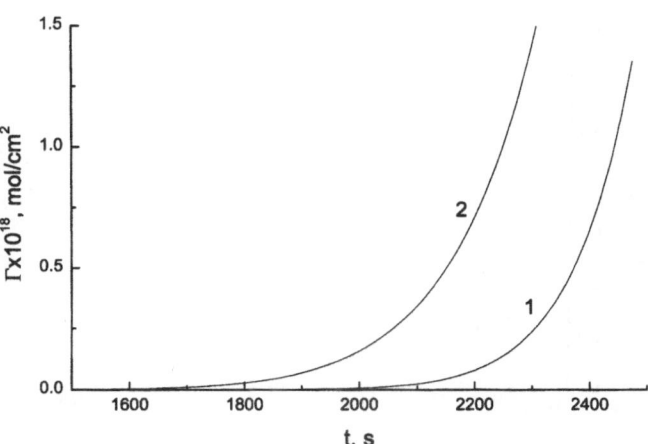

Fig. 4 Adsorption at the free water surface versus time at a distance of 5 mm from the symmetry axis during the induction period: *1* calculated for diffusion surfactant transfer, *2* data of the numerical simulation

obtained which takes into account only diffusion flux by spherical symmetrical distribution of the surfactant in the bulk:

$$\Gamma(t) \approx \frac{4c_0 r_0 h}{\bar{r}^2 (\bar{r} - r_0)^2} \frac{(Dt)^{3/2}}{\pi^{1/2}} e^{-(\bar{r} - r_0)^2 / 4Dt} \qquad (10)$$

where $\bar{r} = \sqrt{h^2 + r^2}$; h is the capillary immersion depth and r is the radial coordinate. Curve 1 in Fig. 4 was calculated from Eq. (10) by using the parameter values mentioned above. It is seen that both curves are similar but curve 2 grows more quickly owing to the contribution of the convective flux and the surface flux.

Figure 5 shows the velocity change with time during the induction period. It is seen that the velocity also grows close to the exponential-like adsorption.

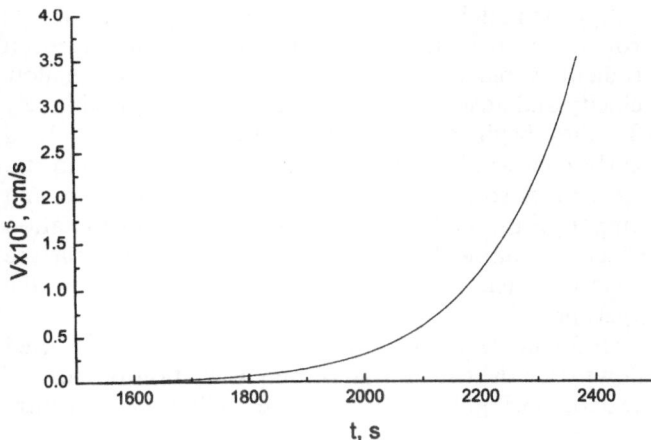

Fig. 5 Velocity on the free water surface versus time at a distance of 2.7 mm from the symmetry axis during the induction period

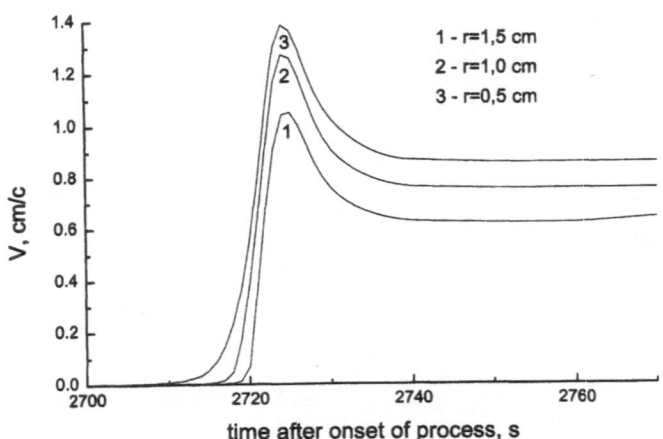

Fig. 6 Velocity at the free water surface versus time after overcoming a threshold at different distances from symmetry axis: *1* 15 mm, *2* 10 mm, *3* 5 mm

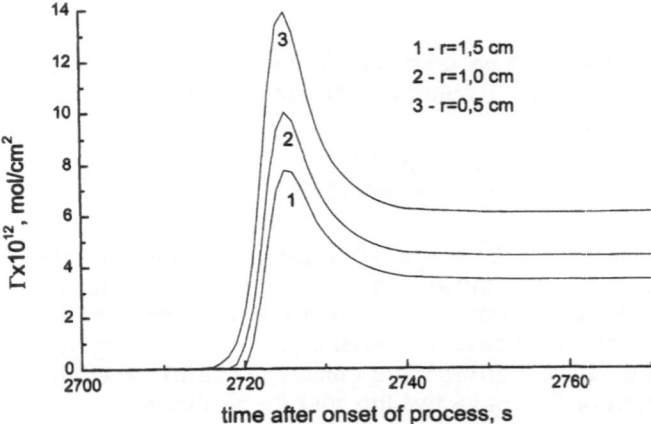

Fig. 7 The dependencies of the adsorption on the free water surface versus time after threshold overcoming at different distances from the symmetry axis: *1* 15 mm, *2* 10 mm, *3* 5 mm

Approximately 45 min after the beginning of the process, when in the system a sufficient concentration gradient is reached, a very rapid growth of the flow velocity and adsorption was found. After approximately 10 s they begin slowly to decrease and then result in nearly constant values (see Fig. 6 for surface velocity, Fig. 7 for adsorption and Fig. 8 for surface tension; and compare Figs. 3 and 8). Thus, the numerical simulation reflects fundamentally the oscillation of the surface tension, which is very similar to that found in the experiment.

Unfortunately the computational procedure used allows only the calculation of one oscillation because of the chosen irregular grid. After the oscillation, the centre

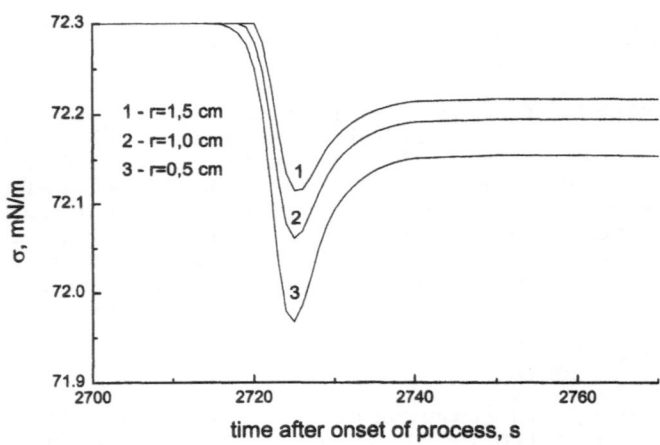

Fig. 8 Surface tension versus time after overcoming a threshold at different distances from the symmetry axis: *1* 15 mm, *2* 10 mm, *3* 5 mm

of the convective motion moves to regions where the grid step is very large and it is thus impossible to provide the necessary precision for further calculations.

Conclusions

The performed numerical simulation provides evidence for a theoretical model which describes the auto-oscillation of surface tension on the basis of equations of fluid mechanics which take account of the convective diffusion and the adsorption of the surfactant. The effect of the surface tension auto-oscillation was found experimentally in the system where a DEP droplet is placed in the bulk of water under a water-air interface.

In the initial stage of the process, only diffusion transfer of surfactant occurs in the liquid. Owing to the system configuration, gradients of the surface active substance and, consequently, of the surface tension increase gradually with time. The surface tension gradient is the driving force for the convective motion in the system. Initially, the velocity of the convection is rather small and also its contribution to the surfactant transfer is small. According to the growth of the adsorption gradient on the surface, the velocity increases with time. At a certain time, the system becomes unstable and all parameters of the system commence to change very fast. Then a strong convective flux spreads the surfactant over the surface, the concentration gradients decrease and the system returns to the stable state.

The results of the numerical simulation presented here confirm the qualitative mechanism of the surface tension auto-oscillation proposed previously [9].

References

1. Zieper J, Oertel H (eds) (1982) Convective transport and instability phenomena, Braun, Karlsruhe
2. Pearson JRA (1958) J Fluid Mech 4:489
3. Scriven LE, Sternling CV (1964) J Fluid Mech 19:321
4. Chen CF, Su TF (1992) Phys Fluids A 4:2360
5. Chu X-L, Velarde MG (1991) Phys Rev A 43:1094
6. Perez-Garsia C, Carneiro G (1991) Phys Fluids A 3:292
7. Smith MK, Davis SH (1983) J Fluid Mech 132:119; (1983) J Fluid Mech 132:145
8. Roze C, Gouesbet G, Darrigo R (1993) J Fluid Mech 250:253
9. Kovalchuk VI, Kamusewitz H, Vollhardt D, Kovalchuk NM (1999) Phys Rev E 60:2029
10. Roache PJ (1976) Computational fluid dynamics. Hermosa, Albuquerque

Progr Colloid Polym Sci (2000) 115:334–341
© Springer-Verlag 2000

M. Mirnik

Electrostatic and chemical interactions of ions in electrolytes and in ionic point-charge double layers. II

M. Mirnik
Laboratory of Physical Chemistry
Faculty of Sciences, University of Zagreb
Marulićev trg 19, P.O.B. 163
10001 Zagreb, Croatia
e-mail: mirnik@zagreb.zoak.pmf.hr
Tel./Fax: +385-1-4828298

Abstract It is described with a figure why the Debye-Hückel theory and its extensions were unable to explain the experimental activity coefficient functions in high concentrations. By fitting the proposed activity coefficient function

$$\gamma_m = 1/[\gamma_{ms}^{-1} + F_+(m^{E+} - 1) + F_-(m^{E-} - 1)]$$

to experimental literature data on the alkali halides and other 1-1 electrolytes, parameters of the same equation are estimated and presented in two tables. Using the parameters and the equation, activity coefficients and activities can be calculated and the molality axes can be transformed into activity axes. In two figures it is demonstrated that the influence of the ionic mass of cations and anions combined with the molar mass of each electrolyte upon the activity coefficient correspond to their position in the Periodic Table. The model of the point-charge double layer is described with a figure. Based on the same model, the function which replaces the Smoluchowski ζ-potential function reads

$$\kappa_p/\kappa_{pm} = 10^{(q_m-q)/(2s)} = 10^{(\zeta_m-\zeta)/(2s)}$$

and the quotient of the Madelung constant potentials, κ/κ_{pm}, replaces the analogous ζ/ζ_m of Smoluchowski. The entities q and q_m are the experimental electrokinetic mobility or electrolyte transport data. Their theoretical dependence on ionic strength, I_c, reads

$$q = q_m - 2s|\log(\kappa_{pm}/\kappa_p)| \text{ and}$$
$$q = q_m - s|\log(I_{cm}/I_c)|$$

The parameters κ_{pm}, q_m and ζ_m are the maximum values defined by I_{cm}. The same equations explain the maxima of the said variable q at I_{cm} experimentally observed by several laboratories.

Key words Activity coefficients · Coagulation · Electrokinetic maxima

Introduction

Since the classical paper of Debye and Hückel [1] was published, no different approaches or extensions of the same theory could satisfactory explain the activity coefficient functions in concentrations $c > \approx 10^{-3}$ mol dm^{-3}. The problem is described in [2]. A convincing explanation why the Debye-Hückel theory (DHT) is unsatisfactory follows from Fig. 1. The abscissa

is the numerical value of the relative distance, $r\kappa = r/(1/\kappa)$, from the reference cation, i.e., it is the multiple of the ionic cloud, $1/\kappa$, which is a decreasing function of the ionic strength (see Fig. 1 in [3]). The reference cation is placed on the abscissa, at the relative distance $r\kappa = 0$. The Coulomb function is $1/r\kappa$. The ion cloud causes the average potential, $e^{-r\kappa}/r\kappa$, i.e., the Coulomb potential minus the potential proportional to the counterion volume charge density. Their difference,

Fig. 1 The Debye-Hückel theory potentials. *Ordinate:* Coulomb, $1/r\kappa$: average Coulomb plus ion cloud, $e^{-r\kappa}r\kappa$, ion cloud, $e^{-r\kappa}r\kappa - 1$, and radial charge density, $r\kappa/e^{-r\kappa}$, as functions of the reduced distance, $r\kappa$ (*abscissa*), from the reference positive charge at $r\kappa = 0$. For a negative reference charge the potentials are of opposite sign. The functions hold also for spherical symmetry

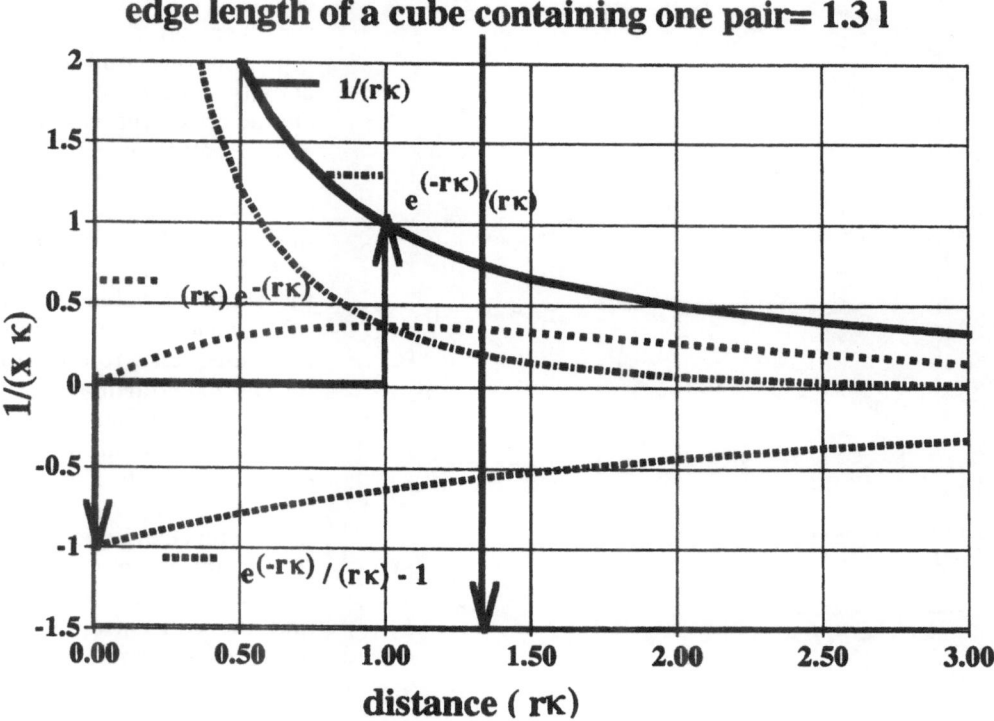

$e^{-r\kappa}/(r\kappa - 1)$, represents the ion cloud potential or, proportional to it, its volume charge density function. The ion cloud exhibits potential -1 upon the reference cation. The functions hold also for the spherical symmetry around the reference cation, or for any space angle around it. The reference cation and its ion cloud form one pair of ions. According to the DHT model, they require a volume of $r\kappa^3 \rightarrow \infty$, while a pair of ions requires the volume of a cube of edge length $r\kappa = \sqrt[3]{2} = 1.26$. The radial charge density function is proportional to $r\kappa\, e^{-r\kappa}$. Its maximum is at $r\kappa \rightarrow \infty$. In an infinite volume, there is an infinite number of ion pairs and not only one pair. The conclusion is that the DHT with its model, even if one introduces the distance of closest approach, is unsatisfactory. In an infinite volume, there is no place for two reference ions, the cation and the anion.

Electrolytes in the aqueous phase

The prerequisite for a satisfactory theory of potential functions in double layers of the interfaces in colloids is a satisfactory theory for electrolytes. The charges on colloid surfaces are adsorbed ions or chemically bound ionic radicals, i.e., fixed ions, also called site-bound ions, around which are electrostatic potential fields with potentials defined by the Coulomb formula. For both, the ions in electrolytes and double layers, the best

model is the point-charge model that was proposed for electrolytes by Debye and Hückel in 1923. There is absolutely no reason why the model for ions fixed on the surface should not be the same as the model for those in the bulk.

In [3], both problems of utmost importance for physical chemistry of electrolytes and for colloid chemistry are solved. They remained unsolved for more than 80 and 75 years, respectively. The cause was the neglect of the different chemical properties of cations and anions and the homogeneous charged, electrical, double-layer model on colloidal surfaces.

The DHT is replaced by the imaginary crystal lattice or Madelung constant theory. Figure 1 in [3] demonstrates the linear dependence of $\log l$, $\log(1/\kappa)$, $\log(1/2\kappa)$ and $\log l_c = \log(2^{1/3}l)$ on $\log I_c$. Figure 2 describes the imaginary lattice element. It consists of eight cubes. In each cube, there are six pairs of $+e$ and $-e$ elementary charges on the spheres of radii $3/\kappa$ and $6/\kappa$. In [3] it is assumed that the same potentials are caused by $\pm e/6$ charges on radii $1/\kappa$ and $1/2\kappa$, respectively. In both cases they cause in the centers either $+\kappa$ or $-\kappa$ potentials. Any finite vessel can be filled up with the necessary number of imaginary crystal lattice elements. The variation of $1/\kappa$ with l_c holds exactly for real electrolytes as well as for the described imaginary crystal.

The proposed function of the dependence of the activity coefficients on molality reads:

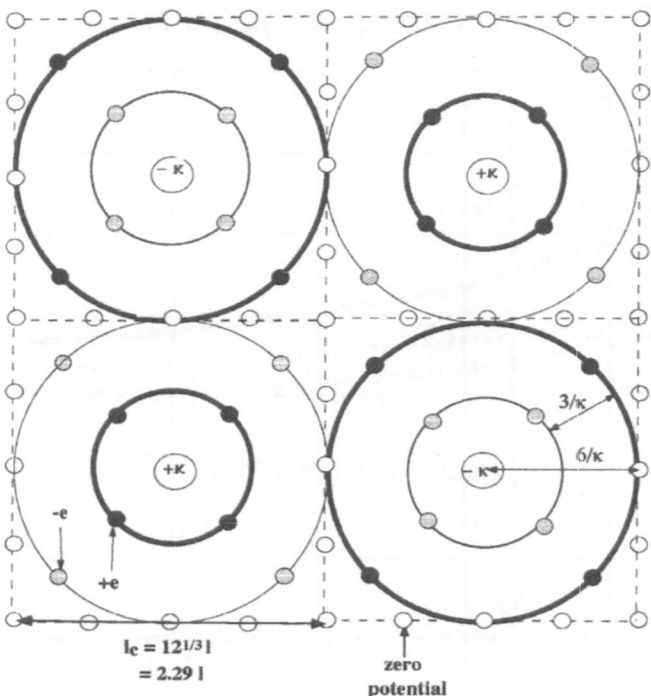

Fig. 2 Cross sections of four cubes containing $6 + e$ and $6 - e$ elementary charges in the designated positions on radii $6/\kappa$ and $3/\kappa$ and causing $+\kappa$ and $-\kappa$ potentials in the centers. Eight cubes represent one lattice element of the imaginary cubic crystal

$$\gamma_m = 1/[\gamma_{ms}^{-1} + F_+(m^{E+} - 1) + F_-(m^{E-} - 1)] \qquad (1)$$

Its deduction is explained in the same reference [3]. The addend $F_+(m^{E+} - 1)$ introduces the influence of the

chemical potential of cations, the addend $F_-(m^{E-} - 1)$ the chemical potential of anions. All three addends represent the equilibrium condition which defines the activity coefficient function γ_m. The parameter γ_{ms}^{-1} characterizes the standard electrostatic free energy of each electrolyte. Of course, all five parameters are necessary for the calculation of activity coefficient functions or activities. The standard value, γ_{ms}^{-1}, is defined in the present special case by $F_+(m^{E+} - 1) + F_-(m^{E-} - 1) = 0$, because $F_+ > 0$ and $F_- < 0$ when $m \leq 1$. If $m > 1$, the standard value is defined by $m = 1$. In Eq. (11) of [3], instead of the present logical γ_{ms}^{-1} there is γ_{ms} with no consequence to the estimated parameter value.

The experimental sets of γ_m values and the corresponding experimental concentrations of many electrolytes can be found in the literature. By fitting Eq. (1) to experimental sets of values, one can estimate the constant parameters γ_{ms}, F_+, E_+, F_- and E_-. Tables of estimated parameters can replace tables of experimental γ_m values. By the use of Eq. (1) and the estimated parameters, one can transform exactly any concentration into its activity and the concentration axes can be transformed into activity axes from experimental data.

Table 1 gives the parameters of alkali halides obtained by fitting Eq. (1) to experimental plots. Note the precision of experiments and theory evidenced by the high $R^2 = 0.999$–0.9999 values of the standard deviation.

In Fig. 3 the influence of the molar mass of the salts (ordinate) upon the reciprocal standard activity coefficient $1/\gamma_{ms}$ (abscissa) is illustrated for alkali halides. The standard $1/\gamma_{ms}$ characterizes each electrolyte, assuming that the remaining four parameters are known. The increase of the cationic mass of Li, Na and K, related to

Table 1 Parameters obtained by fitting Eq. (1) to experimental γ_m values of alkali halides (experimental data from [4])

Salt	Range	R^2	$1/\gamma_{ms}$	F_+	E_+	F_-	E_-
LiCl	0/0.001/1	0.9999	1.2876	1.2902	0.5000	−1.0011	0.8488
LiBr	0/0.001/2	0.9996	1.2473	1.3857	0.5000	−1.1355	0.7968
LiI	0/0.001/2	0.9986	1.1007	1.6052	0.5000	−1.5128	0.6847
NaF	0/0.001/1	1.0000	1.7436	1.7436	0.5000	−0.6268	0.6974
NaCl	0/0.001/1	0.9996	1.5186	1.4049	0.5000	−0.8823	0.7575
NaBr	0/0.001/4	0.9997	1.4538	1.3838	0.5000	−0.9271	0.7708
NaI	0/0.001/4	0.9998	1.3528	1.8458	0.5000	−1.4928	0.6828
KCl	0/0.001/4	1.0000	1.6474	1.2958	0.5000	−0.6480	0.7646
KBr	0/0.001/6	1.0000	1.6207	1.3676	0.5000	−0.7446	0.7424
KI	0/0.001/5	0.9997	1.5474	1.6695	0.5000	−1.1170	0.6746
RbCl	0/0.001/8	0.9997	1.7117	1.3840	0.5000	−0.6685	0.7510
RbBr	0/0.001/5	1.0000	1.7320	1.3050	0.5000	−0.5713	0.7702
RbI	0/0.001/5	1.0000	1.7420	1.2613	0.5000	−0.5182	0.8024
CsF	0/0.001/1	1.0000	1.3772	1.2673	0.5000	−0.8898	0.8095
CsCl	0/0.001/9	0.9999	1.8299	1.1943	0.5000	−0.3639	0.9019
CsBr	0/0.001/5	0.9999	1.8515	1.1267	0.5000	−0.2771	0.9687
CsI	0/0.001/3	1.0000	1.8745	1.3176	0.5000	−0.4412	0.7324

Fig. 3 Interconnection between $1/\gamma_{ms}$ (*abscissa*) and molar mass (*ordinate*) of alkali halides

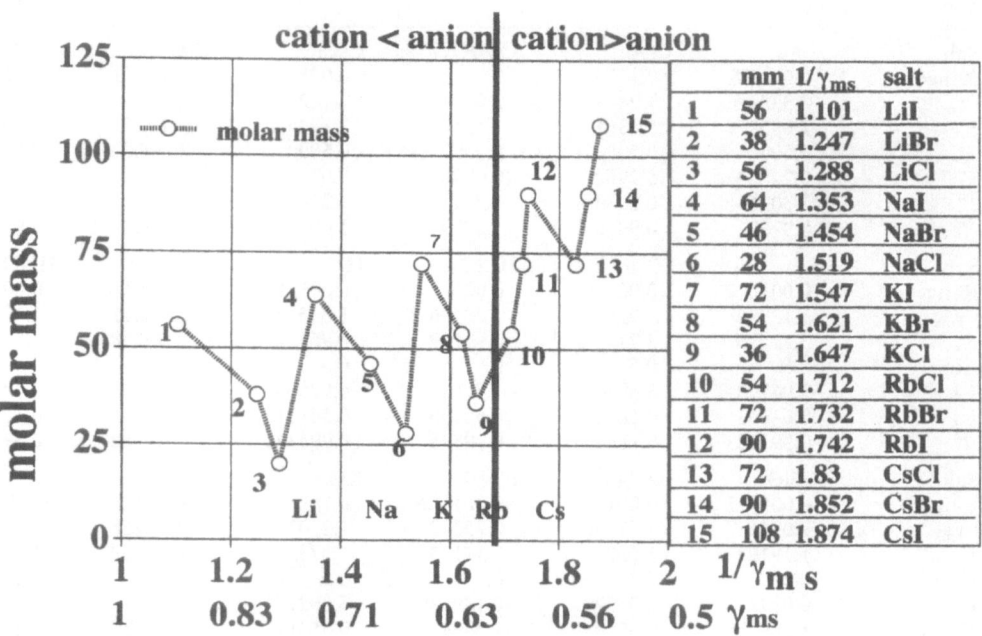

	mm	$1/\gamma_{ms}$	salt
1	56	1.101	LiI
2	38	1.247	LiBr
3	56	1.288	LiCl
4	64	1.353	NaI
5	46	1.454	NaBr
6	28	1.519	NaCl
7	72	1.547	KI
8	54	1.621	KBr
9	36	1.647	KCl
10	54	1.712	RbCl
11	72	1.732	RbBr
12	90	1.742	RbI
13	72	1.83	CsCl
14	90	1.852	CsBr
15	108	1.874	CsI

the decrease of the anionic mass in sequences I, Br and Cl, causes an increase of $1/\gamma_{ms}$. The increase of the cationic mass from K to Rb and Cs causes also the increase of $1/\gamma_{ms}$; however, the sequence of the anionic mass is opposite, i.e., Cl, Br and l. The smallest $1/\gamma_{ms}$ is that of LiI and the greatest is that of CsI. The transition of KCl to RbCl causes a reversal of the sequence of anions. The periodic system holds for the activity coefficients of alkali halides in a very convincing way.

In Table 2 the parameters of Eq. (1) for a large number of 1-1 electrolytes are presented.

In Fig. 4 the interconnection between the molar mass (ordinate) and the standard $1/\gamma_{ms}$ (abscissa) is illustrated for many 1-1 electrolytes. An analogous influence of cationic and anionic masses upon $1/\gamma_{ms}$ can be observed, as for the alkali halides.

The high standard deviation $R^2 = 0.999$–0.9999 indicates the high precision of the experimental values and of the theoretical Eq. (1). Tables 1 and 2 and Figs. 3 and 4 prove exactly that Eq. (1) is correct and thermodynamically justified.

Colloidal ionic solid/liquid systems

Electrokinetics

Plots with increasing and decreasing electrokinetic mobility or with maxima in low ionic strength have been observed experimentally by several prominent laboratories. Functions (2), (3) and (4) (see below) were deduced theoretically in [3] on the basis of the point-charge

double-layer model. The same functions can be fitted to experimental plots. Figure 5 represents schematically the imaginary point-charge double layer in equilibrium with the bulk electrolyte.

In Fig. 5a the fixed $+e$ and the $-e$ charges are drawn. The inter-charge distance between the two fixed $+e$ charges is $l_p = 2l_c = 1/\kappa$. In the plane vertical to the surface across the fixed charges, two full half-circles of $r = 1/\kappa$ are drawn. The $-e$ charge is under the potential κ. The eight pairs of circles of radii $r = 1/\kappa$ and $r = 1/2\kappa$ represent eight elementary lattices. The $\pm e/6$ charges are not drawn. They systematically visualize the first layer of elementary lattices of the bulk electrolyte. The equality of $1/\kappa = l_c = \sqrt[3]{2}\, l$ corresponds to I_c (mol dm^{-3}) $= 1 \times 10^{-5}$.

Figure 5b is the projection on the particle surface of the same lattice charges as described in Fig. 5a. On the surface, around the fixed charges, surfaces are observed inside and outside the DHT circle of $r = 1/\kappa$. In coagulation experiments, owing to Brownian motion, collisions will be efficient, because two surfaces outside the DHT radius can come in direct constant. Then are the attracting van der Waals forces active and cause aggregation or coagulation. If $l_p \leq 1/2\kappa$, the repelling electrostatic forces would prevent direct contact between the surfaces and no coagulation would take place; the particles would remain stable.

From Fig. 5b it follows that the solid phase, in electroosmotic experiments, will be stationary and will not move and the liquid phase will move towards the positive electrode. In electrophoretic experiments, the solid phase will move towards the negative electrode and

Table 2 Parameters obtained by fitting Eq. (1) to experimental γ_m values of various 1-1 electrolytes (experimental data from [4])

Salt	$AgNO_3$	CsAc	$CsBrO_3$	$CsClO_3$	$CsClO_4$	$CsNO_2$	$CsNO_3$	CsOH
Range	0.001/13	0.1/3.5	0.005/0.3	0.005/0.3	0.005/0.3	0.33/36	0.001/1.5	0.001/1.2
$1/\gamma_{ms}$	2.3389	1.2408	2.0919	2.1023	2.4079	1.6408	2.3691	1.2792
F_+	1.0858	5.4365	1.1911	1.1945	0.6712	4.5305	1.1146	1.4383
E_+	0.5000	0.5000	0.5000	0.5000	0.5000	0.5000	0.5000	0.5000
F_-	0.2486	(5.1960)	(0.0990)	(0.0917)	0.7370	(3.9025)	0.2527	(1.1568)
E_-	1.3016	0.5530	1.2913	1.2703	0.5490	0.5115	1.3353	0.7456
R^2	0.9999	0.9978	1.0000	1.0000	1.0000	0.9849	1.0000	0.9998

Salt	HBr	HCl	$HClO_4$	HI	HIO_3	HNO_3	KBO_3	$KClO_3$
Range	0.001/1.7	0.001/3	0.001/3	0.001/3	0.1/17.4	0.001/3	0.001/0.4	0.001/0.7
$1/\gamma_{ms}$	1.13986	1.20529	1.20436	1.02547	5.38411	1.37133	2.25273	2.27615
F_+	3.11488	3.714	1.99159	6.96372	1.99358	1.87949	1.07872	1.0486
E_+	0.5	0.5	0.5	0.5	0.5	0.5	0.5	0.5
F_-	-2.95723	-3.49437	-1.77637	-6.92666	2.39039	-1.50155	0.170903	0.223349
E_-	0.61053	0.58684	0.680668	0.541714	1.05828	0.661586	2.28544	4.42757
R^2	0.99712	0.99618	0.99828	0.995324	0.99988	0.99844	0.99987	0.99987

Salt	$KClO_4$	KCNS	KNO_3	KOAc	KOH	$LiClO_4$	NaAc	NaCNS
Range	0.005/0.3	0.001/5	0.001/3.5	0.1/3.5	0.001/20	001/2	0.1/3.5	0.1/3
$1/\gamma_{ms}$	2.0945	1.67306	2.25348	1.27031	1.48258	1.13041	1.31438	1.69673
F_+	1.19197	1.63217	1.11123	5.57333	0.563097	1.8433	3.83874	1.48919
E_+	0.5	0.5	0.5	0.5	0.5	0.5	0.5	0.5
F_-	-0.0971	-0.95577	-0.139389	-5.30215	-11.6437	-1.70638	-3.52355	-0.7911
E_-	1.389080	0.655035	1.5899	0.552116	0.085872	0.685417	0.576187	0.656282
R^2	0.99999	0.9999	0.99997	0.99726	0.99839	0.99869	0.99836	0.99997

Salt	$NaClO_4$	$NaNO_2$	NaCNS	NH_4Cl	NH_4ClO_4	$TlNO_2$		
Range	0.001/6	0.1/12.25	0.1/3	0.001/4	0.001/2.1	0.001/1.4		
$1/\gamma_{ms}$	1.58873	1.68198	1.4121	1.65927	2.08891	4.3291		
F_+	1.36151	5.99882	1.53830	1.35181	1.2054	0.797486		
E_+	0.5	0.5	0.5	0.5	0.5	0.5		
F_-	-0.77139	-5.3202	-1.1226	-0.69080	-0.11608	2.52446		
E_-	0.730842	0.53373	7.1514	0.737466	0.79882	0.904315		
R^2	0.99998	0.99393	0.99972	0.99998	1.00000	0.99892		

Fig. 4 Interconnection between $1/\gamma_{ms}$ and the molar mass of various 1-1 salts

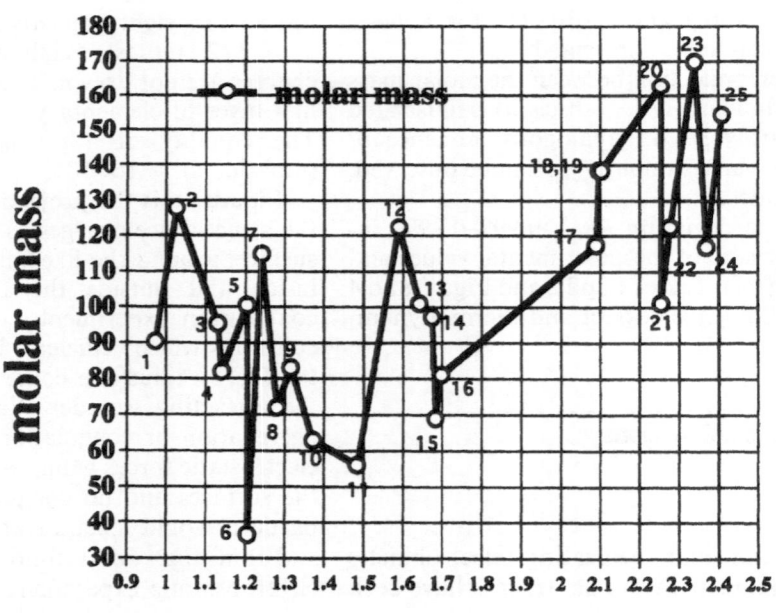

	$1/\gamma_{ms}$	salt
1	0.971	LiClO3
2	1.025	HI
3	1.13	LiClO4
4	1.14	HBr
5	1.204	HClO4
6	1.205	HCl
7	1.241	CsAc
8	1.279	CsOH
9	1.314	NaAc
10	1.371	HNO3
11	1.483	KOH
12	1.589	NaClO4
13	1.641	CsNO2
14	1.673	KCNS
15	1.682	NaNO2
16	1.697	NaCNS
17	2.089	NH4ClO4
18	2.095	KClO4
19	2.102	CsClO3
20	2.253	KBrO3
21	2.253	KNO3
22	2.276	KClO3
23	2.339	AgNO3
24	2.369	CsNO3
25	2.408	CsClO4

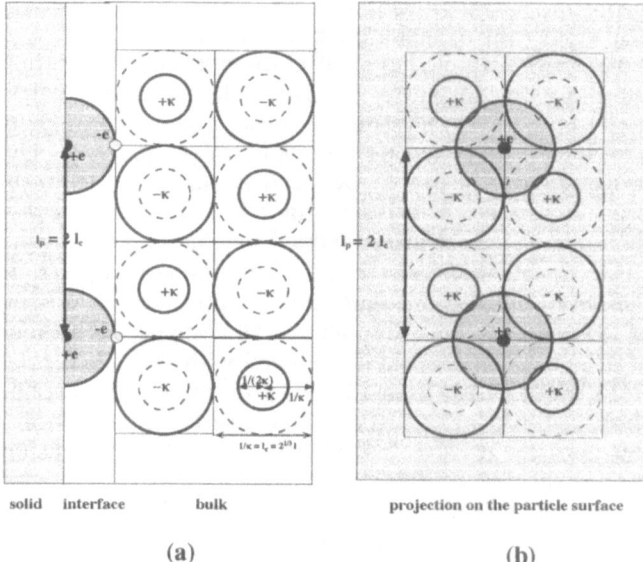

solid   interface   bulk   projection on the particle surface

(a)   (b)

Fig. 5 A schematic presentation of the imaginary point-charge double layer in equilibrium with the bulk electrolyte. **a** Plane vertical to the surface across the fixed $+e$ and counter $-e$ charges and cross section of one layer of lattice elements of the bulk at I_c (mol dm^{-3}) $= 1 \times 10^{-5}$. The 12 elementary charges in each cube of edge length $1/\kappa = l_c = \sqrt[3]{12}l$ of the bulk electrolyte on the sphere surfaces are not drawn. **b** Projection on the particle surface of the same charges as in **a**

the electrolyte phase will be stationary and will not move. The slipping plane is very irregular. Water molecules in the double layer will, depending on the distance from the fixed charges, move at different velocities. The reversal of the sign of assumed fixed charges will cause the reversal of mobility.

In [3], the following two theoretical functions of electrophoretic mobility of stable, not sedimenting particles are proposed:

$$q = q_m - 2s|\log(\kappa_{pm}/\kappa_p)| \tag{2}$$

and

$$q = q_m - s|\log(I_{cm}/I_c)| \tag{3}$$

Here, q is the experimental electrokinetic quotient and q_m is its maximal value at the intersection, $\log(I_{cm})$, of the two lines with positive and negative slopes or of the value of $\log(1/\kappa)$ at the intersection with the $\log(1/\kappa_p)$ line (see Fig. 8 in [3]). Assuming that the inter-charge distance is constant (see Fig. 8 in [8]), the effective thickness of the interface, l_p, is a linear function with $\log(I_c)$.

The electrophoretic experimental q is always measured as the quotient between the length of the particle path in unit time and the applied electrical field strength.

The electoosmotic experimental q is the quotient of the transported volume across the membrane in unit time and the field strength in the electrolyte of the membrane or of the analogous quotients in the streaming potential and streaming current experiments. It is assumed that the

conductivity of the solid phase in comparison with that of the electrolyte is negligible. The proportionality factor or the slope of the line is s. If the experimental plot of Eqs. (2) and (3) is a function with a maximum, the increasing and decreasing parts of the plots can be approximated as straight lines that have the intersection in $\log(I_{cm})$. The present theoretical analysis suggests that the electrokinetic mobility or transport are proportional to the logarithm of the electrostatic potential, κ, of point charges in the double layer, which in turn is proportional to the logarithm concentration or ionic strength, $\log I_c$, and is equal to the same potential in the bulk.

The electrostatic potential quotient, κ_p/κ_{pm}, should be calculated from the experimental electrophoretic mobility or from electroosmotic transport quotient, q, by

$$\kappa_p/\kappa_{pm} = 10^{(q_m-q)/2s} = 10^{(\zeta_m - \zeta)/2s} \tag{4}$$

It replaces the analogous quotient, ζ/ζ_m, of Smoluchovski.

Light scattering experiments of stable, not sedimenting, particles

Dynamic light scattering experiments of coagulation and of formation of stable particles of AgI are described in [5] by Fig. 2. On the ordinate is the intensity of scattered light in relative units, I_{rel}, and on the abscissa the logarithm of time, $\log(t/s)$, after the sol preparation. For all plots [AgI] $= 4 \times 10^{-4}$ moldm^{-3}. The plots number 1 to 11 and plot number 12 is of different shape. Compared with Fig. 1, the plots 1–11, p$I = 0.41$ to 2.25, represent coagulation with Na$^+$ and that of number 12, p$I = 2.44$, represents the formation of stable particles. The I_{rel} does not change for long times and on the 10 min tyndallo-gram its value is minimal. The plots 1–11 are S-shaped and can be interpreted as coagulation plots, which are theoretically described by Eq. (21) in [8]. Plot 12 in Fig. 2 displays linear parts and each linear part should be interpreted as a linear function, i.e.

$$I = I_0 \log t + \log t_{crit} \tag{5}$$

and it holds, for $I = 0$

$$\log t_{crit} = -I_0 \log t \tag{6}$$

Here, I_0 is the slope of the linear plots and the logarithm of the critical time, $\log t_{crit}$, is the intersection with the abscissa of the extended measured linear parts.

Light scattering experiments are always interpreted theoretically by the Rayleigh equation, which can be written in the simplest elementary form (see Eq. (10) in [7]) as:

$$I = BN_p^2 N_a \tag{7}$$

The number of aggregates is N_a and N_p is the number of singlets forming an aggregate. Equation (7) holds for growing scattering aggregates in coagulation, for increasing dense stable particles and for the increase of the number of dense scattering particles of defined constant size that emerge from clear solutions. For a given measuring instrument, the constants $B = const$ have different values for sponge-like aggregates and dense particles.

If dense scattering particles are formed from optically clear solutions and if the size of the scattering dense particles is constant, then $BN_p^2 = B_p = const$, a new constant valid formally also for the dense particles, and the following holds:

$$I_p = B_p N_a \quad \text{or} \quad N_a = I_p/B_p \quad \text{and} \quad N_{a0} = I_{p0}/B_p \quad (8)$$

The intensity of scattered light, I_p, is proportional to the number of scattering particles, N_a.

By insertion of Eq. (8), Eq. (5) is transformed into

$$N_a = N_{a0} \log t + \log t_{crit} \quad (9)$$

The slope of the plots in N_{a0}. One must conclude that the linear parts of the $I_p = I_{rel}$ plot (number 12 of Fig. 2) describe the increase of the number of particles. Unavoidably, after long enough times, the number of scattering particles must become constant, i.e., $N_a(t \gg t_{crit}) = const$, and consequently also $I_{rel}(t \gg t_{crit}) = const$. For each pI plot the emerging particles are of a different constant size because the constant intensities area different for each plot and this means that the particles are of different sizes. The conclusion is based on the fact that the number of all singlets is constant at any time because $N_0 = N_p N_a = const$ for all pI plots. The $I_{rel}(t)$ plot number 12 was not followed for long enough times. It is possible that plots of $pI > 2.44$ I_{rel} $(t \gg t_{crit}) = const$ became constant. In Fig. 1 the dashed line of the plot after 7 days suggests that, at $pI > 2$, the particles are stable and do not sediment. One can conclude that if the same system had been followed for long enough time, I_{rel} would have acquired a constant high value of $I_{rel} > \approx 200$. The plots have different shapes. One should expect that the shape of the plots systematically changes with the increase of pI but it does not. Most probably, the mixing of precipitating solutions is not sufficiently reproducible and a better reproducible technique should be worked out and experiments repeated in the whole range of stable pI values.

In Fig. 1 the thick line of the same [5] shows the 10 min "tyndallogram", i.e., the intensity, I_{rel} on the 10 min plot. In the range $2 > \approx pI < \approx 9.5$ (abscissa), the value $I_{rel} \approx 0.001$ (ordinate) is negligibly small and practically equal to the zero effect. The instrument used was the Pulfrich photometer (measures the intensity of scattered light) combined with the Zeiss tyndallometer (scatters the incident beam). One must conclude that the Brice Phoenix DU2000 SLS used in [5] measures particles of sizes that are too small to be registered by the Pulfrich photometer. In the Brice Phoenix the sols were in cuvettes with parallel windows, while in the former instrument the systems were in test tubes of a 15 mm cross-section. Immediately after mixing the precipitating solutions, the turbidity, I_{rel}, is small; the small singlets present are in very short times transformed into singlets, and later coagulate if a coagulator is present. The processes are illustrated by electron micrographs in [6] (Figs. 1.1, 5.1–5.3, 6.1–6.5 and 7.1–7.3). Theoretically, the same processes are explained by Fig. 6 in [7]. In the latter [7], all intensities I should br replaced by $\sqrt{I}$. It should be pointed out that immediately after the mixing of the precipitating components, all systems, even those of $pI > 2.44$, are turbid if observed with the naked eye.

In times longer than 10 min, after mixing the precipitating solutions for several hours the Pulfrich photometer does not register any increase of I_{rel}. The conclusion is that the Brice Phoenix instrument measures I_{rel} on particles of sizes that are much smaller than the particles registered by the Pulfrich/Zeiss combination.

Instructive light scattering experiments in [9] of the coagulation of $[AgI] = 0.04$ mol dm^{-3} sols, in excess $[KI] = 0.02$ mol dm^{-3} or $pI = 1.2$ and in $[KNO_3] = 0.1$ and 0.2 mol dm^{-3} were analyzed by Figs. 4–6 in [8] on the basis of the Rayleigh equation. Experimental data in [9] describe during 10 min the coagulation of the first-formed dense, small, not sedimenting particles of the sol. Owing to the high $[AgI] = 0.04$ mol dm^{-3}, which is 100 times that of Fig. 2, immediately after mixing, $I_{rel}(t = 0) \approx 0.25 I_\infty$. From the experimental plots, one obtains the estimated I_∞ by fitting. The estimated coagulation half-life or half-period of coagulation is $t_{1/2} = 1$ min.

In [7], [8] and [10], various methods are described to estimate the half-life or half-period of coagulation processes from light scattering experiments.

In [8] (Fig. 1) and [12] (Fig. 1) the formation of small aggregates of from 2 up to 6 singlets is theoretically explained. The theory is experimentally confirmed by the coagulation of monodisperse polystyrene (hard) and poly(vinyl acetate) (soft) particles in [11].

Figures 5 and 14 in [8] are a schematic explanation of counterion and isoelectric coagulation of fresh AgI sols. Both are preceded by coalescence of the smallest particles. If no counterions or Ag$^+$ ions were added, the polydisperse particles would not coalesce. Singlets are formed by coalescence. Their size is equal to the ion cloud radius, which is defined by the ionic strength.

Three papers, [13], [14] and [9], are reviews of the development of colloid chemistry and electrochemistry over more than 50 years, with special emphasis on the controversy between theories based on the point-charge and homogeneous-charge double-layer models. The paper from [3] and the present paper represent the author's theoretical explanation of the latest experiments, published by prominent scientists from several countries, displaying maxima in electrokinetic mobilities and of the activity coefficient function in high concentrations. There remains to be solved the activity coefficient functions of electrolytes of various ion charge numbers, $z_1 \geq 1 - z_2 \geq 1$.

Conclusion

The classical Debye-Hückel theory has been replaced by an imagined crystal lattice or Madelung constant theory. The activity coefficient dependence on concentration has been determined by the electrostatic free energy change and by the chemical potential of cations and anions. Based on this premise, a function is proposed with five constant parameters. Two parameters are defined by the chemical potential of cations, two by the chemical potential of anions and one by their electrostatic free energy change. The same parameters have been estimated for alkali halides and a number of 1-1 electrolytes and are presented in two tables. Using the equation and the parameters, for all 1-1 electrolytes the activity coefficients can be calculated and molality axes can be transformed into activity axes. It is demonstrated graphically that the molar mass of the electrolytes and the ionic masses of the cations and anions define the activity coefficient function in full accordance with their position in the Periodic Table.

The point-charge double-layer model of ions fixed on the surface and in the bulk is described by a scheme. Based on the model, it is demonstrated that the dependence of electrophoretic mobility or electroosmotic transport is defined by the Coulomb equation. A function is proposed that can be fitted to experimental mobility plots with maxima in low ionic strength. It replaces the Smoluchowski ζ-potential equation. No published theory has explained the experimental electrokinetic plots with maxima published by several laboratories.

References

1. Debye P, Hückel E (1923) Phys Z 24:305; (1924) Phys Z 25:97; English translation In: The collected papers of Peter J.W. Debye. Interscience, New York
2. Bockris JO'M, Reddy AKN (1998) Modern Electrochemistry, vol 1, 2nd edn. Plenum Press, New York
3. Mirnik M (1999) Prog Colloid Polym Sci, Springer Verlag pp 188–199
4. Lobo VMM, Quaresma JL (1981) Electrolyte solutions: literature data on thermodynamic and transport properties, vol 2. Coimbra Editora, Coimbra
5. Heimer S, Težak D (1999) Prog Colloid Polym Sci 112:177–182
6. Mirnik M, Strohal P, Wrischer M, Težak B (1958) Kolloid-Z 160:146–156
7. Mirnik M (1988) Croat Chem Acta 61:81–101
8. Mirnik M (1996) Croat Chem Acta 69:125–175; Errata, ibid. (1997) 70:A19–A20
9. Novicki W, Novicka G (1991) J Chem Educ 68:523
10. Mirnik M (1990) Croat Chem Acta 63:113–126
11. Lichtenfeld H, Sonntag H, Dürr Ch (1991) Colloids Surf 54:267
12. Mirnik M (1992) Croat Chem Acta 65:297–307
13. Mirnik M (1970) Croat Chem Acta 42:161–214
14. Mirnik M (1994) Croat Chem Acta 67:493–508

Progr Colloid Polym Sci (2000) 115:342–346
© Springer-Verlag 2000

P. Hansson
S. Schneider
B. Lindman

Macroscopic phase separation in a polyelectrolyte gel interacting with oppositely charged surfactant: correlation between anomalous deswelling and microstructure

P. Hansson (✉)
Department of Physical Chemistry
Uppsala University, P.O. Box 532
75121 Uppsala, Sweden
e-mail: per.hansson@fki.uu.se
Tel.: +46-18-4713650
Fax: +46-18-508542

S. Schneider · B. Lindman
Department of Physical Chemistry 1
Lund University, P.O. Box 124
22100 Lund, Sweden

Abstract The volume transitions in cross-linked sodium polyacrylate gels following the absorption of cetyltrimethylammonium bromide from aqueous solutions is studied. The microstructure of the gels is investigated by means of time-resolved fluorescence quenching, small-angle X-ray scattering, and optical anisotropy. In general, the volume of the pre-swelled gels decreases linearly with the number of absorbed surfactant molecules. The shrinking is due to a collapse of the exterior parts of the gels where essentially all the surfactant is located. The surface phase is very dense and is made up from collapsed micelle/polyelectrolyte complexes arranged on a cubic lattice. The surfactant aggregation number is between 110 and 170. At sufficiently high surfactant-to-polymer ratios the entire gel is in a collapsed single phase state with the same microstructure as the surface phase. It is observed that during the transition from the swollen to the fully collapsed state the gels may be trapped in a semi-swollen state, where the gels have a balloon-like shape due to a pressure across the surface phase. There is a strong correlation between this anomalous behavior and the growth of the micelles from globular to long cylinders and a transition from cubic to hexagonal packing. The observations are compared with the macroscopic surface phases observed during phase transitions in related systems.

Key words Polyelectrolyte gel · Network · Surfactant · Collapse · Phase separation

Introduction

Chemically crosslinked polyelectrolyte gels can absorb large amounts of water. Depending on the degree of crosslinking, the mass of a swollen sodium polyacrylate (NaPA) network may be several hundred times the mass of the dry gel. When the swollen gel is placed in a solution of a surfactant with a charge opposite to that of the network, a deswelling or collapse of the network takes place. The amplitude of the collapse depends on the relative amounts of polymer and surfactant present [1]. This can be explained as a reduction of the osmotic swelling pressure as counterions to the network chains are replaced by surfactant micelles [2]. In the excess of surfactant the gel is transformed into a dense (≈50 wt% water), structurally ordered mesophase [3]. The phenomenon is strongly related to the associative phase separation in mixtures of surfactant and linear polyelectrolytes, where the association is promoted by the release of "trapped" counterions [4]. Similar to the latter case, where a concentrated polyelectrolyte/surfactant phase may separate from an excess polyelectrolyte phase, the collapse of the crosslinked gel is not completed when there is a deficit of surfactant. Under these circumstances a phase separation in the gel takes place. In general, a dense layer of complexed surfactant and polyelectrolyte forms at the exterior part of the gel enclosing a swollen surfactant-free interior part [5, 6], although exceptions to this behavior have been reported [7, 8]. The same behavior, commonly referred to as "skin formation", is observed when the gels absorb oppositely charged linear polyelectrolytes [9] or small proteins [10].

In this paper we investigate the shrinking of NaPA gels placed in aqueous solutions of cetyltrimethylammonium bromide (C_{16}TAB). In particular, we investigate the microstructure of the skins by means of small-angle X-ray scattering (SAXS) and fluorescence quenching experiments and how it relates to the appearance of balloon-like gels. This phenomenon, which was briefly mentioned in a previous study [8], has not been reported before. Gels showing this anomalous behavior are "trapped" in a semi-swollen deformed state at surfactant loading normally giving a much larger collapse.

Materials and methods

C_{16}TAB (Serva) and cetylpyridinium chloride (C_{16}PC) (Merck) were of analytical grade and were used as supplied. Pyrene (99% +) (Janssen), acrylic acid (Aldrich), N,N,N',N'-tetramethylethylenediamine (TEMED), ammonium persulfate, and N,N'-methylenebis(acrylamide) (MBA) (Sigma) were used as received.

Acrylic acid was polymerized in the presence of 1 mol% MBA as the crosslinking agent, using ammonium persulfate and TEMED as radical initiator and accelerator, respectively [11]. The gels were neutralized in 0.5 M NaOH, cut into short cylinders and then repeatedly washed in 10^{-4} M NaOH. The concentration of NaPA in the pure network in equilibrium with 10^{-4} M NaOH was 21 mM, as determined by weighing swollen and freeze dried gels. Surfactant/gel samples were prepared by immersing pre-swollen gel pieces (3–5 g) in C_{16}TAB solutions of initial concentrations: 0.1, 0.5, 1, 3, and 10 mM. All solutions contained 10^{-4} M NaOH. For each C_{16}TAB concentration, samples with different surfactant to polymer ratios were prepared by varying the volume of the solution. After one month the concentration of C_{16}TAB in the aqueous phase $C_{s,aq}$ was determined using a surfactant-sensitive electrode as described elsewhere [11]. The degree of binding β was calculated as $\beta = (C_{s,tot} - C_{s,aq})/C_{p,tot}$, where $C_{s,tot}$ and $C_{p,tot}$ is the total concentration of surfactant and polyelectrolyte monomers, respectively, in the system.

Fluorescence decays were recorded using the single photon counting technique as described in detail elsewhere [8, 12]. Pyrene and C_{16}PC were used as fluorescent probe and quencher, respectively. Quenched decays were analyzed with the relationship [13]:

$$F(t) = F(0)\exp\{-t/\tau_0 + \langle n \rangle(\exp\{-k_q t\} - 1)\} \quad (1)$$

describing the time evolution of the fluorescence intensity $F(t)$ following a short laser pulse. $F(0)$ is the intensity at time zero, τ_0 is the fluorescence lifetime, $\langle n \rangle$ is the average number of quenches per micelle, and k_q is the intra-micellar quenching rate constant. Equation (1), which is a special case of the Infelta-Tachiya equation [14, 15], is valid when the probe and a randomly distributed quencher are stationary in the micelles during the probe lifetime. The average aggregation number N was calculated as $N = \langle n \rangle/X_Q$, where X_Q is the mole fraction of quencher in the $C_{16}P^+/C_{16}TA^+$ mixed micelles. In all samples, X_Q was assumed to be identical to the total mole fraction of quencher in the system. This, as well as the assumption of a random distribution of quencher among the micelles, are excellent approximations in the present system, as discussed elsewhere [8].

SAXS spectra were recorded using a linearly collimated Kratky compact small-angle system equipped with a position-sensitive

detector (OED 50 M, MBraun, Graz, Austria), as described elsewhere [8].

Results

Figure 1 shows the degree of C_{16}TAB binding, β, to the NaPA gels as a function of $C_{s,tot}/C_{p,tot}$. The corresponding variation of the volume V of the gels relative to the volume V_0 of the pre-swollen gels is shown in Fig. 2. We focus first on the case of $C_{s,tot} = 0.5$ mM, which shows the essential features of interest to the present discussion.

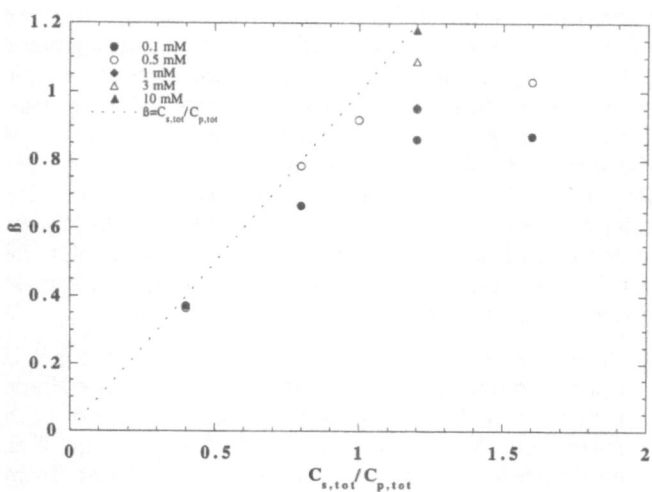

Fig. 1 The variation of β with $C_{s,tot}/C_{p,tot}$ for gels placed in C_{16}TAB solutions of different concentrations; see legend. The *dotted line* corresponds to 100% binding

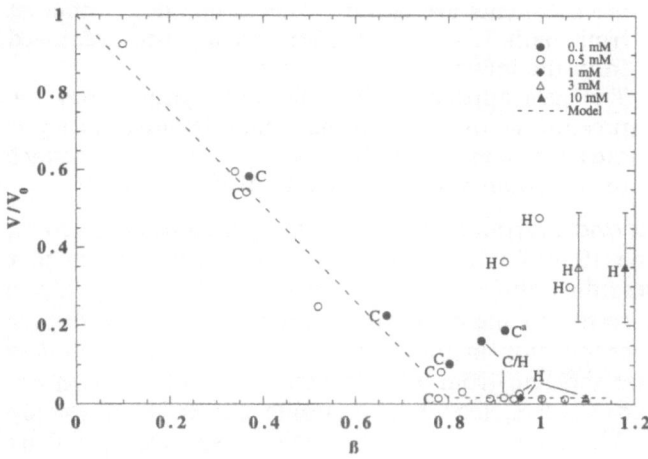

Fig. 2 The relative gel volume (V/V_0) as a function of β for gels placed in C_{16}TAB solutions of different concentrations; see legend. *Letters* indicate the structural order obtained with SAXS: C cubic; H hexagonal; C/H cubic and hexagonal order in different parts of the gel. Superscript a: cubic phase found in a balloon-like gel; see text. The *dashed line* is the prediction of a model described in [11]

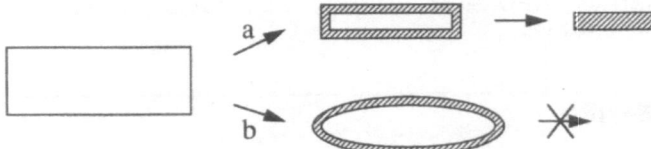

Fig. 3 A schematic presentation of the volume transitions following *a* path 1, and *b* path 2, as described in the text

For $C_{s,tot}/C_{p,tot} < 0.8$, where essentially all surfactant is in the gel, i.e., $\beta \approx C_{s,tot}/C_{p,tot}$, the reduction of the volume is proportional to β. This is expected if the formation of a micelle in the gel is followed by a collapse of a certain number of (swelled-up) polyion segments [11]. An inspection of the gels shows that a macroscopic phase separation takes place in such a way that surfactant-rich skins are formed at the exterior parts of the gels, the thickness of which increases with β. Thus, for $\beta = C_{s,tot}/C_{p,tot} < 0.8$ the skin coexists with a swollen interior phase deficit of surfactant (Fig. 3). At higher ratios there is enough surfactant present to transfer the gel into a single dense phase. However, here we have observed three different paths:

1. "Regular" shrinking: the final volume of the gel is approximately the same as for $\beta \approx 0.8$ and the shape is essentially the same as in the swollen state. For each surfactant bound in excess of $\beta = 0.8$ the free concentration increases strongly, as evident from Fig. 1.
2. "Anomalous" shrinking: despite the high β, the final gel volume is much larger than in path 1 and the entire gel, or parts of it, is balloon-like (Fig. 3). When the skin is punctured, water or a suspension of torn gel fragments are ejected. This is markedly different from path 1, where the skin can be easily removed from the intact swollen gel core.
3. The gels appear to have followed path 2 but the pressure across the skin has made it burst. The gels are either compact with a "contracted" rough surface or just skins resembling empty plastic bags.

As evident from Figs. 1 and 2, the gels placed in 0.1, 1, 3, and 10 mM $C_{16}TAB$ show a similar behavior to that described above. However, at a fixed ratio $C_{s,tot}/C_{p,tot}$, β seems to increase with increasing initial surfactant concentration in the solution. It should be mentioned that the skin from all gels were essentially transparent and colorless, in contrast to results reported by others [6].

The skins from all gel samples were investigated by means of SAXS and/or time-resolved fluorescence quenching (TRFQ). The result is presented in Fig. 4, where the average aggregation number N from TRFQ is given as a function of β. For gels following path 1, N is small but similar to that of globular $C_{16}TAB$ micelles in dilute solutions of the surfactant. Since the gel samples at

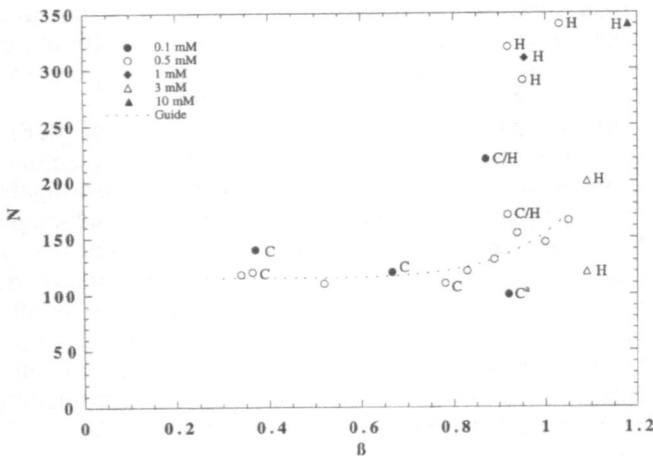

Fig. 4 The surfactant aggregation number N from TRFQ as a function of β from the surface phase of gels placed in $C_{16}TAB$ solutions of different concentration; see legend. $N > 250$ should be regarded as rod-like micelles; see text. The *dashed line* is just a guide to the eye. *Letters* have the same meanings as in Fig. 2. Superscript *a*: cubic phase with small N found in a balloon-like gel; see text

high β are very concentrated, this is perhaps surprising, keeping in mind the formation of worm-like micelles in concentrated $C_{16}TAB$ solutions. It should be noted, however, that the main counterion to the micelle in the gel is the carboxylate group of the network chains. Very little bromide is expected in the gel, at least for $\beta < 0.8$. When the skins were examined with SAXS, several reflections appeared (in some cases 10 distinct peaks), which suggests that the micelles (or, rather, NaPA/$C_{16}TAB$ complexes) are arranged on a cubic lattice. (The investigated samples are denoted "C" in Fig. 4.) In agreement with what was found earlier for gels collapsed in $C_{12}TAB$ solutions [8], and with the $C_{12}TAC$/water micellar cubic phase [16], the peak order is indicative of the space group $Pm3n$.

In contrast, gels following paths 2 and 3 showed TRFQ decays similar to those of quenching in "infinite" rod-like micelles [13], with two exceptions (see Figs. 2 and 4). This type of quenching is qualitatively different from the quenching in discrete micelles described by Eq. (1) [13]. For these gels, N was obtained by forcing Eq. (1) to fit the decay curves, and should therefore be considered just as "large" ($N > 200$). In agreement with the TRFQ results, SAXS measurements revealed that these gels (with one exception) showed hexagonal order (marked "H" in Fig. 4). For two of the samples, one part of the gel was collapsed and seemed to have followed path 1, and the other part was balloon-like. These samples are indicated as "C/H" in Figs. 2 and 4, since the collapsed and balloon-like parts were found to have cubic and hexagonal order, respectively.

The skins from all gel samples were observed between crossed polarizes. The result was unambiguous: all samples with hexagonal order were optically anisotropic; the ones with cubic order were isotropic.

Discussion

As discussed in detail elsewhere [8, 11], the presence of highly charged micelles has a tremendous effect on the network chains in their close vicinity. Owing to the crosslinks, chains (in micelle-free domains) far away from the micelles are also affected, but a collapse of these is prohibited by strong osmotic swelling forces. The situation is thus quite different from changing the solvent quality or the osmotic pressure of the liquid in equilibrium with the gel. The difference is particularly obvious when there is not enough surfactant present to collapse all chains in the network. Under these circumstances a macroscopic phase separation, analogous to that observed in mixtures of linear polyelectrolytes and surfactant [17], may represent the most favorable distribution of the components. However, the crosslinks complicate matters. For instance, the network deformation (i.e., swelling on the microscopic level) must vary continuously between planes parallel to the interface separating the two phases. Thus, the deformation of the network is expected to be anisotropic in both of the two coexisting phases. These problems have been noted by others [18, 19] in connection with the macroscopic surface phases observed during the first-order transitions of temperature-sensitive weakly charged polymer networks following a temperature jump (or quench) [20], and in gels containing carboxylate groups interacting with complexing ions (e.g., Cu^{2+}) [21]. However, in these systems the surface phase is a transient structure which, by preventing the solvent from quickly diffusing out, can arrest the gel for some time in a semi-swollen state [20].

Whether or not the skin formation observed for $\beta < 0.8$ in the present gels is an equilibrium phenomenon is still an open question. It could be a kinetic trap, preventing the micelle/polyelectrolyte complexes from redistributing. However, at higher surfactant-to-polymer ratios the strongly collapsed state as represented by the dotted line in Fig. 2 (for $\beta > 0.8$) is most likely close to the equilibrium. At least, this state can be reached under appropriate conditions. Similar to the temperature-induced collapse of NIPA gels [20], it has been observed that the collapsed state is reached through a continuous growth of the surface phase [6]. Thus, it is likely that the balloon-like gels observed by us appear when the collapse of the exterior parts, i.e., the formation of the skin, is faster than the escape of water. For the collapse of the network to proceed at the interface between the skin and the swollen core, there must be a transport of surfactant through the skin. However, without a concomitant reduction of the core volume, this process cannot proceed unless the core network is destroyed, as was observed in the experiment. Note also that, since it is reasonable to assume that the network chains are in a relaxed state in the fully collapsed gels (following path 1), there must be considerable tension in the skin of the balloon-like gels. The pressure due to the contraction of the skin can thus explain the deformation of the gels.

The strong correlation between hexagonal microstructure and balloon-like gels is very interesting. As yet we can only speculate about its origin. One possibility is that the anisotropic deformation of the surface phase induces the cubic to hexagonal transition, which in turn affects the permeability of the skin to water and/or simple ions.

Conclusions

The absorption of $C_{16}TAB$ by pre-swelled NaPA gels results in a macroscopic phase separation of a water swollen phase and a surfactant-rich surface phase ("skin"). The surface phase is very dense and is made up from collapsed micelle/polyelectrolyte complexes arranged on a cubic lattice. The surfactant aggregation number is between 110 and 170. The volume of the gels decreases with β. At sufficiently high surfactant-to-polymer ratios the gel is in a collapsed single-phase state with the same microstructure as in the skins. This is different from the lamellar structure proposed by Khandurina et al. [22] for the same system. During the transition from the swollen to the fully collapsed state the gels may be trapped in a semi-swollen state, where the gels have a balloon-like shape due to a pressure across the surface phase. There is a strong correlation between this anomalous behavior and the growth of the micelles from globular to long cylinders which is consistent with the observed transition from cubic to hexagonal packing. The type of skin forming gels studied here are expected to be applicable to, for example, drug delivery.

Acknowledgement This work was financially supported by Center for Amphiphilic Polymers (CAP), Lund University, and the Swedish Council for Engineering Sciences (TFR).

References

1. Khokhlov AR, Kramarenko EY, Makhaeva EE, Starodubtzev SG (1992) Macromolecules 25:4779
2. Khokhlov AR, Kramarenko EY, Makhaeva EE, Starodoubtzev SG (1992) Makromol Chem Theory Simul 1:105
3. Khandurina YV, Dembo AT, Rogacheva VB, Zezin AB, Kabanov VA (1994) Polym Sci 36:189
4. Thalberg K, Lindman B, Karlström G (1990) J Phys Chem 94:4289
5. Khandurina YV, Rogacheva VB, Zezin AB, Kabanov VA (1994) Polym Sci 36:184
6. Kabanov VA, Zezin AB, Rogacheva VB, Khandurina YV (1997) Macromol Symp 126:79
7. Filippova OE, Makhaeva EE, Starodubtsev SG (1992) Polym Sci 34:602
8. Hansson P (1998) Langmuir 14:4059
9. Zezin AB, Rogacheva VB, Kabanov VA (1997) Macromol Symp 126:123
10. Skobeleva VV, Rogacheva VB, Zezin AB, Kabanov VA (1996) Dokl Phys Chem 347:52
11. Hansson P (1998) Langmuir 14:2269
12. Almgren M, Hansson P, Mukhtar E, van Stam J (1992) Langmuir 8:2405
13. Almgren M (1992) Adv Colloid Interface Sci 41:9
14. Infelta PP, Grätzel M, Thomas JK (1974) J Phys Chem 78:190
15. Tachiya M (1975) Chem Phys Lett 33:289
16. Balmbra RR, Clunie JS, Goodman JF (1969) Nature 222:1159
17. Piculell L, Lindman B, Karlström G (1998) In: Kwak JCT (ed) Polymer-surfactant systems. Dekker, New York, pp 65
18. Tomari T, Doi M (1995) Macromolecules 28:8334
19. Panyukov S, Rabin Y (1996) Macromolecules 29:8530
20. Matuso ES, Tanaka T (1988) J Chem Phys 89:1695
21. Budtova T, Navard P (1997) Macromolecules 30:6556
22. Khandurina YV, Alexeev VL, Evmenenko GA, Dembo AT, Rogacheva VB, Zezin AB (1995) J Phys II 5:337

Progr Colloid Polym Sci (2000) 115:347–352
© Springer-Verlag 2000

C. Sommer
L. Cannavacciuolo
S. U. Egelhaaf
J. S. Pedersen
P. Schurtenberger

Micelles as model systems for equilibrium polyelectrolytes: a light and neutron scattering study

C. Sommer · P. Schurtenberger (✉)
Department of Physics
University of Fribourg
1700 Fribourg, Switzerland
e-mail: peter.schurtenberger@unifr.ch
Tel.: +41-26-300 9115
Fax: +41-26-300 9747

L. Cannavacciuolo
Polymer Institute, ETH Zürich
CH 8050 Zürich, Switzerland

S. U. Egelhaaf
The University of Edinburgh
Department of Physics & Astronomy
Edinburgh EH9 3JZ, UK

J. S. Pedersen
Condensed Matter Physics and Chemistry
Department, Risø National Laboratory
4000 Roskilde, Denmark

Abstract We demonstrate that aqueous solutions of giant polymer-like non-ionic micelles "doped" with small amounts of ionic surfactants serve as ideal model systems for "equilibrium polyelectrolytes". We report systematic light and neutron scattering investigations of the effect of ionic strength, total concentration and doping level on the static properties of dilute and semi-dilute micellar solutions. In dilute solutions we observe a dramatic influence of (intramicellar) electrostatic interactions on the micellar flexibility, and the results are in close agreement with Monte Carlo simulations. In the semi-dilute regime, strong long-range interactions between micelles induce liquid-like ordering.

Key words Equilibrium polyelectrolytes · Worm-like micelles · Micellar flexibility · Scattering experiments · Electrostatic interactions

Introduction

It has been shown in numerous studies that it is possible to find conditions where micelles grow dramatically, with increasing surfactant concentration, into giant cylindrical aggregates. These giant micelles normally have a high degree of flexibility, and their overall structure is generally well described by polymer theory [1, 2]. The postulate of an analogy between classical polymers and polymer-like micelles has been an important step towards a quantitative understanding of surfactant systems. Several attempts have been made to demonstrate the existence of cylindrical micelles, to characterize the micellar structure using scattering experiments, and to use polymer theory in order to quantitatively describe micellar growth and interactions [3–16].

Despite the considerable experimental and theoretical attention given to the characterization and understanding of polymer-like micelles, we still lack detailed and quantitative information on the flexibility of the micelles as a function of composition, ionic strength or temperature. This is even more important as the Kuhn length b (note that $b = 2l_p$, where l_p is the persistence length) or bending modulus κ (which are related for one-dimensional objects via the thermal energy $k_B T$ through $b = 2\kappa/k_B T$) are key parameters in a more fundamental description of fluid membrane phases provided by the flexible surface model [17]. One of the major problems in the past had been the quantitative determination of b due to the possible influence of the intrinsically high polydispersity of the micelles and the intermicellar interaction effects that could easily interfere with the interpretation of various experimental data [18]. While the scattering data on local length scales are not significantly influenced by interaction effects, the situation is more complex on intermediate length scales. It had been demonstrated quite recently that the analysis of small-angle neutron scattering (SANS) data using newly developed numerical expressions for the scattering function of worm-like chains with excluded volume effects over an extended range of q-values provides a method for obtaining b with high precision, that is independent of polydispersity, and where intermicellar interaction effects can be accounted for [15, 16].

The effect of electrostatic interactions on the flexibility of polyelectrolytes has been the subject of intense experimental and theoretical investigations and resulted in highly controversial results [19–23]. We thus recently proposed to use mixed surfactant systems as suitable models for "equilibrium polyelectrolytes" [16]. The problem with most polyelectrolytes is their rather weak scattering power when performing SANS or small-angle X-ray scattering (SAXS) experiments, which makes it almost impossible to produce data with a sufficient accuracy over the required q range at low concentrations, where single-coil properties can still be resolved. This is clearly not the case for worm-like micelles, and we thus postulated that aqueous solutions of polymer-like micelles "doped" with ionic surfactants can serve as ideal model systems for "equilibrium polyelectrolytes" and help to clarify some of the open questions in the polyelectrolyte literature. Based on the data analysis approach developed recently for the interpretation of scattering data from semi-flexible micelles and polymers, the actual value of the measured Kuhn length can be determined with very high precision for a given system owing to the strong scattering of the micelles. This results in data of remarkable accuracy even at low concentrations, as has been demonstrated recently with solutions of worm-like micelles of the non-ionic surfactant $C_{16}E_6$ "doped" by adding a small amount of ionic surfactants [16].

Here we now present results from a systematic static light and SANS study of non-ionic micelles doped with small amounts of ionic surfactant in order to investigate the influence of the linear charge density and the ionic strength on micellar flexibility, growth, and intermicellar interactions. The static light scattering (SLS) study consisted of a determination of the apparent molar mass and radius of gyration depending on the surfactant concentration at different ionic strengths. This led to qualitative information about the micellar growth and flexibility. The SANS study, corresponding to the investigation of the higher q-range, had as an aim the quantitative evaluation of the effect of intermicellar interactions on the flexibility and local structure. Values of the persistence length were obtained by applying a non-linear least-square fit of a worm-like chain with excluded volume effects to the experimental data. In the semi-dilute regime, the effect of intermicellar electrostatic interactions were studied at different ionic strengths and concentrations.

Materials and methods

The surfactant hexaethylene glycol mono-N-hexadecyl ether ($C_{16}E_6$) was obtained from Nikkol (Tokyo), the ionic surfactant

1-hexadecanesulfonic acid ($C_{16}SO_3Na$) was purchased from TCI, and D_2O (99.9% isotopic purity) was delivered from Cambridge Isotope Laboratories. The samples were obtained by first dissolving both surfactants in D_2O (the $C_{16}E_6$ at 35 °C and the $C_{16}SO_3Na$ at 70 °C) and mixing them together at the required concentrations and ionic strengths. All scattering experiments were carried out at 35 °C. SLS experiments were done and analyzed as described in detail previously [16]. Measurements were performed with a commercial goniometer system (ALV/DLS/SLS-5000F monomode fiber compact goniometer system with ALV-5000 fast correlator). The instrument had been modified to allow for a much larger temperature range (−6 °C to +220 °C) and increased temperature stability (better than ±0.01 °C for several hours). The SANS experiments were performed at the instrument D22 of the ILL in Grenoble, France (dilute solutions), and at the SANS instrument at PSI, Switzerland (semi-dilute solutions). The initial data treatment and the data analysis were performed as described in [16].

Results and discussion

Static light scattering

Figure 1A and B summarizes the light scattering results obtained at three different ionic strengths and a doping level of 3% (weight ratio [$C_{16}SO_3Na$]/[$C_{16}E_6$] = 0.03). We observe two characteristic regimes which exhibit a very different concentration dependence. At low concentrations the dramatic increase of the apparent molar mass M_{app} with increasing values of c primarily reflects the pronounced concentration-induced micellar growth. The resulting micelles are extremely large, and interestingly no measurable ionic strength dependence can be observed for M_{app}, i.e., Fig. 1A clearly shows that the ionic strength has no influence on the micellar growth. Once the micellar size and concentration is large enough, the polymer-like structures overlap and start to entangle. At even higher concentrations the micelles can form an entanglement network, and the solution becomes visco-elastic with properties analogous to semi-dilute polymers. The entanglement threshold c^*, where the transition from the dilute to the semi-dilute regime occurs, is in reality not a sharp boundary but rather a concentration range which starts when the coils touch and ends when the coils are completely entangled [12]. This behavior is also reflected in the scattering data. M_{app} reaches a maximum at approximately the overlap concentration c^* and then decreases with a power law dependence on c. Under these conditions the scattering data are now insensitive to the size of the individual micelles but reflects the osmotic compressibility of the network only.

The results for the concentration dependence of the apparent radius of gyration $R_{g,app}$ shown in Fig. 1B exhibit a similar trend. At low concentrations, $R_{g,app}$ first increases, reflecting the concentration dependence of the micellar aggregation number. At higher concentrations, $R_{g,app}$ reaches a maximum at approximately c^* and decreases at higher values of c. Under these

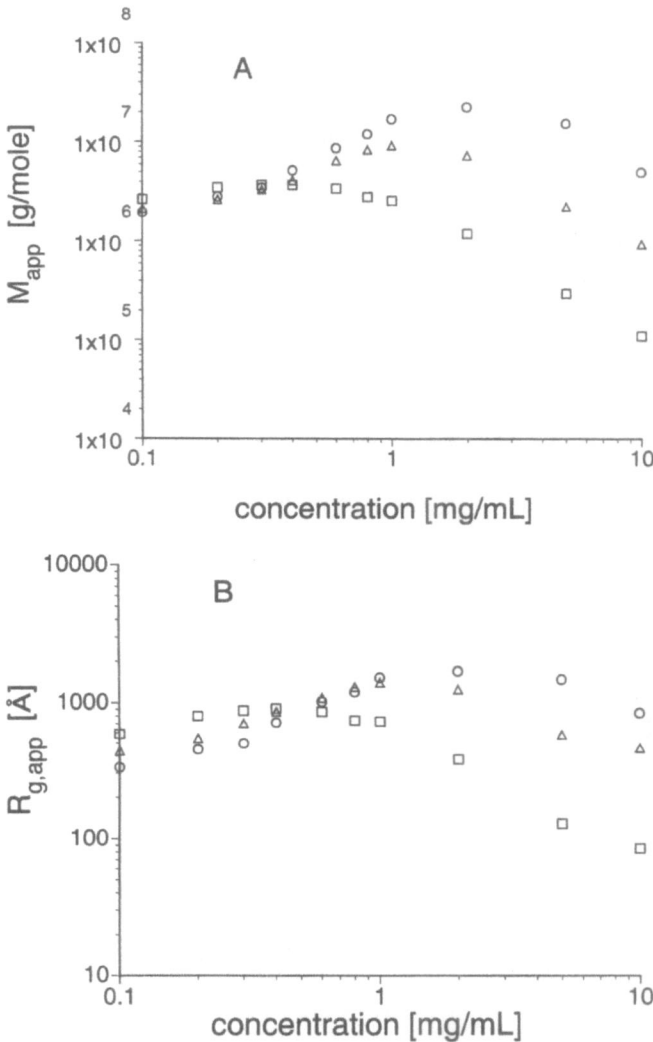

Fig. 1 Apparent molar mass $M_{w,app}$ (**A**) and radius of gyration $R_{g,app}$ (**B**) versus total surfactant concentration for a 3% doping level (weight ratio $[C_{16}SO_3Na]/[C_{16}E_6] = 0.03$)

field models for polyelectrolytes usually divide the total persistence length into an electrostatic $l_{p,el}$ and an intrinsic or "bare" part $l_{p,0}$ such that $l_{p,tot} = l_{p,el} + l_{p,0}$ [24]. Several theoretical models that account for the influence of electrostatic interactions on the persistence length and on the effect of screening from added salt have been presented in the past [19, 22]. A frequently used model to calculate $l_{p,el}$ is the so-called OSF theory which has independently been derived by Odijk, Skolnick and Fixman [19–21]. In this theory, electrostatic contributions to the bending energy are calculated for a rather stiff worm-like chain using a screened Coulomb potential (i.e., a Debye-Hückel approximation) for the electrostatic interactions.

These findings are also consistent with the ionic strength dependence of the overlap concentration. Figure 1A and B clearly shows that $c*$ decreases with decreasing ionic strength. This is due to the fact that the micellar coil size increases at lower ionic strength as a result of the increase in $l_{p,el}$, which correspondingly leads to a shift in $c* \sim M/R_g^3$ [12]. In addition, intermicellar interaction effects become increasingly long-range owing to the additional electrostatic repulsion at low ionic strength, which also leads to a subsequent reduction of the forward scattering intensity and eventually to the formation of a clear peak in the static structure factor $S(q)$ as shown below.

Small-angle neutron scattering

While the data shown in Fig. 1 allow for a qualitative estimate of the dependence of the micellar persistence length upon salt concentration, a quantitative analysis remains difficult owing to the combined effects of micellar polydispersity and intermicellar interactions [12, 16]. This can be done in a much more quantitative way using SANS, where one has access to the micellar structure on the length scale of l_p and is thus only weakly dependent on interaction and polydispersity effects [15, 16].

Figure 2A shows scattering curves at two ionic strengths and 3% doping level. The q dependence of the data is in qualitative and quantitative agreement with the polymer analogy. At low values of q (i.e. $1/q > R_g$), the scattered intensity $I(q)$ is insensitive to structural details and is dominated by the finite overall size of the particles, and one can determine $R_{g,app}$ and M_{app}. At intermediate q (cross section radius $R_c \ll 1/q \ll R_{g,app}$), $I(q)$ becomes much more sensitive to the coil conformation, and polymer theory predicts for flexible polymer coils with excluded volume effects that $I(q)$ should decay with a power law of the form $I(q) \sim q^{-1.66}$. At large values of q, $I(q)$ is controlled by distances over which polymers are rod-like rather than flexible, and we expect a crossover to an asymptotic q^{-1} dependence for

conditions, $R_{g,app}$ does not reflect the size of individual micelles anymore but corresponds to a static correlation length (with $\xi = R_{g,app}/3^{1/2}$), i.e., it reflects the mesh size of the entanglement network which decreases with increasing concentration. However, in contrast to the situation for the micellar mass, the ionic strength seems to play a major role in determining the micellar size. In the dilute regime and at fixed total surfactant concentration, $R_{g,app}$ as a measure of the micellar size strongly increases as the ionic strength decreases. Together with the finding that M_{app} is independent of the ionic strength, this indicates that the micellar flexibility decreases as a result of intramicellar electrostatic interactions. Such an effect is in agreement with the polyelectrolyte analogy, where the commonly used mean

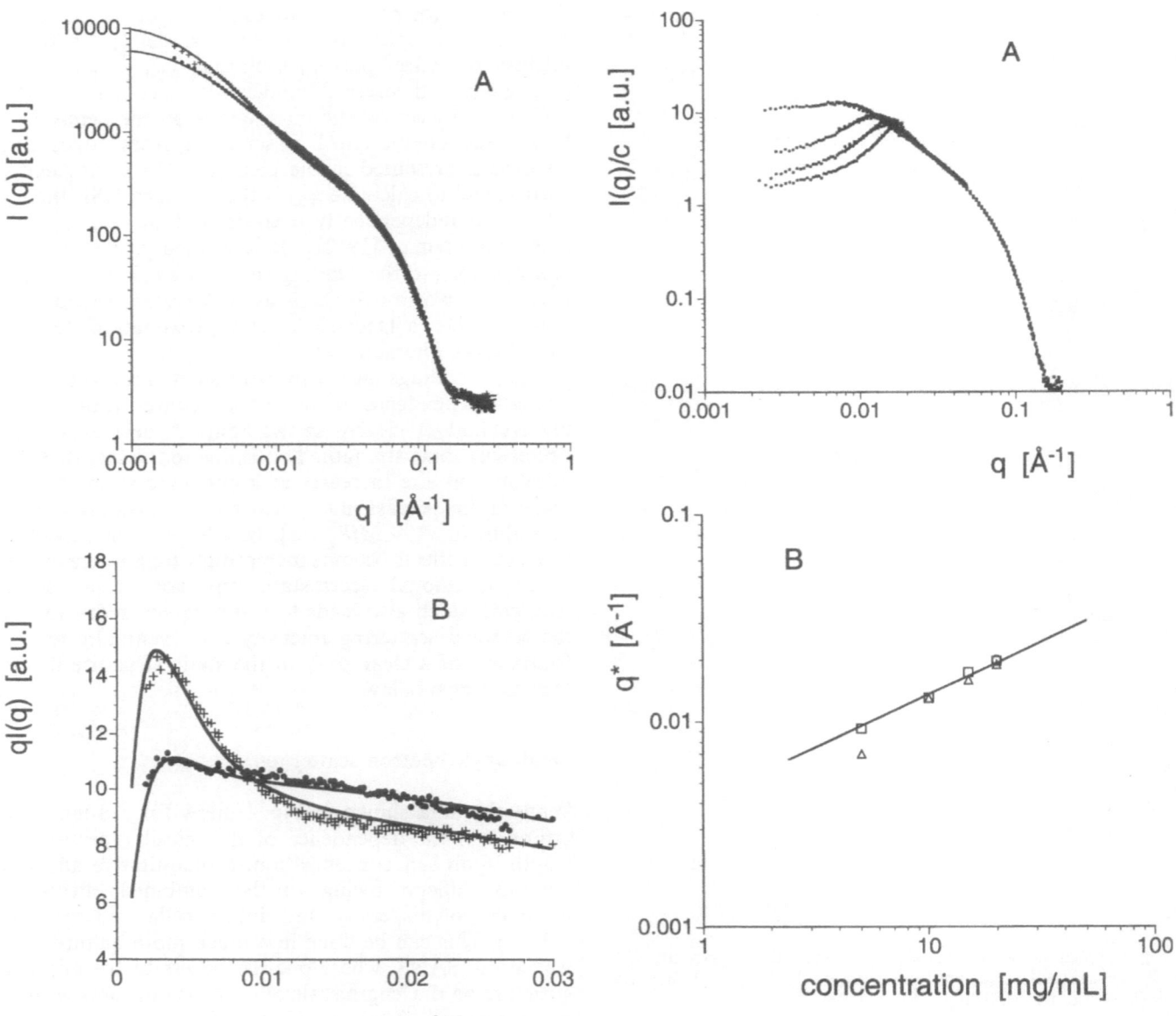

Fig. 2A, B The effect of the ionic strength on the q dependence of the scattered intensity from SANS experiments (+ 0.01 M NaCl; ● 0.001 M NaCl). The *solid lines* are fits based on the scattering function for semi-flexible polymers (see [25]). **A** Data in double logarithmic representation; **B** bending rod or Holtzer plot $qI(q)$ versus q

Fig. 3 A Formation of a peak in $S(q)$ at low salt ([NaCl] = 0.001 M) for doped non-ionic polymer-like micelles at different concentrations c = 5, 10, 15, and 20 mg/ml and **B** location q^* of the structure factor peak versus surfactant concentration for different doping levels [$C_{16}SO_3Na$]/[$C_{16}E_6$] = 0.03 ($\triangle$), 0.06 ($\square$) and 0.09 ($\bigcirc$). Also shown as the *solid line* is a power law of the form $c^{1/2}$

$I(q)$ which is typical for locally cylindrical structures. However, a real polymer is not an infinitely thin chain, and therefore the local cross-section structure of the chains gives rise to a cross-section Guinier behavior and a strong decrease in the scattering intensity at still larger q values. We qualitatively observe these features in the experimental curves shown in Fig. 2A, and the data are very well described by the scattering function for semi-flexible polymers with excluded volume effects given by

the solid lines [25]. In this comparison we have used a recently developed scattering function of the form

$$\frac{d\sigma}{d\Omega}(q) = c\Delta\rho_m^2 S_{wc}(q)S_{cs}(q)M_{app} \tag{1}$$

where $d\sigma/d\Omega$ is the normalized scattering intensity (differential cross section), c is the surfactant concentration (in weight per volume), $\Delta\rho_m$ is the average excess scattering length density per unit mass, and $S_{wc}(q)$ and

$S_{cs}(q)$ are the normalized scattering functions of the infinitely thin worm-like chain and of the cross section.

If we now look closer at the characteristic crossover region, we clearly see the effect of electrostatic interactions on the flexibility of the micelles. At 0.01 M ionic strength the crossover is quite pronounced and occurs at $q \sim 0.01$ nm^{-1}, and a fit of Eq. (1) to the data yields a value of $b_{app} = 42$ nm for the apparent Kuhn length. However, at 0.001 M salt the crossover is now shifted to much lower values of $q \sim 0.006$ nm^{-1}, and we obtain a value of $b_{app} = 81$ nm. The enormous effect of the salt concentration on the micellar structure becomes even more obvious when using a so-called Holtzer or bending rod plot given by $qI(q)$ versus q (Fig. 2B). This representation provides an extremely sensitive way to test the agreement between theoretical and experimental scattered intensity for semi-flexible chains as it significantly amplifies all deviations between data and theoretical curve in the crossover region [16]. Figure 2B clearly shows that at low salt content the data exhibit the typical scattering of rigid cylinders over almost the entire q range, and it is only at very low q where we see a weak crossover due to the much higher value of b_{app} caused by the intramicellar electrostatic interactions. We do in fact obtain very good agreement between measurements and Monte Carlo simulations of a single polyelectrolyte chain with the same linear charge density, which fully supports the polyelectrolyte analogy and indicates that doped worm-like micelles may indeed serve as ideal model systems to shed light on the controversial topic of the effect of electrostatic interactions on flexibility [26].

We have further exploited the micelle-polyelectrolyte analogy and studied the effect of intermicellar interactions on the solution structure. This is shown in Fig. 3A, where SANS data from measurements of four "doped" micellar solutions at low ionic strength ([NaCl] = 0.001 M) are presented. We clearly see the appearance of a well-defined structure factor peak at some finite scattering vector value q^*, which becomes more pronounced

and shifts to higher values of q at higher surfactant concentrations. This peak completely disappears at higher salt concentrations, where the electrostatic interactions are efficiently screened by the salt and the solution exhibits classical polymer behavior [26]. In Fig. 3B we have plotted the location of the peak q^* versus the surfactant concentration for different doping levels. We find that q^* follows a power law of the form $q^* \sim c^{1/2}$, i.e., we observe exactly the same behavior as reported for classical polyelectrolytes [27].

Conclusions

In this work we have further exploited the analogy between partially charged worm-like micelles and "equilibrium polyelectrolytes". Our data show that non-ionic worm-like micelles doped with small amounts of ionic surfactant represent a suitable model for investigating the static properties of polyelectrolytes. We have found that the polyelectrolyte model, and in particular Monte Carlo simulations, quantitatively reproduce the effects of intra- and intermicellar electrostatic interactions. These experiments open up very interesting possibilities to investigate the effect of electrostatic interactions on the micellar flexibility and solution structure. Owing to the high scattering power of the micelles when compared to classical polyelectrolytes, we can perform experiments both in the dilute and semi-dilute regime as a function of parameters such as the linear charge density or the ionic strength. We believe that this should have a significant impact on the current debate about polyelectrolyte properties.

Acknowledgements The support by the Swiss National Science Foundation (grants 20-53381.98, 20-46627.96) is gratefully acknowledged. The neutron scattering experiments were performed at the instrument D22 of the Institute Laue-Langevin in Grenoble, France, and at the SANS instrument of the Paul Scherrer Institute in Switzerland.

References

1. Cates ME, Candau SJ (1990) J Phys Condens Matter 2:6869–6892
2. Schurtenberger P, Cavaco C (1994) J Phys Chem 98:5481–5486
3. Lin T-L, Chen S-H, Gabriel NE, Roberts MF (1987) J Phys Chem 91:406–413
4. Marignan J, Appell J, Bassereau P, Porte G, May RP (1989) J Phys (Paris) 50:3553–3566
5. Hjelm RP, Thiyagarajan P, Alkan H (1988) J Appl Crystallogr 21:858–863
6. Hjelm RP, Thiyagarajan P, Sivia DS, Lindner P, Alkan H, Schwahn D (1990) Prog Colloid Polym Sci 81:225–231
7. Hjelm RP, Thiyagarajan P, Alkan-Onyuksel H (1992) J Phys Chem 96:8653–8661
8. Schurtenberger P, Scartazzini R, Magid LJ, Leser ME, Luisi PL (1990) J Phys Chem 94:3695–3701
9. Schurtenberger P, Magid LJ, King S, Lindner P (1991) J Phys Chem 95:4173–4176
10. Long MA, Kaler EW, Lee SP, Wignall GD (1994) J Phys Chem 98:4402–4410
11. Pedersen JS, Egelhaaf SU, Schurtenberger P (1995) J Phys Chem 99:1299–1305
12. Schurtenberger P, Cavaco C (1994) Langmuir 10:100–108
13. Schurtenberger P, Jerke G, Cavaco C, Pedersen JS (1996) Langmuir 12:2433–2440
14. Schurtenberger P, Cavaco C, Tiberg F, Regev O (1996) Langmuir 12:2894–2899
15. Jerke G, Pedersen JS, Egelhaaf SU, Schurtenberger P (1997) Phys Rev E 56:5772–5788

352

16. Jerke G, Pedersen JS, Egelhaaf SU, Schurtenberger P (1998) Langmuir 14:6013–6024
17. Safran SA (1992) In: Chen SH, Huang JS, Tartaglia P (eds) Structure and dynamics of strongly interacting colloids and supramolecular aggregates in solution. Kluwer, Dordrecht, pp 237–264
18. Schurtenberger P (1996) Curr Opin Colloid Interface Sci 1:773–778
19. Odijk T (1977) J Polym Sci Polym Phys Ed 15:477
20. Skolnick J, Fixmann M (1977) Macromolecules 10:944
21. Fixman M, Skolnick J (1978) Macromolecules 11:863
22. Stevens M, Kremer K (1995) J Chem Phys 103:1669
23. Barrat JL, Joanny JF (1995) Adv Chem Phys 94:1
24. Dautzenberg H et al (1994) Polyelectrolytes: formation, characterization and application. Hanser, München
25. Pedersen JS, Schurtenberger P (1996) Macromolecules 29:7602–7612
26. Sommer C, Cannavacciuolo L, Egelhaaf SU, Pedersen JS, Schurtenberger P (1999) (manuscript in preparation)
27. Morfin I, Reed WF, Rinaudo M, Borsali R (1994) J Phys II 4:1001–1019

Progr Colloid Polym Sci (2000) 115:353–356
© Springer-Verlag 2000

A. Borštnik
H. Stark
S. Žumer

Temperature-induced flocculation of colloidal particles above the nematic-isotropic phase transition

A. Borštnik · S. Žumer
Department of Physics
University of Ljubljana
1000 Ljubljana, Slovenia

H. Stark (✉)
Institut für Theoretische und
Angewandte Physik
Universität Stuttgart, 70550 Germany
e-mail: holger@itap.physik.uni-stuttgart.de
Tel.: +49-711-685 5266
Fax: +49-711-685 5271

Abstract In nematic liquid crystals, rod-like organic molecules align on average parallel to each other, exhibiting a long-range orientational order. Even above the nematic-isotropic phase transition, bounding surfaces induce a liquid crystalline order in the otherwise isotropic liquid. We demonstrate that such an order gives rise to a novel colloidal interaction whose strength can be controlled by temperature. Close to the transition to the nematic phase, the interaction is strongly attractive at a length scale of the order of 10 nm. For electrostatically stabilized colloidal dispersions, we demonstrate that the liquid crystal mediated interaction induces a flocculation of the particles close to the transition temperature reminiscent to the same effect in polymer stabilized dispersions.

Key words Liquid crystals · Colloidal interaction · Flocculation · Wetting

Introduction

The stability of a colloidal dispersion presents a key issue in colloid science since its characteristics change markedly in the transition from the dispersed to the aggregated state [1]. There are always attractive van der Waals forces, which have to be balanced by repulsive interactions to prevent a dispersion of particles from aggregating. This is achieved either by electrostatic repulsion, where the particles carry a surface charge, or by steric stabilization, where they are coated with a soluble polymer brush.

This article introduces a novel colloidal interaction which occurs when particles are dispersed in a nematic liquid crystal slightly above the nematic-isotropic phase transition. In nematic liquid crystals [2], rod-like organic molecules align on average parallel to each other, exhibiting a long-range orientational order. The average direction of the molecules is given by a unit vector n called the director. Furthermore, the Maier-Saupe order parameter S indicates how well the molecules are aligned. It is, for example, proportional to the magnetic anisotropy $\Delta\chi = \chi_\parallel - \chi_\perp$ where $\chi_\parallel$ and $\chi_\perp$ are the respective magnetic susceptibilities parallel and perpendicular to the director. Even above the nematic-isotropic phase transition, bounding surfaces induce a liquid crystalline order ($S \neq 0$) in the otherwise isotropic liquid [3–6]. Its effect on colloidal dispersions is explored in the following (for a treatment in a more general context, see [7]).

Our study was initiated by a recent investigation of nematic emulsions which revealed a long-range colloidal interaction of dipolar type due to the elastic distortion of the director field in the space between the particles [8–10]. Furthermore, a short-range repulsion due to topological defects in the director field was identified [8–10].

A novel colloidal interaction

To treat the surface-induced liquid crystalline order, we employ the Landau-Ginzburg-de Gennes free energy [11]. In the one-constant approximation it takes the form

$$F_{LG} = \int d^3r \left[\frac{3}{4} a_0 (T - T^*) S^2 - \frac{1}{4} bS^3 + \frac{9}{16} cS^4 \right.$$
$$\left. + \frac{3}{4} (\nabla_i S)^2 + \frac{9}{4} (\nabla_i n_j)^2 \right] \qquad (1)$$

The first line is a typical Landau-type free energy. It describes the first-order phase transition from the isotropic to the nematic phase which takes place at a temperature $T_c = T^* + b^2/(27 a_0 c) > T^*$. The second line penalizes distortions in the director field $\boldsymbol{n}(\boldsymbol{r})$ and inhomogeneities in the nematic order paramter $S(\boldsymbol{r})$. To simplify our calculations as much as possible and to illustrate the main features of our problem, we will neglect the non-harmonic terms in Eq. (1). Their effects on the nematic wetting layer is thoroughly studied in [3–6]. Furthermore, we use a surface free energy F_{sur} which forces the liquid crystal molecules to align perpendicular to a bounding surface with $S \approx 0.3$. For a given geometry, the liquid crystalline order then follows from a minimization of the total free energy $F = F_{LG} + F_{sur}$.

A single plate immersed into a nematic liquid crystal above T_c induces a liquid cystalline wetting layer of finite thickness around the plate. The order parameter S decays exponentially into the bulk liquid on a length scale given by the nematic coherence length $\xi_N = \sqrt{L_1/[a_0(T - T^*)]}$. At T_c, $\xi_{NI} = \xi_N(T_c)$ is of the order of 10 nm. Two parallel plates do not interact for separations $d \gg \xi_N$; the interaction energy per unit area, $\Delta F/A = [F(d) - F(d \to \infty)]/A$, decays exponentially in d/ξ_N. However, at separations $d \approx 2\xi_N$, where the nematic layers of the plates start to overlap, a strong attraction sets in (see Fig. 1). It can be understood by a simple argument. Above T_c, the nematic order always possesses higher energy than the isotropic liquid. Therefore, the system can reduce its free energy by reducing the total volume of nematic order when the plates are moved together. The minimum of the interaction energy occurs at $d = 0$ when the liquid with nematic order between the plates is completely removed. This simple argument explains the deep potential well in Fig. 1. It extends to a separation of $2\xi_N$ where the nematic layers start to overlap. Since $\xi_N \propto (T - T^*)^{-1/2}$, the range of the interaction decreases with increasing temperature, and the depth of the potential well becomes smaller.

One spherical particle, suspended in a liquid crystal above the clearing temperature T_c, is again surrounded by a layer of surface-induced nematic order. Its thickness is of the order of the nematic coherence length ξ_N. The director field points radially outward when a perpendicular anchoring of the molecules at the particle surface is assumed. Two particles with a separation $d \gg 2\xi_N$ do not interact. When the separation is reduced to $d \approx 2\xi_N$, a strong attraction sets in for the same reason as in the case of two plates (see Fig. 2). In addition, a repulsion due to the elastic distortion of the director field lines connecting the two particles occurs.

To quantify the two-particle interaction, we modeled the liquid crystalline order around the two particles. This procedure enabled us to calculate the interaction energy U_{LC} analytically [12]. The results for 250 nm particles and the liquid crystal compound 8CB [13] are illustrated in Fig. 3. In the inset we plot the attractive and repulsive contribution to U_{LC} in units of the thermal energy $k_B T$ as a function of the particle separation d. The temperature is T_c. The total interaction energy exhibits a deep potential well with an approximate width of $2\xi_{NI} = 20$nm. At larger separations it is followed by a weak repulsive barrier whose height is approximately $1.5 k_B T$. If $d \gg 2\xi_N$, U_{LC} decays exponentially, $U_{LC} \propto \exp(-d/\xi_N)$. Figure 3 illustrates further that the

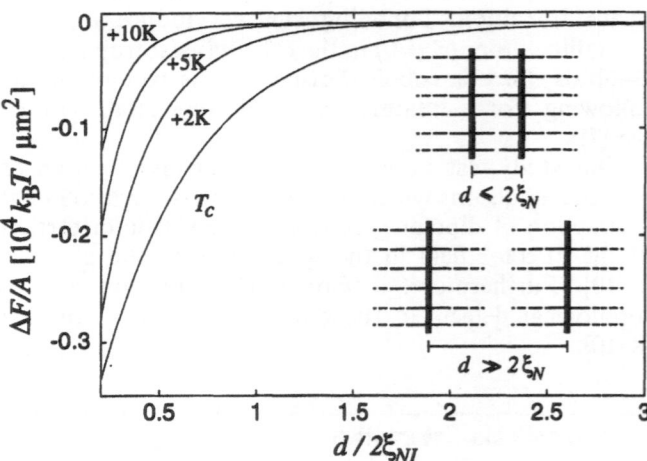

Fig. 1 Interaction energy per unit area, $\Delta F/A$, as a function of the reduced distance $d/2\xi_{NI}$ for various temperatures. For further explanation, see text

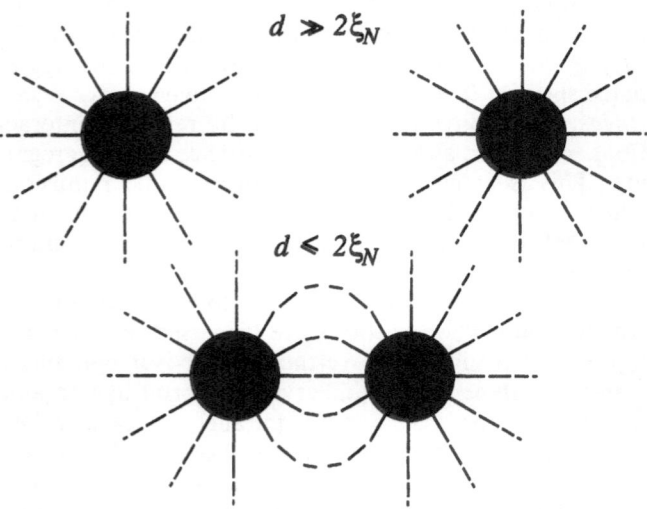

Fig. 2 Two particles at a separation $d \gg 2\xi_N$ do not interact. At $d \approx 2\xi_N$, both a strong attraction and repulsion set in

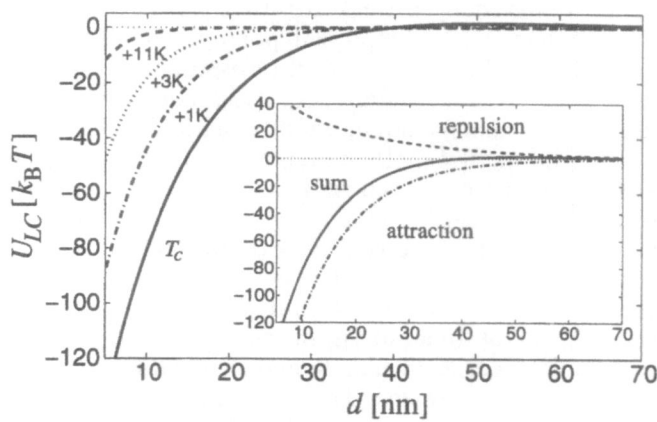

Fig. 3 The liquid crystal mediated interaction U_{LC} in units of $k_B T$ as a function of the particle separation d. The interaction is shown at T_c, $T_c + 1\,K$, $T_c + 3\,K$, and $T_c + 11\,K$. It strongly depends on temperature. *Inset*: U_{LC} is composed of an attractive and repulsive part. A weak repulsive barrier occurs at $d \approx 60$ nm

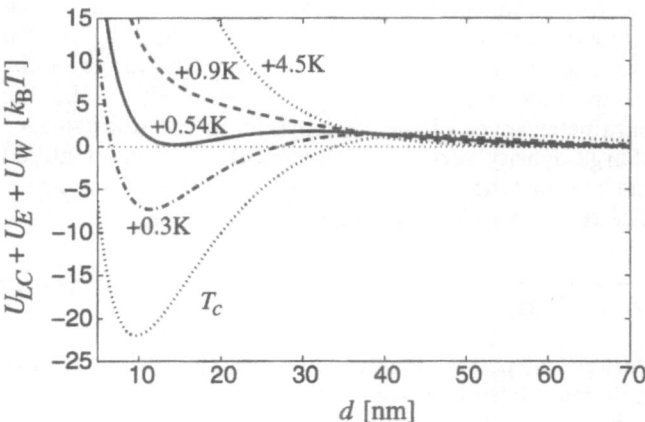

Fig. 4 The total two-particle interaction, $U_{LC} + U_E + U_W$, as a function of particle separation d for various temperatures. A complete flocculation of the particles occurs within a temperature range of about 0.3 K. $q_s = 0.5 \times 10^4\, e_0/\mu m^2$, $\kappa^{-1} = 8.3$ nm, and $A = 1.1\, k_B T$

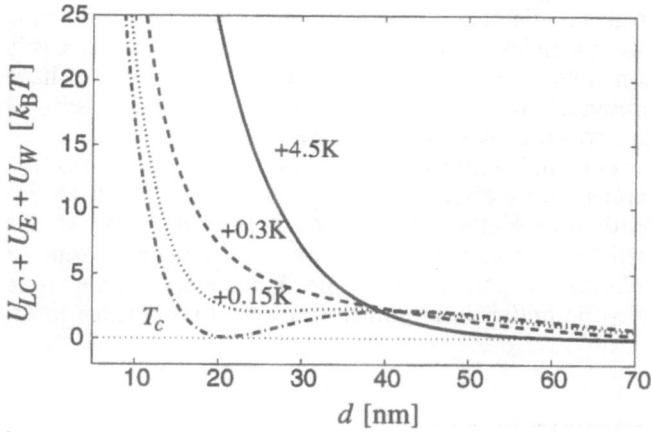

Fig. 5 In comparison to Fig. 4, the surface-charge density is increased to $0.63 \times 10^4\, e_0/\mu m^2$. As a result, flocculation does not occur

depth of the potential well decreases considerably when the dispersion is heated by several Kelvin. That means the liquid crystal mediated interaction can easily be controlled by temperature. It is turned off by heating the dispersion well above T_c.

Flocculation versus dispersion of particles

To study the effect of the liquid crystal mediated interaction on a dispersion of particles, we take into account the van der Waals attraction [1]:

$$U_W = -\frac{A}{6}\left[\frac{2a^2}{d(d+4a)} + \frac{2a^2}{(d+2a)^2} + \ln\frac{d(d+4a)}{(d+2a)^2}\right] \quad (2)$$

where the Hamaker constant for silica particles dispersed in a typical liquid crystal compound amounts to $A = 1.1\, k_B T$ [14]. We stabilize the colloidal dispersion by employing an electrostatic repulsion. The two-particle potential is derived in the Derjaguin approximation for constant surface charge density q_s [1]:

$$U_E = -\pi k_B T \frac{a q_s^2}{z^2 e_0^2 n_p} \ln(1 - e^{-\kappa d}) \quad (3)$$

where e_0 is the fundamental charge, and z is the valence of the ions in the solvent, which have a concentration n_p. The range of the repulsive interaction is determined by the *Debye length* κ^{-1}.

In Fig. 4 we plot the total two-particle interaction $U_{LC} + U_E + U_W$ as a function of particle separation d for various temperatures. We choose $q_s = 0.5 \times 10^4\, e_0/\mu m^2$ and $\kappa^{-1} = 8.3$ nm. At 4.5 K above the transition temperature T_c, the dispersion is stable. With decreasing temperature, a potential minimum at finite separation

develops. At $T_{FD} = T_c + 0.54\,K$, a particle doublet or the aggregated state becomes energetically preferred. We call T_{FD} the temperature of *flocculation transition*. Below T_{FD}, the probability of finding the particles in the aggregated state is larger than the probability that they are dispersed. Already at $T_c + 0.3\,K$ the minimum is $7\,k_B T$ deep, and all particles are condensed in aggregates. This means that within a temperature range of about 0.3 K there is an abrupt change from a completely dispersed to a fully aggregated system, reminiscent to the critical flocculation transition in colloidal dispersions employing polymeric stabilization [1]. Between $d = 30$ nm and 50 nm the two-particle interaction exhibits a small repulsive barrier of about $1.5\,k_B T$. It slows down the doublet formation by a factor of three, i.e., the dynamics of aggregation would not change very dramatically if the barrier were zero. If the surface-charge density q_s is increased to $0.63 \times 10^4\, e_0/\mu m^2$, the dispersed state is thermodynamically stable at all temperatures above T_c, as illustrated in Fig. 5. An increase of the Debye length κ^{-1}, i.e., the range of the electrostatic repulsion, has the same effect. In the parameter space of the electrostatic interaction (surface charge density versus Debye length), we have identified the region where we expect the flocculation to occur. The results will be published elsewhere [14].

Conclusions

Particles dispersed in a liquid crystal above the nematic-isotropic phase transition are surrounded by a surface-induced nematic layer whose thickness is of the order of the nematic coherence length. The particles experience a strong liquid crystal mediated attraction when their nematic layers start to overlap since then the effective volume of liquid crystalline ordering and therefore the free energy is reduced. A repulsive correction results from the distortion of the director field lines connecting two particles. The new colloidal interaction is easily controlled by temperature. In this article we have presented how it can be probed with the help of electrostatically stabilized dispersions.

For sufficiently weak and short-ranged electrostatic repulsion, we observe a sudden flocculation within a few tenth of a Kelvin close to T_c. It is reminiscent to the critical flocculation transition in polymer stabilized colloidal dispersions [1]. The flocculation is due to a deep potential minimum in the total two-particle inter-action followed by a weak repulsive barrier. Thermotropic liquid crystals represent polar organic solvents, and one could wonder if electrostatic repulsion is realizable in such systems. In [15], complex salt is dissolved in nematic liquid crystals and ionic concentrations of up to 10^{-4} mol/l are reported which give rise to Debye lengths employed in this article. Furthermore, when silica spheres are coated with silanamine, the ionogenic group ($= NH_2^+OH^-$) occurs at the particle surface with a density of 3×10^6 molecules/μm^2. It dissociates to a large amount in a liquid crystal compound [16]. In addition, the silane coating provides the required perpendicular boundary condition for the liquid crystal molecules. These two examples illustrate that electrostatic repulsion should be accessible in conventional thermotropic liquid crystals, and we hope to initiate experimental studies which probe the new colloidal force. Our work directly applies to lyotropic liquid crystals [2], i.e., aqueous solutions of non-spherical micelles, when the nematic-isotropic phase transition is controlled by temperature [17, 18]. They are appealing systems since electrostatic stabilization is more easily achieved. When the phase transition is controlled by the micelle concentration ρ_m, as it is usually done, then our diagrams are still valid but with temperature replaced by ρ_m.

In polymer-stabilized dispersions we find that the aggregation of particles sets in gradually when cooling the dispersion down towards T_c. This is in contrast to electrostatic stabilization, where flocculation occurs in a very narrow temperature interval [14].

References

1. Russel WB, Saville DA, Schowalter WR (1995) Colloidal dispersions. Cambridge University Press, Cambridge
2. de Gennes PG, Prost J (1993) The physics of liquid crystals, 2nd edn. Oxford Science Publications, Oxford
3. Sheng P (1976) Phys Rev Lett 37:1059
4. Sheng P (1982) Phys Rev A 26:1610
5. Poniewierski A, Sluckin TJ (1987) Liq Cryst 2:281
6. Borštnik A, Žumer S (1997) Phys Rev E 56:3021
7. Löwen H (1995) Phys Rev Lett 74:1028
8. Poulin P, Stark H, Lubensky T, Weitz D (1997) Science 275:1770
9. Lubensky TC, Pettey D, Currier N, Stark H (1998) Phys Rev E 57:610
10. Stark H (1999) Eur Phys J B 10:311
11. de Gennes PG (1971) Mol Cryst Liq Cryst 12:193
12. Borštnik A, Stark H, Žumer S (1999) Phys Rev E 60:4210
13. Coles HJ (1978) Mol Cryst Liq Cryst Lett 49:67
14. Borštnik A, Stark H, Žumer S (2000) Phys Rev E 61:(in press)
15. Haller I, Young WR, Gladstone G, Teaney DT (1973) Mol Cryst Liq Cryst 24:249
16. Dozov I, personal communication
17. Poulin P, Raghunathan VA, Richetti P, Roux D (1994) J Phys I 4:1557
18. Poulin P, Frances N, Mondain-Monval O (1999) Phys Rev E 59:4384

Progr Colloid Polym Sci (2000) 115 : 357–360
© Springer-Verlag 2000

S. O'Neill
D. Fitzmaurice

Preparation and characterization of size monodisperse TiO$_2$ nanocrystals and their self-assembled two- and three-dimensional arrays

S. O'Neill · D. Fitzmaurice (✉)
Department of Chemistry
University College Dublin, Belfield
Dublin 4, Ireland

Abstract Colloidal TiO$_2$ nanocrystals with average diameter of 6 nm were prepared by hydrolysis of titanium tetraisoproxide in an acidified aqueous solution. Catechol-terminated long-chain alkanes were chemisorbed at the surface of the TiO$_2$ nanocrystals, which were subsequently recovered by centrifugation and redispersed in chloroform. Size-selective precipitation was used to isolate size-monodisperse nanocrystal fractions. Elemental analysis, ^{1}H NMR and FTIR were used to characterize the monolayer of catechol-terminated long-chain alkanes adsorbed at the surface of these nanocrystals. Small-angle X-ray scattering and transmission electron microscopy were used to probe the structure of both the individual nanocrystals and their self-assembled two- and three-dimensional arrays.

Key words Titanium dioxide nanocrystal · Surface modification · Size selective precipitation · Self-assembled aggregates

Introduction

Nanocrystalline metal oxide materials have demonstrated their use, or have potential applications, in many technologies, including solar energy conversion [1–3], batteries [4], catalysis [5, 6], and ductile ceramics [7, 8]. Nanocrystalline titanium dioxide has been the subject of particular scientific and technological attention in these respects.

The performance of nanocrystalline metal oxides in general, and nanocrystalline titanium dioxide in particular, in applications such as these, may be optimized by precisely controlling the morphology of the material, and the properties of the constituent nanocrystals. Precisely controlling the surface chemistry of nanocrystals and their functionalization through the attachment of organic ligands may also provide opportunities to optimize their properties.

One approach to controlling both the morphology and properties of a nanostructured metal oxide material is the following: to prepare size- and shape-monodisperse lyophobic nanocrystals; to disperse these nanocrystals in a volatile lyophilic solvent; and to self-assemble ordered two- and three-dimensional nanocrystal arrays by sol-vent evaporation [9–12]. While this approach has been successfully demonstrated for a variety of materials, including Ag [12], CdSe [10], and γ-Fe$_2$O$_3$ [11], it has not been demonstrated for TiO$_2$. The reason for this is the difficulties encountered to date in preparing size- and shape-monodisperse lyophobized nanocrystals.

In this paper we report the preparation of TiO$_2$ nanocrystals stabilized by chemisorbtion of long-chain alkanes. We also report the size-selective precipitation of these nanocrystals and their self-assembly by solvent evaporation into a two- and three-dimensional array possessing short-range order.

Materials and methods

Preparation of TiO$_2$ nanocrystals

Nitric acid (2.6 mL) was added to distilled, deionized water (375 mL) in a 1 L flask and stirred vigorously. Titanium tetraisopropoxide (62.5 mL) was added dropwise to the solution over 120 s. The mixture was stirred for 2 h at room temperature and 8 h at 80 °C. Analysis was by transmission electron microscopy (TEM) and optical absorption spectroscopy.

Synthesis of capping ligand

Dodecyl 3,4-dihydroxy benzoate (I) was used as a capping ligand in the course of studies reported. Preparation of I is as shown in Scheme 1. Calculated for I ($C_{19}H_{30}O_4$): C 70.77, H 9.38. Found: C 70.77, H 9.35. ^{1}H NMR (chloroform-d): δ 0.83 (t, 3H); δ 1.22 (unresolved multiplet, 18H); δ 1.68 (quintet, 2H); δ 4.18 (t, 2H); δ 6.84 (d, 1H); δ 7.46 (d, 1H); δ 7.52 (s, 1H); δ 7.76 (s, 1H); δ 8.00 (s, 1H).

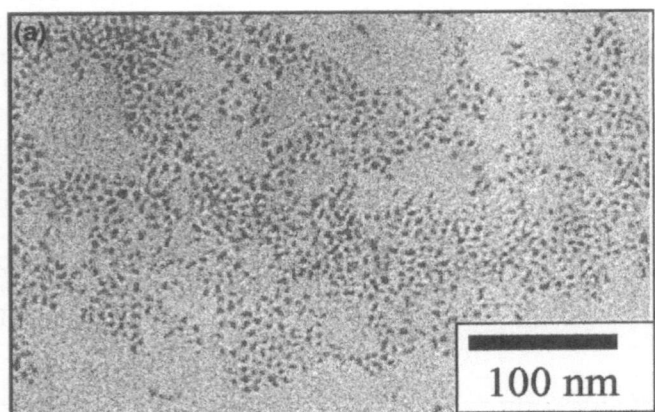

Scheme 1 Synthesis of dodecyl 3,4-dihydroxy benzoate

Capping of titanium dioxide nanocrystals

The above sol was found to contain 52.8 g L^{-1} of TiO_2. The amount of I required to cover the surface of a TiO_2 nanocrystal was calculated assuming each molecule occupies 24 $Å^2$ [13] of the available surface area. A ten-fold excess of capping ligand (700 mg of I in 8 mL EtOH), over that required for monolayer coverage, was added with vigorous stirring, over 1 h to a dispersion of TiO_2 nanocrystals (100 mg of TiO_2 in 1.9 ML). As a consequence of chemisorption of catechol at the surface of a TiO_2 crystal, the initially white-opaque dispersion develops a deep orange color. This orange color is assigned to a ligand-to-metal charge transfer transition. The solution was left to stir for 24 h, when the capped particles precipitated out. The solution was centrifuged at 10,000 rpm for 10 min. The plug of capped particles was redispersed in chloroform. Analysis was by TEM, optical absorption spectroscopy and elemental analysis.

Size-selective precipitation of capped TiO_2 nanocrystals

Capped TiO_2 nanocrystals, prepared as described above, were redispersed in a minimum volume of chloroform, giving a clear orange solution. Size-selective precipitation was performed using ethanol as the non-solvent. The addition was carried out dropwise until the solution turned slightly opaque. At this point the dispersion was centrifuged for 10 min at 6000 rpm. The plug was retained and constituted the first size-selected fraction. More ethanol was added to the recovered supernatant until once again the dispersion looked slightly opaque. It was then centrifuged again. This procedure was repeated to obtain a series of precipitates isolating nanocrystals with decreasing size and increasing monodispersity. Analysis of the precipitates was by TEM, optical absorption spectroscopy and small-angle X-ray scattering (SAXS).

Physical characterization

TEM employing a JEOL JEL-2000 EX electron microscope with a 80 kV accelerating voltage was used to characterize nanocrystal monolayers deposited on carbon-coated copper grids.

SAXS measurements were performed on beamline ID-1 at the European Synchrotron Radiation Facility (ESRF), Grenoble, France. SAXS measurements were collected for titanium dioxide nanocrystal thin films, formed by casting a dispersion of nanocrystals in chloroform on a mica substrate.

Results and discussion

Transmission electron microscopy

Following size-selective precipitation, TiO_2 nanocrystals capped with I were redissolved in a minimum amount of chloroform. The resulting dispersion was diluted 50-fold by addition of chloroform and a drop placed on a carbon-coated copper TEM grid. As may be seen from the corresponding TEM (Fig. 1a), solvent evaporation is accompanied by the self-assembly of a nanocrystal monolayer exhibiting some degree of order. Separation of the nanocrystals from each other is a consequence of the presence of the capping ligand. The particles are prolate in shape. The average nanocrystal dimensions were determined by inspection of the above and similar TEM images; 100 nanocrystals were sized. Successive precipitates were found to decrease in average nanocrys-

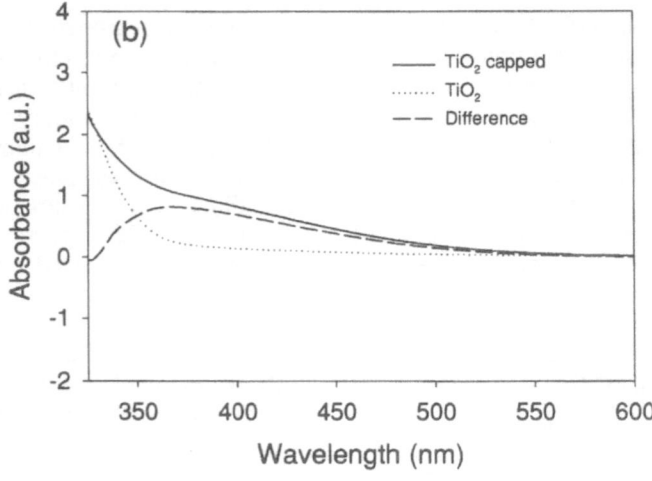

Fig. 1 a Transmission electron micrograph of TiO_2 nanocrystals after size-selective precipitation (13th fraction). **b** Optical absorption spectra of capped and uncapped TiO_2 and the difference of these two spectra

tal size. For the nanocrystals shown (Fig. 1a, 13th fraction) the average lengths determined for the long and short axes were 48 Å and 30 Å, respectively.

Optical absorption spectroscopy

The optical absorption spectrum of the size-selectively precipitated TiO$_2$ nanocrystals (Fig. 1a) shows a characteristic band-edge absorption on which is superimposed an absorption extending to longer wavelengths. This long-wavelength absorption (Fig. 1b), the difference of the spectra measured for capped and uncapped nanocrystals, is similar to that previously reported for TiO$_2$ nanocrystals modified by chemisorption of 1,2-dihdroxybenzene. Specifically, both Moser et al. [14] and Rogriduez et al. [15] observed an absorption centered at 420 nm and assigned this absorption to chemisorbed 1,2-dihydroxybenzene molecules chelated to surface Ti^{4+} atoms. On this basis, the band at 420 nm in Fig. 1b is also assigned to the CT absorption band that results from chelation of a Ti^{4+} site on the surface of the TiO$_2$ nanocrystal by I.

^{1}H Nuclear magnetic resonance

A ^{1}H NMR spectrum of the TiO$_2$ nanocrystals capped by chemisorbed I in chloroform-d confirms that all of I present is adsorbed at the surface of a TiO$_2$ nanocrystal. Specifically, the hydroxyl proton resonances [δ 8.00 (s, 1H) and δ 7.76 (s, 1H)], the aromatic proton resonances [δ 7.52 (s, 1H), δ 7.46 (d, 1H) and δ 6.84 (d, 1H)], and the α-CH$_2$ proton resonance [δ 4.51 (t, 2H)] are not observed. Furthermore, the resonances assigned to the β-CH$_2$ proton resonance [δ 1.68 (quintet, 2H)], the remaining methylene protons [δ 1.25 (unresolved multiplet, 18H)], and the terminal methyl group protons [δ 0.89 (t, 3H)] are all significantly broadened, as shown in Fig. 2. These findings are characteristic of alkanes adsorbed at a nanocrystal surface [16].

Elemental analysis

Elemental analysis gives the mass ratio of organic to metal which, when taken with the average surface area of the nanocrystals determined from TEM, yields the number of I adsorbed at the surface of each TiO$_2$ nanocrystal. On this basis, the area occupied by per adsorbed I is approximately 42 Å^2, implying catechol molecules are not close packed (24 Å^2 [13]) on the TiO$_2$ surface. It is likely the catechol coverage is limited by the number of Ti^{4+} sites on the surface of the nanocrystals at which these molecules are chelated. However, the coverage achieved of 2.36×10^{14} molecules cm^{-2} is sufficient for our purposes.

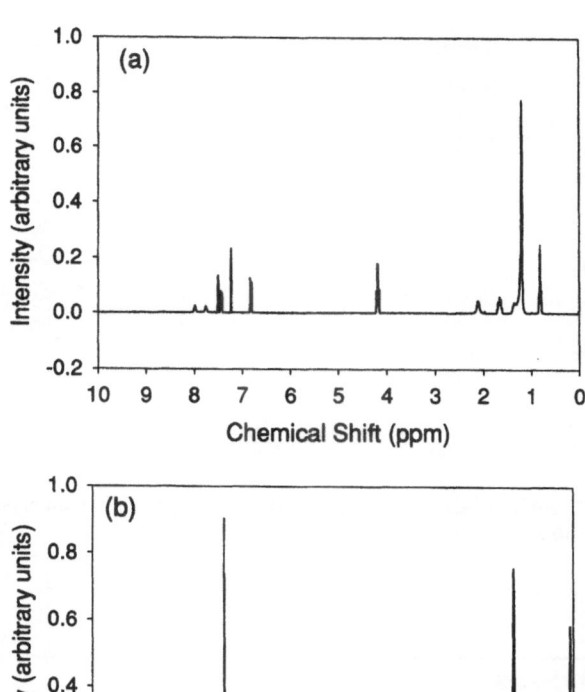

Fig. 2 a ^{1}H NMR of dodecyl 3,4-dihydroxybenzoate (I) in chloroform-d. **b** ^{1}H NMR of TiO$_2$ nanocrystals capped with I and dispersed in chloroform-d

Small-angle X-ray scattering

A series of five capped TiO$_2$ nanocrystal fractions with differing average nanocrystal dimensions were isolated by successive size-selective precipitations. The nanocrystals were cast onto mica substrates. The broad first-order diffraction peak shifts to higher scattering vector (q), implying a decrease in average nanocrystal dimensions (separation) as expected. The average interparticle (center to center) distance determined from this broadened peak for the 13th size selective precipitate was 44 Å (Fig. 3). This agrees well with the dimensions determined for this fraction from TEM, namely 48 Å and 30 Å. The breadth of the diffractive scattering peak further confirms the assertion that ordering is short-range.

Conclusions

Having undertaken a detailed characterization of the titanium dioxide nanocrystals prepared as described above, it is possible to say the following: firstly, the

360

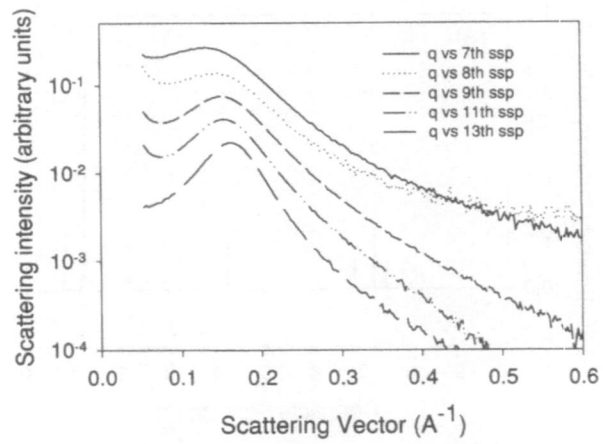

Fig. 3 SAXS of size-selective precipitates of capped TiO_2 nanocrystals (7–13th fractions) cast on a mica surface

TiO_2 particles are elliptical in shape; secondly, the long-chain alkanes terminated with a catechol moiety are chemisorbed at the surface of the TiO_2 nanocrystal; thirdly, coverage of these catechol molecules on the surface of the nanocrystals is approximately 2.36×10^{14} molecules cm^{-2}, indicating that the molecules are not close packed; in fact it is likely coverage is limited by the number of defect sites on the surface of the nanocrystal; fourthly, the nanocrystals assemble in to a monolayer and are separated from each other owing to the presence of the capping ligand; fifthly, size-selective precipitation yields fractions of nanocrystals decreasing in average nanocrystal size which is indicated both by TEM and SAXS.

Acknowledgements The authors thank the following: Dr. H. Rensmo and S. Quinn of the Nanochemistry Group; Drs. D. Cottell and B. Cregg of the Electron Microscopy Centre at University College Dublin; and Dr. P. Boesecke of the European Synchrotron Radiation Facility, Grenoble.

References

1. O'Regan B, Gratzel M (1991) Nature 353:737
2. Li W, Osora H, Otero L, Duncan DC, Fox MA (1998) J Phys Chem 102:5333
3. Bedja I, Kamat PV, Lapin AG, Hotchandani S (1997) Langmuir 13:2398
4. Kavan L, Kratochvilova K, Gratzel M (1995) J Electoanal Chem 394:93
5. Hagfeldt A, Gratzel M (1995) Chem. Rev 95:49
6. Moser WR (ed) (1996) Advanced catalysts and nanostructured materials. Academic Press, San Diego
7. Karch J, Birringer R, Gleiter H (1987) Nature 330:556
8. Mayo MJ, Siegel RW, Narayanasamy A, Nix WD (1990) J Mater Res 5:1073
9. Andres RP, Bielefeld JD, Henderson JI, Janes DB, Kolagunta VR, Kubiack CP, Mahoney WJ, Osifchin RG (1996) Science 273:1690
10. Murray CB, Kagan CR, Bawendi MG (1995) Science 270:1335
11. Bentzon MD, van Wonterghem J, Morup S, Tholen A (1989) Philos Mag B (1989) 60:169
12. Harfenist SA, Wang ZL, Alvarez MA, Vezmar I, Whetten RL (1996) J Phys Chem 100:13904
13. Doyle HJ (1998) PhD Thesis. National University Ireland, Dublin, p 141
14. Moser J, Punchihewa S, Infelta P, Graetzel M (1991) Langmuir 7:3012
15. Rodríguez R, Blesa M, Regazzoni A (1996) J Coll Inter Sci 177:122
16. Badia A, Singh S, Demers L, Cuccia L, Brown G, Lennox R (1996) Chem Eur J 2:359

Progr Colloid Polym Sci (2000) 115:361–366
© Springer-Verlag 2000

F. Mallamace
P. Gambadauro
C. Liao
N. Micali
P. Tartaglia
S.-H. Chen

Effects of the short-range attraction in the kinetic glass transition studied by means of a micellar system

F. Mallamace (✉) · C. Liao · S.-H. Chen
Department of Nuclear Engineering
Massachusetts Institute of Technology
Cambridge MA 02139-4307, USA

F. Mallamace · N. Micali · P. Tartaglia
Istituto Nazionale per la Fisica della
Materia, Italy

P. Gambadauro
Dipartimento di Fisica
Università di Messina, Vill. S. Agata
C.P. 55, 98166 Messina, Italy

N. Micali
Istituto di Tecniche Spettroscopiche del
CNR, Messina, Italy

P. Tartaglia
Dipartimento di Fisica
Università di Roma La Sapienza
P. e A. Moro 5, 00185 Rome, Italy

Abstract Direct experimental evidence is presented, using quasi-elastic light scattering, of the existence of an ergodic to nonergodic transition in a dense micellar system [pluronic L64/water (D_2O)] characterized by a short-ranged attractive interaction. The measured intermediate scattering function (ISF) shows a glass transition on crossing a line which is both temperature and concentration dependent. Detailed analysis of the ISF gives various parameters characterizing the two time-step relaxations predicted by ideal mode coupling theory. Similarities of this kinetic glass transition to the theoretically predicted one in adhesive hard sphere and hard core attractive Yukawa model systems is pointed out.

Key words Micellar systems · Short-range attraction · Scattering function

Introduction

Recently there has been considerable interest in the physics of systems out of equilibrium, particularly in phenomena like the glass transition or the phase transition from an ergodic to a nonergodic state [1]. All of this is related to the properties of many different materials (molecular, ionic, metallic, colloidal, polymeric and some biological materials [1–3]) in which the crystallization process does not occur during cooling, and instead metastable glassy solids form. This is a circumstance of enormous importance in material technology and in different science areas; as a consequence, the quest for understanding supercooling, glass transition and related phenomena has developed an ever increasing level of attention.

Actually it is widely accepted that the crossover from liquid to glass is identified by the so-called "ergodicity breaking" [4], i.e. by a transition from states (liquid) in which the system satisfies the ergodic hypothesis of statistical physics to nonergodic states (glass). In this latter situation, the system during the course of a measurement can revisit only a restricted part of the phase space available to it. Up to now there exists no generally accepted theory in this important field of condensed matter physics; however, the application of mode–mode coupling theories (MCT), developed initially to describe the dynamics of critical fluids, gave an impetus in the study of the glass transition [2, 5, 6]. In connection with MCT, hard sphere colloidal suspensions, having many features in common with atomic fluids, have been used in many experiments as test systems to study such a transition; in particular they can be easily concentrated in dense metastable states without crystallization occurring [7–10].

MCT is a theory for structural relaxation in supercooled liquids, i.e. it relates the molecular motion with molecular trappings in structural cages. Such a cage effect is taken into consideration by including, in the equation of motion, a delayed nonlinear coupling between density fluctuations. The increase of this coupling by increasing the parameters that control the system structure, for example concentration, produces the ergodicity breaking at which a fraction of the fluid structure is arrested. MCT assumes that density fluctuations can be the only relevant slow modes (ideal MCT) or it can also consider current density fluctuations with possible phonon-activated hopping processes (extended MCT) [2]. Colloidal solutions characterized by diffusive dynamics, on all relevant time scales, are entirely described by the ideal MCT in terms of the normalized intermediate scattering function (ISF) [or autocorrelation function of density–density fluctuations $F_q(t)$] [10]. The main prediction of MCT is that the approach of the glass transition, from the fluid, is accompanied by the emergence of two structural relaxation processes, the α and β processes, with critically diverging time scales. Both processes exhibit time scaling properties; the slower α process arrests at the transition while the faster decay, the β process, describes a localized motion that persists into the nonergodic glass phase.

Very recently, to have a correct modelling of the glass transition, using colloidal particles, it has been considered a more realistic situation where the interparticle interaction is due to a hard sphere repulsion plus an attractive contribution [14, 15]; in particular, the glass transition has been investigated using the idealized MCT for systems characterized by short-range attractive interactions, i.e. adhesive hard sphere (Baxter systems [11] or AHS) [14, 15] and hard core attractive Yukawa systems (HCYS) [15], systems with a concentration-temperature phase diagram. By the findings of these studies it turns out to be a more rewarding and complex physics if compared with the results of true hard sphere systems. According to MCT the AHS system shows in the concentration-temperature phase diagram: (1) structural arrest dynamics at high packing fractions due to the competition mechanism of hard core and attractive interactions; (2) two different ideal-glass transition lines; (3) in the region where these two lines meet the slow dynamics changes from a power-law to a logarithmic-law due to the influence of a nearby cusp singularity. For HCYS, MCT predicts: (1) low temperature nonergodic states that extend to the critical and subcritical regions; the percolation (gel transition) line, characteristic of the phase diagram of these systems, is caused by a low temperature extension of the glass transition; (2) the range of the attraction governs the way in which the glass transition line traverses the phase diagram relative to the critical point.

Motivated by these results, we conducted an experimental study by means of quasi-elastic light scattering [more precisely, photon correlation spectroscopy (PCS)] that gives a direct measurement of the ISF in a micellar system (pluronic L64/water). On the basis of previous studies [12] such a system can be considered as a prototype of a Baxter system. In particular its interaction, governed by short ranged sticking, originates a phase diagram that is the same as an AHS system [13]. We studied, for such a micellar suspension, at several high concentrations the temperature dependence of the glass transition with the aim to compare the results obtained with the main findings of the cited MCT studies recently made on systems characterized by analogous interactions. More precisely, similarities and differences of this experimentally determined kinetic glass transition line to recently published results on the sticky hard sphere [14] and attractive Yukawa model systems [15] will be discussed.

Results and discussion

The system we study is a non-ionic triblock copolymer L64 belonging to the pluronic [16] family of compounds. These systems are made of poly(ethylene oxide) (PEO) and poly(propylene oxide) (PPO) arranged symmetrically on each end as PEO_m–PPO_n–PEO_m. The importance of these polymers derives from their properties as temperature-dependent surface active agents in aqueous solutions. The L64 (PEO_{13}–PPO_{30}–PEO_{13}) copolymer has an average molecular weight of 2900 Da and is comprised of 60 wt% of PPO. This copolymer self-aggregates spontaneously, forming monodisperse spherical micelles in an aqueous solution [17] over a wide range of copolymer concentrations c up to more than 50 wt% for temperatures above 30 °C.

We have recently investigated by small angle neutron and light scattering [12] the phase diagram of the disordered micellar phase, the microstructure of the micelles and their mutual interactions. The phase diagram, shown in Fig. 1, similar to Baxter systems is characterized by the existence of an inverted binodal curve with a critical point at 5 wt% and $T = 330.3$ K, and a temperature-concentration dependent percolation locus cutting across the phase diagram starting from near the critical point to at least 50 wt%. The percolation locus is determined by shear moduli jumps (more than two orders of magnitude), by scaling of the frequency dependent viscosity, as shown in Fig. 2, and by a change of the slope of the relaxation time measured by photon correlation spectroscopy (PCS) as a function of temperature. In this work we focus our attention on the concentration range above 45 wt% and temperature above room temperature.

The PCS measurements were performed by using conventional equipment which consisted of a rotating detector assembly that uses a computer driven goniometer and a temperature controlled sample holder. A HeNe laser of 30 mW operating at the wavelength of

$\lambda = 6323$ Å was used as the exciting source. The density–density correlation data have been taken using a Brookhaven digital correlator (model BI-9000AT) with

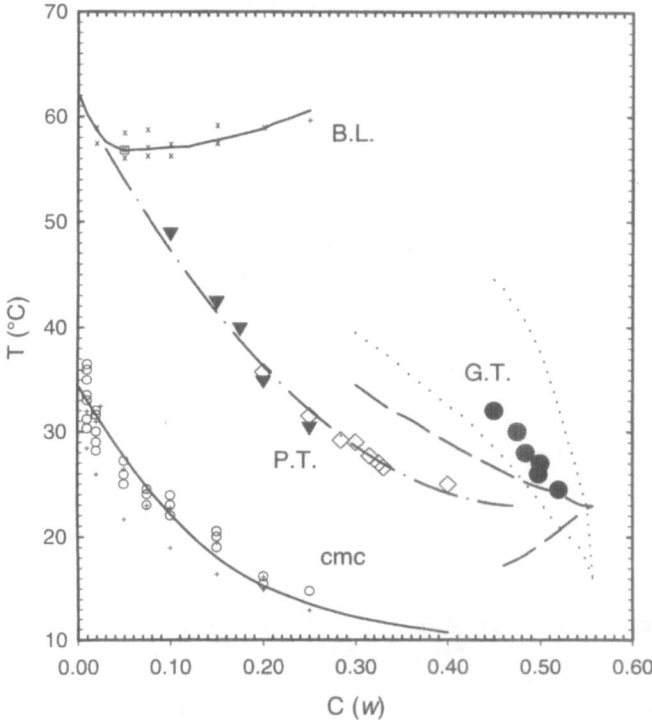

Fig. 1 The phase diagram of the L64/water system showing the binodal line (*BL*), the critical micellar concentrations line (*cmc*), the percolation line measured through viscoelasticity (*full triangles*) and PCS (*empty squares*); model fit of the percolation locus by the AHS (*dash-dotted line*); the measured glass line (*full dots*). The *dot lines* are the computed glass lines for the Yukawa potential with $b = 20$ (*upper curve*) and $b = 30$ (*lower curve*), respectively. The *long dashed curve* represents the glass line computed by means of the AHS

a logarithmic sampling time scale. This feature allows us to describe accurately both the short time region and the very long time one, up to times of the order of seconds. We measure the time and position averaged correlation functions

$$g^{(2)}(q,\tau) = \langle\langle I(q,0)I(q,\tau)\rangle\rangle / \langle\langle I(q)\rangle\rangle^2 \qquad (1)$$

where $I(q, t)$ is the intensity of the light scattered at the wave vector q at time t; the second bracket denotes the positional average by considering different areas of the scattering volume. We follow the method used for colloidal hard sphere systems [10]. Since the system shows structural arrest, and therefore nonergodic behavior, particular care has been taken in order to average over many different samples in the long time region of the density time correlation function. The ISF, $F_q(t)$, has been obtained from $g^{(2)}(q, \tau)$ in a straight forward way. The measurements have been taken at concentrations larger than 40 wt%, where apparently no structural arrest has been observed, although the ISF tends to develop a long time tail which can be described by means of a stretched exponential. When $c > 40$ wt%, as we move at constant concentration and increasing the temperature above the percolation line, we observe a progressive slowing down of the decay of the ISF, up to a point where it becomes flat, indicating the structural arrest. Typical behavior is shown in Fig. 3 at various concentrations and temperatures. The points in the concentration-temperature plane where we observe structural arrest are shown in the phase diagram of the system (Fig. 1) by open circles.

The ISF, as measured by PCS (Fig. 3), shows clearly, by means of the two time-step relaxations, the system evolution towards the ergodic–nonergodic transition; in particular, from the related data the structural arrest can be easily located at a series of loci between concentra-

Fig. 2 *Left*: the real G' and imaginary part G'' (*full symbols*) of the complex shear modulus measured at $\omega = 1.36$ rad/sec; (a) $c = 25$ wt%, (b) $c = 20$ wt%, (c) log-log plot of $|\eta^*|$ vs ω ($c = 20$ wt%, $T = 32$ °C)

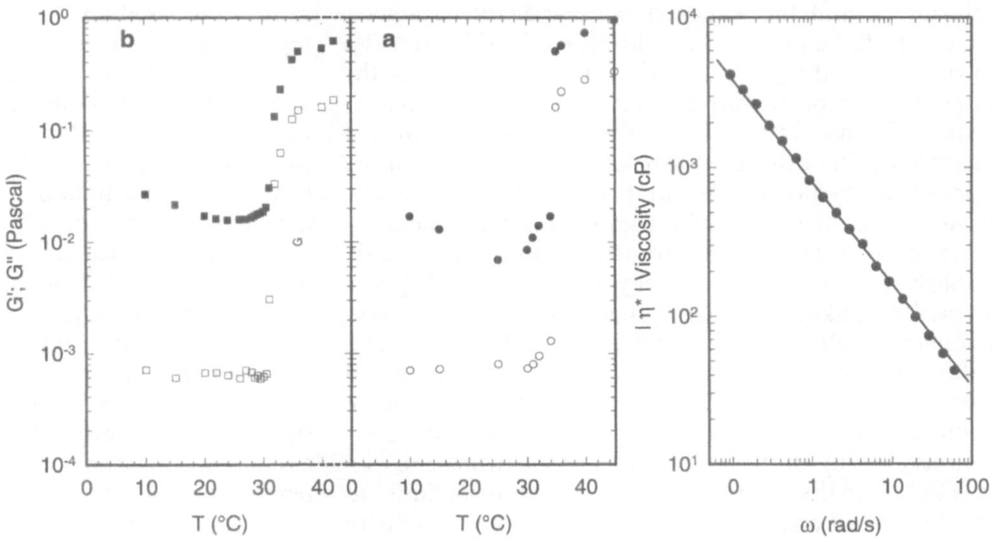

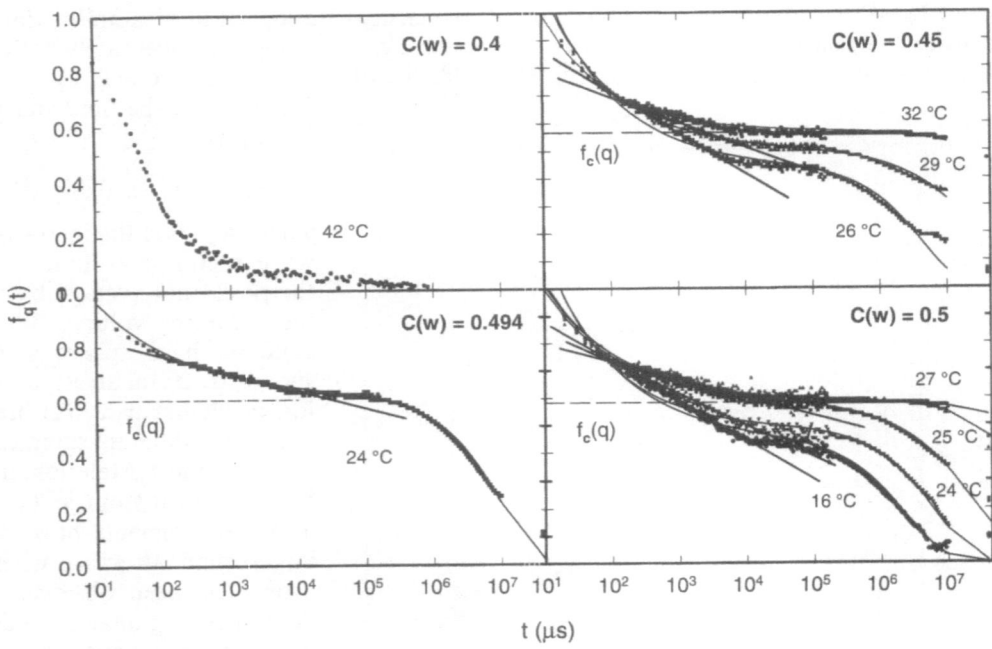

Fig. 3 The normalized ISF at different concentrations and temperatures (above $c = 0.4$ wt%, and in the glass region $c = 0.45$, 0.494 and 0.5 wt%). Continuous lines are the best fit with the MCT, while continuous lines show the logarithmic decay

tions 45.0 wt% to 51.8 wt% and temperatures 24.2 °C to 32 °C. The locus of this ergodic breakdown is clearly both temperature and concentration dependent.

We shall explain these results using the idealized MCT. As previously stated, the MCT has been successfully applied to describe the so-called kinetic glass transition observed for many supercooled liquids [2]. In particular, it has been used to explain dynamic light scattering measurements for the same phenomenon in highly packed colloidal hard spheres [10]. A detailed analysis of the ISF can give the various parameters characterizing the two time-step relaxations predicted by MCT, more precisely the scaling behavior (and the related exponents) that characterizes the α and β relaxations and the so-called nonergodicity parameter f_q^c, i.e. the finite positive value at which the ISF saturates in time above the glass transition. The main point in the interpretation of present data on the structural arrest in terms of glass transition is the introduction of the attractive part of the interparticle potential, which allows nonergodic behavior even at low colloidal concentrations. Similarities and differences of this experimentally determined kinetic glass transition line to recently published results on the sticky hard sphere [14] and attractive Yukawa model systems [15] are discussed. Numerous other experimental studies of hard sphere colloidal systems have shown a dynamical structural arrest; in contrast, there are few experimental studies on colloidal suspensions with short ranged attractive interactions [18–20]. In these latter there is the formation of a gel phase and the corresponding transition from liquid to gel that seems characterized by the same nonergodicity

observed in hard sphere colloidal glasses. The concept of the static percolation transition, and the related clustering effects, has been proposed for the interpretation of the phenomenon of structural arrest in these systems [21], but in some cases this interpretation did not reproduce the experimental results [18, 19].

The general idea, in the study by means of percolation of the gel transition in SHS colloids is that the colloidal particles tend to form polydisperse aggregates with a ramified structure which tend to span the whole sample, thus forming a percolating cluster. The formation of clusters is in turn related to the existence of an attractive tail of the interparticle potential, besides the hard core repulsion. The attractive potential has a short range compared to the particle size, and it is usually represented by an exponential or a square well interaction. Typical examples are the Yukawa potential or sticky spheres potential, introduced by Baxter [11]. In both cases the Ornstein-Zernike equation for the pair correlation function can be solved, in the mean spherical approximation or in the Percus-Yevick approximation, respectively. This type of approach was successful in interpreting percolation phenomena in microemulsions [11] and in the L64/water system [12] we are presently studying. In particular the phase diagram can be fairly well described in terms of the Baxter sticky spheres model, i.e. a square well potential of zero range and infinite depth. In order to take into account the thermodynamic properties of the solvent, one has to assume a functional relation between the measured temperature T and the stickiness parameter $1/\tau$ which describes the model. The good agreement between theory, and the measured binodal percolation lines and critical point for this system is clear from Fig. 1. In addition, the simple inverse proportionality relation $T \approx \tau$ is sufficient to give a fairly good representation of the percolation and binodal lines for both the Yukawa or Baxter models once the critical point given by the theory is assumed to coincide with the experimental one.

On this basis, we considered that from these experimental findings there is the indication that the percolation points, measured through shear viscosity and shear modulus, and the points marking the structural arrest, measured through dynamic light scattering, can be explained only in terms of clustering effects, and not by other different physical phenomena. Therefore, in order to describe the observed structural arrest, we follow the recent literature suggestions [14, 15], i.e. the MCT results for a model potential possessing a hard core and a short range attractive tail.

The MCT equations for the normalized ISF, $f_q(t) = F_q(t)/S_q$ (where S_q is the static structure factor), predict the existence, for a given concentration, of a critical temperature T_{MCT} where the ergodic to nonergodic transition takes place, and $f_q(t)$ tends to the finite value f_q^c for $t \to \infty$. The parameter driving the transition is the separation parameter $\sigma \approx (T_{MCT} - T)/T_{MCT}$. As previously stated, the theory predicts the existence of various regimes in the decay of the density autocorrelator, in terms of different time scales:

1. The short time dynamics region $0 < t < t_0$ dominated by microscopic motions, with t_0 the time scale determined by the short time dynamics (in the case of colloids it is related to the free diffusional coefficient D_0).
2. The β relaxation region $t_0 < t < t_\sigma = t_0|\sigma|^{-1/2a}$, with $\gamma = (1/2a) + (1/2b)$, where the normalized density autocorrelator satisfies the scaling relation

$$f_q(t) = f_q^c + |\sigma|^{1/2} h_q g_\pm(t/t_\sigma) \qquad (2)$$

where the subscript of the scaling function $g_\pm$ refers to the sign of σ and

$$g_\pm(t << t_\sigma) = (t/t_\sigma)^{-a} \qquad (3)$$

$$g_-(t_\sigma << t << t'_\sigma) = -B(t/t_\sigma)^b \qquad (4)$$

where $B > 0$.

3. $t > t'_\sigma = t_0|\sigma|^{-1/2a-1/2b}$, the α relaxation regime, where the scaling law

$$f_q(t) = f_q^c G(t/t'_\sigma) \qquad (5)$$

holds, with $G(q, \tau/\tau_\alpha)$ a nonuniversal function. a ($0 < a < 0.5$) and b ($0 < b < 1$) are nonuniversal exponents determined solely by the so-called exponent parameter λ as

$$\lambda = \Gamma^2(1-a)/\Gamma(1-2a) = \Gamma^2(1+b)/\Gamma(1+2b) \qquad (6)$$

where Γ is the Euler gamma function and λ is in turn determined by the static structure factor S_q.

Analysis of the experimental data allows us to derive the values of the parameters $b = 0.6$ characterizing the von Schweidler law [2] in the β relaxation region and therefore λ for which we obtain the value 0.7. As can be observed in Fig. 3, we have also checked the scaling laws in the same regime with good results. The situation is less satisfactory when trying to describe the experimental data above the plateau of the normalized ISF, which should follow a power law with an exponent a. In that region, in fact, the measured data show a tendency to have a logarithmic dependence on time.

Motivated by our previous experience with percolation phenomena in colloidal systems where many phenomena, as we mentioned in the Introduction, are due to the short range attractive potential that originates clustering effects, we have examined a recent version of the MCT applied to this type of system. We refer to potential functions where besides the hard core a deep and very short ranged potential of the Baxter [14] type or a Yukawa attractive tail [15] are added. A significant and common result, in both cases, is that the glass transition line in the composition temperature plane shows two branches, one essentially corresponding to the glass line of hard sphere systems at high composition, the other extending to much lower compositions. The former is attributed, with the usual mechanism due to the excluded volume, to the repulsive part of the interaction, the latter to the attractive part of the potential. The nonergodicity factors have a qualitatively different behavior on approaching the two lines and in one case [14], the sticky hard spheres system, a new phenomenon arises, a glass–glass transition due to the crossing of the two glass lines. This transition is associated with a higher order glass transition of type A3 [22].

Using the transformation we adopted to map the binodal and percolation lines of Yukawa or Baxter models, we can, at least qualitatively, plot on the phase diagram of the L64/water system the curves corresponding to the glass transitions in the two models. The two lines corresponding to the Yukawa attractive potential $u = K\sigma \exp[-b(r/\sigma - 1)]/r$ for r larger than the hard sphere diameter σ, with different screening parameters b and inverse temperature proportional to K. As we mentioned earlier, the glass line in the models with interparticle attraction has a steep repulsive part around a concentration corresponding to a hard sphere system $c \approx 10c_c$ in the concentration-temperature plane. The attractive part extends instead smoothly in the direction of the critical point. It is this part of the glass line which has a behavior qualitatively similar to the measured structural arrest of our micellar system. The Baxter type glass line is less similar to the measured one. This is probably due to the fact that the A3 model contains a momentum cutoff which should be substituted by a more physical quantity, such as an inverse length typical of a finite square well of finite range. The attractive part of the potential would be in this case much more flexible and gives a better description of the experimental glass line. A final point worth mentioning is the appearance of a time region of the experimental ISF decay which has logarithmic behavior. This feature is characteristic of cusp-like singularities of the A3 type [23], in the vicinity of which this very slow relaxation behavior sets in and tends to influence a large time domain. The logarithmic type behavior that can be seen in the ISF could be a manifestation of such an effect.

Conclusions

In conclusion, we have evidenced that the micellar system made up of a solution in water of the triblock copolymer pluronic L64 shows a structural arrest, or ergodicity breaking (typical of the glass transition). In addition, such a phenomenon is composition-temperature dependent that can be traced to a well-defined glassy transition line in the system phase diagram. In the vicinity of this line the ISF shows the typical two steps decay of supercooled liquids approaching a glass transition. In addition, the measured correlation functions are characterized by a time region where a logarithmic decay exists. This behavior has been related to the existence of an attractive tail of the interparticle potential, which gives rise to a structure factor that, in some approximation, can be analytically evaluated. This in turn allows the application of the MCT of the glass transition, which is significantly different from the analogous calculations for hard spheres. By considering the interparticle attraction results, the glass transition line is in fact composed of two sections that can be traced to the repulsive and the attractive parts of the potential, the latter of which is responsible for the structural arrest we have experimentally observed. Confirmation in this direction is given by the logarithmic decay which is a consequence of the short range attraction, at least in the case of the sticky hard spheres model. Finally, from the reported results on micellar systems there clearly emerges the following general picture: the attractive potential not only is at the origin of the phase separation, but also at the formation of ramified clusters which then originates the percolation line and finally to the structural arrest and the glassy behavior at high concentrations of the polymer.

Acknowledgements The research of F.M. and P.T. is supported by MURST (Prin 97); the research of S.H.C. and C.Y.L. is supported by the Material Science Division of the USDOE.

References

1. Angell CA (1997) In: Mallamace F, Stanley HE (eds) The physics of complex systems. IOS press, Amsterdam, p 571
2. Gotze W (1991) In: Hansen JP, Levesque D, Zinn-Justin J (eds) Liquids freezing and the glass transition. North Holland, Amsterdam, p 287
3. Muschol M, Rosemberger F (1997) J Chem Phys 107:1953
4. Palmer RJ, Stein DC (1985) In: Ngai K, Wright B (eds) Relaxations in complex systems. National Technical Information Service, U.S. Department of Commerce, Springfield, Va. p 253
5. Bengtzelius U, Götze W, Sjölander A (1984) J Phys C 17:5915
6. Fuchs M (1995) Transp Theory Stat Phys 24:855
7. Pusey PN, van Megen W (1987) Phys Rev Lett 59:2083
8. van Megen W, Pusey PN (1991) Phys Rev A 43:5429
9. Krall AH, Weitz DA (1998) Phys Rev Lett 80:778
10. van Megen W, Underwood SM (1993) Phys Rev Lett 70:2766; (1994) Phys Rev E 49:4206
11. Baxter RJ (1968) J Chem Phys 49:2770
12. Lobry L, Micali N, Mallamace F, Liao C, Chen SH (2000) Phys Rev E (in press)
13. Chen SH, Ku C-Y, Liu Y-C (1997) In: Mallamace F, Stanley HE (eds) The physics of complex systems. IOS press, Amsterdam, p 221
14. Fabbian L, Götze W, Sciortino F, Tartaglia P, Thiery F (1999) Phys Rev E 59:R1347
15. Bergenholtz J, Fuchs M (1999) Phys Rev E 59:5706
16. BASF (1989) Pluronic Tetronic Surfactant. Technical Brochure BASF, Parsipanny, NJ
17. Al-Saden AA, Whately TL, Florence AT (1982) J Colloid Interface Sci 90:303; Lindman B, Carlsson A, Kalstrom G, Maltensen M, (1990) Adv Colloid Interface Sci 32:183; Pandya K, Bahadur P, Nagar TN, Bahadur A (1993) Colloids Surf A 70:219
18. Verduin H, Dhont JKG (1995) J Colloid Interface Sci 172:425
19. Bartsch E, Antonietti M, Shupp W, Sillescu H (1992) J Chem Phys 97:3950
20. Poon WCK, Pirie AD, Haw MD, Pusey PN (1997) Physica A 235:110
21. Woutersen ATJM, Mallema J, Blom C, de Kruiff CG (1994) J Chem Phys 101:542
22. Sjögren L (1991) J Phys Condens Matter 3:5023; (1983) J Phys (Paris) IV 3:C1–117
23. Götze W, Sjögren L (1992) Rep Prog Phys 55:241

Progr Colloid Polym Sci (2000) 115:367–370
© Springer-Verlag 2000

H. Löwen
E. Allahyarov
I. D'Amico

Effective interaction between confined colloids: repulsion or attraction?

H. Löwen (✉) · E. Allahyarov
Institut für Theoretische Physik II
Heinrich-Heine-Universität Düsseldorf
40225 Düsseldorf, Germany

E. Allahyarov
Institute for High Temperatures
Russian Academy of Sciences
127412 Moscow, Russia

I. D'Amico
Department of Physics and Astronomy
University of Missouri-Columbia
Columbia, MO 65211, USA

Abstract Using computer simulations of the primitive model, we calculate the effective force acting onto a single macroion and onto a macroion pair in the presence of slit-like confinement between charged plates. For moderate Coulomb coupling, we find that this force is repulsive. Under strong coupling conditions, however, the sign of the force depends on the distance to the plates and on the interparticle distance. In particular, the particle-plate interaction becomes strongly attractive for small distances, which leads to colloidal crystalline layers near the plates as observed in recent experiments.

Key words Charged colloids ·
Effective interaction · Screening ·
Computer simulation address

Introduction

Recent experiments [1–3] have revealed that the effective interaction between like-charged colloidal particles ("macroions") is sensitively affected by a confinement between two parallel charged glass plates. For aqueous polystyrene suspensions studied in experiment, the effective force between two colloidal macroions is found to be repulsive far away from the plates but becomes attractive when the like-charge macroions are located close to an equally charged plate. These findings are surprising as nonlinear Poisson–Boltzmann theory results in a purely repulsive interaction, which was recently shown independently by Neu [4] and Sader and Chan [5]. Assuming an equilibrium situation and neglecting the discrete structure of the solvent and any chemical details, it is tempting to explain the mutual attraction between macroions near plates by counterion fluctuations and correlations which are ignored in the Poisson–Boltzmann approach but fully included in the primitive model of strongly asymmetric electrolytes.

In this paper, we use "exact" computer simulations to calculate the effective interaction between confined charged colloids within the primitive model. We study one or two macroions confined between two parallel charged plates and find that the wall-particle and the interparticle interaction is repulsive for weak Coulomb coupling. For stronger coupling, the behavior of the force changes from repulsive to attractive and back to repulsive as the interparticle distance is varied. In particular, the plate-particle interaction exhibits a short-range attraction for a small distances. This may explain the occurrence of long-lived crystalline colloidal layers on top of the glass plates found in recent experiments [6–8].

The model and target quantities

We consider $N_m = 1, 2$ macroions with bare charge $q_m = Ze > 0$ ($e > 0$ denoting the elementary charge) and mesoscopic diameter d_m confined between two parallel plates that carry a surface charge density $\sigma > 0$. The separation distance between plates is $2L$. For convenience, we choose the z axis to be perpendicular to the plate surface. The origin of the coordinate system is located on the surface of one plate. Image charges are neglected, and there is no added salt. Typically we use a periodically repeated square cell in x and y direction which possesses an area S_p. Hence the macroion number density is $\rho_m = N_m/2LS_p$. The counterions from the plates and the colloids have a microscopic diameter d_c

and carry an opposite charge $q_c = -qe$ where $q > 0$ denotes the valency. Typically, $q = 1, 2$. The total counterion number N_c in the cell is fixed by the condition of global charge neutrality.

The interactions between the particles are described within the framework of the primitive model. We assume the following pair interaction potentials $V_{mm}(r)$, $V_{mc}(r)$, $V_{cc}(r)$ between macroions and counterions, r denoting the corresponding interparticle distance and ϵ the dielectric constant of the solvent:

$$V_{mm}(r) = \begin{cases} \infty & \text{for } r \leq d_m \\ \frac{Z^2 e^2}{\epsilon r} & \text{for } r > d_m \end{cases} \qquad (1)$$

$$V_{mc}(r) = \begin{cases} \infty & \text{for } r \leq (d_m + d_c)/2 \\ -\frac{Zq e^2}{\epsilon r} & \text{for } r > (d_m + d_c)/2 \end{cases} \qquad (2)$$

$$V_{cc}(r) = \begin{cases} \infty & \text{for } r \leq d_c \\ \frac{q^2 e^2}{\epsilon r} & \text{for } r > d_c \end{cases} \qquad (3)$$

The interaction between the particles and the wall is zero as the plates are equally charged.

Our target quantities are the effective forces $\mathbf{F}_j$ acting onto the jth macroion at position $\mathbf{R}_j$ which embody three different parts [9–11]:

$$\mathbf{F}_j = \mathbf{F}_j^{(1)} + \mathbf{F}_j^{(2)} + \mathbf{F}_j^{(3)} \qquad (4)$$

The first term, $\mathbf{F}_j^{(1)}$, is the direct Coulomb repulsion stemming from neighboring macroions and the plates:

$$\mathbf{F}_j^{(1)} = -\nabla_{\mathbf{R}_j} \left(\sum_{i=1; j \neq i}^{N_m} V_{mm}(|\mathbf{R}_i - \mathbf{R}_j|) \right) \qquad (5)$$

The second part $\mathbf{F}_j^{(2)}$ involves the electric part of the counterion-macroion interaction and has the statistical definition:

$$\mathbf{F}_j^{(2)} = \left\langle \sum_{i=1}^{N_c} \nabla_{\mathbf{R}_j} \frac{Zq e^2}{\epsilon |\mathbf{R}_j - \mathbf{r}_i|} \right\rangle_c \qquad (6)$$

where $\langle \cdots \rangle_c$ is the canonical average over the counterion positions $\{\mathbf{r}_i\}$ in the field of fixed macroions. Finally, the third term $\mathbf{F}_j^{(3)}$ describes a depletion (or contact) force arising from the hard-sphere part in $V_{mc}(r)$, which involves a surface integral over the counterion equilibrium number density field $\rho_c^{(0)}(\mathbf{r})$:

$$\mathbf{F}_j^{(3)} = k_B T \int_{S_j} d\mathbf{f} \, \rho_c^{(0)}(\mathbf{r}) \qquad (7)$$

Here S_j is the spherical surface of the jth macroion and $\mathbf{f}$ is a surface vector pointing towards the macroion center and $k_B T$ is the thermal energy.

We define the strength of Coulomb coupling via the dimensionless coupling parameter [11]:

$$\Gamma = \frac{Z}{q} \frac{2\lambda_B}{d_m + d_c} \qquad (8)$$

where the Bjerrum length is $\lambda_B = q^2 e^2 / \epsilon k_B T$.

Results from computer simulation

We have calculated the counterion averages needed to obtain the effective forces between the macroions by computer simulations. We take divalent counterions ($q = 2$) throughout our investigations and fixed the temperature to $T = 293$ K and the width of the slit to $2L = 5d_m$. The dielectric constant is that for water at room temperature ($\epsilon = 78.3$) but we have also investigated the case $\epsilon = 3.9$ in order to enhance the Coulomb coupling Γ formally.

Let us first consider a *single macroion* at position $\mathbf{R}_1 = (0, 0, Z_1)$. The total force $\mathbf{F}_1$ acting on the macroion only depends on the distance Z_1 and points along the unit vector $\mathbf{e}_z$ of the z-axis. Results for $F_1 = \mathbf{F}_1 \cdot \mathbf{e}_z$ are shown in Fig. 1. All data are scaled by the (arbitrary) unit $F_0 = e^2/d_m^2$. For weak Coulomb coupling ($\Gamma = 11$, dashed line), it is repulsive but it exhibits repulsive, attractive and repulsive parts as a function of Z_1 for larger coupling ($\Gamma = 110$, solid line). In the latter case,

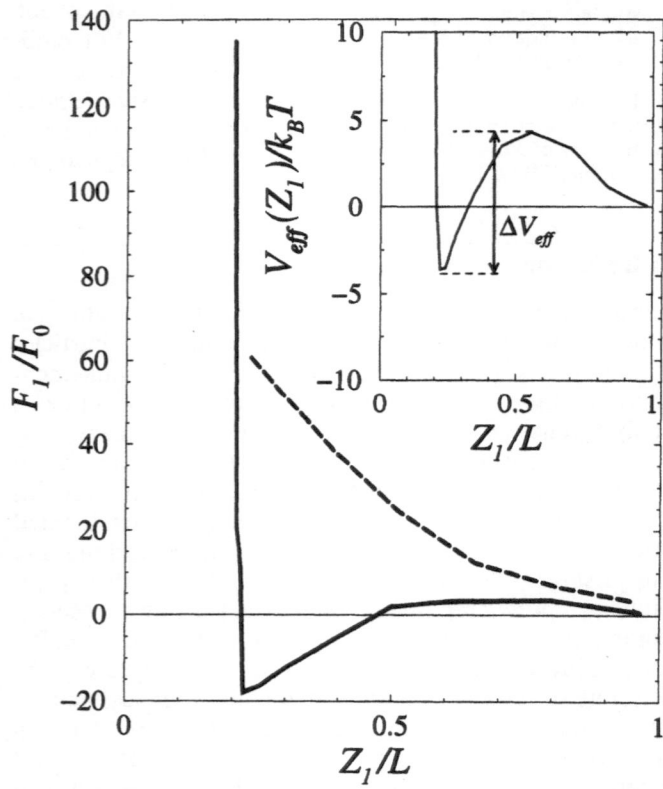

Fig. 1 Force $F_1 = \mathbf{F}_1 \cdot \mathbf{e}_z$ acting on a single macroion versus reduced macroion distance Z_1/L. *Dashed line*: $Z = 200$, $\sigma = 1.24 \times 10^{11}$ (e/cm^2), $\epsilon = 78.3$, $\rho_m = 1.0 \times 10^{12} (1/\text{cm}^3)$, $d_m = 5.32 \times 10^{-6}$ (cm), $d_c = 5.32 \times 10^{-8}$(cm). *Solid line*: $Z = 100$, $\sigma = 1.49 \times 10^{11}$ (e/cm^2), $\epsilon = 3.9$, $\rho_m = 1.7 \times 10^{12}(1/\text{cm}^3)$, $d_m = 5.32 \times 10^{-6}$(cm), $d_c = 5.32 \times 10^{-8}$ (cm). The *inset* shows the effective potential in units of $k_B T$ versus reduced macroion distance Z_1/L together with the energy barrier ΔV_{eff}

the electrostatic part $F_1^{(2)} = \mathbf{F}_1^{(2)} \cdot \mathbf{e}_z$ and the depletion part $F_1^{(3)} = \mathbf{F}_1^{(3)} \cdot \mathbf{e}_z$ behave qualitatively different: $F_1^{(3)}$ is always repulsive and increases with decreasing Z_1, at least if the macroion is not too close to the surface when the counterion depletion between the macroion and the wall induced by the finite counterion core is negligible. This is an expected behavior, since in general there are more counterions close to the walls. The pure electrostatic contribution, $F_1^{(2)}$, on the other hand, exhibits a more subtle behavior. If the macroion is close to the midplane, it is repulsive, then it becomes attractive as the macroion is getting closer to the plates.

For strong coupling, the macroion has three equilibrium positions, two of them are stable, namely the midplane and a position in the vicinity of the plate. In order to extract more information, we have calculated the effective wall-particle potential defined by:

$$V_{\text{eff}}(Z_1) = -\int_0^{Z_1} F_1(h)\mathrm{d}h \qquad (9)$$

by integrating our data with respect to the macroion altitude h. This quantity is shown as an inset in Fig. 1. One first sees that the global minimum is in the vicinity of

the walls. Furthermore, the barrier height ΔV_{eff} to escape from there is about $8k_BT$. This implies that the time for a colloidal particle to escape from the position close to the surface is roughly $\tau_0 \exp(\Delta V_{\text{eff}}/k_BT) = e^8\tau_0 \approx 3000\tau_0$ [12, 13], where τ_0 is a Brownian time scale governing the decay of dynamical correlations of the macroion.

Let us now consider *two equally charged macroions* at positions $\mathbf{R}_1 = (X_1, Y_1, Z_1)$ and $\mathbf{R}_2 = (X_2, Y_2, Z_1)$ with same altitude ($Z_1 = Z_2$). The distance between the macroion centers is $R_{12} = |\mathbf{R}_1 - \mathbf{R}_2|$, where the difference vector $\mathbf{R}_{12} = \mathbf{R}_1 - \mathbf{R}_2$ is the xy-plane. The total force acting on the two macroions can be split into a part pointing in z-direction and another contribution pointing along $\mathbf{R}_{12}$. Hence we write $\mathbf{F}_j = \mathbf{F}_j^{\parallel} + \mathbf{F}_j^{\perp}$ defining $\mathbf{F}_j^{\parallel} = (\mathbf{F}_j \cdot \mathbf{R}_{12}) \cdot \mathbf{R}_{12}/R_{12}^2$ and $\mathbf{F}_j^{\perp} = (\mathbf{F}_j \cdot \mathbf{e}_z) \cdot \mathbf{e}_z$ for $j = 1, 2$. Clearly, $\mathbf{F}_1^{\perp} = \mathbf{F}_2^{\perp}$, and $\mathbf{F}_1^{\parallel} = -\mathbf{F}_2^{\parallel}$.

Simulation results for $F_1^{\parallel} = \mathbf{F}_1^{\parallel} \cdot \mathbf{R}_{12}/R_{12}$ are shown in Fig. 2. For weak coupling ($\Gamma = 11$, dashed line) the force is repulsive, but for larger coupling ($\Gamma = 110$, solid line) there is attraction. In Fig. 3 we fixed the macroion distance and plotted the force perpendicular to the plates, $F_1^{\perp} = \mathbf{F}_1^{\perp} \cdot \mathbf{e}_z$ versus altitude Z_1. The parameters here correspond to strong coupling ($\Gamma = 100$) and again there

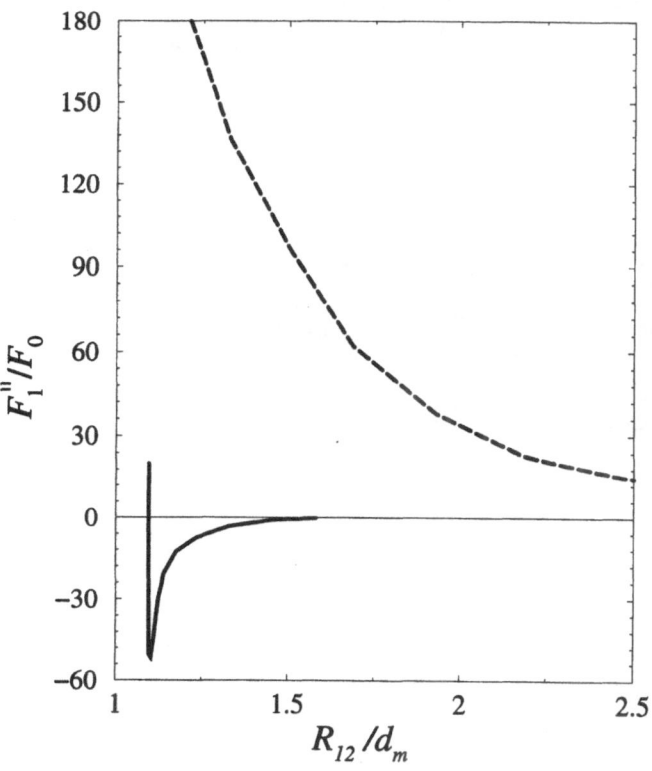

Fig. 2 Parallel part of the effective force acting onto a macroion pair, $F_1^{\parallel}$ versus reduced interparticle distance R_{12}/d_m. The altitude of macroions is $Z_1 = 0.6\,d_m$. *Dashed line*: $Z = 200$, $\sigma = 1.24 \times 10^{11}$ (e/cm^2), $\epsilon = 78.3$, $\rho_m = 2.34 \times 10^{13}(1/\text{cm}^3)$, $d_m = 5.32 \times 10^{-6}$ (cm), $d_c = 5.32 \times 10^{-8}$(cm). *Solid line*: $Z = 100$, $\sigma = 2.98 \times 10^{11}$ (e/cm^2), $\epsilon = 3.9$, $\rho_m = 2.34 \times 10^{13}$ $(1/\text{cm}^3)$, $d_m = 5.32 \times 10^{-6}$(cm), $d_c = 5.32 \times 10^{-8}$(cm)

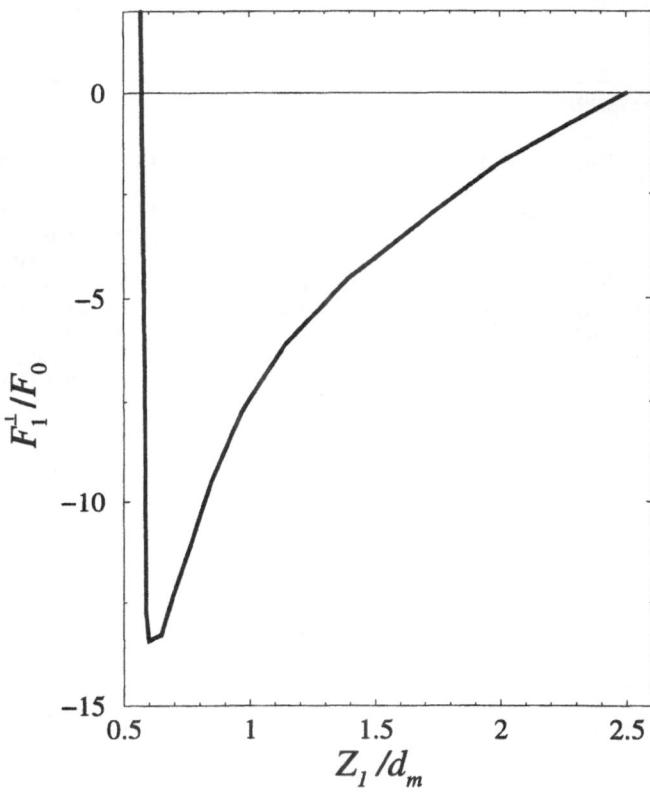

Fig. 3 Perpendicular part of effective force, $F_1^{\perp} = \mathbf{F}_1^{\perp} \cdot \mathbf{e}_z$ versus reduced altitude Z_1/d_m for fixed interparticle spacing $R_{12} = 1.2\,d_m$. System parameters are: $Z = 100$, $\sigma = 2.38 \times 10^{14}(e/\text{cm}^2)$, $\epsilon = 78.3$, $\rho_m = 1.87 \times 10^{17}(1/\text{cm}^3)$, $d_m = 2.66 \times 10^{-7}$(cm), $d_c = 2.66 \times 10^{-8}$(cm)

370

is attraction. Both the interparticle attraction and the wall-particle attraction become stronger in the vicinity of the plate. The depth in the effective wall-particle interaction potential for the perpendicular part is much more than twice as large as in the single macroion case [14]. Thus, a pair of macroions near a planar surface is more stable than a single macroion. Hence, the attraction between the wall and a single macroion is enhanced if more macroions are close to the wall. This gives evidence that the macroions will assemble on top of the surface forming two-dimensional colloidal layers.

Conclusions

In conclusion, we have simulated the effective force between macroions confined in a slit geometry. An effective attraction due to counterion correlations was found for strong Coulomb coupling. In particular, the effective potential of a single macroion confined between two parallel charged plates was found to have two stable minima where the total force vanishes: the first is in the mid-plane, the second close to the walls. This result was confirmed for two macroions. In this case the attraction towards the walls was even stronger than for a single macroion. Our most important conclusion is that the

attractive force will result in two-dimensional colloidal layers on top of the plates. As the depth of the attractive potential is larger than $k_B T$, these layers possess a large life-time with respect to thermal fluctuations. The layers should be crystalline as the interparticle interaction is also attractive. This can explain at least qualitatively the long-lived metastable crystalline layers found in recent experiments on confined samples of charge colloidal suspensions [6, 7].

We remark that our parameters are actually different from those describing the experiments. The main difference is the high surface charge of the glass plates within an area spanned by a typical macroion separation distance. The mechanism of our attraction is similar to that proposed recently by us in the bulk case [11]. A general simple physical picture explaining the attraction is still missing. It only occurs for strong coupling with divalent counterions and is short-ranged. In this respect, it behaves different than in experiment where the attraction was long-ranged. It can be speculated that the attractive part leads to new phenomena relevant, e.g. for adhesion of red blood cells.

Acknowledgements Financial support from the Deutsche Forschungsgemeinschaft (within SFB 237 and Schwerpunkt Benetzung und Strukturbildung an Grenzflächen) is gratefully acknowledged.

References

1. Kepler GM, Fraden S (1994) Phys Rev Lett 73:356
2. Crocker JC, Grier DG (1996) Phys Rev Lett 77:1897
3. Grier DG (1998) Nature 393:621
4. Neu JC (1999) Phys Rev Lett 82:1072
5. Sader JE, Chan DYC (1999) J Colloid Interface Sci 213:268
6. Larsen AE, Grier DG (1996) Phys Rev Lett 76:3862
7. Larsen AE, Grier DG (1997) Nature 385:230
8. Weiss JA, Oxtoby DW, Grier DG (1995) J Chem Phys 103:1180
9. Löwen H, Hansen JP, Madden PA (1993) J Chem Phys 98:3275
10. Allahyarov E, Löwen H, Trigger S (1998) Phys Rev E 57:5818
11. Allahyarov E, D'Amico I, Löwen H (1998) Phys Rev Lett 81:1334
12. Russel WB, Saville DA, Schowalter WR (1989) Colloidal dispersions. Cambridge University Press, Cambridge, p 267
13. Sauer S, Löwen H (1996) J Phys Condens Matter 8:L803
14. Allahyarov E, D'Amico I, Löwen H (1999) Phys Rev E 60: 3199

Progr Colloid Polym Sci (2000) 115:371–375
© Springer-Verlag 2000

E. Zaccarelli
G. Foffi
P. Tartaglia
F. Sciortino
K. A. Dawson

Binary mixtures of sticky spheres using Percus-Yevick theory

E. Zaccarelli (✉) · G. Foffi
K. A. Dawson
Irish Centre for Colloid Science
and Biomaterials
Department of Chemistry
University College Dublin
Belfield, Dublin 4, Ireland
e-mail: emanuela@fiachra.ucd.ie
Tel.: +353-1-7062418
Fax: +353-1-7062415

P. Tartaglia · F. Sciortino
Dipartimento di Fisica
Università di Roma La Sapienza
and Istituto Nazionale di
Fisica della Materia
Unità di Roma La Sapienza
Piazzale Aldo Moro 2
00185 Roma, Italy

Abstract We study a binary mixture of sticky Baxter-like spheres within the Percus-Yevick theory. The equations of liquid-liquid phase equilibrium are solved numerically. Thus, the equations of the derivatives of the chemical potentials, previously derived by Barboy, are integrated along constant-pressure paths, and the phase diagrams for various systems are constructed. The spinodals for the problem are also studied. It transpires that, unlike previous work on Baxter liquid-gas equilibrium within Percus-Yevick, the liquid-liquid separation is well behaved, showing expected features of critical point, symmetric spinodal lying within the binodal, and mean-field exponents. The structure factors for binary mixtures are calculated. A simple prescription for estimating the binary square-well structure factors is also proposed. The work should be of value to those scientists studying mixtures of colloidal particles with an attractive well and a hard core.

Key words Binary mixtures · Square well potential · Baxter model · Phase separation · Spinodal

Introduction

In the early and middle 1970s a series of papers [1–4] was written by a variety of different authors on an interesting model named the sticky sphere or adhesive hard-sphere model. The effort was begun by Baxter, and so the model has since borne his name. One point of the work was to find the simplest possible model that possessed the essential elements of realistic interaction, a hard core, and an attractive potential. The Baxter model may be considered to be such a case where, beginning with a square-well potential, we take the limit where the well depth becomes infinite, and the width zero in such a way to produce non-trivial solutions. One particularly attractive feature of the original Baxter work was that it was possible to find an exact closed solution for the Percus-Yevick (PY) [5] approximation to the model. There followed many subsequent works on the one-component Baxter model, analyzing its phase behaviour, spinodal behaviour, and structure factors [6]. These works lead finally to a very interesting but somewhat different picture to what had been expected [7]. That is, whilst the PY solution to the Baxter model had some peculiar aspects on the low-density side of the phase diagram, the broad picture of a liquid-gas phase separation was confirmed, and studied in detail. However, other works, growing in depth and confidence, began to offer a different picture. These returned to the original model, and attempted to draw some exact conclusions about the sticky sphere model, rather than its PY solution. The outcome was a broad consensus that there was no phase separation, and that likely the only stable condensed phase is the crystal. Subsequent works have largely confirmed these ideas, although recently it has been argued that for all practical purposes even the crystal is irrelevant, and the most likely scenario is that a new type of glass dominated by attractions will emerge [*, 8]. At around the same time as the early PY studies, Barboy [9] wrote out the extension of the PY approach to multicomponent sticky spheres. In an interesting and rather full account dating from the mid-1970s he showed that, whilst there is no simple closed solution for most of the thermodynamic quantities of binary mixtures, there are closed equations from which one can derive all of the relevant quantities. During all this period there was another, almost independent, strand of activity surrounding the PY solution to the sticky sphere model [10–14] (see ref. ∗∗). This was based on scattering experiments, mainly in colloidal and self-assembled systems, where the PY solution was used to fit experimental data. Remarkably good fits were sometimes obtained, and the Baxter model of one-component systems became widely used and well accepted, despite its apparent limitations when viewed from a more fundamental point of view. With hindsight we now understand this success to be connected to the fact that many of the systems of interest were not exactly sticky spheres, but that there was a repulsion, and perhaps a very narrow attractive well [8, 15, 16]. The PY solution to the sticky spheres model was essentially being used as an approximation to a narrow square-well model, which exhibits perfectly reasonable behaviour. Elsewhere [8] we have shown that, apart from the large-q region, the Baxter solution of sticky spheres is an excellent approximation to the narrow square-well problem. To be precise, we have shown that, for q smaller than roughly $\pi/(\epsilon R)$ (ϵR being the well-width, with R the total diameter of the particle including the well), $S(q)$ for the two systems is almost the same, at least up to around $\epsilon = 0.1$. It is a matter of curiosity that, given the success of the one-component model, there was very little attempt to exploit the two-component model PY solution to the Baxter model described in Barboy's original work. The reason may have been that, in the absence of a simple closed formula, there was less incentive from the experimental point of view to pursue matters. Also, as we shall see, the numerical solution of the problem is a little subtle, and this may have been a source of discouragement.

We shall show that it is possible to integrate the partial differential equations for chemical potentials presented by Barboy in a relatively straightforward and elegant manner. Using these, and the equation for pressure, we will be able to construct constant-pressure phase diagrams for any binary Baxter mixture, and to present the accompanying partial structure factors for the components. The results seem quite reasonable, both in terms of phase behaviour and structure factor. A few investigations with molecular dynamics simulations even indicate that the structure factors may be reasonably accurate for very narrow well problems. However, one of the most interesting observations is that the whole phase diagram for liquid-liquid coexistence is entirely well behaved. It exhibits none of the anomalies that the one-component Baxter PY liquid-gas solution does, and both binodals and spinodals are correctly positioned with respect to each other. The critical exponents for the system are entirely normal mean field values, unlike the anomalous behaviour of the one-component model.

Furthermore, there is a region under the binodal where we can identify real-valued solutions which may possibly be interpreted as useful metastable states.

We will argue that given the great speed with which PY solutions may be calculated, and the simplicity of the model, the binary Baxter model is interesting both from a fundamental and practical point of view. We can envisage that the binary system will also be a useful model for experimentalists to fit their binary data to, and that, essentially as before, this remains true not just of sticky spheres, but also of narrow square-well mixtures.

Calculation of phase diagrams

To begin with we remind the reader of the defining equations of the binary mixture. Thus, the direct correlation functions $c_{\alpha,\beta}(\alpha, \beta = 1, 2)$ are defined via the Ornstein-Zernike relation:

$$h_{\alpha\beta}(r) = c_{\alpha\beta}(r) + \sum_{\nu=1}^{2} \int c_{\alpha\nu}(|r - r'|)h_{\nu\beta}(r')\mathrm{d}r' \qquad (1)$$

where $h_{\alpha,\beta}(r) = g(r)_{\alpha,\beta} - 1$ are the total or indirect correlation functions.

The PY approximation amounts to:

$$c_{\alpha\beta} = [1 - \exp(\beta\Phi_{\alpha\beta})]g_{\alpha\beta}(r) \qquad (2)$$

The Baxter potential for the binary mixture is defined as:

$$\frac{\Phi_{\alpha\beta}(r)}{k_\mathrm{B}T} = \begin{cases} +\infty & r < d_{\alpha\beta} \\ \ln\left[\frac{2\tau_{\alpha\beta}(d_{\alpha\beta} - \sigma_{\alpha\beta})}{d_{\alpha\beta}}\right] & d_{\alpha\beta} < \sigma_{\alpha\beta} \\ 0 & \sigma_{\alpha\beta} < r \end{cases} \qquad (3)$$

where $d_{\alpha\alpha}$ is the diameter of a particle of component α, with $d_{\alpha\beta} = (d_{\alpha\alpha} + d_{\beta\beta})/2$ and $\tau_{\alpha\beta}$ determine the stickiness of the interactions, being the effective temperatures of the system. Finally, we arrive at the three equations that define the theory. The equation for the "energy" parameters $\lambda_{\alpha\beta}$ are:

$$\frac{\pi}{6} \sum_\gamma \rho_\gamma d_{\gamma\gamma}^2 (\lambda_{\alpha\gamma}\lambda_{\beta\gamma} - 6\lambda_{\alpha\gamma} - 6\lambda_{\beta\gamma} + 18)$$
$$- 2\frac{(1 - \xi_3)}{d_{\alpha\beta}}\tau_{\alpha\beta}\lambda_{\alpha\beta} = -12(1 - \xi_3)\frac{d_{\alpha\beta}}{d_{\alpha\alpha}d_{\beta\beta}} \qquad (4)$$

where ρ_α is the number density of component α and ξ_3 is the volume fraction of the system, defining in general $\xi_i = \frac{\pi}{6}\sum_\alpha \rho_\alpha d_{\alpha\alpha}^i$. The pressure is given by:

$$\frac{P}{k_\mathrm{B}T} = \frac{6}{\pi}\left\{\frac{\xi_0}{1 - \xi_3} + \frac{3\xi_1\xi_2}{(1 - \xi_3)^2} + 3\left(\frac{\xi_2}{1 - \xi_3}\right)^3\right\}$$
$$- \frac{\pi}{1 - \xi_3}\sum_{\alpha,\beta}\rho_\alpha\rho_\beta t_{\alpha\beta}\left(d_{\alpha\alpha} + d_{\beta\beta} + 3d_{\alpha\alpha}d_{\beta\beta}\frac{\xi_2}{1 - \xi_3}\right)$$
$$+ \frac{4\pi^2}{3}\sum_{\alpha,\beta,\gamma}\rho_\alpha\rho_\beta\rho_\gamma t_{\alpha\gamma}t_{\gamma\beta}t_{\beta\alpha} \qquad (5)$$

where $t_{\alpha\beta} = (\lambda_{\alpha\beta}d_{\alpha\alpha}d_{\beta\beta})/12(1 - \xi_3)$. Finally, the derivatives of the chemical potential are:

$$\frac{1}{k_\mathrm{B}T}\frac{\partial \mu_\beta}{\partial \rho_\alpha} = \frac{\delta_{\alpha\beta}}{\rho_\beta} - 2\pi \int_{s_{\alpha\beta}}^{d_{\alpha\beta}} Q_{\alpha\beta}(r)\mathrm{d}r - 2\pi \int_{s_{\beta\alpha}}^{d_{\beta\alpha}} Q_{\beta\alpha}(r)\mathrm{d}r$$

$$+ 4\pi^2 \sum_\gamma \rho_\gamma \int_{s_{\gamma\alpha}}^{d_{\gamma\alpha}} Q_{\gamma\alpha}(r)\mathrm{d}(r) \int_{s_{\gamma\beta}}^{d_{\gamma\beta}} Q_{\gamma\beta}(t)\mathrm{d}(t) \quad (6)$$

where $Q_{\alpha\beta}(r)$ are known functions. This, then, is the closed theory for mixtures, in general, and binary systems in particular.

To construct a phase diagram we must therefore solve Eq. (4) along a constant pressure and constant temperature cut of the phase diagram, and thereby obtain relevant values of the $\lambda_{\alpha\beta}$. We then must evaluate the chemical potentials $\mu_\alpha(\rho_\alpha, \rho_\beta)$ by integrating their total derivatives with respect of one of the two number densities, i.e.

$$\frac{\mathrm{d}\mu_\beta}{\mathrm{d}\rho_\alpha} = \frac{\partial \mu_\beta}{\partial \rho_\alpha} + \frac{\partial \mu_\beta}{\partial \rho_\beta}\frac{\partial \rho_\beta}{\partial \rho_\alpha} \quad (7)$$

using Eq. (6), where the partial derivative of a density with respect to the other is determined by the chosen constant-pressure value. It is immediately evident that in Eq. (6) a divergence in the diagonal partial derivatives appears when the density goes to zero. If we perform the integration from the starting value $\rho_\alpha = 0$, we also find an unknown constant of integration, that is $\mu_\alpha(0, \rho_\beta^M)$, corresponding to the chemical potential of component α in absence of that component, but the other component density a maximum. This appears to play an unphysical role. Consequently, we have assumed that there is a cancellation of the divergence in the derivative with a divergence in the function $\mu_\alpha(\rho_\alpha, \rho_\beta)$ in the limit $\rho_\alpha \to 0$. Also, to calculate these unknown constants of integration, we integrate Eq. (7) along the whole range of ρ_α, from zero to its maximum value, for which the other density is instead zero; in this way we obtain

$$\int_{\rho_\alpha=0}^{\rho_\alpha=\rho_a^M} \frac{\mathrm{d}\mu_\alpha}{\mathrm{d}\rho_\alpha} = \mu_\alpha(\rho_\alpha^M, 0) - \mu_\alpha\left(0, \rho_\beta^M\right) \quad (8)$$

Since the constant $\mu_\alpha(\rho_\alpha^M, 0)$ is the known chemical potential for the one-component case of density ρ_α^M [3], we can evaluate the unknowns. We point out that this integration is possible because there exists a region inside the binodal where the λ-parameters are real, unlike the case of the one-component model. The integration (8) is still not so straightforward, because we must exclude the spinodal area, that would cause unphysical behaviour for the chemical potentials. The spinodal curve for the liquid-liquid transition is easily calculated numerically, providing that the $\lambda_{\alpha\beta}$ solutions of Eq. (4) also satisfy the condition $S_{cc}(0) = 0$, where S_{cc} is the concentration-concentration structure factor, according to the standard definitions [17].

We present two phase diagrams for comparison. In both cases we have chosen symmetric interactions and we have assumed the parameters $\tau_{\alpha\beta}$ to be proportional to an effective temperature τ, and fixed the proportionality constants equal to 1 for the interactions between like-particles and to 100 between the unlike-particles. These conditions will favour the phase separation in two liquids of types "A-rich" and "B-rich". The first, represented in Fig. 1, is for a mixture that is also symmetric in diameter. We have chosen the parameters in the potential so that the volume fraction, along the phase diagram, is varying between approximately 0.435 and 0.48, being minimum at the critical point. The second of the phase diagrams, represented in Fig. 2, is for a mixture that is asymmetric in diameters by 20%; the range of variation of the volume fraction is approximately 0.41–0.49.

Let us now turn to the partial structure factors $S_{\alpha\beta}(q)$ for the binary PY Baxter model. These have not appeared, in any form easily usable by us, in the literature (although some presentation is made in [18]). The expressions are a little lengthy, but they are worth recording for those who may seek to fit data. Thus:

$$\begin{pmatrix} S_{11}(q) & S_{12}(q) \\ S_{21}(q) & S_{22}(q) \end{pmatrix}^{-1} = \begin{pmatrix} \tilde{Q}_{11}^2(q) + \tilde{Q}_{21}^2(q) & \tilde{Q}_{11}(q)\tilde{Q}_{12}(q) + \tilde{Q}_{21}(q)\tilde{Q}_{22}(q) \\ \tilde{Q}_{11}(q)\tilde{Q}_{12}(q) + \tilde{Q}_{21}(q)\tilde{Q}_{22}(q) & \tilde{Q}_{12}^2(q) + \tilde{Q}_{22}^2(q) \end{pmatrix} \quad (9)$$

where $\tilde{Q}_{\alpha\beta}(q)$ is a complex function:

$$\tilde{Q}_{\alpha\beta}(q) = \delta_{\alpha\beta} - 2\pi(\rho_\alpha\rho_\beta)^{1/2}\left\{ \cos z_\alpha \left[\sin z_\beta \left(\frac{a_\alpha}{2}\left(-\frac{4}{q^3} + \frac{d_{\alpha\alpha} + d_{\beta\beta}}{2q} \right) + \frac{b_\alpha d_{\alpha\alpha}}{q} - \frac{2}{q}\left(\frac{a_\alpha d_{\alpha\beta}^2}{2} + b_\alpha d_{\alpha\beta} - \frac{\lambda_{\alpha\beta} d_{\alpha\alpha} d_{\beta\beta}}{12(1-\xi_3)} \right) \right) \right. \right.$$

$$\left. + \cos z_\beta \frac{a_\alpha d_{\beta\beta}}{q^2} \right] + \sin z_\alpha \left[-\sin z_\beta \frac{a_\alpha d_{\alpha\alpha} + 2b_\alpha}{q^2} + \cos z_\beta d_{\beta\beta} \frac{a_\alpha d_{\alpha\alpha} + 2b_\alpha}{2q} \right] \right\}$$

$$- i2\pi(\rho_\alpha\rho_\beta)^{1/2}\left\{ \left[\sin z_\beta \left(\frac{a_\alpha d_{\alpha\alpha} + 2b_\alpha}{q^2} \right) - \cos z_\beta \left(\frac{a_\alpha d_{\alpha\alpha} + 2b_\alpha}{2q} \right) \right] \cos z_\alpha + \sin z_\alpha \left[\sin z_\beta \left(\frac{a_\alpha}{2}\left(-\frac{4}{q^3} + \frac{d_{\alpha\alpha} + d_{\beta\beta}}{2q} \right) \right. \right. \right.$$

$$\left. \left. \left. + \frac{b_\alpha d_{\alpha\alpha}}{q} - \frac{2}{q}\left(\frac{a_\alpha d_{\alpha\beta}^2}{2} + b_\alpha d_{\alpha\beta} - \frac{\lambda_{\alpha\beta} d_{\alpha\alpha} d_{\beta\beta}}{12(1-\xi_3)} \right) \right) + \cos z_\beta \left(\frac{a_\alpha d_{\beta\beta}}{q^2} \right) \right] \right\} \quad (10)$$

where $z_\alpha = q d_{\alpha\alpha}/2$ and a_α, b_α are,

$$a_\alpha = \frac{1}{(1-\xi_3)^2}\left(1 - \xi_3 + 3\xi_2 d_{\alpha\alpha} - \frac{\pi}{6}d_{\alpha\alpha}\sum_r \rho_r d_{rr}^2 \lambda_{\alpha r}\right)$$

and,

$$b_\alpha = \frac{d_{\alpha\alpha}^2}{2(1-\xi_3)^2}\left(\frac{\pi}{6}d_{\alpha\alpha}\sum_r \rho_r d_{rr}^2 \lambda_{\alpha r} - 3\xi_2\right)$$

We now illustrate in Figs. 3 and 4 the use of the structure factors by exhibiting partial structure factors for a binary mixture for the Baxter model within the PY approximation, in comparison with those obtained via a molecular dynamics simulation for a narrow square-well. To relate a certain temperature T for a narrow square-well potential to the parameters $\tau_{\alpha\beta}$ of the Baxter model it is possible to use the following [15]:

$$\tau_{\alpha\beta} = (12\epsilon_{\alpha\beta})^{-1}\exp(u_{\alpha\beta}/k_B T) \tag{11}$$

where $u_{\alpha\beta}$ are the (negative) well depths. We note that the structure factors here agree reasonably well because the Baxter model is a reasonable approximation to the narrow square-well model. This statement refers only to momenta up to approximately a cut off distance, $q_{\alpha\beta}^* = \pi/(\epsilon_{\alpha\beta}R_{\alpha\beta})$, with $(\epsilon_{\alpha\beta}R_{\alpha\beta})$ the well widths. Beyond this momentum the numerical Baxter model has a very long tail that may, in certain circumstances, give unphysical results. This tail is, of course, not shown in MD calculations, and does not appear in Figs. 3 and 4. Nor it is relevant for many practical purposes such as modelling scattering data. However, if it is intended that

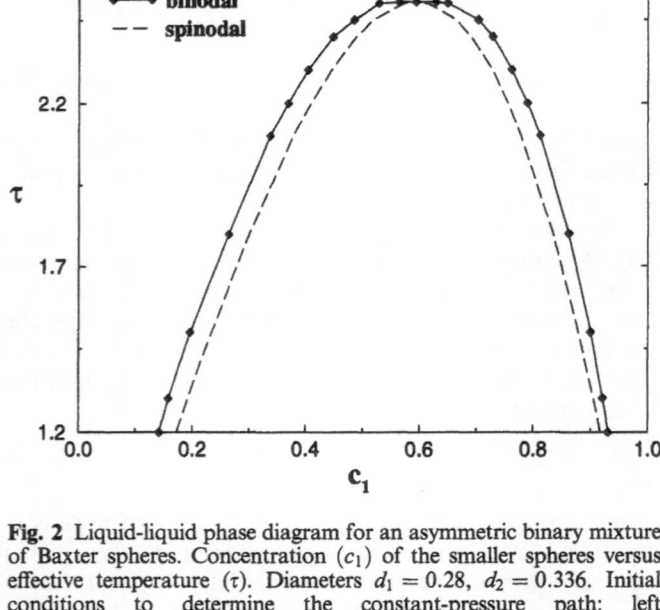

Fig. 2 Liquid-liquid phase diagram for an asymmetric binary mixture of Baxter spheres. Concentration (c_1) of the smaller spheres versus effective temperature (τ). Diameters $d_1 = 0.28$, $d_2 = 0.336$. Initial conditions to determine the constant-pressure path: left ($\rho_1 = 0$, $\rho_2 = 25$), right ($\rho_1 = 36.2$, $\rho_2 = 0$)

the Baxter model for single or multicomponent species be applied to calculate various dynamical data, then more care should be exercised. In particular, viscosity and mode coupling calculations are divergent because of the large momentum tail [*, 8, 13]. We [8] have shown that this can be resolved if the relevant quantities are cut off at momenta $q_{\alpha\beta}^*$. Thus, we now have a reason-

Fig. 1 Liquid-liquid phase diagram for symmetric binary mixtures of Baxter spheres. Concentration (c) versus effective temperature (τ). Diameters $d_1 = d_2 = 0.28$. Initial conditions to determine the constant-pressure path: left ($\rho_1 = 0$, $\rho_2 = 42$), right ($\rho_1 = 42$, $\rho_2 = 0$)

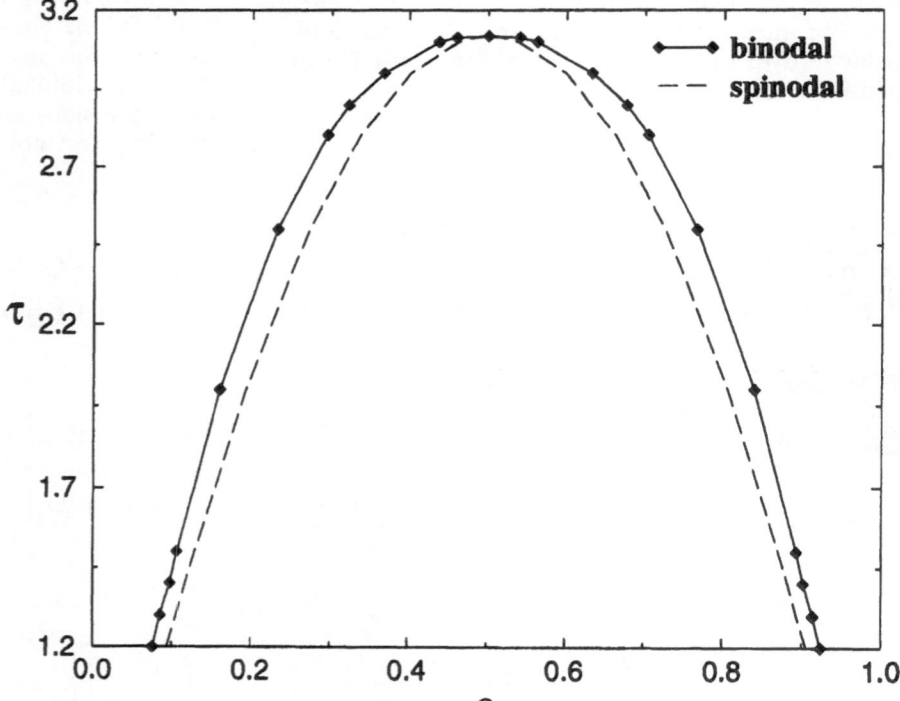

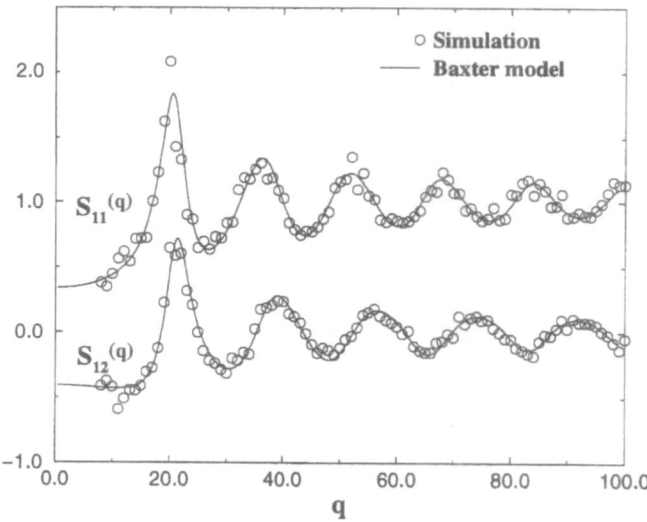

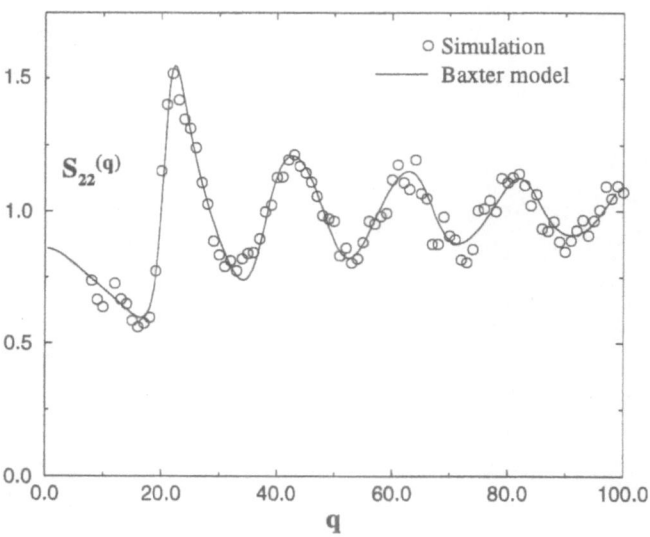

Fig. 3 Comparison of the structure factors calculated via the PY approximation for the Baxter model (*straight line*) and MD simulation (*circles*) of a binary mixture of spheres interacting via a narrow square-well potential. Volume fraction 0.560, composition 50%–50%, temperature $T = 0.1844$, $\tau_{11} = \tau_{12} = \tau_{22} = 0.3715$, $d_{11} = 9.47$, $d_{22} = 11.364$, $R_{11} = 9.5647$, $R_{22} = 11.47764$, $\epsilon_{11} = \epsilon_{12} = \epsilon_{22} = 0.01$. We represent here $S_{11}(q)$ and $S_{12}(q)$

Fig. 4 As in Fig. 3, but $S_{22}(q)$

ably satisfactory model with which to interpret binary mixture data where the well widths are reasonably small, i.e., $\epsilon_{\alpha\beta} < 0.1$.

Conclusions

In this paper we have shown that a long-neglected extension of the Baxter model for single-component spheres to binary mixtures can also be solved to produce a phase diagram and structural information within the same model. Thus Barboy's differential equations have been solved, and liquid-liquid phase equilibria and structure functions for two sample systems have been

presented. The PY solution of a liquid state model is undoubtedly far from satisfactory for many purposes. However, the PY solution is expected to be quite reasonable for many systems. The approach presented here permits one of the only reasonably consistent pictures of binary liquid mixtures with binodals, spinodals, and structural information at constant pressure known for liquid state theory. Superior approximations do exist, but they are usually quite complicated and it not easy to implement the full program of liquid state theory without extensive computations. This has rarely been achieved. Besides the intrinsic theoretic interest, the calculations presented here are sufficiently quick that they may be used in fitting experimental data. This provides those studying colloidal mixtures a simple model with which to interpret their data. We are hopeful that Barboy's work, when treated by the method suggested, may become as useful as the simple single-component solution of Baxter.

References

1. Baxter RJ (1968) J Chem Phys 49:2770
2. Baxter RJ (1967) Phys Rev 154:170
3. Barboy B (1974) J Chem Phys 61:3194
4. Watts RO, Henderson D, Baxter RJ (1971) Adv Chem Phys 21:421
5. Percus JK, Yevick GJ (1958) Phys Rev 110:1
6. Fishman S, Fisher M (1981) Physica A 108:1
7. Stell G (1991) J Stat Phys 63:1203, Sear RP Cond-mat/9805201
* Berghenholtz J, Fuchs M (1999) Phys Rev E 59:5706
8. Foffi G, Zaccarelli E, Sciortino F, Tartaglia P, Dawson KA (1999) J Stat Phys (in press)
9. Barboy B (1975) Chem Phys 11:357
10. Lekkerker HNW, Poon WCK, Pusey PN, Stroobants A, Warren PB (1992) Europhys Lett 20:559
11. Verdun H, Dhont JKG (1995) J Colloid Interface Sci 172:425
12. Tejero CF, Daanoun A, Lekkerker HNW, Baus M (1994) Phys Rev Lett 73:752
13. Liu YC, Chen SH, Huang JS (1996) Phys Rev E 54:1698
14. Menon SVG, Kelkar VK, Manohar C (1991) Phys Rev A 43:1130;
** Tartagua P, Rouch J, Chen SH (1992) Phys Rev A 45:7257; Rouch J, Tartagua P, Chen SH (1993) Phys Rev Lett 40:2643
15. Menon SVG, Manohar C, Rao KS (1991) J Chem Phys 95:9186
16. Regnaut C, Ravey JC (1989) J Chem Phys 91:1211
17. Bhatia AB, Thornton DE (1970) Phys Rev B 2:3004
18. Robertus C, Philippe WH, Joosten JGH, Levine YK (1989) J Chem Phys 90:4482

Progr Colloid Polym Sci (2000) 115 : 377–380
© Springer-Verlag 2000

Progr Colloid Polym Sci (2000) 115: 381–382
© Springer-Verlag 2000

KEY WORD INDEX